T0205722

Lecture Notes in Computer Science 13980

Founding Editors

Gerhard Goos

Juris Hartmanis

Editorial Board Members

Elisa Bertino, *Purdue University, West Lafayette, IN, USA*

Wen Gao, *Peking University, Beijing, China*

Bernhard Steffen ⓘ, *TU Dortmund University, Dortmund, Germany*

Moti Yung ⓘ, *Columbia University, New York, NY, USA*

The series Lecture Notes in Computer Science (LNCS), including its subseries Lecture Notes in Artificial Intelligence (LNAI) and Lecture Notes in Bioinformatics (LNBI), has established itself as a medium for the publication of new developments in computer science and information technology research, teaching, and education.

LNCS enjoys close cooperation with the computer science R & D community, the series counts many renowned academics among its volume editors and paper authors, and collaborates with prestigious societies. Its mission is to serve this international community by providing an invaluable service, mainly focused on the publication of conference and workshop proceedings and postproceedings. LNCS commenced publication in 1973.

Jaap Kamps · Lorraine Goeuriot · Fabio Crestani ·
Maria Maistro · Hideo Joho · Brian Davis ·
Cathal Gurrin · Udo Kruschwitz ·
Annalina Caputo
Editors

Advances in Information Retrieval

45th European Conference on Information Retrieval, ECIR 2023
Dublin, Ireland, April 2–6, 2023
Proceedings, Part I

Springer

Editors
Jaap Kamps 🆔
University of Amsterdam
Amsterdam, Netherlands

Lorraine Goeuriot 🆔
Université Grenoble-Alpes
Saint-Martin-d'Hères, France

Fabio Crestani 🆔
Università della Svizzera Italiana
Lugano, Switzerland

Maria Maistro 🆔
University of Copenhagen
Copenhagen, Denmark

Hideo Joho 🆔
University of Tsukuba
Ibaraki, Japan

Brian Davis 🆔
Dublin City University
Dublin, Ireland

Cathal Gurrin 🆔
Dublin City University
Dublin, Ireland

Udo Kruschwitz 🆔
Universität Regensburg
Regensburg, Germany

Annalina Caputo 🆔
Dublin City University
Dublin, Ireland

ISSN 0302-9743   ISSN 1611-3349 (electronic)
Lecture Notes in Computer Science
ISBN 978-3-031-28243-0   ISBN 978-3-031-28244-7 (eBook)
https://doi.org/10.1007/978-3-031-28244-7

This Springer imprint is published by the registered company Springer Nature Switzerland AG
The registered company address is: Gewerbestrasse 11, 6330 Cham, Switzerland

Preface

The 45th European Conference on Information Retrieval (ECIR 2023) was held in Dublin, Ireland, during April 2–6, 2023, and brought together hundreds of researchers from Europe and abroad. The conference was organized by Dublin City University, in cooperation with the British Computer Society's Information Retrieval Specialist Group (BCS IRSG).

These proceedings contain the papers related to the presentations, workshops, and tutorials given during the conference. This year's ECIR program boasted a variety of novel work from contributors from all around the world. In total, 489 papers from authors in 52 countries were submitted to the different tracks. The final program included 65 full papers (29% acceptance rate), 41 short papers (27% acceptance rate), 19 demonstration papers (66% acceptance rate), 12 reproducibility papers (63% acceptance rate), 10 doctoral consortium papers (56% acceptance rate), and 13 invited CLEF papers. All submissions were peer-reviewed by at least three international Program Committee members to ensure that only submissions of the highest relevance and quality were included in the final program. The acceptance decisions were further informed by discussions among the reviewers for each submitted paper, led by a senior Program Committee member. In a final PC meeting all the final recommendations were discussed, trying to reach a fair and equal outcome for all submissions.

The accepted papers cover the state of the art in information retrieval: user aspects, system and foundational aspects, machine learning, applications, evaluation, new social and technical challenges, and other topics of direct or indirect relevance to search. As in previous years, the ECIR 2023 program contained a high proportion of papers with students as first authors, as well as papers from a variety of universities, research institutes, and commercial organizations.

In addition to the papers, the program also included 3 keynotes, 7 tutorials, 8 workshops, a doctoral consortium, the presentation of selected papers from the 2022 issues of the Information Retrieval Journal, and an industry day. Keynote talks were given by Mounia Lalmas (Spotify), Tetsuya Sakai (Waseda University), and this year's BCS IRSG Karen Spärck Jones Award winner, Yang Wang (UC Santa Barbara). The tutorials covered a range of topics including conversational agents in health; crowdsourcing; gender bias; legal IR and NLP; neuro-symbolic representations; query auto completion; and text classification. The workshops brought together participants to discuss algorithmic bias (BIAS); bibliometrics (BIR); e-discovery (ALTARS); geographic information extraction (GeoExT); legal IR (Legal IR); narrative extraction (Text2story); online misinformation (ROMCIR); and query performance prediction (QPP).

The success of ECIR 2023 would not have been possible without all the help from the team of volunteers and reviewers. We wish to thank all the reviewers and meta-reviewers who helped ensure the high quality of the program. We also wish to thank: the short paper track chairs: Maria Maistro and Hideo Joho; the demo track chairs: Liting Zhou and Frank Hopfgartner; the reproducibility track chair: Leif Azzopardi; the workshop track chairs: Ricardo Campos and Gianmaria Silvello; the tutorial track chairs: Bhaskar Mitra and Debasis Ganguly; the industry track chairs: Nicolas Fiorini and Isabelle Moulinier; the doctoral consortium chair: Gareth Jones; and the awards chair: Suzan Verberne. We thank the students Praveen Acharya, Chinonso Osuji and Kanishk Verma for help with preparing the proceedings. We would like to thank all the student volunteers who helped to create an excellent experience for participants and attendees. ECIR 2023 was sponsored by a range of research institutes and companies. We thank them all for their support.

Finally, we wish to thank all the authors and contributors to the conference.

April 2023

Lorraine Goeuriot
Fabio Crestani
Jaap Kamps
Maria Maistro
Hideo Joho
Annalina Caputo
Udo Kruschwitz
Cathal Gurrin

Organization

General Chairs

Annalina Caputo   Dublin City University, Ireland
Udo Kruschwitz   Universität Regensburg, Germany
Cathal Gurrin   Dublin City University, Ireland

Program Committee Chairs

Jaap Kamps   University of Amsterdam, Netherlands
Lorraine Goeuriot   Université Grenoble Alpes, France
Fabio Crestani   Università della Svizzera Italiana, Switzerland

Short Papers Chairs

Maria Maistro   University of Copenhagen, Denmark
Hideo Joho   University of Tsukuba, Japan

Demo Chairs

Liting Zhou   Dublin City University, Ireland
Frank Hopfgartner   University of Koblenz-Landau, Germany

Reproducibility Track Chair

Leif Azzopardi   University of Strathclyde, UK

Workshop Chairs

Ricardo Campos   Instituto Politécnico de Tomar/INESC TEC,
                            Portugal
Gianmaria Silvello   University of Padua, Italy

Tutorial Chairs

Bhaskar Mitra   Microsoft, Canada
Debasis Ganguly   University of Glasgow, UK

Industry Day Chairs

Nicolas Fiorini   Algolia, France
Isabelle Moulinier   Thomson Reuters, USA

Doctoral Consortium Chair

Gareth Jones   Dublin City University, Ireland

Awards Chair

Suzan Verberne   Leiden University, Netherlands

Publication Chairs

Brian Davis   Dublin City University, Ireland
Joachim Wagner   Dublin City University, Ireland

Local Chairs

Brian Davis   Dublin City University, Ireland
Ly Duyen Tran   Dublin City University, Ireland

Senior Program Committee

Omar Alonso   Amazon, USA
Giambattista Amati   Fondazione Ugo Bordoni, Italy
Ioannis Arapakis   Telefonica Research, Spain
Jaime Arguello   University of North Carolina at Chapel Hill, USA
Javed Aslam   Northeastern University, USA

Krisztian Balog	University of Stavanger & Google Research, Norway
Patrice Bellot	Aix-Marseille Université - CNRS (LSIS), France
Michael Bendersky	Google, USA
Mohand Boughanem	IRIT University Paul Sabatier Toulouse, France
Jamie Callan	Carnegie Mellon University, USA
Ben Carterette	Spotify, USA
Charles Clarke	University of Waterloo, Canada
Bruce Croft	University of Massachusetts Amherst, USA
Maarten de Rijke	University of Amsterdam, Netherlands
Arjen de Vries	Radboud University, Netherlands
Giorgio Maria Di Nunzio	University of Padua, Italy
Laura Dietz	University of New Hampshire, USA
Shiri Dori-Hacohen	University of Connecticut, USA
Carsten Eickhoff	Brown University, USA
Tamer Elsayed	Qatar University, Qatar
Liana Ermakova	HCTI, Université de Bretagne Occidentale, France
Hui Fang	University of Delaware, USA
Nicola Ferro	University of Padova, Italy
Ingo Frommholz	University of Wolverhampton, UK
Norbert Fuhr	University of Duisburg-Essen, Germany
Debasis Ganguly	University of Glasgow, UK
Nazli Goharian	Georgetown University, USA
Marcos Goncalves	Federal University of Minas Gerais, Brazil
Julio Gonzalo	UNED, Spain
Jiafeng Guo	Institute of Computing Technology, China
Matthias Hagen	Friedrich-Schiller-Universität Jena, Germany
Martin Halvey	University of Strathclyde, UK
Allan Hanbury	TU Wien, Austria
Donna Harman	NIST, USA
Faegheh Hasibi	Radboud University, Netherlands
Claudia Hauff	Spotify, Netherlands
Ben He	University of Chinese Academy of Sciences, China
Jiyin He	Signal AI, UK
Dietmar Jannach	University of Klagenfurt, Austria
Adam Jatowt	University of Innsbruck, Austria
Hideo Joho	University of Tsukuba, Japan
Gareth Jones	Dublin City University, Ireland
Joemon Jose	University of Glasgow, UK
Jaap Kamps	University of Amsterdam, Netherlands

Paul Thomas	Microsoft, Australia
Nicola Tonellotto	University of Pisa, Italy
Theodora Tsikrika	Information Technologies Institute, CERTH, Greece
Julián Urbano	Delft University of Technology, Netherlands
Suzan Verberne	LIACS, Leiden University, Netherlands
Gerhard Weikum	Max Planck Institute for Informatics, Germany
Marcel Worring	University of Amsterdam, Netherlands
Andrew Yates	University of Amsterdam, Netherlands
Jakub Zavrel	Zeta Alpha, Netherlands
Min Zhang	Tsinghua University, China
Shuo Zhang	Bloomberg, Norway
Justin Zobel	University of Melbourne, Australia
Guido Zuccon	University of Queensland, Australia

Program Committee

Shilpi Agrawal	Linkedin, India
Qingyao Ai	Tsinghua University, China
Dyaa Albakour	Signal AI, UK
Mohammad Aliannejadi	University of Amsterdam, Netherlands
Satya Almasian	Heidelberg University, Germany
Omar Alonso	Amazon, USA
Ismail Sengor Altingovde	Middle East Technical University, Turkey
Giuseppe Amato	ISTI-CNR, Italy
Enrique Amigó	UNED, Spain
Sophia Ananiadou	University of Manchester, UK
Linda Andersson	Artificial Researcher IT GmbH, TU Wien, Austria
Vito Walter Anelli	Politecnico di Bari, Italy
Negar Arabzadeh	University of Waterloo, Canada
Arian Askari	Leiden Institute of Advanced Computer Science, Leiden University, Netherlands
Giuseppe Attardi	Università di Pisa, Italy
Maurizio Atzori	University of Cagliari, Italy
Sandeep Avula	Amazon, USA
Mossaab Bagdouri	Walmart Global Tech, USA
Ebrahim Bagheri	Ryerson University, Canada
Georgios Balikas	Salesforce Inc, France
Krisztian Balog	University of Stavanger & Google Research, Norway
Alvaro Barreiro	University of A Coruña, Spain

Célia da Costa Pereira	Université Côte d'Azur, France
Duc Tien Dang Nguyen	University of Bergen, Norway
Maarten de Rijke	University of Amsterdam, Netherlands
Arjen de Vries	Radboud University, Netherlands
Yashar Deldjoo	Polytechnic University of Bari, Italy
Gianluca Demartini	University of Queensland, Australia
Amey Dharwadker	Meta, USA
Emanuele Di Buccio	University of Padua, Italy
Giorgio Maria Di Nunzio	University of Padua, Italy
Gaël Dias	Normandie University, France
Laura Dietz	University of New Hampshire, USA
Vlastislav Dohnal	Faculty of Informatics, Masaryk University, Czechia
Zhicheng Dou	Renmin University of China, China
Antoine Doucet	University of La Rochelle, France
Pan Du	Thomson Reuters Labs, Canada
Tomislav Duricic	Graz University of Technology, Austria
Liana Ermakova	HCTI, Université de Bretagne Occidentale, France
Ralph Ewerth	L3S Research Center, Leibniz Universität Hannover, Germany
Guglielmo Faggioli	University of Padova, Italy
Anjie Fang	Amazon.com, USA
Hossein Fani	University of Windsor, Canada
Yue Feng	UCL, UK
Marcos Fernández Pichel	Universidade de Santiago de Compostela, Spain
Juan M. Fernández-Luna	University of Granada, Spain
Nicola Ferro	University of Padua, Italy
Komal Florio	Università di Torino, Italy
Thibault Formal	Naver Labs Europe, France
Ophir Frieder	Georgetown University, USA
Ingo Frommholz	University of Wolverhampton, UK
Maik Fröbe	Friedrich-Schiller-Universität Jena, Germany
Norbert Fuhr	University of Duisburg-Essen, Germany
Michael Färber	Karlsruhe Institute of Technology, Germany
Petra Galuščáková	Université Grenoble Alpes, France
Debasis Ganguly	University of Glasgow, UK
Dario Garigliotti	No affiliation, Norway
Eric Gaussier	LIG-UJF, France
Kripabandhu Ghosh	Indian Institute of Science Education and Research (IISER) Kolkata, India
Anastasia Giachanou	Utrecht University, Netherlands

Manoj Kesavulu	Dublin City University, Ireland
Atsushi Keyaki	Hitotsubashi University, Japan
Johannes Kiesel	Bauhaus-Universität Weimar, Germany
Benjamin Kille	Norwegian University of Science and Technology, Norway
Tracy Holloway King	Adobe, USA
Udo Kruschwitz	University of Regensburg, Germany
Lakhotia Kushal	Outreach, USA
Mucahid Kutlu	TOBB University of Economics and Technology, Turkey
Saar Kuzi	Amazon, USA
Wai Lam	The Chinese University of Hong Kong, China
Birger Larsen	Aalborg University, Denmark
Dawn Lawrie	Johns Hopkins University, USA
Jochen L. Leidner	Coburg University of Applied Sciences/University of Sheffield/KnowledgeSpaces, Germany
Mark Levene	Birkbeck, University of London, UK
Qiuchi Li	University of Padua, Italy
Wei Li	University of Roehampton, UK
Xiangsheng Li	Tsinghua University, China
Shangsong Liang	Sun Yat-sen University, China
Siyu Liao	Amazon, USA
Trond Linjordet	University of Stavanger, Norway
Matteo Lissandrini	Aalborg University, Denmark
Suzanne Little	Dublin City University, Ireland
Haiming Liu	University of Southampton, UK
Benedikt Loepp	University of Duisburg-Essen, Germany
Andreas Lommatzsch	TU Berlin, Germany
Chenyang Lyu	Dublin City University, Ireland
Weizhi Ma	Tsinghua University, China
Sean MacAvaney	University of Glasgow, UK
Andrew Macfarlane	City, University of London, UK
Joel Mackenzie	University of Queensland, Australia
Khushhall Chandra Mahajan	Meta Inc., USA
Maria Maistro	University of Copenhagen, Denmark
Antonio Mallia	New York University, Italy
Thomas Mandl	University of Hildesheim, Germany
Behrooz Mansouri	University of Southern Maine, USA
Jiaxin Mao	Renmin University of China, China
Stefano Marchesin	University of Padova, Italy
Mirko Marras	University of Cagliari, Italy
Monica Marrero	Europeana Foundation, Netherlands

Miguel Martinez	Signal AI, UK
Bruno Martins	IST and INESC-ID - Instituto Superior Técnico, University of Lisbon, Portugal
Flavio Martins	INESC-ID, Instituto Superior Técnico, Universidade de Lisboa, Portugal
Maarten Marx	University of Amsterdam, Netherlands
Yosi Mass	IBM Haifa Research Lab, Israel
David Maxwell	Delft University of Technology, UK
Richard McCreadie	University of Glasgow, UK
Graham McDonald	University of Glasgow, UK
Dana McKay	RMIT University, Australia
Paul McNamee	Johns Hopkins University, USA
Parth Mehta	IRSI, India
Florian Meier	Aalborg University, Copenhagen, Denmark
Ida Mele	IASI-CNR, Italy
Zaiqiao Meng	University of Glasgow, UK
Donald Metzler	Google, USA
Tomasz Miksa	TU Wien, Austria
Ashlee Milton	University of Minnesota, USA
Alistair Moffat	University of Melbourne, Australia
Ali Montazeralghaem	University of Massachusetts Amherst, USA
Jose Moreno	IRIT/UPS, France
Alejandro Moreo Fernández	Istituto di Scienza e Tecnologie dell'Informazione "A. Faedo", Italy
Philippe Mulhem	LIG-CNRS, France
Cristina Ioana Muntean	ISTI CNR, Italy
Henning Müller	HES-SO, Switzerland
Suraj Nair	University of Maryland, USA
Franco Maria Nardini	ISTI-CNR, Italy
Fedelucio Narducci	Politecnico di Bari, Italy
Wolfgang Nejdl	L3S and University of Hannover, Germany
Manh Duy Nguyen	DCU, Ireland
Thong Nguyen	University of Amsterdam, Netherlands
Jian-Yun Nie	Université de Montreal, Canada
Sérgio Nunes	University of Porto, Portugal
Diana Nurbakova	National Institute of Applied Sciences of Lyon (INSA Lyon), France
Kjetil Nørvåg	Norwegian University of Science and Technology, Norway
Michael Oakes	University of Wolverhampton, UK
Hiroaki Ohshima	Graduate School of Applied Informatics, University of Hyogo, Japan

Salvatore Orlando	Università Ca' Foscari Venezia, Italy
Iadh Ounis	University of Glasgow, UK
Pooja Oza	University of New Hampshire, USA
Özlem Özgöbek	Norwegian University of Science and Technology, Norway
Deepak P.	Queen's University Belfast, UK
Panagiotis Papadakos	Information Systems Laboratory - FORTH-ICS, Greece
Javier Parapar	IRLab, University of A Coruña, Spain
Pavel Pecina	Charles University, Czechia
Gustavo Penha	Delft University of Technology, Brazil
Maria Soledad Pera	TU Delft, Netherlands
Vivien Petras	Humboldt-Universität zu Berlin, Germany
Giulio Ermanno Pibiri	Ca' Foscari University of Venice, Italy
Francesco Piccialli	University of Naples Federico II, Italy
Karen Pinel-Sauvagnat	IRIT, France
Florina Piroi	TU Wien, Institue of Information Systems Engineering, Austria
Marco Polignano	Università degli Studi di Bari Aldo Moro, Italy
Martin Potthast	Leipzig University, Germany
Ronak Pradeep	University of Waterloo, Canada
Xin Qian	University of Maryland, USA
Fiana Raiber	Yahoo Research, Israel
David Rau	University of Amsterdam, Netherlands
Andreas Rauber	Vienna University of Technology, Austria
Gábor Recski	TU Wien, Austria
Weilong Ren	Shenzhen Institute of Computiing Sciences, China
Zhaochun Ren	Shandong University, China
Chiara Renso	ISTI-CNR, Pisa, Italy, Italy
Thomas Roelleke	Queen Mary University of London, UK
Kevin Roitero	University of Udine, Italy
Haggai Roitman	eBay Research, Israel
Paolo Rosso	Universitat Politècnica de València, Spain
Stevan Rudinac	University of Amsterdam, Netherlands
Anna Ruggero	Sease Ltd., Italy
Tony Russell-Rose	Goldsmiths, University of London, UK
Ian Ruthven	University of Strathclyde, UK
Sriparna Saha	IIT Patna, India
Tetsuya Sakai	Waseda University, Japan
Eric Sanjuan	Laboratoire Informatique d'Avignon—Université d'Avignon, France
Maya Sappelli	HAN University of Applied Sciences, Netherlands

Yannis Tzitzikas	University of Crete and FORTH-ICS, Greece
Md Zia Ullah	Edinburgh Napier University, UK
Kazutoshi Umemoto	University of Tokyo, Japan
Julián Urbano	Delft University of Technology, Netherlands
Ruben van Heusden	University of Amsterdam, Netherlands
Aparna Varde	Montclair State University, USA
Suzan Verberne	LIACS, Leiden University, Netherlands
Manisha Verma	Amazon, USA
Vishwa Vinay	Adobe Research, India
Marco Viviani	Università degli Studi di Milano-Bicocca - DISCo, Italy
Ellen Voorhees	NIST, USA
Xi Wang	University College London, UK
Zhihong Wang	Tsinghua University, China
Wouter Weerkamp	TomTom, Netherlands
Gerhard Weikum	Max Planck Institute for Informatics, Germany
Xiaohui Xie	Tsinghua University, China
Takehiro Yamamoto	University of Hyogo, Japan
Eugene Yang	Human Language Technology Center of Excellence, Johns Hopkins University, USA
Andrew Yates	University of Amsterdam, Netherlands
Elad Yom-Tov	Microsoft, Israel
Ran Yu	University of Bonn, Germany
Hamed Zamani	University of Massachusetts Amherst, USA
Eva Zangerle	University of Innsbruck, Austria
Richard Zanibbi	Rochester Institute of Technology, USA
Fattane Zarrinkalam	University of Guelph, Canada
Sergej Zerr	Rhenish Friedrich Wilhelm University of Bonn, Germany
Fan Zhang	Wuhan University, China
Haixian Zhang	Sichuan University, China
Min Zhang	Tsinghua University, China
Rongting Zhang	Amazon, USA
Ruqing Zhang	Institute of Computing Technology, Chinese Academy of Sciences, China
Mengyisong Zhao	University of Sheffield, UK
Wayne Xin Zhao	Renmin University of China, China
Jiang Zhou	Dublin City University, Ireland
Liting Zhou	Dublin City University, Ireland
Steven Zimmerman	University of Essex, UK
Justin Zobel	University of Melbourne, Australia
Lixin Zou	Tsinghua University, China

Guido Zuccon   University of Queensland, Australia

Additional Reviewers

Ashkan Alinejad
Evelin Amorim
Negar Arabzadeh
Dennis Aumiller
Mohammad Bahrani
Mehdi Ben Amor
Giovanni Maria Biancofiore
Ramraj Chandradevan
Qianli Chen
Dhivya Chinnappa
Isabel Coutinho
Washington Cunha
Xiang Dai
Marco de Gemmis
Alaa El-Ebshihy
Gloria Feher
Yasin Ghafourian
Wolfgang Gritz
Abul Hasan
Phuong Hoang
Eszter Iklodi
Andrea Iovine
Tania Jimenez
Pierre Jourlin

Anoop K.
Tuomas Ketola
Adam Kovacs
Zhao Liu
Daniele Malitesta
Cataldo Musto
Evelyn Navarrete
Zhan Qu
Saed Rezayi
Ratan Sebastian
Dawn Sepehr
Simra Shahid
Chen Shao
Mohammad Sharif
Stanley Simoes
Matthias Springstein
Ting Su
Wenyi Tay
Alberto Veneri
Chenyang Wang
Lorenz Wendlinger
Mengnong Xu
Shuzhou Yuan

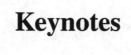

Keynotes

Personalization at Spotify

Mounia Lalmas

Spotify

Abstract. One of Spotify's missions is "to match fans and creators in a personal and relevant way". This talk will share some of the research work aimed at achieving this, from using machine learning to metric validation, and illustrated through examples within the context of Spotify's home and search. An important aspect will focus on illustrating that when aiming to personalize for both recommendation and search, it is important to consider the heterogeneity of both listener and content. One way to do this is to consider the following three angles when developing machine learning solutions for personalization: (1) Understanding user journey; (2) Optimizing for the right metric; and (3) Thinking about diversity.

On A Few Responsibilities of (IR) Researchers: Fairness, Awareness, and Sustainability

Tetsuya Sakai

Waseda University, Tokyo, Japan
tetsuyasakai@acm.org

Abstract. I would like to discuss with the audience a few keywords which I believe should be considered as foundation pillars of modern research practices, namely, *fairness*, *awareness*, and *sustainability*. Other important pillars such as *ethics* are beyond the scope of this keynote.

Fairness. By this I mean fairness in terms of exposure etc. for the items being ranked or recommended. As an example, I will describe the ongoing NTCIR-17 Fair Web Task, which is about ensuring group fairness of web search results.[1] More specifically, I will explain the model behind the Group Fairness and Relevance evaluation measure, which can handle ordinal groups (e.g. *high* h-index researchers vs. *medium* h-index researchers vs. others) as well as intersectional group fairness.

Awareness. What I mean by this word is that researchers should always try to see "both sides" and make informed decisions instead of just blindly accepting recommendations from a few particular researchers, even if they are great people. Conference PC chairs and journal editors should also be aware of both sides and provide appropriate guidance to authors and reviewers.[2]

Sustainability. From this year until 2025, Paul Thomas and I will be working as the first sustainability chairs of SIGIR. So I would like to discuss how IR researchers may want to minimise and/or compensate for the negative impact of our activities on earth and on society. As a related development, I will mention SIGIR-AP (Asia/Pacific), a regional SIGIR conference which will be launched this year. I will also solicit ideas from the audience for the IR community to go greener and beyond.

Keywords: Awareness · Diversity and inclusion · Fairness · Sustainability

---

[1] http://sakailab.com/fairweb1/.
[2] https://twitter.com/tetsuyasakai/status/1596298720316129280.

On A few Responsibilities of (IR) Researchers: Fairness, Awareness and Sustainability

2022 KAREN SPÄRCK JONES AWARD LECTURE Large Language Models for Question Answering: Challenges and Opportunities

William Yang Wang ⓘ

University of California, Santa Barbara, Santa Barbara, CA 93117, USA
william@cs.ucsb.edu

Abstract. A key goal for Artificial Intelligence is to design intelligent agents that can reason with heterogeneous representations and answer open-domain questions. The advances in large language models (LLMs) bring exciting opportunities to create disruptive technologies for question answering (QA). In this talk, I will demonstrate that major challenges for open-domain QA with LLMs include the capability to reason seamlessly between textual and tabular data, to understand and reason with numerical data, and to adapt to specialized domains. To do this, I will describe our recent work on teaching machines to reason in semi-structured tables and unstructured text data. More specifically, I will introduce (1) Open Question Answering over Tables and Text (OTT-QA), a new large-scale open-domain benchmark that combines information retrieval and language understanding for multihop reasoning over tabular and textual data; (2) FinQA and ConFinQA, two challenging benchmarks for exploring the chain of numerical reasoning in conversational finance question answering. I will also describe other exciting research directions in open-domain question answering.

Keywords: Large language models · Question answering · Reasoning

Contents – Part I

Full Papers

Self-supervised Contrastive BERT Fine-tuning for Fusion-Based
Reviewed-Item Retrieval 3
  Mohammad Mahdi Abdollah Pour, Parsa Farinneya, Armin Toroghi,
  Anton Korikov, Ali Pesaranghader, Touqir Sajed, Manasa Bharadwaj,
  Borislav Mavrin, and Scott Sanner

User Requirement Analysis for a Real-Time NLP-Based Open Information
Retrieval Meeting Assistant 18
  Benoît Alcaraz, Nina Hosseini-Kivanani, Amro Najjar,
  and Kerstin Bongard-Blanchy

Auditing Consumer- and Producer-Fairness in Graph Collaborative
Filtering 33
  Vito Walter Anelli, Yashar Deldjoo, Tommaso Di Noia,
  Daniele Malitesta, Vincenzo Paparella, and Claudio Pomo

Exploiting Graph Structured Cross-Domain Representation
for Multi-domain Recommendation 49
  Alejandro Ariza-Casabona, Bartlomiej Twardowski,
  and Tri Kurniawan Wijaya

Injecting the BM25 Score as Text Improves BERT-Based Re-rankers 66
  Arian Askari, Amin Abolghasemi, Gabriella Pasi, Wessel Kraaij,
  and Suzan Verberne

Quantifying Valence and Arousal in Text with Multilingual Pre-trained
Transformers 84
  Gonçalo Azevedo Mendes and Bruno Martins

A Knowledge Infusion Based Multitasking System for Sarcasm Detection
in Meme 101
  Dibyanayan Bandyopadhyay, Gitanjali Kumari, Asif Ekbal,
  Santanu Pal, Arindam Chatterjee, and Vinutha BN

Multilingual Detection of Check-Worthy Claims Using World Languages
and Adapter Fusion 118
  Ipek Baris Schlicht, Lucie Flek, and Paolo Rosso

Market-Aware Models for Efficient Cross-Market Recommendation 134
  Samarth Bhargav, Mohammad Aliannejadi, and Evangelos Kanoulas

TourismNLG: A Multi-lingual Generative Benchmark for the Tourism
Domain 150
  *Sahil Manoj Bhatt, Sahaj Agarwal, Omkar Gurjar, Manish Gupta,
  and Manish Shrivastava*

An Interpretable Knowledge Representation Framework for Natural
Language Processing with Cross-Domain Application 167
  Bimal Bhattarai, Ole-Christoffer Granmo, and Lei Jiao

Graph-Based Recommendation for Sparse and Heterogeneous User
Interactions 182
  *Simone Borg Bruun, Kacper Kenji Leśniak, Mirko Biasini,
  Vittorio Carmignani, Panagiotis Filianos, Christina Lioma,
  and Maria Maistro*

It's Just a Matter of Time: Detecting Depression with Time-Enriched
Multimodal Transformers 200
  Ana-Maria Bucur, Adrian Cosma, Paolo Rosso, and Liviu P. Dinu

Recommendation Algorithm Based on Deep Light Graph Convolution
Network in Knowledge Graph 216
  Xiaobin Chen and Nanfeng Xiao

Query Performance Prediction for Neural IR: Are We There Yet? 232
  *Guglielmo Faggioli, Thibault Formal, Stefano Marchesin,
  Stéphane Clinchant, Nicola Ferro, and Benjamin Piwowarski*

Item Graph Convolution Collaborative Filtering for Inductive
Recommendations 249
  *Edoardo D'Amico, Khalil Muhammad, Elias Tragos, Barry Smyth,
  Neil Hurley, and Aonghus Lawlor*

CoLISA: Inner Interaction via Contrastive Learning for Multi-choice
Reading Comprehension 264
  Mengxing Dong, Bowei Zou, Yanling Li, and Yu Hong

Viewpoint Diversity in Search Results 279
  *Tim Draws, Nirmal Roy, Oana Inel, Alisa Rieger, Rishav Hada,
  Mehmet Orcun Yalcin, Benjamin Timmermans, and Nava Tintarev*

COILCR: Efficient Semantic Matching in Contextualized Exact Match
Retrieval 298
   Zhen Fan, Luyu Gao, Rohan Jha, and Jamie Callan

Bootstrapped nDCG Estimation in the Presence of Unjudged Documents 313
   Maik Fröbe, Lukas Gienapp, Martin Potthast, and Matthias Hagen

Predicting the Listening Contexts of Music Playlists Using Knowledge
Graphs 330
   Giovanni Gabbolini and Derek Bridge

Keyword Embeddings for Query Suggestion 346
   Jorge Gabín, M. Eduardo Ares, and Javier Parapar

Domain-Driven and Discourse-Guided Scientific Summarisation 361
   Tomas Goldsack, Zhihao Zhang, Chenghua Lin, and Carolina Scarton

Injecting Temporal-Aware Knowledge in Historical Named Entity
Recognition 377
   Carlos-Emiliano González-Gallardo, Emanuela Boros,
   Edward Giamphy, Ahmed Hamdi, José G. Moreno, and Antoine Doucet

A Mask-Based Logic Rules Dissemination Method for Sentiment
Classifiers 394
   Shashank Gupta, Mohamed Reda Bouadjenek, and Antonio Robles-Kelly

Contrasting Neural Click Models and Pointwise IPS Rankers 409
   Philipp Hager, Maarten de Rijke, and Onno Zoeter

Sentence Retrieval for Open-Ended Dialogue Using Dual Contextual
Modeling 426
   Itay Harel, Hagai Taitelbaum, Idan Szpektor, and Oren Kurland

Temporal Natural Language Inference: Evidence-Based Evaluation
of Temporal Text Validity 441
   Taishi Hosokawa, Adam Jatowt, and Kazunari Sugiyama

Theoretical Analysis on the Efficiency of Interleaved Comparisons 459
   Kojiro Iizuka, Hajime Morita, and Makoto P. Kato

Intention-Aware Neural Networks for Question Paraphrase Identification 474
   Zhiling Jin, Yu Hong, Rui Peng, Jianmin Yao, and Guodong Zhou

Automatic and Analytical Field Weighting for Structured Document
Retrieval 489
   Tuomas Ketola and Thomas Roelleke

An Experimental Study on Pretraining Transformers from Scratch for IR 504
   Carlos Lassance, Hervé Dejean, and Stéphane Clinchant

Neural Approaches to Multilingual Information Retrieval 521
   Dawn Lawrie, Eugene Yang, Douglas W. Oard, and James Mayfield

CoSPLADE: Contextualizing SPLADE for Conversational Information
Retrieval 537
   Nam Hai Le, Thomas Gerald, Thibault Formal, Jian-Yun Nie,
   Benjamin Piwowarski, and Laure Soulier

SR-CoMbEr: Heterogeneous Network Embedding Using Community
Multi-view Enhanced Graph Convolutional Network for Automating
Systematic Reviews 553
   Eric W. Lee and Joyce C. Ho

Multimodal Inverse Cloze Task for Knowledge-Based Visual Question
Answering 569
   Paul Lerner, Olivier Ferret, and Camille Guinaudeau

A Transformer-Based Framework for POI-Level Social Post Geolocation 588
   Menglin Li, Kwan Hui Lim, Teng Guo, and Junhua Liu

Document-Level Relation Extraction with Distance-Dependent Bias
Network and Neighbors Enhanced Loss 605
   Hao Liang and Qifeng Zhou

Investigating Conversational Agent Action in Legal Case Retrieval 622
   Bulou Liu, Yiran Hu, Yueyue Wu, Yiqun Liu, Fan Zhang, Chenliang Li,
   Min Zhang, Shaoping Ma, and Weixing Shen

MS-Shift: An Analysis of MS MARCO Distribution Shifts on Neural
Retrieval 636
   Simon Lupart, Thibault Formal, and Stéphane Clinchant

Listwise Explanations for Ranking Models Using Multiple Explainers 653
   Lijun Lyu and Avishek Anand

Improving Video Retrieval Using Multilingual Knowledge Transfer 669
   Avinash Madasu, Estelle Aflalo, Gabriela Ben Melech Stan,
   Shao-Yen Tseng, Gedas Bertasius, and Vasudev Lal

Service Is Good, Very Good or Excellent? Towards Aspect Based
Sentiment Intensity Analysis 685
    Mamta and Asif Ekbal

Effective Hierarchical Information Threading Using Network Community
Detection 701
    Hitarth Narvala, Graham McDonald, and Iadh Ounis

HADA: A Graph-Based Amalgamation Framework in Image-text Retrieval 717
    Manh-Duy Nguyen, Binh T. Nguyen, and Cathal Gurrin

Author Index 733

Contents – Part II

Full Papers

Extractive Summarization of Financial Earnings Call Transcripts: Or:
When GREP Beat BERT 3
   Tim Nugent, George Gkotsis, and Jochen L. Leidner

Parameter-Efficient Sparse Retrievers and Rerankers Using Adapters 16
   Vaishali Pal, Carlos Lassance, Hervé Déjean, and Stéphane Clinchant

Feature Differentiation and Fusion for Semantic Text Matching 32
   Rui Peng, Yu Hong, Zhiling Jin, Jianmin Yao, and Guodong Zhou

Multivariate Powered Dirichlet-Hawkes Process 47
   Gaël Poux-Médard, Julien Velcin, and Sabine Loudcher

Fragmented Visual Attention in Web Browsing: Weibull Analysis of Item
Visit Times 62
   Aini Putkonen, Aurélien Nioche, Markku Laine, Crista Kuuramo,
   and Antti Oulasvirta

Topic-Enhanced Personalized Retrieval-Based Chatbot 79
   Hongjin Qian and Zhicheng Dou

Improving the Generalizability of the Dense Passage Retriever Using
Generated Datasets 94
   Thilina C. Rajapakse and Maarten de Rijke

SegmentCodeList: Unsupervised Representation Learning for Human
Skeleton Data Retrieval 110
   Jan Sedmidubsky, Fabio Carrara, and Giuseppe Amato

Knowing What and How: A Multi-modal Aspect-Based Framework
for Complaint Detection 125
   Apoorva Singh, Vivek Gangwar, Shubham Sharma, and Sriparna Saha

What Is Your Cause for Concern? Towards Interpretable Complaint Cause
Analysis 141
   Apoorva Singh, Prince Jha, Rohan Bhatia, and Sriparna Saha

DeCoDE: Detection of Cognitive Distortion and Emotion Cause
Extraction in Clinical Conversations 156
  Gopendra Vikram Singh, Soumitra Ghosh, Asif Ekbal,
  and Pushpak Bhattacharyya

Domain-Aligned Data Augmentation for Low-Resource and Imbalanced
Text Classification 172
  Nikolaos Stylianou, Despoina Chatzakou, Theodora Tsikrika,
  Stefanos Vrochidis, and Ioannis Kompatsiaris

Privacy-Preserving Fair Item Ranking 188
  Jia Ao Sun, Sikha Pentyala, Martine De Cock, and Golnoosh Farnadi

Multimodal Geolocation Estimation of News Photos 204
  Golsa Tahmasebzadeh, Sherzod Hakimov, Ralph Ewerth,
  and Eric Müller-Budack

Topics in Contextualised Attention Embeddings 221
  Mozhgan Talebpour, Alba García Seco de Herrera, and Shoaib Jameel

New Metrics to Encourage Innovation and Diversity in Information
Retrieval Approaches 239
  Mehmet Deniz Türkmen, Matthew Lease, and Mucahid Kutlu

Probing BERT for Ranking Abilities 255
  Jonas Wallat, Fabian Beringer, Abhijit Anand, and Avishek Anand

Clustering of Bandit with Frequency-Dependent Information Sharing 274
  Shen Yang, Qifeng Zhou, and Qing Wang

Graph Contrastive Learning with Positional Representation
for Recommendation 288
  Zixuan Yi, Iadh Ounis, and Craig Macdonald

Domain Adaptation for Anomaly Detection on Heterogeneous Graphs
in E-Commerce 304
  Li Zheng, Zhao Li, Jun Gao, Zhenpeng Li, Jia Wu, and Chuan Zhou

Short Papers

Improving Neural Topic Models with Wasserstein Knowledge Distillation 321
  Suman Adhya and Debarshi Kumar Sanyal

Towards Effective Paraphrasing for Information Disguise 331
   Anmol Agarwal, Shrey Gupta, Vamshi Bonagiri, Manas Gaur,
   Joseph Reagle, and Ponnurangam Kumaraguru

Generating Topic Pages for Scientific Concepts Using Scientific
Publications 341
   Hosein Azarbonyad, Zubair Afzal, and George Tsatsaronis

De-biasing Relevance Judgements for Fair Ranking 350
   Amin Bigdeli, Negar Arabzadeh, Shirin Seyedsalehi, Bhaskar Mitra,
   Morteza Zihayat, and Ebrahim Bagheri

A Study of Term-Topic Embeddings for Ranking 359
   Lila Boualili and Andrew Yates

Topic Refinement in Multi-level Hate Speech Detection 367
   Tom Bourgeade, Patricia Chiril, Farah Benamara,
   and Véronique Moriceau

Is Cross-Modal Information Retrieval Possible Without Training? 377
   Hyunjin Choi, Hyunjae Lee, Seongho Joe, and Youngjune Gwon

Adversarial Adaptation for French Named Entity Recognition 386
   Arjun Choudhry, Inder Khatri, Pankaj Gupta, Aaryan Gupta,
   Maxime Nicol, Marie-Jean Meurs, and Dinesh Kumar Vishwakarma

Exploring Fake News Detection with Heterogeneous Social Media
Context Graphs 396
   Gregor Donabauer and Udo Kruschwitz

Justifying Multi-label Text Classifications for Healthcare Applications 406
   João Figueira, Gonçalo M. Correia, Michalina Strzyz,
   and Afonso Mendes

Doc2Query–: When Less is More 414
   Mitko Gospodinov, Sean MacAvaney, and Craig Macdonald

Towards Quantifying the Privacy of Redacted Text 423
   Vaibhav Gusain and Douglas Leith

Detecting Stance of Authorities Towards Rumors in Arabic Tweets:
A Preliminary Study 430
   Fatima Haouari and Tamer Elsayed

Leveraging Comment Retrieval for Code Summarization 439
   Shifu Hou, Lingwei Chen, Mingxuan Ju, and Yanfang Ye

CPR: Cross-Domain Preference Ranking with User Transformation 448
   Yu-Ting Huang, Hsien-Hao Chen, Tung-Lin Wu, Chia-Yu Yeh,
   Jing-Kai Lou, Ming-Feng Tsai, and Chuan-Ju Wang

ColBERT-FairPRF: Towards Fair Pseudo-Relevance Feedback in Dense
Retrieval 457
   Thomas Jaenich, Graham McDonald, and Iadh Ounis

C^2LIR: Continual Cross-Lingual Transfer for Low-Resource Information
Retrieval 466
   Jaeseong Lee, Dohyeon Lee, Jongho Kim, and Seung-won Hwang

Joint Extraction and Classification of Danish Competences for Job
Matching 475
   Qiuchi Li and Christina Lioma

A Study on FGSM Adversarial Training for Neural Retrieval 484
   Simon Lupart and Stéphane Clinchant

Dialogue-to-Video Retrieval 493
   Chenyang Lyu, Manh-Duy Nguyen, Van-Tu Ninh, Liting Zhou,
   Cathal Gurrin, and Jennifer Foster

Time-Dependent Next-Basket Recommendations 502
   Sergey Naumov, Marina Ananyeva, Oleg Lashinin, Sergey Kolesnikov,
   and Dmitry I. Ignatov

Investigating the Impact of Query Representation on Medical Information
Retrieval 512
   Georgios Peikos, Daria Alexander, Gabriella Pasi, and Arjen P. de Vries

Where a Little Change Makes a Big Difference: A Preliminary Exploration
of Children's Queries 522
   Maria Soledad Pera, Emiliana Murgia, Monica Landoni,
   Theo Huibers, and Mohammad Aliannejadi

Visconde: Multi-document QA with GPT-3 and Neural Reranking 534
   Jayr Pereira, Robson Fidalgo, Roberto Lotufo, and Rodrigo Nogueira

Towards Detecting Interesting Ideas Expressed in Text 544
   Bela Pfahl and Adam Jatowt

Towards Linguistically Informed Multi-objective Transformer Pre-training
for Natural Language Inference 553
    Maren Pielka, Svetlana Schmidt, Lisa Pucknat, and Rafet Sifa

Dirichlet-Survival Process: Scalable Inference of Topic-Dependent
Diffusion Networks 562
    Gaël Poux-Médard, Julien Velcin, and Sabine Loudcher

Consumer Health Question Answering Using Off-the-Shelf Components 571
    Alexander Pugachev, Ekaterina Artemova, Alexander Bondarenko,
    and Pavel Braslavski

MOO-CMDS+NER: Named Entity Recognition-Based Extractive
Comment-Oriented Multi-document Summarization 580
    Vishal Singh Roha, Naveen Saini, Sriparna Saha, and Jose G. Moreno

Don't Raise Your Voice, Improve Your Argument: Learning to Retrieve
Convincing Arguments 589
    Sara Salamat, Negar Arabzadeh, Amin Bigdeli, Shirin Seyedsalehi,
    Morteza Zihayat, and Ebrahim Bagheri

Learning Query-Space Document Representations for High-Recall
Retrieval 599
    Sara Salamat, Negar Arabzadeh, Fattane Zarrinkalam,
    Morteza Zihayat, and Ebrahim Bagheri

Investigating Conversational Search Behavior for Domain Exploration 608
    Phillip Schneider, Anum Afzal, Juraj Vladika, Daniel Braun,
    and Florian Matthes

Evaluating Humorous Response Generation to Playful Shopping Requests 617
    Natalie Shapira, Oren Kalinsky, Alex Libov, Chen Shani,
    and Sofia Tolmach

Joint Span Segmentation and Rhetorical Role Labeling with Data
Augmentation for Legal Documents 627
    T. Y. S. S Santosh, Philipp Bock, and Matthias Grabmair

Trigger or not Trigger: Dynamic Thresholding for Few Shot Event
Detection 637
    Aboubacar Tuo, Romaric Besançon, Olivier Ferret, and Julien Tourille

The Impact of a Popularity Punishing Hyperparameter on ItemKNN
Recommendation Performance 646
    Robin Verachtert, Jeroen Craps, Lien Michiels, and Bart Goethals

Neural Ad-Hoc Retrieval Meets Open Information Extraction 655
  Duc-Thuan Vo, Fattane Zarrinkalam, Ba Pham, Negar Arabzadeh,
  Sara Salamat, and Ebrahim Bagheri

Augmenting Graph Convolutional Networks with Textual Data
for Recommendations 664
  Sergey Volokhin, Marcus D. Collins, Oleg Rokhlenko,
  and Eugene Agichtein

Utilising Twitter Metadata for Hate Classification 676
  Oliver Warke, Joemon M. Jose, and Jan Breitsohl

Evolution of Filter Bubbles and Polarization in News Recommendation 685
  Han Zhang, Ziwei Zhu, and James Caverlee

Capturing Cross-Platform Interaction for Identifying Coordinated
Accounts of Misinformation Campaigns 694
  Yizhou Zhang, Karishma Sharma, and Yan Liu

Author Index 703

Contents – Part III

Reproducibility Papers

Knowledge is Power, Understanding is Impact: Utility and Beyond Goals,
Explanation Quality, and Fairness in Path Reasoning Recommendation   3
    Giacomo Balloccu, Ludovico Boratto, Christian Cancedda,
    Gianni Fenu, and Mirko Marras

Stat-Weight: Improving the Estimator of Interleaved Methods Outcomes
with Statistical Hypothesis Testing   20
    Alessandro Benedetti and Anna Ruggero

A Reproducibility Study of Question Retrieval for Clarifying Questions   35
    Sebastian Cross, Guido Zuccon, and Ahmed Mourad

The Impact of Cross-Lingual Adjustment of Contextual Word
Representations on Zero-Shot Transfer   51
    Pavel Efimov, Leonid Boytsov, Elena Arslanova, and Pavel Braslavski

Scene-Centric vs. Object-Centric Image-Text Cross-Modal Retrieval:
A Reproducibility Study   68
    Mariya Hendriksen, Svitlana Vakulenko, Ernst Kuiper,
    and Maarten de Rijke

Index-Based Batch Query Processing Revisited   86
    Joel Mackenzie and Alistair Moffat

A Unified Framework for Learned Sparse Retrieval 101
    Thong Nguyen, Sean MacAvaney, and Andrew Yates

Entity Embeddings for Entity Ranking: A Replicability Study 117
    Pooja Oza and Laura Dietz

Do the Findings of Document and Passage Retrieval Generalize
to the Retrieval of Responses for Dialogues? 132
    Gustavo Penha and Claudia Hauff

PyGaggle: A Gaggle of Resources for Open-Domain Question Answering 148
    Ronak Pradeep, Haonan Chen, Lingwei Gu, Manveer Singh Tamber,
    and Jimmy Lin

Pre-processing Matters! Improved Wikipedia Corpora for Open-Domain
Question Answering 163
    Manveer Singh Tamber, Ronak Pradeep, and Jimmy Lin

From Baseline to Top Performer: A Reproducibility Study of Approaches
at the TREC 2021 Conversational Assistance Track 177
    Weronika Lajewska and Krisztian Balog

Demonstration Papers

Exploring Tabular Data Through Networks 195
    Aleksandar Bobic, Jean-Marie Le Goff, and Christian Gütl

InfEval: Application for Object Detection Analysis 201
    Kirill Bogomasov, Tim Geuer, and Stefan Conrad

The System for Efficient Indexing and Search in the Large Archives
of Scanned Historical Documents 206
    Martin Bulín, Jan Švec, and Pavel Ircing

Public News Archive: A Searchable Sub-archive to Portuguese Past News
Articles 211
    Ricardo Campos, Diogo Correia, and Adam Jatowt

TweetStream2Story: Narrative Extraction from Tweets in Real Time 217
    Mafalda Castro, Alípio Jorge, and Ricardo Campos

SimpleRad: Patient-Friendly Dutch Radiology Reports 224
    Koen Dercksen, Arjen P. de Vries, and Bram van Ginneken

Automated Extraction of Fine-Grained Standardized Product Information
from Unstructured Multilingual Web Data 230
    Alexander Flick, Sebastian Jäger, Ivana Trajanovska,
    and Felix Biessmann

Continuous Integration for Reproducible Shared Tasks with TIRA.io 236
    Maik Fröbe, Matti Wiegmann, Nikolay Kolyada, Bastian Grahm,
    Theresa Elstner, Frank Loebe, Matthias Hagen, Benno Stein,
    and Martin Potthast

Dynamic Exploratory Search for the Information Retrieval Anthology 242
    Tim Gollub, Jason Brockmeyer, Benno Stein, and Martin Potthast

Text2Storyline: Generating Enriched Storylines from Text 248
    Francisco Gonçalves, Ricardo Campos, and Alípio Jorge

Uptrendz: API-Centric Real-Time Recommendations in Multi-domain
Settings 255
    Emanuel Lacic, Tomislav Duricic, Leon Fadljevic, Dieter Theiler,
    and Dominik Kowald

Clustering Without Knowing How To: Application and Evaluation 262
    Daniil Likhobaba, Daniil Fedulov, and Dmitry Ustalov

Enticing Local Governments to Produce FAIR Freedom of Information
Act Dossiers 269
    Maarten Marx, Maik Larooij, Filipp Perasedillo, and Jaap Kamps

Which Country Is This? Automatic Country Ranking of Street View Photos 275
    Tim Menzner, Florian Mittag, and Jochen L. Leidner

Automatic Videography Generation from Audio Tracks 281
    Debasis Ganguly, Andrew Parker, and Stergious Aji

Ablesbarkeitsmesser: A System for Assessing the Readability of German
Text 288
    Florian Pickelmann, Michael Färber, and Adam Jatowt

FACADE: Fake Articles Classification and Decision Explanation 294
    Erasmo Purificato, Saijal Shahania, Marcus Thiel,
    and Ernesto William De Luca

PsyProf: A Platform for Assisted Screening of Depression in Social Media 300
    Anxo Pérez, Paloma Piot-Pérez-Abadín, Javier Parapar,
    and Álvaro Barreiro

SOPalign: A Tool for Automatic Estimation of Compliance with Medical
Guidelines 307
    Luke van Leijenhorst, Arjen P. de Vries, Thera Habben Jansen,
    and Heiman Wertheim

Tutorials

Understanding and Mitigating Gender Bias in Information Retrieval
Systems 315
    Amin Bigdeli, Negar Arabzadeh, Shirin Seyedsalehi, Morteza Zihayat,
    and Ebrahim Bagheri

ECIR 23 Tutorial: Neuro-Symbolic Approaches for Information Retrieval 324
    Laura Dietz, Hannah Bast, Shubham Chatterjee, Jeff Dalton,
    Edgar Meij, and Arjen de Vries

Legal IR and NLP: The History, Challenges, and State-of-the-Art 331
  Debasis Ganguly, Jack G. Conrad, Kripabandhu Ghosh,
  Saptarshi Ghosh, Pawan Goyal, Paheli Bhattacharya,
  Shubham Kumar Nigam, and Shounak Paul

Deep Learning Methods for Query Auto Completion 341
  Manish Gupta, Meghana Joshi, and Puneet Agrawal

Trends and Overview: The Potential of Conversational Agents in Digital
Health 349
  Tulika Saha, Abhishek Tiwari, and Sriparna Saha

Crowdsourcing for Information Retrieval 357
  Dmitry Ustalov, Alisa Smirnova, Natalia Fedorova,
  and Nikita Pavlichenko

Uncertainty Quantification for Text Classification 362
  Dell Zhang, Murat Sensoy, Masoud Makrehchi,
  and Bilyana Taneva-Popova

Workshops

Fourth International Workshop on Algorithmic Bias in Search
and Recommendation (Bias 2023) 373
  Ludovico Boratto, Stefano Faralli, Mirko Marras, and Giovanni Stilo

The 6th International Workshop on Narrative Extraction from Texts:
Text2Story 2023 377
  Ricardo Campos, Alípio Jorge, Adam Jatowt, Sumit Bhatia,
  and Marina Litvak

2nd Workshop on Augmented Intelligence in Technology-Assisted Review
Systems (ALTARS) 384
  Giorgio Maria Di Nunzio, Evangelos Kanoulas, and Prasenjit Majumder

QPP++ 2023: Query-Performance Prediction and Its Evaluation in New
Tasks 388
  Guglielmo Faggioli, Nicola Ferro, Josiane Mothe, and Fiana Raiber

Bibliometric-Enhanced Information Retrieval: 13th International BIR
Workshop (BIR 2023) 392
  Ingo Frommholz, Philipp Mayr, Guillaume Cabanac, and Suzan Verberne

Geographic Information Extraction from Texts (GeoExT) 398
  Xuke Hu, Yingjie Hu, Bernd Resch, and Jens Kersten

ROMCIR 2023: Overview of the 3rd Workshop on Reducing Online
Misinformation Through Credible Information Retrieval 405
    Marinella Petrocchi and Marco Viviani

ECIR 2023 Workshop: Legal Information Retrieval 412
    Suzan Verberne, Evangelos Kanoulas, Gineke Wiggers, Florina Piroi,
    and Arjen P. de Vries

Doctoral Consoritum

Building Safe and Reliable AI Systems for Safety Critical Tasks
with Vision-Language Processing 423
    Shuang Ao

Text Information Retrieval in Tetun 429
    Gabriel de Jesus

Identifying and Representing Knowledge Delta in Scientific Literature 436
    Alaa El-Ebshihy

Investigation of Bias in Web Search Queries 443
    Fabian Haak

Monitoring Online Discussions and Responses to Support the Identification
of Misinformation 450
    Xin Yu Liew

User Privacy in Recommender Systems 456
    Peter Müllner

Conversational Search for Multimedia Archives 462
    Anastasia Potyagalova

Disinformation Detection: Knowledge Infusion with Transfer Learning
and Visualizations 468
    Mina Schütz

A Comprehensive Overview of Consumer Conflicts on Social Media 476
    Oliver Warke

Designing Useful Conversational Interfaces for Information Retrieval
in Career Decision-Making Support 482
    Marianne Wilson

CLEF Lab Descriptions

iDPP@CLEF 2023: The Intelligent Disease Progression Prediction
Challenge 491
*Helena Aidos, Roberto Bergamaschi, Paola Cavalla, Adriano Chiò,
Arianna Dagliati, Barbara Di Camillo, Mamede Alves de Carvalho,
Nicola Ferro, Piero Fariselli, Jose Manuel García Dominguez,
Sara C. Madeira, and Eleonora Tavazzi*

LongEval: Longitudinal Evaluation of Model Performance at CLEF 2023 499
*Rabab Alkhalifa, Iman Bilal, Hsuvas Borkakoty,
Jose Camacho-Collados, Romain Deveaud, Alaa El-Ebshihy,
Luis Espinosa-Anke, Gabriela Gonzalez-Saez, Petra Galuščáková,
Lorraine Goeuriot, Elena Kochkina, Maria Liakata, Daniel Loureiro,
Harish Tayyar Madabushi, Philippe Mulhem, Florina Piroi,
Martin Popel, Christophe Servan, and Arkaitz Zubiaga*

The CLEF-2023 CheckThat! Lab: Checkworthiness, Subjectivity, Political
Bias, Factuality, and Authority 506
*Alberto Barrón-Cedeño, Firoj Alam, Tommaso Caselli,
Giovanni Da San Martino, Tamer Elsayed, Andrea Galassi,
Fatima Haouari, Federico Ruggeri, Julia Maria Struß,
Rabindra Nath Nandi, Gullal S. Cheema, Dilshod Azizov,
and Preslav Nakov*

Overview of PAN 2023: Authorship Verification, Multi-author Writing
Style Analysis, Profiling Cryptocurrency Influencers, and Trigger
Detection: Extended Abstract 518
*Janek Bevendorff, Mara Chinea-Ríos, Marc Franco-Salvador,
Annina Heini, Erik Körner, Krzysztof Kredens, Maximilian Mayerl,
Piotr Pęzik, Martin Potthast, Francisco Rangel, Paolo Rosso,
Efstathios Stamatatos, Benno Stein, Matti Wiegmann,
Magdalena Wolska, and Eva Zangerle*

Overview of Touché 2023: Argument and Causal Retrieval: Extended
Abstract 527
*Alexander Bondarenko, Maik Fröbe, Johannes Kiesel,
Ferdinand Schlatt, Valentin Barriere, Brian Ravenet, Léo Hemamou,
Simon Luck, Jan Heinrich Reimer, Benno Stein, Martin Potthast,
and Matthias Hagen*

CLEF 2023 SimpleText Track: What Happens if General Users Search
Scientific Texts? 536
*Liana Ermakova, Eric SanJuan, Stéphane Huet, Olivier Augereau,
Hosein Azarbonyad, and Jaap Kamps*

Science for Fun: The CLEF 2023 JOKER Track on Automatic Wordplay
Analysis 546
Liana Ermakova, Tristan Miller, Anne-Gwenn Bosser,
Victor Manuel Palma Preciado, Grigori Sidorov, and Adam Jatowt

ImageCLEF 2023 Highlight: Multimedia Retrieval in Medical, Social
Media and Content Recommendation Applications 557
Bogdan Ionescu, Henning Müller, Ana Maria Drăgulinescu,
Adrian Popescu, Ahmad Idrissi-Yaghir, Alba García Seco de Herrera,
Alexandra Andrei, Alexandru Stan, Andrea M. Storås,
Asma Ben Abacha, Christoph M. Friedrich, George Ioannidis,
Griffin Adams, Henning Schäfer, Hugo Manguinhas, Ihar Filipovich,
Ioan Coman, Jérôme Deshayes, Johanna Schöler, Johannes Rückert,
Liviu-Daniel Ştefan, Louise Bloch, Meliha Yetisgen, Michael A. Riegler,
Mihai Dogariu, Mihai Gabriel Constantin, Neal Snider,
Nikolaos Papachrysos, Pål Halvorsen, Raphael Brüngel,
Serge Kozlovski, Steven Hicks, Thomas de Lange, Vajira Thambawita,
Vassili Kovalev, and Wen-Wai Yim

LifeCLEF 2023 Teaser: Species Identification and Prediction Challenges 568
Alexis Joly, Hervé Goëau, Stefan Kahl, Lukáš Picek,
Christophe Botella, Diego Marcos, Milan Šulc, Marek Hrúz,
Titouan Lorieul, Sara Si Moussi, Maximilien Servajean,
Benjamin Kellenberger, Elijah Cole, Andrew Durso, Hervé Glotin,
Robert Planqué, Willem-Pier Vellinga, Holger Klinck, Tom Denton,
Ivan Eggel, Pierre Bonnet, and Henning Müller

BioASQ at CLEF2023: The Eleventh Edition of the Large-Scale
Biomedical Semantic Indexing and Question Answering Challenge 577
Anastasios Nentidis, Anastasia Krithara, Georgios Paliouras,
Eulalia Farre-Maduell, Sulvador Lima-Lopez, and Martin Krallinger

eRisk 2023: Depression, Pathological Gambling, and Eating Disorder
Challenges 585
Javier Parapar, Patricia Martín-Rodilla, David E. Losada,
and Fabio Crestani

Overview of EXIST 2023: sEXism Identification in Social NeTworks 593
Laura Plaza, Jorge Carrillo-de-Albornoz, Roser Morante,
Enrique Amigó, Julio Gonzalo, Damiano Spina, and Paolo Rosso

DocILE 2023 Teaser: Document Information Localization and Extraction 600
Štěpán Šimsa, Milan Šulc, Matyáš Skalický, Yash Patel, and Ahmed Hamdi

Author Index 609

Full Papers

Self-supervised Contrastive BERT Fine-tuning for Fusion-Based Reviewed-Item Retrieval

Mohammad Mahdi Abdollah Pour[1]([✉]), Parsa Farinneya[1], Armin Toroghi[1],
Anton Korikov[1], Ali Pesaranghader[2], Touqir Sajed[2], Manasa Bharadwaj[2],
Borislav Mavrin[2], and Scott Sanner[1]

[1] University of Toronto, Toronto, Canada
{m.abdollahpour,parsa.farinneya,armin.toroghi,
anton.korikov}@mail.utoronto.ca, ssanner@mie.utoronto.ca
[2] LG Electronics, Toronto AI Lab, Toronto, Canada
{ali.pesaranghader,touqir.sajed,manasa.bharadwaj,borislav.mavrin}@lge.com

Abstract. As natural language interfaces enable users to express increasingly complex natural language queries, there is a parallel explosion of user review content that can allow users to better find items such as restaurants, books, or movies that match these expressive queries. While Neural Information Retrieval (IR) methods have provided state-of-the-art results for matching queries to documents, they have not been extended to the task of Reviewed-Item Retrieval (RIR), where query-review scores must be aggregated (or fused) into item-level scores for ranking. In the absence of labeled RIR datasets, we extend Neural IR methodology to RIR by leveraging self-supervised methods for contrastive learning of BERT embeddings for both queries and reviews. Specifically, contrastive learning requires a choice of positive and negative samples, where the unique two-level structure of our item-review data combined with meta-data affords us a rich structure for the selection of these samples. For contrastive learning in a Late Fusion scenario (where we aggregate query-review scores into item-level scores), we investigate the use of positive review samples from the same item and/or with the same rating, selection of hard positive samples by choosing the least similar reviews from the same anchor item, and selection of hard negative samples by choosing the most similar reviews from different items. We also explore anchor sub-sampling and augmenting with meta-data. For a more end-to-end Early Fusion approach, we introduce contrastive item embedding learning to fuse reviews into single item embeddings. Experimental results show that Late Fusion contrastive learning for Neural RIR outperforms all other contrastive IR configurations, Neural IR, and sparse retrieval baselines, thus demonstrating the power of exploiting the two-level structure in Neural RIR approaches as well as the importance of preserving the nuance of individual review content via Late Fusion methods.

Keywords: Neural information retrieval · Natural language processing · Contrastive learning · Language models

---

M. M. Abdollah Porur and P. Farinneya—Equal contribution

J. Kamps et al. (Eds.): ECIR 2023, LNCS 13980, pp. 3–17, 2023.
https://doi.org/10.1007/978-3-031-28244-7_1

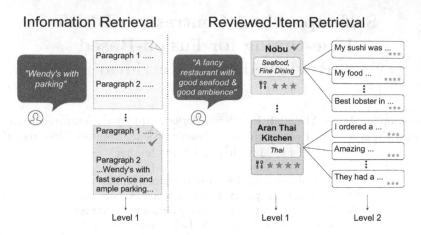

Fig. 1. Structural difference between IR (left) and RIR (right). In RIR, items have reviews covering different aspects. In contrast, documents in the IR task do not have this **two-level item-review structure**.

1   Introduction

The rise of expressive natural language interfaces coupled with the prevalence of user-generated review content provide novel opportunities for query-based retrieval of reviewed items. While Neural Information Retrieval (IR) methods have provided state-of-the-art results for query-based document retrieval ([16]), these methods do not directly extend to review-based data that provides a unique two-level structure in which an item has several reviews along with ratings and other meta-data. Due to differences between standard IR document retrieval ([21]) and the task of retrieving reviewed-items indirectly through their reviews, we coin the term Reviewed-Item Retrieval (RIR) for this task. Figure 1 illustrates the structural difference between IR and RIR. In RIR, each item includes a set of reviews and each review expresses different perspectives about that item.

Unlike standard Neural IR, which produces query-document scores for document ranking, RIR requires query-item scores to be obtained indirectly using review text. This can be done both via Late Fusion (LF) methods that simply aggregate query-review scores into item scores or Early Fusion (EF) methods that build an item representation for direct query-item scoring. Given the absence of labeled data for RIR, we explore self-supervised contrastive learning methods for Neural RIR that exploit the two-level structure of the data in both the LF and EF frameworks with the following contributions:

1. For LF, we propose different positive and negative sampling methods that exploit the two-level structure of RIR data as well as data augmentation methods for contrastive fine-tuning of a BERT [4] Language Model (LM).
2. For EF, we propose end-to-end contrastive learning of item embeddings to fuse reviews of each item into a single embedding vector.
3. We experimentally show that LF-based Neural RIR methods outperform EF and other contrastive IR methods, Neural IR, and sparse retrieval baselines.

Overall, these results demonstrate the power of exploiting the two-level structure in Neural RIR approaches. By showing the superiority of LF over EF, we also demonstrate the benefit of aggregating scores for individual reviews *after* similarity with a query is computed, thus preserving the nuances of each review during scoring. Ultimately, this work develops a foundation for the application and extension of Neural IR techniques to RIR.

2 Background

2.1 IR

Given a set of documents $\mathcal{D}$ and a query $q \in \mathcal{Q}$, an IR task $\mathcal{IR}\langle \mathcal{D}, q \rangle$ is to assign a similarity score $S_{q,d} \in \mathbb{R}$ between the query and each document $d \in \mathcal{D}$ and return a list of top-scoring documents.

Before the advent of Neural IR methods, most methods depended on sparse models such as TF-IDF ([20]) and its variants such as BM25 [19], which heavily relied on exact term matches and measures of term informativeness. However, the need for exact term matches, the availability of large datasets, and increases in computational power led to a shift from traditional models to deep neural networks for document ranking [8,10]. Recently, Nogueira and Cho [18] have initiated a line of research on Neural IR by fine-tuning BERT [4] for ranking candidate documents with respect to queries.

Recent works have substantially extended BERT-based Neural IR methods. CoCondenser [6] is a resource-efficient model with excellent performance on the MS-MARCO benchmark ([17]). This model is based on Condenser ([5]), which alters the BERT architecture to emphasize more attention to the classification embedding (i.e., the so-called "[CLS]" output of BERT). Contriever ([12]) is a state-of-the-art self-supervised contrastive Neural IR method that does not rely on query-document annotations to train.

Although our work on Neural RIR is influenced by Neural IR, the structure of IR and RIR differ as illustrated by Fig. 1. This requires methods specifically for working with the two-level structure of data in both training and inference in the RIR task. To the best of our knowledge, these methods have not been explored in the literature.

2.2 Fusion

Information retrieval from two-level data structures has previously been studied by Zhang and Balog [27], though they did not study *neural* techniques, which are the focus of our work. Specifically, Zhang and Balog define the *object retrieval* problem, where (high-level) objects are described by multiple (low-level) documents and the task is to retrieve objects given a query. This task requires *fusing* information from the document level to the object level, which can be done before query scoring, called Early Fusion, or after query scoring, called Late Fusion. Our contributions include extending Early and Late Fusion methods to self-supervised contrastive Neural IR.

Formally, let $i \in \mathcal{I}$ be an object described by a set of documents $\mathcal{D}_i \subset \mathcal{D}$ and let $r_{i,k}$ denote the k'th document describing object i. Given a query $q \in \mathcal{Q}$, fusion is used to aggregate document information to the object level and obtain a query-object similarity score $S_{q,i} \in \mathbb{R}$.

Late Fusion. In Late Fusion, similarity scores are computed between documents and a query and then aggregated into a query-object score. Given an embedding space $\mathbb{R}^m$, let $g : \mathcal{D} \cup \mathcal{Q} \to \mathbb{R}^m$ map $r_{i,k}$ and q to their embeddings $g(r_{i,k}) = r_{i,k}$ and $g(q) = q$, respectively. Given a similarity function $f(\cdot, \cdot) : \mathbb{R}^m \times \mathbb{R}^m \to \mathbb{R}$, a query-document score $S_{q,r_{i,k}}$ for document $r_{i,k}$ is computed as

$$S_{q,r_{i,k}} = f(\boldsymbol{q}, \boldsymbol{r_{i,k}}) \tag{1}$$

To aggregate such scores into a similarity score for object i, the top-K query-document scores for that object are averaged:

$$S_{q,i} = \frac{1}{K} \sum_{j=1}^{K} S_{q,r_{i,j}} \tag{2}$$

where $S_{q,r_{i,j}}$ is the j'th top query-document score for object i. The Late Fusion process is illustrated on the left side of Fig. 2.

Previously, Bursztyn *et al.* [3] introduced a method (one of our baselines) which can be interpreted as neural late fusion with $K = 1$, in which query and document embeddings are obtained from an LM fine-tuned using conventional Neural IR techniques. In contrast, we develop neural late fusion methods where $K \neq 1$ and introduce novel contrastive LM fine-tuning techniques for neural fusion.

Early Fusion. In Early Fusion, document information is aggregated to the object level before query scoring takes place. Various aggregation approaches are possible and we discuss the details of our proposed methods in Sect. 3.2; however, the purpose of aggregation is to produce an object embedding $\boldsymbol{i} \in \mathbb{R}^m$. Query-item similarity is then computed directly as

$$S_{q,i} = f(\boldsymbol{q}, \boldsymbol{i}) \tag{3}$$

The Early Fusion process is illustrated on the right side of Fig. 2.

Interestingly, related work [22,25] which uses hierarchical neural networks to classify hierarchical textual structures can be interpreted as Supervised Early Fusion, since the hierarchical networks learn to aggregate low-level text information into high-level representations. In contrast to these works, we study retrieval, use self-supervised contrastive learning, and explore both Early and Late Fusion.

2.3 Contrastive Representation Learning

To generate embeddings, our methods rely on contrastive representation learning. Given data samples $x \in \mathcal{X}$ where each x is associated with a label, the goal

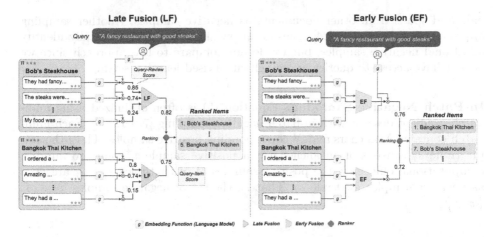

Fig. 2. (Left) Late Fusion demonstrates how embedding of queries and reviews are used to get a similarity score with Eq. (1) and how items are ranked according to the query-item score. (Right) Early Fusion demonstrates how reviews are fused together to build a vector representation for comparing the item with the query.

of contrastive learning is to train an embedding function $g : \mathcal{X} \rightarrow \mathbb{R}^m$ which maximizes the similarity between two embeddings $f(g(x^a), g(x^b)) = f(\boldsymbol{x}^a, \boldsymbol{x}^b)$ if x^a and x^b are associated with the same label, and minimizes this similarity if x^a and x^b are associated with different labels. During training, these similarities are evaluated between some anchor $x^A \in \mathcal{X}$, positive samples $x^+ \in \mathcal{X}$ associated with the same label as x^A, and negative samples $x^- \in \mathcal{X}$ associated with different labels than x^A. Specifically, given a tuple of $N + 1$ inputs $(x^A, x^+, x_1^-, ..., x_{N-1}^-)$, a similarity function f, and an embedding function g, the N-pair loss [23] is defined as:

$$L_{\text{con}}^g (x^A, x^+, \{x_i^-\}_{i=1}^{N-1}) = -\log \frac{e^{f(g(x^A), g(x^+))}}{e^{f(g(x^A), g(x^+))} + \sum_{i=1}^{N-1} e^{f(g(x^A), g(x_i^-))}} \quad (4)$$

This equation is equivalent to the softmax loss for multi-class classification of N classes. Letting $s = (x^A, x^+, \{x_i^-\}_{i=1}^{N-1})$ denote a sample tuple and $\mathcal{S}$ be the set of samples used for training, the objective of contrastive learning is:

$$\min_g L^g(\mathcal{S}) = \min_g \sum_{s \in \mathcal{S}} L_{\text{con}}^g(s) \quad (5)$$

Sampling Methods for Neural IR. For contrastive learning to be used for Neural IR, a set of samples $\mathcal{S}$ must be created from the documents. We focus on two sampling methods for this task as baselines. The *Inverse Cloze Task* (ICT) ([12]) uses two mutually exclusive spans of a document as a positive pair (x^A, x^+) and spans from other documents as negative samples. *Independent Cropping* (IC) ([6,12]) takes two random spans of a document as a positive pair and spans from other documents as negative samples. Another sampling approach used in ANCE [24] relies on query similarity to dynamically identify additional negative samples, but we do not compare to this approach since we do not have access to queries in our self-supervised learning setting.

In-Batch Negative Sampling. Equation (5) is often optimized using mini-batch gradient descent with in-batch negative sampling [7,9,13,26]. Each mini-batch contains N anchors and N corresponding positive samples. The j'th sample tuple s_j consists of an anchor x_j^A, a positive sample x_j^+, and, to improve computational efficiency, the positive samples of other tuples $s_{j',j' \neq j}$ are used as negative samples for tuple s_j. That is, the set of negative samples for s_j is $\{x_{j'}^+\}_{j'=1, j' \neq j}^N$.

3   Proposed Fusion-based Methods for RIR

We now define the Reviewed-Item Retrieval problem as a specific and highly applicable case of object retrieval (Sect. 2.2). We then demonstrate how the two-level structure of reviewed-item data can be exploited by our novel contrastive fine-tuning methods for late and early fusion. In the Reviewed-Item Retrieval problem $\mathcal{RIR}\langle \mathcal{I}, \mathcal{D}, q \rangle$, we are given a set of n items $\mathcal{I}$ where each item $i \in \mathcal{I}$ is described by a set of reviews $\mathcal{D}_i \subset \mathcal{D}$, and where the k'th review of item i is denoted by $r_{i,k}$. A review $r_{i,k}$ can only describe one item i, and this requirement makes RIR a special case of object retrieval since, in object-retrieval, a document can be associated with more than one object. Given a query $q \in \mathcal{Q}$, the goal is to rank the items based on the $S_{q,i}$ score for each item-query pair (q, i).

3.1   CLFR: Contrastive Learning for Late Fusion RIR

We now present our novel contrastive fine-tuning method for late fusion, in which review nuance is preserved during query scoring. In Contrastive Learning for Late Fusion RIR (CLFR), we fine-tune a language model g (the embedding function) using the contrastive loss function in Eq. (5) to produce embeddings of queries and reviews. The similarities of query and review embeddings are then evaluated using the dot product and aggregated into query-item scores by Eq. (2).

As opposed to single-level Neural IR contrastive sampling, RIR enables us to use the item-review structure in our sampling methods. Specifically, a positive pair (x^A, x^+) is constructed from two reviews of the same item, while negative samples $\{x_i^-\}_{i=1}^{N-1}$ are obtained from reviews of different items. We explore several variations of sampling, including the use of item ratings, item keywords, and review embedding similarity, described below. While most of our methods use full reviews as samples, we also experiment with the use of review sub-spans for anchors; the goal of these variants is to reduce the discrepancy between anchor length and query length, since queries are typically much shorter than reviews. To keep training balanced, the same number of samples is used for all items.

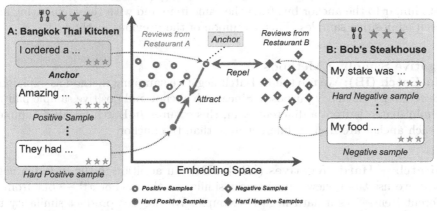

Fig. 3. The objective of contrastive fine-tuning is to fine-tune the LM so that positive sample embeddings are closer to the anchor while negative sample embeddings are further from the anchor. Hard positive and hard negative samples are selected based on their distance to the anchor. A hard positive sample (e.g., the green filled circle) is a review from the same item which is furthest from the anchor. A hard negative sample (e.g., the red filled diamond) is the closest review to the anchor but from a different item. (Color figure online)

Positive Sampling Methods

I. Same Item (SI): We want two reviews from the same item to be close to each other in the embedding space. Therefore, if we have a review from item i as an anchor x^A, the positive sample is a different review from item i, $x^+ \in \mathcal{D}_i$, sampled randomly unless otherwise mentioned.

II. Same Item, Same Rating (SI, SR): Building on SI, we further filter the positive pair reviews from the same item to have the same user rating as well. The motivation for this method is that reviews with the same rating are from people that had the same level of satisfaction from their experience. They are likely expressing a different aspect of their experience or a different phrasing of the same experience. This may be helpful for better embeddings of reviews according not only to language similarity but also user experience similarity.

III. Least Similar, Same Item (LS, SI): Choosing a set of the hardest positive samples for the model to train on may help to further boost the performance. For the positive samples, we choose the samples that are from the same item and are furthest away from anchor in the BERT embedding space. Figure 3 (cf. green rings) shows this sampling method.

IV. Least Similar, Same Item, Same Rating (LS, SI, SR): The reviews least similar to the anchor but from the same item and with the same rating are chosen as positive samples. This is a union of the previous methods.

Negative Sampling Methods
I. In-Batch (IB): We use the In-Batch negative sampling method as explained in Sect. 2.3 for efficiency purposes. Since each anchor and positive sample pair in the mini-batch is from a different item, this ensures In-Batch negative samples for each anchor are from a different item than the anchor.

In-Batch + Hard Negatives (IB + HN): In addition to the In-Batch negatives, we use the review that is most similar to the anchor x^A — but from a different item — as a hard negative sample. We use dot product similarity to find the most similar review to each anchor in the (BERT) embedding space, as illustrated in Fig. 3, and hard negatives are cached from the fine-tuning procedure. By adding the hard negatives to the In-Batch negatives we aim to make the model more robust.

Data Augmentation. We propose two data augmentation methods for RIR. Our first method exploits the meta-data tags that each item has, and our second method aims to mitigate the gap between self-supervised training and inference by shortening the anchor text.

I. Prepending meta-data (PPMD): Many item-review structures contain meta-data alongside textual and structural information. In our inference dataset, the items are restaurants that have categorical labels such as "Pizza, Salad, Fast Food" and usually include the cuisine type such as "Mediterranean, Middle Eastern". In order to use this meta-data without changing the model architecture, we prepend this data in text format to the review text during inference. This augmentation is done with the goal of better working with queries referring to categories and cuisine types.

II. Anchor sub-sampling: User queries are usually shorter than reviews. Thus, instead of taking a review as an anchor (which are the pseudo queries during the self-supervised fine-tuning), we take a shorter span from the review as the anchor in what we call sub-sampling anchor to a span (**SASP**). In an alternative method, we take a single sentence from the review as the anchor, and call this sub-sampling anchor to a sentence (**SASN**). For both cases, the span or sentence is chosen randomly, and we still use full reviews as positive and negative samples.

3.2   CEFR: Contrastive Learning for Early Fusion RIR

We now present our contrastive learning approaches for Early Fusion, in which item embeddings are learned before query-scoring. In Early Fusion, the query-item score $S_{q,i}$ is computed by Eq. (3) once we have an embedding i for each item i. A naive approach for obtaining the item embedding is to take the average of the review embeddings $r_{i,k}$'s of item i. We call this naive approach Average EF.

We also introduce a new method which we call Contrastive Learning for Early Fusion RIR (CEFR), where we use contrastive learning to simultaneously fine-tune g and learn an item embedding i in an end-to-end style. In particular, we let i be a learnable vector which we use instead of an anchor embedding $g(x^A)$ and initialize i with the naive item embedding obtained from Average EF. For a tuple $(i, x^+, \{x_j^-\}_{j=1}^{N-1})$, the loss given by Eq. (4) thus becomes

$$L_{\text{con}}^g(i, x^+, \{x_j^-\}_{j=1}^{N-1}) = -\log \frac{e^{f(i,g(x^+))}}{e^{f(i,g(x^+))} + \sum_{j=1}^{N-1} e^{f(i,g(x_j^-))}} \qquad (6)$$

Given a set of training samples $\mathcal{S}$, with $s_i = (i, x^+, \{x_j^-\}_{j=1}^{N-1})$ representing the sample tuple for item i, and denoting $\mathcal{I}$ as the set of learnable item embeddings, the objective of contrastive learning in CEFR is

$$\min_{\mathcal{I},g} L^g(\mathcal{S}) = \min_{\mathcal{I},g} \sum_{i=1}^{N} L_{\text{con}}^g(s_i) \qquad (7)$$

Note that CEFR simultaneously learns item embeddings and fine-tunes the language model g (the embedding function), which is intended to allow flexibility in learning i. For each item i, the positive sample $x_i^+ \in \mathcal{D}_i$ is a review from item i, and the negative samples are obtained via In-Batch negative sampling.

4   Experiments

4.1   Reviewed-Item Retrieval Dataset (RIRD)

To address the lack of existing RIR datasets, we curated the *Reviewed-Item Retrieval Dataset (RIRD)*[1] to support our analysis. We used reviews related to 50 popular restaurants in Toronto, Canada obtained from the Yelp dataset.[2] We selected restaurants with a minimum average rating of 3 and at least 400 reviews that were not franchises (e.g., McDonalds) since we presume franchises are well-known and do not need recommendations. We created 20 queries for 5 different conceptual categories highlighted in Table 1 (with examples). These 5 groups capture various types of natural language preference statements that occur in this task. We then had a group of annotators assess the binary relevance of each of these 100 queries to the 50 restaurants in our candidate set. Each review was

---

[1] RIRD: https://github.com/D3Mlab/rir_data.
[2] Yelp Dataset: https://www.yelp.com/dataset.

labeled by 5 annotators and the annotations showed a kappa agreement score of 0.528, demonstrating moderate agreement according to Landis and Koch [15], which is expected given the subjectivity of the queries in this task ([1]). There is a total number of 29k reviews in this dataset.

Table 1. Categories of queries and their examples

Query category	Example
Indirect queries	*I am on a budget*
Queries with negation	*Not Sushi but Asian*
General queries	*Nice place with nice drinks*
Detailed queries	*A good cafe for a date that has live music*
Contradictory queries	*A fancy, but affordable place*

4.2   Experimental Setup

All experiments for fine-tuning models and obtaining semantic representations for reviews and queries were done with 8 Google TPUs. Each fine-tuning experiment took between 5–7 h depending on the experimental setup. We use an Adam optimizer [14] with a learning rate of 10^{-5} and use a validation set of 20% of reviews for hyperparameter tuning. Each experiment was repeated 5 times with 5 different seeds to mitigate possible noise in dataset. We use a batch size of $N = 48$, and add one hard negative per tuple if hard negatives are used. The similarity function f is the dot product. For the choice of LM, we use uncased base BERT ([11]), and we select $K \in \{1, 10, |\mathcal{D}_j|\}$ for Eq. 2.2, where $|\mathcal{D}_i|$ is the number of reviews for item i. Our code for reproducing all results is publicly available.[3]

4.3   Baselines

Traditional IR: We use TF-IDF and BM25 with LF as baselines. First, all query-review scores are computed by TF-IDF and BM25 ($b = 0.75$, $k_1 = 1.6$) ranking functions. These scores are then fused in the same manner as in Sect. 2.2, where query-review level information is aggregated to the query-item level using Eq. (2). Stopword removal and Lemmatization were also applied for these models using NLTK ([2]).

Masked Language Modeling (MLM): Masked Language Modeling (MLM) is a primary self-supervised fine-tuning objective for many LMs such as BERT [4]. We use off-the-shelf BERT as a baseline and also fine-tune it with an MLM objective on our review data as another baseline. We train for 100 epochs with early stopping and a learning rate of 10^{-5}.

---

[3] Code: https://github.com/D3Mlab/rir.

Neural IR Models: In order to compare state-of-the-art Neural IR models with our proposed models in the self-supervised regime, we use the following contrastively trained Neural IR language models: Contriever ([12]), Condenser ([5]) and CoCondenser ([6]). We use publicly released pre-trained models for these baselines. Since only one model is available, confidence intervals are not applicable.

IR-Based Contrastive: We use contrastive learning with self-supervised IR, explained in Sect. 2.3, to fine-tune the LM as a baseline. The positive samples are created by ICT and IC from a single review, and In-Batch negative sampling is used. Since these methods are agnostic to the item associated with a review, as long as a review is not used for the positive pair, negative samples could come from reviews of the same item or reviews of different items. We use these baselines to examine the importance of using the two-level item-review structure in addition to the review text in our contrastive learning methods for RIR, since IR-based contrastive learning (ICT, IC) does not leverage that structure.

4.4 Evaluation Metrics

We use Mean R-Precision (R-Prec) and Mean Average Precision (MAP) with 90% confidence intervals to evaluate the performance of RIR in our experiments. We note that R-Precision is regarded as a stable measure for averaging when the number of relevant items varies widely per query.

4.5 Results and Discussion

This section studies four Research Questions (RQ) and discusses our experimental results for each. Specifically, we compare the effects of various sampling and data augmentation methods, evaluate Late Fusion against Early Fusion, and examine how our proposed methods compare to baselines.

RQ1: Which sampling method from Section 3.1 gives the best performance for LM fine-tuning for LF? The choice of sampling method for contrastive fine-tuning has an important effect on retrieval performance, but must be analysed jointly with the effect of K. In Table 2, we explore all the sampling options outlined in Sect. 3.1 with our base CLFR method using Same Item (SI) and In-Batch Negatives (IB) sampling. Regardless of K, there is no significant improvement from using Same Item (SI) Same Rating (SR) sampling over Same Item (SI) sampling. However, adding Least Similar (LS) to the base CLFR and having Same Rating (SR) provides an absolute and significant improvement of 0.028 and 0.036 for $K = 1$ and $K = |\mathcal{D}_j|$ in R-Prec but no significant improvement for $K = 10$. Adding hard negative samples fails to provide significant improvement regardless of the positive sampling method.

RQ2: What is the impact of data augmentation methods from Section 3.1 for LM fine-tuning in LF setting? The results in Table 2 show that prepending meta-data (i.e., cuisine type) to review text gives no significant improvement. We see that sub-sampling the anchors, by span or by sentence, achieves the best performance. On average, SASP and SASN improve R-Prec on the base CLFR method by a significant amount for every K. This shows that being robust to the length of the anchor through sub-sampling plays a crucial role for contrastive learning in Neural RIR. Due to space limitations, full comparative evaluation results are made available in online supplementary material.[4]

RQ3: Does Early Fusion or Late Fusion work better for RIR? Table 3 shows the performance of our Early and Late Fusion methods. We can see that our Late Fusion methods outperform Early Fusion significantly. We conjecture that since Early Fusion fuses all of an item's reviews into a single embedding vector, it may lose the nuance of individual review expressions or differences that averaged out in the Early Fusion process. In contrast, Late Fusion preserves the nuances of individual reviews until query-scoring.

Table 2. Results for exploring different techniques for CLFR with 90% confidence intervals. Positive and negative sampling methods are explained in Sect. 3.1. All rows have the base Same Item (SI) positive and In-Batch (IB) negative sampling. Additional sampling methods are specified in the first two columns.

Sampling		$K=1$		$K=10$		$K=|\mathcal{D}_j|$ (Avg)	
Pos.	Neg.	R-Prec	MAP	R-Prec	MAP	R-Prec	MAP
–	–	0.497 ± 0.010	0.569 ± 0.021	0.514 ± 0.011	0.587 ± 0.013	0.495 ± 0.017	0.575 ± 0.018
SR	–	0.499 ± 0.016	0.563 ± 0.022	0.509 ± 0.011	0.583 ± 0.010	0.501 ± 0.029	0.572 ± 0.030
LS	–	0.513 ± 0.014	0.589 ± 0.018	0.500 ± 0.014	0.584 ± 0.015	0.487 ± 0.008	0.569 ± 0.013
SR, LS	–	0.525 ± 0.004	0.596 ± 0.014	0.515 ± 0.013	0.590 ± 0.016	$\mathbf{0.531\pm0.006}$	$\mathbf{0.607\pm0.014}$
–	HN	0.511 ± 0.012	0.589 ± 0.011	0.512 ± 0.016	0.590 ± 0.012	0.516 ± 0.014	0.589 ± 0.018
SR, LS	HN	0.504 ± 0.015	0.571 ± 0.006	0.516 ± 0.016	0.594 ± 0.012	0.517 ± 0.012	0.596 ± 0.013
PPMD	–	0.506 ± 0.004	0.576 ± 0.012	0.503 ± 0.007	0.579 ± 0.010	0.507 ± 0.010	0.573 ± 0.009
SASP	–	0.523 ± 0.013	0.595 ± 0.016	0.531 ± 0.006	0.611 ± 0.005	0.525 ± 0.008	0.599 ± 0.013
SASN	–	$\mathbf{0.532\pm0.019}$	$\mathbf{0.609\pm0.020}$	$\mathbf{0.545\pm0.009}$	$\mathbf{0.626\pm0.009}$	0.530 ± 0.012	0.610 ± 0.011

Table 3. Comparing our non-contrastive and contrastive Early Fusion (EF) methods with our base and best Late Fusion (LF) methods. Average EF takes an average BERT embedding of reviews as the item embedding. Base LF is CLFR with Same Item and In-Batch negatives. Best LF is CLFR with Same item and SASN and In-Batch negatives. We can see the noticeable improvement of using LF over EF. We conjecture this is due to the preservation of nuance in individual reviews during query scoring.

EF Model	R-Prec			MAP		
Average	0.297			0.364		
CEFR	0.438 ± 0.003			0.519 ± 0.003		
	$K=1$		$K=10$		$K=|\mathcal{D}_j|$ **(Avg)**	
LF Model	**R-Prec**	**MAP**	**R-Prec**	**MAP**	**R-Prec**	**MAP**
Base LF	0.497 ± 0.010	0.569 ± 0.021	0.514 ± 0.011	0.587 ± 0.013	0.495 ± 0.017	0.575 ± 0.018
Best LF	$\mathbf{0.532\pm0.019}$	$\mathbf{0.609\pm0.020}$	$\mathbf{0.545\pm0.009}$	$\mathbf{0.626\pm0.009}$	$\mathbf{0.530\pm0.012}$	$\mathbf{0.610\pm0.011}$

---

[4] Supplementary Materials: https://ssanner.github.io/papers/ecir23_rir.pdf.

RQ4: How effective is our (best) method of using the structure of the review data for Late Fusion in RIR (CLFR) compared to existing baselines? Table 4 compares the performance of state-of-the-art unsupervised IR methods with our best contrastive fine-tuning method for RIR. The out-of-domain Neural IR pre-trained language models CoCondenser and Contriever perform noticeably better than base BERT, with CoCondenser performing the best among them. The traditional sparse retrieval models (TF-IDF and BM25) both outperform non-fine-tuned BERT, which emphasizes the importance of fine-tuning BERT for the downstream task of RIR. We also see that the self-supervised contrastive fine-tuning methods for IR (ICT and IRC) outperform non-fine-tuned BERT, but fall far behind the CoCondenser model. MLM fine-tuning of BERT also does not improve the performance of RIR, which is expected since it neither utilizes the structure of the data nor does this training objective directly support IR or RIR tasks. In contrast, by using the item-review data structure, our best method (CLFR with Same Item, SASN, and In-Batch Negatives) outperforms all contrastive IR, Neural IR, and sparse retrieval baselines.

Table 4. CLFR (our best model) results on RIRD dataset versus baselines from Sect. 4.3. IR-based methods (Sect. 2.3) and CLFR use In-Batch negatives (IB), and CLFR uses Same Item (SI) positive samples and SASN augmentation (Sect. 3.1).

Model	$K=1$		$K=10$		$K=|\mathcal{D}_j|$ (Avg)	
	R-Prec	MAP	R-Prec	MAP	R-Prec	MAP
TF-IDF	0.345	0.406	0.378	0.442	0.425	0.489
BM25	0.393	0.450	0.417	0.490	0.421	0.495
BERT	0.295	0.343	0.296	0.360	0.297	0.364
MLM	0.289	0.347	0.303	0.366	0.298	0.353
Condenser	0.358	0.410	0.390	0.449	0.378	0.428
CoCondenser	0.445	0.505	0.481	0.553	0.482	0.570
Contriever	0.375	0.427	0.418	0.482	0.458	0.519
IR-based, IRC	0.355 ± 0.024	0.422 ± 0.033	0.355 ± 0.022	0.424 ± 0.028	0.398 ± 0.031	0.464 ± 0.026
IR-based, ICT	0.331 ± 0.011	0.395 ± 0.016	0.339 ± 0.023	0.405 ± 0.025	0.328 ± 0.009	0.384 ± 0.014
CLFR	$\mathbf{0.532 \pm 0.019}$	$\mathbf{0.609 \pm 0.020}$	$\mathbf{0.545 \pm 0.009}$	$\mathbf{0.626 \pm 0.009}$	$\mathbf{0.530 \pm 0.012}$	$\mathbf{0.610 \pm 0.011}$

5   Conclusion

In this paper, we proposed and explored novel self-supervised contrastive learning methods for both Late Fusion and Early Fusion methods that exploit the two-level item-review structure of RIR. We empirically observed that Same Item (SI) positive sampling and In-Batch negative sampling with sub-sampling anchor reviews to a sentence for Late Fusion achieved the best performance. This model significantly outperformed state-of-the-art Neural IR models, showing the importance of using the structure of item-review data for the RIR task. We also showed Late Fusion outperforms Early Fusion, which we hypothesize is due to the preservation of review nuance during query-scoring. Most importantly, this work opens new frontiers for the extension of self-supervised contrastive Neural-IR techniques to leverage multi-level textual structures for retrieval tasks.

References

1. Balog, K., Radlinski, F., Karatzoglou, A.: On interpretation and measurement of soft attributes for recommendation. In: Proceedings of the 44th International ACM SIGIR Conference on Research and Development in Information Retrieval, pp. 890–899 (2021)
2. Bird, S., Klein, E., Loper, E.: Natural language processing with Python: analyzing text with the natural language toolkit. O'Reilly Media, Inc. (2009)
3. Bursztyn, V., Healey, J., Lipka, N., Koh, E., Downey, D., Birnbaum, L.: "It doesn't look good for a date": Transforming critiques into preferences for conversational recommendation systems. In: Proceedings of the 2021 Conference on Empirical Methods in Natural Language Processing, pp. 1913–1918, Association for Computational Linguistics (Nov 2021). https://doi.org/10.18653/v1/2021.emnlp-main.145
4. Devlin, J., Chang, M.W., Lee, K., Toutanova, K.: Bert: Pre-training of deep bidirectional transformers for language understanding pp. 4171–4186 (Jun 2019). https://doi.org/10.18653/v1/N19-1423
5. Gao, L., Callan, J.: Condenser: a pre-training architecture for dense retrieval. In: Proceedings of the 2021 Conference on Empirical Methods in Natural Language Processing, pp. 981–993 (2021)
6. Gao, L., Callan, J.: Unsupervised corpus aware language model pre-training for dense passage retrieval. In: Proceedings of the 60th Annual Meeting of the Association for Computational Linguistics (Volume 1: Long Papers), pp. 2843–2853 (2022)
7. Gillick, D., Kulkarni, S., Lansing, L., Presta, A., Baldridge, J., Ie, E., Garcia-Olano, D.: Learning dense representations for entity retrieval. In: Proceedings of the 23rd Conference on Computational Natural Language Learning (CoNLL), pp. 528–537 (2019)
8. Gupta, P., Bali, K., Banchs, R.E., Choudhury, M., Rosso, P.: Query expansion for mixed-script information retrieval. In: Proceedings of the 37th International ACM SIGIR Conference on Research & Development in Information Retrieval, pp. 677–686, SIGIR '14, Association for Computing Machinery, New York, NY, USA (2014), ISBN 9781450322577, https://doi.org/10.1145/2600428.2609622
9. Henderson, M., et al.: Efficient natural language response suggestion for smart reply (2017)
10. Huang, P.S., He, X., Gao, J., Deng, L., Acero, A., Heck, L.: Learning deep structured semantic models for web search using clickthrough data. In: Proceedings of the 22nd ACM International Conference on Information & Knowledge Management, pp. 2333–2338, CIKM '13, Association for Computing Machinery, New York, NY, USA (2013), ISBN 9781450322638. https://doi.org/10.1145/2505515.2505665
11. Huggingface: Bert base model (uncased). https://huggingface.co/bert-base-uncasedd (2022)
12. Izacard, G., et al.: Towards unsupervised dense information retrieval with contrastive learning. arXiv preprint arXiv:2112.09118 (2021)
13. Karpukhin, V., et al.: Dense passage retrieval for open-domain question answering. arXiv preprint arXiv:2004.04906 (2020)
14. Kingma, D.P., Ba, J.: Adam: A method for stochastic optimization. arXiv preprint arXiv:1412.6980 (ICLR 2015)
15. Landis, J.R., Koch, G.G.: The measurement of observer agreement for categorical data. biometrics, pp. 159–174 (1977)

16. Lin, J., Nogueira, R., Yates, A.: Pretrained transformers for text ranking: Bert and beyond. Synth. Lect. Human Lang. Technol. **14**(4), 1–325 (2021)
17. Nguyen, T., et al.: MS MARCO: A human generated machine reading comprehension dataset. In: CoCo@ NIPS (2016)
18. Nogueira, R., Cho, K.: Passage re-ranking with bert. arXiv preprint arXiv:1901.04085 (2019)
19. Robertson, S., Zaragoza, H.: The probabilistic relevance framework: BM25 and beyond. Now Publishers Inc (2009)
20. Salton, G., Wong, A., Yang, C.S.: A vector space model for automatic indexing. Commun. ACM **18**(11), 613–620 (1975)
21. Schütze, H., Manning, C.D., Raghavan, P.: Introduction to information retrieval, vol. 39. Cambridge University Press Cambridge (2008)
22. Shing, H.C., Resnik, P., Oard, D.W.: A prioritization model for suicidality risk assessment. In: Proceedings of the 58th annual meeting of the association for computational linguistics, pp. 8124–8137 (2020)
23. Sohn, K.: Improved deep metric learning with multi-class n-pair loss objective. In: Advances in Neural Information Processing Systems, vol. 29 (2016)
24. Xiong, L., et al.: Approximate nearest neighbor negative contrastive learning for dense text retrieval. arXiv preprint arXiv:2007.00808 (2020)
25. Yang, Z., Yang, D., Dyer, C., He, X., Smola, A., Hovy, E.: Hierarchical attention networks for document classification. In: Proceedings of the 2016 Cconference of the North American Chapter of the Association for Computational Linguistics: Human Language Technologies, pp. 1480–1489 (2016)
26. Yih, W.t., Toutanova, K., Platt, J.C., Meek, C.: Learning discriminative projections for text similarity measures. In: Proceedings of the Fifteenth Conference on Computational Natural Language Learning, pp. 247–256 (2011)
27. Zhang, S., Balog, K.: Design patterns for fusion-based object retrieval. In: Jose, J.M., et al. (eds.) ECIR 2017. LNCS, vol. 10193, pp. 684–690. Springer, Cham (2017). https://doi.org/10.1007/978-3-319-56608-5_66

User Requirement Analysis for a Real-Time NLP-Based Open Information Retrieval Meeting Assistant

Benoît Alcaraz[1]([⊠]) [iD], Nina Hosseini-Kivanani[1] [iD], Amro Najjar[2] [iD],
and Kerstin Bongard-Blanchy[1] [iD]

[1] University of Luxembourg, Esch-sur-Alzette, Luxembourg
{benoitalcaraz,ninahosseini-kivanani,kerstinbongard-blanchy}@uni.lu
[2] Luxembourg Institute of Science and Technology (LIST),
Esch-sur-Alzette, Luxembourg
amro.najjar@list.lu

Abstract. Meetings are recurrent organizational tasks intended to drive progress in an interdisciplinary and collaborative manner. They are, however, prone to inefficiency due to factors such as differing knowledge among participants. The research goal of this paper is to design a recommendation-based meeting assistant that can improve the efficiency of meetings by helping to contextualize the information being discussed and reduce distractions for listeners. Following a Wizard-of-Oz setup, we gathered user feedback by thematically analyzing focus group discussions and identifying this kind of system's key challenges and requirements. The findings point to shortcomings in contextualization and raise concerns about distracting listeners from the main content. Based on the findings, we have developed a set of design recommendations that address context, interactivity and personalization issues. These recommendations could be useful for developing a meeting assistant that is tailored to the needs of meeting participants, thereby helping to optimize the meeting experience.

Keywords: Mediated human-human interaction · Recommender systems · User-centered design · User experience · Natural language

1 Introduction

Meetings are intended to drive progress in an interdisciplinary and collaborative manner and have become a recurrent task within companies. A meeting context entails many variables. Speech may be unevenly distributed among participants [3]. Multiple topics may be discussed in parallel [34], with a possible but not systematic display of visuals. Finally, a meeting can be held entirely remotely or partially so (i.e., hybrid).

Discussions in meetings frequently stray into unplanned topics. In such cases, no visual support will have been prepared, so other participants cannot evaluate

J. Kamps et al. (Eds.): ECIR 2023, LNCS 13980, pp. 18–32, 2023.
https://doi.org/10.1007/978-3-031-28244-7_2

the reliability of the verbal contribution. The other participants may also be less familiar with the topic, making it difficult for them to follow the discussion. With the Internet easily accessible via connected devices, it is tempting to quickly conduct a web search about the topic under discussion. This, however, distracts participants during the meeting [25]. Consequently, a tool for quick information retrieval connected to a display system could help them understand in real time what another participant is saying (e.g., a participant is talking about a particular country, and the system displays a map of that country). In addition, combining this tool with the presentation slides would serve to enhance them with additional information. The potential benefits of such a tool include shortening the speaker's preparation time, facilitating the audience's comprehension of the topic under consideration, and providing a summary of what was said during the meeting.

However, to optimize the utility and user adoption of a recommendation-based meeting assistant, it is essential that such a system should meet user needs since system accuracy is not the only performance criteria [36]. To this end, this paper presents a user study which took place during the discovery phase at the beginning of the user-centered design process [4]. We derive actionable design recommendations from the study's findings, including contextual and interactive aspects. These contribute to global understanding of user requirements for recommender systems in a meeting context. Design recommendations for the meeting assistant are discussed. together with limitations and future perspectives respectively.

2   Related Work

This section on the related work is divided into three subtopics. First, in Sect. 2.1, we discuss existing meeting assistants, their use, and their design. Second, in Sect. 2.2, we mention general guidelines about recommender systems that cover different ways of increasing trust in such systems. The last part (Sect. 2.3) elaborates on the importance of user experience in recommender systems.

2.1   Meeting Assistants

In the past few years, various meeting assistant solutions have been developed to improve collaboration in meetings. These solutions fall into several categories: *(i)* those that provide automated meeting assistant features such as automatic transcription and annotation. Cognitive Assistant that Learns and Organizes (CALO) [34] is one of the most famous examples of meeting assistants that belong to this category. Another more recent example, from industry, is the WebEx meeting assistant developed by Cisco[1], *(ii)* ontology-based assistants like that of [33], whose aim is to provide means to manage virtual meetings in a way that allows the participants to add contributions, consult the contributions

---

[1] https://www.webex.com/webex-assistant.html.

of others, and obtain supporting information and meetings history, *(iii)* natural language processing (NLP) focused meeting assistants such as the Intel Meeting Assistant [3], whose main innovative features are online automatic speech recognition, speaker diarization, keyphrase extraction and sentiment detection. *(iv)* information retrieval and user interface management user assistants (the topic of this paper).

This paper concerns the final category. The goal of an information retrieval and user interface management user assistant is to retrieve and present relevant information about what is being discussed by participants in a meeting. Only few works in the literature fall into this category. The first concerns Augmented Multi-party Interaction with Distance Access (AMIDA) [25] with its tangible interface [13], a helper agent [19]. AMIDA is a recommender system that, based on speech, presents documents from previous meetings to participants. It is based on the *Automatic Content-Linking Device* (ACLD) recommender system [26]. The documents retrieved are primarily textual. Then, participants can manipulate documents by sliding their fingers on tables with the help of the tangible interface. Yet, this research is limited to the corpus available within an offline database. Moreover, it presupposes that there is a room available that is especially equipped to display the presented material, thereby shifting the focus of the work to the front end. The second work describes Image Recommender Robot Meeting Assistant (IRRMA), a recommender system embodied in a QTRobot[2] that, based on speech, presents image documents retrieved from web databases [1]. It comes with the drawback that its databases are not limited to documents about previous meetings, making it prone to recommending non-related content accidentally.

This paper aims to address the above limitations. To do so, we ran an empirical study in order to determine which features should be included and avoided in a future system.

2.2   General Guidelines Concerning Recommender Systems

Amershi et al. [2] have provided a total of eighteen guidelines (G) on how to design recommender systems for various contexts (e.g., commercial, web search, voice agents). Some of these guidelines apply to our meeting assistant. The applicable guidelines are: G3. Time of service based on context; G4. Show contextually relevant information; G7. Support efficient invocation; G8. Support efficient dismissal; G11. Make clear why the system did what it did; G17. Provide global controls. The user study described in this paper looked into the understanding of users who were domain experts of these guidelines.

In addition, some work has been done on understanding how to increase user trust in recommender systems. Indeed, trust is a key point since it directly impacts how users perceive recommendation materials [6,15]. It has been shown that the system setup impacts user trust too [27]. The study presented in [22] denotes the impact of embodiment over time on users' trust in an agent. It shows

---

[2] https://luxai.com/humanoid-social-robot-for-research-and-teaching/.

that users initially trust physical agents more than virtual ones. Nevertheless, the difference in trust tends to decrease with the amount of time passed with the agent. Other studies [17,32] tell us that trust comes mainly from the preference elicitation feature and explanations from the recommender system, even when the recommendation material is not optimal. It has also been shown that high transparency regarding the origin of the information provided increases users' trust in the recommendations [29]. These different results show that what affects user trust remains unclear and merits further investigation.

Recent research on AI ethics has contributed to defining and understanding ethical challenges associated with recommender systems. Milano et al. [24] identified six major ethical challenges, including opacity, privacy, and inappropriate content. They prescribed a set of guidelines to overcome these challenges ranging from user-specified filters, and explainability with factual explanations, to the adoption of multi-sided recommendation material to improve the fairness of such systems.

2.3   User Experience of Recommender Systems

Recommender systems pose various technical challenges, which have been partially addressed in existing research [37]. However, we note a lack of regard for eliciting users' needs in the design and development of solutions. Recently, user experience (UX) of recommender systems has received growing attention because of its impact on the ability of such systems to help users make better choices, for example, when selecting from a long list of available options (see e.g., [8]). According to [16], UX is the first feeling the user has regarding the product, which emphasizes its importance. Even though a lot of UX research has been conducted on recommender systems providing personalized and user-centered recommendation material (e.g., [9,12,23,28]), only a few exist when it comes to group and user-agnostic recommendations. Even if the features of most of the existing recommender systems were designed according to market studies and experimental results, only a minority of them address our topic, as shown in [7,35]. Moving forward, the available knowledge is insufficient as a basis for designing a recommender system that takes into account end-user needs.

To close this gap and gather user requirements, we conducted a user study as a preliminary step to developing a system intended to be used by people collaborating in a work context.

3   Method

3.1   User-Centered Design Approach

Our research project concerns the design and development of tools that support collaborative work. In this paper, we focus on meetings. Our system will be a meeting assistant to help people follow the contents of a meeting better. The system will potentially serve educators and people working in business contexts.

A user-centered approach is indispensable to devising a system that people will adopt into their work routines. The project supports the creation of products and services tailored to address end-users' needs, wants and limitations by considering their perspective throughout the design process and ensuring that the final products and services meet those needs and are easy to use [30]. The user-centered design process starts with a 'discover and define phase' during which user requirements are gathered and defined before heading into an ideation phase to produce concept ideas and generate solutions. The process is iterative and includes regular evaluations of the ideas and solutions [4]. As a first step, we presented a qualitative experiential study that served to identify user requirements by simulating the recommendation system during a live presentation with small user groups. The simulation was carried out via a Wizard-of-Oz setup [11] (in such a setup, an administrator hides in an adjacent room). The administrator simulated the behavior of an intelligent computer application by intercepting communications between the participants and the system.

3.2 Participants

To gather user requirements for the meeting assistant, we conducted a focus group study at the University of Luxembourg's user lab. Seventeen professionals participated. Fifteen audience participants were divided into four focus groups (three groups had four participants and one group had three). Six participants identified as women and nine as men. Five were Bachelor's students in Computer Science, two were Bachelor's students in Psychology, two were Master's students in Civil Engineering, and five were human-computer interaction (HCI) professionals. Their ages ranged from 20 to 41 years (an average of 27 years). In addition, two presenters (male Computer Science researchers aged 28 and 42) introduced a different machine learning topic in two sessions (topic A: genetic algorithms; topic B: natural language processing NLP). All the participants received a compensation of 30 EUR, which was deemed fair by the Ethics Committee of the University of Luxembourg.

3.3 Study Material

The study took place during the early design phase of the meeting assistant, where user requirements are gathered. Due to the lack of a functional prototype, we had to pre-select the recommendation material and could not carry out the test in an actual meeting scenario in which people discuss freely. We chose a meeting scenario in which one person presents to a group. The recommendation material was pre-selected from Wikimedia by the first author, using a minimalist procedure that selected only the first search results based on three keywords from the speaker's discourse. Simulating the system instead of relying on existing systems reduces the risk of detecting local rather than global design issues and enables exploration of the wider design space.

Three to four attendees per session followed a presentation on one of the two machine learning topics during which a recommender system suggested visual content. The audience was seated around a table facing the presenter. Each individual had an iPad lying on the table in front of them. The recommender system under the Wizard of Oz setup sent information to individual iPads via a Microsoft Teams video call in screen-sharing mode (see Fig. 1). The system was placed before the participants without instructions to prevent priming on the topic of recommender systems. After the presentation, iPads were also used to conduct a user experience (UX) questionnaire.

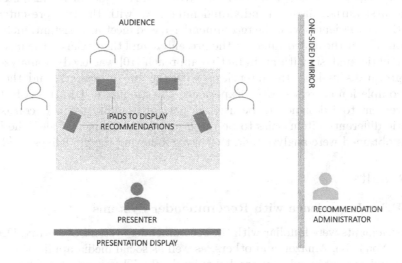

Fig. 1. Room layout for the Wizard-of-Oz setup.

3.4 Study Procedure

We obtained the participants' consent to record them during the session. Each session took 60 min, starting with a 15-min presentation where they were observed via a one-sided mirror to follow how their gazes alternated between the presentation and the material presented on the iPad. Following the presentation, they filled in a questionnaire providing demographic information and feedback about their experience with the recommender system-based meeting assistant during the presentation. The questionnaire used the standardized User Experience Questionnaire (UEQ) scale [21] and included sentence completion exercises [20]. Finally, the first and last authors (over three and one sessions respectively) moderated a semi-structured focus group discussion [14] with the participants. Each session took about 30 min and addressed the shortcomings and potential benefits of the system's design along the following topics:

1. their previous experience with existing recommender systems;
2. their experience with the tested recommender system;
3. their experience with the medium (individual iPad) of the tested recommender system.

The focus group format was based on a semi-structured interview [14] and chosen because it naturally follows from a group activity setting. Individual interviews would have been a valid alternative, but this would have imposed waiting times on the participants. Moreover, focus groups allow participants to build on each other's responses. Together, they can recall experiences in greater detail, which makes it a suitable research method for extracting rich qualitative insights.

The first author also ran individual interviews with the two presenters to probe their experiences with the recommender-based meeting assistant, including questions about the added value for the presenters and their take on the medium.

Thematic analysis with an inductive approach [10] was used to analyze the focus group discussions, the interviews with the two presenters, and the sentence completion exercises. User Experience Questionnaire (UEQ) is a fast UX measurement tool designed to be used as part of usability tests. It consists of semantic differential item pairs to be rated on a 7-point Likert scale. The UEQ ratings obtained were analyzed via that scale following the guidelines at [21].

4   Results

4.1   Prior Experience with Recommender Systems

The participants were familiar with the recommender systems of Netflix, Deezer, Spotify, YouTube, Amazon and others, as well as social media applications that recommend content based on users' behavioral data. They were ambivalent about their experiences with those systems. They appreciate the convenience of being proposed suitable options without the need to search. However, they regularly encounter irrelevant recommendations. Furthermore, they expressed mistrust in AI, fears about being manipulated, and loss of agency (*"Sometimes we're scared about the technology that makes our life so easy. I don't like to be recommended everything by AI technology. Sometimes I prefer to search for something useless for a lot of time, trying to find a solution by myself instead of receiving [it]."* G3-P4). Their trust in such systems is closely linked to 1) the pertinence of the recommendations and 2) the discretion with which user data is gathered and handled (*"[T]hen it's the time they help us, and we see the help is trustworthy. [...] But if you don't want them at [that] moment or they give us the information that they [have] just gathered, it's untrustworthy. Because we, as humans, don't want to be manipulated. And if we see [...] somebody knows some information about us [...] we feel insecure."* G3-P2).

4.2   Participants' Experience with the Meeting Assistant

Through this study, we sought to understand how the audience experienced the tested recommender system-based meeting assistant in a presentation setting.

Looking at the UEQ data, we see that the overall experience under the setup tested was not very satisfying. The system scores (Table 1) were approximately neutral on all five criteria: attractiveness, perspicuity, efficiency, dependability, and stimulation. The score for novelty was positive, but the data obtained for this scale was not sufficiently consistent (Guttman's Lambda-2 Coefficient 0.54).

Table 1. Scores for UEQ scales

UEQ Scales	Mean	Variance	Lambda2
Attractiveness	0.067	1.73	0.89
Perspicuity	−0.100	2.39	0.90
Efficiency	0.133	1.41	0.76
Dependability	−0.100	1.22	0.67
Stimulation	0.217	1.78	0.88
Novelty	0.867	0.61	0.54

Looking at the single items (Fig. 2), we see that the attributes experienced most strongly indicate the system's innovative, inventive and creative character. The attributes experienced most weakly indicate that the participants perceived the system as unpredictable and rather annoying. These scores align with comments from the focus group that the participants found it challenging to make connections between the recommended material and the presentation. This was either caused by a lack of contextual information or timing issues (*"I didn't trust it much since we didn't get much context and [...] the interval seems kind of random. Like each time you change the slides, it would change, but sometimes it wouldn't."* G2-P1). The lack of contextual hints hindered participants who were not familiar with the topic being presented from understanding the material provided by the recommender system better. By contrast, participants who were familiar with the topic were critical that many of the visuals provided by the recommender system did not add anything to enhance their understanding (*"It's a concept that I'm very familiar with [...] I didn't get any value out of the information displayed here. It was supposed to be supportive material, but it wasn't concrete enough for me to help."* PG4-P3). Furthermore, all the participants expressed uncertainty about where the material was coming from—whether it was prepared by the presenter or was the result of an algorithm (*"Where does it come up with these recommendations? Does it come from the slides directly? Does the presenter have to give keywords? Does the presenter give the images?"* G2-P3).

Due to the lack of introduction to the system, about a quarter of the participants did not see all the material provided by the recommender system while following the presentation. All feared that the meeting assistant was potentially a source of distraction. They felt uneasy about looking at the material on the iPad instead of looking at the presenter (*"I cared [about] him—if we are all on the tablet and then he has the impression that we are not listening to him."* G1-P2).

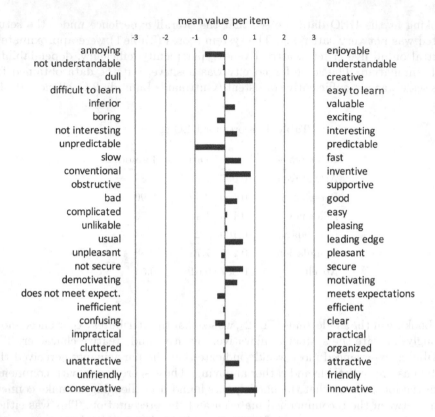

Fig. 2. Scores for UEQ items

Nevertheless, some also said that the system was stimulating in moments when their thoughts started to drift away from the presentation (*"If your attention starts [to stray], you can maybe look here and keep your [mind on] the subject."* G3-P3). Moreover, participants saw that material from the recommender system has the potential to bring them back when they have lost the thread (*"I think that we cannot concentrate for a whole presentation. [...] I think the brain needs to [get] distract[ed] somewhere. That's why it goes [over] there and then goes to the presentation. So I think it's a good distraction. It's a distraction [that] is not very distracting."* G2-P2). The content of the recommended material could signal the start of a new part of the presentation so that the listener would be motivated to try and follow it again. Multiple comments also underlined the wish for more user agency over the recommender system, in particular, the ability to navigate between the recommended material (*"[S]ome interaction would go a long way. Maybe [being] able to go back to previous [material] and then be able to come back to [...]. [A] bit more control would be good."* G3-P3).

Regarding system setup, all participants appreciated the individual setup with the tablet close by. However, some believed it would be easier to make connections between the presentation and the recommended material if they were displayed side by side. The presenter could then also refer to the recommended material (*"Maybe it's easier for the presenter to make the link between what is happening with the recommended system if it is on the same level."* G1-P3).

An individual device offers desirable possibilities for personalizing the recommendations. The participants envisioned two different approaches to personalized recommendation content. These would either help to simplify the narrative by making explanatory material more accessible or augment the presentation with stimulating content for people who want more information. Ultimately, most agree that such a system would probably work better as a personalized system as this would allow recommendation material to be tailored to the background of each listener (*"It would be nice to have these things personalized for each person. So me, I'm a little bit stupid with graphs. So all the graphs would be shown on this device for me, but for him, he will have some texts because he is very good with text. He can read and learn with text. So we will have the main content [presented differently]. That would be awesome."* G3-P2). The participants generally preferred personalized recommendations derived in accordance with their parameters instead of recommendations based on their behavioral data ("To some extent, it's not intelligent, but it's more like user-driven." G4-P2).

The tested setup was non-humanoid, with a simple tablet displaying the recommended visual material. Discussion about the potential character of such a system was non-conclusive. Many participants thought this very neutral setup was suitable for a meeting assistant ("[User-]friendliness doesn't really play a big role. I think [I would like] several buttons to call back and that's it." G2-P2). A more humanoid setup similar to Amazon Alexa was met with interest by some. However, there was consensus that users should be able to select a humanoid or non-humanoid setup according to their own preference, which some mentioned could also depend on the context of use ("You know, I may or may not use it for class, depending also on how well the lecturer uses the system." G2-P3).

4.3   The Presenters' Experience with the Meeting Assistant

We gathered the following points about the presenters' point of view. It seems that they quickly forgot about the presence of the meeting assistant even though they were initially aware of it. This contrasts with some participants' remarks that it might irritate presenters.

After showing the presenters the material displayed by the meeting assistant during the presentation, one considered the recommended material to be not very useful, but the other did see the potential value of the system. Both presenters agreed that they would appreciate having control over the material displayed by the meeting assistant, especially when they are inappropriate, even if that rarely happens. Regarding the system setup, the second presenter would be willing to try a setup that involves a robot acting as a meeting assistant, but like most participants, he was not convinced of its suitability either.

4.4   Envisioned Use Cases for the Meeting Assistant

Despite the shortcomings found in user experience during the test session, most participants thought that the meeting assistant had great potential. They imagined a wide range of use cases from education to business settings. In education, they envisaged the system as an individual assistant that knows to what extent the lecture falls into the listener's field of knowledge (*"[p]resentations where the people watching are not the target audience and need additional information for them to better follow and understand what is being shown."* P2-1). It would then display (animated or non-animated) visual material, definitions, essential keywords, even additional information to keep the listener's interest (*"Useful for kids [who] are distracted most of the time, so they have something [to address] distraction, and [...] lead them to the main subject."* P3-2). Combining this with interactivity was seen as highly desirable. They imagined being able to combine the recommended material with the lecture slides and their own annotations. A similar use scenario mentioned by the participants was conferences, where a personalized system would allow more and less experienced listeners to follow the presentations better (*"For some fields that you are not secure about, never heard of, it's kind of probably useful to have like little graphs or some recommendations that help you understand the topic."* P3-1). The same goes for business meetings with people from different backgrounds, where the assistant might save people time googling technical terms (*"[m]eetings where discussions really happen and people come together from different backgrounds for one objective because not everyone knows the same information."* P4-1). They also see it having utility as a fact-checking tool when people have diverging opinions (*"...like corporate meetings, because they also have different levels of complexity in their opinions. And [...] you should look at the screen, and you make sure."* P4-2). Finally, the participants also saw potential use cases in entertainment settings like museums.

5   Discussion and Recommendations

The focus group discussions brought forth manifold insights into how a recommender system-based meeting assistant should be designed. This section outlines our recommendations.

5.1   Introduce the System to Users

We had not introduced the system to the users before because this might not happen in a real-world scenario. The participants' reactions demonstrated the need to manage user expectations and introduce the system, detailing what it can do and where the information comes from. This can be done with simple stand-by text on the first screen. It would help participants become aware of incoming recommendation material and judge to what extent they can trust them.

5.2   Annotate Visual Material with Titles and Keywords

To help users understand the recommendations shown, it is useful to display the keywords used for the search. This will help them know what the displayed visual represents and determine whether it is still relevant to what the presenter is currently saying. The display of small animations could be treated as even more suitable than static visuals, as it is stated that they enhance understanding of topics [18]. However, the feasibility of providing animations depends on the data available in the databases searched. They could be hard to find consistently.

5.3   Cater to User Preferences

Suppose the participants' profiles are known in advance, and they consent to sharing some of their data. In that case, the system could adapt its search criteria across different databases, taking account of the confidence level of each participant. Doing this would improve the relevance and usefulness of the recommended material for each participant, avoiding material that is too hard to understand for non-experts and too basic for experts. On a different matter, users should also be able to indicate what type of material they prefer, ranging from text to images or even animations.

5.4   Add Interactivity Options

Our experiment showed that users require a cue to let them know when new recommendation material is available. An audio signal would be too intrusive, but one option could be to add an overlay icon to the presentation display to let people know there is new material to view from the recommender system. Another option is to let people click a button to trigger a search for material about what is currently being presented. However, according to the literature, people prefer a proactive system to a reactive or human-initiated one [5]. Beyond a recommendation cue, our results indicate a strong user need to navigate back and forth between the recommended material.

5.5   Companion Style

Unlike some of the literature, which states that using a human-like robot increases users' trust in such systems [31], we cannot give a clear recommendation regarding a specific style of meeting assistant. Whether users prefer neutral or humanoid systems appears to be a rather personal and context-dependent choice.

6   Limitations and Future Work

The study could not employ an existing prototype and had to simulate a recommender system via a Wizard-of-Oz setup [11]. The choice to test during a presentation rather than a "meeting-style" discussion among participants was motivated by the impossibility of simulating recommendations in real time regarding any kind of dialogue. In addition, the pre-selected recommendations did not always perfectly match the discourse when the presenters strayed from their initial script. The setup did not allow for the simulation of personalized recommendations at this early concept stage. In our future work, we plan to pursue the development of a meeting assistant based on the design recommendations. Early prototypes will be user tested on various features, including different setups, personalized recommendations, and presenter control.

7   Conclusion

In this study, we simulated a meeting assistant system that displays recommendation material to an audience. Our user study aimed to investigate how users perceive recommender systems employed in real-world meeting situations and to identify user requirements ranging from contextual triggers to navigation options. Our findings suggest that there are some challenges involved with contextualizing the recommendation material and that listeners may be distracted by the recommendations if they are not engaged in the meeting.

Acknowledgement. This research is supported by the Luxembourg National Research Fund (FNR): IPBG2020/IS/14839977/C21.

This work has been partially supported by the Chist-Era grant CHIST-ERA19-XAI-005, by Swiss National Science Foundation (G.A. 20CH21 195530), Italian Ministry for Universities and Research, Luxembourg National Research Fund (G.A. INTER/CHIST/19/14589586), Scientific and Research Council of Turkey (TÜBİTAK, G.A. 120N680).

References

1. Alcaraz, B., Hosseini-Kivanani, N., Najjar, A.: IRRMA: an image recommender robotic meeting assistant. In: Dignum, F., Mathieu, P., Corchado, J.M., De La Prieta, F. (eds.) Advances in Practical Applications of Agents, Multi-Agent Systems, and Complex Systems Simulation. The PAAMS Collection. PAAMS 2022. LNCS, vol. 13616. Springer, Cham (2022). https://doi.org/10.1007/978-3-031-18192-4_36
2. Amershi, S., et al.: Guidelines for human-AI interaction. In: Proceedings of the 2019 Chi Conference on Human Factors in Computing Systems, pp. 1–13 (2019)
3. Assayag, M., et al.: Meeting assistant application. In: Sixteenth Annual Conference of the International Speech Communication Association (2015)
4. Ball, J.: The double diamond: a universally accepted depiction of the design process (2022). www.designcouncil.org.uk/our-work/news-opinion/double-diamond-universally-accepted-depiction-design-process
5. Baraglia, J., Cakmak, M., Nagai, Y., Rao, R.P., Asada, M.: Efficient human-robot collaboration: when should a robot take initiative? Int. J. Robot. Res. **36**(5–7), 563–579 (2017)
6. Billings, D.R., Schaefer, K.E., Chen, J.Y., Hancock, P.A.: Human-robot interaction: developing trust in robots. In: Proceedings of the Seventh Annual ACM/IEEE International Conference on Human-Robot Interaction, pp. 109–110 (2012)

7. Çano, E., Morisio, M.: Hybrid recommender systems: a systematic literature review. Intell. Data Anal. **21**(6), 1487–1524 (2017)
8. Champiri, Z.D., Mujtaba, G., Salim, S.S., Yong Chong, C.: User experience and recommender systems. In: 2019 2nd International Conference on Computing, Mathematics and Engineering Technologies (iCoMET), pp. 1–5 (2019). https://doi.org/10.1109/ICOMET.2019.8673410
9. Chiu, M.C., Huang, J.H., Gupta, S., Akman, G.: Developing a personalized recommendation system in a smart product service system based on unsupervised learning model. Comput. Ind. **128**, 103421 (2021)
10. Clarke, V., Braun, V., Hayfield, N.: Thematic analysis. Qualit. Psychol. a practical guide to research methods **222**, 248 (2015)
11. Dahlbäck, N., Jönsson, A., Ahrenberg, L.: Wizard of OZ studies-why and how. Knowl.-Based Syst. **6**(4), 258–266 (1993)
12. Do, V., Corbett-Davies, S., Atif, J., Usunier, N.: Online certification of preference-based fairness for personalized recommender systems. In: Proceedings of the AAAI Conference on Artificial Intelligence, vol. 36, pp. 6532–6540 (2022)
13. Ehnes, J.: A tangible interface for the AMI content linking device-the automated meeting assistant. In: 2009 2nd Conference on Human System Interactions, pp. 306–313. IEEE (2009)
14. Fessenden, T.: Focus groups 101 (2022). www.nngroup.com/articles/focus-groups-definition/
15. Ge, Y., et al.: A survey on trustworthy recommender systems. arXiv preprint arXiv:2207.12515 (2022)
16. Hassenzahl, M.: User experience (UX) towards an experiential perspective on product quality. In: Proceedings of the 20th Conference on l'Interaction Homme-Machine, pp. 11–15 (2008)
17. Herse, S., et al.: Do you trust me, blindly? factors influencing trust towards a robot recommender system. In: 2018 27th IEEE International Symposium on Robot and Human Interactive Communication (RO-MAN), pp. 7–14. IEEE (2018)
18. Höffler, T.N., Leutner, D.: Instructional animation versus static pictures: a meta-analysis. Learn. Instr. **17**(6), 722–738 (2007)
19. Isbister, K., Nakanishi, H., Ishida, T., Nass, C.: Helper agent: designing an assistant for human-human interaction in a virtual meeting space. In: Proceedings of the SIGCHI Conference on Human Factors in Computing Systems, pp. 57–64 (2000)
20. Lallemand, C., Mercier, E.: Optimizing the use of the sentence completion survey technique in user research: a case study on the experience of e-reading. In: CHI Conference on Human Factors in Computing Systems, pp. 1–18 (2022)
21. Laugwitz, B., Held, T., Schrepp, M.: Construction and evaluation of a user experience questionnaire. In: Holzinger, A. (ed.) USAB 2008. LNCS, vol. 5298, pp. 63–76. Springer, Heidelberg (2008). https://doi.org/10.1007/978-3-540-89350-9_6
22. van Maris, A., Lehmann, H., Natale, L., Grzyb, B.: The influence of a robot's embodiment on trust: a longitudinal study. In: Proceedings of the Companion of the 2017 ACM/IEEE International Conference on Human-Robot Interaction, pp. 313–314 (2017)
23. Martijn, M., Conati, C., Verbert, K.: "knowing me, knowing you": personalized explanations for a music recommender system. User Model. User-Adapt. Inter. **32**(1), 215–252 (2022). https://doi.org/10.1007/s11257-021-09304-9
24. Milano, S., Taddeo, M., Floridi, L.: Recommender systems and their ethical challenges. AI Soc. **35**(4), 957–967 (2020)

25. Popescu-Belis, A., et al.: The AMIDA automatic content linking device: just-in-time document retrieval in meetings. In: Popescu-Belis, A., Stiefelhagen, R. (eds.) MLMI 2008. LNCS, vol. 5237, pp. 272–283. Springer, Heidelberg (2008). https://doi.org/10.1007/978-3-540-85853-9_25

26. Popescu-Belis, A., Kilgour, J., Nanchen, A., Poller, P.: The ACLD: speech-based just-in-time retrieval of meeting transcripts, documents and websites. In: Proceedings of the 2010 International Workshop On Searching Spontaneous Conversational Speech, pp. 45–48 (2010)

27. Rae, I., Takayama, L., Mutlu, B.: In-body experiences: embodiment, control, and trust in robot-mediated communication. In: Proceedings of the SIGCHI Conference on Human Factors in Computing Systems, pp. 1921–1930 (2013)

28. Renjith, S., Sreekumar, A., Jathavedan, M.: An extensive study on the evolution of context-aware personalized travel recommender systems. Inf. Process. Manage. **57**(1), 102078 (2020)

29. Sinha, R., Swearingen, K.: The role of transparency in recommender systems. In: CHI2002 Extended Abstracts On Human Factors In Computing Systems, pp. 830–831 (2002)

30. Still, B., Crane, K.: Fundamentals of user-centered design: a practical approach. CRC Press (2017)

31. Stroessner, S.J., Benitez, J.: The social perception of humanoid and non-humanoid robots: effects of gendered and machinelike features. Int. J. Soc. Robot. **11**(2), 305–315 (2019)

32. Symeonidis, P., Nanopoulos, A., Manolopoulos, Y.: Moviexplain: a recommender system with explanations. In: Proceedings of the Third ACM Conference on Recommender systems, pp. 317–320 (2009)

33. Thompson, P., James, A., Stanciu, E.: Agent based ontology driven virtual meeting assistant. In: Kim, T., Lee, Y., Kang, B.-H., Ślęzak, D. (eds.) FGIT 2010. LNCS, vol. 6485, pp. 51–62. Springer, Heidelberg (2010). https://doi.org/10.1007/978-3-642-17569-5_8

34. Tur, G., Tur, G., et al.: The CALO meeting assistant system. IEEE Trans. Audio Speech Lang. Process. **18**(6), 1601–1611 (2010)

35. Villegas, N.M., Sánchez, C., Díaz-Cely, J., Tamura, G.: Characterizing context-aware recommender systems: a systematic literature review. Knowl.-Based Syst. **140**, 173–200 (2018)

36. Wu, W., He, L., Yang, J.: Evaluating recommender systems. In: Seventh International Conference on Digital Information Management (ICDIM 2012), pp. 56–61. IEEE (2012)

37. Zhang, S., Yao, L., Sun, A., Tay, Y.: Deep learning based recommender system: a survey and new perspectives. ACM Comput. Surv. (CSUR) **52**(1), 1–38 (2019)

Auditing Consumer- and Producer-Fairness in Graph Collaborative Filtering

Vito Walter Anelli, Yashar Deldjoo, Tommaso Di Noia, Daniele Malitesta[(✉)],
Vincenzo Paparella, and Claudio Pomo[(✉)]

Politecnico di Bari, Bari, Italy
{vitowalter.anelli,yashar.deldjoo,tommaso.dinoia,daniele.malitesta,
vincenzo.paparella,claudio.pomo}@poliba.it

Abstract. To date, *graph* collaborative filtering (CF) strategies have been shown to outperform pure CF models in generating accurate recommendations. Nevertheless, recent works have raised concerns about fairness and potential biases in the recommendation landscape since unfair recommendations may harm the interests of Consumers and Producers (CP). Acknowledging that the literature lacks a careful evaluation of graph CF on CP-aware fairness measures, we initially evaluated the effects on CP-aware fairness measures of eight state-of-the-art graph models with four pure CF recommenders. Unexpectedly, the observed trends show that graph CF solutions do not ensure a large item exposure and user fairness. To disentangle this performance puzzle, we formalize a taxonomy for graph CF based on the mathematical foundations of the different approaches. The proposed taxonomy shows differences in node representation and neighbourhood exploration as dimensions characterizing graph CF. Under this lens, the experimental outcomes become clear and open the doors to a multi-objective CP-fairness analysis (Codes are available at: https://github.com/sisinflab/ECIR2023-Graph-CF.).

Keywords: Graph collaborative filtering · Fairness · Multi-objective analysis

1 Introduction and Motivations

Recommender systems (RSs) are ubiquitous and utilized in a wide range of domains from e-commerce and retail to media streaming and online advertising. Personalization, or the system's ability to suggest relevant and engaging products to users, has long served as a key indicator for gauging the success of RSs. In recent decades, collaborative filtering (CF) [10], the predominant modeling paradigm in RSs, has shifted from neighborhood techniques [10,30,31] to frameworks based on the learning of users' and items' latent factors [16,29,49].

Authors are listed in alphabetical order.

J. Kamps et al. (Eds.): ECIR 2023, LNCS 13980, pp. 33–48, 2023.
https://doi.org/10.1007/978-3-031-28244-7_3

More recently, deep learning (DL) models have been proposed to overcome the linearity of traditional latent factors approaches.

Among these DL algorithms, graph-based methods view the data in RSs from the perspective of graphs. By modeling users and items as nodes with latent representations and their interactions as edges, the data can be naturally represented as a user-item bipartite graph. By iteratively aggregating contributions from near- and long-distance neighborhoods, the so-called message-passing schema updates nodes' initial representations and effectively distills the collaborative signal [43]. Early works [5,50] adopted the vanilla graph convolutional network (GCN) [15] architecture and paved the way to advanced algorithms lightening the message-passing schema [8,14] and exploring different graph sampling strategies [47]. Recent approaches propose simplified formulations [21,26] that optionally transfer the graph CF paradigm to different spaces [33,34]. As some graph edges may provide noisy contributions to the message-passing schema [39], a research line focuses on meaningful user-item interactions [36,42,45]. In this context, explainability is the natural next step [18] towards the disentanglement of user-item connections into a set of user intents [44,46].

On the other side, the adoption of DL (and, often, black-box) approaches to the recommendation task has raised issues regarding the fairness of RSs. The concept of fairness in recommendation is multifaceted. Specifically, the two core aspects to categorize recommendation fairness may be summarized as (1) the primary parties engaged (consumers vs. producers) and (2) the type of benefit provided (exposure vs. relevance). Item suppliers are more concerned about exposure fairness than customers because they want to make their products better known and visible (**P**roducer fairness). However, from the customer's perspective, relevance fairness is of utmost importance, and hence system designers must ensure that exposure of items is equally effective across user groups (**C**onsumer fairness). A recent study highlights that nine out of ten publications on recommendation fairness concentrated on either C-fairness or P-fairness [22], disregarding the joint evaluation between C-fairness, P-fairness, and the accuracy.

The various graph CF *strategies* described above have historically centered on the enhancement of system accuracy, but, actually, never focused on the recommendation fairness dimensions. Despite some recent graph-based approaches have specifically been designed to address C-fairness [11,17,27,40,41,48] and P-fairness [6,19,20,35,51,52], there is a notable *knowledge gap* in the literature about the effects of the state-of-the-art graph *strategies* on the three objectives of C-fairness, P-fairness, and system accuracy. This work intends to complement the previous research and provide answers to pending research problems such as how different graph models perform for the three evaluation objectives. By measuring these dimensions in terms of **overall accuracy**, **user fairness**, and **item exposure**, we observe these aspects in detail[1].

Motivating Example. A preliminary comparison of the leading graph and classical CF models is carried out to provide context for our study. The graph-based

---

[1] In the rest of the paper, when no confusion arises, we will refer to C-fairness with user fairness, to P-fairness with item exposure, and to their combination as CP-fairness.

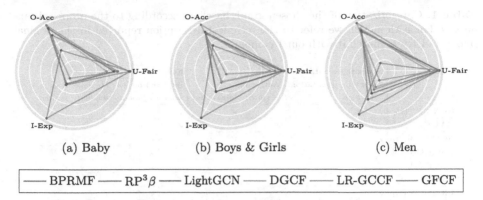

(a) Baby   (b) Boys & Girls   (c) Men

—— BPRMF —— RP$^3\beta$ —— LightGCN —— DGCF —— LR-GCCF —— GFCF

Fig. 1. Kiviat diagrams indicating the performance of selected pure and graph CF recommenders on overall accuracy (i.e., O-Acc, calculated with the *nDCG@20*), item exposure (i.e., I-Exp, calculated with the *APLT@20* [1]), and user fairness (U-Fair, calculated with the *UMADrat@20* [9]). Higher means better.

models include LightGCN [14], DGCF [44], LR-GCCF [8], and GFCF [33], which are tested against two classical CF baselines, namely BPRMF [28] and RP$^3\beta$ [25], on the Baby, Boys & Girls, and Men datasets from the Amazon catalog [23]. We train each baseline using a total of 48 unique hyper-parameter settings and select the optimal configuration for each baseline as the one achieving the highest accuracy on the validation set (as in the original papers). Overall accuracy, user fairness, and item exposure (as introduced above) are evaluated. Figure 1 displays the performance of the selected baselines on the three considered recommendation objectives. For better visualization, all values are scaled between 0 and 1 using min-max normalization, and, when needed, they are replaced by their 1's complement to adhere to the "higher numbers are better" semantics. As a result, in each of the three dimensions, the values lay in [0, 1] with higher values indicating the better. Please, note that such an experimental evaluation is not the main focus of this work but it is the motivating example for the more extensive analysis we present later. The interested reader may refer to **Appendix** A for a presentation of the full experimental settings to reproduce these results and the ones reported in the following sections of the paper.

First, according to Fig. 1, graph CF models are significantly more accurate than the classical CF ones, even if the latter perform far better in terms of item exposure. Moreover, the displayed trends suggest there is no clear winner on the user fairness dimension: classical CF models show promising performance, while some graph CF models do not achieve remarkable results. As a final observation, an underlying trade-off between the three evaluation goals seems to exist, and it might be worth investigating it in-depth. Such outcomes open to a more complete study on how **different strategy patterns** recognized in graph CF may affect the three recommendation objectives, which is the scope of this work.

Research Questions and Contributions. In the remainder of this paper, we therefore attempt to answer the following two research questions (RQs):

Table 1. Categorization of the chosen graph baselines according to the proposed taxonomy. For each model, we refer to the technical description reported in the original paper and try to match it with our taxonomy.

Models	Nodes representation				Neighborhood exploration			
	Latent representation		Weighting		Explored nodes		Message passing	
	low	high	weighted	unweighted	same	different	implicit	explicit
GCN-CF* [15]		✓		✓	✓			✓
GAT-CF* [39]		✓	✓		✓			✓
NGCF [43]	✓			✓		✓		✓
LightGCN [14]	✓			✓		✓		✓
DGCF [44]	✓		✓			✓		✓
LR-GCCF [8]	✓			✓	✓	✓		✓
UltraGCN [21]	✓				✓	✓	✓	
GFCF [33]						✓	✓	

*The postfix -CF indicates that we re-adapted the original implementations (tailored for the task of node classification) to the task of personalized recommendation.

RQ1. Given the different graph CF strategies, the raising question is *"Can we explain the variations observed when testing several graph models on overall accuracy, item exposure, and user fairness separately?"* According to a recent benchmark that identifies some state-of-the-art graph techniques [54], the suggested graph CF taxonomy (Table 1) extends the set of graph-based models introduced in the motivating example by examining eight state-of-the-art graph CF baselines through their strategies for *nodes representation* and *neighborhood exploration*. We present a more nuanced view of prior findings by analyzing the impact of each taxonomy dimension on overall accuracy and CP-fairness.

RQ2. The demonstrated performance prompts the questions: *"How and why nodes representation and neighborhood exploration algorithms can strike a trade-off between overall accuracy, item exposure, and user fairness?"* We employ the Pareto optimality to determine the influence of such dimensions in two-objective scenarios, where the objectives include overall accuracy, item exposure, and user fairness. The Pareto frontier is computed for three 2-dimensional spaces: accuracy/item exposure, accuracy/user fairness, and item exposure/user fairness.

2 Nodes Representation and Neighborhood Exploration in Graph Collaborative Filtering: A Formal Taxonomy

2.1 Preliminaries

Let $\mathcal{U}$ be the set of N users, and $\mathcal{I}$ the set of M items in the system, respectively. We represent the observed interactions between users and items in a binary format (i.e., implicit feedback). Specifically, let $\mathbf{R} \in \mathbb{R}^{N \times M}$ be the user-item feedback matrix, where $r_{u,i} = 1$ if user $u \in \mathcal{U}$ and item $i \in \mathcal{I}$ have a recorded interaction, $r_{u,i} = 0$ otherwise. Following the above preliminaries, we introduce

$\mathcal{G} = (\mathcal{U}, \mathcal{I}, \mathbf{R})$ as the bipartite and undirected graph connecting users and items (the graph nodes) when there exists a recorded bi-directional interaction among them (the graph edges). Nodes features for user $u \in \mathcal{U}$ and $i \in \mathcal{I}$ are suitably encoded as the embeddings $\mathbf{e}_u \in \mathbb{R}^d$ and $\mathbf{e}_i \in \mathbb{R}^d$, with $d << N, M$. Given the dual nature of user and item derivations, we only report user-side formulas.

2.2 Updating Node Representation Through Message-Passing

The representation of users' and items' nodes are updated by leveraging the graph topology from $\mathcal{G}$. In this respect, the message-passing schema has recently gained attention in the literature. The algorithm works by aggregating the information (i.e., the *messages*) from the *neighbor* nodes into the *ego* node, and the process is recursively performed for multiple hops thus exploring wider neighborhood portions. In general, the message-passing for l hops is:

$$\mathbf{e}_u^{(l)} = \omega\left(\left\{\mathbf{e}_{i'}^{(l-1)}, \forall i' \in \mathcal{N}(u)\right\}\right),\tag{1}$$

where $\omega(\cdot)$ and $\mathcal{N}(\cdot)$ are the aggregation function and neighborhood node set, respectively, while l is in $1 \le l \le L$, where L is a hyper-parameter. Note that the following statements hold: $\mathbf{e}_u^{(0)} = \mathbf{e}_u$ and $\mathbf{e}_i^{(0)} = \mathbf{e}_i$. A reworking of Eq. (1) for $l \in \{2, 3\}$ allows *same-* and *different*-type node representation emerge [3]:

$$
\begin{cases}
\textit{Same-type} & \underbrace{\mathbf{e}_u^{(2)}}_{(\text{user})} = \omega\left(\left\{\omega\left(\left\{\underbrace{\mathbf{e}_{u''}^{(0)}}_{(\text{user})}, \forall u'' \in \mathcal{N}(i') \setminus \{u\}\right\}\right), \forall i' \in \mathcal{N}(u)\right\}\right) \\
\textit{node} \\
\textit{representation} \\[2em]
\textit{Different-type} & \underbrace{\mathbf{e}_u^{(3)}}_{(\text{user})} = \omega\left(\left\{\omega\left(\left\{\omega\left(\left\{\underbrace{\mathbf{e}_{i'''}^{(0)}}_{(\text{item})}, \forall i''' \in \mathcal{N}(u'') \setminus \{i''\}\right\}\right),\right.\right.\right. \\
\textit{node} & \hspace{2em}\left.\left.\left. \forall u'' \in \mathcal{N}(i') \setminus \{u''\}\right\}\right), \forall i' \in \mathcal{N}(u)\right\}\right). \\
\textit{representation}
\end{cases}
\tag{2}
$$

To better clarify the extent of Eq. (2), after an **even** and an **odd** number of explored hops, *ego* node updates leverage by design *same-* and *different*-type node connections, i.e., user-user/item-item and user-item/item-user as evident from Eq. (2). While the existing literature does not always consider the two scenarios as distinct, we underline the importance of investigating the influence of different node-node connections explored during the message-passing. In light of the above, we will count the number of explored hops as follows: $\mathbf{e}_*^{(2l)}, \forall l \in \{1, 2, \dots, \frac{L}{2}\}$ as obtained through l **same**-type node connections (denoted as *same-l*), and $\mathbf{e}_*^{(2l-1)}, \forall l \in \{1, 2, \dots, \frac{L}{2}\}$ as obtained through l **different**-type node connections (denoted as *different-l*). In the following, we introduce the graph convolutional network (GCN) and its recent CF applications.

The Baseline: Graph Convolutional Network (GCN). The standard graph convolutional network from Kipf and Welling [15] performs feature transformation, message aggregation, application of a one-layer neural network, element-wise addition, and ReLU activation, respectively. Let us consider

$\mathbf{W}^{(l)} \in \mathbb{R}^{d_{l-1} \times d_l}$ and $\mathbf{b}^{(l)} \in \mathbb{R}^{d_l}$ as the weight matrix and the bias for the l-th explored hop. The message-passing for user u is:

$$\mathbf{e}_u^{(l)} = \text{ReLU} \left(\sum_{i' \in \mathcal{N}(u)} \left(\mathbf{W}^{(l)} \mathbf{e}_{i'}^{(l-1)} + \mathbf{b}^{(l)} \right) \right). \tag{3}$$

GCN for Collaborative Filtering. Inspired by the GCN message-passing approach, the authors from Wang et al. [43] propose neural graph collaborative filtering (NGCF). At each hop exploration, the model aggregates the neighborhood information and the inter-dependencies among the *ego* and the neighborhood nodes. Formally, the aggregation could be formulated as follows:

$$\mathbf{e}_u^{(l)} = \text{LeakyReLU} \left(\sum_{i' \in \mathcal{N}(u)} \left(\mathbf{W}_{\text{neigh}}^{(l)} \mathbf{e}_{i'}^{(l-1)} + \mathbf{W}_{\text{inter}}^{(l)} \left(\mathbf{e}_{i'}^{(l-1)} \odot \mathbf{e}_u^{(l-1)} \right) + \mathbf{b}^{(l)} \right) \right), \tag{4}$$

where LeakyReLU is the activation function, $\mathbf{W}_{\text{neigh}}^{(l)} \in \mathbb{R}^{d_{l-1} \times d_l}$ and $\mathbf{W}_{\text{inter}}^{(l)} \in \mathbb{R}^{d_{l-1} \times d_l}$ are the neighborhood and inter-dependencies weight matrices, respectively, while $\odot$ is the Hadamard product.

He et al. [14] propose a light convolutional network, namely LightGCN, with the rationale to simplify the message-passing schema from GCN and NGCF by dropping feature transformations (i.e., the weight matrices and biases) and the non-linearity applied after the message aggregation. Specifically, they implement:

$$\mathbf{e}_u^{(l)} = \sum_{i' \in \mathcal{N}(u)} \mathbf{e}_{i'}^{(l-1)}. \tag{5}$$

The variation shows superior accuracy to the state-of-the-art. A slightly different solution [8] can outperform LightGCN regarding the accuracy level.

2.3   Weighting the Importance of Graph Edges

The message-passing schema is inherently designed to aggregate into the *ego* node all messages coming from its neighborhood. Nevertheless, the *binary* nature of the user-item feedback (i.e., 0/1) would suggest that not all recorded user-item interactions necessarily hide the same importance to the nodes they involve.

In general, let $a_{y \to x}^{(l)}$ be the importance of the neighbor node y on its ego node x after l explored hops. We re-write the formulation of the message-passing after l explored hops (presented in Eq. (1)) as:

$$\mathbf{e}_u^{(l)} = \omega \left(\left\{ a_{i' \to u}^{(l)} \mathbf{e}_{i'}^{(l-1)}, \forall i' \in \mathcal{N}(u) \right\} \right). \tag{6}$$

The Baseline: Graph Attention Network (GAT). Attention mechanisms have reached considerable success in the GCN-related literature to weight the contribution of neighbor messages before aggregation. The original study [39] proposes the following message-passing formulation:

$$
\begin{aligned}
\mathbf{e}_u^{(l)} &= \sum_{i' \in \mathcal{N}(u)} \left(a_{i' \to u}^{(l)} \mathbf{W}_{\text{neigh}}^{(l)} \mathbf{e}_{i'}^{(l-1)} + \mathbf{b}^{(l)} \right) \\
&= \sum_{i' \in \mathcal{N}(u)} \left(\alpha \left(\mathbf{e}_{i'}^{(l-1)}, \mathbf{e}_u^{(l-1)} \right) \mathbf{W}_{\text{neigh}}^{(l)} \mathbf{e}_{i'}^{(l-1)} + \mathbf{b}^{(l)} \right),
\end{aligned}
\tag{7}
$$

where $\alpha(\cdot)$ is the importance function depending on the lastly-calculated embeddings of the neighbor and the ego nodes, e.g., $a_{i' \to u}^{(l)} = \alpha \left(\mathbf{e}_{i'}^{(l-1)}, \mathbf{e}_u^{(l-1)} \right)$.

GAT for Collaborative Filtering. The authors from Wang et al. [44] design a message-passing schema that calculates the importance of neighborhood nodes for *ego* nodes by disentangling the intents underlying each user-item interaction. Similarly to He et al. [14] and Chen et al. [8], they therefore propose the following embedding update formulation:

$$
\begin{aligned}
\mathbf{e}_u^{(l)} &= \sum_{i' \in \mathcal{N}(u)} a_{i' \to u}^{(l)} \mathbf{e}_{i'}^{(l-1)} \\
&= \sum_{i' \in \mathcal{N}(u)} \alpha \left(\mathbf{e}_{i'}^{(l-1)}, \mathbf{e}_u^{(l-1)}, K, T \right) \mathbf{e}_{i'}^{(l-1)},
\end{aligned}
\tag{8}
$$

where $\alpha(\cdot, K, T)$ is the importance function of the lastly-calculated embeddings from the neighbor and the *ego* nodes, e.g., $a_{i' \to u}^{(l)} = \alpha \left(\mathbf{e}_{i'}^{(l-1)}, \mathbf{e}_u^{(l-1)}, K, T \right)$, K is the total number of intents, and T is the total number of routing iterations to repeat the disentangling procedure.

2.4 Going Beyond Message-Passing

The recent graph learning literature [7,53] has outlined the phenomenon of *over-smoothing*, that leads node representations to become more similar as more hops are explored. The issue is generally tackled by limiting the neighborhood exploration to (maximum) three hops, and to two hops when attention mechanisms are introduced. However, the idea of improving accuracy by restricting the number of explored neighborhoods is counter-intuitive and "conflicts" with the rationale behind collaborative filtering [4]. This awareness led works such as Mao et al. [21] and Shen et al. [33] to surpass and simplify the traditional concept of message-passing. UltraGCN [21] adopts negative sampling to contrast over-smoothing and additional objective terms to (i) approximate the infinite neighborhood exploration and (ii) mine relevant "unexpected" node-node interactions such as the item-item ones. Conversely, GFCF [33] translates the graph-based recommendation task into the graph signal processing domain to obtain a closed-form formulation for approximating the infinite neighborhood exploration. Given that such recent strategies do not *explicitly* perform the message-passing schema as presented above, in the remaining sections of this paper, we will adopt the terms *explicit* and *implicit* message-passing as shorthands to denote the two model families, respectively.

2.5   A Taxonomy of Graph CF Approaches

We propose (see Table 1) a taxonomy to classify the state-of-the-art graph models. The taxonomy considers the recurrent **strategy patterns** as emerged by conducting an in-depth review and analyzing the different graph CF approaches.

- **Node representation** indicates the representation strategy to model users' and items' nodes. It involves the *dimensionality* of node embeddings, and the possibility of *weighting* the neighbor node contributions.
- **Neighborhood exploration** refers to the procedure for exploring the multi-hop neighborhoods of each node to update the node latent representation. It involves the type of *node-node connections* which are explored, and the *message-passing* schema (i.e., *explicit* or *implicit* as previously defined).

In the next two sections, we will assess the performance of the graph CF models from the taxonomy in Table 1. Thus, we consider GCN-CF [15], GAT-CF [39], NGCF [43], LightGCN [14], DGCF [44], LR-GCCF [8], UltraGCN [21], and GFCF [33] for a total of eight graph CF solutions.

3   Taxonomy-aware Evaluation

This section aims to answer RQ1 (*"Can we explain the variations observed when testing several graph models on overall accuracy, item exposure, and user fairness separately?"*) by showing how the proposed taxonomy of graph strategies can explain the recommendation evaluation on CP-Fairness and overall accuracy. We experiment with 48 hyper-parameter configurations to investigate various combinations of graph CF techniques for *message-passing*, *explored nodes*, *edge weighting*, and *latent representations*. Results refer to the Amazon Men dataset and top-20 lists (Table 2). Please note that we report the **best** metric result for each <dimension, value> pair (the corresponding best graph recommendation model is displayed below each metric result) to ease the interpretation of results and provide meaningful insights.

- **Message-passing.** We investigate the two widely-recognized message-passing strategies: *implicit* and *explicit*. The most obvious pattern indicates that both sets have almost the same number of top-performing models in each of the evaluation criteria. *Explicit* graph approaches perform better on item exposure, where they outperform *implicit* techniques (i.e., on *Gini* and *APLT*) two out of three times by a significant margin. On the one hand, this tendency may be due to the absence of a direct message (information) propagating along the user-item graph in *implicit* techniques, which prevents the user node from exploring vast item segments. On the other hand, it appears that models from both families perform similarly on accuracy and user fairness, indicating that there is no obvious reason to favor *implicit* over *explicit* or vice versa.
- **Explored nodes.** Here, we examine four methods to explore nodes (adopting the message-passing re-formulation from Eq. (2)): *same* and *different*, with 1

Table 2. Best metric results (and corresponding graph CF model) for each <dimension, value> pair, on the Amazon Men dataset for top-20 lists. **Bold** is used to indicate the best result in the pairs having a two-valued dimension, while † is used only for the "explored nodes" dimension to indicate also the best results on *same* and *different*. The symbols ↑ and ↓ indicate whether better stands for high or low values. We use "*rank*" and "*rat*" as the *UMADrank@k* and *UMADrat@k*.

Dimensions	Values	Overall accuracy		Item exposure			User fairness	
		Recall↑	nDCG↑	EFD↑	Gini↑	APLT↑	rank↓	rat↓
Message passing	implicit	**0.1222** (GFCF)	**0.0911** (GFCF)	**0.2615** (GFCF)	0.2871 (UltraGCN)	0.1808 (UltraGCN)	0.0123 (UltraGCN)	**0.0022** (UltraGCN)
	explicit	**0.1223** (LR-GCCF)	0.0884 (LR-GCCF)	0.2536 (LR-GCCF)	**0.5090** (LR-GCCF)	**0.3823** (GAT-CF)	**0.0002** (DGCF)	0.0169 (LightGCN)
Explored nodes	same-1	0.1221† (LR-GCCF)	0.0884† (LR-GCCF)	0.2500† (LR-GCCF)	0.4377 (LR-GCCF)	0.3433 (GAT-CF)	0.0002† (DGCF)	0.0022† (UltraGCN)
	same-2	0.1184 (LightGCN)	0.0841 (LightGCN)	0.2380 (LightGCN)	0.5090† (LR-GCCF)	0.3823† (GAT-CF)	0.0002† (DGCF)	0.0209 (NGCF)
	different-1	0.1222† (GFCF)	0.0911† (GFCF)	0.2615† (GFCF)	0.4093 (NGCF)	0.3424 (GAT-CF)	0.0002† (DGCF)	0.0022† (UltraGCN)
	different-2	0.1210 (DGCF)	0.0850 (DGCF)	0.2407 (LightGCN)	0.4934† (LR-GCCF)	0.3438† (LR-GCCF)	0.0002† (DGCF)	0.0388 (LightGCN)
Weighting	weighted	0.1210 (DGCF)	0.0857 (DGCF)	0.2428 (DGCF)	0.3240 (DGCF)	**0.3823** (GAT-CF)	**0.0002** (DGCF)	0.0301 (DGCF)
	unweighted	**0.1223** (LR-GCCF)	**0.0884** (LR-GCCF)	**0.2536** (LR-GCCF)	**0.5090** (LR-GCCF)	0.3438 (LR-GCCF)	0.0101 (GCN-CF)	**0.0169** (LightGCN)
Latent representations	emb-64	0.1193 (LR-GCCF)	0.0871 (LR-GCCF)	0.2479 (LR-GCCF)	**0.5090** (LR-GCCF)	0.3627 (GAT-CF)	**0.0002** (DGCF)	0.0054 (UltraGCN)
	emb-128	0.1221 (LR-GCCF)	0.0883 (LR-GCCF)	**0.2536** (LR-GCCF)	**0.5090** (LR-GCCF)	0.3644 (GAT-CF)	**0.0002** (DGCF)	0.0111 (UltraGCN)
	emb-256	**0.1223** (LR-GCCF)	**0.0884** (LR-GCCF)	0.2532 (LR-GCCF)	0.5038 (LR-GCCF)	**0.3823** (GAT-CF)	**0.0002** (DGCF)	**0.0022** (UltraGCN)

and 2 hops. Similarly to the trend found for the message-passing dimension, the results demonstrate that the two primary categories (*same* and *different*) are nearly equally performing across all measurements, with *same-2* and *different-1* being the prominent ones. In detail, the *different-1* exploration outperforms the *same-2* on the overall accuracy level (GFCF is the leading model here). Conversely, *same-2* is the best strategy for item exposure (with LR-GCCF and GAT-CF leading). As observed for the message-passing, user fairness does not give a reason to choose between *same* and *different*. The exploration of 1 hop in *same* and *different* settings is the preferable technique, even if 2 hops connections lead to a better item exposure.

- **Weighted.** This study examines *weighted* and *unweighted* graph CF techniques. Differently from above, we observe that *unweighted* solutions provide the best performance on almost all CP-fairness metrics, with LR-GCCF steadily being the superior approach. The only trend deviation refers to GAT-CF (i.e., a *weighted* method) surpassing *unweighted* solutions on the *APLT* level, that is, recommending items from the long-tail. The behavior is likely attributable to the design of *weighted* techniques, which can investigate farther neighbors of the *ego* node (observe the performance of GAT-CF on the *same-2* dimension), leading user profiles to match distant (and possibly niche) products in the catalog. On the contrary, it is interesting to notice how the other two metrics accounting for item exposure (i.e., *EFD* as item novelty measure and *Gini* as item diversity measure) seem to privilege *unweighted* graph techniques (i.e., LR-GCCF). The observed behaviors differ as the three metrics provide completely different perspectives of the *item exposure*, and thus they are uncorrelated.

- **Latent representations.** We compare the performance of graph CF techniques adopting latent representations with *64*, *128*, and *256* features, respectively. It is worth noticing that higher latent representations (i.e., *128* and *256*) result in better performance on all measurements. Specifically, it appears that the *128* dimension is the turning point after which the trend becomes stable (i.e., the metric values for *128* and *256* are frequently comparable). This may be an important insight since the majority of research works in recent literature tend to employ *64*-embedded representations of nodes without exploring further dimensionalities (see Table 1 as a reference).

4   Trade-off Analysis

This section analyses how the graph CF baselines balance the trade-off among accuracy, item exposure, and user fairness, and aims to answer RQ2 (*"How and why nodes representation and neighborhood exploration algorithms can strike a trade-off between overall accuracy, item exposure, and user fairness?"*). Due to space constraints, we report the results only for the Amazon Men dataset. The negative Pearson correlation values for accuracy/item exposure (*nDCG/APLT*) and accuracy/user fairness (*nDCG/UMADrank*) suggest that a trade-off may be necessary, and desirable. In addition, the same correlation metric indicates the necessity of a trade-off for item exposure/user fairness (*APLT/UMADrank*). Among the strategy patterns identified in the proposed taxonomy (see Table 1), we select the most important architectural dimensions, **message-passing** and **weighting** of graph edges, to conduct this study. In detail, the analysis studies three combined categories: (1) models with implicit message-passing (denoted as *implicit*); (2) models with explicit message-passing and neighborhood weighting (denoted as *explicit/weighted*); (3) models with explicit message-passing without neighborhood weighting (denoted as *explicit/unweighted*). For each analyzed trade-off, we select the Pareto optimal solutions[2] of the baselines laying on the model-specific Pareto frontier [24]. Figure 2 plots graph models Pareto frontiers in the common *objective function spaces* related to the considered trade-offs. The careful reader may notice the different axis' scales across the graphics due to the metric values. The colors of Pareto optimal solutions are model-specific, while the line style is used to distinguish the categories: dotted lines for *implicit*, dash-dot lines for *explicit/weighted*, and dashed lines for *explicit/unweighted*.

---

[2] A solution is Pareto optimal if no other solution can improve an objective without hurting the other one.

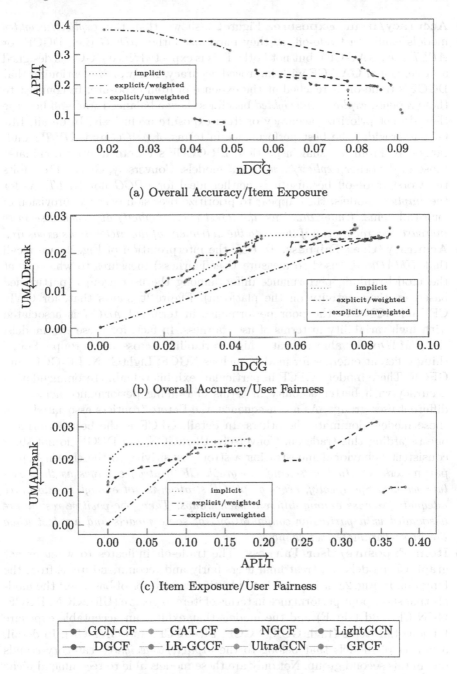

(a) Overall Accuracy/Item Exposure

(b) Overall Accuracy/User Fairness

(c) Item Exposure/User Fairness

Fig. 2. Overall Accuracy/Item Exposure, Overall Accuracy/User Fairness, and Item Exposure/User Fairness trade-offs on Amazon Men, assessed through *nDCG/APLT*, *nDCG/UMADrank*, and *APLT/UMADrank*, respectively. Each point depicts a model hyper-parameter configuration set belonging to the corresponding Pareto frontier. Colors refer to a particular baseline, while lines styles discern their technical strategies based on the proposed taxonomy. Arrows indicates the optimization direction for each metric on x and y axes.

- **Accuracy/Item Exposure.** Figure 2a shows that the *explicit/weighted* models exhibit a trade-off, as they maximize either $nDCG$ (i.e., DGCF) or $APLT$ (i.e., GAT-CF), but not both. This is expected since DGCF is designed as a version of GAT-CF with improved accuracy. It is worth mentioning that DGCF's trade-off is reached at the expense of item exposure. In contrast to these models, *explicit/unweighted* baselines show a balanced trade-off because they do not prioritize accuracy or item exposure exclusively. In detail, LR-GCCF provides the best performance in terms of $nDCG$ and $APLT$ simultaneously. From a visual inspection, LR-GCCF's Pareto frontier dominates those of the other *explicit/unweighted* models. Conversely, GCN-CF exhibits the worst trade-off because it is neither ideal for $nDCG$ nor $APLT$. As for the *implicit* models, they appear to prioritize precision over the provision of long-tail items. *Under this lens, the latest (i.e., implicit) approaches seem to increase accuracy, even if this is to the detriment of the niche items exposure.*
- **Accuracy/User Fairness.** To ease the interpretation of Fig. 2b, we recall that *UMADrank* (used to measure User Fairness) measures to what extent the model ranking performance differs among the user groups (partitioned based on their activity on the platform). Figure 2b shows that, for GAT-CF and GCN-CF, the poor performance in terms of $nDCG$ is associated with high variability in terms of user fairness. In fact, for these two models, the *UMADrank* value indicates high variability across user groups. Something different emerges for models such as NGCF, LightGCN, LR-GCF, and GFCF. These models, GFCF in particular, exhibit valuable recommendation accuracy with better stability in terms of ranking performance across the different user groups. As a consequence, the Pareto frontiers associated with these models dominate the others. In detail, GFCF is the best-performing one regarding this trade-off. Conversely, UltraGCN and DCGF do not show consistent behavior demonstrating a strong sensitivity to the chosen hyper-parameters set. *In this setting, no graph CF strategy emerges as the absolute winner. Specifically, every graph CF strategy is not enough to guarantee adequate fairness among different user groups. Then, the positive results are associated with particular configurations of some models and are lost when the hyper-parameter set changes.*
- **Item Exposure/User Fairness.** The trade-off indicates to what extent graph CF models can treat final users fairly and recommend items from the long tail. In Fig. 2c, it is possible to identify two groups of baselines: the models that show poor performance in terms of item exposure (UltraGCN, DGCF, GCN-CF, and GFCF) and the models that exhibit an acceptable exposure for long-tail items (LightGCN, NGCF, LR-GCCF, and GAT-CF). In detail, a cluster of models that belong to the *explicit/unweighted* category stands out in this second group. Not only are these models able to recommend niche items, but also they are stable (among the user groups) in terms of accuracy. On the contrary, although GAT-CF lies close to the *utopia point*[3], it exhibits greater variability regarding the accuracy metric. Indeed, comparing Fig. 2c

---

[3] The point that simultaneously minimizes (maximizes) all the metrics.

with Fig. 2a, GAT-CF demonstrates to achieve adequate user fairness, but its performance is still very poor in terms of accuracy. *To summarize, even if a system designer could be more interested in promoting models solely guaranteeing the best value for APLT (Producer Fairness), the explicit/unweighted strategies can generally ensure a satisfactory (for Consumers and Producers) trade-off between user fairness and item exposure.*

5   Conclusion and Future Work

We assess the performance of graph CF models on Consumer and Producer (CP)-fairness metrics showing that their superior accuracy capabilities is reached at the expense of user fairness, item exposure, and their combination. By recognizing nodes representation and neighborhood exploration as the two main dimensions of a novel graph CF taxonomy, we study their influence on CP-fairness and overall accuracy separately and simultaneously. The outcomes raise concerns about the effective application of recent approaches in graph CF (e.g., implicit message-passing techniques). On such basis, we are performing further investigations on other datasets and algorithms, and we are working on new graph models balancing accuracy and CP-Fairness.

Acknowledgment. This work was partially supported by the following projects: IPZS-PRJ4_IA_NORMATIVO, Codice Pratica VHRWPD7 - CUP B97I19000980007 - COR 1462424 ERP 4.0, Grant Agreement Number 101016956 H2020 PASSEPARTOUT, Secure Safe Apulia, Codice Pratica 3PDW2R7 SERVIZI LOCALI 2.0, MISE CUP: I14E20000020001 CTEMT - Casa delle Tecnologie Emergenti Comune di Matera, PON ARS01_00876 BIO-D, CT_FINCONS_II.

A   Experimental Settings and Protocols

Datasets. As a pre-processing stage, for each dataset, we randomly sample 60k interactions and drop users and items with less than five interactions to avoid the cold-start effect [12,13]. The final dataset statistics are: (1) Baby has 5,842 users, 7,925 items, 35,475 interactions; (2) Boys & Girls has 3,042 users, 12,912 items, 35,762 interactions; (3) Men has 3,909 users, 27,656 items, 51,519 interactions.

Reproducibility. Datasets are split using the 70/10/20 train/validation/test hold-out strategy. Baselines are trained through grid search (48 explored configurations), with a batch size of 256 and 400 epochs. Datasets and codes (implemented with Elliot [2]) are available at this *link*.

Evaluation. As for the *overall accuracy*, we use the recall (*Recall@k*) and the normalized discounted cumulative gain (*nDCG@k*). Concerning the *item exposure*, we focus on: (1) item novelty [37,38] through the expected free discovery (*EFD@k*) measuring the expected portion of relevantly-recommended items that have already been seen by the users; (2) item diversity [32] with the 1's complement of the Gini index ($Gini@k$), a statistical dispersion measure which estimates how a model suggests heterogeneous items to users; (3) the average percentage of items from the long-tail ($APLT@k$) which are recommended in users' lists [1] to calculate recommendation's bias towards popular items. *User fairness* indicates how equally each user group receives accurate recommendations. Users are split into quartiles based on the number of items they interacted with. We then measure $UMADrat@k$ and the $UMADrank@k$ [9], where the former stands for the average deviation in the predicted ratings among users groups, while the latter represents the average deviation in the recommendation accuracy (calculated in terms of $nDCG@k$) among users groups. The best hyper-parameter configurations are found by considering *Recall@20* on the validation.

References

1. Abdollahpouri, H., Burke, R., Mobasher, B.: Controlling popularity bias in learning-to-rank recommendation. In: RecSys, pp. 42–46, ACM (2017)
2. Anelli, V.W., et al.: Elliot: A comprehensive and rigorous framework for reproducible recommender systems evaluation. In: SIGIR, pp. 2405–2414, ACM (2021)
3. Anelli, V.W., et al.: How neighborhood exploration influences novelty and diversity in graph collaborative filtering. In: MORS@RecSys, CEUR Workshop Proceedings, vol. 3268, CEUR-WS.org (2022)
4. Anelli, V.W., et al.: Reshaping graph recommendation with edge graph collaborative filtering and customer reviews. In: DL4SR@CIKM, CEUR Workshop Proceedings, vol. 3317, CEUR-WS.org (2022)
5. van den Berg, R., Kipf, T.N., Welling, M.: Graph convolutional matrix completion. CoRR abs/1706.02263 (2017)
6. Boltsis, G., Pitoura, E.: Bias disparity in graph-based collaborative filtering recommenders. In: SAC, pp. 1403–1409, ACM (2022)
7. Chen, D., Lin, Y., Li, W., Li, P., Zhou, J., Sun, X.: Measuring and relieving the over-smoothing problem for graph neural networks from the topological view. In: AAAI, pp. 3438–3445, AAAI Press (2020)
8. Chen, L., Wu, L., Hong, R., Zhang, K., Wang, M.: Revisiting graph based collaborative filtering: A linear residual graph convolutional network approach. In: AAAI, pp. 27–34, AAAI Press (2020)
9. Deldjoo, Y., Anelli, V.W., Zamani, H., Bellogín, A., Noia, T.D.: A flexible framework for evaluating user and item fairness in recommender systems. User Model. User Adapt. Interact. **31**(3), 457–511 (2021)
10. Ekstrand, M.D., Riedl, J., Konstan, J.A.: Collaborative filtering recommender systems. Found. Trends Hum. Comput. Interact. **4**(2), 175–243 (2011)
11. Fu, Z., et al.: Fairness-aware explainable recommendation over knowledge graphs. In: SIGIR, pp. 69–78, ACM (2020)
12. He, R., McAuley, J.J.: Ups and downs: Modeling the visual evolution of fashion trends with one-class collaborative filtering. In: WWW, pp. 507–517, ACM (2016)
13. He, R., McAuley, J.J.: VBPR: visual bayesian personalized ranking from implicit feedback. In: AAAI, pp. 144–150, AAAI Press (2016)
14. He, X., Deng, K., Wang, X., Li, Y., Zhang, Y., Wang, M.: Lightgcn: Simplifying and powering graph convolution network for recommendation. In: SIGIR, pp. 639–648, ACM (2020)

15. Kipf, T.N., Welling, M.: Semi-supervised classification with graph convolutional networks. In: ICLR (Poster), OpenReview.net (2017)
16. Koren, Y., Bell, R.M., Volinsky, C.: Matrix factorization techniques for recommender systems. Computer **42**(8), 30–37 (2009)
17. Li, C., Hsu, C., Zhang, Y.: Fairsr: Fairness-aware sequential recommendation through multi-task learning with preference graph embeddings. ACM Trans. Intell. Syst. Technol. **13**(1), 16:1–16:21 (2022)
18. Ma, J., Cui, P., Kuang, K., Wang, X., Zhu, W.: Disentangled graph convolutional networks. In: ICML, Proceedings of Machine Learning Research, vol. 97, pp. 4212–4221, PMLR (2019)
19. Mansoury, M., Abdollahpouri, H., Pechenizkiy, M., Mobasher, B., Burke, R.: Fairmatch: A graph-based approach for improving aggregate diversity in recommender systems. In: UMAP, pp. 154–162, ACM (2020)
20. Mansoury, M., Abdollahpouri, H., Pechenizkiy, M., Mobasher, B., Burke, R.: A graph-based approach for mitigating multi-sided exposure bias in recommender systems. ACM Trans. Inf. Syst. **40**(2), 32:1–32:31 (2022)
21. Mao, K., Zhu, J., Xiao, X., Lu, B., Wang, Z., He, X.: Ultragcn: Ultra simplification of graph convolutional networks for recommendation. In: CIKM, pp. 1253–1262, ACM (2021)
22. Naghiaei, M., Rahmani, H.A., Deldjoo, Y.: Cpfair: Personalized consumer and producer fairness re-ranking for recommender systems. In: SIGIR, pp. 770–779, ACM (2022)
23. Ni, J., Li, J., McAuley, J.J.: Justifying recommendations using distantly-labeled reviews and fine-grained aspects. In: EMNLP/IJCNLP (1), pp. 188–197, Association for Computational Linguistics (2019)
24. Paparella, V.: Pursuing optimal trade-off solutions in multi-objective recommender systems. In: RecSys, pp. 727–729, ACM (2022)
25. Paudel, B., Christoffel, F., Newell, C., Bernstein, A.: Updatable, accurate, diverse, and scalable recommendations for interactive applications. ACM Trans. Interact. Intell. Syst. **7**(1), 1:1–1:34 (2017)
26. Peng, S., Sugiyama, K., Mine, T.: SVD-GCN: A simplified graph convolution paradigm for recommendation. In: CIKM, pp. 1625–1634, ACM (2022)
27. Rahman, T.A., Surma, B., Backes, M., Zhang, Y.: Fairwalk: Towards fair graph embedding. In: IJCAI, pp. 3289–3295, ijcai.org (2019)
28. Rendle, S., Freudenthaler, C., Gantner, Z., Schmidt-Thieme, L.: BPR: bayesian personalized ranking from implicit feedback. In: UAI, pp. 452–461, AUAI Press (2009)
29. Rendle, S., Krichene, W., Zhang, L., Anderson, J.R.: Neural collaborative filtering vs. matrix factorization revisited. In: RecSys, pp. 240–248, ACM (2020)
30. Resnick, P., Iacovou, N., Suchak, M., Bergstrom, P., Riedl, J.: Grouplens: An open architecture for collaborative filtering of netnews. In: CSCW, pp. 175–186, ACM (1994)
31. Sarwar, B.M., Karypis, G., Konstan, J.A., Riedl, J.: Item-based collaborative filtering recommendation algorithms. In: WWW, pp. 285–295, ACM (2001)
32. Shani, G., Gunawardana, A.: Evaluating Recommendation Systems. In: Ricci, F., Rokach, L., Shapira, B., Kantor, P.B. (eds.) Recommender Systems Handbook, pp. 257–297. Springer, Boston, MA (2011). https://doi.org/10.1007/978-0-387-85820-3_8
33. Shen, Y., et al.: How powerful is graph convolution for recommendation? In: CIKM, pp. 1619–1629, ACM (2021)

34. Sun, J., Cheng, Z., Zuberi, S., Pérez, F., Volkovs, M.: HGCF: hyperbolic graph convolution networks for collaborative filtering. In: WWW, pp. 593–601, ACM / IW3C2 (2021)
35. Sun, J., et al.:A framework for recommending accurate and diverse items using bayesian graph convolutional neural networks. In: KDD, pp. 2030–2039, ACM (2020)
36. Tao, Z., Wei, Y., Wang, X., He, X., Huang, X., Chua, T.: MGAT: multimodal graph attention network for recommendation. Inf. Process. Manag. **57**(5), 102277 (2020)
37. Vargas, S.: Novelty and diversity enhancement and evaluation in recommender systems and information retrieval. In: SIGIR, pp. 1281, ACM (2014)
38. Vargas, S., Castells, P.: Rank and relevance in novelty and diversity metrics for recommender systems. In: RecSys, pp. 109–116, ACM (2011)
39. Velickovic, P., Cucurull, G., Casanova, A., Romero, A., Liò, P., Bengio, Y.: Graph attention networks. In: ICLR (Poster), OpenReview.net (2018)
40. Voit, M.M., Paulheim, H.: Bias in knowledge graphs - an empirical study with movie recommendation and different language editions of dbpedia. In: LDK, OASIcs, vol. 93, pp. 14:1–14:13, Schloss Dagstuhl - Leibniz-Zentrum für Informatik (2021)
41. Wang, N., Lin, L., Li, J., Wang, H.: Unbiased graph embedding with biased graph observations. In: WWW, pp. 1423–1433, ACM (2022)
42. Wang, X., He, X., Cao, Y., Liu, M., Chua, T.: KGAT: knowledge graph attention network for recommendation. In: KDD, pp. 950–958, ACM (2019)
43. Wang, X., He, X., Wang, M., Feng, F., Chua, T.: Neural graph collaborative filtering. In: SIGIR, pp. 165–174, ACM (2019)
44. Wang, X., Jin, H., Zhang, A., He, X., Xu, T., Chua, T.: Disentangled graph collaborative filtering. In: SIGIR, pp. 1001–1010, ACM (2020)
45. Wang, Y., Tang, S., Lei, Y., Song, W., Wang, S., Zhang, M.: Disenhan: Disentangled heterogeneous graph attention network for recommendation. In: CIKM, pp. 1605–1614, ACM (2020)
46. Wu, J., et al.: Disenkgat: Knowledge graph embedding with disentangled graph attention network. In: CIKM, pp. 2140–2149, ACM (2021)
47. Wu, J., et al.: Self-supervised graph learning for recommendation. In: SIGIR, pp. 726–735, ACM (2021)
48. Wu, L., Chen, L., Shao, P., Hong, R., Wang, X., Wang, M.: Learning fair representations for recommendation: A graph-based perspective. In: WWW, pp. 2198–2208, ACM / IW3C2 (2021)
49. Wu, Y., DuBois, C., Zheng, A.X., Ester, M.: Collaborative denoising auto-encoders for top-n recommender systems. In: WSDM, pp. 153–162, ACM (2016)
50. Ying, R., He, R., Chen, K., Eksombatchai, P., Hamilton, W.L., Leskovec, J.: Graph convolutional neural networks for web-scale recommender systems. In: KDD, pp. 974–983, ACM (2018)
51. Zhao, M., et al.: Investigating accuracy-novelty performance for graph-based collaborative filtering. In: SIGIR, pp. 50–59, ACM (2022)
52. Zheng, Y., Gao, C., Chen, L., Jin, D., Li, Y.: DGCN: diversified recommendation with graph convolutional networks. In: WWW, pp. 401–412, ACM / IW3C2 (2021)
53. Zhou, K., Huang, X., Li, Y., Zha, D., Chen, R., Hu, X.: Towards deeper graph neural networks with differentiable group normalization. In: NeurIPS (2020)
54. Zhu, J., et al.: BARS: towards open benchmarking for recommender systems. In: SIGIR, pp. 2912–2923, ACM (2022)

Exploiting Graph Structured Cross-Domain Representation for Multi-domain Recommendation

Alejandro Ariza-Casabona[1]([✉]) [iD], Bartlomiej Twardowski[2,3] [iD], and Tri Kurniawan Wijaya[3] [iD]

[1] Universitat de Barcelona, Barcelona, Spain
`alejandro.ariza14@ub.edu`
[2] Computer Vision Center, UAB, Barcelona, Spain
`btwardowski@cvc.uab.es`
[3] Huawei Ireland Research Center, Dublin, Ireland
{`bartlomiej.twardowski,tri.kurniawan.wijaya`}`@huawei.com`

Abstract. Multi-domain recommender systems benefit from cross-domain representation learning and positive knowledge transfer. Both can be achieved by introducing a specific modeling of input data (i.e. disjoint history) or trying dedicated training regimes. At the same time, treating domains as separate input sources becomes a limitation as it does not capture the interplay that naturally exists between domains. In this work, we efficiently learn multi-domain representation of sequential users' interactions using graph neural networks. We use temporal intra- and inter-domain interactions as contextual information for our method called MAGRec (short for *Multi-domAin Graph-based Recommender*). To better capture all relations in a multi-domain setting, we learn two graph-based sequential representations simultaneously: domain-guided for recent user interest, and general for long-term interest. This approach helps to mitigate the negative knowledge transfer problem from multiple domains and improve overall representation. We perform experiments on publicly available datasets in different scenarios where MAGRec consistently outperforms state-of-the-art methods. Furthermore, we provide an ablation study and discuss further extensions of our method.

Keywords: Multi-domain recommendation · Graph neural networks · Sequence-aware recommender system

1 Introduction

Recommender systems are introduced to solve the task of quickly retrieving the most suitable items from large catalogs to the corresponding users. The complexity of the task comes not only from the huge amount of information that already subdivides the recommendation problem into three stages (matching, ranking and re-ranking) [8], but also from the multi-objective minimization interest in

J. Kamps et al. (Eds.): ECIR 2023, LNCS 13980, pp. 49–65, 2023.
https://doi.org/10.1007/978-3-031-28244-7_4

multi-stakeholder platforms (accuracy, novelty, fairness, explainability, business income, etc.) [1] and the different types of recommendation scenarios (sequential, session-based, social, cross-domain, multi-domain, etc.).

Whilst most recommender systems focus on a single domain recommendation (SDR) problem, it is increasingly common to find large-scale commercial platforms containing products from multiple domains in which the user/item set reaches a certain overlapping degree. These multiple domains may correspond to different advertisement types or product categories that present certain domain commonalities but still require specialized promotion strategies or models for effective profitability. Unfortunately, having a single "naive" model to serve all domains may achieve subpar results. As a consequence, several models were proposed to tackle cross-domain recommendation (CDR) with the aim of transferring knowledge from a source domain to one or multiple target domains [11,19,22].

The main problem that appears with CDR is the fact that each domain requires separate training or fine-tuning of a different model per each set of source-target domains, but large-scale commercial platforms may consider a single model to improve performance for all domains simultaneously. This challenge has been recently introduced as Multi-Domain Recommendation (MDR) [27] in which the distinct domain data distributions caused by different user behavioral patterns are modelled altogether. An advantage of MDR systems is that they make use of all available data for training unlike previous alternatives, resulting in positive outcomes for low-resource domains and optimal performance if shared information among domains is properly exploited.

As previously explored, one approach to improve learning on multiple domains is to use multi-task learning (MTL) models by considering each domain as a different task [4,18,25] . However, MTL models are built to model different label spaces rather than partially distinct input data distributions that characterize the MDR problem. That is the reason why these models incorporate separate task-specific output layers and share the bottom input representational layers, leading to a poor ability to represent domain commonalities in the label space. To overcome MTL limitations, existing studies [16,27] have focused on adding domain-shared knowledge to the classification block of the recommender system, and defining new training strategies to avoid the domain conflict and domain overfitting problems. However, little to no effort has been made on properly modeling the available multi-domain structural information from the past user behavior history. Existing approaches ignore this information and rely on a simple domain indicator. Therefore, the input modeling capabilities are limited and excessive time-consuming training strategies are necessary to achieve a successful convergence.

In order to fully exploit domain interest fluctuations and specific data distributions, we combine the power of Sequence-Aware Recommender Systems and Graph Neural Networks (GNN) into a new model named MAGRec for MDR. MAGRec receives multi-domain graph representations of historical user interactions performed in multiple domains, and processes them using a two-branch network architecture. On the one hand, one branch focuses on the user's most recent interest via domain-aware message passing through the sequential graph. On the other hand, the second branch tries to create a contextualized global user representation via graph structure learning and local pooling operations. Finally, this rich representation of the input is combined and fed to the classification network that models the MDR task. Given that our focus lies on the input representation modeling and its ignored presence in the MDR literature, unlike previous works that adapt the classification network and training strategies, our model only uses a single fully connected network (FCN) as the classification block. This approach also benefits those cases where the number of domains is very large and both training times and model parameters reach unacceptable limits. Nonetheless, it is important to note that our domain-aware graph modeling network could be combined with previous state-of-the-art adopted strategies, which is left for future work.

The main contributions of this work can be summarized as follows:

- We propose a new MDR model: MAGRec. The use of graph neural networks to capture domain relationships from past user interactions in multiple domains, together with the integration of global and recent user interest with domain-shared Graph Structure Learning, provides a faster alternative to MDR that puts the focus on the input modeling stage of the model.
- We explore different input representations from a "naive" to a more complex multi-domain sequential representation and test them on multiple sequential and graph-based recommender systems.
- Extensive experiments on different MDR scenarios and models, including: single-domain sequential CTR prediction models, MTL recommendation models, and MDR strategies show the consistent viability of our model, and it clearly opens new directions in MDR research such as the combination of input representation capabilities with task modeling strategies.

2   Related Work

Sequence-aware and graph recommender systems are commonly applied for the SDR problem. In our work, we propose an approach based on recent advancement from latest research in those fields to tackle the MDR problem, which we describe below.

Sequence-Aware CTR Prediction. The path that single-domain CTR prediction models have followed over the past few years go from shallow models [20] to complex deep learning architectures. The later are trying to extract more complex patterns from users' behavior by adding feature interaction modeling strategies such as cross-connections [30], improving numerical features representation [9] or exploiting sequential patterns in the past user's behavior [34,35], just to name a few. However, it was a sequential modeling that improved recommender systems in many domains with the introduction of recurrent architectures [10], convolutional networks [29] and attention modules [12,13]. Therefore, they have been adopted for CTR prediction task as well. However, these techniques remain unexplored in MDR where domain sequential interest fluctuations must be correctly accounted for.

An important remark from recent sequence-aware models is the necessity to properly integrate the long- and short-term user interest representations [5, 17,23]. In this work, we integrate domain contextualization for both short- and long-term user modeling. The former is achieved via a weighted edge message passing mechanism, and the latter via domain-aware graph structure learning and local pooling strategies.

Graph Neural Networks for Recommendation. The use of GNNs has notably improved sequential recommendation [5,24,31]. The importance of structure modeling has even been addressed with hypergraph representations to account for higher-order node relations. [32] proved the beneficial effects of combining different input graph representations such as hypergraphs and reduced line graphs for session-based user representation. Furthermore, [5] support previous findings regarding the necessity of extracting global user interest with graph clustering (for noise removal) and properly fuse it with the most recent user interests for more accurate recommendations.

In our work, we use GNNs to model intra- and inter-domain connectivity patterns from a graph-based multi-domain user history representation by exploiting item node features and domain-related edge features.

Multi-Domain Recommendation. Unlike CDR, where source and target domains are clearly defined and most benchmarks assess a knowledge transfer between domains, MDR aims at improving performance on all domains simultaneously with, preferably, a single model to reduce computational complexity.

On the one hand, having a single traditional model without domain knowledge and sharing all parameters to serve all domains may put a huge burden on the generalizability-specialty trade-off of the downstream task. On the other hand, having a separate model per domain is unacceptable in platforms with many domains, due to the number of trainable parameters and lack of training data for certain domains. A middle ground for CDR and MDR can be found in [7,14,33]. [14] proposed a sequential modeling of the input with dual embedding and dual attention mechanisms. However, generalizing this work for more than two domains is not trivial. [7,33] created a cross-domain graph representation but ignored the sequential connections that are important for inter-domain interest flow. Moreover, generalization to an arbitrary number of domains is barely explored.

Alternatively, other approaches to MDR included several MTL architectures in order to model each domain as a separate task [3,4,18,28]. All of them have task-shared input representation network i.e., expert networks, and multiple task-specific networks for a final prediction denoted as tower networks by MMoE. Nonetheless, as previously mentioned, the MDR problem differs from MTL in the fact that the task/label space is the same and face different input distributions from multiple domains. Consequently, [27] focused their efforts on building a task-modeling network with star topology that could model all domains as a shared task called STAR. It proved the importance of domain contextualization with the use of a partitioned normalization per domain and an auxiliary network. Despite improving MDR performance, STAR still suffers from the domain conflict problem and it is prone to overfit on sparse domains as stated in [16]. To overcome those limitations, [16] introduced MAMDR as an attempt to adapt the MDR training strategy to include a domain negotiation and a domain regularization approach. Nevertheless, as these MDR approaches were built on top of MTL strategies and despite STAR supporting the evidence that a simple domain indicator is important contextual information in these scenarios, they pay little attention on how to model domain interconnections and structural information from the input.

In this work, we aim to prove the relevance of structured multi-domain representation in MDR problems. Note that [16,27] are complementary to this work and could potentially be used in combination with our model to overcome task modelling limitations on many-domain scenarios, thus, boosting the performance.

3   Methodology

In this section, we provide a brief introduction to the MDR problem and our proposed two-branch architecture. Furthermore, we detail how domain information is exploited by each branch to leverage recent and global user interest representations.

Problem Formulation. In traditional recommendation scenarios, it is common to have a set of users and a set of items, denoted by $\mathcal{U}$ and $\mathcal{I}$ respectively. In addition, we have a set of user interactions denoted by $\mathcal{B}$. In the single domain setting, each interaction $b \in \mathcal{B}$ is a tuple (u, i, t, y), where $u \in \mathcal{U}$, $i \in \mathcal{I}$, t is the timestamp of the interaction and y is the user action (e.g. was there a click or not in CTR prediction). In the MDR setting, each tuple must further contain the domain $d \in \mathcal{D}$ in which that interaction took place, i.e. $b = (d, u, i, t, y)$.

The goal in a conventional single-domain setting is to predict the action y a user u will take for a candidate next item i_k in a domain d, given their previous behaviors as a sequence in that particular domain $\mathcal{B}_u^d = [b_{u,1}^d, b_{u,2}^d, ..., b_{u,N}^d]$. However, when there are $|\mathcal{D}|$ different domains that need to be taken into account, some domains may encounter partial or total overlap in their user and/or item sets. Consequently, user history $\mathcal{B}_u$ can be interpreted in multiple ways depending on how domain data is aggregated and which interactions are considered. In order to account for more complex representations to be fed into the model, $\mathcal{B}_u$ is transformed into the corresponding graph representation $\mathcal{G}_u = \{\mathcal{V}_u, \mathcal{E}_u\}$, where $\mathcal{V}_u$ is the set of vertices/items, and $\mathcal{E}_u$ is the set of edges containing intra- and, potentially, inter-domain sequential connections. This transformation is data dependent, see *Experiments* section.

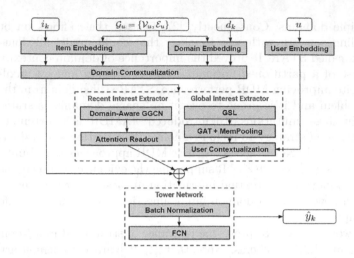

Fig. 1. Illustration of MAGRec. MAGRec extracts both recent and global user interests from a contextualized multi-domain user history graph representation. The recent interest extractor learns cross-domain sequential patterns from intra- and inter-domain interactions focused on the last user-item interaction. The global interest extractor learns item correlations and their affinity to long-term user core preferences via graph structure learning and local graph pooling techniques.

Proposed Architecture. Our proposed MAGRec model is primarily built to maximize input representation capabilities for the multi-domain recommendation problem. Its architecture is illustrated in Fig. 1. As previously mentioned, the input includes the user u, the selected user history representation $\mathcal{G}_u$, the candidate item i_k and the domain to which the candidate item belongs d_k. Note that the graph contains both node and edge features. Specifically, node features are represented by item embeddings and edge attributes correspond to the source and target domains in the sense of a directional connection. Next, a brief explanation of each block of our architecture is presented:

Embedding. An initial embedding layer is applied to the items, users and domains in order to obtain their latent representations, i.e. $\vec{i_j} \in \mathbb{R}^{m_i}$, $\vec{u_j} \in \mathbb{R}^{m_u}$ and $\vec{d_j} \in \mathbb{R}^{m_d}$ respectively.

Domain Contextualization. In order to transform item embeddings from multiple domains to a shared latent space, we apply a dense transformation layer to the concatenation of the item embedding and its associated domain embedding as follows:

$$\vec{i_j}' = \mathbf{W_d} \cdot (\vec{i_j} \parallel \vec{d_j}) \tag{1}$$

where $\parallel$ is the concatenation operator and $\mathbf{W_d}$ is a matrix of learnable parameters. Note that other aggregation operators could also be used, such as element-wise summation or Hadamard product.

Recent Interest Extractor. This block receives the selected graph representation of the user behavior history $\mathcal{G}_u$, already contextualized by the previous module, and it is responsible for temporally aggregating implicit signals into a strong recent user interest representation. This process is performed in two steps:

Domain-Aware Gated Graph Convolution. To learn inter-domain sequential patterns and intra-domain structural constraints, it is important to model domain commonalities by considering how much information a domain is able to pass onto another domain. Therefore, an edge weight is computed using the source and target domains as edge attributes:

$$e_{z,j} = \langle \mathbf{W_{src}} \cdot \vec{d_z}, \mathbf{W_{trg}} \cdot \vec{d_j} \rangle \tag{2}$$

where $\mathbf{W_{src}}$ and $\mathbf{W_{trg}}$ are learnable parameters and $\langle \vec{x}, \vec{y} \rangle$ is the dot product operator. The next step is to learn smooth node representations with a proper sequence modeling technique applied to graphs. We opted for stacking L Gated Graph Convolutional (GGCN) layers [15] with the previously computed cross-domain edge weights:

$$\vec{h}_j^{(l+1)} = GGCN(e_{z,j}, \vec{h}_z^{(l)}) \tag{3}$$

where $\vec{h}_z^{(0)} = \vec{i_j}'$.

Attention Graph Readout. A graph level embedding of the short-term user interests is obtained by a weighted aggregation of the node embeddings. To compute those weights, a local attention mechanism is applied to each node j and the node corresponding to the last interaction of $\mathcal{B}_u$:

$$\alpha_j = softmax_j \left(\mathbf{W_{l_2}} \cdot \sigma(\mathbf{W_{l_1}} \cdot (\vec{h}_j^{(L)} \parallel \vec{h}_N^{(L)}))^T \right)$$

where $\mathbf{W_{l_1}}$ and $\mathbf{W_{l_2}}$ are learnable parameters and σ is a non-linear activation function. Finally, a weighted aggregation with the computed coefficients generates the graph embedding:

$$\vec{r}_u = \sum_{j \in \mathcal{V}_u} \alpha_j \cdot \vec{h}_j^{(l+1)} \tag{4}$$

Global Interest Extractor. A graph built from user history is useful for extracting strong recent interests but the history may also contain noisy interactions that confound long-term core user preferences. The global interest extractor branch is tasked with obtaining a global time independent user interest representation. It consists of the following modules:

Graph Structure Learning. A new graph representation $\mathcal{G}_u' = \{\mathcal{V}_u, \mathcal{E}_u'\}$ whose adjacency matrix $\mathcal{A}'$ is created based on a relative ranking of multi-head kernel similarity scores between any two vertices in $\mathcal{V}_u$:

$$\mathcal{A}_{j,z}^h = Kernel(\mathbf{W_{gsl}^h} \cdot \vec{i_j}', \mathbf{W_{gsl}^h} \cdot \vec{i_z}') \tag{5}$$

$$\mathcal{A}_{j,z} = \frac{1}{H} \sum_{h=1}^{H} \mathcal{A}_{j,z}^h \qquad (6)$$

where H denotes the number of kernel heads to extract specific node similarities and $\mathbf{W}_{gsl}^h$ are learnable parameters. We select the cosine similarity as the kernel function. Following [5], a relative ranking with a threshold γ equal to 0.5 is applied for graph sparsification:

$$\mathcal{A}'_{j,z} = \begin{cases} 1, & \text{if } \mathcal{A}_{j,z} \geq \gamma; \\ 0, & \text{otherwise}; \end{cases} \qquad (7)$$

Local Graph Pooling. MemPooling [2], a hierarchical graph representation learning technique based on multiple memory layers for local clustering, is applied on top of a Graph Attention (GAT) query network to obtain a global user interest embedding of the last N interactions:

$$\vec{g}_u = MemPool \left(GAT(\vec{i_j}' | j \in \mathcal{G}'_u) \right) \qquad (8)$$

User Contextualization. Given that a user can have more than N past interactions, a "residual" user contextualization is aimed at a more global interest representation:

$$\vec{g_u}' = \mathbf{W_u} \cdot (\vec{g}_u \parallel \vec{u}) \qquad (9)$$

where $\mathbf{W_u}$ are learnable parameters.

Note that learning a new graph structure based on item similarities increases the model cross-domain representation capabilities as it helps to pull similar items from different domains together in the shared space.

Classification Network. Previous MDR work has focused their efforts on this block by extracting ideas from MTL strategies combining domain-shared and domain-specific parameters. However, given the aim of this paper on multi-domain data modeling, we decided to implement a single tower network with global batch normalization and FCN. This tower network receives the concatenation of the contextualized candidate item embedding and both, recent and global user interest representations $(\vec{i}_k, \vec{g_u}', \vec{r_u})$, and it outputs the probability of the candidate item being clicked next by the current user $(\hat{y}_k)$. It is important to note that we leave the combination of MDR-MTL strategies and our MDR graph-based representational model for future work.

4   Experiments

Dataset[1]. Following previous work [11,14,16], we use the Amazon 5-core review dataset [21] by combining data corresponding to different domains (product categories) with varying degrees of users overlap, dataset size and click-trough-rates (CTR). In order to bridge CDR and MDR evaluation scenarios, we form eight

---
[1] Code and dataset partitions are available at https://github.com/alarca94/magrec.

sub-datasets: four 2-domain, one 3-domain, one 6-domain and one 13-domain partition. Similar to previous studies [16], for each partition, we keep the existing user-item reviews as positive samples and perform negative sampling on the set of items that the user has not interacted with, using a randomly generated CTR per domain to simulate domain distinction. For sequential recommendation, we perform a sliding window approach on the user histories and set the minimum and maximum window lengths to 5 and 80 respectively. Finally, for each user, a temporal split of ratio 6:2:2 is used to create the train, validation and test sets. Table 1 summarizes the basic statistics for all datasets.

Table 1. Dataset statistics.

Dataset	# User	# Item	# Train	# Val	# Test
Amazon-2a	153,658	55,089	1,766,478	267,805	267,884
Amazon-2b	290,631	105,609	3,593,621	610,585	610,537
Amazon-2c	188,539	59,978	2,245,655	348,101	347,917
Amazon-2d	254,736	95,394	3,009,960	555,399	555,701
Amazon-3	113,144	33,193	1,443,804	220,155	220,109
Amazon-6	444,737	170,815	7,580,825	1,397,290	1,395,688
Amazon-13	500,569	212,241	7,105,872	1,380,084	1,379,630

Baselines. To demonstrate the sequential modeling effectiveness of MAGRec, we compare it against competitive single-domain recommenders, including non-sequential, sequential and graph-based alternatives. Furthermore, we consider several MTL and MDR strong baselines to determine how well our model performs in multi-domain settings:

- WDL [6]: non-sequential model based on the combination of wide linear and deep neural models with cross-product feature transformations.
- DIN [35]: non-sequential model that aggregates the user behavior history using a softmax attention pooling based on the candidate item.
- DIEN [34]: sequential model with a two-layer GRU that implements an attentional update gate for a proper interest evolution modeling.
- FGNN [24]: competitive session-based recommender model adapted to predict CTR. It uses several weighted attention graph layers followed by a GRU set-to-set readout function.
- Shared-Bottom (SB) [4]: MTL model with shared parameters in the bottom layers and $|\mathcal{D}|$ domain-specific tower networks.
- MMoE [18]: MTL model that adopts a Mixture of Experts layout with $|\mathcal{D}|$ experts and a gating mechanism per domain connecting experts to the respective tower networks.
- STAR [27]: MDR model with partitioned normalization and star topology network to leverage shared and specific domain knowledge. They also use an auxiliary domain-aware network.

– MAMDR [16]: MDR learning method that uses domain regularization and domain negotiation on top of a star topology network.

To give a fair comparison, all non-sequential models are fed with a mean pooled representation of the user behavior sequence. Some of the methods can be used as a complementary to ours.

Evaluation Metrics. We use logloss and area under the ROC curve (AUC) as it is the most common metrics used to evaluate the performance of CTR prediction. We report overall metrics for all domains and, wherever possible, domain-specific results.

Implementation Details. Both graph-based models (FGNN[2] and MAGRec) are implemented using Pytorch Geometric. MAMDR and STAR implementations correspond to the ones provided by MAMDR authors[3]. For the other methods, they have been implemented by DeepCTR [26]. To make a fair comparison, all FCNs are set to [128, 64] parameters; the user and item embedding sizes are 64; the domain embedding size is 128; the initial learning rate is set to $1e^{-3}$ and the batch size is 512; the selected optimizer is Adam and Binary Cross Entropy is the loss function. For the model specific parameters, in the case of DeepCTR and graph-based models, a TPE hyperparameter search is performed on the Amazon 2d dataset and the best hyperparameters are used for the rest of the datasets. MAGRec best hyperparameters are 2 GGCN layers, 1 GAT layer with 2 attention heads and 3 MemPooling layers with [32, 10, 1] centroids respectively. Regarding MTL and MDR model specific parameters, we keep the ones selected in [16]. Experiments were conducted on a Tesla P100-PCIE-16GB GPU and 500GB RAM.

Two Domains Results. The results for two domain datasets are presented in Table 2. Overall and per-domain AUC and logloss values are presented. MAGRec outperforms all other methods in a meaning of single and combined domain performance for all scenarios. Amazon-2c is the scenario with the biggest logloss difference between two domains. The second-best method in this setting is an MTL approach. MMoE is consistently better than others (on the second place after MAGRec). Single-domain recommenders in this scenario can still be strong baselines, i.e. DIN and DIEN outperform STAR in all datasets and MAMDR for Amazon-2b,d. Lastly, session-based FGNN performs the worst on all partitions.

Input Representation Analysis. As described in the *Methodology* section, an input graph for MAGRec can be prepared in a few different ways. Consequently, we have experimented with three different graph representations i.e. *Disjoint,*

---

[2] https://github.com/RuihongQiu/FGNN.
[3] https://github.com/RManLuo/MAMDR.

Table 2. Results on 2-domain datasets and different methods. Bold numbers indicate the leading results, while underlined numbers represent the second-best scores. Mean and std. from seven runs are presented.

Dataset	Model	Domain 1		Domain 2		Both	
		Logloss	AUC	Logloss	AUC	Logloss	AUC
Amazon-2a	DIEN	0.510 ± 0.007	0.786 ± 0.003	0.393 ± 0.003	0.853 ± 0.003	0.452 ± 0.004	0.819 ± 0.002
	DIN	0.496 ± 0.003	0.806 ± 0.002	0.398 ± 0.004	0.842 ± 0.002	0.447 ± 0.004	0.824 ± 0.001
	FGNN	0.530 ± 0.003	0.766 ± 0.002	0.451 ± 0.003	0.792 ± 0.004	0.491 ± 0.003	0.779 ± 0.003
	MAMDR	0.481 ± 0.007	0.819 ± 0.005	0.377 ± 0.003	0.865 ± 0.002	0.429 ± 0.004	0.842 ± 0.002
	MMoE	0.474 ± 0.013	0.823 ± 0.007	0.372 ± 0.009	0.869 ± 0.005	0.423 ± 0.003	0.846 ± 0.001
	SB	0.470 ± 0.013	0.823 ± 0.010	0.375 ± 0.006	0.865 ± 0.004	0.423 ± 0.005	0.844 ± 0.004
	STAR	0.543 ± 0.032	0.777 ± 0.005	0.419 ± 0.021	0.830 ± 0.008	0.481 ± 0.022	0.804 ± 0.003
	WDL	0.481 ± 0.013	0.816 ± 0.008	0.378 ± 0.005	0.863 ± 0.002	0.430 ± 0.004	0.840 ± 0.003
	MAGRec	**0.466 ± 0.004**	**0.831 ± 0.001**	**0.363 ± 0.003**	**0.875 ± 0.001**	**0.415 ± 0.003**	**0.853 ± 0.001**
Amazon-2b	DIEN	**0.407 ± 0.055**	0.849 ± 0.004	0.405 ± 0.022	0.841 ± 0.001	0.406 ± 0.033	0.845 ± 0.003
	DIN	0.451 ± 0.000	0.845 ± 0.000	0.400 ± 0.000	0.838 ± 0.000	0.425 ± 0.000	0.841 ± 0.000
	FGNN	0.485 ± 0.006	0.820 ± 0.005	0.421 ± 0.003	0.822 ± 0.003	0.453 ± 0.005	0.821 ± 0.004
	MAMDR	0.467 ± 0.007	0.834 ± 0.004	0.408 ± 0.004	0.836 ± 0.002	0.437 ± 0.005	0.835 ± 0.003
	MMoE	0.443 ± 0.027	0.851 ± 0.016	0.386 ± 0.017	0.849 ± 0.011	0.414 ± 0.015	0.850 ± 0.010
	SB	0.472 ± 0.031	0.830 ± 0.018	0.393 ± 0.010	0.841 ± 0.010	0.433 ± 0.020	0.836 ± 0.013
	STAR	0.467 ± 0.010	0.832 ± 0.003	0.417 ± 0.019	0.828 ± 0.004	0.442 ± 0.007	0.830 ± 0.001
	WDL	0.461 ± 0.019	0.838 ± 0.014	0.396 ± 0.013	0.842 ± 0.008	0.428 ± 0.009	0.840 ± 0.008
	MAGRec	0.416 ± 0.002	**0.869 ± 0.002**	**0.378 ± 0.003**	**0.856 ± 0.002**	**0.397 ± 0.002**	**0.862 ± 0.002**
Amazon-2c	DIEN	**0.446 ± 0.060**	0.805 ± 0.006	0.411 ± 0.023	0.835 ± 0.002	0.429 ± 0.036	0.820 ± 0.004
	DIN	0.496 ± 0.001	0.805 ± 0.001	0.399 ± 0.001	0.834 ± 0.001	0.448 ± 0.001	0.819 ± 0.001
	FGNN	0.525 ± 0.002	0.773 ± 0.000	0.432 ± 0.001	0.808 ± 0.001	0.478 ± 0.001	0.791 ± 0.001
	MAMDR	0.484 ± 0.005	0.816 ± 0.004	0.396 ± 0.002	0.844 ± 0.001	0.440 ± 0.002	0.830 ± 0.002
	MMoE	0.462 ± 0.009	0.829 ± 0.006	0.393 ± 0.007	0.845 ± 0.005	0.428 ± 0.005	0.837 ± 0.003
	SB	0.468 ± 0.011	0.825 ± 0.007	0.394 ± 0.007	0.845 ± 0.006	0.431 ± 0.007	0.835 ± 0.005
	STAR	0.528 ± 0.017	0.782 ± 0.004	0.411 ± 0.008	0.827 ± 0.003	0.469 ± 0.006	0.805 ± 0.001
	WDL	0.469 ± 0.004	0.822 ± 0.004	0.401 ± 0.007	0.841 ± 0.006	0.435 ± 0.005	0.831 ± 0.004
	MAGRec	0.462 ± 0.007	**0.830 ± 0.006**	**0.383 ± 0.004**	**0.852 ± 0.004**	0.423 ± 0.006	**0.841 ± 0.005**
Amazon-2d	DIEN	0.427 ± 0.026	0.857 ± 0.005	0.413 ± 0.029	0.844 ± 0.005	0.420 ± 0.023	0.851 ± 0.004
	DIN	0.413 ± 0.008	0.872 ± 0.005	0.398 ± 0.006	0.842 ± 0.002	0.405 ± 0.007	0.857 ± 0.003
	FGNN	0.468 ± 0.004	0.837 ± 0.002	0.427 ± 0.003	0.825 ± 0.002	0.448 ± 0.003	0.831 ± 0.002
	MAMDR	0.473 ± 0.017	0.857 ± 0.005	0.391 ± 0.001	0.852 ± 0.001	0.432 ± 0.008	0.855 ± 0.002
	MMoE	0.393 ± 0.014	0.885 ± 0.009	0.394 ± 0.016	0.849 ± 0.012	0.393 ± 0.014	0.867 ± 0.010
	SB	0.416 ± 0.021	0.870 ± 0.015	0.408 ± 0.015	0.837 ± 0.013	0.412 ± 0.017	0.853 ± 0.013
	STAR	0.455 ± 0.006	0.842 ± 0.006	0.431 ± 0.014	0.824 ± 0.005	0.443 ± 0.006	0.833 ± 0.002
	WDL	0.414 ± 0.011	0.871 ± 0.008	0.418 ± 0.019	0.830 ± 0.014	0.416 ± 0.015	0.850 ± 0.011
	MAGRec	**0.375 ± 0.002**	**0.897 ± 0.001**	**0.363 ± 0.001**	**0.873 ± 0.001**	**0.369 ± 0.001**	**0.885 ± 0.001**

Flattened and *Interacting History.* In *Disjoint History*, domains are considered as independent data sources, meaning that for a candidate item i_k from domain d_k, the graph is constructed from the past user history in that particular domain. This representation is inline with the way MDR models [16,27] handle alternate domain training and, therefore, it is not able to exploit the existing cross-domain behaviors in the overall user history. A complete but naive cross-domain representation is the sequence of user-item interactions in all domains as a single timeline, thus, filling the gap between specific domain user sessions with the user sessions occurring in other domains. This representation is what we refer to as *Flattened History.* The downside of it is that user interests are assumed to evolve

Table 3. Comparison of different methods for graph $\mathcal{G}_u$ preparation on Amazon-2a dataset for selected methods. DIN and DIEN cannot be combined with *Interacting*.

Input	DIN		DIEN		FGNN		MAGRec	
	Logloss	AUC	Logloss	AUC	Logloss	AUC	Logloss	AUC
Disjoint	0.486 ± 0.012	0.778 ± 0.017	0.493 ± 0.013	0.790 ± 0.003	0.496 ± 0.001	0.774 ± 0.004	0.565 ± 0.209	0.827 ± 0.021
Flattened	**0.447 ± 0.004**	**0.824 ± 0.001**	**0.453 ± 0.004**	**0.816 ± 0.002**	0.495 ± 0.003	**0.779 ± 0.003**	0.417 ± 0.003	0.849 ± 0.002
Interacting	–	–	–	–	0.497 ± 0.002	0.778 ± 0.002	**0.415 ± 0.003**	**0.853 ± 0.001**

Table 4. Average logloss and AUC values for different methods and many domains scenarios

Model	3 domains		6 domains		13 domains	
	Logloss	AUC	Logloss	AUC	Logloss	AUC
SB	0.482 ± 0.007	0.802 ± 0.007	0.516 ± 0.017	0.736 ± 0.023	0.574 ± 0.022	0.661 ± 0.016
MMoE	0.489 ± 0.037	0.809 ± 0.003	0.514 ± 0.029	0.754 ± 0.022	0.572 ± 0.026	0.683 ± 0.009
WDL	0.478 ± 0.008	0.804 ± 0.006	0.505 ± 0.007	0.748 ± 0.013	0.553 ± 0.011	0.663 ± 0.014
STAR	0.528 ± 0.020	0.766 ± 0.003	0.488 ± 0.014	0.810 ± 0.002	0.545 ± 0.024	0.760 ± 0.003
FGNN	0.512 ± 0.001	0.766 ± 0.001	0.453 ± 0.001	0.808 ± 0.002	0.486 ± 0.002	0.775 ± 0.000
MAMDR	0.472 ± 0.002	0.812 ± 0.002	0.430 ± 0.004	0.835 ± 0.001	0.483 ± 0.036	0.772 ± 0.042
DIN	0.473 ± 0.000	0.804 ± 0.000	0.427 ± 0.000	0.818 ± 0.000	0.461 ± 0.000	0.791 ± 0.000
DIEN	0.475 ± 0.004	0.807 ± 0.002	0.427 ± 0.001	0.825 ± 0.001	0.457 ± 0.005	0.797 ± 0.001
MAGRec	**0.461 ± 0.003**	**0.819 ± 0.003**	**0.416 ± 0.001**	**0.836 ± 0.001**	**0.436 ± 0.001**	**0.815 ± 0.002**

smoothly across all domains and unrelated domains could potentially create an interest bottleneck in a k-hop message passing for user recent interest modelling. To overcome this limitation, we propose an *Interacting History* representation in which the *Flattened History* is enriched with domain skip connections to enable uninterrupted intra-domain paths and proper signal propagation in cases where two domains share little to no commonalities. Table 3 compares all three representations for two domain setting on Amazon-2a dataset and different methods. It is clear that *Interacting History* gives the best results with 26.5% and 0.5% improvement over *Disjoint* and *Flattened History* representations respectively for MAGRec. Results are consistent for sequential (DIN, DIEN) and graph-based models (FGNN, MAGRec) proving the importance of multi-domain context as well as intra- and inter-domain paths. FGNN cannot benefit from the *Interacting History*, as it lacks a dedicated mechanism for Domain Contextualization during message propagation.

More Domains Scenarios. In order to evaluate our method performance beyond two domain scenarios, we use three, six and 13 domains. The results are in Table 4. MAGRec receives the best logloss and AUC in all scenarios. Interestingly, SDR models (i.e., DIN, DIEN) achieve second-best performance on 13-domains where both, MAMDR and STAR, struggle to model all domains well at the same time. As stated in [16], the incremental drop in performance is even more noticeable for STAR. MTL-based models such as MMoE also cope to handle larger number of domains. Figure 2 presents variability across all domains

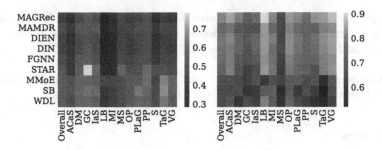

Fig. 2. Results for all 13 domains and methods presented as a heatmap. Logloss (left) and AUC (right).

Table 5. Results for MAGRec and Amazon-2d dataset where Recent Interest Extractor (RIE), Global Interest Extraction (GIE) and Domain Contextualization (DC) are ablated.

RIE	GIE	DC	Domain 1		Domain 2		Both	
			Logloss	AUC	Logloss	AUC	Logloss↓	AUC
	✓	✓	0.434 ± 0.001	0.805 ± 0.002	0.487 ± 0.001	0.812 ± 0.003	0.460 ± 0.001	0.808 ± 0.002
		✓	0.429 ± 0.001	0.810 ± 0.001	0.477 ± 0.003	0.820 ± 0.002	0.453 ± 0.002	0.815 ± 0.001
✓			0.365 ± 0.001	0.872 ± 0.001	0.381 ± 0.002	0.893 ± 0.001	0.373 ± 0.001	0.882 ± 0.000
✓	✓		0.364 ± 0.000	0.873 ± 0.001	0.378 ± 0.000	0.895 ± 0.000	0.371 ± 0.000	0.884 ± 0.001
✓		✓	0.364 ± 0.002	0.874 ± 0.002	0.376 ± 0.003	0.897 ± 0.001	0.370 ± 0.002	0.885 ± 0.001
✓	✓	✓	0.364 ± 0.001	0.873 ± 0.001	0.374 ± 0.002	0.897 ± 0.001	$\mathbf{0.369 \pm 0.001}$	$\mathbf{0.885 \pm 0.001}$

and used methods. MAGRec and DIEN are clear winners here. While MMoE, SB, WDL are worse, with a high variability between domains.

Ablation Study. In Table 5, we present the ablation study results for MAGRec network, where particular modules of the network are disabled. The biggest improvement in logloss 0.08 is when Recent Interest Extractor (RIE) is used. Then, Global Interest Extractor (GIE) gives a smaller boost of 0.002. When combining all, with Domain Contextualization (DC), we observe the best outcome of 0.369 logloss value. Combining DC with GIE alone does not help to get better overall performance.

5 Conclusions and Future Work

This paper presents a new method, MAGRec, for multi-domain recommendation that uses GNNs to model intra- and inter-domain sequential relations. In our experiments on publicly available datasets, our method shows better performance compared to state-of-the-art methods in multiple different settings: from two up to 13 different domain combinations. Additionally, we performed a series of experiments that proved the usefulness of *Interacting History* representation as well as *Recent* and *Global Interest Extractors*. As future work, higher-order graph representations could be explored as input for sparse multi-domain representation. Additionally, aforementioned integration with other MDR training strategies can further improve the results.

Acknowledgements. This work was partially supported by the FairTransNLP-Language Project (MCIN/AEI/10.13039/501100011033/FEDER,UE).

References

1. Abdollahpouri, H., et al.: Multistakeholder recommendation: survey and research directions. User Model. User-Adapt. Inter. **30**(1), 127–158 (2020)
2. Ahmadi, A.H.K., Hassani, K., Moradi, P., Lee, L., Morris, Q.: Memory-based graph networks. In: 8th International Conference on Learning Representations, ICLR 2020, Addis Ababa, Ethiopia, 26–30 April 2020. OpenReview.net (2020). https://openreview.net/forum?id=r1laNeBYPB
3. Aoki, R., Tung, F., Oliveira, G.L.: Heterogeneous multi-task learning with expert diversity. IEEE/ACM Transactions on Computational Biology and Bioinformatics, vol. PP (2022)
4. Caruana, R.: Multitask learning, pp. 95–133. Kluwer Academic Publishers, USA (1998)
5. Chang, J., et al.: Sequential recommendation with graph neural networks. In: Proceedings of the 44th International ACM SIGIR Conference on Research and Development in Information Retrieval, pp. 378–387. SIGIR 2021, Association for Computing Machinery, New York, NY, USA (2021). https://doi.org/10.1145/3404835.3462968
6. Cheng, H.T., et al.: Wide & deep learning for recommender systems. In: Proceedings of the 1st Workshop on Deep Learning for Recommender Systems, pp. 7–10. DLRS 2016, Association for Computing Machinery, New York, NY, USA (2016). https://doi.org/10.1145/2988450.2988454
7. Cui, Q., Wei, T., Zhang, Y., Zhang, Q.: HeroGRAPH: a heterogeneous graph framework for multi-target cross-domain recommendation. In: Vinagre, J., Jorge, A.M., Al-Ghossein, M., Bifet, A. (eds.) Proceedings of the 3rd Workshop on Online Recommender Systems and User Modeling co-located with the 14th ACM Conference on Recommender Systems (RecSys 2020), Virtual Event, 25 Sept 2020. CEUR Workshop Proceedings, vol. 2715. CEUR-WS.org (2020). https://ceur-ws.org/Vol-2715/paper6.pdf
8. Fang, H., Zhang, D., Shu, Y., Guo, G.: Deep learning for sequential recommendation: Algorithms, influential factors, and evaluations. ACM Trans. Inf. Syst. **39**(1), 1–42 (2020). https://doi.org/10.1145/3426723
9. Guo, H., Chen, B., Tang, R., Zhang, W., Li, Z., He, X.: An embedding learning framework for numerical features in CTR prediction. In: Zhu, F., Ooi, B.C., Miao, C. (eds.) KDD 2021: The 27th ACM SIGKDD Conference on Knowledge Discovery and Data Mining, Virtual Event, Singapore, 14–18 August 2021, pp. 2910–2918. ACM (2021). https://doi.org/10.1145/3447548.3467077
10. Hidasi, B., Karatzoglou, A., Baltrunas, L., Tikk, D.: Session-based recommendations with recurrent neural networks. In: Bengio, Y., LeCun, Y. (eds.) 4th International Conference on Learning Representations, ICLR 2016, San Juan, Puerto Rico, 2-4 May 2016, Conference Track Proceedings (2016). arxiv.org/abs/1511.06939

11. Hu, G., Zhang, Y., Yang, Q.: CoNet: collaborative cross networks for cross-domain recommendation. In: Proceedings of the 27th ACM International Conference on Information and Knowledge Management, CIKM 2018, Torino, Italy, 22–26 Oct 2018, pp. 667–676. ACM (2018). https://doi.org/10.1145/3269206.3271684

12. Kang, W.C., McAuley, J.: Self-attentive sequential recommendation. In: 2018 IEEE International Conference on Data Mining (ICDM), pp. 197–206 (2018). https://doi.org/10.1109/ICDM.2018.00035

13. Li, J., Ren, P., Chen, Z., Ren, Z., Lian, T., Ma, J.: Neural attentive session-based recommendation. In: Proceedings of the 2017 ACM on Conference on Information and Knowledge Management, pp. 1419–1428. CIKM 2017, Association for Computing Machinery, New York, NY, USA (2017). https://doi.org/10.1145/3132847.3132926

14. Li, P., Jiang, Z., Que, M., Hu, Y., Tuzhilin, A.: Dual attentive sequential learning for cross-domain click-through rate prediction. In: Proceedings of the 27th ACM SIGKDD Conference on Knowledge Discovery & Data Mining, pp. 3172–3180. KDD 2021, Association for Computing Machinery, New York, NY, USA (2021). https://doi.org/10.1145/3447548.3467140

15. Li, Y., Tarlow, D., Brockschmidt, M., Zemel, R.S.: Gated graph sequence neural networks. In: Bengio, Y., LeCun, Y. (eds.) 4th International Conference on Learning Representations, ICLR 2016, San Juan, Puerto Rico, 2–4 May 2016, Conference Track Proceedings (2016). arxiv.org/abs/1511.05493

16. Luo, L., et al.: MAMDR: a model agnostic learning method for multi-domain recommendation. CoRR abs/2202.12524 (2022). arxiv.org/abs/2202.12524

17. Lv, F., et al.: SDM: sequential deep matching model for online large-scale recommender system. In: Proceedings of the 28th ACM International Conference on Information and Knowledge Management, pp. 2635–2643. CIKM 2019, Association for Computing Machinery, New York, NY, USA (2019). https://doi.org/10.1145/3357384.3357818

18. Ma, J., Zhao, Z., Yi, X., Chen, J., Hong, L., Chi, E.H.: Modeling task relationships in multi-task learning with multi-gate mixture-of-experts. In: Proceedings of the 24th ACM SIGKDD International Conference on Knowledge Discovery & Data Mining, pp. 1930–1939. KDD 2018, Association for Computing Machinery, New York, NY, USA (2018). https://doi.org/10.1145/3219819.3220007

19. Man, T., Shen, H., Jin, X., Cheng, X.: Cross-domain recommendation: an embedding and mapping approach. In: Proceedings of the 26th International Joint Conference on Artificial Intelligence, pp. 2464–2470. IJCAI2017, AAAI Press (2017)

20. McMahan, H.B., et al.: Ad click prediction: a view from the trenches. In: Proceedings of the 19th ACM SIGKDD International Conference on Knowledge Discovery and Data Mining, pp. 1222–1230 (2013)

21. Ni, J., Li, J., McAuley, J.: Justifying recommendations using distantly-labeled reviews and fine-grained aspects. In: Proceedings of the 2019 Conference on Empirical Methods in Natural Language Processing and the 9th International Joint Conference on Natural Language Processing (EMNLP-IJCNLP), pp. 188–197. Association for Computational Linguistics, Hong Kong, China (2019). https://doi.org/10.18653/v1/D19-1018

22. Ouyang, W., et al.: MiNet: mixed interest network for cross-domain click-through rate prediction. In: Proceedings of the 29th ACM International Conference on Information & Knowledge Management, pp. 2669–2676. CIKM 2020, Association for Computing Machinery, New York, NY, USA (2020). https://doi.org/10.1145/3340531.3412728

23. Pi, Q., et al.: Search-based user interest modeling with lifelong sequential behavior data for click-through rate prediction. In: Proceedings of the 29th ACM International Conference on Information & Knowledge Management, pp. 2685–2692. CIKM 2020, Association for Computing Machinery, New York, NY, USA (2020). https://doi.org/10.1145/3340531.3412744
24. Qiu, R., Li, J., Huang, Z., YIn, H.: Rethinking the item order in session-based recommendation with graph neural networks. In: Proceedings of the 28th ACM International Conference on Information and Knowledge Management. p. 579–588. CIKM 2019, Association for Computing Machinery, New York, NY, USA (2019). https://doi.org/10.1145/3357384.3358010
25. Schoenauer-Sebag, A., Heinrich, L., Schoenauer, M., Sebag, M., Wu, L., Altschuler, S.: Multi-domain adversarial learning. In: International Conference on Learning Representations (2019). https://openreview.net/forum?id=Sklv5iRqYX
26. Shen, W.: Deepctr: Easy-to-use, modular and extendible package of deep-learning based CTR models. https://github.com/shenweichen/deepctr (2017)
27. Sheng, X.R., et al.: One model to serve all: Star topology adaptive recommender for multi-domain CTR prediction. In: Proceedings of the 30th ACM International Conference on Information & Knowledge Management, pp. 4104–4113. CIKM 2021, Association for Computing Machinery, New York, NY, USA (2021). https://doi.org/10.1145/3459637.3481941
28. Tang, H., Liu, J., Zhao, M., Gong, X.: Progressive layered extraction (PLE): a novel multi-task learning (MTL) model for personalized recommendations. In: Fourteenth ACM Conference on Recommender Systems, pp. 269–278. RecSys 2020, Association for Computing Machinery, New York, NY, USA (2020). https://doi.org/10.1145/3383313.3412236
29. Tang, J., Wang, K.: Personalized top-n sequential recommendation via convolutional sequence embedding. In: Proceedings of the Eleventh ACM International Conference on Web Search and Data Mining, pp. 565–573. WSDM 2018, Association for Computing Machinery, New York, NY, USA (2018). https://doi.org/10.1145/3159652.3159656
30. Wang, R., Fu, B., Fu, G., Wang, M.: Deep & cross network for ad click predictions. In: Proceedings of the ADKDD2017. ADKDD2017, Association for Computing Machinery, New York, NY, USA (2017). https://doi.org/10.1145/3124749.3124754
31. Wu, S., Tang, Y., Zhu, Y., Wang, L., Xie, X., Tan, T.: Session-based recommendation with graph neural networks. In: Proceedings of the Thirty-Third AAAI Conference on Artificial Intelligence and Thirty-First Innovative Applications of Artificial Intelligence Conference and Ninth AAAI Symposium on Educational Advances in Artificial Intelligence. AAAI2019/IAAI2019/EAAI2019, AAAI Press (2019). https://doi.org/10.1609/aaai.v33i01.3301346
32. Xia, X., Yin, H., Yu, J., Wang, Q., Cui, L., Zhang, X.: Self-supervised hypergraph convolutional networks for session-based recommendation. In: Thirty-Fifth AAAI Conference on Artificial Intelligence, AAAI 2021, Thirty-Third Conference on Innovative Applications of Artificial Intelligence, IAAI 2021, The Eleventh Symposium on Educational Advances in Artificial Intelligence, EAAI 2021, Virtual Event, 2–9 Feb 2021, pp. 4503–4511. AAAI Press (2021). https://ojs.aaai.org/index.php/AAAI/article/view/16578

33. Zhao, C., Li, C., Fu, C.: Cross-domain recommendation via preference propagation graphNet. In: Proceedings of the 28th ACM International Conference on Information and Knowledge Management, pp. 2165–2168. CIKM 2019, Association for Computing Machinery, New York, NY, USA (2019). https://doi.org/10.1145/3357384.3358166

34. Zhou, G., et al.: Deep interest evolution network for click-through rate prediction. In: The Thirty-Third AAAI Conference on Artificial Intelligence, AAAI 2019, The Thirty-First Innovative Applications of Artificial Intelligence Conference, IAAI 2019, The Ninth AAAI Symposium on Educational Advances in Artificial Intelligence, EAAI 2019, Honolulu, Hawaii, USA, 27 Jan - 1 Feb 2019, pp. 5941–5948. AAAI Press (2019). https://doi.org/10.1609/aaai.v33i01.33015941

35. Zhou, G., et al.: Deep interest network for click-through rate prediction. In: Guo, Y., Farooq, F. (eds.) Proceedings of the 24th ACM SIGKDD International Conference on Knowledge Discovery & Data Mining, KDD 2018, London, UK, 19–23 Aug 2018, pp. 1059–1068. ACM (2018). https://doi.org/10.1145/3219819.3219823

Injecting the BM25 Score as Text Improves BERT-Based Re-rankers

Arian Askari[1(✉)], Amin Abolghasemi[1], Gabriella Pasi[2], Wessel Kraaij[1], and Suzan Verberne[1]

[1] Leiden Institute of Advanced Computer Science,
Leiden University, Leiden, The Netherlands
{a.askari,m.a.abolghasemi,w.kraaij,s.verberne}@liacs.leidenuniv.nl
[2] Department of Informatics, Systems and Communication,
University of Milano-Bicocca, Milan, Italy
gabriella.pasi@unimib.it

Abstract. In this paper we propose a novel approach for combining first-stage lexical retrieval models and Transformer-based re-rankers: we inject the relevance score of the lexical model as a token in the middle of the input of the cross-encoder re-ranker. It was shown in prior work that interpolation between the relevance score of lexical and BERT-based re-rankers may not consistently result in higher effectiveness. Our idea is motivated by the finding that BERT models can capture numeric information. We compare several representations of the BM25 score and inject them as text in the input of four different cross-encoders. We additionally analyze the effect for different query types, and investigate the effectiveness of our method for capturing exact matching relevance. Evaluation on the MSMARCO Passage collection and the TREC DL collections shows that the proposed method significantly improves over all cross-encoder re-rankers as well as the common interpolation methods. We show that the improvement is consistent for all query types. We also find an improvement in exact matching capabilities over both BM25 and the cross-encoders. Our findings indicate that cross-encoder re-rankers can efficiently be improved without additional computational burden and extra steps in the pipeline by explicitly adding the output of the first-stage ranker to the model input, and this effect is robust for different models and query types.

Keywords: Injecting BM25 · Two-stage retrieval · Transformer-based rankers · BM25 · Combining lexical and neural rankers

1 Introduction

The commonly used ranking pipeline consists of a first-stage retriever, e.g. BM25 [47], that efficiently retrieves a set of documents from the full document collection, followed by one or more re-rankers [40,59] that improve the initial ranking. Currently, the most effective re-rankers are BERT-based rankers with a

J. Kamps et al. (Eds.): ECIR 2023, LNCS 13980, pp. 66–83, 2023.
https://doi.org/10.1007/978-3-031-28244-7_5

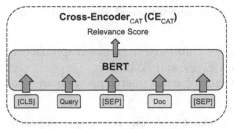
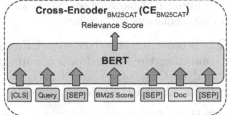

Fig. 1. Regular cross-encoder input   **Fig. 2.** Injection of BM25 in input

cross-encoder architecture, concatenating the query and the candidate document in the input [2,25,40,44]. In this paper, we refer to these re-rankers as Cross-Encoder$_{CAT}$ (CE$_{CAT}$). In the common re-ranking set-up, BM25 [47] is widely leveraged [7,20,27] for finding the top-k documents to be re-ranked; however, the relevance score produced by BM25 based on exact lexical matching is not explicitly taken into account in the second stage. Besides, although cross-encoder re-rankers substantially improve the retrieval effectiveness compared to BM25 alone [34], Rau et al. [43] show that BM25 is a more effective *exact lexical matcher* than CE$_{CAT}$ rankers; in their exact-matching experiment they only use the words from the passage that also appear in the query as the input of the CE$_{CAT}$. This suggests that CE$_{CAT}$ re-rankers can be further improved by a better exact word matching, as the presence of query words in the document is one of the strongest signals for relevance in ranking [48,50]. Moreover, obtaining improvement in effectiveness by interpolating the scores (score fusion [58]) of BM25 and CE$_{CAT}$ is challenging: a linear combination of the two scores has shown to decrease effectiveness on the MSMARCO Passage collection compared to only using the CE$_{CAT}$ re-ranker in the second stage retrieval [34].

To tackle this problem, in this work, we propose a method to enhance CE$_{CAT}$ re-rankers by directly injecting the BM25 score as a string to the input of the Transformer. Figure 2 show our method for the injection of BM25 in the input of the CE re-ranker. We refer to our method as CE$_{BM25CAT}$. Our idea is inspired by the finding by Wallace et al. [54] that BERT models can capture numeracy. In this regard, we address the following research questions:

RQ1: What is the effectiveness of BM25 score injection in addition to the query and document text in the input of CE re-rankers?

To answer this question we setup two experiments on three datasets: MSMARCO, TREC DL'19 and '20. First, since the BM25 score has no defined range, we investigate the effect of different representations of the BM25 score by applying various normalization methods. We also analyze the effect of converting the normalized scores of BM25 to integers. Second, we evaluate the best representation of BM25 – based on our empricial study – on four cross-encoders: BERT-base, BERT-large [53], DistillBERT [49], and MiniLM [56], comparing CE$_{BM25CAT}$ to CE$_{CAT}$ across different Transformer models with a smaller and larger number of parameters. Next, we compare our proposed approach to common interpolation approaches:

RQ2: What is the effectiveness of $CE_{BM25CAT}$ compared to common approaches for combining the final relevance scores of CE_{CAT} and BM25?
To analyze $CE_{BM25CAT}$ and CE_{CAT} in terms of exact matching compared to BM25 we address the following question:

RQ3: How effective can $CE_{BM25CAT}$ capture exact matching relevance compared to BM25 and CE_{CAT}?
Furthermore, to provide an explanation on the improvement of $CE_{BM25CAT}$, we perform a qualitative analysis of a case where CE_{CAT} fails to identify the relevant document that is found using $CE_{BM25CAT}$ with the help of the BM25 score.[1]

To the best of our knowledge, there is no prior work on the effectiveness of cross-encoder re-rankers by injecting a retrieval model's score into their input. Our main contributions in this work are four-fold:

1. We provide a strategy for efficiently utilizing BM25 in cross-encoder re-rankers, which yields statistically significant improvements on all official metrics and is verified by thorough experiments and analysis.
2. We find that our method is more effective than the approaches which linearly interpolate the scores of BM25 and CE_{CAT}.
3. We analyze the exact matching effectiveness of CE_{CAT} and $CE_{BM25CAT}$ in comparison to BM25. We show that $CE_{BM25CAT}$ is a more powerful exact matcher than BM25 while CE_{CAT} is less effective than BM25.
4. We analyze the effectiveness of CE_{CAT} and $CE_{BM25CAT}$ on different query types. We show that $CE_{BM25CAT}$ consistently outperforms CE_{CAT} over all type of queries.

After a discussion of related work in Sect. 2, we describe the retrieval models employed in Sect. 3 and the specifics of our experiments and methods in Sect. 4. The results are examined and the research questions are addressed in Sect. 5. Finally, the conclusion is described in Sect. 6.

2   Related Work

Modifying the Input of Re-rankers. Boualili et al. [12,13] propose a method for highlighting exact matching signals by marking the start and the end of each occurrence of the query terms by adding markers to the input. In addition, they modify original passages and expand each passage with a set of generated queries using Doc2query [41] to overcome the vocabulary mismatch problem. This strategy is different from ours in two aspects: (1) the type of information added to the input: they add four tokens as markers for each occurrence of query terms, adding a burden to the limited input length of 512 tokens for query

---

[1] In this work, we interchangeably use the words document and passage to refer to unit that should be retrieved.

and document together, while we only add the BM25 score. (2) The need for data augmentation: they need to train a Doc2query model to provide the exact matching signal for improving the BERT re-ranker while our strategy does not need any extra overhead in terms of data augmentation. A few recent, but less related examples are Al-Hajj et al. [4], who experiment with the use of different supervised signals into the input of the cross-encoder to emphasize target words in context and Li et al. [30], who insert boundary markers into the input between contiguous words for Chinese named entity recognition.

Numerical Information in Transformer Models. Thawani et al. [52] provide an extensive overview of numeracy in NLP models up to 2021. Wallace et al. [54] analyze the ability of BERT models to work with numbers and come to the conclusion that the models capture numeracy and are able to do numerical reasoning; however the models appeared to struggle with interpreting floats. Moreover, Zhang et al. [63] show that BERT models capture a significant amount of information about numerical scale except for general common-sense reasoning. There are various studies that are inspired by the fact that Transformer models can correctly process numbers [11,15,21,22,26,38]. Gu et al. [23] incorporate text, categorical and numerical data as different modalities with Transformers using a combining module accross different classification tasks. They discover that adding tabular features increases the effectiveness while using only text is insufficient and results in the worst performance.

Methods for Combining Rankers. Linearly interpolating different rankers' scores has been studied extensively in the literature [8–10,34,58]. In this paper, we investigate multiple linear and non-linear interpolation ensemble methods to analyze the performance of them for combining BM25 and CE_{CAT} scores in comparison to $CE_{BM25CAT}$. For the sake of a fair analysis, we do not compare $CE_{BM25CAT}$ with a Learning-to-rank approach that is trained on 87 features by [65]. The use of ensemble methods brings additional overhead in terms of efficiency because it adds one more extra step to the re-ranking pipeline. It is noteworthy to mention that in this paper, we concentrate on analyzing the improvement by combining the first-stage retriever and a BERT-based re-ranker: BM25 and CE_{CAT} respectively. However, we are aware that combining scores of BM25 and Dense Retrievers that both are first-stage retrievers has also shown improvements [1,6,55] that are outside the scope of our study. In particular, CLEAR [20] proposes an approach to train the dense retrievers to encode semantics that BM25 fails to capture for first stage retrieval. However, in this study, our aim is to improve re-ranking in the second stage of two-stage retrieval setting.

3 Methods

3.1 First Stage Ranker: BM25

Lexical retrievers estimate the relevance of a document to a query based on word overlap [46]. Many lexical methods, including vector space models, Okapi BM25, and query likelihood, have been developed in previous decades. We use BM25 because of its popularity as first-stage ranker in current systems. Based on the statistics of the words that overlap between the query and the document, BM25 calculates a score for the pair:

$$s_{lex}(q,d) = BM25(q,d) = \sum_{t \in q \cap d} rsj_t \cdot \frac{tf_{t,d}}{tf_{t,d} + k_1\{(1-b) + b\frac{|d|}{l}\}} \quad (1)$$

where t is a term, $tf_{t,d}$ is the frequency of t in document d, rsj_t is the Robertson-Spärck Jones weight [47] of t, and l is the average document length. k_1 and b are parameters [32,33].

3.2   CE$_{CAT}$: Cross-Encoder Re-rankers Without BM25 Injection

Concatenating query and passage input sequences is the typical method for using cross-encoder (e.g., BERT) architectures with pre-trained Transformer models in a re-ranking setup [25,36,40,60]. This basic design is referred to as CE$_{CAT}$ and shown in Fig. 1. The query $q_{1:m}$ and passage $p_{1:n}$ sequences are concatenated with the $[SEP]$ token, and the $[CLS]$ token representation computed by CE is scored with a single linear layer W_s in the CE$_{CAT}$ ranking model:

$$CE_{CAT}(q_{1:m}, p_{1:n}) = CE([CLS]\, q\, [SEP]\, p\, [SEP]) * W_s \quad (2)$$

We use CE$_{CAT}$ as our baseline re-ranker architecture. We evaluate different cross-encoder models in our experiments and all of them follow the above design.

3.3   CE$_{BM25CAT}$: Cross-Encoder Re-rankers with BM25 Injection

To study the effectiveness of injecting the BM25 score into the input, we modify the input of the basic input format as follows and call it CE$_{BM25CAT}$:

$$CE_{BM25CAT}(q_{1:m}, p_{1:n}) = CE([CLS]\, q\, [SEP]\, BM25\, [SEP]\, p\, [SEP]) * W_s \quad (3)$$

where BM25 represent the relevance score produced by BM25 between query and passage.

We study different representations of BM25 to find the optimal approach for injecting BM25 into the cross-encoders. The reasons are: (1) BM25 scores do not have an upper bound and should be normalized for having an interpretable score given a query and passage; (2) BERT-based models can process integers better than floating point numbers [54] so we analyze if converting the normalized score to an integer is more effective than injecting the floating point score. For normalizing BM25 scores, we compare three different normalization methods: Min-Max, Standardization (Z-score), and Sum:

$$Min\text{-}Max(s_{BM25}) = \frac{s_{BM25} - s_{min}}{s_{max} - s_{min}} \quad (4)$$

$$Standard(s_{BM25}) = \frac{s_{BM25} - \mu(S)}{\sigma(S)} \quad (5)$$

$$Sum(s_{BM25}) = \frac{s_{BM25}}{sum(S)} \tag{6}$$

where s_{BM25} is the original score, and s_{max} and s_{min} are the maximum and minimum scores respectively, in the ranked list. $Sum(S)$, $\mu(S)$, and $\sigma(S)$ refer to sum, average and standard deviation over the scores of all passages retrieved for a query. The anticipated effect of the Sum normalizer is that the sum of the scores of all passages in the ranked list will be 1; thus, if the top-n passages receive much higher scores than the rest, their normalized scores will have a larger difference with the rest of passages' scores in the ranked list; this distance could give a good signal to $CE_{BM25CAT}$. We experiment with Min-Max and Standardization in a local and a global setting. In the local setting, we get the minimum or maximum (for Min-Max) and mean and standard deviation (for Standard) from the ranked list of scores per query. In the global setting, we use $\{0, 50, 42, 6\}$ as {minimum, maximum, mean, standard deviation} as they have been empirically suggested in prior work to be used as default values across different queries to globally normalize BM25 scores [37]. In our data, the {minimum, maximum, mean, standard deviation} values are $\{0, 98, 7, 5\}$ across all queries. Because of the differences between the recommended defaults and the statistics of our collections, we explore other global values for Min-Max, using $25, 50, 75, 100$ as maximum and 0 as minimum. However, we got the best result using default values of [37]. To convert the float numbers to integers we multiply the normalized score to 100 and discard decimals. Finally, we store the number as a string.

3.4   Linear Interpolation Ensembles of BM25 and CE$_{CAT}$

We compare our approach to common ensemble methods [34,64] for interpolating BM25 and BERT re-rankers. We combine the scores linearly using the following methods: (1) Sum: compute sum over BM25 and CE$_{CAT}$ scores, (2) Max: select maximum between BM25 and CE$_{CAT}$ scores, and (3) Weighted-Sum:

$$s_i = \alpha \cdot s_{BM25} + (1 - \alpha) \cdot s_{CE_{CAT}} \tag{7}$$

where s_i is the weighted sum produced by the interpolation, s_{BM25} is the normalized BM25 score, $s_{CE_{CAT}}$ is the CE$_{CAT}$ score, and $\alpha \in [0..1]$ is a weight that indicates the relative importance. Since CE$_{CAT}$ score $\in [0, 1]$, we also normalize BM25 score using Min-Max normalization. Furthermore, we train ensemble models that take s_{BM25} and $s_{CE_{CAT}}$ as features. We experiment with three different classifiers for this purpose: SVM with a linear kernel, SVM with an RBFkernel, Naive Bayes, and Multi Layer Perceptron (MLP) as a non-linear method and report the best classifier performance in Sect. 5.1.

4   Experimental Design

Dataset and Metrics. We conduct our experiments on the MSMARCO-passage collection [39] and the two TREC Deep Learning tracks (TREC-DL'19 and TREC-DL'20) [17,19]. The MSMARCO-passage dataset contains about 8.8 million passages (average length: 73.1 words) and about 1 million natural language queries (average length: 7.5 words) and has been extensively used to train deep language models for ranking because of the large number of queries. Following prior work on MSMARCO [28,34,35,67,68], we use the dev set ($\sim 7k$ queries) for our empirical evaluation. $MAP@1000$ and $nDCG@10$ are calculated in addition to the official evaluation metric $MRR@10$. The passage corpus of MSMARCO is shared with TREC DL'19 and DL'20 collections with 43 and 54 queries respectively. We evaluate our experiments on these collections using $nDCG@10$ and $MAP@1000$, as is standard practice in TREC DL [17,19] to make our results comparable to previously published and upcoming research. We cap the query length at 30 tokens and the passage length at 200 tokens following prior work [25].

Training Configuration and Model Parameters. We use the Huggingface library [57], Cross-encoder package of Sentence-transformers library [45], and PyTorch [42] for the cross-encoder re-ranking training and inference. For injecting the BM25 score as text, we pass the BM25 score in string format into the BERT tokenizer in a similar way to passing query and document. Please note that the integer numbers are already included in the BERT tokenizer's vocabulary, allowing for appropriate tokenization. Following prior work [25] we use the Adam [29] optimizer with a learning rate of $7 * 10^{-6}$ for all cross-encoder layers, regardless of the number of layers trained. To train cross-encoder re-rankers for each TREC DL collection, we use the other TREC DL query set as the validation set and we select both TREC DL ('19 and '20) query sets as the validation set to train CEs for the MSMARCO Passage collection. We employ early stopping, based on the nDCG@10 value of the validation set. We use a training batch size of 32. For all cross-Encoder re-rankers, we use Cross-Entropy loss [66] . For the lexical retrieval with BM25 we employ the tuned parameters from the Anserini documentation [32,33].[2]

5   Results

5.1   Main Results: Addressing Our Research Questions

Choice of BM25 Score Representation. As introduced in Sect. 3.3, we compare different representations of the BM25 score in Table 1 for injection into $CE_{BM25CAT}$. We chose MiniLM [56] for this study as it has shown competitive results in comparison to BERT-based models while it is 3 times smaller and 6 times faster.[3] Our first interesting observation is that injecting the original float score rounded down to 2 decimal points (row b) of BM25 into the input seems to slightly improve the effectiveness of re-ranker. We assume this is due to the fact that the average query and passage length is relatively small in the MSMARCO

---

[2] The code is available on https://github.com/arian-askari/injecting_bm25_score_bert.
[3] https://huggingface.co/microsoft/MiniLM-L12-H384-uncased.

Passage collection, which prevents from getting high numbers – with low interpretability for BERT – as BM25 score. Second, we find that the normalized BM25 score with Min-Max in the global normalization setting converted to integer (row f) is the most significant effective[4] representation for injecting BM25.

Table 1. Effectiveness results. Lines b-n refer to the MiniLM$_{BMCAT}$ re-ranker using different representations of the BM25 score as text. Significance is shown with † for the best result (row f) compared to MiniLM$_{CAT}$ (row a). Statistical significance was measured with a paired t-test ($p < 0.05$) with Bonferroni correction for multiple testing.

	Normalization	Local/Global	Float/Integer	MSMARCO DEV nDCG@10	MAP	MRR@10
(a)	MiniLM$_{CAT}$ (without injecting BM25 score)			.419	.363	.360
(b)	Original score	—	—	.420	.364	.362
(c)	Min-Max	Local	Float	.411	.359	.354
(d)	Min-Max	Local	Integer	.414	.361	.355
(e)	Min-Max	Global	Float	.422	.365	.363
(f)	Min-Max	Global	Integer	**.424**†	**.368**†	**.367**†
(g)	Standard	Local	Float	.407	.355	.352
(h)	Standard	Local	Integer	.410	.358	.354
(i)	Standard	Global	Float	.420	.363	.361
(j)	Standard	Global	Integer	.421	.365	.363
(k)	Sum	–	Float	.402	.349	.338
(l)	Sum	–	Integer	.405	.350	.342

The global normalization setting gives better results for both Min-Max (rows e, f) and Standardization (rows i, j) than local normalization (rows c, d and g, h).[5] The reason is probably that in the global setting a candidate document obtains a high normalized score (close to 1 in the floating point representation) if its original score is close to default maximum (for Min-Max normalization) so the normalized score could be more interpretable across different queries. On the other hand, in the local setting, the passages ranked at position 1 always receive 1 as normalized score with Min-Max even if its original score is not high and it does not have a big difference with the last passage in the ranked list.

Moreover, converting the normalized float score to integers gives better results for both Min-Max (rows d, f) and Standardization (rows h, j) than the float representation (rows c, e and g, i). We find that Min-Max normalization is a better representation for injecting BM25 than Standardization, which could

---

[4] Although the evaluation metrics are not in an interval scale, Craswell et al. [18] show that they are mostly reliable in practice on MSMARCO for statistical testing.

[5] The range of normalized integer scores using the best normalizer (row f) are from 0 to 196 as the maximum BM25 score in the collection is 98.

Table 2. Effectiveness results. Fine-tuned cross-encoders are used for re-ranking over BM25 first stage retrieval with a re-ranking depth of 1000. † indicates a statistically significant improvement of a cross-encoder with BM25 score injection as text into the input (Cross-encoder$_{BM25CAT}$) over the same cross-encoder without BM25 score injection (Cross-encoder$_{CAT}$). Statistical significance was measured with a paired t-test ($p < 0.05$) with Bonferroni correction for multiple testing.

Model	TREC DL 20		TREC DL 19		MSMARCO DEV		
	nDCG@10	MAP	nDCG@10	MAP	nDCG@10	MAP	MRR@10
BM25	.480	.286	.506	.377	.234	.195	.187
Re-rankers							
BERT-Base$_{CAT}$	.689	.447	.713	.441	.399	.346	.342
BERT-Base$_{BM25CAT}$	.705†	.475†	.723†	.453†	.422†	.367†	.364†
BERT-Large$_{CAT}$	.695	.464	.714	.467	.401	.344	.360
BERT-Large$_{BM25CAT}$	**.728†**	**.482†**	**.731†**	**.477†**	**.424†**	.367†	**.369†**
DistilBERT$_{CAT}$	.670	.442	.679	.440	.383	.310	.325
DistilBERT$_{BM25CAT}$	.682†	.456†	.699†	.451†	.390†	.323†	.339†
MiniLM$_{CAT}$	.681	.448	.704	.452	.419	.363	.360
MiniLM$_{BM25CAT}$	.710†	.473†	.711†	.463†	**.424†**	**.368†**	.367†

be due to the fact that in Min-Max the normalized score could not be negative, and, as a result, interpreting the injected score is easier for CE$_{BM25CAT}$. We find that the Sum normalizer (rows k and l) decreases effectiveness. Apparently, our expectation that Sum would help distinguish between the top-n passages and the remaining passages in the ranked list (see Sect. 5.1) is not true.

Impact of BM25 Injection for Various Cross-encoders (RQ1). Table 2 shows that injecting the BM25 score – using the best normalizer which is Min-Max in the global normalization setting converted to integer – into all four cross-encoders improves their effectiveness in all of the metrics compared to using them without injecting BM25. This shows that injecting the BM25 score into the input as a small modification to the current re-ranking pipeline improves the re-ranking effectiveness. This is without any additional computational burden as we train CE$_{CAT}$ and CE$_{BM25CAT}$ in a completely equal setting in terms of number of epochs, batch size, etc. We receive the highest result by BERT-Large$_{BM25CAT}$ for cross-encoder with BM25 injection, which could be due to the higher number of parameters of the model. We find that the results of MiniLM are similar to those for BERT-Base on MSMARCO-DEV while the former is more efficient.

Comparing BM25 Injection with Ensemble Methods (RQ2). Table 3 shows that while injecting BM25 leads to improvement, regular ensemble methods and Naive Bayes classifier fail to do so; combining the scores of BM25 and BERT$_{CAT}$ in a linear and non-linear (MLP) interpolation ensemble setting even leads to lower effectiveness than using the cross-encoder as sole re-ranker. Therefore, our strategy is a better solution than linear interpolation. We only report results for Naive Bayes – having BM25 and BERT$_{CAT}$ score as features – as

Table 3. The effectiveness of injecting BM25 score into the input (Bert-Base$_{BM25CAT}$) compared to interpolation performance of BM25 and Bert-Base$_{CAT}$ using common ensemble methods.

Model	Ensemble	MSMARCO DEV		
		nDCG@10	MAP	MRR@10
BM25	—	.234	.195	.187
BERT-Base$_{CAT}$	—	.399	.346	.342
BM25 and BERT-Base$_{CAT}$	Sum	.270	.225	.218
BM25 and BERT-Base$_{CAT}$	Max	.237	.197	.190
BM25 and BERT-Base$_{CAT}$	Weighted-Sum (tuned)	.353	.295	.290
BM25 and BERT-Base$_{CAT}$	Naive Bayes	.314	.260	.254
BERT-Base$_{BM25CAT}$	BM25 Score Injection	**.422**	**.367**	**.364**

it had the highest effectiveness of the four estimators. Still, the effectiveness is much lower than BERT$_{BM25CAT}$ and also lower than a simple Weighted-Sum. Weighted-Sum (tuned) in Table 3 is tuned on the validation set, for which $\alpha = 0.1$ was found to be optimal. We analyze the effect of different α values in a weighted linear interpolation (Weighted-Sum) to draw a more complete picture on the impact of combining scores on the DEV set. Figure 3 shows that by increasing the weight of BM25, the effectiveness decreases. The figure also shows that the tuned alpha which was found on the validation set in Table 3 is not the most optimal possible alpha value for the DEV set. The highest effectiveness for $\alpha = 0.0$ in Fig. 3 confirms we should not combine the scores by current interpolation methods and only using scores of Bert-Base$_{CAT}$ is better, at least for the MSMARCO passage collection.

Exact Matching Relevance Results (RQ3). To conduct exact matching analysis, we replace the passage words that do not appear in the query with the $[MASK]$ token, leaving the model only with a skeleton of the original passage and force it to rely on the exact word matches between query and passage [43]. We do not train models on this input but use our models that were fine-tuned on the original data. Table 4 shows that BERT-Base$_{BM25CAT}$ performs better than both BM25 and BERT-Base$_{CAT}$ in the exact matching setting on all metrics. Moreover, we found that the percentage of relevant passages ranked in top-10 that are common between BM25 and BERT$_{BM25CAT}$ is 40%, which is higher than the percentage of relevant passages between BM25 and BERT$_{CAT}$ (37%). Therefore, the higher effectiveness of BERT$_{BM25CAT}$ in exact matching setting could be at least partly because it mimics BM25 more than BERT$_{CAT}$. In comparison, this percentage is 57 between BERT$_{BM25CAT}$ and BERT$_{CAT}$.

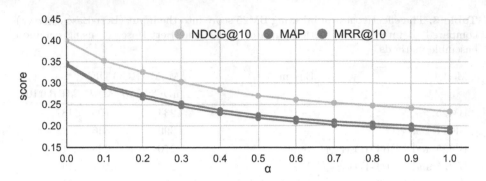

Fig. 3. Effectiveness on MSMARCO DEV with varying the interpolation weight of BM25 and BERT-Base$_{CAT}$ scores. $\alpha = 0$ means only BERT$_{CAT}$ scores are used.

Table 4. Comparing exact matching effectiveness of BERT-Base$_{BM25CAT}$ and BERT-Base$_{CAT}$ by keeping only the query words in each passage for re-ranking. The increase and decrease of effectiveness compared to BM25 is indicated with ↑ and ↓.

Model	Input	MSMARCO DEV		
		nDCG@10	MAP	MRR@10
BM25	Full text	.234	.195	.187
BERT-Base$_{CAT}$	Only query words	.218 (↓1.6)	.186 (↓0.9)	.180 (↓0.7)
BERT-Base$_{BM25CAT}$	Only query words	**.243** (↑.9)	**.209** (↑1.4)	**.202** (↑1.5)

5.2   Analysis of the Results

Query Types. In order to analyze the effectiveness of BERT-base$_{CAT}$ and BERT-base$_{BM25CAT}$ across different types of questions, we classify questions based on the lexical answer type. We use the rule-based answer type classifier[6] inspired by [31] to extract answer types. We classify MSMARCO queries into 6 answer types: abbreviation, location, description, human, numerical and entity. 4105 queries have a valid answer type and at least one relevant passage in the top-1000. We perform our analysis in two different settings: normal (full-text) and exact-matching (keeping only query words and replacing non-query words with $[MASK]$). The average $MRR@10$ per query type is shown in Table 5. The table shows that BERT$_{BM25CAT}$ is more effective than BERT$_{CAT}$ consistently on all types of queries.

---

[6] https://github.com/superscriptjs/qtypes.

Table 5. MRR@10 on MSMARCO-DEV per query type for comparing BERT-Base$_{BM25CAT}$ and BERT-Base$_{CAT}$ on different query types in full-text and exact-matching (only keeping query words) settings.

Model	Input	ABBR	LOC	DESC	HUM	NUM	ENTY
# queries		9	493	1887	455	933	328
BERT-BaseCAT	Full text	.574	.477	.397	.435	.361	.399
BERT-BaseBM25CAT	Full text	**.592**	**.503**	**.428**	**.457**	**.405**	**.411**
BM25	Only query words	.184	.256	.215	.238	.200	**.221**
BERT-BaseCAT	Only query words	.404	.204	.224	.240	.177	.200
BERT-BaseBM25CAT	Only query words	**.438**	**.278**	**.245**	**.258**	**.215**	.216

Query [SEP] BM25 [SEP] Passage	Label	Model: Rank
[CLS] what is the shingles jab ? [SEP] 22 [SEP] the shingles vaccine . the vaccine , called zostavax , is given as a single injection under the skin (subcutaneously) . it can be given at any time in the year . unlike with the flu jab	R	BM25: 3 BERT$_{BM25CAT}$: 1 BERT$_{CAT}$: 104
[CLS] what is the shingles jab ? [SEP] 11 [SEP] shingle is a shingle corruption of german schindle (schindel) meaning a roofing slate . shingles historically were called tiles and shingle was a term applied to wood shingles , as is still mostly the case outside the us [SEP]	N	BM25: 146 BERT$_{BM25CAT}$: 69 BERT$_{CAT}$: 1

Fig. 4. Example query and two passages in the input of BERT$_{BM25CAT}$. The color of each word indicates the word-level attribution value according to Integrated Gradient (IG) [51], where red is positive, blue is negative, and white is neutral. We use the brightness of different colors to indicate the values of these gradients. (Color figure online)

Qualitative Analysis. We show a qualitative analysis of one particular case in Fig. 4 to analyze more in-depth what the effect of BM25 injection is and why it works. In the top row, while BERT$_{CAT}$ mistakenly ranked the relevant passage at position 104, BM25 ranked that passage at position 3 and BERT$_{BM25CAT}$ – apparently helped by BM25 – ranked that relevant passage at position 1. In the bottom row, BERT$_{CAT}$ mistakenly ranked the irrelevant passage at position 1 and informed by the low BM25 score, BERT$_{BM25CAT}$ ranked it much lower, at 69. In order to interpret the importance of the injected BM25 score in the input of CE$_{BM25CAT}$ and show its contributions to the matching score in comparison to other words in the query and passage, we use Integrated Gradient (IG) [51] which has been proven to be a stable and reliable interpretation method in many different applications including Information Retrieval [16,61,62].[7] On both rows of Fig. 4, we see that the BM25 score ('22' in the top row and '11' in the bottom row) is a highly attributed term in comparison to other terms. This shows that

---

[7] We refer readers to [51] for a detailed explanation.

injecting the BM25 score assists $BERT_{BM25CAT}$ to identify relevant or non-relevant passages better than $BERT_{CAT}$.

As a more general analysis, we randomly sampled 100 queries from MSMARCO-DEV. For each query, we took the top-1000 passages retrieved by BM25, we fed all pairs of query and their corresponding retrieved passages ($100k$ pairs) into $BERT_{BM25CAT}$, and computed the attribution scores over the input at the word-level. We ranked tokens based on their importance using the absolute value of their attribution score and found the mode of the rank of the BM25 token over all samples is 3. This shows that $BERT_{BM25CAT}$ highly attributes the BM25 token for ranking.

6   Conclusion and Future Work

In this paper we have proposed an efficient and effective way of combining BM25 and cross-encoder re-rankers: injecting the BM25 score as text in the input of the cross-encoder. We find that the resulting model, $CE_{BM25CAT}$, achieves a statistically significant improvement for all evaluated cross-encoders. Additionally, we find that our injection approach is much more effective than linearly interpolating the initial ranker and re-ranker scores. In addition, we show that $CE_{BM25CAT}$ performs significantly better in an exact matching setting than both BM25 and CE_{CAT} individually. This suggests that injecting the BM25 score into the input could modify the current paradigm for training cross-encoder re-rankers.

While it is crystal clear that our focus is not on chasing the state-of-the-art, we believe that as future work, our method could be applied into any cross-encoder in the current multi-stage ranking pipelines which are state-of-the-art for the MSMARCO Passage benchmark [24]. Moreover, previous studies show that combining BM25 and BERT re-rankers on *Robust04* [5] leads to improvement [3]. It is interesting to study the effect of injecting BM25 for this task because documents often have to be truncated to fit the maximum model input length [14]; injecting the BM25 score might give information to the cross-encoder re-ranker about the lexical relevance of the whole text of the document. Another interesting direction is to study how Dense Retrievers can benefit from injecting lexical ranker scores. Moreover, injecting scores of several lexical rankers and adding more traditional Learning-to-Rank features could be also interesting.

Acknowledgments. This work was supported by the EU Horizon 2020 ITN/ETN on Domain Specific Systems for Information Extraction and Retrieval (H2020-EU.1.3.1., ID: 860721).

References

1. Abolghasemi, A., Askari, A., Verberne, S.: On the interpolation of contextualized term-based ranking with bm25 for query-by-example retrieval. In: Proceedings of the 2022 ACM SIGIR International Conference on Theory of Information Retrieval. ICTIR 2022, pp. 161–170. Association for Computing Machinery, New York (2022). https://doi.org/10.1145/3539813.3545133

2. Abolghasemi, A., Verberne, S., Azzopardi, L.: Improving BERT-based query-by-document retrieval with multi-task optimization. In: Hagen, M., et al. (eds.) ECIR 2022. LNCS, vol. 13186, pp. 3–12. Springer, Cham (2022). https://doi.org/10.1007/978-3-030-99739-7_1

3. Akkalyoncu Yilmaz, Z., Wang, S., Yang, W., Zhang, H., Lin, J.: Applying BERT to document retrieval with birch. In: Proceedings of the 2019 Conference on Empirical Methods in Natural Language Processing and the 9th International Joint Conference on Natural Language Processing (EMNLP-IJCNLP): System Demonstrations, pp. 19–24. Association for Computational Linguistics, Hong Kong, November 2019. https://doi.org/10.18653/v1/D19-3004, https://aclanthology.org/D19-3004

4. Al-Hajj, M., Jarrar, M.: ArabglossBERT: fine-tuning BERT on context-gloss pairs for WSD. arXiv preprint arXiv:2205.09685 (2022)

5. Allan, J.: Overview of the TREC 2004 robust retrieval track. In: Proceedings of TREC, vol. 13 (2004)

6. Althammer, S., Askari, A., Verberne, S., Hanbury, A.: DoSSIER@ COLIEE 2021: leveraging dense retrieval and summarization-based re-ranking for case law retrieval. arXiv preprint arXiv:2108.03937 (2021)

7. Anand, M., Zhang, J., Ding, S., Xin, J., Lin, J.: Serverless BM25 search and BERT reranking. In: DESIRES, pp. 3–9 (2021)

8. Askari, A., Verberne, S.: Combining lexical and neural retrieval with longformer-based summarization for effective case law retrieva. In: Proceedings of the Second International Conference on Design of Experimental Search and Information REtrieval Systems, pp. 162–170. CEUR (2021)

9. Askari, A., Verberne, S., Pasi, G.: Expert finding in legal community question answering. In: Hagen, M., et al. (eds.) ECIR 2022. LNCS, vol. 13186, pp. 22–30. Springer, Cham (2022). https://doi.org/10.1007/978-3-030-99739-7_3

10. Bartell, B.T., Cottrell, G.W., Belew, R.K.: Automatic combination of multiple ranked retrieval systems. In: SIGIR 1994, pp. 173–181. Springer, London (1994). https://doi.org/10.1007/978-1-4471-2099-5_18

11. Berg-Kirkpatrick, T., Spokoyny, D.: An empirical investigation of contextualized number prediction. In: Proceedings of the 2020 Conference on Empirical Methods in Natural Language Processing (EMNLP), pp. 4754–4764 (2020)

12. Boualili, L., Moreno, J.G., Boughanem, M.: MarkedBERT: integrating traditional IR cues in pre-trained language models for passage retrieval. In: Proceedings of the 43rd International ACM SIGIR Conference on Research and Development in Information Retrieval, pp. 1977–1980 (2020)

13. Boualili, L., Moreno, J.G., Boughanem, M.: Highlighting exact matching via marking strategies for ad hoc document ranking with pretrained contextualized language models. Inf. Retrieval J. 1–47 (2022)

14. Boytsov, L., Lin, T., Gao, F., Zhao, Y., Huang, J., Nyberg, E.: Understanding performance of long-document ranking models through comprehensive evaluation and leaderboarding. arXiv preprint arXiv:2207.01262 (2022)

15. Chen, C.C., Huang, H.H., Chen, H.H.: Numclaim: investor's fine-grained claim detection. In: Proceedings of the 29th ACM International Conference on Information and Knowledge Management, pp. 1973–1976 (2020)

16. Chen, L., Lan, Y., Pang, L., Guo, J., Cheng, X.: Toward the understanding of deep text matching models for information retrieval. arXiv preprint arXiv:2108.07081 (2021)

17. Craswell, N., Mitra, B., Yilmaz, E., Campos, D.: Overview of the TREC 2020 deep learning track. arXiv preprint arXiv:2102.07662 (2021)

18. Craswell, N., Mitra, B., Yilmaz, E., Campos, D., Lin, J.: MS marco: benchmarking ranking models in the large-data regime. In: Proceedings of the 44th International ACM SIGIR Conference on Research and Development in Information Retrieval, pp. 1566–1576 (2021)
19. Craswell, N., Mitra, B., Yilmaz, E., Campos, D., Voorhees, E.M.: Overview of the TREC 2019 deep learning track. arXiv preprint arXiv:2003.07820 (2020)
20. Gao, L., Dai, Z., Chen, T., Fan, Z., Van Durme, B., Callan, J.: Complement lexical retrieval model with semantic residual embeddings. In: Hiemstra, D., Moens, M.-F., Mothe, J., Perego, R., Potthast, M., Sebastiani, F. (eds.) ECIR 2021. LNCS, vol. 12656, pp. 146–160. Springer, Cham (2021). https://doi.org/10.1007/978-3-030-72113-8_10
21. Geva, M., Gupta, A., Berant, J.: Injecting numerical reasoning skills into language models. arXiv preprint arXiv:2004.04487 (2020)
22. Gretkowski, A., Wiśniewski, D., Ławrynowicz, A.: Should we afford affordances? Injecting conceptnet knowledge into BERT-based models to improve commonsense reasoning ability. In: Corcho, O., Hollink, L., Kutz, O., Troquard, N., Ekaputra, F.J. (eds.) EKAW 2022. LNCS, pp. 97–104. Springer, Cham (2022). https://doi.org/10.1007/978-3-031-17105-5_7
23. Gu, K., Budhkar, A.: A package for learning on tabular and text data with transformers. In: Proceedings of the Third Workshop on Multimodal Artificial Intelligence, pp. 69–73. Association for Computational Linguistics, Mexico City, June 2021. https://doi.org/10.18653/v1/2021.maiworkshop-1.10, https://www.aclweb.org/anthology/2021.maiworkshop-1.10
24. Han, S., Wang, X., Bendersky, M., Najork, M.: Learning-to-rank with BERT in TF-ranking. arXiv preprint arXiv:2004.08476 (2020)
25. Hofstätter, S., Althammer, S., Schröder, M., Sertkan, M., Hanbury, A.: Improving efficient neural ranking models with cross-architecture knowledge distillation. arXiv preprint arXiv:2010.02666 (2020)
26. Johnson, D., Mak, D., Barker, D., Loessberg-Zahl, L.: Probing for multilingual numerical understanding in transformer-based language models. arXiv preprint arXiv:2010.06666 (2020)
27. Kamphuis, C., de Vries, A.P., Boytsov, L., Lin, J.: Which BM25 do you mean? A large-scale reproducibility study of scoring variants. In: Jose, J.M., et al. (eds.) ECIR 2020. LNCS, vol. 12036, pp. 28–34. Springer, Cham (2020). https://doi.org/10.1007/978-3-030-45442-5_4
28. Khattab, O., Zaharia, M.: ColBERT: efficient and effective passage search via contextualized late interaction over BERT. In: Proceedings of the 43rd International ACM SIGIR conference on research and development in Information Retrieval, pp. 39–48 (2020)
29. Kingma, D.P., Ba, J.: Adam: a method for stochastic optimization. arXiv preprint arXiv:1412.6980 (2014)
30. Li, L., et al.: MarkBERT: marking word boundaries improves Chinese BERT. arXiv preprint arXiv:2203.06378 (2022)
31. Li, X., Roth, D.: Learning question classifiers. In: COLING 2002: The 19th International Conference on Computational Linguistics (2002)
32. Lin, J., Ma, X., Lin, S.C., Yang, J.H., Pradeep, R., Nogueira, R.: Pyserini: a Python toolkit for reproducible information retrieval research with sparse and dense representations. In: Proceedings of the 44th Annual International ACM SIGIR Conference on Research and Development in Information Retrieval (SIGIR 2021), pp. 2356–2362 (2021)

33. Lin, J., Ma, X., Lin, S.C., Yang, J.H., Pradeep, R., Nogueira, R.: Pyserini: BM25 baseline for MS marco document retrieval, August 2021. https://github.com/castorini/pyserini/blob/master/docs/experiments-msmarco-doc.md
34. Lin, J., Nogueira, R., Yates, A.: Pretrained transformers for text ranking: BERT and beyond. Synth. Lect. Hum. Lang. Technol. **14**(4), 1–325 (2021)
35. MacAvaney, S., Nardini, F.M., Perego, R., Tonellotto, N., Goharian, N., Frieder, O.: Expansion via prediction of importance with contextualization. In: Proceedings of the 43rd International ACM SIGIR Conference on Research and Development in Information Retrieval, pp. 1573–1576 (2020)
36. MacAvaney, S., Yates, A., Cohan, A., Goharian, N.: CEDR: contextualized embeddings for document ranking. In: Proceedings of the 42nd International ACM SIGIR Conference on Research and Development in Information Retrieval, pp. 1101–1104 (2019)
37. Michael, N., Diego, C., Joshua, P., LP, B.: Learning to rank, May 2022. https://solr.apache.org/guide/solr/latest/query-guide/learning-to-rank.html#feature-engineering
38. Muffo, M., Cocco, A., Bertino, E.: Evaluating transformer language models on arithmetic operations using number decomposition. In: Proceedings of the Thirteenth Language Resources and Evaluation Conference, pp. 291–297. European Language Resources Association, Marseille, June 2022. https://aclanthology.org/2022.lrec-1.30
39. Nguyen, T., et al.: MS marco: a human generated machine reading comprehension dataset. In: CoCo@ NIPs (2016)
40. Nogueira, R., Cho, K.: Passage re-ranking with BERT. arXiv preprint arXiv:1901.04085 (2019)
41. Nogueira, R., Yang, W., Lin, J., Cho, K.: Document expansion by query prediction. arXiv preprint arXiv:1904.08375 (2019)
42. Paszke, A., et al.: Automatic differentiation in PyTorch (2017)
43. Rau, D., Kamps, J.: How different are pre-trained transformers for text ranking? In: Hagen, M., et al. (eds.) ECIR 2022. LNCS, vol. 13186, pp. 207–214. Springer, Cham (2022). https://doi.org/10.1007/978-3-030-99739-7_24
44. Rau, D., Kamps, J.: The role of complex NLP in transformers for text ranking. In: Proceedings of the 2022 ACM SIGIR International Conference on Theory of Information Retrieval, pp. 153–160 (2022)
45. Reimers, N., Gurevych, I.: Sentence-BERT: sentence embeddings using Siamese BERT-networks. In: Proceedings of the 2019 Conference on Empirical Methods in Natural Language Processing. Association for Computational Linguistics, November 2019. https://arxiv.org/abs/1908.10084
46. Robertson, S., Zaragoza, H., et al.: The probabilistic relevance framework: Bm25 and beyond. Found. Trends® Inf. Retrieval **3**(4), 333–389 (2009)
47. Robertson, S.E., Walker, S.: Some simple effective approximations to the 2-Poisson model for probabilistic weighted retrieval. In: Croft, B.W., van Rijsbergen, C.J. (eds.) SIGIR 1994, pp. 232–241. Springer, London (1994). https://doi.org/10.1007/978-1-4471-2099-5_24
48. Salton, G., McGill, M.J.: Introduction to Modern Information Retrieval. McGraw-Hill, New York (1983)
49. Sanh, V., Debut, L., Chaumond, J., Wolf, T.: DistilBERT, a distilled version of BERT: smaller, faster, cheaper and lighter. arXiv preprint arXiv:1910.01108 (2019)

50. SARACEVIC, T.: A review of an a framework for the thinking on the notion in information science. J. Am. Soc. Inf. Sci. **26**
51. Sundararajan, M., Taly, A., Yan, Q.: Axiomatic attribution for deep networks. In: International Conference on Machine Learning, pp. 3319–3328. PMLR (2017)
52. Thawani, A., Pujara, J., Szekely, P.A., Ilievski, F.: Representing numbers in NLP: a survey and a vision. arXiv preprint arXiv:2103.13136 (2021)
53. Vaswani, A., et al.: Attention is all you need. In: Advances in Neural Information Processing Systems, pp. 5998–6008 (2017)
54. Wallace, E., Wang, Y., Li, S., Singh, S., Gardner, M.: Do NLP models know numbers? Probing numeracy in embeddings. arXiv preprint arXiv:1909.07940 (2019)
55. Wang, S., Zhuang, S., Zuccon, G.: BERT-based dense retrievers require interpolation with BM25 for effective passage retrieval. In: Proceedings of the 2021 ACM SIGIR International Conference on Theory of Information Retrieval. ICTIR 2021, pp. 317–324. Association for Computing Machinery, New York (2021). https://doi.org/10.1145/3471158.3472233
56. Wang, W., Wei, F., Dong, L., Bao, H., Yang, N., Zhou, M.: Minilm: deep self-attention distillation for task-agnostic compression of pre-trained transformers. Adv. Neural. Inf. Process. Syst. **33**, 5776–5788 (2020)
57. Wolf, T., et al.: Huggingface's transformers: state-of-the-art natural language processing. arXiv preprint arXiv:1910.03771 (2019)
58. Wu, S.: Applying statistical principles to data fusion in information retrieval. Expert Syst. Appl. **36**(2), 2997–3006 (2009)
59. Yan, M., Li, C., Wu, C., Xia, J., Wang, W.: IDST at TREC 2019 deep learning track: deep cascade ranking with generation-based document expansion and pre-trained language modeling. In: TREC (2019)
60. Yilmaz, Z.A., Yang, W., Zhang, H., Lin, J.: Cross-domain modeling of sentence-level evidence for document retrieval. In: Proceedings of the 2019 Conference on Empirical Methods in Natural Language Processing and the 9th International Joint Conference on Natural Language Processing (EMNLP-IJCNLP), pp. 3490–3496 (2019)
61. Zhan, J., Mao, J., Liu, Y., Guo, J., Zhang, M., Ma, S.: Interpreting dense retrieval as mixture of topics. arXiv preprint arXiv:2111.13957 (2021)
62. Zhan, J., Mao, J., Liu, Y., Zhang, M., Ma, S.: An analysis of BERT in document ranking. In: Proceedings of the 43rd International ACM SIGIR Conference on Research and Development in Information Retrieval, pp. 1941–1944 (2020)
63. Zhang, X., Ramachandran, D., Tenney, I., Elazar, Y., Roth, D.: Do language embeddings capture scales? arXiv preprint arXiv:2010.05345 (2020)
64. Zhang, X., Yates, A., Lin, J.: Comparing score aggregation approaches for document retrieval with pretrained transformers. In: Hiemstra, D., Moens, M.-F., Mothe, J., Perego, R., Potthast, M., Sebastiani, F. (eds.) ECIR 2021. LNCS, vol. 12657, pp. 150–163. Springer, Cham (2021). https://doi.org/10.1007/978-3-030-72240-1_11
65. Zhang, Y., Hu, C., Liu, Y., Fang, H., Lin, J.: Learning to rank in the age of muppets: effectiveness-efficiency tradeoffs in multi-stage ranking. In: Proceedings of the Second Workshop on Simple and Efficient Natural Language Processing, pp. 64–73 (2021)
66. Zhang, Z., Sabuncu, M.: Generalized cross entropy loss for training deep neural networks with noisy labels. In: Advances in Neural Information Processing Systems, vol. 31 (2018)

67. Zhuang, S., Li, H., Zuccon, G.: Deep query likelihood model for information retrieval. In: Hiemstra, D., Moens, M.-F., Mothe, J., Perego, R., Potthast, M., Sebastiani, F. (eds.) ECIR 2021. LNCS, vol. 12657, pp. 463–470. Springer, Cham (2021). https://doi.org/10.1007/978-3-030-72240-1_49
68. Zhuang, S., Zuccon, G.: Tilde: term independent likelihood model for passage re-ranking. In: Proceedings of the 44th International ACM SIGIR Conference on Research and Development in Information Retrieval, pp. 1483–1492 (2021)

Quantifying Valence and Arousal in Text with Multilingual Pre-trained Transformers

Gonçalo Azevedo Mendes[1,2] and Bruno Martins[1,2(✉)]

[1] Instituto Superior Técnico, Universidade de Lisboa, Lisbon, Portugal
{goncalo.a.mendes,bruno.g.martins}@tecnico.ulisboa.pt
[2] INESC-ID, Lisbon, Portugal

Abstract. The analysis of emotions expressed in text has numerous applications. In contrast to categorical analysis, focused on classifying emotions according to a pre-defined set of common classes, dimensional approaches can offer a more nuanced way to distinguish between different emotions. Still, dimensional methods have been less studied in the literature. Considering a valence-arousal dimensional space, this work assesses the use of pre-trained Transformers to predict these two dimensions on a continuous scale, with input texts from multiple languages and domains. We specifically combined multiple annotated datasets from previous studies, corresponding to either emotional lexica or short text documents, and evaluated models of multiple sizes and trained under different settings. Our results show that model size can have a significant impact on the quality of predictions, and that by fine-tuning a large model we can confidently predict valence and arousal in multiple languages. We make available the code, models, and supporting data.

Keywords: Transformer-based multilingual language models ·
Emotion analysis in text · Predicting valence and arousal

1 Introduction

The task of analyzing emotions expressed in text is commonly modeled as a classification problem, representing affective states (e.g., Ekman's six basic emotions [22]) as specific classes. The alternative approach of dimensional emotion analysis focuses on rating emotions according to a pre-defined set of dimensions, offering a more nuanced way to distinguish between different emotions [7]. Emotional states are represented on a continuous numerical space, with the most common dimensions defined as valence and arousal. In particular, valence describes the pleasantness of a stimulus, ranging from negative to positive feelings. Arousal represents the degree of excitement provoked by a stimulus, from calm to excited. The Valence-Arousal (VA) space [4] corresponds to a 2-dimensional space to which a text sequence can be mapped.

This study proposes using pre-trained multilingual Transformer models to predict valence and arousal ratings in text from different languages and domains.

J. Kamps et al. (Eds.): ECIR 2023, LNCS 13980, pp. 84–100, 2023.
https://doi.org/10.1007/978-3-031-28244-7_6

Models pre-trained on huge amounts of data from multiple languages can be fine-tuned to different types of downstream tasks with relatively small datasets in one or few languages, and still obtain reliable results on different languages [43]. While previous research focused on monolingual VA prediction as regression from text, this study compiled 34 publicly available psycho-linguistic datasets, from different languages, into a single uniform dataset. We then evaluated multilingual DistilBERT [48] and XLM-RoBERTa [17] models, to understand the impact of model size and training conditions on the ability to correctly predict affective ratings from textual contents.

Experimental results show that multilingual VA prediction is possible with a single Transformer model, particularly when considering the larger XLM-RoBERTa model. Even if performance differs across languages, most results improve or stay in line with the results from previous research focused on predicting these affective ratings on a single language. The code, models, and data used in this study are available on a GitHub repository[1].

The rest of the paper is organized as follows: Sect. 2 presents related work, while Sect. 3 describes the models considered for predicting valence and arousal. Section 4 describes the corpora used for model training and evaluation. Section 5 presents our findings and compares the results. Finally, Sect. 6 summarizes the main findings and discusses possibilities for future work.

2   Related Work

Since Russel [47] first proposed a two-dimensional model of emotions, based on valence and arousal, much research has been done on dimensional emotion analysis. Most relevant to this study are the main lexicons [5,38,49,55,60] and corpora [6,8,42,66] annotated according to these dimensions, used in previous work. Still, while several NLP and IR studies have addressed dimensional emotion extraction, most previous work has focused on categorical approaches [1].

Trying to predict valence and arousal has long been a relevant topic, both at the word-level [11,20,26,45,51,62,67] and at the sentence/text-level [7,9,32, 36,41,44,52,59,63]. Recchia et al. used pointwise mutual information coupled with k-NN regression to estimate valence and arousal for words [45]. Hollis et al. resorted to linear regression modelling [26]. Sedoc et al. combined distributional approaches with signed spectral clustering [51]. Du and Zhang explored the use of CNNs [20]. Wu et al. used a densely connected LSTM network and word features to identify emotions on the VA space for words and phrases [62]. More recently, Buechel et al. proposed a method for creating arbitrarily large emotion lexicons in 91 languages, using a translation model, a target language embedding model, and a multitask learning feed-forward neural network [11]. This last work is interesting when compared to ours, as it is one of the few attempts to predict VA at a multilingual level, if only for individual words.

Paltoglou et al. attempted text-level VA prediction by resorting to affective dictionaries, as supervised machine learning techniques were inadequate for the

---

[1] https://www.github.com/gmendes9/multilingual_va_prediction.

small dataset used in their tests [41]. Preoţiuc-Pietro et al. compiled a corpus of Facebook posts and built a bag-of-words (BoW) linear regression prediction model [44]. Similarly, Buechel and Hahn used BoW representations in conjunction with TF-IDF weights [7,9]. More recently, several studies have compared CNNs and RNNs, amongst other neural architectures [32,52,59,63]. For instance, Lee et al. explored different methods for prediction, ranging from linear regression to multiple neural network architectures [36]. This last study explored the use of a BERT model, but differs from our work as the data is not multilingual. The present work follows the steps of some of the aforementioned studies leveraging deep learning, aiming to build a single multilingual model capable of predicting affective ratings for valence and arousal.

3   Models for Predicting Valence and Arousal from Text

We address the prediction of valence and arousal scores as text-based regression, using pre-trained multilingual models adapted from the Huggingface library [61]. In particular, we use DistilBERT [48] and XLM-RoBERTa [17] models.

The multilingual DistilBERT model, consisting of 134M parameters, is based on a 6 layer Transformer encoder, with 12 attention heads and a hidden state size of 768. The model can train two times faster with only a slight performance decrease (approx. 5%), compared to a multilingual BERT-base model with 25% more parameters. As for XLM-RoBERTa, we used both the base (270M parameters) and large (550M parameters) versions. The base version is a 12 layer Transformer, with 12 attention heads and a hidden state size of 768. The large version uses 24 layers, 16 attention heads, and a hidden state size of 1024.

Both these models are pre-trained on circa 100 different languages, which will likely enable the generalization to languages for which there are no annotated data in terms of valence and arousal ratings. These models are fine-tuned for the task at hand with a regression head on top, consisting of a linear layer on top of the pooled representation from the Transformer (i.e., the representation of the first token in the input sequence).

The regression head produces two outputs, which are processed through a hard sigmoid activation function, forcing the predicted values on both dimensions to respect the target interval between zero and one.

Three loss functions were initially compared for model training, namely the Mean Squared Error (MSE), the Concordance Correlation Coefficient Loss (CCCL), and a recently proposed Robust Loss (RL) function [3]. In all these cases, the models are trained with the sum of the loss for the valence and arousal predictions, equally weighting both affective dimensions.

MSE is the most used loss function in regression problems and can be defined as the mean of the squared differences between predicted ($\hat{y}$) and ground-truth (y) values, as shown in Eq. 1.

$$\text{MSE} = \frac{1}{N} \sum_{i=0}^{N} (y_i - \hat{y}_i)^2. \tag{1}$$

The CCCL corresponds to a correlation-based function, evaluating the ranking agreement between the true and predicted values, within a batch of instances. It varies from the Pearson correlation by penalizing the score in proportion to the deviation if the predictions shift in value. Atmaja and Akagi [56] compared this function to the MSE and Mean Absolute Error (MAE) loss functions for the task of predicting emotional ratings from speech signals using LSTM neural networks, suggesting that this loss yields a better performance than error-based functions. The CCCL follows Eq. 2, where $\rho_{y\hat{y}}$ represents the Pearson correlation coefficient between y and $\hat{y}$, σ represents the standard deviation, and μ the mean value. Notice that the correlation ranges from -1 to 1, and thus we use one minus the correlation as the loss.

$$\text{CCCL} = 1 - \frac{2\rho_{y\hat{y}}\sigma_y\sigma_{\hat{y}}}{\sigma_{y^2} + \sigma_{\hat{y}^2} + (\mu_y - \mu_{\hat{y}})^2}. \tag{2}$$

The RL function generalizes some of the most common robust loss functions (e.g., the Huber loss), that reduce the influence of outliers [3], being described in its general form through Eq. 3. In this function, x is the variable being minimized, corresponding to the difference between true and predicted values (i.e., $x_i = y_i - \hat{y}_i$). The function involves two parameters that tune its shape, namely $\alpha \in \mathbb{R}$ that controls the robustness, and a scale parameter $c > 0$ which controls the size of its quadratic bowl.

$$\text{RL} = \frac{1}{N}\sum_{i=0}^{N}\begin{cases} \frac{1}{2}(x_i/c)^2 & \text{if } \alpha = 2 \\ \log\left(\frac{1}{2}(x_i/c)^2 + 1\right) & \text{if } \alpha = 0 \\ 1 - \exp\left(-\frac{1}{2}(x_i/c)^2\right) & \text{if } \alpha = \infty \\ \frac{|\alpha-2|}{\alpha}\left(\left(\frac{(x_i/c)^2}{|\alpha-2|} + 1\right)^{\alpha/2} - 1\right) & \text{otherwise.} \end{cases} \tag{3}$$

A lower value of α implies penalizing minor errors at the expense of larger ones, while a higher value of α allows more inliers while increasing the penalty for outliers. We used the adaptive form of this robust loss function, where the parameter α is optimized and tuned during model training via stochastic gradient descent, as explained in the original paper [3].

We also tested two hybrid loss functions derived from the previous ones, combining their different properties and merits. While the MSE and the RL functions analyze results at the instance level, the CCCL function does the same at the batch level. With this in mind, one hybrid loss function combines the CCCL and the MSE functions, while the other combines the CCCL with the RL function, in both cases through a simple addition.

4   Resources

We collected 34 different public datasets to form a large corpus of annotated data for the emotional dimensions of valence and arousal, with the intent to build the largest possible multilingual dataset. The original datasets comprise 13 different languages, which represent up to 2.5 billion native speakers worldwide[2,3]. There are two types of datasets described on Table 1, namely word and short text datasets, respectively associating valence and arousal ratings to either individual words or short text sequences. All of these datasets were manually annotated by humans, either via crowdsourcing or by experienced linguists/psychologists, according to the Self-Assessment Manikin (SAM) method [4]. In addition, several lexicons relate to the Affective Norms for English Words (ANEW) resource, corresponding to either adaptations to other languages or extensions in terms of the number of words [5]. ANEW was the first lexicon providing real-valued scores for the emotional dimensions of valence and arousal. It is important to note that this lexicon is excluded from our corpus for being part of larger datasets that were included, such as the one from Warriner et al. [60].

Overall, merging the 34 datasets allowed us to build a large multilingual VA dataset, consisting of 128,987 independently annotated instances (i.e., 30,657 short texts and 98,330 words). The textual sequences were left unchanged from the source datasets. As for the valence and arousal ratings, we took the mean annotated values when ratings were obtained from multiple individuals, and normalized the scores between 0 and 1. The normalization was performed according to the equation $z_i = (x_i - \min(x))/(\max(x) - \min(x))$, in which z_i denotes the normalized value, x_i the original value, and min and max denote the extremes of the scales in which the original scores were rated on.

Table 1 presents a statistical characterization for the short text datasets in its first half, followed by the word datasets. Each entry describes the dataset source language, the dataset size, and the mean number of words (this last variable in the case of the short texts). An exploratory analysis of the VA ratings supports a better understanding of the score distributions. In turn, Fig. 1 presents the distribution of the ratings for the entire merged dataset, as well as for its two subsets (i.e., words and short texts). The ratings were plotted on the two-dimensional valence-arousal space, and they are visualized with the help of a kernel density estimate. The individual distributions of the two dimensions are displayed on the margins. The analysis of the resulting merged dataset leads to the conclusion that there is a quadratic relationship between the two emotional dimensions, with a tendency for increased arousal on high and low valence values, and abundant low arousal scores in the middle of the valence scale. A similar pattern was previously observed in several different studies in Psychology, such as in the original ANEW study and its extensions [5,18,28,33,39,42,64].

5   Experimental Evaluation

Each of the individual original datasets were randomly split in half and combined with the others to form two subsets of data equally representative of all the datasets, later used for 2-fold cross-validation. For each configuration, two models were separately trained on each fold, and then separately used to make

---

[2] https://www.cia.gov/the-world-factbook/countries/world/people-and-society.
[3] https://www.ethnologue.com/.

Table 1. Dataset characterization. μ_{length} represents the mean text length of each instance, in terms of the number of words. μ and σ represent the mean and standard deviation, in the emotional ratings, respectively.

Dataset	Language	Items	μ_{length}	Arousal μ	σ		Valence μ	σ	
EmoBank [8,10]	English	10062	23.27	0.51	0.06		0.49	0.09	
IEMOCAP [12]	English	10039	19.22	0.56	0.22		0.48	0.17	
Facebook Posts [44]	English	2894	28.15	0.29	0.25		0.53	0.15	
EmoTales [24]	English	1395	17.91	0.55	0.12		0.49	0.15	
ANET [6]	English	120	31.96	0.66	0.22		0.52	0.33	
PANIG [16]	German	619	9.12	0.47	0.12		0.40	0.22	
COMETA sentences [15]	German	120	16.75	0.48	0.15		0.50	0.20	
COMETA stories [15]	German	64	90.17	0.53	0.15		0.56	0.21	
CVAT [66]	Mandarin	2969	58.00	0.48	0.13		0.48	0.17	
CVAI [63]	Mandarin	1465	29.53	0.51	0.12		0.32	0.06	
ANPST [28]	Polish	718	28.16	0.48	0.13		0.47	0.22	
MAS [42]	Portuguese	192	8.94	0.52	0.17		0.49	0.28	
Yee [65]	Cantonese	292		0.40	0.15		0.58	0.17	
Ćoso et al. [18]	Croatian	3022		0.45	0.15		0.51	0.21	
Moors et al. [40]	Dutch	4299		0.52	0.14		0.49	0.18	
Verheyen et al. [57]	Dutch	1000		0.52	0.17		0.50	0.20	
NRC-VAD [38]	English	19971		0.50	0.17		0.50	0.22	
Warriner et al. [60]	English	13915		0.40	0.11		0.51	0.16	
Scott et al. [50]	English	5553		0.45	0.14		0.51	0.19	
Söderholm et al. [54]	Finnish	420		0.50	0.13		0.50	0.25	
Eilola et al. [21]	Finnish	210		0.36	0.19		0.44	0.26	
FAN [39]	French	1031		0.41	0.13		0.56	0.17	
FEEL [25]	French	835		0.56	0.17		0.43	0.20	
BAWL-R [58]	German	2902		0.44	0.17		0.51	0.21	
ANGST [49]	German	1034		0.52	0.16		0.51	0.24	
LANG [29]	German	1000		0.39	0.20		0.50	0.13	
Italian ANEW [23]	Italian	1121		0.52	0.19		0.51	0.26	
Xu et al. [64]	Mandarin	11310		0.52	0.14		0.52	0.16	
CVAW [36,66]	Mandarin	5512		0.50	0.18		0.44	0.21	
ANPW_R [27]	Polish	4905		0.39	0.11		0.50	0.16	
NAWL [46]	Polish	2902		0.34	0.13		0.53	0.20	
Portuguese ANEW [53]	Portuguese	1034		0.49	0.14		0.50	0.23	
S.-Gonzalez et al. [55]	Spanish	14031		0.70	0.22		0.72	0.16	
Kapucu et al. [30]	Turkish	2031		0.50	0.11		0.49	0.20	

predictions for the instances in the other fold (containing instances not seen during training), with final evaluation metrics computed on the complete set of results (the predictions from the models trained on each fold were joined, and the metrics were computed over the full set of predictions). Hyperparameters were defined through an initial set of tests and kept constant for all models. The batch size was fixed at 16, and models were trained during 10 epochs. AdamW was the chosen optimizer, and we used it together with a linear learning rate schedule with warm-up. The learning rate was set at $6 \cdot 10^{-6}$, with a warm-up ratio of $1 \cdot 10^{-1}$. We experimented with various model and loss function combinations, namely by using the three differently-sized pre-trained Transformer models, as well as the loss functions presented in Sect. 3.

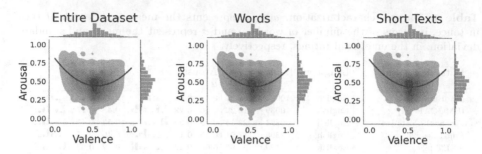

Fig. 1. Distribution of dataset instances in the valence-arousal space. Each dimensions' distribution is shown with a histogram on the corresponding axis. An orange trend line shows the quadratic relation between valence and arousal. (Color figure online)

Three different evaluation metrics were used to assess and compare model performance, namely the Mean Absolute Error (MAE), the Root Mean Squared Error (RMSE), and the Pearson correlation coefficient (ρ). The MAE, as detailed by Eq. 4, corresponds to the sum of absolute errors between observations x_i and predictions y_i.

$$\text{MAE} = \frac{1}{N} \sum_{i=1}^{N} |x_i - y_i|. \tag{4}$$

The RMSE, as shown by Eq. 5, is the square root of the mean square of the differences between observations x_i and predictions y_i.

$$\text{RMSE} = \sqrt{\frac{1}{N} \sum_{i=1}^{N} (x_i - y_i)^2}. \tag{5}$$

Finally, the Pearson correlation coefficient, given by Eq. 6, is used to assess the presence of a linear relationship between the ground truth x and the predicted results given by y.

$$\rho = \frac{\sum_{i=1}^{N}(x_i - \bar{x})(y_i - \bar{y})}{\sqrt{\sum_{i=1}^{N}(x_i - \bar{x})^2 (y_i - \bar{y})^2}}. \tag{6}$$

While the first two metrics should be minimized, the latter is best when it is closer to one, i.e., the value denoting a perfect correlation.

5.1 Results with Different Models and Loss Functions

Table 2 summarizes the results for the different combinations of model size and loss function. The single thing that affects evaluation metrics the most is the size of the pre-trained Transformer model being used. The best performing model was the large version of XLM-RoBERTa, respectively returning on average 9% and 20% better correlation results than XLM-RoBERTa-base and DistilBERT. For each model, we compared the five loss functions, highlighting in bold the

Table 2. Comparison between different models and loss functions.

Model	Loss	ρ_V	ρ_A	$RMSE_V$	$RMSE_A$	MAE_V	MAE_A
DistilBERT	MSE	0.663	0.594	0.138	0.132	0.102	0.101
	CCCL	0.657	0.590	0.150	0.146	0.111	0.111
	RL	**0.668**	**0.598**	**0.138**	**0.132**	**0.101**	**0.101**
	MSE+CCCL	0.657	0.590	0.149	0.145	0.110	0.111
	RL+CCCL	0.664	0.591	0.147	0.144	0.109	0.110
XLM RoBERTa base	MSE	0.757	0.646	**0.121**	0.125	**0.088**	0.095
	CCCL	0.757	0.653	0.136	0.144	0.101	0.110
	RL	0.757	**0.657**	0.122	**0.125**	0.088	**0.095**
	MSE+CCCL	0.757	0.655	0.135	0.141	0.099	0.108
	RL+CCCL	**0.757**	0.657	0.134	0.141	0.099	0.107
XLM RoBERTa large	MSE	0.810	0.695	**0.109**	**0.120**	**0.079**	**0.091**
	CCCL	**0.817**	0.698	0.117	0.132	0.085	0.099
	RL	0.802	0.689	0.114	0.122	0.083	0.092
	MSE+CCCL	0.815	**0.699**	0.121	0.135	0.089	0.103
	RL+CCCL	0.813	0.694	0.119	0.133	0.087	0.100

best performing one for each metric, and evaluating separately for valence and arousal. In short, the choice of loss function has less impact on the quality of the results. For the best model, we see differences in correlation of up to 2% between best and worst performing loss functions. Comparatively, in the error metrics, these differences can be of up to 12%. As such, looking to identify the best model/loss-function combination, we gave more relevance to the error metrics. We identified the MSE loss function as the best performing one, adding to the fact that this loss function is also the simplest of the set of functions that were tested. Consequently, further results are presented for that model/loss pair.

When analyzing the results, it is possible to break them down into two categories: predicting valence and arousal for individual words or, on the other hand, for short texts (see Table 3). Our models are more accurate at predicting word-level scores, although this is also a more straightforward problem with less ambiguity. An essential fact to take from the results is the greater difficulty in predicting the affective dimension of arousal. Previous research has also stated that human ratings themselves varied much more in annotating arousal when compared to the valence dimension [41].

5.2   Results per Language and Dataset

Further analysis focused on the results of predicting ratings for each of the original datasets, with results summarized on Table 4.

Table 3. Comparing VA prediction on words or short texts using the XLM-RoBERTa-large model and considering the MSE loss function for training.

Dataset	ρ_V	ρ_A	RMSE$_V$	RMSE$_A$	MAE$_V$	MAE$_A$
All data	0.810	0.695	0.109	0.120	0.079	0.091
Words	0.833	0.686	0.107	0.116	0.078	0.090
Short texts	0.682	0.711	0.115	0.132	0.082	0.093

For most word datasets, compared in the bottom half of Table 4, our best model performed to high standards, showing a correlation between predicted values and the ground truth of around 0.8 for valence and 0.7 for arousal. As a comparison, when evaluating correlation on Warriner's dataset [60], our work achieved $\rho_V = 0.84$ and $\rho_A = 0.65$, while Hollis [26] achieved $\rho_V = 0.80$ and $\rho_A = 0.63$. Although good scores are observed for most datasets, we can also identify some outliers, like in the case of the dataset from Kapucu et al. [30].

As for the short text datasets, compared in the top half of Table 4, performance varies more significantly, with an overall lower correlation and a higher error. A particular case is the COMETA stories dataset [15], which shows a correlation close to zero. The COMETA dataset is a database of conceptual metaphors, in which half of the text instances contain metaphors while the other half corresponds to their literal counterparts. The obtained results indicate that even the best model does not cope well with metaphorical phrasing. Comparing our model to the method from Preoţiuc-Pietro et al. [44], the correlation values we obtained for the Facebook Posts dataset were $\rho_V = 0.80$ and $\rho_A = 0.78$, while they achieved $\rho_V = 0.65$ and $\rho_A = 0.85$ (i.e., we have better results for valence, and worse for arousal). In [66], Yu et al. predict VA on the CVAT dataset using the ratings obtained for the CVAW words. They obtained correlation results of $\rho_V = 0.54$ and $\rho_A = 0.16$, while our approach obtained $\rho_V = 0.89$ and $\rho_A = 0.62$. In subsequent research, the same team tried to predict VA ratings with different neural network approaches, including a model based on BERT, for which they obtained $\rho_V = 0.87$ and $\rho_A = 0.58$ on the same dataset [36].

It should be noted that all previous comparisons against other studies are merely indicative, given that the experimental conditions (e.g., the data splits used for training and evaluation) were very different.

We performed a similar comparison to evaluate the result quality in distinct languages, grouping prediction results by language. It was possible to conclude that our best model yields good results in most languages. The most challenging languages in terms of word prediction are Finnish and Turkish, with the model seemingly excelling at Portuguese, Mandarin, and English, to name a few. The lower scores observed for Finnish and Turkish can be explained by the small sample of training data in those languages, respectively 0.48% and 1.57% of the entire dataset, as well as by the complex morphology and productive compounding associated with these languages, as found by Buechel et al. [11].

As for the short texts, compared in detail in Table 5, the most challenging language was German. On this subject, we note that the German training sample contains the metaphorical instances of the COMETA dataset, which can explain the gap in the results for this language. Predicting valence in English also proved demanding. If analyzed in detail, the results are heavily influenced by the IEMOCAP dataset, which makes up for 46% of the English short text corpus. IEMOCAP is a particular dataset, created through the video recording of actors performing scripts designed to contain select emotions [12]. We used the transcriptions of the audio, which is annotated for valence and arousal in the dataset. Contrarily to all other datasets, these instances were annotated from videos, which can portray a large range of sentiments for the same textual script, depending on aspects such as posture and intonation of the actors. This implies that annotations range over a broader scope too, which likely affects the quality of the prediction results.

Table 4. Evaluation results for the short texts (top) and words (bottom) datasets, using the XLM-RoBERTa-large model and considering the MSE loss.

Dataset	Language	ρ_V	ρ_A	$RMSE_V$	$RMSE_A$	MAE_V	MAE_A
EmoBank	English	0.736	0.440	0.061	0.071	0.044	0.052
IEMOCAP	English	0.469	0.656	0.159	0.173	0.126	0.132
Facebook Posts	English	0.797	0.776	0.098	0.176	0.075	0.124
EmoTales	English	0.560	0.405	0.127	0.123	0.095	0.091
ANET	English	0.920	0.859	0.135	0.111	0.095	0.087
PANIG	German	0.597	0.563	0.181	0.111	0.137	0.085
COMETA sent	German	0.853	0.598	0.103	0.120	0.074	0.096
COMETA stories	German	0.072	0.042	0.254	0.160	0.206	0.130
CVAT	Mandarin	0.890	0.623	0.082	0.105	0.062	0.085
CVAI	Mandarin	0.517	0.720	0.068	0.089	0.053	0.071
ANPST	Polish	0.868	0.607	0.113	0.111	0.082	0.089
MAS	Portuguese	0.935	0.694	0.115	0.124	0.082	0.100
Yee	Cantonese	0.875	0.718	0.090	0.121	0.069	0.099
Ćoso et al.	Croatian	0.784	0.646	0.133	0.120	0.096	0.093
Moors et al.	Dutch	0.776	0.653	0.116	0.125	0.081	0.098
Verheyen et al.	Dutch	0.791	0.637	0.130	0.137	0.096	0.109
NRC-VAD	English	0.858	0.754	0.111	0.124	0.086	0.097
Warriner et al.	English	0.843	0.655	0.101	0.114	0.078	0.090
Scott et al.	English	0.884	0.636	0.095	0.117	0.067	0.092
Söderholm et al.	Finnish	0.645	0.492	0.188	0.138	0.147	0.109
Eilola et al.	Finnish	0.807	0.534	0.164	0.191	0.117	0.161
FAN	French	0.755	0.605	0.116	0.112	0.086	0.087
FEEL	French	0.823	0.664	0.131	0.131	0.096	0.103
BAWL-R	German	0.749	0.629	0.139	0.133	0.101	0.105
ANGST	German	0.837	0.738	0.135	0.117	0.092	0.089
LANG	German	0.802	0.696	0.100	0.144	0.074	0.115
Italian ANEW	Italian	0.846	0.644	0.138	0.148	0.099	0.118
Xu et al.	Mandarin	0.882	0.754	0.078	0.098	0.055	0.077
CVAW	Mandarin	0.904	0.666	0.094	0.136	0.071	0.108
ANPW_R	Polish	0.846	0.689	0.093	0.088	0.065	0.069
NAWL	Polish	0.828	0.581	0.111	0.122	0.081	0.096
Portuguese ANEW	Portuguese	0.893	0.779	0.106	0.103	0.074	0.081
S.-Gonzalez et al.	Spanish	0.808	0.689	0.100	0.095	0.074	0.072
Kapucu et al.	Turkish	0.571	0.373	0.165	0.127	0.125	0.101

Table 5. Evaluation results for individual languages on the short text datasets, using the XLM-RoBERTa-large model and considering the MSE loss function.

Language	ρ_V	ρ_A	$RMSE_V$	$RMSE_A$	MAE_V	MAE_A
English	0.592	0.719	0.118	0.138	0.085	0.096
Mandarin	0.892	0.657	0.077	0.100	0.059	0.080
German	0.619	0.533	0.179	0.117	0.133	0.090
Portuguese	0.935	0.694	0.115	0.124	0.082	0.100
Polish	0.868	0.607	0.113	0.111	0.082	0.089

Stemming from these last conclusions, we performed one more separate experiment. Considering the same training setting, we trained the model with a combined dataset not containing the two seemingly troublesome datasets, COMETA stories and IEMOCAP. Compared to previous results, the Pearson's ρ for valence increased from 0.8095 to 0.8423, and arousal's correlation increased from 0.6974 to 0.7107. Performance gains were observed for all tested languages. In particular, valence and arousal correlation values for German short texts increased 13% and 7%, and most noticeably for English they increased 31% and 11%, respectively. This took the scores obtained for these two languages, which are well represented in the training instances, to levels akin to most other languages, and explained the previously noticed discrepancy in the evaluations.

5.3   Results in Zero-Shot Settings

With the previous results in mind, a question remained on whether our best model could generalize well to other languages in which it was not trained on. For that purpose, two other XLM-RoBERTa-large models were fine-tuned under the same training setup. Specifically, these models were trained with all the data from the merged dataset except for either the Polish or the Portuguese instances. These instances were saved for subsequent zero-shot evaluations, separately focusing on each of these languages. This trial aimed to assert whether the proposed approach can generalize to a language not used for training. Polish and Portuguese were chosen for this purpose, as both these languages are represented in our dataset, simultaneously with word and short text instances. Despite being reasonably popular languages, they are not as extensively present as English, and thus they allow us to adequately simulate the scenario of testing the proposed model on a new language not seen during training, and also not seen extensively during the model pre-training stage (i.e., the DiltilBERT and XML-RoBERTa models, despite being multilingual, have seen much more English training data in comparison to other languages).

We can compare the results of these zero-shot experiments, presented in Table 6, with the results obtained for the Polish and Portuguese subsets of predictions presented previously in Table 4. When comparing correlation and error metrics, we found overall worse results. However, the difference is not significant, and the results are in fact higher than some of the observed results for other languages on which the model was fine-tuned on. The zero-shot performance for

Table 6. Zero-shot evaluation for Polish (PL) and Portuguese (PT) data, using the XLM-RoBERTa-large model and considering the MSE loss function.

Training on	Predicting on	ρ_V	ρ_A	RMSE$_V$	RMSE$_A$	MAE$_V$	MAE$_A$
All	Any PL input	0.839	0.648	0.101	0.103	0.072	0.080
All excl. PL		0.818	0.618	0.111	0.135	0.080	0.108
All	Any PT input	0.895	0.756	0.108	0.107	0.075	0.084
All excl. PT		0.886	0.735	0.112	0.112	0.079	0.088
All	PL words	0.833	0.631	0.100	0.102	0.071	0.079
All excl. PL		0.814	0.647	0.111	0.135	0.079	0.108
All	PT words	0.893	0.779	0.106	0.103	0.074	0.081
All excl. PT		0.906	0.777	0.102	0.107	0.071	0.084
All	PL short texts	0.868	0.607	0.113	0.111	0.082	0.089
All excl. PL		0.860	0.487	0.113	0.135	0.085	0.108
All	PT short texts	0.935	0.694	0.115	0.124	0.082	0.100
All excl. PT		0.923	0.627	0.155	0.135	0.121	0.109

both languages shows promising prospects for the application of the proposed approach to different languages without available emotion corpora.

6 Conclusions and Future Work

This paper presented a bi-dimensional and multilingual model to predict real-valued emotion ratings from instances of text. First, a multi-language emotion corpus of words and short texts was assembled. This goes in contrast to most previous studies, which focused solely on words or texts in a single language. The corpus, consisting of 128,987 instances, features annotations for the psycho-linguistic dimensions of Valence and Arousal (VA), spanning 13 different languages. Subsequently, DistilBERT and XLM-RoBERTa models were trained for VA prediction using the multilingual corpus. The evaluation methodology used Pearson's ρ and two error metrics to assess the results. Overall, the predicted ratings showed a high correlation with human ratings, and the results are in line with those of previous monolingual predictive approaches. Additionally, this research highlights the challenge of predicting arousal to the same degree of confidence of predicting valence from text. In sum, the evaluation of our best model showed competitive results against previous approaches, having the advantage of generalization to different languages and different types of text.

An interesting idea to explore in future work concerns applying uncertainty quantification[4] to the predicted ratings, for instance as explained by Angelopoulos and Bates [2]. Instead of predicting a single pair of values for the valence and arousal ratings, the aim would be to predict a high confidence interval of values in which valence and arousal are contained. Future work can also address the

---

[4] https://mapie.readthedocs.io/en/latest/.

study of data augmentation methods (e.g., based on machine translation), in an attempt to further improve result quality and certainty.

Another interesting direction for future work concerns extending the work reported in this paper to consider multimodal emotion estimation. Instead of the models considered here, we can consider fine-tuning a large multilingual vision-and-language model[5] such as CLIP [13], combining the textual datasets together with affective image datasets like the International Affective Picture System (IAPS) [35], the Geneva Affective PpicturE Database (GAPED) [19], the Nencki Affective Picture System (NAPS) [37], the Open Affective Standardized Image Set (OASIS) [34], or others [14,31].

Acknowledgements. This research was supported by the European Union's H2020 research and innovation programme, under grant agreement No. 874850 (MOOD), as well as by the Portuguese Recovery and Resilience Plan (RRP) through project C645008882-00000055 (Responsible.AI), and by Fundação para a Ciência e Tecnologia (FCT), through the INESC-ID multi-annual funding with reference UIDB/50021/2020, and through the projects with references DSAIPA/DS/0102/2019 (DEBAQI) and PTDC/CCI-CIF/32607/2017 (MIMU).

References

1. Alm, C.O., Roth, D., Sproat, R.: Emotions from text: machine learning for text-based emotion prediction. In: Proceedings of the Conference on Empirical Methods in Natural Language Processing (2005)
2. Angelopoulos, A.N., Bates, S.: A gentle introduction to conformal prediction and distribution-free uncertainty quantification. arXiv preprint arXiv:2107.07511 (2021)
3. Barron, J.T.: A general and adaptive robust loss function. In: Proceedings of the IEEE Conference on Computer Vision and Pattern Recognition (2019)
4. Bradley, M.M., Lang, P.J.: Measuring emotion: the self-assessment manikin and the semantic differential. J. Behav. Ther. Exp. Psychiatry **25**(1), 49–59 (1994)
5. Bradley, M.M., Lang, P.J.: Affective Norms for English Words (ANEW): instruction manual and affective ratings. Technical report C-1, The Center for Research in Psychophysiology, University of Florida (1999)
6. Bradley, M.M., Lang, P.J.: Affective Norms for English Text (ANET): affective ratings of text and instruction manual. Technical report D-1, University of Florida (2007)
7. Buechel, S., Hahn, U.: Emotion analysis as a regression problem - dimensional models and their implications on emotion representation and metrical evaluation. In: Proceedings of the European Conference on Artificial Intelligence (2016)
8. Buechel, S., Hahn, U.: EmoBank: studying the impact of annotation perspective and representation format on dimensional emotion analysis. In: Proceedings of the Conference of the European Chapter of the Association for Computational Linguistics (2017)

---

[5] https://huggingface.co/laion/CLIP-ViT-H-14-frozen-xlm-roberta-large-laion5B-s13B-b90k.

9. Buechel, S., Hahn, U.: A flexible mapping scheme for discrete and dimensional emotion representations: evidence from textual stimuli. In: Proceedings of the Annual Meeting of the Cognitive Science Society (2017)
10. Buechel, S., Hahn, U.: Readers vs. writers vs. texts: coping with different perspectives of text understanding in emotion annotation. In: Proceedings of the Linguistic Annotation Workshop (2017)
11. Buechel, S., Rücker, S., Hahn, U.: Learning and evaluating emotion lexicons for 91 languages. In: Proceedings of the Annual Meeting of the Association for Computational Linguistics (2020)
12. Busso, C., et al.: IEMOCAP: interactive emotional dyadic motion capture database. Lang. Resour. Eval. **42**(4), 335–359 (2008)
13. Carlsson, F., Eisen, P., Rekathati, F., Sahlgren, M.: Cross-lingual and multilingual CLIP. In: Proceedings of the Language Resources and Evaluation Conference (2022)
14. Carretié, L., Tapia, M., López-Martín, S., Albert, J.: EmoMadrid: an emotional pictures database for affect research. Motiv. Emotion **43**(6) (2019)
15. Citron, F.M.M., Lee, M., Michaelis, N.: Affective and psycholinguistic norms for German conceptual metaphors (COMETA). Behav. Res. Methods **52**(3), 1056–1072 (2020)
16. Citron, F.M., Cacciari, C., Kucharski, M., Beck, L., Conrad, M., Jacobs, A.M.: When emotions are expressed figuratively: psycholinguistic and Affective Norms of 619 Idioms for German (PANIG). Behav. Res. Methods **48**(1), 91–111 (2016)
17. Conneau, A., et al.: Unsupervised cross-lingual representation learning at scale. In: Proceedings of the Annual Meeting of the Association for Computational Linguistics (2020)
18. Ćoso, B., Guasch, M., Ferré, P., Hinojosa, J.A.: Affective and concreteness norms for 3,022 Croatian words. Q. J. Exp. Psychol. **72**(9), 2302–2312 (2019)
19. Dan-Glauser, E.S., Scherer, K.R.: The Geneva Affective Picture Database (GAPED): a new 730-picture database focusing on valence and normative significance. Behav. Res. Methods **43**(2) (2011)
20. Du, S., Zhang, X.: Aicyber's system for IALP 2016 shared task: character-enhanced word vectors and boosted neural networks. In: Proceedings of the International Conference on Asian Language Processing (2016)
21. Eilola, T.M., Havelka, J.: Affective norms for 210 British English and Finnish nouns. Behav. Res. Methods **42**(1), 134–140 (2010)
22. Ekman, P.: An argument for basic emotions. Cogn. Emotion **6**(3–4), 169–200 (1992)
23. Fairfield, B., Ambrosini, E., Mammarella, N., Montefinese, M.: Affective norms for Italian words in older adults: age differences in ratings of valence, arousal and dominance. PLoS ONE **12**(1), e0169472 (2017)
24. Francisco, V., Hervás, R., Peinado, F., Gervás, P.: EmoTales: creating a corpus of folk tales with emotional annotations. Lang. Resour. Eval. **46**(3), 341–381 (2012)
25. Gilet, A.L., Grühn, D., Studer, J., Labouvie-Vief, G.: Valence, arousal, and imagery ratings for 835 French attributes by young, middle-aged, and older adults: the French Emotional Evaluation List (FEEL). Eur. Rev. Appl. Psychol. **62**(3), 173–181 (2012)
26. Hollis, G., Westbury, C., Lefsrud, L.: Extrapolating human judgments from skip-gram vector representations of word meaning. Q. J. Exp. Psychol. **70**, 1–45 (2016)
27. Imbir, K.K.: Affective Norms for 4900 Polish Words Reload (ANPW_R): assessments for valence, arousal, dominance, origin, significance, concreteness, imageability and age of acquisition. Front. Psychol. **7**, 1081 (2016)

28. Imbir, K.K.: The Affective Norms for Polish Short Texts (ANPST) database properties and impact of participants' population and sex on affective ratings. Front. Psychol. **8**, 855 (2017)

29. Kanske, P., Kotz, S.A.: Leipzig affective norms for German: a reliability study. Behav. Res. Methods **42**(4), 987–991 (2010)

30. Kapucu, A., Kılıç, A., Özkılıç, Y., Sarıbaz, B.: Turkish emotional word norms for arousal, valence, and discrete emotion categories. Psychol. Rep. **124**(1), 188–209 (2021)

31. Kim, H.R., Kim, Y.S., Kim, S.J., Lee, I.K.: Building emotional machines: recognizing image emotions through deep neural networks. IEEE Trans. Multimedia **20**(11) (2018)

32. Kratzwald, B., Ilić, S., Kraus, M., Feuerriegel, S., Prendinger, H.: Deep learning for affective computing: text-based emotion recognition in decision support. Decis. Support Syst. **115**, 24–35 (2018)

33. Kron, A., Goldstein, A., Lee, D., Gardhouse, K., Anderson, A.: How are you feeling? Revisiting the quantification of emotional qualia. Psychol. Sci. **24**(8), 1503–1511 (2013)

34. Kurdi, B., Lozano, S., Banaji, M.R.: Introducing the open affective standardized image set (OASIS). Behav. Res. Methods **49**(2) (2017)

35. Lang, P.J., Bradley, M.M., Cuthbert, B.N.: International Affective Picture System (IAPS): affective ratings of pictures and instruction manual. Technical report, NIMH Center for the Study of Emotion and Attention (2005)

36. Lee, L.H., Li, J.H., Yu, L.C.: Chinese EmoBank: building valence-arousal resources for dimensional sentiment analysis. Trans. Asian Low-Resource Lang. Inf. Process. **21**(4), 1–18 (2022)

37. Marchewka, A., Żurawski, Ł., Jednoróg, K., Grabowska, A.: The Nencki Affective Picture System (NAPS): introduction to a novel, standardized, wide-range, high-quality, realistic picture database. Behav. Res. Methods **46**(2) (2014)

38. Mohammad, S.: Obtaining reliable human ratings of valence, arousal, and dominance for 20,000 English words. In: Proceedings of the Annual Meeting of the Association for Computational Linguistics (2018)

39. Monnier, C., Syssau, A.: Affective Norms for French words (FAN). Behav. Res. Methods **46**(4), 1128–1137 (2014)

40. Moors, A., et al.: Norms of valence, arousal, dominance, and age of acquisition for 4,300 Dutch words. Behav. Res. Methods **45**(1), 169–177 (2013)

41. Paltoglou, G., Theunis, M., Kappas, A., Thelwall, M.: Predicting emotional responses to long informal text. IEEE Trans. Affect. Comput. **4**(1), 106–115 (2013)

42. Pinheiro, A.P., Dias, M., Pedrosa, J., Soares, A.P.: Minho Affective Sentences (MAS): probing the roles of sex, mood, and empathy in affective ratings of verbal stimuli. Behav. Res. Methods **49**(2), 698–716 (2017)

43. Pires, T., Schlinger, E., Garrette, D.: How multilingual is multilingual BERT? In: Proceedings of the Annual Meeting of the Association for Computational Linguistics (2019)

44. Preoţiuc-Pietro, D., et al.: Modelling valence and arousal in Facebook posts. In: Proceedings of the Workshop on Computational Approaches to Subjectivity, Sentiment and Social Media Analysis (2016)

45. Recchia, G., Louwerse, M.: Reproducing affective norms with lexical co-occurrence statistics: predicting valence, arousal, and dominance. Qu. J. Exp. Psychol. **68**, 1–41 (2014)

46. Riegel, M., et al.: Nencki Affective Word List (NAWL): the cultural adaptation of the Berlin Affective Word List-Reloaded (BAWL-R) for Polish. Behav. Res. Methods **47**(4), 1222–1236 (2015)
47. Russell, J.: A circumplex model of affect. J. Pers. Soc. Psychol. **39**, 1161–1178 (1980)
48. Sanh, V., Debut, L., Chaumond, J., Wolf, T.: DistilBERT, a distilled version of BERT: smaller, faster, cheaper and lighter. arXiv preprint arXiv:1910.01108 (2019)
49. Schmidtke, D.S., Schröder, T., Jacobs, A.M., Conrad, M.: ANGST: affective norms for German sentiment terms derived from the affective norms for English words. Behav. Res. Methods **46**(4), 1108–1118 (2014)
50. Scott, G.G., Keitel, A., Becirspahic, M., Yao, B., Sereno, S.C.: The Glasgow norms: ratings of 5,500 words on nine scales. Behav. Res. Methods **51**(3), 1258–1270 (2019)
51. Sedoc, J., Preoţiuc-Pietro, D., Ungar, L.: Predicting emotional word ratings using distributional representations and signed clustering. In: Proceedings of the Conference of the European Chapter of the Association for Computational Linguistics (2017)
52. Shad Akhtar, M., Ghosal, D., Ekbal, A., Bhattacharyya, P., Kurohashi, S.: A multi-task ensemble framework for emotion, sentiment and intensity prediction. arXiv preprint arXiv:1808.01216 (2018)
53. Soares, A., Comesaña, M., Pinheiro, A., Simões, A., Frade, S.: The adaptation of the Affective Norms for English Words (ANEW) for European Portuguese. Behav. Res. Methods **44**(1), 256–269 (2012)
54. Söderholm, C., Häyry, E., Laine, M., Karrasch, M.: Valence and arousal ratings for 420 Finnish nouns by age and gender. PLoS ONE **8**(8), e72859 (2013)
55. Stadthagen-Gonzalez, H., Imbault, C., Pérez Sánchez, M.A., Brysbaert, M.: Norms of valence and arousal for 14,031 Spanish words. Behav. Res. Methods **49**(1), 111–123 (2017)
56. Tris Atmaja, B., Akagi, M.: Evaluation of error- and correlation-based loss functions for multitask learning dimensional speech emotion recognition. J. Phys: Conf. Ser. **1896**, 012004 (2021)
57. Verheyen, S., De Deyne, S., Linsen, S., Storms, G.: Lexicosemantic, affective, and distributional norms for 1,000 Dutch adjectives. Behav. Res. Methods **52**(3), 1108–1121 (2020)
58. Vo, M.L.H., Conrad, M., Kuchinke, L., Urton, K., Hofmann, M.J., Jacobs, A.M.: The Berlin Affective Word List Reloaded (BAWL-R). Behav. Res. Methods **41**(2), 534–538 (2009)
59. Wang, J., Yu, L.C., Lai, K., Zhang, X.: Tree-structured regional CNN-LSTM model for dimensional sentiment analysis. IEEE/ACM Trans. Audio Speech Lang. Process. **28**, 581–591 (2020)
60. Warriner, A.B., Kuperman, V., Brysbaert, M.: Norms of valence, arousal, and dominance for 13,915 English lemmas. Behav. Res. Methods **45**(4), 1191–1207 (2013)
61. Wolf, T., et al.: Transformers: state-of-the-art natural language processing. In: Proceedings of the Conference on Empirical Methods in Natural Language Processing (2020)
62. Wu, C., Wu, F., Huang, Y., Wu, S., Yuan, Z.: THU_NGN at IJCNLP-2017 task 2: dimensional sentiment analysis for Chinese phrases with deep LSTM. In: Proceedings of the International Joint Conference on Natural Language Processing (2017)
63. Xie, H., Lin, W., Lin, S., Wang, J., Yu, L.C.: A multi-dimensional relation model for dimensional sentiment analysis. Inf. Sci. **579**, 832–844 (2021)

64. Xu, X., Li, J., Chen, H.: Valence and arousal ratings for 11,310 simplified Chinese words. Behav. Res. Methods **54**(1), 26–41 (2022)
65. Yee, L.: Valence, arousal, familiarity, concreteness, and imageability ratings for 292 two-character Chinese nouns in Cantonese speakers in Hong Kong. PLoS ONE **12**(3), e0174569 (2017)
66. Yu, L.C., et al.: Building Chinese affective resources in valence-arousal dimensions. In: Proceedings of the Conference of the North American Chapter of the Association for Computational Linguistics (2016)
67. Yu, L.C., Wang, J., Lai, K.R., Zhang, X.: Pipelined neural networks for phrase-level sentiment intensity prediction. IEEE Trans. Affect. Comput. **11**(3), 447–458 (2020)

A Knowledge Infusion Based Multitasking System for Sarcasm Detection in Meme

Dibyanayan Bandyopadhyay[1]([ID]), Gitanjali Kumari[1][ID], Asif Ekbal[1][ID],
Santanu Pal[2][ID], Arindam Chatterjee[2][ID], and Vinutha BN[2]

[1] Department of Computer Science and Engineering, Indian Institute of Technology
Patna, Bihta, India
{dibyanayan_2111cs02,gitanjali_2021cs03,asif}@iitp.ac.in
[2] Wipro AI Labs, Bengaluru, India
{santanu.pal2,arindam.chatterjee4,vinutha.narayanmurthy}@wipro.com

Abstract. In this paper, we hypothesize that sarcasm detection is closely associated with the emotion present in memes. Thereafter, we propose a deep multitask model to perform these two tasks in parallel, where sarcasm detection is treated as the primary task, and emotion recognition is considered an auxiliary task. We create a large-scale dataset consisting of 7416 memes in Hindi, one of the widely spoken languages. We collect the memes from various domains, such as *politics*, *religious*, *racist*, and *sexist*, and manually annotate each instance with three sarcasm categories, i.e., *i) Not Sarcastic, ii) Mildly Sarcastic* or *iii) Highly Sarcastic* and 13 fine-grained emotion classes. Furthermore, we propose a novel Knowledge Infusion (KI) based module which captures sentiment-aware representation from a pre-trained model using the *Memotion* dataset. Detailed empirical evaluation shows that the multitasking model performs better than the single-task model. We also show that using this KI module on top of our model can boost the performance of sarcasm detection in both single-task and multi-task settings even further. Code and dataset are available at this link: https://www.iitp.ac.in/ ai-nlp-ml/resources.html#Sarcastic-Meme-Detection.

Keywords: Sarcasm detection in meme · Emotion recognition · Knowledge infusion · Multitasking

1 Introduction

Recently, there has been a growing interest in the domain of computational social science for detecting various attributes (e.g., Hateful Tweets, Fake news, Offensive posts) of social media articles to curb their circulation and thereby reduce social harm. Often, these articles use sarcasm to veil offensive content with a humorous tone. This is due to the very nature of sarcasm, where the article utilizes superficial

D. Bandyopadhyay and G. Kumari—Equal Contribution.

humor to mask the intended offensive meaning. This makes detecting harmful content more difficult. Also, sarcasm detection, in itself, is a challenging task because detecting it often requires prior context about the article creators, audience, and social constructs, and these are hard to determine from the article itself [34].

Memes are a form of multimodal (Image+Text) media that is becoming increasingly popular on the internet. Though it was initially created to spread humor, some memes help users to spread negativity in society in the form of sarcasm [18,35,37]. In the context of memes, detecting sarcasm is very difficult, as memes are multimodal in nature. In meme, just like offensiveness detection [15], we cannot uncover the complex meaning of sarcasm until we know all the modalities and their contributions in sarcastic content. Also, a sarcastic sentence always has an implied negative sentiment because it intends to express contempt and derogation [16].

Refer to example meme 3 of Fig. 1, which is taken from the political domain. It says, *"While selling mangoes on a handcart, I asked a man, "brother, this mango is not ripe by giving chemicals." The vendor replied, "No, brother, it has been ripened after listening to Person-A's inner thoughts."* When we observe this meme from an outer perspective (by considering both text and image), it is seen that the meme was formed solely to spread humor with no apparent twist.

	Test Sample 1	Test Sample 2	Test Sample 3
Sarcasm	Highly Sarcastic	Non-sarcastic	Highly sarcastic
Emotions	insult,joy	joy	joy,insult

Fig. 1. Some samples from our dataset. We have masked individual faces with white color and their name with Person-A throughout the paper to maintain anonymization

But, after carefully analyzing the intended meaning, we observe that the meme creator is using sarcasm to offend Person-A. The meme creator wants to convey two conflicting emotions with the help of this sarcastic meme, i.e., both *insult* and *joy* up to varying degrees. Additionally, the meme creator is implicitly 'annoyed' (which may be considered as a negative sentiment) with Person-A. This demonstrates a clear interplay between emotions and sentiment in a meme and their associations with sarcasm.

Armed with the above analysis, we hypothesize through the help of the associated emotions and the overall sentiment, we can detect sarcasm and vice-versa. Multimodality helps us to understand the intent of the meme creator with more certainty. We list the contributions of this work as follows:

- We create a high-quality and large-scale multimodal meme dataset annotated with three labels (non-sarcastic, mildly sarcastic, and highly sarcastic) to detect and also quantify the sarcasm given in a meme and 13 fine-grained emotion classes.
- We propose a deep neural model which simultaneously detects *sarcasm* and recognizes *emotions* in a given meme. Multitasking ensures that we exploit the emotion of the meme, which aids in detecting sarcasm more easily. We also propose a module denoted as knowledge infusion (KI) by which we leverage pre-trained sentiment-aware representation in our model.
- Empirical results show that the proposed KI module significantly outperforms the *naive* multimodal model.

2   Related Work

The multimodal approach to sarcasm detection in memes is a relatively recent trend rather than just text-based classification [4,24]. A semi-supervised framework [39] for the recognition of sarcasm proposed a robust algorithm that utilizes features specific to (Amazon) product reviews. Poria et al. [30] developed pre-trained sentiment, emotion, and personality models to predict sarcasm on a text corpus through a Convolutional Neural Network (CNN), which effectively detects sarcasm. Bouazizi et al. [4] proposed four sets of features, i.e., sentiment-related features, punctuation-related features, syntactic and semantic features, and pattern-related features that cover the different types of sarcasm. Then, they used these features to classify tweets as sarcastic/non-sarcastic.

The use of multi-modal sources of information has recently gained significant attention from researchers in affective computing. Ghosal et al. [12] proposed a recurrent neural network-based attention framework that leverages contextual information for multi-modal sentiment prediction. Kamrul et al. [14] presented a new multi-modal dataset for humor detection called UR-FUNNY that contains three modalities, *viz.* text, vision, and acoustic.

Researchers have also put their efforts towards sarcasm detection in the direction of conversational Artificial Intelligence (AI) [8,13,17]. For multimodal sarcasm detection in conversational AI, Castro et al. [5] created a new dataset, *MUStARD*, with high-quality annotations by including both multimodal and conversational context features. Majumder et al. [26] demonstrated that sarcasm detection could also be beneficial to sentiment analysis and designed a multitask learning framework to enhance the performance of both tasks simultaneously. Similarly, Chauhan et al. [6] have also shown that sarcasm can be detected with better accuracy when we know the *sarcasm* and *sentiment* of the speaker. This work differs from conversational sarcasm detection owing to the presence of contextual and background information in memes. Specifically, in this paper, we try to demonstrate that these multitasking approaches hold true in the domain of memes as well when sentiment-aware representation is used as contextual information via the proposed knowledge infusion (KI) scheme.

3 Dataset

Data Collection and Preprocessing: Our data collection process is inspired by previous studies done on meme analysis [18,35]. We collect 7416 freely available public domain memes to keep a strategic distance from any copyright issues. For retrieval, we consider domains like politics, religion, and social issues like terrorism, racism, sexism, etc., following a list of a total of 126 keywords (e.g. Terrorism, Politics, Entertainment) in Hindi for retrieval.

To save the data annotation effort, we manually discard the collected memes which are (i) noisy such as background pictures are not clear, (ii) non-Hindi, i.e., meme texts are written in other languages except Hindi, and (iii) non-multimodal, i.e., memes contain either text or visual content. Next, we extract the textual part of each meme using an open source Optical Character Recognition(OCR) tool: Tesseract[1]. The OCR errors are manually post-corrected by annotators. We finally consider 7,416 memes for data annotation.

Data Annotation:

Sarcasm. Previous works for sarcasm detection use a dataset that is labeled with two classes of sarcasm, i.e., i) Sarcastic, ii) Non Sarcastic [6,35]. To make the problem even more challenging for a system, we annotate each sample in the dataset with three labels of sarcasm *viz. Non-sarcastic, Mildly sarcastic* and *Highly Sarcastic*. Although it is easy for a system to detect either non-sarcastic or highly sarcastic memes, it becomes harder when the system has to disambiguate a non-sarcastic meme from a mildly sarcastic one and vice-versa.

Emotion. Most psycho-linguistic theories [11,41] claim that there are only a few primary emotions, like Anger, Fear, that form the basis of all other emotions. However, Merely these primary emotions could not adequately represent the diverse emotional states that humans are capable of [20]. Taking inspiration from their work, we construct a list of 120 affective keywords collected from our pre-defined four domains (i.e., *politics, religious, racist, and sexist*). After mapping these affective keywords to their respective emotions and depending on the specific affective keyword present in a meme, we annotate every sample of the dataset with 13 fine-grained categories of emotions, *viz. Disappointment, Disgust, Envy, Fear, Irritation, Joy, Neglect, Nervousness, Pride, Rage, Sadness, Shame, and Suffering*.

Annotation Guidelines. We employed two annotators with an expert-level understanding of Hindi for annotation purposes. At first, we provided expert-level training based on 100 sample memes. After automatic Data collection and Data preprocessing steps, the annotators were then asked to annotate the memes with three classes of sarcasm and thirteen classes of fine-grained emotions. The annotation process is significantly prone to the creation of a racially and politically biased dataset as illustrated in [7]. To counter bias, i) we made sure that the terms included were inclusive of all conceivable politicians, political organizations, young politicians, extreme groups, and religions and were not prejudiced

---

[1] github.com/tesseract-ocr/tesseract.

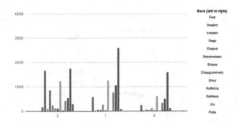

classes	instance	% distribution
Non-Sarcastic(0)	1798	24.25
Mildly Sarcastic(1)	2770	37.35
Highly Sarcastic(2)	2848	38.40

Fig. 2. Distribution of fine-grained emotion categories for each sarcasm class

Fig. 3. Data statistics of our annotated corpus for sarcasm

against any one group, and ii) Annotators were instructed not to make decisions based on what they believe but what the social media user wants to transmit through that meme. In the case of disagreements between the annotators during the annotation process, we resolved them by agreeing on a common point after thorough discussions. The annotators were asked to annotate each sample meme with as many emotion classes as possible. This is done to ensure that we do not have a severe class imbalance in emotion categories for each sarcasm class. Despite that, there is an overall small class imbalance for emotion categories in our proposed dataset. Note that while learning and evaluating our models on top of our dataset, this small class imbalance is ignored. To assess inter-rater agreement, we utilized Cohen's Kappa coefficient [2]. For the sarcasm label, we observed Cohen's Kappa coefficient score of 0.7197, which is considered to be a reliable score. Similarly, for 13 fine-grained emotion labels, the reported Krippendorff's Alpha Coefficient [21] is 0.6174 in a multilabel scenario, which is relatively low. But previous annotation tasks [1,3,28] have shown that even with binary or ternary classification schemes, human annotators agree only about 70-80% of the time and the more categories there are, the harder it becomes for annotators to agree. Based on this point, 0.6174 can be considered a good score for inter-annotator agreement. In Fig. 3, we show the data statistics of sarcasm classes and the respective distribution of emotion classes in Fig. 2.

4 Methods

This section presents the details of our proposed multitasking model architecture by which we perform two tasks in parallel, *viz.* Sarcasm detection and Emotion recognition. We also describe the knowledge infusion (KI) mechanism, which is a novel addition to the multitasking model.

The basic diagram of the proposed model is shown in Fig. 4. The following section discusses our method in detail:

4.1 Feature Extraction Layer

We use memes (M) as input to our model, which is comprised of an image (V) and an associated text (T). These are then fed into pre-trained and frozen CLIP

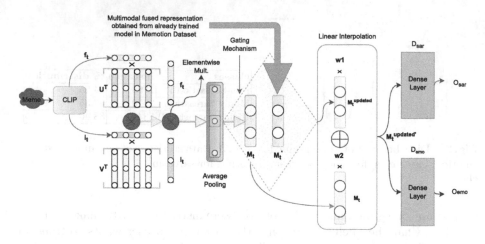

Fig. 4. A generic architecture of our proposed model

model [31]² to obtain the text representation (f_t) and visual representation (i_t), respectively.

We summarize the above steps by the following equation:

$$T, V \in M$$
$$f_t, i_t = CLIP(T, V) \tag{1}$$

4.2 Multimodal Fusion

Text (f_t) and visual representation (i_t) obtained from the feature extraction layer are then fed into a Fusion Module to prepare a fused multimodal representation.

We have CLIP extracted text feature (f_t) and visual features (i_t) having dimensions $\mathbb{R}^{m \times 1}$ and $\mathbb{R}^{n \times 1}$ respectively. Further, assume we need a multimodal representation M_t having dimension $\mathbb{R}^{o \times 1}$. The fusion module is comprised of two trainable weight matrices U and V having dimensions $\mathbb{R}^{m \times ko}$ such that the following projection followed by the average-pooling operation is performed.

$$M_t = AveragePool(U^T f_t \circ V^T i_t, k) \tag{2}$$

where $\circ$ denotes element-wise multiplication operation and k denotes the stride of the overlapped window (similar to convolutional stride) to perform the pooling operation.

4.3 Knowledge Infusion (KI)

We devise a simple knowledge infusion (KI) technique to enrich multimodal representation (M_t) for better performance in our downstream classification tasks.

---

² https://github.com/FreddeFrallan/Multilingual-CLIP.

Our KI method consists of two steps: (i) Obtaining a learned representation from an already trained model, (ii) Utilizing the learned representation via a gating mechanism to 'enrich' M_t. The following subsections deal with the aforementioned steps in detail.

KI Learned Representation. We fine-tune a copy of our model (parent model) until convergence. We use *Memotion* dataset[3] for training. We perform multitasking by classifying each meme instance into (i) one of three classes for sarcasm and (ii) one of the three classes of sentiment.[4] This is done using two task specific classification layers, D'_{sar} and D'_{sent}, respectively, on top of the shared layers.

After the model is completely trained, we freeze its layers. We provide an input meme to extract multimodal representation M'_t from its trained fusion module. Subsequently, M'_t is used to enrich M_t via the proposed gating mechanism described below.

Gating Mechanism. Firstly, we obtain multimodal representation (M_t) following Eq. 2. Instead of feeding M_t directly into the subsequent classifier layers, we use a gating mechanism by which we pass extra information (M'_t) as needed and update M_t according to the following equation:

$$M_t^{updated} = f(M_t, M'_t) \tag{3}$$

where f is a generic function used to show the 'gating' mechanism.

Importance Weighting. Firstly an importance score (z) is obtained via Eq. 4 to measure the importance of M_t and M'_t for the downstream classification task.

$$z = \sigma(U^T M_t + V^T M'_t + b) \tag{4}$$

where σ refers to sigmoid operation and U and V are learnable weight matrices and b is a bias term.

Update Scheme. We linearly interpolate M_t and M'_t via z to obtain KI infused representation ($M_t^{updated}$) following Eq. 5.

$$M_t^{updated} = z \times M_t + (1 - z) \times M'_t \tag{5}$$

Linear Interpolation. The parent model was never trained to detect emotion in memes, and the obtained representation from the gating mechanism ($M_t^{updated}$) thus conflicts with the employed multitasking objective (simultaneously detecting sarcasm and recognizing emotion). To compensate for this issue, we tweak our training objective by replacing $M_t^{updated}$ with $M_t^{updated'}$, which is given by:

$$M_t^{updated'} = w1 \times M_t^{updated} + w2 \times M_t \tag{6}$$

where, $w1$ and $w2$ are scalar weight parameters initialized to 0.5.

---

[3] https://competitions.codalab.org/competitions/35688.

[4] Each meme in *Memotion.* dataset is annotated with both sarcasm and sentiment classes.

4.4 Classification

Our model performs multitasking to detect sarcasm and recognize emotion in an input meme.

For both of these tasks, task-specific dense classification layers are used, and both of the task-specific layers get the same multimodal representation from the previous 'shared' layers. Specifically, the fusion layer acts as a shared layer, and the updated multimodal representation (M_t') is then used as an input to two separate dense layers. We denote the dense layer as D_{sar} for sarcasm detection and D_{emo} for emotion recognition, respectively.

Previous operations can be described as follows:

$$O_{sar} = D_{sar}(M_t^{updated'}, activation = softmax)$$
$$O_{emo} = D_{emo}(M_t^{updated'}, activation = sigmoid) \qquad (7)$$
$$O_{sar} \in \mathbb{R}^{1 \times 3}; O_{emo} \in \mathbb{R}^{1 \times 13}$$

O_{sar} and O_{emo} are respectively the logit outputs associated to the D_{sar} and D_{emo} classifier heads. These output vectors are then used to calculate the respective cross-entropy loss to optimize the model.

Experimental Setups: We train our proposed model with a batch size of 32 for 50 epochs with an early-stopping callback. We use Adam optimizer [19] (i.e. $lr = 0.001$, $\beta 1 = 0.9$, $\beta 2 = 0.999$, $\epsilon = 1e - 8$) with standard hyperparameters throughout the experiments.

5   Results

We evaluate our proposed architecture with unimodal inputs, i.e., Text only (T) and Vision only (V), and compare their performance with multimodal inputs (T+V). For all of the input combinations (T, V, T+V), we perform our experiments for both Single Task Learning (STL) and Multitask learning (MTL) setups. In the STL setup, we only consider the model to learn to detect sarcasm in a given meme, whereas in the MTL setup, the model learns from the mutual interaction of two similar tasks, *viz.* sarcasm detection and emotion recognition. For each of the STL and MTL setups, we also show the effect of knowledge infusion by training our proposed model with the KI component.

STL Setup: In STL setup, we train the models to detect sarcasm in a meme by only training its D_{sar} classifier head. Furthermore, we train two separate models based on whether or not we use the KI component.

1. M_{sar}: This model is trained by only optimizing its D_{sar} head. Also we set $M_t^{updated'} = M_t$ to disable Knowledge infusion.

2. M_{emo}: This model is trained by only optimizing its D_{emo} head. This is done to measure the emotion recognition performance of our model without multitasking.

3. M_{sar}^{KI}: This is same as M_{sar} except KI is enabled here.

Table 1. *Sarcasm detection performance.* For both text-only (T) and vision-only (V) unimodal architectures, we show the performance of our proposed model for sarcasm detection. For comparison purposes, we also show multimodal (T+V) system performance. Here, Knowledge Infusion (KI) is disabled.

Setup	Model	T+V				T				V			
		R	P	F1	Acc	R	P	F1	Acc	R	P	F1	Acc
STL	M_{sar}	59.88	**63.28**	59.88	63.87	**53.18**	53.79	**53.24**	**55.88**	**55.94**	58.69	56.00	59.13
MTL	$M_{sar+emo}$	**61.07**	62.43	**61.11**	**64.61**	53.04	**54.48**	53.14	55.81	56.75	**62.03**	**56.28**	**60.75**

Table 2. *Sarcasm detection performance.* Here, Knowledge Infusion (KI) is enabled. $M_{sar+emo}^{KI}$ is statistically significant to M_{sar} ($p < 0.05$). *McNemar's test* [27] is performed to determine statistical significance level.

Setup	Model	T+V				T				V			
		R	P	F1	Acc	R	P	F1	Acc	R	P	F1	Acc
STL	M_{sar}^{KI}	**63.15**	64.01	63.29	65.89	**58.15**	58.32	**58.19**	60.14	56.89	57.63	57.01	59.81
MTL	$M_{sar+emo}^{KI}$	63.11	**65.01**	**63.37**	**66.64**	58.14	**60.01**	57.80	**62.31**	**57.79**	**60.73**	**57.25**	**62.24**

MTL Setup: In MTL setup, we simultaneously train D_{sar} and D_{emo} classifier heads of the model to perform multitasking by detecting both sarcasm and emotion in a meme (shown by $sar+emo$ subscript). Mutual interaction between two similar tasks helps in learning both. Similar to the STL setup, two models are trained for MTL setup.

1. $M_{sar+emo}$: It is trained by optimizing its D_{sar} head for detecting sarcasm and D_{emo} for recognizing emotion. We set $M_t^{updated'} = M_t$ to disable Knowledge infusion.

2. $M_{sar+emo}^{KI}$: It is same as $M_{sar+emo}$, except KI is enabled here.

In this section, we show the results that outline the comparison between the single-task (STL) and multi-task (MTL) learning frameworks. We use 7416 data points with a train-test split of $80 - 20$. 15% of the train set is used for validation purposes. For the evaluation of sarcasm classification in Table 1 and Table 2, we use macro-F1 score (F1), precision (P), and recall score (R), and accuracy (Acc) as the preferred metrics. In the STL setup, we observe that the M_{sar}^{KI} performs better than M_{sar}. This shows enabling knowledge infusion aids the model in detecting sarcasm. We observe that even the MTL setup benefits by enabling knowledge infusion (KI). This is evident from the increased performance of +2.26 F1-score when $M_{sar+emo}^{KI}$ compared to $M_{sar+emo}$. This improvement could be attributed to the sentiment-aware hidden representation (M_t'), which helps our model perform better by transferring knowledge via the proposed gating mechanism.

We also observe that for both STL and MTL setups, the multimodal input settings (T+V) show better performance than unimodal input settings (T or V). Our best performing model ($M_{sar+emo}^{KI}$) obtains an F1 score of 63.37, surpassing all the baselines. This performance gain is also statistically significant to M_{sar} ($p < 0.05$). We separately show the performance of our model on the emotion

Table 3. *Emotion Recognition performance* for multimodal (T+V) setting. *hloss* refers to Hamming Loss [40].

Task	M_{emo}				$M_{sar+emo}$			
	R	P	F1	hloss	R	P	F1	hloss
Emo. recognition	46.93	75.36	57.84	12.88	51.07	71.11	59.46	13.11

recognition task in Table 3. We observe increased performance of $M_{sar+emo}$ compared to M_{emo}, which supports the proposed MTL objective that training the model with both sarcasm detection and emotion recognition helps each other.

Ablation Study. Performance of both M_{sar}^{KI} and $M_{sar+emo}$ depend on the objective by which the parent model is trained. We can train the parent model with (i) *sar* objective (only detecting sarcasm) by only training it's D'_{sar} classifier head; or (ii) *sar+sent* objective (detecting both sarcasm and sentiment via multitasking) by training its D'_{sar} head and D'_{sent} simultaneously. We also observe that besides the proposed gating mechanism, the generic gating mechanism shown in Eq. 3 can be implemented by the following methodologies as (i) Concatenation followed by projection (*cat+proj*) to combine M_t and M'_t and (ii) Minimize KL divergence (*KL_div*) between M_t and M'_t. For all the above combinations, we show their performance in Table 4. We observed that our proposed KI fusion with *sar+sent* pretraining of the parent model performs better than other techniques, as it can be inferred intuitively. Furthermore, to demonstrate the superior performance of our proposed KI method, we sequentially finetune our model on our dataset after being trained on Memotion dataset using two setups:

Table 4. Ablation 1 Results of two models *viz sar* only and *sar+sent* pretraining objective of parent model with different KI fusion methods.

Obj.	KI fusion	M_{sar}^{KI}				$M_{sar+emo}^{KI}$			
		R	P	F1	Acc	R	P	F1	Acc
sar	*proposed*	62.07	62.82	62.31	64.34	62.05	**65.05**	61.89	66.37
	KL_div	61.85	64.11	62.06	65.29	61.14	64.25	61.00	65.30
	cat+proj	60.70	61.87	60.89	62.31	59.63	64.08	59.24	64.07
sar+sent	*proposed*	**63.15**	64.01	63.29	65.89	63.12	65.00	**63.37**	**66.64**
	KL_div	61.75	64.33	62.00	65.15	62.34	64.67	62.49	66.00
	cat+proj	61.12	62.28	61.31	64.20	60.86	63.58	61.20	63.59

i) KI-enabled fine-tuning: Instead of only fine-tuning sequentially, we also enable the KI mechanism using the proposed gating scheme.

ii) KI-disabled fine-tuning: We only fine-tune sequentially without enabling the KI mechanism. In Table 5, we observe that KI-enabled finetuning performs better than KI-disabled fine-tuning, and the proposed KI setup performs better than sequential finetuning (by comparing Table 2 and Table 5). Also, we observe that, (i) KI helps the sequential fine-tuning procedure and (ii) Combining KI with sequential fine-tuning is not effective as only performing KI.

Table 5. Ablation 2 Performance of sequential fine-tuning process.

Obj.	Process	STL				MTL			
		R	P	F1	Acc	R	P	F1	Acc
KI enabled	*seq. finetuning*	**63.32**	**63.53**	**62.11**	**65.08**	**63.10**	63.92	**62.08**	**65.47**
KI disabled	*seq. finetuning*	60.25	61.78	60.34	63.73	61.5	**64.25**	61.74	65.02

Table 6. Baseline performance. Note that our proposed model ($M^{KI}_{sar+emo}$) outperforms the developed baselines. *hloss* refers to Hamming Loss.

Task	STL				MTL			
	R	P	F1	hloss	R	P	F1	hloss
CNN+VGG19 [36]	48.71	44.91	45.12	–	53.32	58.01	51.52	29.16
BiLSTM+VGG19 [32]	46.41	46.12	42.85	–	51.52	52.67	51.62	30.81
mBERT+VGG19 [29]	45.82	49.61	45.23	–	48.66	48.79	47.80	31.30
mBERT+ViT [9]	62.88	64.12	60.69	–	63.29	64.63	61.24	14.64
VisualBERT [23]	64.74	63.66	60.64	–	62.44	63.09	60.38	14.10
LXMERT [38][1]	60.38	60.66	60.47	–	60.09	60.31	60.10	14.75
Ours $M^{KI}_{sar+emo}$	**63.15**	**64.01**	**63.29**	–	**63.11**	**65.01**	**63.37**	**13.11**

[1] For training both VisualBERT and LXMERT, we automatically translated the Hindi textual part of our dataset to English using MBart model [25]

Baselines. In Table 6, we show the baseline results for both STL and MTL setups. Majorly the developed baselines are categorized into two categories: i) *Late fusion models:* Here, visual representations are separately encoded by CNN or ViT, and then they are fused together (the first four baselines in Table 6),

ii) *Early fusion models:* Here the visual and textual information are fused in the beginning before encoding them via a transformer architecture (last two baselines in Table 6). We observe that our proposed method outperforms all the baselines both in sarcasm detection (by F1 score) and emotion recognition tasks (by Hamming Loss). To determine if these performance gains are statistically significant, we used *McNemar's test* [27], which is a widely used hypothesis testing method used in NLP [10,22,42]. We observe that the performance gains are statistically significant with p-values < 0.05.

6   Analysis

We perform a detailed quantitative and qualitative analysis of some samples from the test set. In Table 8, we show 3 examples with true labels of the sarcasm class. We compare models for both STL and MTL setups by comparing their predicted labels with actual labels. We observe that the MTL model with KI objective $(M_{sar+emo}^{KI})$ helps to capture related information from the meme to correctly predict the associated sarcasm class. To analyze whether the multi-modality helps in the context of detecting sarcasm, we also analyze two predicted examples in the left portion of Fig. 6. In the first example, we see that the text-only (T) model fails to detect sarcasm, whereas the multimodal (T+V) model correctly classifies it. The text *'Come, brother, beat me'* alone is not sarcastic, but whenever we add Mahatma Gandhi's picture as a context, the meme becomes sarcastic. This is correctly classified by the multimodal (T+V) M_{sar} model. Similarly, in the second example, without textual context, the image part is non-sarcastic, and thus, the vision only (V) M_{sar} model wrongly classifies this meme as non-sarcastic. Adding textual context helps the multimodal model to correctly classify this meme as a sarcastic meme. To explain the prediction behavior of our model, we use a well-known model-agnostic interpretability method known as LIME (Locally Interpretable Model-Agnostic Explanations) [33]. In the right portion of Fig. 6, we observe the parts of the meme and texts where the model focuses on making predictions, which aligns with our intuition. We also observe that despite the strong performance of our proposed model, it still fails to predict the sarcasm class correctly in a few cases. In Table 7, we show some of the memes with actual and predicted sarcasm labels from the multimodal (T+V) framework $(M_{sar}, M_{sar}^{KI}, M_{sar+emo}, M_{sar+emo}^{KI},)$. We show the six most common reasons why the models are failing to predict the actual class associated with the memes (Fig. 5).

Table 7. Error analysis: error cases and the possible reasons frequently occurring with each of them

Meme name	Ssarcasm class					Possible reason
	Act	M_{sar}	M_{sar}^{KI}	$M_{sar+emo}$	$M_{sar+emo}^{KI}$	
meme1	0	2	2	2	2	Hazy picture
meme2	0	2	1	2	2	Uninformative picture
meme3	0	2	2	2	2	Background knowledge
meme4	0	1	1	1	1	Common sense
meme5	1	2	2	2	2	Hindi words in English font
meme6	2	1	1	0	1	Code mixing

Fig. 5. Associated memes with the previous Table 7

Table 8. Sample test examples with predicted sarcasm label for STL and MTL models. Label definition: **2**: Highly Sarcastic, **1**: Mildly Sarcastic, **0**: Not Sarcastic.

		Sample 1	Sample 2	Sample 3
	True Label	2	1	0
STL	M_{sar}	0	2	1
	M_{sar}^{KI}	2	0	1
MTL	$M_{sar+emo}$	1	2	2
	$M_{sar+emo}^{KI}$	2	1	0

T	T+V	Image	Text
×	✓		ओ भई मारो मुझे मारो। Come brother, Beat me

V	T+V	Image	Text
×	✓		भारत माता की जय बोलूँ तो जीतने दोगे ? Will you let me win, if I say "Long Live Mother India"

(a) Multimodal (T+V) vs
Unimodal (V) M_{sar}

(b) LIME Visualization for
explainability

Fig. 6. Left: Two examples where we show multimodal (T+V) M_{sar} model performs better than unimodal (T and V only) M_{sar} models., **Right:** LIME outputs for the model explanation.

7 Conclusion

In this paper, we propose a *multitask knowledge-infused (KI)* model that leverages emotions and sentiment to identify the presence of sarcasm in a meme. To this end, we manually create a large-scale benchmark dataset by annotating 7,416 memes for sarcasm and emotion. Detailed error analyses and ablation studies show the efficacy of our proposed model, which produces promising results compared to the baseline models. Our analysis has found that the model could not perform well enough in a few cases due to the lack of contextual knowledge. In the future, we intend to explore about including background context to solve this problem more effectively.

Acknowledgement. The research reported in this paper is an outcome of the project "**HELIOS-Hate, Hyperpartisan, and Hyperpluralism Elicitation and Observer System**", sponsored by Wipro.

References

1. Bayerl, P.S., Paul, K.I.: What determines inter-coder agreement in manual annotations? A meta-analytic investigation. Comput. Linguist. **37**(4), 699–725 (2011). https://doi.org/10.1162/COLI_a_00074, https://aclanthology.org/J11-4004
2. Bernadt, M., Emmanuel, J.: Diagnostic agreement in psychiatry. Br. J. Psychiatry: J. Mental Sci. **163**, 549–50 (1993). https://doi.org/10.1192/S0007125000034012
3. Boland, K., Wira-Alam, A., Messerschmidt, R.: Creating an annotated corpus for sentiment analysis of German product reviews. GESIS-Technical reports, 2013/05 (2013)
4. Bouazizi, O.: Tomoaki: a pattern-based approach for sarcasm detection on twitter. IEEE Access **4**, 5477–5488 (2016). https://doi.org/10.1109/ACCESS.2016.2594194
5. Castro, S., Hazarika, D., Pérez-Rosas, V., Zimmermann, R., Mihalcea, R., Poria, S.: Towards multimodal sarcasm detection (an _obviously_ perfect paper). CoRR abs/1906.01815 (2019). http://arxiv.org/abs/1906.01815

6. Chauhan, D.S., Dhanush, S.R., Ekbal, A., Bhattacharyya, P.: Sentiment and emotion help sarcasm? A multi-task learning framework for multi-modal sarcasm, sentiment and emotion analysis. In: Proceedings of the 58th Annual Meeting of the Association for Computational Linguistics, pp. 4351–4360. Association for Computational Linguistics, Online, July 2020. https://doi.org/10.18653/v1/2020.acl-main.401, https://aclanthology.org/2020.acl-main.401

7. Davidson, T., Bhattacharya, D., Weber, I.: Racial bias in hate speech and abusive language detection datasets. In: Proceedings of the Third Workshop on Abusive Language Online, pp. 25–35. Association for Computational Linguistics, Florence, August 2019. https://doi.org/10.18653/v1/W19-3504, https://aclanthology.org/W19-3504

8. Dong, X., Li, C., Choi, J.D.: Transformer-based context-aware sarcasm detection in conversation threads from social media. CoRR abs/2005.11424 (2020). https://arxiv.org/abs/2005.11424

9. Dosovitskiy, A., et al.: An image is worth 16x16 words: transformers for image recognition at scale (2020). https://doi.org/10.48550/ARXIV.2010.11929, https://arxiv.org/abs/2010.11929

10. Dror, R., Baumer, G., Shlomov, S., Reichart, R.: The Hitchhiker's guide to testing statistical significance in natural language processing. In: Proceedings of the 56th Annual Meeting of the Association for Computational Linguistics (Volume 1: Long Papers). pp. 1383–1392. Association for Computational Linguistics, Melbourne, Australia, July 2018. https://doi.org/10.18653/v1/P18-1128, https://aclanthology.org/P18-1128

11. Ekman, P., Cordaro, D.T.: What is meant by calling emotions basic. Emot. Rev. **3**, 364–370 (2011)

12. Ghosal, D., Akhtar, M.S., Chauhan, D., Poria, S., Ekbal, A., Bhattacharyya, P.: Contextual inter-modal attention for multi-modal sentiment analysis. In: Proceedings of the 2018 Conference on Empirical Methods in Natural Language Processing, pp. 3454–3466. Association for Computational Linguistics, Brussels, October–November 2018. https://doi.org/10.18653/v1/D18-1382, https://aclanthology.org/D18-1382

13. Ghosh, D., Fabbri, A.R., Muresan, S.: The role of conversation context for sarcasm detection in online interactions. CoRR abs/1707.06226 (2017). http://arxiv.org/abs/1707.06226

14. Hasan, M.K., et al.: UR-FUNNY: a multimodal language dataset for understanding humor. CoRR abs/1904.06618 (2019). http://arxiv.org/abs/1904.06618

15. He, S., Zheng, X., Wang, J., Chang, Z., Luo, Y., Zeng, D.: Meme extraction and tracing in crisis events. In: 2016 IEEE Conference on Intelligence and Security Informatics (ISI), pp. 61–66 (2016). https://doi.org/10.1109/ISI.2016.7745444

16. Joshi, A., Bhattacharyya, P., Carman, M.J.: Automatic sarcasm detection: a survey. ACM Comput. Surv. **50**(5), 73:1–73:22 (2017). https://doi.org/10.1145/3124420

17. Joshi, A., Tripathi, V., Bhattacharyya, P., Carman, M.J.: Harnessing sequence labeling for sarcasm detection in dialogue from TV series 'Friends'. In: Proceedings of The 20th SIGNLL Conference on Computational Natural Language Learning, pp. 146–155. Association for Computational Linguistics, Berlin, August 2016. https://doi.org/10.18653/v1/K16-1015, https://aclanthology.org/K16-1015

18. Kiela, D., Firooz, H., Mohan, A., Goswami, V., Singh, A., Ringshia, P., Testuggine, D.: The hateful memes challenge: detecting hate speech in multimodal memes. CoRR abs/2005.04790 (2020). https://arxiv.org/abs/2005.04790

19. Kingma, D.P., Ba, J.: Adam: a method for stochastic optimization. CoRR abs/1412.6980 (2015)
20. Kosti, R., Alvarez, J.M., Recasens, A., Lapedriza, A.: Emotic: emotions in context dataset. In: 2017 IEEE Conference on Computer Vision and Pattern Recognition Workshops (CVPRW), pp. 2309–2317 (2017). https://doi.org/10.1109/CVPRW.2017.285
21. krippendorff, k.: Computing Krippendorff's alpha-reliability, January 2011
22. Kumar, A., Joshi, A.: Striking a balance: alleviating inconsistency in pre-trained models for symmetric classification tasks. In: Findings of the Association for Computational Linguistics: ACL 2022, pp. 1887–1895. Association for Computational Linguistics, Dublin, May 2022. https://doi.org/10.18653/v1/2022.findings-acl.148, https://aclanthology.org/2022.findings-acl.148
23. Li, L.H., Yatskar, M., Yin, D., Hsieh, C.J., Chang, K.W.: VisualBERT: a simple and performant baseline for vision and language (2019). https://doi.org/10.48550/ARXIV.1908.03557, https://arxiv.org/abs/1908.03557
24. Liu, L., Priestley, J.L., Zhou, Y., Ray, H.E., Han, M.: A2text-net: a novel deep neural network for sarcasm detection. In: 2019 IEEE First International Conference on Cognitive Machine Intelligence (CogMI), pp. 118–126 (2019). https://doi.org/10.1109/CogMI48466.2019.00025
25. Liu, Y., et al.: Multilingual denoising pre-training for neural machine translation (2020). https://doi.org/10.48550/ARXIV.2001.08210, https://arxiv.org/abs/2001.08210
26. Majumder, N., Poria, S., Peng, H., Chhaya, N., Cambria, E., Gelbukh, A.: Sentiment and sarcasm classification with multitask learning. IEEE Intell. Syst. **34**, 38–43 (2019). https://doi.org/10.1109/MIS.2019.2904691
27. McNemar, Q.: Note on the sampling error of the difference between correlated proportions or percentages. Psychometrika **12**(2), 153–157 (1947). https://doi.org/10.1007/bf02295996
28. Öhman, E.: Emotion annotation: rethinking emotion categorization. In: DHN Post-Proceedings (2020)
29. Pires, T., Schlinger, E., Garrette, D.: How multilingual is multilingual BERT? (2019). https://doi.org/10.48550/ARXIV.1906.01502, https://arxiv.org/abs/1906.01502
30. Poria, S., Cambria, E., Hazarika, D., Vij, P.: A deeper look into sarcastic tweets using deep convolutional neural networks. CoRR abs/1610.08815 (2016). http://arxiv.org/abs/1610.08815
31. Radford, A., et al.: Learning transferable visual models from natural language supervision. CoRR abs/2103.00020 (2021). https://arxiv.org/abs/2103.00020
32. Rhanoui, M., Mikram, M., Yousfi, S., Barzali, S.: A CNN-BILSTM model for document-level sentiment analysis. Mach. Learn. Knowl. Extract. **1**(3), 832–847 (2019). https://doi.org/10.3390/make1030048, https://www.mdpi.com/2504-4990/1/3/48
33. Ribeiro, M.T., Singh, S., Guestrin, C.: "why should I trust you?": Explaining the predictions of any classifier. In: Proceedings of the 22nd ACM SIGKDD International Conference on Knowledge Discovery and Data Mining, San Francisco, CA, USA, 13–17 August 2016, pp. 1135–1144 (2016)
34. Rockwell, P., Theriot, E.M.: Culture, gender, and gender mix in encoders of sarcasm: a self-assessment analysis. Commun. Res. Rep. **18**(1), 44–52 (2001). https://doi.org/10.1080/08824090109384781, https://doi.org/10.1080/08824090109384781
35. Sharma, C., et al.: SemEval-2020 task 8: memotion analysis - the visuo-lingual metaphor! CoRR abs/2008.03781 (2020). https://arxiv.org/abs/2008.03781

36. Simonyan, K., Zisserman, A.: Very deep convolutional networks for large-scale image recognition. CoRR abs/1409.1556 (2015)
37. Suryawanshi, S., Chakravarthi, B.R., Arcan, M., Buitelaar, P.: Multimodal meme dataset (MultiOFF) for identifying offensive content in image and text. In: Proceedings of the Second Workshop on Trolling, Aggression and Cyberbullying, pp. 32–41. European Language Resources Association (ELRA), Marseille, France, May 2020. https://aclanthology.org/2020.trac-1.6
38. Tan, H., Bansal, M.: LXMERT: learning cross-modality encoder representations from transformers (2019). https://doi.org/10.48550/ARXIV.1908.07490, https://arxiv.org/abs/1908.07490
39. Tsur, O., Rappoport, A.: RevRank: a fully unsupervised algorithm for selecting the most helpful book reviews. In: ICWSM (2009)
40. Venkatesan, R., Er, M.J.: Multi-label classification method based on extreme learning machines. In: 2014 13th International Conference on Control Automation Robotics Vision (ICARCV), pp. 619–624 (2014). https://doi.org/10.1109/ICARCV.2014.7064375
41. Wilson, P.A., Lewandowska, B.: The Nature of Emotions. In: Cambridge University Press (2012)
42. Zhu, H., Mak, D., Gioannini, J., Xia, F.: NLPStatTest: a toolkit for comparing NLP system performance. In: Proceedings of the 1st Conference of the Asia-Pacific Chapter of the Association for Computational Linguistics and the 10th International Joint Conference on Natural Language Processing: System Demonstrations, pp. 40–46. Association for Computational Linguistics, Suzhou, China, December 2020. https://aclanthology.org/2020.aacl-demo.7

Multilingual Detection of Check-Worthy Claims Using World Languages and Adapter Fusion

Ipek Baris Schlicht[1,2]([✉]) [iD], Lucie Flek[3] [iD], and Paolo Rosso[1] [iD]

[1] PRHLT Research Center, Universitat Politècnica de València, Valencia, Spain
ibarsch@doctor.upv.es, prosso@dsic.upv.es
[2] DW Innovation, Essen, Germany
[3] CAISA Lab, University of Marburg, Frankfurt, Germany
lucie.flek@uni-marburg.de

Abstract. Check-worthiness detection is the task of identifying claims, worthy to be investigated by fact-checkers. Resource scarcity for non-world languages and model learning costs remain major challenges for the creation of models supporting multilingual check-worthiness detection.

This paper proposes cross-training adapters on a subset of world languages, combined by adapter fusion, to detect claims emerging globally in multiple languages. (1) With a vast number of annotators available for world languages and the storage-efficient adapter models, this approach is more cost efficient. Models can be updated more frequently and thus stay up-to-date. (2) Adapter fusion provides insights and allows for interpretation regarding the influence of each adapter model on a particular language.

The proposed solution often outperformed the top multilingual approaches in our benchmark tasks.

Keywords: Fact-checking · Checkworthiness detection · Mutilingual · Adapters

1 Introduction

There is an increasing demand for automated tools that support fact-checkers and investigative journalists, especially in the event of breaking or controversial news [11,19]. Identifying and prioritizing claims for fact-checking, aka. check-worthy (CW) claim detection, is the first task of such automated systems [9]. This task helps guiding fact-checkers to potentially harmful claims for further investigation. CW claims, as shown in Table 1, are verifiable, of public interest and may invoke emotional responses [20]. Most studies in this area focus on monolingual approaches, predominantly using English datasets for learning the model. Support for multilingualism has become an essential feature for fact-checkers who investigate non-English resources [8].

This paper is accepted at ECIR 2023.

© The Author(s), under exclusive license to Springer Nature Switzerland AG 2023
J. Kamps et al. (Eds.): ECIR 2023, LNCS 13980, pp. 118–133, 2023.
https://doi.org/10.1007/978-3-031-28244-7_8

Table 1. An example of a check-worthy claim from CT21.

Those who falsely claim Vaccines are 100% safe and who laugh at anti-vaxxers will be dependent on those who refuse to tell the official lies to say that wide-spread vaccination *is* critical. 23 die in Norway after receiving COVID-19 vaccine: [URL] via [MENTION]

Although transformers achieved competitive results on several multilingual applications [3,4] including CW detection [18,21], there are still two main challenges in the CW task: (1) Because of the task complexity, there are a few publicly available datasets in multiple languages. Updating a multilingual model to detect even recently emerged claims requires data annotation and timely retraining. Finding fact-checking experts to annotate samples would be hard, especially in low-resourced languages. (2) Storing standalone copies of each fine-tuned model for every supported language requires vast storage capacities. Because of the limited budgets of non-profit organizations and media institutes, it would be infeasible to update the models frequently.

World Languages (WLs) are languages spoken in multiple countries, while non-native speakers can still communicate with each other using the WL as a foreign language. English, Arabic and Spanish are examples of WLs[1]. It would appear that finding expert annotators for collecting samples in a WL is easier than finding expert annotators for low-resourced languages.

As a resource-efficient alternative to fully fine-tuning transformer models, adapters [14,31] have been recently proposed. Adapters are lightweight and modular neural networks, learning tasks with fewer parameters and transferring knowledge across tasks [14,22,31] as well as languages [24]. Fine-tuned adapters require less storage than fully fine-tuned pre-trained models.

In this paper, we propose cross-lingual training of datasets in WLs with adapters to mitigate resource scarcity and provide a cost-efficient solution. We first train Task Adapters (TAs) [23] for each WL and incorporate an interpretable Adapter Fusion (AF) [22] to combine WL TAs for an effective knowledge transfer among the heterogeneous sources.

Our contributions for this paper are summarized as follows[2]:

- We extensively analyze the WL AF models on the multilingual CW claim detection task and evaluate the models on zero-shot learning (i.e., the target languages were unseen during training). We show that the models could perform better than monolingual TAs and fully fine-tuned models. They also outperformed the related best performing methods in some languages. In addition, zero-shot learning is possible with the WL AF models.
- We construct an evaluation to quantify the performance of the models on claims about global/local topics across the languages. Our approach for curating the evaluation set could be reused for the assessment of other multilingual, social tasks.

---

[1] https://bit.ly/3eMIZ9q.
[2] We share our source code at https://bit.ly/3rH6yXu.

– We present a detailed ablation study for understanding the limitations of AF models and their behavior on the CW task.

2 Related Work

2.1 Identifying Check-Worthy Claims

Early studies applied feature engineering to machine learning models and neural network models to identify CW claims in political debates and speeches. Claim-Buster [12,13] was one of the first approaches using a Support Vector Machine (SVM) classifier trained on lexical, affective feature sets. Gencheva et al. [7] combined sentence-level and contextual information from political debates for training neural network architectures. Lespagnol et al. [17] employed information nutritional labels [6] and word embeddings as features. Vasileva et al. [35] applied multitask learning from different fact-checking organizations to decide whether a statement is CW. Jaradat et al. [15] used MUSE word embeddings to support the CW detection task in English and Arabic.

CheckThat! (CT) organized multilingual CW detection tasks since 2020 [2, 18,21]. They support more languages every year and an increasing number of multilingual systems have been submitted. Schlicht et al. [28] proposed a model supporting all languages in the dataset. They used a multilingual sentence transformer [26] and then fine-tuned the model jointly on a language identification task. Similarly, Uyangodage et al. [34] fine-tuned mBERT on a dataset containing balanced samples for each language. Recently, Du et al. [5] fine-tuned mT5 [37] on the CW detection task, jointly with multiple related tasks and languages by inserting prompts. All of the listed methods were limited to the languages they were trained on.

Kartal and Kutlu [16] evaluated mBERT for zero-shot learning in English, Arabic and Turkish, and observed the performance drop in cross-training. A more effective method would be required for cross-lingual training. Furthermore, none of the approaches tackled the resource efficiency issue. In this paper, we use adapters and train them only on WLs for resource efficiency and evaluate unseen languages, leveraging AF to understand which target language in the dataset benefits from transferring knowledge from WLs.

2.2 Adapters

Adapters have been successfully applied to pre-trained models for efficient fine-tuning to various applications. Early studies [14,22] used adapters for task adaptation in English. Pfeiffer et al. [22] propose an AF module for learning to combine TAs as an alternative to multi-task learning. Some researchers [24,25,33] exploit language-specific adapters for cross-lingual transfer. This paper builds upon the works of [22,25]. We exploit a cross-lingual training set to learn TAs and then use AF to combine them effectively and provide interpretability on which source TAs are efficient on the target language.

Table 2. Statistics of CT21 and CT22.

	Split	CT21 Total	CT21 %CW	CT22 Total	CT22 %CW
ar	Train	3444	22.15	2513	38.28
	Dev	661	40.09	235	42.55
	Test	600	40.33	682	35.28
es	Train	2495	8.02	4990	38.14
	Dev	1247	8.74	2500	12.20
	Test	1248	9.62	4998	14.09
en	Train	822	35.28	2122	21.07
	Dev	140	42.86	195	22.56
	Test	350	5.43	149	26.17
tr	Test	1013	18.07	303	4.62
bg	Test	357	21.29	130	43.85
nl	Test	-	-	666	47.45

Table 3. Statistics of Global and Local Topics

	Split	CT21 Total	CT21 %CW	CT22 Total	CT22 %CW
ar	Global	269	43.12	116	46.55
	Local	40	42.5	-	-
es	Global	1208	9.93	148	22.97
	Local	-	-	917	12.43
nl	Global	-	-	103	56.31
	Local	-	-	15	46.67
en	Global	349	5.44	14	28.57
tr	Global	887	8.91	-	-
bg	Global	356	21.35	25	32

3 Datasets

3.1 Task Datasets

We looked for multilingual datasets in WLs and other languages for the experiments. CT21 [29] and CT22 [18] are the only publicly available datasets that meet this requirement. CT21 includes English, Arabic, Turkish, Spanish, and Bulgarian samples. It deals mainly with Covid-19 events, except for the Spanish samples, which focus only on politics. Compared with CT21, CT22 also includes samples in Dutch. The English, Arabic, Bulgarian, and Dutch samples in the dataset build on the corpora on Covid-19 [1]. The researchers collected new samples in Turkish and Spanish. The Spanish samples were augmented with CT21.

The statistics of the datasets are shown in Table 2. English, Spanish and Arabic are the WLs contained in the datasets. Both datasets are imbalanced, i,e. CW samples are under-represented across the languages. Although English is considered a high-resource language, there are considerably fewer English samples than samples in other languages. Since some samples of the datasets could overlap, we conducted our research experiments per dataset.

3.2 Topical Evaluation Dataset

Some countries are culturally dissimilar or might have different political agendas, thus CW topics might differ among countries. For example, while vaccination was a globally CW topic throughout the COVID-19 pandemic, some COVID-19 myths were believed only by a few communities [30]. An ideal multilingual system should perform well on global as well as local topics.

Global topics are the topics present in all languages in all datasets, while local topics are present in only one language. We created dedicated datasets to evaluate the performance of the WL models for identifying CW claims across global

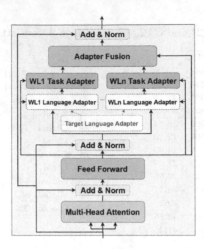

Fig. 1. The architecture of the WL+AF models within a transformer. WL+AF+LA is the setup when the LAs (blocks with the dashed lines) are stacked into the task adapters.

and local topics. This evaluation dataset contains solely global and local topics and was created as follows. To learn topics across datasets, we first translated the datasets into English and encoded them with a sentence transformer [26]. Second, we learned a topic modeling on the training datasets in the WLs by using BERTopic [10] to assign topics to test samples in all languages. Test samples with topics present in all WL languages were added as presumably global topics to the evaluation set. BERTopic labeled topics that are unrelated to the learnt topics with -1. To select samples with local topics, we chose the samples labeled with -1 from the evaluation and applied a new topic modeling on them. Test samples with topics that were not present in the test dataset of any other language (i.e., local topics) were added to the evaluation set for local topics. The statistics of the local and global evaluation sets are presented in Table 3.

4   Methodology

4.1   World Language Adapter Fusion

This section describes the WL+AF models. The WL+AF models are transformers containing additional modules called adapters. During training, only the adapters are trained and the weights of pre-trained model are frozen. Therefore, it is a parameter-efficient approach. We experiment with two types of WL+AF models. WL+AF is a standard setup for combining WL task adapters. WL+AF+LA has additional language adapters (LAs) [24] by stacking task adapters. We illustrate the architecture of the models in Fig. 1. The input of the architecture is a text while the output is a probability score describing the check-worthiness of the input.

Transformer Encoder. To provide cross-lingual representations, we use multilingual BERT (mBERT) [4] and XLM-Roberta (XLM-R) [3][3]. Because the transformers were previously used by the related studies [1,34]. Furthermore, there are publicly available LAs for the transformers. Before encoding text with transformers, we first clean it from URLs, mentions and retweets, then truncate or pad it to 128 tokens.

Task Adapter. The annotations might be affected by the different cultural backgrounds and events across the countries, like the CW claims. Additionally, some may have been created from journalists following manual fact-checking procedures while others stem from crowd-sourcing [18,21] For these reasons, we treat each WL dataset as a separate task. Then, we obtain TAs by optimizing on their corresponding WL dataset. TAs consist of one down projection layer followed by a ReLU activation and one up projection layer [23]. The TA of each WL is fine-tuned on its corresponding dataset.

Adapter Fusion. To share knowledge from the different TAs for predicting samples in an unseen language or new topics, we need to combine them effectively. AF is a method for combining multiple TAs, and mitigating the common issues of multi-task learning, such as catastrophic forgetting and complete model retraining for supporting new tasks [22]. AF consists of an attention module which learns how to combine knowledge from the TAs dynamically. By analyzing the attention weights of the AF, we could learn which source task adapter contributes more to test predictions. We combine the TAs trained on the WLs with an AF. Then, the AF is fine-tuned on mixed datasets of the WLs to use for the target languages.

Language Adapter. To learn CW claim detection from various cross-lingual sources, we should ensure that the model does not learn a language classification task but learns how to identify CW claims. Therefore, we need to separate language-specific knowledge from task-specific knowledge. A recent study [24] demonstrated that LAs achieved better results when transferring the knowledge of a task performed in one source language into another language. To encode language-specific information into the transformer model and to see the LAs impact on the performance of the AF models, we use the LAs from Adapter Hub [23]. The architecture of the LAs is analog to those of the TAs. However, these LAs were pre-trained on the unlabeled Wikipedia [24] by using a masked language model objective. While learning task-specific transformations with TAs, each world LA is stacked on its corresponding TA. The weights of the LAs are kept frozen, so they are not learned again. During inference of the target task, the source LA is replaced with the target LA.

4.2 Implementation Details

We download the pre-trained transformers from Huggingface [36]. We fine-tune the TAs on English, Arabic, and Spanish training samples (WLs), and evaluate

---

[3] We use the base version of the model, which consists of 12 layers.

the models for their capability of zero-shot learning by testing on the other languages. As LAs for Arabic, English, Spanish and Turkish samples in the dataset [24], we download the pre-trained LAs from Adapter Hub [23] for both transformers. However, a LA for Bulgarian and Dutch is not available in Adapter Hub, therefore, we use the English LA.

We use the trainers of AdapterHub [23] for both full fine-tuning and adapter tuning. We set the number of epochs as 10 and train the models with a learning rate of 1e-4 for CT21 and 2e-5 for CT22[4], and batch size of 16. We use the best models on the development set according to the dataset metrics. We repeat the experiments ten times using different random seeds using an NVIDIA GeForce RTX 2080.

5 Baselines

We compare the performance of the WL AF models against various baselines:

- **Top performing systems:** We chose the top-performing systems on CT21 and CT22 which used a single model for the multilingual CW detection instead of containing language-specific models. Schlicht et al. [28] is a runner-up[5] system on CT21. The original implementation uses a sentence transformer and the model was fine-tuned on all the language datasets. For a fair comparison, we set the language identification task to WLs and replaced the sentence transformer with mBERT and XLM-R. The state-of-the-art model on CT22 is based on mT5-xlarge [5], a multi-task text-to-text transformer. Like Schlicht et al., the model is fine-tuned on all of the corpora by using multi-task learning. Due to the limited computing resources, we couldn't fine-tune this model on WLs alone. We report the results from [5].
- **Fully fine tuned (FFT) Transformers (on Single Language):** We fine-tune the datasets on a single language of the WLs to evaluate the efficiency of cross-lingual learning: AR+FFT, EN+FFT, and ES+FFT. We also add BG+FFT, ES+FFT, and NL+FFT as the baseline for zero-shot learning.
- **Task Adapters (on Single Language):** These baselines contain a task adapter followed by a LA, a widely used setup for cross-lingual transfer learning with adapters [24]: AR+TA+LA, EN+TA+LA, ES+TA+LA. Comparing our model to these baselines can help understand whether cross-training with WLs is efficient.
- **Other WL Models:** We evaluate the AF models against a model containing a task adapter trained on WLs (WL+TA). In addition to this baseline, to see if we need a complex fusion method, we use WL+TA+LA+Mean, which takes the average of the predictions by AR+TA, EN+TA and ES+TA. Finally, we analyze the adapter tuning on multiple WLs against the fully fine tuning of mBERT: WL+FFT.

---

[4] 2e-5 gives better results on the development set of CT22.

[5] BigIR is the state of art approach, but there is no associated paper/code describing the system.

Table 4. MAP scores for CT21 and F1 scores for CT22 of the CW detection in WLs. The **bold** indicates the best score and <u>underline</u> indicates the second best score. Overall the AF models performed well on multiple languages while the performance of other models are sensitive to the characteristics of the training set.

		CT21			CT22			
		ar	es	en	ar	es	en	avg
	Du et al. [5]	-	-	-	**62.8**	57.1	**51.9**	-
mBERT	AR+FFT	50.17	15.78	6.03	<u>55.52</u>	17.85	42.92	31.38
	ES+FFT	41.22	20.30	6.80	37.30	54.05	45.28	34.16
	EN+FFT	51.63	15.90	10.80	40.97	22.49	44.29	31.01
	AR+TA+LA	58.20	19.97	8.62	18.13	17.61	45.73	28.04
	ES+TA+LA	48.07	18.93	11.06	37.30	54.05	45.73	35.86
	EN+TA+LA	50.81	40.81	**21.21**	56.39	24.63	12.93	34.46
	WL+FFT	47.93	51.50	13.85	51.50	63.20	39.94	<u>44.65</u>
	Schlicht et al. [28]	51.51	31.04	7.87	45.93	<u>66.48</u>	34.18	39.50
	WL+TA	53.77	46.58	14.44	39.54	62.69	37.19	42.37
	WL+TA+LA+Mean	54.89	35.72	12.96	0.00	33.21	51.03	31.30
	WL+AF	55.13	46.29	16.05	36.45	64.32	39.73	42.96
	WL+AF+LA	<u>55.32</u>	46.58	15.66	39.87	<u>64.64</u>	37.27	43.22
XLM-R	AR+FFT	43.55	12.80	5.88	38.72	6.83	41.29	24.85
	ES+FFT	41.22	15.90	6.80	40.25	21.69	43.51	28.23
	EN+FFT	43.64	10.49	5.64	31.46	64.66	45.69	33.60
	AR+TA+LA	58.16	25.87	7.64	6.93	0.06	1.40	16.68
	ES+TA+LA	50.27	**52.51**	11.53	41.24	65.45	38.79	43.30
	EN+TA+LA	56.39	24.63	12.93	12.82	0.06	28.74	22.60
	WL+FFT	47.93	13.85	6.37	44.53	64.75	<u>50.93</u>	38.06
	Schlicht et al. [28]	51.56	21.61	7.32	42.59	**67.13**	36.40	37.77
	WL+TA	58.02	<u>50.76</u>	11.53	42.91	63.36	31.69	43.05
	WL+TA+LA+Mean	**59.32**	46.11	10.49	25.32	35.55	35.28	35.35
	WL+AF	58.39	49.42	13.29	39.84	65.66	46.96	**45.59**
	WL+AF+LA	<u>58.83</u>	47.26	<u>16.06</u>	35.17	65.80	43.46	44.43

6   Results and Discussion

In this section, we present and analyze the results of the WL AF(+LA) models. We compare the models performance at CW detection for (1) WLs (2) zero-shot languages and (3) local and global topics. Lastly, we compare the performance of WL+AF and WL+AF+LA to investigate whether LA is effective in model performance.

As seen in Table 4, the models trained on single languages are able to perform well for other WLs if only provided with training sets of considerable size, or language of the training and test sets are same. Additionally, Schlicht et al. [28] and WL+FFT were performing well only on CT22, overall, the AF models, perform well for various languages. Du et al. [5] outperformed the AF models for Arabic and English samples of CT22, but it underperformed for the Spanish samples.

As shown in Table 5, the AF models achieve good results on target sets in zero-shot languages. It shows that the fusion of multiple sources with adapters could be beneficial in knowledge transfer and is better than the other fusion

Table 5. MAP for **CT21** and F1 for **CT22** of the CW detection in zero-shot languages. The **bold** indicates the best score and underline indicates the second best score. The AF models performed well, even outperformed WL+FFT and Schlicht et al. [28] and some of the monolingual approaches in terms of average score.

		CT21		CT22			
		tr	bg	tr	bg	nl	avg
	Du et al. [5]	-	-	17.3	61.7	**64.2**	-
mBERT	BG+FFT	-	35.64	-	57.16	-	-
	TR+FFT	28.47	-	14.66	-	-	-
	NL+FFT	-	-	-	-	49.80	-
	AR+FFT	30.14	22.12	10.71	54.84	45.55	32.67
	ES+FFT	23.17	24.15	8.19	47.67	33.82	27.4
	EN+FFT	42.80	33.20	13.57	54.58	54.32	39.69
	AR+TA+LA	58.08	26.76	8.04	57.43	58.99	41.86
	ES+TA+LA	49.09	28.09	8.58	47.67	42.16	35.12
	EN+TA+LA	54.16	44.98	8.19	33.92	33.82	35.01
	WL+FFT	27.61	24.29	10.90	58.53	29.03	30.07
	Schlicht et al. [28]	27.81	24.41	7.86	51.94	29.33	28.27
	WL+TA	54.11	37.04	9.73	48.02	36.95	37.17
	WL+TA+LA+Mean	62.32	34.52	12.02	47.17	39.91	39.19
	WL+AF	50.46	39.80	9.63	53.55	38.76	38.44
	WL+AF+LA	50.94	40.27	9.73	52.75	43.07	39.35
XLM-R	BG+FFT	-	24.68	-	43.47	-	-
	TR+FFT	24.57	-	**18.90**	-	-	-
	NL+FFT	-	-	-	-	58.61	-
	AR+FFT	23.40	21.86	11.62	45.10	38.73	28.14
	ES+FFT	23.17	24.15	9.07	62.72	31.24	30.07
	EN+FFT	22.63	21.76	18.43	49.25	45.62	31.54
	AR+TA+LA	56.19	21.37	0.48	9.75	5.22	18.60
	ES+TA+LA	44.98	23.86	8.04	46.72	30.52	30.82
	EN+TA+LA	58.38	41.61	15.19	8.42	25.83	29.89
	WL+FFT	27.61	24.29	14.02	63.79	36.34	33.21
	Schlicht et al. [28]	25.86	22.19	9.55	53.75	24.44	27.16
	WL+TA	59.37	39.72	15.60	**66.14**	35.98	43.36
	WL+TA+LA+Mean	63.65	32.28	9.90	31.27	19.91	31.40
	WL+AF	57.46	46.86	12.73	59.12	40.83	43.4
	WL+AF+LA	**61.74**	41.78	17.77	63.88	37.59	**44.55**

Table 6. F1 scores of the models on global topics for each dataset. Adapter training is more effective than fully fine-tuning. Although WL+TA outperformed the AF models in particular languages, at average the AF models performed better.

	CT21					CT22					avg
	tr	es	ar	en	bg	es	ar	en	bg	nl	
WL+FFT	0.00	0.00	2.60	0.59	0.10	77.59	**52.13**	**52.05**	**48.52**	38.53	27.21
Schlicht et al. [28]	18.15	17.92	58.84	7.18	19.75	**81.81**	48.12	28.33	36.55	30.34	34.70
WL+TA	**52.37**	38.09	61.01	**13.58**	**38.44**	77.28	45.47	22.57	41.31	**45.09**	38.52
WL+AF	45.02	**41.57**	61.51	13.36	36.23	79.14	44.61	50.96	40.58	43.19	**45.62**
WL+AF+LA	48.52	37.44	**61.79**	13.38	28.72	77.95	42.84	44.23	41.09	43.99	44.00

Table 7. F1 scores of the models on local topics for each dataset. The AF models show similar results to WL+TA and outperformed WL+FFT.

	CT21	CT21		
	ar	es	nl	avg
WL+FFT	0.93	60.64	31.11	30.89
Schlicht et al. [28]	46.80	**64.65**	14.44	41.96
WL+TA	**50.88**	61.65	38.47	**50.33**
WL+AF	48.15	62.01	34.38	48.18
WL+AF+LA	46.41	63.07	**41.16**	50.21

Table 8. Number of training parameters and file size comparisons for the models. mT5 is larger than mBERT and XLM-R.

Model	Base model	Parameters	Model size
WL+FFT	mBERT	178 M	711.5 MB
	XLM-R	278 M	1.1 GB
Schlicht et al. [28]	mBERT	179 M	716.3 MB
	XLM-R	279 M	1.1 GB
TA & WL+TA	mBERT	1.5 M	6 MB
	XLM-R	1.5 M	6 MB
AF	mBERT	22 M	87.4 MB
	XLM-R	22 M	87.4 MB
LA	mBERT	-	147.78 MB
	XLM-R	-	147.78 MB
	mT5	3.7 B	15 GB

method WL+TA+LA+Mean. It is noteworthy that Du et al [5] was trained on all samples of the training datasets and hence has no zero-shot learning capacity. Although Du et al. achieved a better performance on the Dutch samples, the AF models could obtain similar results in other languages. In terms of resource efficiency, the AF models required less space than WL+FFT and mT5 for storing new weights, as shown in Table 8, which make them more suitable than updating mT5 for newsrooms with a limited budget.

We compare the performance of models trained on multiple WLs for identifying CW claims about global or local topics. We tested this experiment with the evaluation set described in Sect. 3.2 in terms of F1 score. We take the average of the scores of the models coded with mBERT and XLM-R and present them in Tables 6 and 7, respectively, for global and local topics. Overall, the AF models performed better than WL+FFT and Schlicht et al. [28] for both types. However, WL+TA performed similarly to WL+AF+LA in predicting local statements in Arabic samples in CT21.

Last, we compare the performance of WL+AF and WL+AF+LA to investigate whether LA is effective in model performance. We computed the Fleiss Kappa scores of the AF models for each experiment and language. The overall score is 0.63, which is a moderate agreement. We further investigate the disagreements where the kappa is below 0.5. The conflicts mainly occurred in the zero shot languages and English, with the lowest CW samples on both datasets. Since sometimes WL+AF+LA is better than WL+AF and vice versa, we conclude that LA is not effective in our experiments. The pre-trained LAs were trained on the

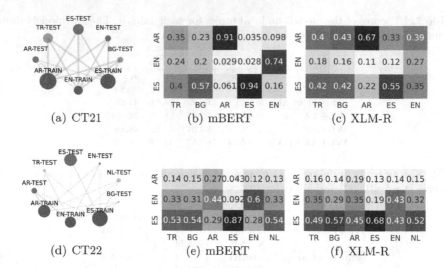

Fig. 2. The left images are topical relation graphs of CT21 (top left) and CT22 (bottom left). In the graphs, the size of the nodes varies by the number of samples, and the edge thickness depends on the overlapping topics. x-axis of heatmaps shows task adapters, y-axis shows the test samples in the different languages. (b) (c) attention heatmaps of mBERT (e) (f) attention heatmaps of XLM-R. Topical distribution, and the sample sizes of the training datasets impact the task adapters' activations. Especially XLM-R TAs are more sensitive than mBERT.

Wikipedia texts [24]. Thus, they might miss the properties of social texts, which are mostly noisy.

7   Further Analysis

In this section, we present further analysis of the AF models. We investigate AF attentions and then apply an error analysis on the models' predictions.

Interpretation of the Fusion Attentions. The AF models can provide an interpretation of which source task adapter might be useful when transferring the knowledge into the target dataset. This kind of analysis would help a data scientist at a newsroom on a decision on which WL should be collected for updating model and managing new resources. To check the AF behavior on WL+TA+LA+AF, we took the average of the softmax probabilities of the layer of each task adapter in the fusion layer. The higher probability means the more useful the task for determining the label [22]. In addition, to correlate the attention with the source datasets, we created a graph displaying the topical relationship between source and target sets. In the graph, the nodes are the monolingual datasets; the edges are the overlapped topics between the source and target dataset, weighted by the percentage of the samples about the overlapped topic. The size of nodes are scaled according to sample size. Figure 2 shows

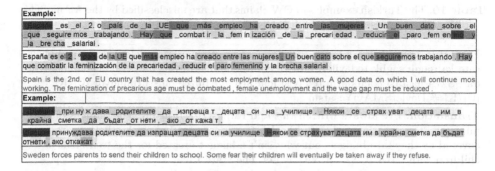

Fig. 3. The CW claims predicted correctly by the WL+TA+AF, the examples are in Spanish and Bulgarian. The order of the texts for each example: 1) Visualizations on mBERT, 2) Visualizations on XLM-R 3) English translation. The models focus more on GPE (e.g. country names) than the other entity types. We colorized the claims based on their integrated gradients [32].

Table 9. The performance of the AF model at predicting entity types in terms of average F1 and the standard deviation. The models could predict GPE more accurately than the others.

	Geo-political Entity	Organization	Number	People
F1	46.83 ± 17.37	40.44 ± 18.62	40.99 ± 22.52	37.88 ± 18.83

the graph for both datasets and the attention weights of mBERT and XLM-R task adapters. Topical distributions and source datasets' size affect which task adapter activates. XLM-R TAs are more sensitive to the source data size and topical relationship. For instance, the Spanish tests in CT22 are weakly connected with the Arabic and English source datasets, and the Spanish TA of XLM-R has less activation than the mBERT TA.

Error Analysis. Last, we analyze the misclassified/correctly classified samples by both AF models. As shown in Fig. 3, we noted that the AF model focuses on geo-political entities (GPE). The models could categorize the claims with GPE better than claims containing other type of entities as shown in Table 9. The importance of the GPE could be learned from the WL corpus whose CW samples have no negligible amount of these entities (e.g. %76 of CT21 and %77 of CT22 Arabic source datasets are GPE). However, the models cannot predict the claims requiring local context, especially in the zero-shot languages. Moreover, the models cannot identify the claims whose veracity can be changed to not CW by time. Some examples are shown in Table 10.

Training Efficiency. We measured the models' training time for one epoch on the datasets. The TA training is on average 4 min less than the fully fine tuning. However, the AF training without LAs lasts 3 min more, and the training with LAs 9 min more than the training time of WL+FFT which was approx. 22 min. The methods such as AdapterDrop [27] could speed up the AF training.

Table 10. The Turkish examples of CW claims that are misclassified by the AF models. Additionally, we present their English translations and claim explanations.

Example 1: Bunu hep yazdım yine yazacağım , Bakanların aileleri , annesi - babası tam kapsamlı sağlık tedavileri buna (Estetik dahil) devlet bütcesinden karşılanıyor da , SMA hastası çocukların tedavisi için niye bir bütçe oluşturulmuyor [UNK] [UNK] # DevletSMAyıYaşatsın

Translation: I've always written this and I will write it again. The families of the ministers, their mother-father full health treatments (including Aesthetics) are covered by the state budget, but why isn't a budget created for the treatment of children with SMA [UNK] [UNK] # Let The State Live

Explanation: Example of a CW claim that requires local context. SMA is a disease that affects children, and the treatment of SMA is a controversial issue in Turkey

---

Example 2: Koronavirüs salgınında vaka sayısı 30 bin 021 [UNK] e ulaştı # Corona # COVID # coronavirus

Translation: The number of cases in the coronavirus epidemic reached 30 thousand 021 [UNK] # Corona # COVID # coronavirus

Explanation: An example of a CW claim whose veracity could be changed by time

---

8   Conclusion and Future Work

In this paper, we investigated the cost efficient cross-training of adapter fusion models on world languages to detect check-worthiness in multiple languages. The proposed solution performs well on multiple languages, even on zero-shot learning. Thanks to adapter fusion, the effectiveness of the adapters on particular languages was possible.

The attention of some task adapters seems to depend on the topic and sample distribution in the source dataset. Ensuring a topical balance across world languages appears to be important. Our error analysis results indicate that local context is required to detect local claims. We recommend the usage of background knowledge injection to detect local claims.

In the future, we would like to investigate the injection of background knowledge in adapters and verify our results in additional domains (e.g. war), employing more languages such as German and focusing on zero-shot learning.

Acknowledgements. We would like to thank the anonymous reviewers, Joan Plepi, Flora Sakketou, Akbar Karimi and Nico Para for their constructive feedback. The work of Ipek Schlicht was part of the KID2 project (led by DW Innovation and co-funded by BKM). The work of Lucie Flek was part of the BMBF projects DeFaktS and DynSoDA. The work of Paolo Rosso was carried out in the framework of IBERIFIER (INEA/CEF/ICT/A202072381931 n.2020-EU-IA-0252), XAI Disinfodemics (PLEC2021-007681) and MARTINI (PCI2022-134990–2).

References

1. Alam, F., et al.: Fighting the covid-19 infodemic: modeling the perspective of journalists, fact-checkers, social media platforms, policy makers, and the society. In: Findings of the Association for Computational Linguistics: EMNLP 2021, pp. 611–649 (2021)
2. Barrón-Cedeño, A., et al.: Overview of CheckThat! 2020: Automatic Identification and Verification of Claims in Social Media. In: Arampatzisa, A., et al. (eds.) CLEF 2020. LNCS, vol. 12260, pp. 215–236. Springer, Cham (2020). https://doi.org/10.1007/978-3-030-58219-7_17
3. Conneau, A., et al.: Unsupervised cross-lingual representation learning at scale. In: Proceedings of the 58th Annual Meeting of the Association for Computational Linguistics, pp. 8440–8451 (2020)
4. Devlin, J., Chang, M., Lee, K., Toutanova, K.: BERT: pre-training of deep bidirectional transformers for language understanding. In: Burstein, J., Doran, C., Solorio, T. (eds.) In: Proceedings of the 2019 Conference of the North American Chapter of the Association for Computational Linguistics: Human Language Technologies, NAACL-HLT 2019, Minneapolis, MN, USA, June 2–7, 2019, (Long and Short Papers) 1, pp. 4171–4186. Association for Computational Linguistics (2019)
5. Du, S.M., Gollapalli, S.D., Ng, S.K.: Nus-ids at checkthat! 2022: identifying check-worthiness of tweets using checkthat5. Working Notes of CLEF (2022)
6. Fuhr, N.: An information nutritional label for online documents. SIGIR Forum **51**(3), 46–66 (2017)
7. Gencheva, P., Nakov, P., Màrquez, L., Barrón-Cedeño, A., Koychev, I.: A context-aware approach for detecting worth-checking claims in political debates. In: Mitkov, R., Angelova, G. (eds.) Proceedings of the International Conference Recent Advances in Natural Language Processing, RANLP 2017, Varna, Bulgaria, September 2–8, 2017, pp. 267–276. INCOMA Ltd. (2017)
8. Ginsborg, L., Gori, P.: Report on a survey for fact checkers on covid-19 vaccines and disinformation. Tech. rep., European Digital Media Observatory (EDMO) (2021). https://cadmus.eui.eu/handle/1814/70917
9. Graves, L.: Understanding the promise and limits of automated fact-checking (2018)
10. Grootendorst, M.: Bertopic: Neural topic modeling with a class-based TF-IDF procedure. arXiv preprint arXiv:2203.05794 (2022)
11. Hassan, N., Adair, B., Hamilton, J.T., Li, C., Tremayne, M., Yang, J., Yu, C.: The quest to automate fact-checking. In: Proceedings of the 2015 Computation Journalism Symposium (2015)
12. Hassan, N., Arslan, F., Li, C., Tremayne, M.: Toward automated fact-checking: Detecting check-worthy factual claims by claimbuster. In: Proceedings of the 23rd ACM SIGKDD International Conference on Knowledge Discovery and Data Mining, Halifax, NS, Canada, August 13–17, 2017, pp. 1803–1812. ACM (2017)
13. Hassan, N., Li, C., Tremayne, M.: Detecting check-worthy factual claims in presidential debates. In: Bailey, J., et al., (eds.) Proceedings of the 24th ACM International Conference on Information and Knowledge Management, CIKM 2015, Melbourne, VIC, Australia, October 19–23, 2015, pp. 1835–1838. ACM (2015)
14. Houlsby, N., et al.: Parameter-efficient transfer learning for NLP. In: ICML. In: Proceedings of Machine Learning Research, vol. 97, pp. 2790–2799. PMLR (2019)

15. Jaradat, I., Gencheva, P., Barrón-Cedeño, A., Màrquez, L., Nakov, P.: ClaimRank: detecting check-worthy claims in Arabic and English. In: Proceedings of the 2018 Conference of the North American Chapter of the Association for Computational Linguistics: Demonstrations, pp. 26–30. Association for Computational Linguistics, New Orleans, Louisiana (2018)

16. Kartal, Y.S., Kutlu, M.: Re-think before you share: A comprehensive study on prioritizing check-worthy claims. In: IEEE Trans. Comput. Soci. Syst. (2022)

17. Lespagnol, C., Mothe, J., Ullah, M.Z.: Information nutritional label and word embedding to estimate information check-worthiness. In: Proceedings of the 42nd International ACM SIGIR Conference on Research and Development in Information Retrieval. SIGIR'19, Assoc. Comput. Mach., New York, NY, USA, pp. 941–944. (2019)

18. Nakov, P., et al.: The clef-2022 checkthat! lab on fighting the COVID-19 infodemic and fake news detection. In: European Conference on Information Retrieval (2022)

19. Nakov, P., et al.: Automated fact-checking for assisting human fact-checkers. In: IJCAI. pp. 4551–4558. https://ijcai.org

20. Nakov, P., et al.: The CLEF-2021 CheckThat! Lab on Detecting Check-Worthy Claims, Previously Fact-Checked Claims, and Fake News. In: Hiemstra, D., et al. (eds.) ECIR 2021. LNCS, vol. 12657, pp. 639–649. Springer, Cham (2021). https://doi.org/10.1007/978-3-030-72240-1_75

21. Nakov, P., et al.: Overview of the CLEF–2021 CheckThat! Lab on Detecting Check-Worthy Claims, Previously Fact-Checked Claims, and Fake News. In: Candan, K., et al. (eds.) CLEF 2021. LNCS, vol. 12880, pp. 264–291. Springer, Cham (2021). https://doi.org/10.1007/978-3-030-85251-1_19

22. Pfeiffer, J., Kamath, A., Rücklé, A., Cho, K., Gurevych, I.: Adapterfusion: Non-destructive task composition for transfer learning. In: Merlo, P., Tiedemann, J., Tsarfaty, R. (eds.) In: Proceedings of the 16th Conference of the European Chapter of the Association for Computational Linguistics: Main Volume, EACL 2021, Online, April 19–23, 2021, pp. 487–503. Association for Computational Linguistics (2021)

23. Pfeiffer, J., et al.: AdapterHub: a framework for adapting transformers. In: Proceedings of the 2020 Conference on Empirical Methods in Natural Language Processing: System Demonstrations, pp. 46–54. Association for Computational Linguistics, Online (2020)

24. Pfeiffer, J., Vulić, I., Gurevych, I., Ruder, S.: MAD-X: an Adapter-Based Framework for Multi-Task Cross-Lingual Transfer. In: Proceedings of the 2020 Conference on Empirical Methods in Natural Language Processing (EMNLP), pp. 7654–7673. Association for Computational Linguistics, (2020)

25. Pfeiffer, J., Vulić, I., Gurevych, I., Ruder, S.: UNKs everywhere: adapting multilingual language models to new scripts. In: Proceedings of the 2021 Conference on Empirical Methods in Natural Language Processing. Association for Computational Linguistics, Online and Punta Cana, Dominican Republic, pp. 10186–10203 (2021)

26. Reimers, N., Gurevych, I.: Sentence-bert: Sentence embeddings using siamese bert-networks. In: Proceedings of the 2019 Conference on Empirical Methods in Natural Language Processing and the 9th International Joint Conference on Natural Language Processing (EMNLP-IJCNLP), pp. 3982–3992 (2019)

27. Rücklé, A., et al.: AdapterDrop: On the efficiency of adapters in transformers. In: Proceedings of the 2021 Conference on Empirical Methods in Natural Language Processing, pp. 7930–7946. Association for Computational Linguistics, Online and Punta Cana, Dominican Republic (2021)

28. Schlicht, I.B., de Paula, A.F.M., Rosso, P.: UPV at checkthat! 2021: mitigating cultural differences for identifying multilingual check-worthy claims. In: CLEF (Working Notes). CEUR Workshop Proceedings, 2936, pp. 465–475. CEUR-WS.org (2021)

29. Shaar, S., et al.: Overview of the CLEF-2021 checkthat! lab task 1 on check-worthiness estimation in tweets and political debates. In: CLEF (Working Notes). CEUR Workshop Proceedings, **2936**, pp. 369–392. CEUR-WS.org (2021)

30. Singh, k, et al.: Misinformation, believability, and vaccine acceptance over 40 countries: Takeaways from the initial phase of the covid-19 infodemic. PLoS ONE **17**(2), e0263381 (2022)

31. Stickland, A.C., Murray, I.: BERT and pals: Projected attention layers for efficient adaptation in multi-task learning. In: ICML. Proc. Mach. Learn. Res. **97**, pp. 5986–5995. PMLR (2019)

32. Sundararajan, M., Taly, A., Yan, Q.: Axiomatic attribution for deep networks. In: ICML. Proc. Mach. Learn. Res. vol. 70, pp. 3319–3328. PMLR (2017)

33. Üstün, A., Bisazza, A., Bouma, G., van Noord, G.: UDapter: Language adaptation for truly universal dependency parsing. In: Proceedings of the 2020 Conference on Empirical Methods in Natural Language Processing (Empirical Methods in Natural Language Processing) Association for Computational Linguistics, pp. 2302–2315 (2020)

34. Uyangodage, L., Ranasinghe, T., Hettiarachchi, H.: can multilingual transformers fight the COVID-19 infodemic? In: Proceedings of the International Conference on Recent Advances in Natural Language Processing (RANLP 2021), pp. 1432–1437. INCOMA Ltd., Held Online (2021)

35. Vasileva, S., Atanasova, P., Màrquez, L., Barrón-Cedeño, A., Nakov, P.: It takes nine to smell a rat: Neural multi-task learning for check-worthiness prediction. In: Mitkov, R., Angelova, G. (eds.) Proceedings of the International Conference on Recent Advances in Natural Language Processing, RANLP 2019, Varna, Bulgaria, September 2–4, 2019, pp. 1229–1239. INCOMA Ltd. (2019)

36. Wolf, T., et al.: Transformers: State-of-the-art natural language processing. In: Proceedings of the 2020 Conference on Empirical Methods in Natural Language Processing: System Demonstrations, pp. 38–45. Association for Computational Linguistics, Online (2020)

37. Xue, L., et al.: mt5: a massively multilingual pre-trained text-to-text transformer. In: Proceedings of the 2021 Conference of the North American Chapter of the Association for Computational Linguistics: Human Language Technologies, pp. 483–498 (2021)

Market-Aware Models for Efficient Cross-Market Recommendation

Samarth Bhargav(✉), Mohammad Aliannejadi, and Evangelos Kanoulas

University of Amsterdam, Amsterdam, The Netherlands
{s.bhargav,m.aliannejadi,e.kanoulas}@uva.nl

Abstract. We consider the cross-market recommendation (CMR) task, which involves recommendation in a low-resource *target* market using data from a richer, auxiliary *source* market. Prior work in CMR utilised meta-learning to improve recommendation performance in target markets; meta-learning however can be complex and resource intensive. In this paper, we propose market-aware (MA) models, which directly model a market via *market embeddings* instead of meta-learning across markets. These embeddings transform item representations into market-specific representations. Our experiments highlight the effectiveness and efficiency of MA models both in a pairwise setting with a single target-source market, as well as a global model trained on all markets in unison. In the former pairwise setting, MA models on average outperform market-unaware models in 85% of cases on nDCG@10, while being time-efficient—compared to meta-learning models, MA models require only 15% of the training time. In the global setting, MA models outperform market-unaware models consistently for some markets, while outperforming meta-learning-based methods for all but one market. We conclude that MA models are an efficient and effective alternative to meta-learning, especially in the global setting.

Keywords: Cross-market recommendation · Domain adaptation · Market adaptation

1 Introduction

Cross-market recommendation (CMR) involves improving recommendation performance in a target market using data from one or multiple auxiliary source markets. Data from *source* markets, which have rich- and numerous interactions, are leveraged to aid performance in a *target* market with fewer interactions. For instance, an e-commerce company well-established in Germany may want to start selling its products in Denmark. Using CMR methods, data from the German market can be utilised to augment recommender performance in the Danish market. This task is challenging since target market data can be scarce or otherwise unavailable, and user behaviours may differ across markets [2,7,24].

Research in CMR tackles multiple challenges. One challenge is to select the best source market, which is crucial since user behaviours across markets may vary [2, 24], which may harm performance instead of bolstering it. Furthermore, effectively utilising data from multiple markets at the same time without harming performance can be challenging [2]. Another key obstacle is effectively modelling a market, in addition to users and items. Bonab et al. [2] treat recommendation in each market as a task in a multi-task learning (MTL) framework, using meta-learning to learn model parameters. This is followed by a fine-tuning step per market. These two steps enable models to learn both common behaviours across markets as well as market-specific behaviours. However, meta-learning can be resource intensive compared to other methods. In addition to this, utilising new data from source markets requires re-running the meta-learning step.

We propose market-aware (MA) models to address these limitations. We aim to *explicitly* model each market as an embedding, using which an item representation can be transformed and 'customised' for the given market. Compared to meta-learning models, we show that MA models are far more efficient to train. Furthermore, they are trained in one go, enabling easier model updates when new data is collected. MA models are built on the hypothesis that explicit modelling of markets allows better generalisation. In essence, an item representation is a product of (i) an *across-market* item embedding and (ii) a market embedding. The former is learnt from data across markets, and aims to capture an item representation applicable across markets; the latter enables market-specific behaviours to be captured.

In our experiments, we compare MA models with market-unaware baselines as well as meta-learning models. We do so in multiple settings, utilising data from several markets: the *pairwise* setting, which deals with a single target-source pair, and the *global* setting which trains one model for recommendation in all markets. In the *pairwise* setting, we show that *MA* models improve over market-unaware models for many markets, and match or beat meta-learning methods. This is significant since we show that training MA models require approximately the same time as market-unaware models and only 15% of the time required to train meta-learning models. We show that MA models especially excel in the *global* setting, outperforming meta-learning methods for nearly every market. We examine the following research questions[1]:

RQ1. *Given a single source and target market, does explicitly modelling markets with embeddings lead to effective performance in the target market?* We compare MA models against market-unaware as well as meta-learning models. We show MA models achieve the best performance for most markets, and when a single, best source is available they match or outperform baselines for all markets.

RQ2. *How computationally expensive are MA models compared to market-unaware and meta-learning models?* We show that MA models require similar training times as market-unaware models, and require fewer computational resources to train compared to meta-learning models while achieving similar or better performance.

---

[1] https://github.com/samarthbhargav/efficient-xmrec

RQ3. *How do MA models compare against market-unaware models and meta-learning models when a* global *model is trained on all markets in unison?* We show that MA models outperform or match market-unaware baselines, outperforming meta-learning models for all but one market.

2 Related Work

While both cross-domain recommendation (CDR) and CMR focus on improving recommender effectiveness using data from other domains (i.e. item categories) or markets, they present different challenges: CDR involves recommending items in a different domain for the same set of users, with the general assumption that the model learns from interactions of overlapping users. In CMR, *items* are instead shared across different markets, with each market having a different set of users. Interactions from auxiliary markets are leveraged to boost performance for users in the target market for a similar set of items.

Cross-domain Recommendation. CDR has been researched extensively [6,12,14,17,18,20,22,23]. Prior approaches involve clustering-based algorithms [21] and weighing the influence of user preferences based on the domain [23]. Lu et al. [20] show that domain transfer may sometimes harm performance in the target domain. Neural approaches using similarity networks like DSSM [13] or transfer learning [6,12] can be effective. DDCTR [18] utilises iterative training across domains. Augmenting data with 'virtual' data [4,22], as well as considering additional sources [27] have been shown to help. Other approaches leverage domain adaptation [9] for leveraging content for full cold-start [15], utilising adversarial approaches [19,25] or formulating it as an extreme classification problem [26]. Our approach is inspired by contextual invariants [17], which are behaviours that are consistent across domains, similar to our hypothesis that there are behaviours common across markets.

Cross-market Recommendation. CMR is relatively new and understudied compared to CDR. Ferwerda et al. [7] studied CMR from the perspective of country based diversity. Roitero et al. [24] focus on CMR for music, investigating trade-offs between learning from local/single markets vs. a global model, proposing multiple training strategies. [2] release a new dataset for the Cross Market Product recommendation problem, which we utilise in our experiments. They design a meta-learning approach to transfer knowledge from a source market to a target market by freezing and forking specific layers in their models. The WSDM Cup 2022 challenge also dealt with this dataset, where most top teams utilised an ensemble of models based on different data pairs. Cao et al. [3] builds on the XMRec dataset and proposes multi-market recommendation, training a model to learn intra- and inter-market item similarities. In this work, we show that meta-learning methods are expensive to train. Instead, we show that market embeddings can encode and effectively transfer market knowledge, beating or matching the performance of complex models while being much more efficient to train.

3   Methodology

We outline market-unaware models in Sect. 3.1, followed by market-aware models as well as meta-learning models in Sect. 3.2.

Notation. Given a set of markets $\{\mathbb{M}_0, \mathbb{M}_1, \ldots, \mathbb{M}_t\}$, such that market l has a items $\mathcal{I}_l$ and z_l users $\mathcal{U}_l = \{U_l^1 \ldots U_l^{z_l}\}$. We assume the *base* market $\mathbb{M}_0$ has $\mathcal{I}_0$ s.t. $\mathcal{I}_0 \supset \mathcal{I}_l$ for all $1 \leq l \leq m$. The task is to adapt a given market $\mathbb{M}_l$ using data from other markets $\mathbb{M}_{m \neq l}$ as well as data from the target market. We use $\mathbf{p}_u$ for the user embedding for user u, $\mathbf{q}_i$ for the item embedding for item i, and finally $\mathbf{o}_l$ for the market embedding for market l. y_{ui} and $\hat{y}_{ui}$ is the actual and predicted rating respectively. $\odot$ denotes an element-wise product.

3.1   Market-Unaware Models

These models do not differentiate between users and items from different markets and are termed *market-unaware* since they do not explicitly model the market. We first outline three such models previously employed for CMR [2, 11]:

- **GMF**: The generalized matrix factorization (GMF) model computes the predicted rating $\hat{y}_{ui}$ given $\mathbf{p}_u$, $\mathbf{q}_i$ and parameters $\mathbf{h}$:

$$\hat{y}_{ui} = \text{sigmoid}(\mathbf{h}^T(\mathbf{p}_u \odot \mathbf{q}_i))$$

- **MLP**: An multi-layer perceptron (MLP) uses a L layer fully-connected network, such that:

$$\mathbf{m}_0 = \begin{bmatrix} \mathbf{p}_u \\ \mathbf{q}_i \end{bmatrix}$$
$$\mathbf{m}_{L-1} = \text{ReLU}(\mathbf{W}_{L-1}^T \text{ReLU}(\ldots \text{ReLU}(\mathbf{W}_1^T \mathbf{m}_0 + \mathbf{b}_1)) + \mathbf{b}_{L-1})$$
$$\hat{y}_{ui} = \text{sigmoid}(\mathbf{h}^T \mathbf{m}_{L-1})$$

- **NMF**: neural matrix factorization (NMF) combines both MLP and GMF. Given $\mathbf{p}_u^1$, $\mathbf{q}_i^1$ for the MLP, and $\mathbf{p}_i^2$, $\mathbf{q}_u^2$ for GMF, the NMF model computes the score as follows:

$$\mathbf{m}_0 = \begin{bmatrix} \mathbf{p}_u^1 \\ \mathbf{q}_i^1 \end{bmatrix}$$
$$\mathbf{m}_{MLP} = \text{ReLU}(\mathbf{W}_L^T \text{ReLU}(\ldots \text{ReLU}(\mathbf{W}_1^T \mathbf{m}_0 + \mathbf{b}_1))) + \mathbf{b}_L)$$
$$\mathbf{m}_{GMF} = \mathbf{p}_u^2 \odot \mathbf{q}_i^2$$
$$\hat{y}_{ui} = \text{sigmoid}(\mathbf{h}^T \begin{bmatrix} \mathbf{m}_{GMF} \\ \mathbf{m}_{MLP} \end{bmatrix})$$

For adapting to CMR, different sets of users from different markets are treated similarly, and training is performed on a combined item pool resulting in a single model. During inference for a user, however, only items from that market are ranked.

3.2   Market-Aware Models

We first discuss models proposed by Bonab et al. [2], followed by our proposed methods.

Meta-learning Baselines. Bonab et al. [2] propose using meta-learning in an MTL setting where each market is treated as a 'task'. model-agnostic metalearning (MAML) [8] is employed to train the base NMF model across markets. MAML employs two loops for training, an inner loop that optimises a particular market, and an outer loop that optimises across markets. This makes training expensive, as we will show in our experiments. Once a MAML model is trained, the FOREC model is obtained as follows for a given source/target market: (a) the MAML model weights are copied over to a new model, 'forking' it, (b) parts of the weights of the model are frozen and finally (c) the frozen model is fine-tuned on the given market.

Both MAML and FOREC are market aware but do not *explicitly* model the market i.e. a single item embedding is learned in MAML models for all markets, and while market adaptation is achieved through fine-tuning for FOREC, it requires maintaining separate sets of parameters, unlike the proposed MA models.

Market Aware Models. Markets here are explicitly modelled by learning embeddings for each of them, in addition to user and item embeddings. A market embedding *adapts* an item to the current market, which we argue is crucial for items that may be perceived differently in different markets. This aspect should be reflected in the latent representation of the item, motivating our approach. Both meta-learning and MA models learn item representations across markets, but MA models this explicitly via an element-wise product between a representation for an item and a market embedding. This produces item embeddings adapted to a given market. We augment the market-unaware baselines with market embeddings, producing MA models. We leave more complex methods, for instance—a neural network that models item/market interactions instead of an element-wise produce—for future work.

To obtain a *market-adapted* item embedding, we first (one-hot) encode a market l, to obtain a market embedding $\mathbf{o}_l$; the dimensionality of $\mathbf{o}_l$ is the same as $\mathbf{p}_u$ and $\mathbf{q}_i$. The scores are computed as follows for the three proposed models:

– **MA-GMF**: For a user u in market l, and item i, we have embeddings $\mathbf{p}_u$, $\mathbf{o}_l$ and $\mathbf{q}_i$:

$$\hat{y}_{ui} = \text{sigmoid}(\mathbf{h}^T(\mathbf{p}_u \odot (\mathbf{o}_l \odot \mathbf{q}_i)))$$

- **MA-MLP**: This is the same as the MLP, with the initial embedding $\mathbf{m}_0$ augmented with market information: $\mathbf{m}_0 = \begin{bmatrix} \mathbf{p}_u \\ \mathbf{q}_i \odot \mathbf{o}_l \end{bmatrix}$
- **MA-NMF**: The NMF model utilises both modifications listed above. That is:

$$\mathbf{m}_{GMF} = \mathbf{p}_u^2 \odot (\mathbf{o}_l \odot \mathbf{q}_i^2)$$

$$\mathbf{m}_0 = \begin{bmatrix} \mathbf{p}_u^1 \\ \mathbf{q}_i^1 \odot \mathbf{o}_l \end{bmatrix}$$

These models are trained similarly to the market-unaware models, except the market is taken into consideration when making recommendations. Market embeddings are learned via backpropagation, similar to how user and item embeddings are learned, using a binary cross entropy loss [11].

Our proposed technique adds market awareness to all the models. Besides this, the proposed models are easier to update with new interactions compared to MAML/FOREC. While FOREC requires the expensive MAML pre-training followed by the fork and fine-tune step, MA models simply can be trained with new interaction data. In spite of this simplicity, MA models achieve similar performance compared to meta-learning models while requiring far lesser time to train, which we demonstrate in the following section.

4   Experimental Setup

We conduct two sets of experiments. The first set of experiments trains models with a single auxiliary source market for improving recommendation performance in a given target market. We term these *pairwise* experiments since one model is trained for a given source-target market pair. The second set of experiments deals with a *global* model trained on all markets in unison, with the goal of improving overall performance. We outline the dataset, evaluation, baselines, hyperparameters and training followed by a description of the experiments.

Dataset. We use the XMarket dataset [2] for all experiments. XMarket is an CMR dataset gathered from a leading e-commerce website with multiple markets. We utilise the largest subset, 'Electronics', considering the following markets (# users, # items, # interactions): *de* (2373/ 2210/ 22247), *jp* (487/ 955 /4485), *in* (239/ 470/ 2015), *fr* (2396/ 1911/ 22905), *ca* (5675/ 5772/ 55045), *mx* (1878/ 1645/ 17095), *uk* (4847/ 3302/ 44515), *us* (35916/ 31125/ 364339). We consider all markets except *us* as a target market, with all markets (including *us*) as possible source markets. Experiments are limited to XMarket as it is the only public dataset for research in CMR.

Evaluation. The data (per market) is split into a train/validation/test set, where one left-out item from the user history is used in the validation and test set. This follows the leave-one-out strategy [5,10–12,16]. We extract 99 negatives per user for evaluating recommender performance in the validation/test set, following Bonab et al. [2]. In the pairwise experiments, the best-source market is picked based on the validation set performance. We report nDCG@10 on the test set in all results, with significance tests using a paired two-sided t-test with the Bonferroni correction. While we report only nDCG@10, we note that we observed similar trends for HR@10.

Compared methods. Market-aware models are denoted with an 'MA-' prefix, and are compared with the following models:

- Single-market models: These are models trained only on the target market data without any auxiliary source data, see Sect. 3.1. We train all three models GMF, NMF and MLP.
- Cross-market models: In addition to target market data, these models are trained with either one source market (for *pairwise* experiments), or all source markets (for *global* experiments). Models trained with at least one source market have a '++' suffix e.g. GMF++ and MA-GMF++.
- Meta-learning models (see Sect. 3.2) similarly utilise data from one or more auxiliary markets:
  - MAML [2,8]: These are models trained using MAML, with weights initialised from a trained NMF++ model [2].
  - FOREC [2]: This model uses the trained MAML model to first freeze certain parts of the network, followed by a fine-tuning step on the target market.

Model hyperparameters. We set model parameters from [2][2]: the dimensionality of the user, item, and market embeddings are set to 8, with a 3-Layer [12,23,23] network for MLP/NMF models. For MAML models, we set the fast learning rate $\beta = 0.1$ with 20 shots.

Training. All models are trained for 25 epochs using the Adam optimiser with a batch size of 1024. We use learning rates from [2], for GMF we use 0.005, for MLP and NMF we use 0.01. All models also utilise an L-2 regularisation loss with $\lambda = 1e - 7$. The NMF model is initialised with weights from trained GMF and MLP models. MAML models are trained on top of the resulting NMF models, and FOREC models utilise the trained MAML models for the fork-and-fine-tune procedure [2]. MA variants use the same hyperparameters as the market-unaware models. The objective function for all models is binary cross-entropy, given positive items and 4 sampled negatives [2,11]. For pairwise experiments, data from the source market is (randomly) down-sampled to the target market [2], which ensures that models are comparable across different-sized source markets. For global models, all data is concatenated together without any down-sampling[3].

---

[2] https://github.com/hamedrab/FOREC.
[3] We observed that this greatly improved performance for almost all markets.

Table 1. AVG results: Models are first trained on a single target-source pair and performance across sources are averaged. We report the nDCG@10 on the test set, with best performance in **bold**. Significance test ($p < \frac{0.05}{9}$) results are also reported comparing MA models with market-unaware (‡), MAML (∗) and FOREC (+).

Method	de	jp	in	fr	ca	mx	uk
GMF++	0.2045	0.0916	0.1891	0.2026	0.1937	0.4204	0.3222
MA-GMF++	0.2148‡	0.1079	0.2013	0.2022	0.2203‡	0.4283‡	0.3327‡
MLP++	0.2836	0.1653	0.4376	0.2704	0.2905	0.5274	0.4346
MA-MLP++	0.2909‡+∗	0.1741	**0.4502**	0.2805‡	**0.3073‡+∗**	0.5311	0.4349∗
NMF++	0.2927	**0.1826**	0.4403	0.2844	0.2844	0.5367	**0.4379**
MA-NMF++	**0.3055‡+∗**	0.1824	0.4471	**0.2893+∗**	0.3002‡+∗	**0.5387+∗**	0.4370∗
MAML	0.2808	0.1770	0.4320	0.2785	0.2794	0.5288	0.4296
FOREC	0.2835	0.1758	0.4345	0.2816	0.2772	0.5302	0.4330

Pairwise Experiments. The first set of experiments dealing with **RQ1** and **RQ2**, which we call *pairwise* (Sect. 5.1), assumes a single auxiliary market is available for a given target market. Since there are multiple source markets, we report both the average performance in the target market *across source markets* — termed AVG — as well as performance in the target market using the *best source market*, termed BST . The two tables relay different results: the average performance indicates the *expected* performance of a method since the 'best' source market might be unknown, or only a single source may exist; whereas the best-source results are indicative of the maximum achievable performance *if* a good source market is already known (this is typically unknown [24]).

Global Experiments. The second set of experiments corresponding to **RQ3** utilises data from multiple auxiliary markets at once to train a *global* recommender, with the goal to achieve good performance for all markets. We term these experiments *Global* (Sect. 5.2). We describe the results of the two sets of experiments in the following section.

5 Results and Discussion

5.1 Pairwise Experiments

Tables 1 and 2 report the results of the *pairwise* experiments, where the models only use one auxiliary market at a time. We report both AVG , the average performance of models using different auxiliary markets for the same target market (Table 1), as well as BST , the best auxiliary market (Table 2). The best auxiliary market is determined based on the validation set performance. Moreover, the results of the single-market baseline models are only reported in Table 2. We first examine **RQ1**, comparing the performance of MA models against baselines in both the AVG and BST settings. We end with discussion of **RQ2**, which compares training times across models.

Do MA models improve over market unaware models on average?
Using Table 1, we first examine if MA models outperform market-unaware models in the AVG setting e.g. GMF++ against MA-GMF++. We see that the MA-GMF++ outperforms GMF++ for every market except *fr*. MA-MLP++ outperforms MLP++ for all markets, and MA-NMF++ outperforms NMF++ on all markets except *jp* and *uk*. For the *de* and *ca* markets, we see that MA models always outperform their non-MA variant. In addition, for the *uk* and *mx* markets, MA-GMF++ significantly outperforms GMF++; and for *fr* we see that MA-MLP++ significantly outperforms MLP++. Despite large improvements in some markets e.g. MA-MLP++ improves nDCG@10 by 0.12 points over MLP++ for *in*, we do not see a significant result, which may be due to the conservative Bonferroni correction, or fewer test users for *in* (requiring larger effect sizes). Overall, MA models outperform their market unaware equivalent in 18 of 21 settings. In summary, we can conclude that *in the AVG setting, the proposed market-aware models outperform market-unaware baselines for nearly all markets.* This demonstrates the robustness of *MA* models since these improvements are across multiple source markets.

How do MA models compare against meta-learning models in the AVG setting? We compare MA models against MAML and FOREC considering AVG , in Table 1. MA-GMF++ never outperforms MAML/ FOREC, but the differences in model sizes render this comparison unfair. A fairer comparison would be with MA-NMF++: we see that it outperforms MAML for 5 of 7 markets: *de, fr, ca, mx* and *uk*. Additionally, FOREC is significantly outperformed by MA-NMF++ for 4 of 7 markets: *de, fr, ca* and *mx*. We note, however, that at least one MA model outperforms both MAML/FOREC for all markets, and at least one MA model *significantly* outperforms MAML/FOREC for *de* (both), *fr* (both), *ca* (both), *mx* (MAML only) and *uk* (MAML only). Therefore, *we can thus conclude that market-aware models either match or outperform meta-learning models for many markets in AVG setting.*

Do MA models outperform market-unaware models when trained with the best available source? Viewing Table 2, we first note that MA models outperform all single market variants, highlighting the utility of selecting a good source market, consistent with prior research [2,24]. MA models *significantly* outperform single-market variants depending on the market and model, with more significant improvements seen for MA-GMF++ (5 of 7 markets) than MA-MLP++ (3 of 7) or MA-NMF++ (3 of 7). Consistent improvements over the single-market models are surprisingly seen for some larger markets i.e. *ca* and *de* (but not for *uk*), showing larger markets can sometimes benefit from auxiliary market data. However, the results are less consistent when comparing the MA models with their augmented but market-unaware models, especially as model size increases. MA-GMF++ improves over GMF++ in 4 of 7 markets, MA-MLP++ improves over MLP++ in 3 of 7 markets, and finally, MA-NMF++ improves over NMF++ only in 2 markets. In fact, for *in, fr, mx* and *uk*, we see that NMF++ outperforms MA-NMF++. Furthermore, only MA-NMF++ on *ca*

Table 2. BST : Models are trained on all source markets, the best source is selected based on validation set performance. We report nDCG@10 on the test set, along with significance test results ($p < \frac{0.05}{12}$) comparing MA models with single market (†), market unaware (‡), MAML (∗) and FOREC (+).

Method	de	jp	in	fr	ca	mx	uk
GMF	0.2574	0.0823	0.0511	0.2502	0.2566	0.5066	0.4136
GMF++	0.2670	0.1093	0.2838	0.2708	0.2818	0.5338	0.4399
MA-GMF++	0.2831†	0.1453†	0.3338†	0.2654	0.2907†	0.5145	0.4336†
MLP	0.2986	0.1340	0.4506	0.2869	0.2934	0.5367	0.4465
MLP++	0.3170	0.1865	0.4470	0.3016	0.3100	0.5455	0.4585
MA-MLP++	0.3167†	0.1806†	0.4584	0.3026	0.3105†+∗	0.5419	0.4544
NMF	0.3214	0.1717	0.4265	0.3014	0.2848	0.5430	0.4488
NMF++	0.3332	0.1921	**0.4595**	**0.3271**	0.3008	**0.5590**	**0.4702**
MA-NMF++	**0.3415**†+∗	0.1896	0.4433	0.3228†	**0.3158**†‡+∗	0.5573	0.4578
MAML	0.3168	**0.2083**	0.4491	0.3152	0.2989	0.5463	0.4671
FOREC	0.3040	0.1983	0.4458	0.3191	0.2927	0.5442	0.4683

significantly outperforms NMF++. We can thus conclude that while MA models improve over market unaware models in some cases, *selecting a source market remains an important factor for improving performance given a target market*. While this conclusion holds, we note that in general, data from multiple source markets may be unavailable, or otherwise data from target markets might be unavailable—making best source selection unviable [24]. In such cases, results from the average-source experiments have to be considered.

How Do MA models compare against meta-learning models when trained on the best source? We now compare MA models against MAML/ FOREC. We first note that at least one MA model beats MAML/FOREC for all markets but *jp* and *uk*. MA-NMF++, in particular, outperforms both MAML and FOREC for 4 of 7 markets. We see MA-NMF++ significantly outperforms both MAML/FOREC for *de* and *ca*. MAML achieves the best performance for *jp*, beating other models by a large margin. In conclusion, we observe similar performance of our MA models compared to meta-learning models, while outperforming them in some cases. This again indicates the effectiveness of our market embedding layer, especially when the training times are considered, which we discuss next.

How do training times compare across models? Are MA models time-efficient? We plot the time taken to train all models for a given target market (distributed across the seven different source markets) in Fig. 1, where the time taken is on a log scale. From this, we can see that the meta-learning models take far longer to train compared to MA models. We note that MA models

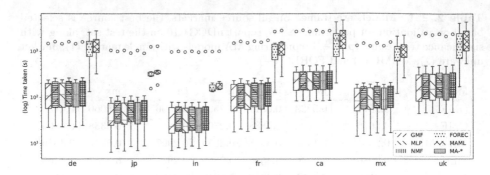

Fig. 1. Time taken to train a model for a target market across all source markets, where time is on a log scale. MA and market-unaware models require similar training times, while meta-learning models require significantly more.

require *only* 15% of the time taken to train meta-learning models, with MA models requiring about the same time to train as market-unaware models. This is unsurprising, since MAML requires an inner and outer loop, as well as requiring the expensive computation of second-order derivatives [1,8]. FOREC uses MAML in addition to fine-tuning the target market, so training FOREC takes up even more training time. In conclusion, MA models achieve better or similar performance to MAML/FOREC while requiring much less training time.

Discussion. Overall, we can conclude for AVG that MA models outperform both market-unaware baselines as well as meta-learning models, demonstrating the effectiveness of MA models across multiple sources. For BST i.e. when best-source selection is viable, the results are mixed: MA models always outperform single model variants; they outperform market-unaware models for many, but not all, markets; and an MA model either matches or outperforms meta-learning models for all markets.

A fair question to ask is whether an increase in performance of MA over market-unaware models can be attributed to the increase in the number of parameters from the market embeddings. However, this increase is minuscule compared to model sizes, especially for NMF and MLP i.e. for t markets and D dimensional user/item/market embeddings, the increase is just tD parameters. In the pairwise experiments, this difference is just $16(= 2 * 8)$, much fewer than 19929, the number of parameters of a MLP model for the smallest target/source pair (in/jp).

While meta-learning models implicitly model the market during training, MA models show that this may be insufficient. We attribute the success of MA models to this explicit modelling of the markets: by adapting item representations depending on the market, the model may be better able to distinguish between recommendation in different markets more than market-unaware and meta-learning models. As we observe a better performance on AVG , we can *conclude for **RQ1**, that market-aware models exhibit a more robust performance*

Table 3. Global experiments: All markets are trained in unison. Best model for a market is in **bold**. Significance test ($p < \frac{0.05}{9}$) results are also reported comparing MA models with market unaware ($\ddagger$), MAML ($*$) and FOREC ($+$).

Method	de	jp	in	fr	ca	mx	uk
GMF++	0.3166	0.1781	0.4535	0.2884	0.2921	0.5245	0.4481
MA-GMF++	0.3073	0.1817	0.4554	0.2836	0.3015*	0.5262	0.4504
MLP++	0.3268	0.2127	0.4479	0.2953	0.3048	0.5376	0.4491
MA-MLP++	0.3158	0.2195	0.4398	0.2958	**0.3178**$^{\ddagger+*}$	0.5258	0.4535
NMF++	0.3262	0.1930	**0.4796**	0.3030	0.2851	0.5340	0.4476
MA-NMF++	**0.3442**$^{\ddagger+}$	**0.2212**	0.4602	**0.3052**	0.3112$^{\ddagger+*}$	**0.5536**‡*	**0.4604**$^{\ddagger+*}$
MAML	0.3281	0.1860	0.4736	0.3022	0.2836	0.5317	0.4474
FOREC	0.3249	0.1956	0.4778	0.3033	0.2947	0.5409	0.4474

compared to other models either matching or outperforming baselines in many settings. While this indicates that market-aware models are more effective models in general, in some cases meta-learning models seem to learn better from the most suitable market: in these cases, MA models achieve similar performance. However, it is critical to note that MA models achieve this while requiring far less computational power. Moreover, it is evident that MA models do not add much complexity to non-MA models, while empowering the model to capture the market's attributes more effectively, resulting in an efficient and effective model.

5.2 Global Experiments

Table 3 reports the results of training one global recommendation model for all markets. We see that MA models outperform baselines in many cases, even beating meta-learning models for almost all markets.

How do MA models compare with market-unaware models? MA-variant models outperform market-unaware models in 15 of 21 settings, but results differ across models: MA-GMF++ (5 of 7), MA-MLP++ (4 of 7) and MA-NMF++ (6 of 7). MA-MLP++ significantly outperforms MLP++ for *ca* whereas MA-NMF++ significantly outperforms NMF++ for four markets. We also note that MA models for the largest markets, *uk* and *ca*, outperform both market-unaware and meta-learning models. We observe mixed results for smaller markets: for *jp*, MA consistently improves over market-unaware variants, but for *in*, only MA-GMF++ outperforms GMF++. Overall, we can conclude that *MA models outperform market-aware models in several settings, especially for larger markets and models.*

How do MA models compare with meta-learning-based models? We first note that an MA model (typically MA-NMF++) beats MAML/FOREC for all markets except *in*. Indeed, MA-NMF++ beats *both* MAML and FOREC for all markets except *in*. It *significantly* outperforms MAML for *ca, mx* and *uk* markets, and FOREC for *de, ca* and *uk*—the larger markets. For *ca*, we see all three MA models significantly outperform MAML, with MA-MLP++ and MA-GMF++ significantly outperforming FOREC. On the whole, we see that in a global setting, MA models outperform meta-learning methods in nearly all markets, and in particular the larger markets.

Discussion. We can conclude for **RQ3** that MA models are more suitable than market unaware or meta-learning models if a global model is used for recommendation across all markets. This is critical for cases where various markets exist, empowering the model to take advantage of various user behaviours across different markets to improve recommendation in the target market. Moreover, it also leaves the problem of selecting the 'best source' to the model (i.e. the market embedding), as the model consumes the whole data and synthesises knowledge from multiple markets. MA models seem to have an advantage over market-unaware and meta-learning models, especially for larger markets. This is likely due to the market embedding, allowing markets to distinguish source- and target-market behaviours. As more data is collected, MA models, which perform better in the global setting for larger markets, are likely to have a clear advantage.

6   Conclusions and Future Work

In this work, we proposed simple yet effective MA models for the CMR task. In a *pairwise* setting where models are trained with a single source market, MA models on average outperform baselines in most settings, showcasing their robustness. Considering the best source market, we showed that MA models match or outperform baselines for many markets. We showed that they require far less time to train compared to meta-learning models. Next, we trained a global model for all markets and showed that MA models match or outperform market-unaware models in nearly all settings, and outperform meta-learning models for all but one market. For future work, we plan to experiment with more complex MA models in a limited data setting. We also plan to investigate the utility of MA models in a zero-shot setting, substituting the market-embedding of the new market with a similar market. In addition, we want to consider data selection techniques, since we speculate that not all data from a given source market will be useful for a given target market.

Acknowledgements. The authors would like to thank Hamed Bonab for help with code and data, and Ruben van Heusden for help with statistical testing. The authors also thank the reviewers for their valuable feedback. This research was supported by the NWO Innovational Research Incentives Scheme Vidi (016.Vidi.189.039), the NWO Smart Culture - Big Data/Digital Humanities (314-99-301), and the H2020-EU.3.4. - SOCIETAL CHALLENGES - Smart, Green And Integrated Transport (814961). All content represents the opinion of the authors, which is not necessarily shared or endorsed by their respective employers and/or sponsors.

References

1. Antoniou, A., Edwards, H., Storkey, A.J.: How to train your MAML. In 7th International Conference on Learning Representations, ICLR 2019, New Orleans, LA, USA, May 6–9, (2019). OpenReview.net (2019). https://openreview.net/forum?id=HJGven05Y7

2. Bonab, H., Aliannejadi, M., Vardasbi, A., Kanoulas, E., Allan, J.: Cross-Market Product Recommendation, pp. 110–119. Association for Computing Machinery, New York, NY, USA (2021). ISBN 9781450384469, https://doi.org/10.1145/3459637.3482493

3. Cao, J., Cong, X., Liu, T., Wang, B.: Item similarity mining for multi-market recommendation. In: Proceedings of the 45th International ACM SIGIR Conference on Research and Development in Information Retrieval, SIGIR '22, pp. 2249–2254, New York, NY, USA, 2022. Association for Computing Machinery. ISBN 9781450387323. https://doi.org/10.1145/3477495.3531839

4. Chae, D.-K., Kim, J., Chau, D.H., Kim, S.-W.: AR-CF: augmenting virtual users and items in collaborative filtering for addressing cold-start problems. In: Proceedings of the 43rd International ACM SIGIR Conference on Research and Development in Information Retrieval, SIGIR 2020, pp. 1251–1260, New York, NY, USA, 2020. Association for Computing Machinery. ISBN 9781450380164. https://doi.org/10.1145/3397271.3401038

5. Cheng, H.-T., et al.: Wide & deep learning for recommender systems. In: Proceedings of the 1st Workshop on Deep Learning for Recommender Systems, DLRS 2016, pp. 7–10, New York, NY, USA, (2016). Association for Computing Machinery. ISBN 9781450347952. https://doi.org/10.1145/2988450.2988454

6. Elkahky, A.M., Song, Y., He, X.: A multi-view deep learning approach for cross domain user modeling in recommendation systems. In: Proceedings of the 24th International Conference on World Wide Web, WWW 2015, pp. 278–288, Republic and Canton of Geneva, CHE (2015). International World Wide Web Conferences Steering Committee. ISBN 9781450334693. https://doi.org/10.1145/2736277.2741667

7. Ferwerda, B., Vall, A., Tkalcic, M., Schedl, M.: Exploring music diversity needs across countries. In: Proceedings of the 2016 Conference on User Modeling Adaptation and Personalization, UMAP (2016), pp. 287–288, New York, NY, USA (2016). Association for Computing Machinery. ISBN 9781450343688. https://doi.org/10.1145/2930238.2930262

8. Finn, C., Abbeel, P., Levine, S.: Model-agnostic meta-learning for fast adaptation of deep networks. In: Proceedings of the 34th International Conference on Machine Learning - Volume 70, ICML 2017, pp. 1126–1135. JMLR.org (2017)

9. Ganin, Y., Lempitsky, V.: Unsupervised domain adaptation by backpropagation. In: Bach, F., Blei, D., (eds) Proceedings of the 32nd International Conference on Machine Learning, volume 37 of Proceedings of Machine Learning Research, pp. 1180–1189, Lille, France, 07–09 Jul 2015. PMLR. https://proceedings.mlr.press/v37/ganin15.html

10. Ge, Y., Xu, S., Liu, S., Fu, Z., Sun, F., Zhang, Y.: Learning personalized risk preferences for recommendation. In: Proceedings of the 43rd International ACM SIGIR Conference on Research and Development in Information Retrieval, SIGIR 2020, pp. 409–418, New York, NY, USA (2020). Association for Computing Machinery. ISBN 9781450380164. https://doi.org/10.1145/3397271.3401056

11. He, X., Liao, L., Zhang, H., Nie, L., Hu, X., Chua, T.-S.: Neural collaborative fil-
    tering. In: Proceedings of the 26th International Conference on World Wide Web,
    WWW 2017, pp. 173–182, Republic and Canton of Geneva, CHE (2017). Interna-
    tional World Wide Web Conferences Steering Committee. ISBN 9781450349130.
    https://doi.org/10.1145/3038912.3052569
12. Hu, G., Zhang, Y., Yang, Q.: CoNet: collaborative cross networks for cross-domain
    recommendation. In: Proceedings of the 27th ACM International Conference on
    Information and Knowledge Management, CIKM 2018, pp. 667–676, New York,
    NY, USA (2018). Association for Computing Machinery. ISBN 9781450360142.
    https://doi.org/10.1145/3269206.3271684
13. Huang, P.-S., He, X., Gao, J., Deng, L., Acero, A., Heck, L.: Learning deep
    structured semantic models for web search using clickthrough data. In: Proceed-
    ings of the 22nd ACM International Conference on Information & Knowledge
    Management, CIKM 2013, pp. 2333–2338, New York, NY, USA (2013). Associ-
    ation for Computing Machinery. ISBN 9781450322638. https://doi.org/10.1145/
    2505515.2505665
14. Im, I., Hars, A.: Does a one-size recommendation system fit all? the effectiveness
    of collaborative filtering based recommendation systems across different domains
    and search modes. ACM Trans. Inf. Syst., 26(1), 4-es, Nov 2007. ISSN 1046–8188.
    https://doi.org/10.1145/1292591.1292595
15. Kanagawa, H., Kobayashi, H., Shimizu, N., Tagami, Y., Suzuki, T.: Cross-domain
    recommendation via deep domain adaptation. In: Azzopardi, L., Stein, B., Fuhr,
    N., Mayr, P., Hauff, C., Hiemstra, D. (eds.) ECIR 2019. LNCS, vol. 11438, pp.
    20–29. Springer, Cham (2019). https://doi.org/10.1007/978-3-030-15719-7_3
16. Kang, W.-C., McAuley, J.: Self-attentive sequential recommendation. In: ICDM,
    pp. 197–206 (2018)
17. Krishnan, A., Das, M., Bendre, M., Yang, H., Sundaram, H.: Transfer learning via
    contextual invariants for one-to-many cross-domain recommendation. In: Proceed-
    ings of the 43rd International ACM SIGIR Conference on Research and Develop-
    ment in Information Retrieval, SIGIR 2020, pp. 1081–1090, New York, NY, USA
    (2020). Association for Computing Machinery. ISBN 9781450380164. https://doi.
    org/10.1145/3397271.3401078
18. Li, P., Tuzhilin, A.: DDTCDR: deep dual transfer cross domain recommenda-
    tion. In: Proceedings of the 13th International Conference on Web Search and
    Data Mining, WSDM 2020, pp. 331–339, New York, NY, USA (2020). Associ-
    ation for Computing Machinery. ISBN 9781450368223. https://doi.org/10.1145/
    3336191.3371793
19. Li, Y., Xu, J.-J., Zhao, P.-P., Fang, J.-H., Chen, W., Zhao, L.: ATLRec: An atten-
    tional adversarial transfer learning network for cross-domain recommendation. J.
    Comput. Sci. Technol. 35(4), 794–808 (2020). https://doi.org/10.1007/s11390-020-
    0314-8
20. Lu, Z., Zhong, E., Zhao, L., Xiang, E.W., Pan, W., Yang, Q.: selective transfer
    learning for cross domain recommendation, pp. 641–649. https://doi.org/10.1137/
    1.9781611972832.71
21. Mirbakhsh, N., Ling, C.X.: Improving top-n recommendation for cold-start users
    via cross-domain information. ACM Trans. Knowl. Discov. Data, 9(4), Jun 2015.
    ISSN 1556–4681. https://doi.org/10.1145/2724720
22. Perera, D., Zimmermann, R.: CNGAN: generative adversarial networks for cross-
    network user preference generation for non-overlapped users. In: The World Wide
    Web Conference, WWW 2019, pp. 3144–3150, New York, NY, USA (2019). Asso-

ciation for Computing Machinery. ISBN 9781450366748. https://doi.org/10.1145/3308558.3313733

23. Rafailidis, D., Crestani, F.: A collaborative ranking model for cross-domain recommendations. In: Proceedings of the 2017 ACM on Conference on Information and Knowledge Management, CIKM 2017, pp. 2263–2266, New York, NY, USA (2017). Association for Computing Machinery. ISBN 9781450349185. https://doi.org/10.1145/3132847.3133107

24. Roitero, K., Carterrete, B., Mehrotra, R., Lalmas, M.: Leveraging behavioral heterogeneity across markets for cross-market training of recommender systems. In: Companion Proceedings of the Web Conference 2020, WWW 2020, pp. 694–702, New York, NY, USA (2020). Association for Computing Machinery. ISBN 9781450370240. https://doi.org/10.1145/3366424.3384362

25. Wang, C., Niepert, M., Li, H.: RecSys-DAN: discriminative adversarial networks for cross-domain recommender systems. IEEE Trans. Neural Netw. Learn. Syst. **31**(8), 2731–2740 (2020). https://doi.org/10.1109/TNNLS.2019.2907430

26. Yuan, F., Yao, L., Benatallah, B.: DARec: deep domain adaptation for cross-domain recommendation via transferring rating patterns. In: Proceedings of the 28th International Joint Conference on Artificial Intelligence, IJCAI'19, pp. 4227–4233. AAAI Press (2019). ISBN 9780999241141

27. Zhao, C., Li, C., Xiao, R., Deng, H., Sun, A.: CATN: cross-domain recommendation for cold-start users via aspect transfer network. In: Proceedings of the 43rd International ACM SIGIR Conference on Research and Development in Information Retrieval, SIGIR 2020, pp. 229–238, New York, NY, USA (2020). Association for Computing Machinery. ISBN 9781450380164. https://doi.org/10.1145/3397271.3401169

TourismNLG: A Multi-lingual Generative Benchmark for the Tourism Domain

Sahil Manoj Bhatt[1(✉)], Sahaj Agarwal[2], Omkar Gurjar[1], Manish Gupta[1,2],
and Manish Shrivastava[1]

[1] IIIT-Hyderabad, Hyderabad, India
sahil.bhatt@research.iiit.ac.in, omkar.gurjar@students.iiit.ac.in,
m.shrivastava@iiit.ac.in
[2] Microsoft, Hyderabad, India
{sahagar,gmanish}@microsoft.com

Abstract. The tourism industry is important for the benefits it brings
and due to its role as a commercial activity that creates demand and
growth for many more industries. Yet there is not much work on data
science problems in tourism. Unfortunately, there is not even a standard
benchmark for evaluation of tourism-specific data science tasks and models. In this paper, we propose a benchmark, TourismNLG, of five natural language generation (NLG) tasks for the tourism domain and release
corresponding datasets with standard train, validation and test splits.
Further, previously proposed data science solutions for tourism problems do not leverage the recent benefits of transfer learning. Hence, we
also contribute the first rigorously pretrained mT5 and mBART model
checkpoints for the tourism domain. The models have been pretrained
on four tourism-specific datasets covering different aspects of tourism.
Using these models, we present initial baseline results on the benchmark
tasks. We hope that the dataset will promote active research for natural
language generation for travel and tourism. (https://drive.google.com/
file/d/1tux19cLoXc1gz9Jwj9VebXmoRvF9MF6B/.)

Keywords: NLG for Tourism · Long QA · Blog-title generation ·
Forum-title generation · Paragraph generation · Short QA

1 Introduction

According to World Travel and Tourism Council, in 2019, travel and tourism
accounted for (1) 10.3% of global GDP, (2) 333 million jobs, which is 1 in 10
jobs around the world, and (3) US$1.7 trillion visitor exports (6.8% of total
exports, 27.4% of global services exports)[1]. Tourism boosts the revenue of the
economy, creates thousands of jobs, develops the infrastructure of a country,
and plants a sense of cultural exchange between foreigners and citizens. This
commercially important industry has resulted in a lot of online data.

---

[1] https://wttc.org/research/economic-impact.

© The Author(s), under exclusive license to Springer Nature Switzerland AG 2023
J. Kamps et al. (Eds.): ECIR 2023, LNCS 13980, pp. 150–166, 2023.
https://doi.org/10.1007/978-3-031-28244-7_10

Data in the tourism domain exists in the form of (1) public web pages (blogs, forums, wiki pages, general information, reviews), or (2) travel booking information owned by travel portals which includes customer travel history, schedules, optimized itineraries, pricing, customer-agent conversations, etc. Accordingly, work on tourism data mining has mostly focused on structured extraction of trip related information [26], mining reviews (personalized sentiment analysis of tourist reviews [25], establishing review credibility [5,14]), and automatic itinerary generation [9–11,15]. Most of this work has however relied on traditional ways of performing natural language processing (NLP). However, recently transfer learning using pretrained models has shown immense success across almost all NLP tasks. Transformer [32] based models like Bidirectional Encoder Representations from Transformers (BERT) [12], Generative Pre-trained Transformer (GPT-2) [27], Extra-Long Network (XLNet) [34], Text-to-Text Transfer Transformer (T5) [28] have been major contributors to this success. These models have been pretrained on generic corpora like Books Corpus or Wikipedia pages. To maximize benefits, researchers across various domains have come up with domain specific pretrained models like BioBERT (biomedical literature corpus) [22], SciBERT (biomedical and computer science literature corpus) [6], ClinicalBERT (clinical notes corpus) [18], FinBERT (financial services corpus) [4], PatentBERT (patent corpus) [21], LegalBERT (law webpages) [8], etc. However, there are no models specifically pretrained for the tourism task. Further, there is no standard benchmark for tourism related tasks.

Travel text is very different from usual text across domains. Skibitska [30] investigated the degree of specialization of language of tourism in different kinds of tourism-related texts. They group tourism vocabulary into groups like types of tours and tourism (e.g. agro tourism, incentive tour, rural tourism, week-end tour, day trip etc.), industry professionals (e.g. guide, event organizer, travel agent, tourist information center assistant, etc.), accommodation (e.g. standard room, daily average rate, reservation, cancellation, room facilities, spa, check in, prepaid room etc.), catering (e.g. full board, white glove service, buffet, a la carte, coffee shop, tip, bev nap, etc.), transportation (e.g. charge, refund, non-refundable, actual passenger car hours, excess baggage, scheduled flight, frequent flyer, etc.), excursion (e.g., itinerary, overnight, local venue, sightseeing, city guide, departure point, meeting point, hop on hop off etc.), abbreviations (e.g. IATA, AAA, WTO, NTA, etc.). Compared to usual blogs, travel and tourism blog titles often (a) include a destination or type of travel experience to emphasize the appeal of the location, (b) emphasize the "adventure" aspect of traveling, (c) include the words "explore" or "discover" to emphasize the discovery of new places, (d) include words like "journey," "voyage," or "road trip" to emphasize the journey aspect of traveling, and (e) use vivid adjectives or descriptive phrases to emphasize the beauty and uniqueness of the destination. Compared to generic answers, answers on travel forums are (a) more focused on specific destinations, (b) typically more concise and to-the-point, (c) include advice or tips and less opinion-based, (d) written in a more conversational tone, (e) frequently include personal stories and anecdotes, (f) often written in the first person, (g) written in a positive or helpful manner. Paragraphs in travel webpages tend to describe culture and event sequences (in blogs), temporal facts and planning (on forums), etc. For factual short question answering, the answer types are rather restricted in tourism domain to architectural types, geographic names, population, timings, cost, directions etc.

Hence, we propose a benchmark, TOURISMNLG, consisting of five novel natural language generation (NLG) tasks in the travel and tourism domain. Overall, the number of instances across these five tasks add up to 4.2M instances. We make the datasets corresponding to these tasks, along with their train, validation and test splits publicly available. We also make the code and all our pretrained models publicly available.

Given this benchmark of five tourism NLG tasks and four different tourism-specific multi-lingual pretraining datasets, we also perform domain-adaptive pretraining of mT5 [33] and mBART [23] models for the tourism domain. Since all our tasks are generative, we chose mT5 and mBART as our primary model architectures. We show the efficacy of our models by finetuning them on the proposed TOURISMNLG benchmark tasks both individually as well as in a multi-task setup. This sets a good baseline for further researchers to compare their results on the TOURISMNLG benchmark.

Overall, in this paper, we make the following contributions. (1) We propose a benchmark of five novel and diverse tourism NLG tasks called TOURISMNLG. As part of this benchmark, we also contribute four datasets along with standard splits to the research community. (2) We pretrain multiple tourism-domain specific models. We also make the pretrained models publicly available. (3) We experiment with multiple pretraining and finetuning setups, and present initial baseline results on the TOURISMNLG benchmark.

2   Related Work

2.1   Data Science in Tourism

Published work on data science in the tourism domain has been very sparse. It has been mainly focused on structured extraction of trip related information, mining reviews, and automatic itinerary generation. Popescu et al. [26] use tagged photos uploaded by tourists on Flickr to deduce trip related information such as visit times for a tourist spot, while Pantano et al. [25] build a model using online tourist reviews to predict tourists' future preferences. Generation of automatic travel itineraries is another well-explored problem, with papers like [9,15] and [11] that use travel histories and online data generated by tourists to automatically generate itineraries for future travellers under monetary and time constraints. Specifically in NLP, Gurjar et al. [17] study aspect extraction for tourism domain for eleven factors, Kapoor et al. [20] identify travel-blog-worthy sentences from Wikipedia articles, and Iinuma et al. [19] propose a methodology to summarise multiple blog entries by finding important sentences and images using a graph-based approach. Finally, [3] proposed a large dataset from the hotel domain, for hotel recommendation task. Unfortunately, there is hardly any work on natural language generation for the tourism domain. We attempt to fill this gap in this paper.

2.2 Domain-specific Pretrained Models

Several previous works have presented models trained for various domains and their respective domain specific tasks. Transformer models like BERT [12] have been adapted to create pre-trained models like BioBERT [22] for Bio-Medical Domain, SciBERT [6] for scientific data domain and two models proposed by [2], one for generic clinical text and other for discharge summaries specifically, among others such as FinBERT [4] for NLP tasks in Financial Domain, Covid-Twitter-BERT [24] trained on Covid related Twitter content, and PatentBERT [21]. There are also models in the legal domain (LegalBERT [8]) and even specialized ones to model conversational dialogues (DialoGPT [35]). Since there are no domain-specific pretrained models for the tourism domain, we propose the TOURISMNLG benchmark and the associated initial models.

3 TOURISMNLG Benchmark

In this section, we present details of the four datasets and five NLG tasks which form the TOURISMNLG benchmark.

3.1 TOURISMNLG Datasets

The TOURISMNLG benchmark consists of four datasets: TravelWeb, TravelBlog, TripAdvisorQnA, and TravelWiki. These datasets were carefully chosen to cover diverse online content in the public domain.

TravelWeb: Given a large web crawl, we use a proprietary domain classifier based on [7] to choose webpages broadly in the travel and tourism domain. Further, to be able to reuse cleaned text for these URLs, from this set of URLs, we retain those URLs which are also present in mC4 dataset[2]. We use Marisa trie [13] for computing the intersection efficiently. We gather the webpage contents from the clean and pre-processed mC4 chunks. This leads to a dataset containing the URL, body text and publication timestamp. We remove instances where body text is empty. Overall, the dataset contains 454553 documents published from 2013 to 2020. The webpages belong to 80157 unique websites. The top few websites include wikipedia, tripadvisor, britannica, rome2rio, lonelyplanet, maplandia, expedia and theculturetrip.

TravelBlog: We collected travel blogs from travelblog.org. The dataset contains blog title, publication date, body text. The dataset contains 491276 blogs from 2009 to 2020. The dataset is divided into ten geo-categories with the following data split: Africa (33226), Antarctica (376), Asia (91626), Central America

---

[2] https://huggingface.co/datasets/mc4.

and Caribbean (21505), Europe (119202), Middle East (12093), North America (85270), Oceania (69783), Oceans and Seas (1802) and South America (56393). We use this dataset for the blog title generation task. We remove instances where blog titles and/or body text are empty.

TripAdvisorQnA: We prepare a tourism-centred Question-Answering dataset by collecting questions asked on TripAdvisor's forums[3] The dataset comprises of 217352 questions across several tourism-related categories such as Air Travel, Road Trips, Solo Travel, Cruises, Family Travel, etc. For each question, we collect the question title, the question description and all the responses to the query by other forum members. We also collect public user details for the user who posted the question as well as details of users who posted answers. In addition to this, we store the number of "helpful votes" for every response denoting the usefulness of the responses as voted by others. We use this to judge the quality of response for a particular question, and use the most voted response as the golden answer for our tasks.

We identified a list of standard messages from the TripAdvisor staff that were very common throughout the dataset. We identified at least three phrases in each of the language subsets, to remove such comments. For example, in English, common messages included the phrases: "this post was determined to be inappropriate", "this post has been removed", "message from tripadvisor staff". Similarly, in Spanish, we removed examples where answers contained "el personal de tripadvisor ha eliminado", "esta publicación ha sido eliminada por su autor", "no es posible responder a este tema ya que ha sido cerrado por inactividad". Since the dataset is large, we use only one response per question, but we have included all responses in the dataset. We remove instances where page titles are empty or where there is no answer.

TravelWiki: We gathered a list of top 1000 tourism spots worldwide from websites like lonelyplanet. Next, we discovered their Wikipedia pages (basic string match with typical normalizations) and gathered a histogram of Infobox template names for those Wikipedia pages. Further, we manually looked at the top 100 templates and identified a list of 29 Infobox templates like "nrhp", "uk place", "mountain", "river", etc. which seemed relevant to travel and tourism. Subsequently we collect the list of English Wikipedia pages containing these templates, and finally their counterparts in other languages using WikiData[4] mappings. For each page in our data we collect the first paragraph of that page along with key attributes (and their values) from the corresponding Infoboxes. Each Infobox is associated with 5.76 key-value pairs on average. TravelWiki contains content from about 3077404 tourism-related Wikipedia pages gathered from Wikipedia websites for 31 languages. Although the dataset contains the entire Infobox per page, we sample one random key-value pair per webpage and use it for the Paragraph generation and Short Question Answering (QA) tasks.

---

[3] https://www.tripadvisor.<countryCode>/ListForums-g1-World.html.   We   used these country codes: in, it, es, fr, de, pt, jp and ru.

[4] https://www.wikidata.org/wiki/.

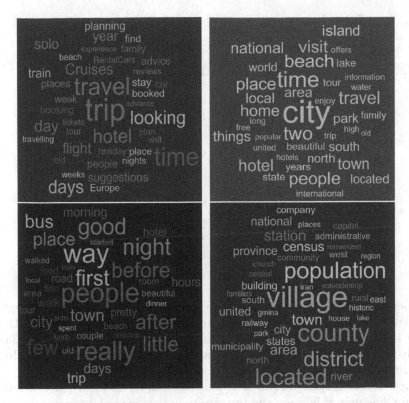

Fig. 1. Word cloud for top few words in text associated with TripAdvisorQnA (top-left), TravelWeb (top-right), TravelBlog (bottom-left) and TravelWiki (bottom-right) datasets.

The dataset consists of many key-value pairs that are present disproportionately. For example, "Country: United States" occurs 50948 times in the dataset. In order to prevent the model from getting biased into predicting the most commonly occurring value corresponding to a given key, while sampling, we ensure that the value appears in a max of 2000 instances across the entire dataset. Note that this sampling is done only for the Short QA task, since we are trying to predict a value corresponding to a key. In the case of paragraph generation task, all valid key-value pairs in the instance are used. We remove examples where the text is empty or if there are no key-value pairs in the infobox.

Figure 1 shows word clouds for top few words in text associated with TripAdvisorQnA, TravelWeb, TravelBlog and TravelWiki datasets. We manually removed stop words and create word clouds using English documents. We observe that formal words like village, country, population, district are frequent in TravelWiki. On the other hand, informal language with words like people, way, really, night are common in TravelBlog. Further, TripAdvisor has a lot of frequent words like trip, travel, suggestions, time, hotel, etc. which are related to advice around travel planning.

Table 1. Language distribution across the four datasets.

TripAdvisorQnA		TravelWeb		TravelBlog		TravelWiki	
Language	Docs (%)	Language	Docs (%)	Language	Docs (%)	Language	Docs (%)
en	58.84	en	70.93	en	88.84	en	27.28
it	21.84	de	4.91	de	2.26	pl	6.85
es	10.48	es	4.32	fr	2.14	fr	6.31
fr	4.87	fr	4.21	nl	1.19	fa	5.85
de	1.74	it	2.32	es	0.79	it	5.29
pt	1.07	fa	1.85	it	0.51	es	5.09
ru	0.88	nl	1.81	da	0.46	uk	4.71
ja	0.16	pt	1.45	fi	0.36	nl	4.66
ca	0.02	pl	0.93	no	0.29	sv	4.02
nl	0.01	ru	0.85	sk	0.26	de	3.81
Others	0.09	Others	6.42	Others	2.90	Others	26.13

Document language is not explicitly known for these datasets except for TravelWiki. Hence, we predict the same using the langdetect library. Table 1 shows language distribution across languages for these datasets. TravelBlog is heavily skewed towards English but TravelWiki and TripAdvisorQnA have higher representation from other languages.

Each dataset is split (stratified by language count) into four parts as follows: pretrain, finetune, validation and test. Pretrain part is used for pretraining, finetune part is used for task-specific finetuning, validation part is used for early stopping as well as hyper-parameter tuning and test part is used for reporting metrics. We allocate 7500 instances each for validation and test, remaining instances are divided equally into pretrain and finetune. We do not use Travel-Web dataset for any specific downstream task, thus for this dataset, we allocate 7500 instances for validation and the remaining for pretraining. Table 2 shows basic statistics of the datasets in TOURISMNLG.

Table 2. Characteristics of the datasets in TOURISMNLG.

Dataset	Domain	|Pretrain|	|Finetune|	|Dev|	|Test|	Tasks
TravelWeb	General	447053	–	7500	–	–
TravelBlog	Social Media	238143	238133	7500	7500	Blog-Title Generation
TripAdvisorQnA	Community Question Answering	101210	101142	7500	7500	Forum-Title Generation, Long Question Answering
TravelWiki	Encyclopedia	1531208	1531196	7500	7500	Paragraph Generation, Short Question Answering

3.2 TourismNLG Tasks

Our goal is to provide an accessible benchmark for standard evaluation of models in the tourism domain. We select the tasks in the benchmark based on the following principles.

- Tourism specific: Tasks should be specific to tourism or defined on tourism-specific datasets.
- Task difficulty: Tasks should be sufficiently challenging.
- Task diversity: The generated output is of different sizes. Short QA generates very short answers. Blog-title generation as well as forum-title generation generate
  sentence-sized outputs. Paragraph generation as well as Long QA tasks expect much longer outputs.
- Training efficiency: Tasks should be trainable on a single GPU for less than a day. This is to make the benchmark accessible, in particular to practitioners working with low resource languages under resource constraints.

TourismNLG consists of five generative NLP tasks. We give an overview of all tasks, including the average input and output sequence length for each task, in Table 3, and describe the tasks briefly as follows.

Paragraph Generation: Given the (key, value) pairs from Wikipedia Infobox, the goal is to generate the first paragraph of the Wikipedia page. For example, Input: "official_name=skegness; population=18910; population_date=2001; post _town=skegness; shire_district=east lindsey; shire_county=lincolnshire". Output: "skegness ist eine kleinstadt im district east lindsey der englischen grafschaft lincolnshire. sie hatte 18910 einwohner im jahre 2001. die stadt ist ein beliebtes touristenziel und ist auch bekannt als skeg, skeggy, costa del skeg oder das blackpool der ostküste und hat ein berühmtes maskottchen den jolly fisherman".

Short QA: Given the first paragraph of the Wikipedia page, and a key from the Wikipedia Infobox, the goal is to generate the correct value for that key. For example, Input: "Dinbaghan (also Romanized as Dīnbāghān; also known as Deym-e Bāghān) is a village in Jowzar Rural District, in the Central District of Mamasani County, Fars Province, Iran. At the 2006 census, its population was 115, in 23 families" and key is "County". Output: "Mamasani".

Blog-Title Generation: This is essentially the page title or headline generation task. Given body text of a travel blog article, the goal is to generate its title. For example, Input: "spent 10 days in el salvador, a country a similar size to wales. one of the first things we noticed was how polite and courteous everyone was. no-one batted an eyelid at the monstrously tall, ice-cream eating, lost europeans showing up. started off in alegria, a tiny town which is the highest in the country. a cooler mountain climate was very welcome after escaping the furnace that was san miguel...". Output: "Trip to El Salvador".

Forum-Title Generation: Given a description and an answer on a TripAdvisor Question-Answer forum page, the goal is to generate the title associated with the

Table 3. Characteristics of the tasks in TOURISMNLG. |I|= Avg Input Sequence Lengths (in words). |O|= Avg Output Sequence Lengths (in words).

Task	Dataset	Input	Output	Metrics	|I|	|O|
Paragraph Generation	TravelWiki	(key, value)	paragraph	ROUGE-1, ROUGE-L, METEOR, MRR	2.29	49.34
Short QA	TravelWiki	paragraph, key	value	F1, Accuracy, MRR	50.34	1.29
Blog-Title Generation	TravelBlog	body text	title	ROUGE-1, ROUGE-L, METEOR, MRR	793.9	4.41
Forum-Title Generation	TripAdvisorQnA	description, answer	title	ROUGE-1, ROUGE-L, METEOR, MRR	156.3	5.32
Long QA	TripAdvisorQnA	question title and description	answer	ROUGE-1, ROUGE-L, METEOR, MRR	103.2	58.46

(description, answer) pair. For example, Input: "hoping to go on a mediterranean [/cruises] cruise on september 2016. when is the best time to book to get a good price. we looked at a [/cruises] cruise to the baltics back in may 2015 and were on the point of booking it 2 months beforehand.the price changed on a daily basis,one day it was £1199,the following day it dropped to £999 and when we sorted out payment for a few days later,it went back up to £1299." Output: "when is it best to book a cruise"

Long QA: Given a question (title+description) on a TripAdvisor Question-Answer forum page, the goal is to generate the answer associated with the question. For example, Input: "trenhotel granada to barcelona 13 september. hi there i can see on renfe that i can book the trenhotel on the 12th and 14th of september but not the 13th. at first i thought it may be a saturday thing, but i checked and the saturdays before all have trenhotel.am i missing something? is it possible that this train will appear at a later date? i know its silly but we were really hoping to get this date so we didnt have to cut granada or barcelona short and spend a ful day on a train". Output: "it should run daily.just wait a bit and keep checking, it may well appear shortly. renfe are really bad about loading reservations the full 60 days out."

4   Baseline Models for TOURISMNLG

In this work, our goal is to build generic pretrained models for the tourism domain which can be finetuned for individual tasks.

4.1   Model Selection

All of our tasks contain multi-lingual data and are generative in nature. mT5 [33] and mBART [23] are both multilingual encoder-decoder Transformer models and have been shown to be very effective across multiple NLP tasks like question answering, natural language inference, named entity recognition, etc. Thus, mT5 and mBART were natural choices for our purpose. mT5 [33] was pretrained on the mC4 dataset[5] comprising of web data in 101 different languages and leverages a unified text-to-text format. mBART [23] was pretrained on the CommonCrawl corpus using the BART objective where the input texts are noised by masking phrases and permuting sentences, and a single Transformer model is learned to recover the texts.

Specifically, our mT5-base model is an encoder-decoder model with 12 layers each for encoder as well as decoder. It has 12 heads per layer, feed forward size of 2048, keys and values are 64 dimensional, d_{model}=768, and a vocabulary size of 250112. Overall the model has 582.40M parameters. Our mBART-large-50 model [31] also has 12 layers each for encoder as well as decoder. It has 16 heads per layer, feed forward size of 4096, d_{model}=1024, and a vocabulary size of 250054. Overall the model has 610.87M parameters. Note that the two models have almost the same size.

We use the mC4-pretrained mT5-base and CommonCrawl-pretrained mBART-large models, and perform domain adaptive pretraining to adapt them to the tourism domain. These are then further finetuned using task-specific labeled data. We discuss pretraining and finetuning in detail later in this section.

mT5 requires every task to be modelled as sequence-to-sequence generation task preceded by a task prompt specifying the type of task. Thus, we use this format both while pretraining as well as finetuning. For the language modeling task while domain-specific pretraining, we use the task prefix "language-modeling". For the downstream TourismNLG tasks, we use the following task prefixes: infobox-2-para, para-2-infobox, blog-title-generation, forum-title-generation, and answer-generation. Further, mBART also requires a language code to be passed as input. Thus, for mBART, we pass language code, task prefix and task-specific text as input[6].

4.2   Pre-Training and Finetuning

For domain adaptive pretraining, we leverage our four datasets described in detail in the previous section. We pretrain mT5 as well as mBART using two different approaches: MLM and MLM+Tasks. MLM models have been pretrained only on masked language modeling (MLM) loss; MLM+Tasks models are pretrained using a combination of the MLM and task-specific losses. Pretraining tasks include masked language modeling on all the four datasets, paragraph generation and Short QA on TravelWiki, blog-title generation on TravelBlog and forum-title generation and Long QA on TripAdvisorQnA.

For MLM, the goal was to reconstruct the original text across all positions on the decoder side. The decoder input is the original text with one position offset. MLM uses text combined across pretrain parts of all datasets; large input

---

[5] https://www.tensorflow.org/datasets/catalog/c4#c4multilingual_nights_stay.
[6] If the language of current instance was not among the 50 supported by the mBART model, we passed language=English.

sequences were chunked and masked to create training instances. All the other pretraining tasks are sequence generation tasks. Thus, the input was fed to the encoder and loss was defined with respect to tokens sampled at the decoder.

For pretraining, we use the standard categorical cross entropy (CCE) loss. For MLM, CCE is computed for masked words. For other tasks, CCE is computed for task-specific output words.

We finetune the pretrained models in two ways: (1) Single-task finetune (2) Multi-task finetune. Finetuning on individual tasks leads to one finetuned model per task. Managing so many models might be cumbersome. Thus, we also finetune one single model across all tasks. Another benefit of multi-task finetuning is that it can benefit from cross-task correlations.

4.3   Metrics

We evaluate our models using standard NLG metrics like ROUGE-1, ROUGE-L and METEOR for four tasks except Short QA where we report F1 and accuracy (exact match). These metrics are syntactic match-based and hence cannot appropriately evaluate predictions against the ground truth from a semantic perspective. For example a blog title like "Trip to Bombay" is semantically very similar to "Five days of fun in Mumbai, Maharashtra, India" but has no word overlap.

One approach is to create a set with the prediction and K hard negative candidates and check if the predicted output is most similar to the ground truth. Hence, we use the popular mean reciprocal rank metric (as also done in [29] for dialog quality evaluation) which is computed as follows. For every instance in the test set, we gather 10 negative candidates. Given the predicted output, we rank the 11 candidates (1 ground-truth and 10 negatives) and return the reciprocal of the rank of the ground-truth. Ranking is done in the descending order of similarity between the prediction output and candidate text using paraphrase-multilingual-MiniLM-L12-v2 model from Huggingface[7].

Negative candidates are sampled as follows. Given a test instance and its ground-truth output, we compute 20 most similar outputs from the train and dev sets of the same language. For the Short QA task, negatives are sampled from instances such that the "key" in the input also matches. Amongst most similar 20 candidates, top 10 are rejected since they could be very similar to ground-truth and hence may not be negative. The remaining ten are used as negative candidates. Note that these are fairly hard negatives and help differentiate clearly between strongly competing approaches.

4.4   Implementation Details for Reproducibility

We use a machine with 4 A100 GPUs with CUDA 11.0 and PyTorch 1.7.1. We use a batch size of 16 with AdamW optimizer. We pretrain as well as finetune

---

[7] https://huggingface.co/sentence-transformers/paraphrase-multilingual-MiniLM-L12-v2.

for 3 epochs. Maximum sequence length is set to 256 for both input as well as output. We perform greedy decoding.

Pretraining: We initialize using google/mt5-base and facebook/mbart-large-50 checkpoints. We use a learning rate of 1e-5. Pretraining takes approximately 12, 26, 14 and 37 h for mT5 MLM, mT5 MLM+Tasks, mBART MLM and mBART MLM+Tasks models respectively. We use a dropout of 0.1.

Finetuning: We use a learning rate of 5e-6 and 3e-6 for single-task finetune and multi-task finetune respectively.

Table 4. Results on TourismNLG Tasks: Long QA, Blog-title Generation, Forum-title Generation, Paragraph Generation and Short QA. For finetuning, STF=Single Task Finetune and MTF=Multi-task Finetune. For pretraining, (-) means no pretraining, (A) means MLM, (B) means MLM+Tasks. Best results in each block are highlighted.

	Model	Long QA				Blog-title generation				Forum-title generation			
		R-1	R-L	METEOR	MRR	R-1	R-L	METEOR	MRR	R-1	R-L	METEOR	MRR
MTF	mT5 (-)	9.31	7.51	4.79	62.93	16.18	15.99	9.32	20.45	21.58	21.12	13.74	27.77
	mT5 (A)	10.55	8.18	5.35	64.20	15.86	15.75	8.93	20.86	22.09	21.50	14.13	27.53
	mT5 (B)	**13.73**	**9.71**	**8.55**	68.17	17.49	17.40	9.81	22.01	25.78	25.10	16.26	31.81
	mBART (-)	10.46	7.82	6.09	70.97	19.30	19.14	11.24	23.81	27.22	26.48	17.47	33.58
	mBART (A)	11.00	7.86	7.13	72.93	20.07	19.95	11.77	24.23	29.00	28.08	19.10	35.59
	mBART (B)	12.47	8.79	8.28	**75.14**	**21.01**	**20.84**	**12.28**	**25.03**	**30.77**	**29.72**	**20.53**	**37.99**
STF	mT5 (-)	10.22	7.36	7.57	60.61	12.48	12.31	7.69	16.11	15.70	15.33	10.10	22.21
	mT5 (A)	11.80	8.46	9.11	69.39	14.90	14.74	8.72	18.99	23.92	23.12	15.96	30.80
	mT5 (B)	11.03	8.37	7.62	72.49	17.98	17.85	10.41	21.96	28.59	27.57	18.96	35.07
	mBART (-)	13.17	9.20	9.22	75.94	20.44	20.21	12.99	23.98	31.55	30.21	21.61	37.73
	mBART (A)	12.39	8.81	9.17	75.63	20.99	20.70	13.21	24.00	31.65	30.38	21.56	38.16
	mBART (B)	**13.82**	**9.88**	**9.65**	**76.12**	**21.86**	**21.59**	**13.74**	**24.95**	**33.00**	**31.56**	**22.30**	**39.42**

	Model	Paragraph generation				Short QA		
		R-1	R-L	METEOR	MRR	F1	Accuracy	MRR
MTF	mT5 (-)	23.90	21.15	17.53	48.01	48.80	64.93	75.70
	mT5 (A)	32.27	28.85	22.22	54.61	58.98	73.68	82.83
	mT5 (B)	34.26	30.63	24.27	**56.79**	61.76	75.77	84.28
	mBART (-)	33.16	29.80	24.76	54.45	62.86	76.56	85.25
	mBART (A)	**35.11**	**31.71**	**26.84**	56.64	63.89	77.45	85.98
	mBART (B)	33.87	30.46	25.75	56.16	**64.60**	**77.96**	**86.38**
STF	mT5 (-)	19.73	17.48	12.87	33.93	48.80	65.00	75.40
	mT5 (A)	25.70	22.72	17.46	43.63	62.69	76.36	84.49
	mT5 (B)	26.08	22.88	19.14	44.84	64.17	77.60	85.51
	mBART (-)	35.23	31.31	**28.04**	53.16	69.66	81.37	88.36
	mBART (A)	31.73	28.26	24.18	50.07	71.07	82.39	89.10
	mBART (B)	**35.62**	**31.43**	27.30	**55.59**	**71.17**	**82.40**	**89.16**

5 Experiments and Results

In this section, we first present the main TourismNLG benchmark results using various proposed models. Next, we briefly present notes on pretraining stability, qualitative analysis of model outputs, human evaluation and detailed error analysis.

TourismNLG Benchmark Results: Table 4 shows results obtained using our models under various pretraining and finetuning setups for the five TourismNLG tasks on the test set. From the two tables we make the following observations: (1) STF models lead to better results compared to MTF models.

But MTF models are very close across all metrics. Thus, rather than retaining individual STF models, deploying just one MTF model is recommended. (2) Domain-pretraining helps. Domain-pretrained models are better than standard models. (3) Pretraining using MLM+Tasks is better than just MLM-based pretraining. (4) Lastly, mBART models are significantly better than mT5 models except for the MTF Long QA setting.

Pretraining Stability: Figure 2 shows the variation in loss with epochs for the mT5 MLM, mT5 MLM+Tasks, mBART MLM and mBART MLM+Tasks models respectively.

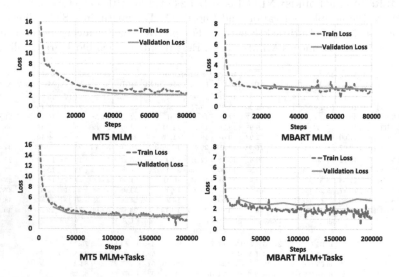

Fig. 2. Variation in Pretraining Loss on Training as well as Validation Data for our mT5 and mBART models under the MLM and MLM+Tasks settings.

Qualitative Analysis: Table 5 shows an example of generated output using our best model for each of the five tasks. Due to lack of space, we show shorter examples. We observe that the generated results are very relevant and well-formed.

Human Evaluation and Error Analysis: Finally, to check fluency and relevance of the generated outputs for various tasks, one of the authors manually labeled 50 English samples per task on a 5-point scale. Note that we did not do such an evaluation for the Short QA task since the output is just the value (and not expected to be a well-formed sentence). Table 6 shows that our model generates human consumable output with high quality. Fluency measures the degree to which a text 'flows well', is coherent [1] and is not a sequence of unconnected parts. Relevance measures correctness and the overall factual quality of the generated answer.

Table 5. Examples of predictions using our best model

Task	Input	Output
Paragraph Generation	title=amorebieta-etxano; Nome=amorebieta-etxano; divisione amm grado 1=paesi baschi; abitanti=16182	amorebieta-etxano è un comune spagnolo di 16182 abitanti situato nella comunità autonoma dei paesi baschi
Short QA	Seion Cwmaman is a Welsh Baptist church originally established in 1859. The chapel closed in 2013 but the church still meets at another location in the village. Guess "architectural type"	chapel
Blog-title Generation	Text from https://www.travelblog.org/Asia/ Malaysia/Johor/Johor-Bahru/blog-526949.html	johor bahru - day 1
Forum-title Generation	Please suggest Honeymoon destination starting the end of april for 2 weeks,and we want beach activities Thank you :) Maldives and Bali offer great honeymoon stay. Some hotels even have wedding vow renewal on the beach. Bali high end resorts even have private beach to you only!!!	honeymoon end of april
Long QA	With an extended amount of time off, my family and I are looking for a small epic adventure. Would like to include activities for the 4 year old son and relaxing for the parents. We have done the Caribbean all inclusive vacations and want something more adventurous than laying on a beach for 2 weeks. Traveling for 14 days anywhere from December to February and son is a great traveler. Any climate will do, but like the warmth and no skiing. What memorable vacations have you experienced and would suggest?	i would look at the cancun area. it is a great place to visit

Table 6. Human evaluation results

Task	Fluency	Relevance
Paragraph Generation	4.18	3.36
Blog-title Generation	4.88	4.00
Forum-title Generation	4.82	4.08
Long QA	4.28	3.84

Table 7. Error Analysis: # errors across categories for each task (out of 50 judged samples).

Error Category	Paragraph Generation	Blog-title Generation	Forum-title Generation	Long QA
Less Creative Response	5	3	0	5
Hallucination	15	5	6	3
Grammatical error	11	1	1	13
Incomplete	0	12	5	8

We also performed analysis of the kinds of errors in the generated outputs. Table 7 shows distribution of errors across major categories. Less creative responses include cases where bland responses were generated, e.g., simply using city name as blog title, repeating the question as answer for long QA task or asking the user to post on another forum, or simply concatenating the key-value pairs as output for paragraph generation task. Incomplete category includes cases like blog/forum titles that do not take into account the entire context or output that does not answer the user's question completely in long QA task. Hallucination includes cases where forum/blog titles have nothing to do with the input text, or unseen irrelevant information is added in the generated output.

6   Conclusions

In this paper, we propose the first benchmark for NLG tasks in the tourism domain. The TOURISMNLG benchmark consists of five novel natural language generation tasks. We also pretrained mT5 and mBART models using various tourism domain-specific pretraining tasks and datasets. Our models lead to encouraging results on these novel tasks. We make our code, data and pretrained models publicly available. We hope that our work will help further research on natural language generation in the travel and tourism domain. We expect that such models will help in writing automated tour guides, travel reviews and blogs, trip planning, travel advisories, multi-destination itinerary creation, and travel question answering. In the future, we plan to extend this work by including multi-modal tasks and datasets like [16].

References

1. Abhishek, T., Rawat, D., Gupta, M., Varma, V.: Fact aware multi-task learning for text coherence modeling. In: Gama, J., Li, T., Yu, Y., Chen, E., Zheng, Y., Teng, F. (eds.) Advances in Knowledge Discovery and Data Mining. PAKDD 2022. LNCS, vol. 13281, pp. 340–353. Springer, Cham (2022). https://doi.org/10.1007/978-3-031-05936-0_27
2. Alsentzer, E., et al.: Publicly available clinical BERT embeddings. arXiv preprint arXiv:1904.03323 (2019)
3. Antognini, D., Faltings, B.: HotelRec: a novel very large-scale hotel recommendation dataset. arXiv preprint arXiv:2002.06854 (2020)
4. Araci, D.: Finbert: Financial sentiment analysis with pre-trained language models. arXiv preprint arXiv:1908.10063 (2019)
5. Ayeh, J.K., Au, N., Law, R.: "Do we believe in TripAdvisor?" examining credibility perceptions and online travelers' attitude toward using user-generated content. J. Travel Res. **52**(4), 437–452 (2013)
6. Beltagy, I., Lo, K., Cohan, A.: SciBERT: a pretrained language model for scientific text. arXiv preprint arXiv:1903.10676 (2019)
7. Bennett, P., Svore, K., Dumais, S.: Classification-enhanced ranking. In: World Wide Web Conference (WWW), pp. 111–120 (2010)
8. Chalkidis, I., Fergadiotis, M., Malakasiotis, P., Aletras, N., Androutsopoulos, I.: LEGAL-BERT: the muppets straight out of law school. arXiv preprint arXiv:2010.02559 (2020)

9. Chang, H.T., Chang, Y.M., Tsai, M.T.: ATIPS: automatic travel itinerary planning system for domestic areas. Computational Intelligence and Neuroscience 2016 (2016)

10. Chen, G., Wu, S., Zhou, J., Tung, A.K.: Automatic itinerary planning for traveling services. IEEE Trans. Knowl. Data Eng. **26**(3), 514–527 (2013)

11. De Choudhury, M., Feldman, M., Amer-Yahia, S., Golbandi, N., Lempel, R., Yu, C.: Automatic construction of travel itineraries using social breadcrumbs. In: Proceedings of the 21st ACM Conference on Hypertext and Hypermedia, pp. 35–44 (2010)

12. Devlin, J., Chang, M.W., Lee, K., Toutanova, K.: BERT: pre-training of deep bidirectional transformers for language understanding. arXiv preprint arXiv:1810.04805 (2018)

13. Ercoli, S., Bertini, M., Del Bimbo, A.: Compact hash codes and data structures for efficient mobile visual search. In: 2015 IEEE International Conference on Multimedia & Expo Workshops (ICMEW), pp. 1–6. IEEE (2015)

14. Filieri, R., Alguezaui, S., McLeay, F.: Why do travelers trust tripadvisor? antecedents of trust towards consumer-generated media and its influence on recommendation adoption and word of mouth. Tour. Manage. **51**, 174–185 (2015)

15. Friggstad, Z., Gollapudi, S., Kollias, K., Sarlos, T., Swamy, C., Tomkins, A.: Orienteering algorithms for generating travel itineraries. In: Proceedings of the Eleventh ACM International Conference on Web Search and Data Mining, pp. 180–188 (2018)

16. Gatti, P., Mishra, A., Gupta, M., Gupta, M.D.: VisToT: vision-augmented table-to-text generation. In: EMNLP (2022)

17. Gurjar, O., Gupta, M.: Should i visit this place? inclusion and exclusion phrase mining from reviews. In: Hiemstra, D., Moens, M.-F., Mothe, J., Perego, R., Potthast, M., Sebastiani, F. (eds.) ECIR 2021. LNCS, vol. 12657, pp. 287–294. Springer, Cham (2021). https://doi.org/10.1007/978-3-030-72240-1_27

18. Huang, K., Altosaar, J., Ranganath, R.: ClinicalBERT: modeling clinical notes and predicting hospital readmission. arXiv preprint arXiv:1904.05342 (2019)

19. Iinuma, S., Nanba, H., Takezawa, T.: Automatic summarization of multiple travel blog entries focusing on travelers' behavior. In: Stangl, B., Pesonen, J. (eds.) Information and Communication Technologies in Tourism 2018, pp. 129–142. Springer, Cham (2018). https://doi.org/10.1007/978-3-319-72923-7_11

20. Kapoor, A., Gupta, M.: Identifying relevant sentences for travel blogs from wikipedia articles. In: Database Systems for Advanced Applications. DASFAA 2022. Lecture Notes in Computer Science, vol. 13247, pp. 532–536. Springer, Cham (2022). https://doi.org/10.1007/978-3-031-00129-1_50

21. Lee, J.S., Hsiang, J.: PatentBERT: patent classification with fine-tuning a pretrained bert model. arXiv preprint arXiv:1906.02124 (2019)

22. Lee, J., et al.: BioBERT: a pre-trained biomedical language representation model for biomedical text mining. Bioinformatics **36**(4), 1234–1240 (2020)

23. Liu, Y., et al.: Multilingual denoising pre-training for neural machine translation. Trans. Assoc. Comput. Linguist. **8**, 726–742 (2020)

24. Müller, M., Salathé, M., Kummervold, P.E.: Covid-twitter-BERT: a natural language processing model to analyse covid-19 content on twitter. arXiv preprint arXiv:2005.07503 (2020)

25. Pantano, E., Priporas, C.V., Stylos, N.: 'you will like it!'using open data to predict tourists' response to a tourist attraction. Tour. Manage. **60**, 430–438 (2017)

26. Popescu, A., Grefenstette, G.: Deducing trip related information from flickr. In: Proceedings of the 18th international conference on World Wide Web, pp. 1183–1184 (2009)

27. Radford, A., Wu, J., Child, R., Luan, D., Amodei, D., Sutskever, I.: Language models are unsupervised multitask learners. OpenAI Blog **1**(8), 1–24 (2019)

28. Raffel, C., et al.: Exploring the limits of transfer learning with a unified text-to-text transformer. arXiv:1910.10683 (2019)

29. Santra, B., et al.: Representation learning for conversational data using discourse mutual information maximization. In: Proceedings of the 2022 Annual Conference of the North American Chapter of the Association for Computational Linguistics (NAACL) (2022)

30. Skibitska, O.: The language of tourism: translating terms in tourist texts. Transl. J. **18**(4) (2015)

31. Tang, Y., et al.: Multilingual translation with extensible multilingual pretraining and finetuning. ArXiv abs/2008.00401 (2020)

32. Vaswani, A., et al.: Attention is all you need. In: NIPS, pp. 5998–6008 (2017)

33. Xue, L., et al.: mt5: a massively multilingual pre-trained text-to-text transformer. In: Proceedings of the 2021 Conference of the North American Chapter of the Association for Computational Linguistics: Human Language Technologies, pp. 483–498 (2021)

34. Yang, Z., Dai, Z., Yang, Y., Carbonell, J., Salakhutdinov, R., Le, Q.V.: XLNet: generalized autoregressive pretraining for language understanding. arXiv:1906.08237 (2019)

35. Zhang, Y., et al.: DIALOGPT:lLarge-scale generative pre-training for conversational response generation. In: Proceedings of the 58th Annual Meeting of the Association for Computational Linguistics: System Demonstration, pp. 270–278 (2020)

An Interpretable Knowledge Representation Framework for Natural Language Processing with Cross-Domain Application

Bimal Bhattarai[✉], Ole-Christoffer Granmo, and Lei Jiao

Centre for AI Research, University of Agder, Grimstad, Norway
bobsbimal58@gmail.com, {bimal.bhattarai,ole.granmo,lei.jiao}@uia.no

Abstract. Data representation plays a crucial role in natural language processing (NLP), forming the foundation for most NLP tasks. Indeed, NLP performance highly depends upon the effectiveness of the preprocessing pipeline that builds the data representation. Many representation learning frameworks, such as Word2Vec, encode input data based on local contextual information that interconnects words. Such approaches can be computationally intensive, and their encoding is hard to explain. We here propose an interpretable representation learning framework utilizing Tsetlin Machine (TM). The TM is an interpretable logic-based algorithm that has exhibited competitive performance in numerous NLP tasks. We employ the TM clauses to build a sparse propositional (boolean) representation of natural language text. Each clause is a class-specific propositional rule that links words semantically and contextually. Through visualization, we illustrate how the resulting data representation provides semantically more distinct features, better separating the underlying classes. As a result, the following classification task becomes less demanding, benefiting simple machine learning classifiers such as Support Vector Machine (SVM). We evaluate our approach using six NLP classification tasks and twelve domain adaptation tasks. Our main finding is that the accuracy of our proposed technique significantly outperforms the vanilla TM, approaching the competitive accuracy of deep neural network (DNN) baselines. Furthermore, we present a case study showing how the representations derived from our framework are interpretable. (We use an asynchronous and parallel version of Tsetlin Machine: available at https://github.com/cair/PyTsetlinMachineCUDA).

Keywords: Natural language processing (NLP) · Tsetlin machine (TM) · Propositional logic · Knowledge representation · Domain adaptation · Interpretable representation

1 Introduction

The performance of machine- and deep learning in NLP heavily relies on the representation of natural language text. Therefore, succeeding with such models

© The Author(s), under exclusive license to Springer Nature Switzerland AG 2023
J. Kamps et al. (Eds.): ECIR 2023, LNCS 13980, pp. 167–181, 2023.
https://doi.org/10.1007/978-3-031-28244-7_11

requires efficient preprocessing pipelines that produce effective data representations. Firstly, data representation influences the accuracy of the classifier by determining how much helpful information it can extract from raw data. Secondly, dense and high-dimensional representation models can be more costly to compute. Indeed, recent advances in deep neural networks (DNNs) have brought forward both the accuracy benefits and the complexity of NLP data representation.

Since natural language data is unstructured, encompassing multiple granularities, tasks, and domains, achieving sufficient natural language understanding is still challenging. Simultaneously, state-of-the-art language models like BERT and GPT-3 struggle with high computational complexity and lack of explainability [2, 35]. One might argue that attention is an explanation. However, attention merely highlights which part of the input the model used to produce its output. It does not break down the focus area into semantically meaningful units and cannot explain the ensuing reasoning process leading to an output decision [36]. Further, computation-wise, the complexity of attention is quadratic.

DNN NLP models usually represent words in vector space. Word2Vec is one early and widely used vector-based representation approach introduced by Mikolov et al. in 2013 [29]. In Word2Vec, a single-layer neural network learns the context of words and relates the words based on the inner product of context vectors. Similarly, GloVe is a popular unsupervised model incorporating corpus-wide word co-occurrence statistics [31]. The cornerstone of the latter two approaches is the distributional hypothesis, which states that words with similar contexts have similar meanings. While boosting generalization ability by co-locating similar words in vector space, the dense vectors are expensive to compute and difficult to interpret.

The Tsetlin Machine (TM) is a logic-based pattern recognition approach that blends summation-based (cf. logistic regression) and rule-based approaches (cf. decision trees). Recent studies on TMs report promising performance in NLP, including sentiment analysis [44], novelty detection [6,9], fake news detection [8], semantic relation analysis [34], and robustness toward counterfactual data [46].

The TM leverages propositional logic for interpretable modeling and bitwise operation for efficiency. Yet, recent TM research reports increasingly competitive NLP accuracy compared to deep learning at reduced complexity and increased efficiency. Simple AND rules, referred to as clauses, give these properties, employing set-of-words (SOW) as features. The clauses are self-contained, hence parallelizable [1]. Simultaneously, they can capture discriminative patterns that are interpretable [10].

Contributions: In this paper, we propose a representation learning framework for NLP classification utilizing TM. We use the TM clauses for supervised pre-training, building an abstract logical representation of the training data. We then show that the logical representation may be effective already after three epochs of training for six NLP classification tasks. We also evaluate our logic-based approach on twelve domain adaptation tasks from the Amazon dataset.

Furthermore, as the learning of TM is human-interpretable, we provide a case study to explore the explainability of our representation.

2   Related Work

Conventional representation learning mostly focuses on feature engineering for data representation. For example, [23] introduced distributed representation for symbolic data, further developed in the context of statistical language modelling [4] and in neural net language models [3]. The neural language models are based on learning a distributed representation for each word, termed as a *word embedding.* One of the most common techniques in NLP is the bag of words (BOW) representation [49], extended to n-grams, topic modelling [42], and fuzzy BOW [50]. Other techniques include representing text as graphs [28]. However, because these models lack pre-trained knowledge, the representations produced are in general not robust, and consequently, they have degraded performance [47].

In recent years, there has been tremendous progress in NLP models employing pretrained language models [19,24,33]. Most of the state-of-the-art NLP solutions are today initialized using various pre-trained input data representations such as word2vec [29], GloVe [31], and FastText [12]. These word embeddings map words into informative low-dimensional vectors, which aid neural networks in computing and understanding languages. While the initialization of input using such embeddings has demonstrated improved performance in NLP tasks, adopting these sophisticated pretrained language models for data representation comes with a cost. First, the models are intrinsically complicated, being trained on immense amounts of data through fine-tuning of a very large number of parameters [2]. Second, as complexity rises, the interpretability of the input representation becomes more ambiguous [21]. One interpretation of such models is based on the attention mechanism, which assigns weights to input features. However, a more extensive investigation demonstrates that attention weights do not in general provide useful explanations [36].

TMs [22] are a recent rule-based machine learning approach that demonstrates competitive performance with DNN, providing human-interpretable rules using propositional reasoning [5]. Several studies have demonstrated the interpretability of TM, with competitive accuracy in comparison with other deep learning approaches. Examples of applications for TM include regression [17] , natural language understanding [6–9,34,44,45], and speech understanding [26]. Furthermore, [10] analyzed the local and global interpretability of TM clauses, showing how the TM discrimination capability can be interpreted by inspecting each clause. However, these studies generally employ TM as a classifier. In this work, we exploit the data representations created by a TM to train computationally simple machine learning classifiers such as Support Vector Machine (SVM) and Logistic Regression (LR). Our intent is to develop rich context-specific language representations by using the clauses of a TM to capture the patterns and sub-patterns of data, utilized for later classification and domain adaptation tasks.

3   Data Representation Framework

3.1   Tsetlin Machine

A TM consists of dedicated teams of two-action Tsetlin Automata (TA) [41] that operate with boolean input and its negations. Each TA has $2N$ states (i.e., N states per action) and performs an action depending on its current state, which is either an "Include" action (in state 1 to N) or "Exclude" action (in state $N + 1$ to $2N$). The TA states update based on reinforcement feedback in the form of rewards and penalties. Rewards strengthen the TA action, enhancing the confidence of the TA in the current action, whereas a penalty suppresses the action. The feedback helps the TA reach the optimal action, which is the one that maximizes the expected number of rewards. The learning of TM comes from multiple teams of TAs that build conjunctive clauses in propositional logic. During learning, each TM clause captures a specific sub-pattern, comprising negated and non-negated inputs. The output is decided by counting the number of matching sub-patterns recognized by the clauses.

The TM accepts a vector $\mathbf{x} = [x_1, \ldots, x_o]$ of o propositional features as Boolean input, to be categorized into one of Cl classes, $Y = (y_1, y_2, \ldots, y_{Cl})$, where Cl is the total number of classes. These features are then turned into a set of literals that comprises of the features themselves as well as their negated counterparts: $L = \{x_1, \ldots, x_o, \neg x_1, \ldots, \neg x_o\}$.

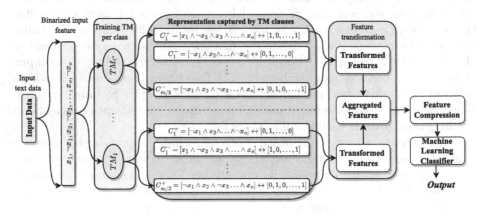

Fig. 1. Knowledge representation framework.

If there are Cl classes and m sub-patterns per class, a TM employs $Cl \times m$ conjunctive clauses to express the sub-patterns. For a given class[1], we index its clauses by j, $1 \leq j \leq m/2$, each clause being a conjunction of literals. In general, half of the clauses are assigned positive polarity, i.e., C_j^+, and the other half are

---

[1] Without loss of generality, we consider only one of the classes, thereby simplifying the notation. Any TM class is modeled and processed in the same way.

assigned negative polarity, i.e., C_j^-. A clause C_j^ψ, $\psi \in \{-, +\}$, is produced by ANDing a subset of the literals, $L_j^\psi \subseteq L$:

$$C_j^\psi(\mathbf{x}) = \bigwedge_{l_k \in L_j^\psi} l_k. \tag{1}$$

Here, $l_k, 1 \le k \le 2o$, is a feature or its negation. L_j^ψ is a subset of the literal set L. For example, a particular clause $C_j^+(\mathbf{x}) = x_1 \wedge \neg x_2$ consists of literals $L_j^+ = \{x_1, \neg x_2\}$ and it outputs 1 if $x_1 = 1$ and $x_2 = 0$.

The number of clauses m assigned to each class is user-configurable. The clause outputs are merged into a classification decision by summation and thresholding using the unit step function $u(v) = 1$ **if** $v \ge 0$ **else** 0:

$$\hat{y} = u \left(\sum_{j=1}^{m/2} C_j^+(\mathbf{x}) - \sum_{j=1}^{m/2} C_j^-(\mathbf{x}) \right). \tag{2}$$

From Eq. (2), we can see that the classification is accomplished based on a majority vote, with the positive clauses voting for the class and the negative ones voting against it. For example, the classifier $\hat{y} = u \left(x_1 \bar{x}_2 + \bar{x}_1 x_2 - x_1 x_2 - \bar{x}_1 \bar{x}_2 \right)$ captures the XOR-relation. The TM learning involves guiding the TAs to take optimal actions. Each clause receives feedback for each round of training, which is transmitted to its individual TAs. The TM utilizes Type I and Type II feedback that governs the rewards and penalties distributed to the TAs. In short, Type I feedback is designed to develop frequent patterns, eliminate false negatives, and make clauses evaluate to 1. Type II feedback, on the other hand, enhances pattern discrimination, suppresses false positives, and makes clauses evaluate to 0. Both types of feedback allow clauses to learn numerous sub-patterns from data. The details of the learning process can be found in [22].

3.2 Data Representation

The trained TM is comprised of clauses that express sub-patterns in the data. In our NLP tasks, sub-patterns typically contain contextual combinations of words that explicitly characterize a specific class. In essence, the operation of TM in NLP consists of building rules by *ANDing* groups of word literals that occur together in similar contexts. As such, the clauses (rules) are contextually rich and specific, which we here exploit to build accurate representations. By being modular and decomposable, our representation also signifies which clauses are relevant for a given input. Since our representations are based on logical rules, we will refer to them as *knowledge representations* (cf. knowledge-based systems).

Our overall procedure is illustrated in Fig. 1 and can be detailed as follows. In brief, consider a trained TM with m clauses. Given the input text $\mathbf{x} = [x_1, \ldots, x_o]$, we transform it into a representation consisting of logical rules:

$$TM_{trans}^{\mathbf{x}} = \zeta^{\mathbf{x}} = \|_{y=1}^{Cl} [C_1^{y,+}(\mathbf{x}), \ldots, C_{m/2}^{y,+}(\mathbf{x}), C_1^{y,-}(\mathbf{x}), \ldots, C_{m/2}^{y,-}(\mathbf{x})]. \tag{3}$$

Here, $\|_{y=1}^{Cl}$ denotes the array concatenation of the positive- and negative polarity clauses for class 1 to Cl. Each clause can be computed using Eq. (1).

Next, we perform feature compression since the transformed feature array produced in Eq. (3) can be too sparse for many machine learning algorithms. Assume that the total number of input examples is E and that each example is converted to a vector $\zeta^{\mathbf{x}} = (x_1, x_2, \ldots, x_{d_\zeta})$, $\zeta^{\mathbf{x}} \in R^{d_\zeta}$ of dimensionality $d_\zeta = Cl \cdot m$. The dimensionality of the input matrix then becomes $E \times d_\zeta$. We transform this input matrix further by centering the data: $A = [\Omega_1, \Omega_2, \ldots, \Omega_E]$. The center can be determined as follows:

$$\mathbf{x}_r = \frac{\sum_{e=1}^{E} \mathbf{x}_e}{E}, \tag{4}$$

$$\Omega_e = \mathbf{x}_e - \mathbf{x}_r. \tag{5}$$

The covariance matrix of A is $Cov(A, A) = E[(A - E[A])(A - E[A])^T]$ and it contains eigenvalues arranged in decreasing orders i.e., $\gamma_1 > \gamma_2 > \ldots > \gamma_E$ with corresponding eigenvectors $v_1, v_2, \ldots, v_E$. The set of original vectors can then be presented in the eigen space as follows:

$$\Omega = \alpha_1 v_1 + \alpha_2 v_2 + \ldots + \alpha_E v_E = \sum_{i=1}^{E} \alpha_i v_i. \tag{6}$$

After picking the top $\mathcal{P}$ eigenvectors v_i and corresponding eigenvalues γ_i, we have:

$$\Omega = \alpha_1 v_1 + \alpha_2 v_2 + \ldots + \alpha_{\mathcal{P}} v_{\mathcal{P}} = \sum_{i=1}^{\mathcal{P}} \alpha_i v_i, \tag{7}$$

where $\mathcal{P} << \mathcal{E}$. In the above equation, a vector of coefficients $[\alpha_1, \alpha_2, \ldots, \alpha_{\mathcal{P}}]$ represents the final representation formed in a Principal Component Analysis (PCA) space. We can observe that the number of dimensions is reduced while the most important features are retained by eigenvectors corresponding to large eigenvalues. Also, $\mathcal{P}$ eigenvalues in $\mathcal{E}$ are selected as follows:

$$\frac{\sum_{i=1}^{\mathcal{P}} \gamma_i}{\sum_{i=1}^{\mathcal{E}} \gamma_i} \geqslant \theta. \tag{8}$$

Here, θ can be 0.9 or 0.95. Now, each input example Ω_i can be expressed by a linear combination of $\mathcal{P}$ eigen vectors $\alpha_i = v_j^T \Omega_i$, where $j = 1, 2, \ldots, \mathcal{P}$, which is a compressed representation of the input given to the attached classifier, such as SVM or LR.

4   Experiments and Results

In this section, we evaluate the performance of the logical data representation created by transforming the input using the trained TMs[2].

---

[2] Classification is done using SVM from scikit-learn with default parameters.

4.1   Datasets

We conduct our experiments on six publicly accessible benchmark text classification datasets.

- TREC-6 [13] is an open-domain, fact-based question classification dataset.
- WebKB [16] comprises manually classified web pages gathered from the computer science departments of 4 universities and classified into 5 categories.
- MPQA [43] is a dataset for detecting opinion polarity.
- CR [20] is a customer review dataset with each sample labeled as positive or negative.
- SUBJ [30] is a review classification into subjective or objective.
- R8 [18] is a subset of the Reuters-21578 with 8 classes.

4.2   Implementation Details

We utilize the publicly accessible predefined train and test splits for all the datasets. The TM model is first initialized with three parameters: the number of clauses m, threshold T, and specificity s. We generate the TM representation under 3 settings: early stopping, mid stopping, and best stopping. The early stopping, mid stopping, and best stopping correspond to running our experiment with 3, 10, and 250 epochs. This enables us to observe the effect of quick TM convergence on the representation and classification performance. Following that, the TM model is trained for the different settings, and the training and testing input are transformed into respective representations using the trained model. The representations obtained at this point are uncompressed and in sparse Boolean form. Thereafter, the representation compression is done using PCA and Linear Discriminant Analysis (LDA)[3]. Finally, the compressed representation is fed into a simple machine learning classifier such as linear SVM and LR, where the only features are the transformed vectors from TM. We repeat this procedure for each setting, and the results are reported in Subsect. 4.4.

Table 1. Performance comparison (in accuracy) of vanilla TM with and without Knowledge representation in three stopping settings.

Datasets	$TM_{vanilla}$	$TM_{representation}$		
		TM_{best}	TM_{mid}	TM_{early}
TREC	91.6	95	**95.6**	92.2
MPQA	74.55	**87.3**	82.75	81.33
SUBJ	86.8	88.4	89.9	**90.1**
WebKB	91.69	**93.05**	92.19	92.47
CR	80.55	**83.06**	77.76	81.48
R8	95.93	**96.84**	96.71	96.29

---

[3] We use the default scikit-learn parameters for PCA and LDA for feature compression.

4.3   Baselines

We compared the performance of our framework with deep learning and general pre-trained models. We adopted BERT [14] as a general pre-trained baseline. These models achieve state-of-the-art performance on a variety of NLP tasks. For deep learning models, we also included long short-term memory (LSTM) and convolutional neural network (CNN). We present the results of all the baseline models from the original papers. The LSTM model in our work is from [27], which represents the entire text based on the last hidden state. We use a BiL-STM model from [15,39,48]. The CNN models are taken from [15,25], which employ pretrained word embedding for initialization. The result for DiSAN, which adopts directional self-attention, is taken from [38]. The BERT model is from [14].

Table 2. Performance comparison of our model with baseline algorithms. We reproduce the results with the same hyperparameter configurations for all baselines for a fair comparison and report average accuracy across 10 different random seeds.

Models	TREC	MPQA	SUBJ	WebKB	CR	R8
LSTM	87.19	89.43	85.66	85.32	80.06	96.09
BiLSTM	91.0	89.5	92.3	-	-	96.31
CNN-non-static	93.6	89.05	93.4	-	84.3	95.71
CNN-static	92.0	89.06	93.0	-	84.7	94.02
CNN-multichannel	92.2	89.4	93.2	-	85.0	-
DiSAN	94.2	90.1	94.2	-	84.8	-
BERT	**97.6**	**90.66**	**97.0**	79.0	**86.58**	96.02
TM	91.6	74.55	86.8	91.69	80.55	95.93
TM_{rep}	95.6	87.3	90.1	**93.05**	83.06	**96.84**

4.4   Results and Analysis

The performance of our representation is compared to vanilla TM under 3 different settings (as explained in Subsect. 4.2) shown in Table 1. The $TM_{vanilla}$ is the text classification using legacy TM. And TM_{rep} makes use of the representation generated by TM under various settings. On all datasets, we observe that the classification accuracy using the representation outperforms vanilla TM. For MPQA, we can see a massive improvement of around 13% followed by approximately 4% for TREC. With only 3 epochs, we can observe that the TM representation performs well, with the highest accuracy in the SUBJ dataset. This also demonstrates how the quick convergence of TM enables the generation of richer representation within a small number of epochs, hence benefiting representation production time.

Table 3. Computation for TM vs BERT in TREC dataset.

Parameter	$BERT$	TM
10 Epochs	297s	96s
Memory in (MB)	4637	1131

The performance result of our representation framework is compared with the other baselines (from Subsect. 4.3) in Table 2. We observe that our framework performs competitively with other baselines. The model beats all other baselines in WebKB and R8, with an accuracy of 93.05% and 96.84% respectively. WebKB largely entails classifying personal attributes captured by sparse data, such as categorizing individual students and professors from academia. The sparseness of the data may explain the poor performance of BERT according to a study that reports that BERT completely ignores the minority class at test time for low-resource tasks such as few-shot learning and rare entity recognition [40]. For TREC, MPQA, and CR, BERT outperforms all other models. Our model, on the other hand, achieves 95.6% on TREC, placing it as the second-best model in terms of performance. LSTM performs the worst of all models, whereas BiLSTM performs competitively. Surprisingly, a basic CNN model with static vectors gives competitive results against the more sophisticated attention-based DiSAN model. We see that the performance of vanilla TM falls short when compared with models initialized with pre-trained word embeddings. Overall, the TM representation performs competitively compared with computationally intensive models such as BERT, which is trained on a big text corpus. Training time and memory consumption of TM and BERT are shown in Table 3. To calculate training time, we run both TM and BERT for 10 epochs. Note that the TM is trained entirely from scratch, whereas the BERT is simply fine-tuned. We observe that the TM outperforms BERT by 3× in terms of training time and memory utilization.

4.5 Visualization

To investigate how our representation enhances the performance on NLP tasks, we plot the learned representation using t-SNE. We visualize the input with and without the TM representation in Fig. 3. Additionally, the figure contains the corresponding BERT hidden layer representation[4]. We observe that both representations provide richer and more precise information because the clusters get more separated and clear-cut. Further, notice how the TM clauses compress the data, significantly reducing the number of distinct data points (Fig. 3b). Each data point relates important features, formulated in propositional logic. Additionally, we demonstrate in Fig. 2 how the number of epochs influences the

---

[4] For BERT representation, the pretrained "BERT Base Uncased" model is fine-tuned with 3 epochs, and hidden states from the 11th layer are visualized.

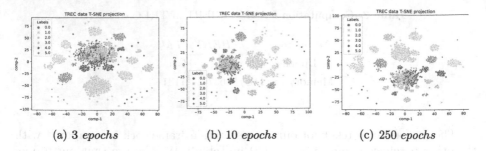

(a) 3 *epochs*   (b) 10 *epochs*   (c) 250 *epochs*

Fig. 2. Visualization of t-SNE projection of representation produced in 3 training settings on TREC dataset.

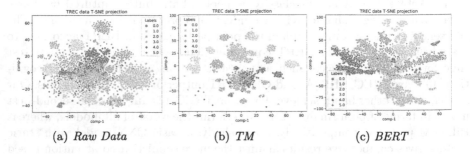

(a) *Raw Data*   (b) *TM*   (c) *BERT*

Fig. 3. Visualization of t-SNE projection for raw data, TM, and BERT representation on TREC dataset.

representation. We observe that as the number of epochs increases, the cluster becomes increasingly compact and distinct.

4.6   Concluding Remarks

From the above empirical results and visual analysis, our conclusion is that the TM representation considerably enhances the input feature space, resulting in enhanced performance. As explored further below in the case study on interpretability, the advantage of using TM can at least partially be explained by its ability to capture both semantic and structural word representations from the input. Additionally, unlike DNNs, our model provides a reasonable trade-off between performance and explainability. That is, the TM representation is computationally simple and explainable through the logic-based propositional rules composed by the clauses.

5   A Case Study: Interpretability

In this section, we demonstrate a case study showing how the representation produced by our framework is interpretable. Let us assume the following input sentence from the TREC dataset: "what is the highest waterfall in the united states?" with the label "Location" and "what is the date of boxing day?" with the label "Entity". After tokenization, the vocabulary will consist of the following tokens: ["what", "highest", "waterfall", "united", "states", "date", "boxing", "day"]. During training, the clauses capture the distinctive pattern to designate each label. Figure 4 contains some sample clauses that support "Location".

$$C_1 = [``highest", ``waterfall'"] \rightarrow C_1 = 1$$

$$C_2 = [``highest'", ``\neg boxing", ``\neg date"] \rightarrow C_2 = 0$$

$$\vdots$$

$$C_m = [``highest", ``waterfall", ``united", ``states"] \rightarrow C_m = 1$$

Fig. 4. Literals captured by clauses.

Referring to Fig. 1, the given input can be represented using TM clauses from a trained model. The representation can be written in an array: $[C_1, C_2, \ldots, C_m] \rightarrow [1, 0, \ldots, 1]$. For a given input, the representation consists of an array of clauses that are activated. And the clauses that are activated encapsulate the propositional rules necessary to make the correct classification decision. As a result, the representation is dense with information and can be completely interpretable. For example, for the above input, C_1 and C_m in Fig. 4 are activated in the representation. The vocabulary encompassed by these clauses are ["highest", "waterfall", "united", "states"]. That is, these clauses encapsulate the propositional rules associated with the label "Location".

Table 4. Domain adaptation performance (accuracy %) on Amazon review dataset.

	S-only	MMD	DANN	CORAL	WDGRL	ACAN	BERT	TM_{rep}
B → D	81.09	82.57	82.07	82.74	83.05	83.45	**86.75**	84.94
B → E	75.23	80.95	78.98	82.93	83.28	81.20	82.80	**86.21**
B → K	77.78	83.55	82.76	84.81	85.45	83.05	86.20	**87.57**
D → B	76.46	79.93	79.35	80.81	80.72	82.35	81.55	**85.06**
D → E	76.24	82.59	81.64	83.49	83.58	82.80	80.60	**86.81**
D → K	79.68	84.15	83.41	85.35	86.24	78.60	83.00	**87.75**
E → B	73.37	75.72	75.95	76.91	77.22	79.75	81.85	**84.83**
E → D	73.79	77.69	77.58	78.08	78.28	81.75	**83.85**	83.43
E → K	86.64	87.37	86.63	87.87	88.16	83.35	**90.80**	87.88
K → B	72.12	75.83	75.81	76.95	77.16	80.80	82.10	**82.30**
K → D	75.79	78.05	78.53	79.11	79.89	82.10	82.05	**83.07**
K → E	85.92	86.27	86.11	86.83	86.29	86.60	**88.35**	88.31
AVG	77.84	81.22	80.74	82.16	82.43	82.15	84.13	**85.68**

6   Application: Domain Adaptation

We here demonstrate that the input representations produced from our framework can be used in domain adaptation tasks. These results thus reinforce our previous conclusion that the representations are rich, and also applicable as contexts in cross-domain applications. We employ Amazon reviews datasets [11], which comprises 4 domains: Books (B), DVD (D), Electronic (E), and Kitchen & Housewares (K), with 12 adaptation scenarios. Each domain has around 2000 labeled and approximately 4000 unlabeled reviews. We follow the transductive setting in [32] to train in the source domain and test in the target domains. For a fair comparison, the results for the baseline algorithms are obtained directly from [37,51]. The results are summarized in Table 4. As shown, the new approach can outperform baseline algorithms in 9 out of 12 tasks. And on average, our model beats all the other algorithms.

7   Conclusion

In this paper, we propose a data representation framework that enhances the performance of Tsetlin Machines (TMs). Our approach is capable of producing more sophisticated data representation through the utilization of semantic and contextual patterns captured by clauses in TMs. We conduct extensive experiments on NLP classification and domain adaptation using publicly available datasets. In NLP classification, our experimental findings suggest that our method is competitively equal to complicated and non-transparent DNNs, including BERT. In domain adaptation, we outperform all other baselines, illustrating that the representation produced from our framework can be employed in cross-domain applications. Additionally, using a t-SNE plot, we visualize how the representation can enhance input features by utilizing distinctive decision boundaries for each class. Finally, we present a case study demonstrating the interpretability of TM-generated representation.

References

1. Abeyrathna, K.D., et al.: Massively Parallel and Asynchronous tsetlin Machine Architecture Supporting Almost Constant-Time Scaling. In: The Thirty-eighth International Conference on Machine Learning (ICML), pp. 10–20 (2021)
2. Bender, E.M., Gebru, T., McMillan-Major, A., Shmitchell, S.: On the dangers of stochastic parrots: can language models be too big? In: Proceedings of the 2021 ACM Conference on Fairness, Accountability, and Transparency, pp. 610–623 (2021)
3. Bengio, Y.: Neural net language models. Scholarpedia 3(1), 3881 (2008)
4. Bengio, Y., Ducharme, R., Vincent, P.: A neural probabilistic language model. Adv. Neural Inf. Proc. Syst. 13 (2000)
5. Berge, G.T., et al.: Using the tsetlin machine to learn human-interpretable rules for high-accuracy text categorization with medical applications. IEEE Access 7, 115134–115146 (2019)

6. Bhattarai, B., Granmo, O.C., Jiao, L.: Measuring the novelty of natural language text using the conjunctive clauses of a tsetlin machine text classifier. In: Proceedings of ICAART (2021)
7. Bhattarai, B., Granmo, O.C., Jiao, L.: Convtexttm: an explainable convolutional tsetlin machine framework for text classification. In: Proceedings of the Thirteenth Language Resources and Evaluation Conference, pp. 3761–3770 (2022)
8. Bhattarai, B., Granmo, O.C., Jiao, L.: Explainable tsetlin machine framework for fake news detection with credibility score assessment. In: Proceedings of the Thirteenth Language Resources and Evaluation Conference (2022)
9. Bhattarai, B., Granmo, O.C., Jiao, L.: Word-level human interpretable scoring mechanism for novel text detection using tsetlin machines. Appl. Intell. (2022)
10. Blakely, C., Granmo, O.: Closed-Form Expressions for Global and Local Interpretation of tsetlin Machines. In: Fujita, H., Selamat, A., Lin, J., Ali, M. (eds.) IEA/AIE 2021. LNCS (LNAI), vol. 12798, pp. 158–172. Springer, Cham (2021). https://doi.org/10.1007/978-3-030-79457-6_14
11. Blitzer, J., Dredze, M., Pereira, F.: Biographies, bollywood, boom-boxes and blenders: Domain adaptation for sentiment classification. In: Proceedings of the 45th annual meeting of the association of computational linguistics, pp. 440–447 (2007)
12. Bojanowski, P., Grave, E., Joulin, A., Mikolov, T.: Enriching word vectors with subword information. Trans. Assoc. Comput. Linguist. 5, 135–146 (2017)
13. Chang, E., Seide, F., Meng, H.M., Chen, Z., Shi, Y., Li, Y.C.: A system for spoken query information retrieval on mobile devices. IEEE Trans. Speech Audio proc. 10(8), 531–541 (2002)
14. Chen, Q., Zhang, R., Zheng, Y., Mao, Y.: Dual contrastive learning: Text classification via label-aware data augmentation. arXiv preprint arXiv:2201.08702 (2022)
15. Chen, T., Xu, R., He, Y., Wang, X.: Improving sentiment analysis via sentence type classification using bilstm-CRF and CNN. Expert Syst. Appl. 72, 221–230 (2017)
16. Craven, M.W., et al.: Learning to extract symbolic knowledge from the world wide web. In: AAAI/IAAI (1998)
17. Darshana Abeyrathna, K., Granmo, O.C., Zhang, X., Jiao, L., Goodwin, M.: The regression tsetlin machine: a novel approach to interpretable nonlinear regression. Phil. Trans. R. Soc. A 378(2164), 20190165 (2020)
18. Debole, F., Sebastiani, F.: An analysis of the relative hardness of reuters-21578 subsets. J. Am. Soc. Inform. Sci. Technol. 56(6), 584–596 (2005)
19. Devlin, J., Chang, M.W., Lee, K., Toutanova, K.: Bert: pre-training of deep bidirectional transformers for language understanding. In: NAACL (2019)
20. Ding, X., Liu, B., Yu, P.S.: A holistic lexicon-based approach to opinion mining. In: Proceedings of the 2008 international conference on web search and data mining, pp. 231–240 (2008)
21. Gilpin, L.H., Bau, D., Yuan, B.Z., Bajwa, A., Specter, M., Kagal, L.: Explaining explanations: An overview of interpretability of machine learning. In: 2018 IEEE 5th International Conference on data science and advanced analytics (DSAA), pp. 80–89. IEEE (2018)
22. Granmo, O.C.: The tsetlin machine-a game theoretic bandit driven approach to optimal pattern recognition with propositional logic. arXiv preprint arXiv:1804.01508 (2018)
23. Hinton, G., McClelland, J., Rumelhart, D.: Distributed representations. In: The Philosophy of Artificial Intelligence. Oxford University Press, pp. 248–280.(1990)

24. Ilic, S., Marrese-Taylor, E., Balazs, J.A., Matsuo, Y.: Deep contextualized word representations for detecting sarcasm and irony. In: WASSA EMNLP (2018)
25. Kim, Y.: Convolutional neural networks for sentence classification. In: Proceedings of the 2014 Conference on Empirical Methods in Natural Language Processing (EMNLP), pp. 1746–1751 (2014)
26. Lei, J., Rahman, T., Shafik, R., Wheeldon, A., Yakovlev, A., Granmo, O.C., Kawsar, F., Mathur, A.: Low-power audio keyword spotting using tsetlin machines. J. Low Power Electron. Appl. **11**(2), 18 (2021)
27. Liu, P., Qiu, X., Huang, X.: Recurrent neural network for text classification with multi-task learning. In: Proceedings of the Twenty-Fifth. International Joint Conference on Artificial Intelligence, pp. 2873–2879 (2016)
28. Luo, Y., Uzuner, Ö., Szolovits, P.: Bridging semantics and syntax with graph algorithms-state-of-the-art of extracting biomedical relations. Brief. Bio inform. **18**(1), 160–178 (2017)
29. Mikolov, T., Sutskever, I., Chen, K., Corrado, G.S., Dean, J.: Distributed representations of words and phrases and their compositionality. Adv. Neural Inf. Proc. Syst. **26** (2013)
30. Pang, B., Lee, L.: A sentimental education: Sentiment analysis using subjectivity summarization based on minimum cuts. In: Proceedings of the 42nd Annual Meeting of the Association for Computational Linguistics (ACL-04), pp. 271–278 (2004)
31. Pennington, J., Socher, R., Manning, C.D.: Glove: global vectors for word representation. In: Proceedings of the 2014 conference on empirical methods in natural language processing (EMNLP), pp. 1532–1543 (2014)
32. Qu, X., Zou, Z., Cheng, Y., Yang, Y., Zhou, P.: Adversarial category alignment network for cross-domain sentiment classification. In: Proceedings of the 2019 Conference of the North American Chapter of the Association for Computational Linguistics: Human Language Technologies, (Long and Short Papers), **1**, pp. 2496–2508 (2019)
33. Radford, A., Narasimhan, K., Salimans, T., Sutskever, I.: Improving language understanding by generative pre-training (2018)
34. Saha, R., Granmo, O.C., Goodwin, M.: Mining interpretable rules for sentiment and semantic relation analysis using tsetlin machines. In: Proceedings of International Conference on Innovative Techniques and Applications of Artificial Intelligence (2020)
35. Green, A.I.: Schwartz, R., Dodge, J., Smith, N., Etzioni, O. Commun. ACM **63**, 54–63 (2020)
36. Serrano, S., Smith, N.A.: Is attention interpretable? In: Proceedings of the 57th Annual Meeting of the Association for Computational Linguistics, pp. 2931–2951 (2019)
37. Shen, J., Qu, Y., Zhang, W., Yu, Y.: Wasserstein distance guided representation learning for domain adaptation. In: Thirty-second AAAI conference on artificial intelligence, pp. 4058–4065 (2018)
38. Shen, T., Zhou, T., Long, G., Jiang, J., Pan, S., Zhang, C.: Disan: Directional self-attention network for RNN/CNN-free language understanding. In: Proceedings of the AAAI conference on artificial intelligence (2018)
39. Tai, K.S., Socher, R., Manning, C.D.: Improved semantic representations from tree-structured long short-term memory networks. In: Proceedings of the 53rd Annual Meeting of the Association for Computational Linguistics and the 7th International Joint Conference on Natural Language Processing (Long Papers), **1**, pp. 1556–1566 (2015)

40. Tänzer, M., Ruder, S., Rei, M.: Memorisation versus generalisation in pre-trained language models. In: Proceedings of the 60th Annual Meeting of the Association for Computational Linguistics (Long Papers)**1**, pp. 7564–7578 (2022)

41. Tsetlin, M.L.: On behaviour of finite automata in random medium. Avtomat. i Telemekh **22**(10), 1345–1354 (1961)

42. Wallach, H.M.: Topic modeling: beyond bag-of-words. In: Proceedings of the 23rd international conference on Machine learning, pp. 977–984 (2006)

43. Wilson, T., Wiebe, J., Hoffmann, P.: Recognizing contextual polarity in phrase-level sentiment analysis. In: Proceedings of human language technology conference and conference on empirical methods in natural language processing, pp. 347–354 (2005)

44. Yadav, R., Jiao, L., Granmo, O.C., Goodwin, M.: Human-level interpretable learning for aspect-based sentiment analysis. In: Proceedings of AAAI (2021)

45. Yadav, R.K., Jiao, L., Granmo, O.C., Goodwin, M.: Interpretability in word sense disambiguation using tsetlin machine. In: Proceedings of ICAART, pp. 402–409 (2021)

46. Yadav, R.K., Jiao, L., Granmo, O.C., Goodwin, M.: Robust Interpretable Text Classification against Spurious Correlations Using and-rules with Negation. In: The 31st International Joint Conference on Artificial Intelligence (IJCAI) (2022)

47. Yang, J., et al.: A survey of knowledge enhanced pre-trained models. arXiv preprint arXiv:2110.00269 (2021)

48. Zhang, T., Huang, M., Zhao, L.: Learning structured representation for text classification via reinforcement learning. In: Proceedings of the AAAI Conference on Artificial Intelligence (2018)

49. Zhang, Y., Jin, R., Zhou, Z.H.: Understanding bag-of-words model: a statistical framework. Int. J. Mach. Learn. Cybern. **1**(1), 43–52 (2010)

50. Zhao, R., Mao, K.: Fuzzy bag-of-words model for document representation. IEEE Trans. Fuzzy Syst. **26**(2), 794–804 (2017)

51. Zhou, J., Tian, J., Wang, R., Wu, Y., Xiao, W., He, L.: Sentix: A sentiment-aware pre-trained model for cross-domain sentiment analysis. In: Proceedings of the 28th International Conference on Computational Linguistics, pp. 568–579 (2020)

Graph-Based Recommendation for Sparse and Heterogeneous User Interactions

Simone Borg Bruun[1]([✉])(iD), Kacper Kenji Leśniak[1], Mirko Biasini[2],
Vittorio Carmignani[2], Panagiotis Filianos[2], Christina Lioma[1](iD),
and Maria Maistro[1](iD)

[1] Department of Computer Science, University of Copenhagen,
Copenhagen, Denmark
{simoneborgbruun,kkl,c.lioma,mm}@di.ku.dk
[2] FullBrain, Copenhagen, Denmark
{mirko,vitto,panos}@fullbrain.org

Abstract. Recommender system research has oftentimes focused on approaches that operate on large-scale datasets containing millions of user interactions. However, many small businesses struggle to apply state-of-the-art models due to their very limited availability of data. We propose a graph-based recommender model which utilizes heterogeneous interactions between users and content of different types and is able to operate well on small-scale datasets. A genetic algorithm is used to find optimal weights that represent the strength of the relationship between users and content. Experiments on two real-world datasets (which we make available to the research community) show promising results (up to 7% improvement), in comparison with other state-of-the-art methods for low-data environments. These improvements are statistically significant and consistent across different data samples.

Keywords: Personalized page rank · Genetic algorithm · Collaborative filtering

1 Introduction

With the advent of the internet, huge amounts of data have become available. This allows to design and develop novel Recommender Systems (RSs) based on complex Machine Learning (ML) and Deep Learning (DL) approaches, often characterized as data-hungry approaches. Many recent recommender models belong to this category, so a recommender dataset of size 100 K might already be considered small [23]. Moreover, when using such datasets, a pre-processing step is often applied to remove all users with less than a certain number of interactions, e.g., 5, because several models are not able to learn with only few data points per user [14,24,28,42].

S. B. Bruun and K. K. Leśniak—Contributed equally.

In the era of big data, Small and Medium Enterprises (SMEs) struggle to find their way, given that they might not have access to such a huge amount of data. However, SMEs are fundamental actors in the global economy, as they represent about 90% of businesses and more than 50% of employment worldwide [11]. In these cases, RSs able to cope with low data scenarios are necessary [13].

In ML and DL, small data problems are notoriously hard and are usually solved with a number of well-studied techniques [29]: (1) data augmentation, where synthetic samples are generated from the training set [27,44,51]; (2) transfer learning, where models learn from a related task and transfer the knowledge [25,50]; (3) self supervision, where models learns from pseudo or weak labels [37,49]; (4) few-shot learning, i.e., (meta-)learning from many related tasks with the aim of improving the performance on the problem of interest [35,39,40]; (5) exploiting prior knowledge manually encoded, for example external side information and Knowledge Graphs (KGs) [3,47]. However, except for hand-coded knowledge, the above approaches still require a considerable amount of initial data or access to a different, but similar domain, where plenty of data is available. On the other side, knowledge bases are application dependent, require access to expert knowledge, and are not always available.

In this paper, we **contribute** a novel recommender approach able to operate in small data scenarios: our model does not require large volumes of initial data and is not application dependent. We use a heterogeneous graph, where vertices denote entities, e.g., users and different types of content, and edges represent interactions between users and content, e.g., a user posting a message on a social media. Then we use Personalized PageRank (PPR) to recommend items. Note that, edges represent any interaction with users and content, not only interaction with recommendable items. We assign weights to edges in the graph, which represent the strength of the relationship between users and content. In previous work within RSs [26,52,53], such weights are usually pre-defined depending on the application. We do not make any assumption on the values of such weights and optimize them with a genetic algorithm [17]. To the best of our knowledge, heuristic algorithms have never been applied to learn edge weights in the context of RSs. Our approach is evaluated on two real-world use cases: (1) an emergent educational social network, where there are few user interactions due to the initial stage of the platform, but a large number of items; (2) an insurance e-commerce platform, where there are many users but few user interactions, because users do not interact often with insurance products, and few items by nature of the insurance domain. Experimental results are promising, showing up to a 7% improvement over state-of-the-art baselines.

2   Related Work

Recommendation with small data has been tackled heuristically, i.e., by recommending items based on a set of specific rules [18]. Such rules have to be designed for each use case, making these models application dependent. Hybrid RSs have also been proposed for small data, for instance by merging Content

Based (CB) and association rules [19,20]. Note that the datasets in [19,20] are not publicly available. Item-to-item recommendation is addressed in [36] with a CF approach as a counterfactual problem, where a small collection of explicit user preferences is used to improve propensity estimation. We cannot use this in our work because: (1) our task is not item-to-item recommendations; (2) we do not have access to explicit user preferences; (3) a large dataset (MovieLens 25M [15]) is still needed to estimate propensity (the small annotated dataset is only used to debias the propensity estimate). In [38], a hybrid user-based model combines CF, rule-based recommendation, and the top popular recommender with domain-specific and contextual information in the area of a small online educational community. The dataset is not publicly available and the approach is domain-dependent, hence not applicable to our work. Finally, conversion rate prediction for small-scale recommendation is used in [32], with an ensemble of deep neural networks that are trained and evaluated on a non-public dataset of millions of users, impressions, and clicks. Our small data scenario does not include enough data to train this ensemble model.

Solutions for cold start cases (where users or items have few or no interactions) are hybrid combinations of CF, CB, demographic and contextual information [5,12,33], or ML methods such as data augmentation [51], transfer learning [2,50], etc. (see Sect. 1). Data augmentation is used in [27], where a CF model creates synthetic user ratings and is then combined with a CB model. We cannot use this in our task because we do not have explicit ratings (we use any user interaction as implicit feedback). In [49], self-supervision and data augmentation are combined on the user-item graph, and in [37], self-supervision on the user-item graph is enhanced with features extracted from user reviews. Few shot learning and meta-learning have also been used. In [35], a neural recommender is trained over head items with frequent interactions, and this meta-knowledge is transferred to learn prototypes for long-tail items. In [39], recommendations for cold users are generated with a meta-learner that accounts for interest drift and geographical preferences. In [3], knowledge bases (KG) are used to enrich feature representations, and in [47] a neural attention mechanism learns the high order relation in the user-item graph and the KG.

All above approaches [2,3,35,37,39,47,49–51] are evaluated on popular publicly available datasets, e.g., MovieLens [15], Yelp, Amazon, CiteULike [43], Weeplaces, etc. These datasets are much larger than those in our case (see Tables 1 and 2) and allow using self-supervision, few-shot learning, attention mechanisms, and other neural models that we cannot use due to the extremely low amount of data. Transfer learning and domain adaptation require large amounts of training data from a similar task or a related domain, which are not (publicly) available for our use cases.

Lastly, graph-based RSs can be robust as they enable information to propagate through vertices, unlike matrix completion which is affected by data sparsity [45]. This motivates recent approaches using GNN [34,37,46,49,51]. However, these are not applicable to small data problems because there are not enough samples to train GNN models. PathRank [26] uses a heterogeneous user-item graph with additional vertices that are attributes of items, e.g., movie genre, director, etc. Recommendations are generated with a random walk similar to PPR, but constraints are used to ensure that the random walks follow certain predefined paths. These are application dependent. In [54], the user-item graph is extended with item attributes, and meta-paths are defined to determine how two entities in the graph (vertices of different types) are connected (this encodes entity similarity). A preference diffusion score is defined for specific meta-paths, based on user implicit feedback and co-occurrences of entities, and used to recommend items. Unlike [26,54], we build the heterogeneous graph from all user interactions, not only interactions with items. We also do not include item attributes in the graph and we do not use predefined paths or meta-paths. We assume any possible path and optimize edge weights with a genetic algorithm.

Injected Preference Fusion (IPF) [52] extends PPR with a session-based temporal graph (STG) that includes both long- and short-term user preferences. STG is a bipartite graph where users, items, and sessions are vertices. Nonnegative weights are associated with edges, which control the balance between long- and short-term preferences. Multi-Layer Context Graph (MLCG) [53] is a three-layer graph, where each layer represents a different type of context: (1) user context, e.g., gender and age; (2) item context, e.g., similarity between items; and (3) decision context, e.g., location and time. Different weights are associated with intra- and inter-layer edges, defined as functions of vertice co-occurrence. Unlike [52,53], (1) we represent in the graph all user interactions (not only those with recommendable items); (2) we do not consider temporal, contextual, or demographic features, which may not be available; (3) we do not use predefined weights, but we optimize them with a genetic algorithm.

3 Approach

Usually, graph-based recommendation consists of 2 steps: (1) building the graph structure (Sect. 3.1), and (2) recommending items (Sect. 3.2). We introduce an intermediate step between (1) and (2), where we optimize weights associated to different edge types (Sect. 3.3).

3.1 Heterogeneous Graph Representation of User Interactions

Let us consider a set of users who interact with content of various types, for example posts and comments in social networks or items and services in an online store. We represent users and content (vertices), and their relationships (edges) with a heterogeneous graph [41]. Vertices belong to different types, e.g., users, items, posts, etc. Edges have different types depending on the action that they represent, e.g., the edge with the type "like" can connect a user with a post. Moreover, edges have a direction because some actions are not symmetric, e.g., a user can follow another user, but not be followed by the same user. Note that all types of user interactions are included in the graph, i.e., also interactions

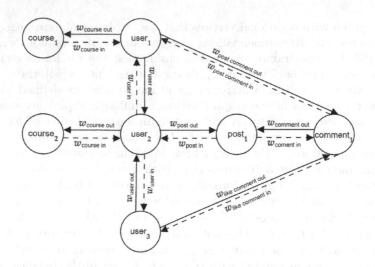

Fig. 1. Example of the heterogeneous graph in a social network.

with objects that are not recommendable items, for example, a user creating a post or reporting a claim.

A heterogeneous graph, or heterogeneous information network, is a type of directed graph $G = (\mathcal{V}, \mathcal{E}, \mathcal{A}, \mathcal{R})$, where vertices and edges represent different types of entities and relationships among them. Each vertex $v \in \mathcal{V}$ and edge $e \in \mathcal{E}$ is associated to its type through a mapping function $\tau \colon \mathcal{V} \to \mathcal{A}$ and $\phi \colon \mathcal{E} \to \mathcal{R}$ respectively. $\mathcal{A}$ is the set of vertex types, or tags, e.g., users or various type of content. $\mathcal{R}$ is the set of edge types, e.g., a user liking a post.

We denote edges as $e = (i,j)$, where i and $j \in \mathcal{V}$ and $i \neq j$. Edge types are mapped to positive weights representing the strength of the relationship between two vertices. Formally, given an edge (i,j), we define a weight function $W \colon \mathcal{R} \to \mathbb{R}^{+}$ such that $W(\phi((i,j))) = w_{i,j}$, with the constraint that $w_{i,j} > 0$ (Sect. 3.3 explains how to compute optimal weights $w_{i,j}$).

Since G is a directed graph, $\phi((i,j)) \neq \phi((j,i))$, i.e., $\mathcal{R}$ contains distinct types for ingoing and outgoing edges. Therefore, each user interaction is represented as two weighted edges[1], $w_{i,j}$ from the user to another vertex and $w_{j,i}$ from that vertex to the user vertex, e.g., a user liking a post and a post being liked by a user. The weights for those outgoing and ingoing edges might differ. Edges can also exist between two content vertices when two entities are related (for example a comment that was created under a post). In case of multiple interactions between 2 vertices, e.g., a user can both create a post and like it, we define a different type of edge, i.e., a new value in $\mathcal{R}$, which represents the two actions. The weight of such edge corresponds to the sum of the weights of each individual type, e.g., the creation and liking of a post. In practice, this happens only for a few actions and does not affect the size of $\mathcal{R}$ significantly.

---

[1] Except for non-symmetric actions, e.g., following a user.

Figure 1 illustrates an example of a heterogeneous graph in an educational social network, where user 1 follows course 1 (note that course means university course in an educational setting), user 2 follows course 2, and user 1 and 3 follow user 2. Furthermore, user 2 has created post 1, user 1 has created comment 1 under post 1 and user 3 has liked comment 1.

3.2 Generating Recommendations Using Random Walks

Given a user, the recommendation task consists in ranking vertices from the heterogeneous graph. To do this, we use PPR [31]. Starting at a source vertex s, the PPR value of a target vertex, t, is defined as the probability that an α-discounted random walk from vertex s terminates at t. An α-discounted random walk represents a random walker that, at each step, either terminates at the current vertex with probability α, or moves to a random out-neighbor with probability $1 - \alpha$. Formally, let G be a graph of n vertices, let $O(i)$ denote the set of end vertices of the outgoing edges of vertex i, and let the edge (i, j) be weighted by $w_{ij} > 0$. The steady-state distribution of an α-discounted random walk in G, starting from vertex s, is defined as follows:

$$\pi = (1-\alpha)P^T\pi + \alpha e_s \quad \text{where} \quad P = (p_{i,j})_{i,j\in V} = \frac{w_{i,j}}{\sum_{k\in O(i)} w_{i,k}} \cdot \mathbb{1}_{\{j\in O(i)\}} \quad (1)$$

$\alpha \in (0,1)$, P is the transition matrix, e_s is a one-hot vector of length n with $e_s(s) = 1$, and $\mathbb{1}$ is the indicator function, equal to one when $j \in O(i)$. Eq. (1) is a linear system that can be solved using the power-iteration method [31].

Solving Eq. (1) returns a π for each user containing the PPR values (i.e., the ranks) of all the content vertices with respect to that user. A recommendation list is then generated by either ordering the content vertices by their ranks and selecting the top-k, or by selecting the most similar neighbors by their ranks, then ordering the content by the neighbors' interaction frequency with the content. We implement both methods (see Sect. 4.3).

3.3 Optimizing Edge Weights Using Genetic Algorithm

Next, we explain how to compute the weight function W, which assigns the optimal weights for outgoing and ingoing edges of each interaction type. In our data (which is presented in Sect. 4.1), the number of interaction types was 11 and 9, which required to optimize respectively 22 and 18 parameters (see Table 3). With such a large search space, using grid search and similar methods would be very inefficient. Instead, we use a heuristic algorithm to find the optimal weights. Heuristic methods can be used to solve optimization problems that are not well suited for standard optimization algorithms, for instance, if the objective function is discontinuous or non-differentiable. In particular, we use a genetic algorithm [17] as our optimization algorithm, as it is a widely known and used algorithm, which is relatively straightforward to get started with and has been shown to serve as a strong baseline in many use cases [30]. The algorithm in [17] consists of 5 components, which in our case are specified as follows: **1. Initial population:** A population consisting of a set of gene vectors is initialized. In our case, each gene vector is a vector of weights with size $|\mathcal{R}|$, and each gene is uniformly initialized from a predefined range. **2. Fitness function:** Each of the initialized gene vectors is evaluated. In our case, recommendations are generated with PPR as described in Sect. 3.2 where the graph is weighted by the genes. The fitness function can be any evaluation measure that evaluates the quality of the ranked list of recommendations, e.g., normalized Discounted Cumulative Gain (nDCG), Mean Average Precision (MAP), etc. **3. Selection:** Based on the fitness score, the best gene vectors are selected to be parents for the next population. **4. Crossover:** Pairs of parents are mated with a uniform crossover type, i.e., offspring vectors are created where each gene in the vector is selected uniformly at random from one of the two mating parents. **5. Mutation:** Each gene in an offspring vector has a probability of being mutated, meaning that the value is modified by a small fraction. This is to maintain diversity and prevent local optima. Finally, new offspring vectors are added to the population, and step 2 to 5 are repeated until the best solution converges.

4   Experiments

Next we describe the experimental evaluation: the use cases and datasets in Sect. 4.1; training and evaluation in Sect. 4.2; baselines and hyperparameters in Sect. 4.3; and results in Sect. 4.4. The code is publicly available[2].

4.1   Use Cases and Datasets

To evaluate our model, we need a dataset that satisfies the following criteria: (1) interaction scarcity; (2) different types of actions which might not be directly associated with items. To the best of our knowledge, most publicly available datasets include only clicks and/or purchases or ratings, so they do not satisfy the second criterion. We use two real-world datasets described next. The **educational social network** dataset was collected from a social platform for students between March 17, 2020 to April 6, 2022. We make this dataset public available[3]. The users can, among others, follow courses from different universities, create and rate learning resources, and create, comment and like posts. The content vertices are: courses, universities, resources, posts, and comments and the goal is to recommend courses. The platform is maintained by a SME in the very early stage of growth and the dataset from it contains 5088 interactions, made by 878 different users with 1605 different content objects resulting in a data sparsity of 0.996. Dataset statistics are reported in Table 1.

The second dataset is an **insurance** dataset [7] collected from an insurance vendor between October 1, 2018 to September 30, 2020[4]. The content vertices

---

[2] https://github.com/simonebbruun/genetically_optimized_graph_RS.
[3] https://github.com/carmignanivittorio/ai_denmark_data.
[4] https://github.com/simonebbruun/cross-sessions_RS.

Table 1. Dataset statistics: educational social network.

Type of interaction	Follow		Post		Comment		Source		Join University
	Course	User	Create	Like	Create	Like	Create	Rate	
Training	1578 (28.12%)	842 (15%)	92 (1.64%)	339 (6.04%)	116 (2.07%)	96 (1.71%)	75 (1.34%)	113 (2.01%)	1400 (24.95%)
Validation	415 (7.39%)								
Test	546 (9.73%)								

Table 2. Dataset statistics: insurance dataset.

Type of interaction	Purchase items	E-commerce		Personal account		Claims reporting		Information	
		Items	Services	Items	Services	Items	Services	Items	Services
Training	4853 (13.65%)	6897 (19.4%)	1775 (4.99%)	287 (0.81%)	17050 (47.96%)	154 (0.43%)	6 (0.02%)	1129 (3.18%)	2118 (5.96%)
Validation	601 (1.69%)								
Test	680 (1.91%)								

are items and services within a specified section of the insurance website being either e-commerce, claims reporting, information or personal account. Items are insurance products (e.g., car insurance) and additional coverages of insurance products (e.g., roadside assistance). Services can, among others, be specification of "employment" (required if you have an accident insurance), and information about "the insurance process when moving home". User interactions are purchases and clicks on the insurance website. The goal is to recommend items. The dataset contains 432249 interactions, made by 53569 different users with 55 different item and service objects resulting in a data sparsity of 0.853. Dataset statistics are reported in Table 2.

4.2 Evaluation Procedure

We split the datasets into training and test set as follows. As test set for the educational social network dataset, we use the last course interaction (leave-one-out) for each user who has more than one course interaction. The remaining is used as training set. All interactions occurring after the left-out course interaction in the test set are removed to prevent data leakage. As test set for the insurance dataset, we use the latest 10% of purchase events (can be one or more purchases made by the same user). The remaining interactions (occurring before the purchases in the test set) are used as training set.

For each user in the test set, the RS generates a ranked list of content vertices to be recommended. For the educational social network dataset, courses that the user already follows are filtered out from the ranked list. For the insurance dataset, it is only possible for a user to buy an additional coverage if the user has the corresponding base insurance product, therefore we filter out additional coverages if this is not the case, as per [1].

As evaluation measures, we use Hit Rate (HR) and Mean Reciprocal Rank (MRR). Since in most cases, we only have one true content object for each user in the test set (leave-one-out), MRR corresponds to MAP and is somehow proportional to nDCG (they differ in the discount factor). For the educational social network, we use standard cutoffs, i.e., 5 and 10. For the insurance dataset, we use a cutoff of 3 because the total number of items is 16, therefore with higher cut-offs all measures will reach high values, which will not inform on the actual quality of the RSs.

4.3   Baselines, Implementation, and Hyperparameters

The focus of this work is to improve the quality of recommendations on small data problems, such as the educational social network dataset. Therefore, we consider both simple collaborative filtering baselines that are robust on small datasets as well as state-of-the-art neural baselines: *Most Popular* recommends the content with most interactions across users; *UB-KNN* is a user-based nearest neighbor model that computes similarities between users, then ranks the content by the interaction frequency of the top-k neighbors. Similarity is defined as the cosine similarity between the binarized vectors of user interactions; *SVD* is a latent factor model that factorizes the matrix of user interactions by singular value decomposition [8]; *NeuMF* factorizes the matrix of user interactions and replaces the user-item inner product with a neural architecture [16]; *NGCF* represents user interactions in a bipartite graph and uses a graph neural network to learn user and item embeddings [48]; *Uniform Graph* is a graph-based model that ranks the vertices using PPR [31] with all edge weights equal to 1; *User Study Graph* is the same as uniform, but the weights are based on a recent user study conducted on the same educational social network [6]. Users assigned 2 scores to each action type: the effort required to perform the action, and the value that the performed action brings to the user. We normalized the scores and used effort scores for outgoing edges and value scores for ingoing edges. The exact values are in our source code. A similar user study is not available for the insurance domain.

All implementation is in `Python 3.9`. Hyperparameters are tuned on a validation set, created from the training set in the same way as the test set (see Sect. 4.2). For the educational social network, optimal hyperparameters are the following: damping factor $\alpha = 0.3$; PPR best predictions are obtained by ranking vertices; 30 latent factors for SVD; and number of neighbors $k = 60$ for UB-KNN. Optimal hyperparametrs in the insurance dataset are: damping factor $\alpha = 0.4$; PPR best predictions are obtained by ranking user vertices and select the closest 90 users; 10 latent factors for SVD; and number of neighbors $k = 80$ for UB-KNN.

The genetic algorithm is implemented with `PyGAD 2.16.3` with MRR as fitness function and the following parameters: initial population: 10; gene range: $[0.01, 2]$, parents mating: 4; genes to mutate: 10%; mutation range: $[-0.3, 0.3]$. We optimize the edge weights using the training set to build the graph and the validation set to evaluate the fitness function. The optimal weights are reported

Table 3. Optimized interaction weights averaged over five runs of the genetic algorithm.

(a) Educational social network dataset.

Trained for	Course follows	
Direction of edge	Out	In
User follows user	0.95	0.86
User follows course	0.73	1.88
User creates post	1.11	1.09
User creates resource	1.27	0.55
User creates comment	0.91	1.03
User likes post	1.27	0.61
User likes comment	0.42	0.99
User rates resource	0.77	0.84
Comment under post	0.91	1.17
User joins university	0.28	1.06

(b) Insurance dataset.

Trained for	Purchase items	
Direction of edge	Out	In
Purchase items	0.24	1.05
E-commerce items	0.64	1.49
E-commerce services	1.21	1.18
Personal account items	1.06	0.36
Personal account services	0.92	0.66
Claims reporting items	0.93	0.58
Claims reporting services	0.90	1.32
Information items	1.60	0.79
Information services	0.64	0.51

in Table 3. In order to provide stability of the optimal weights, we report the average weights obtained by five runs of the genetic algorithm.

4.4   Results

Table 4 reports experimental results. On both datasets, UB-KNN outperforms SVD and NeuMF, and the best-performing baseline is the uniform graph-based model on the educational social network dataset and the NGCF model on the insurance dataset. This corroborates previous findings, showing that graph-based RSs are more robust than matrix factorization when data is sparse [45] and neural models need a considerable amount of data to perform well. Graph-based methods account for indirect connectivity among content vertices and users, thus outperforming also UB-KNN, which defines similar users on subsets of commonly interacted items. Our genetically optimized graph-based model outperforms all baseline models on the educational social network dataset and obtains competing results with the NGCF model on the insurance dataset, showing that the genetic algorithm can successfully find the best weights, which results in improved effectiveness. In order to account for randomness of the genetic algorithm, we run the optimization of weights five times and report the mean and standard deviation of the results in Table 4. The standard deviation is lowest on the insurance dataset, but even on the very small educational dataset, the standard deviation is relatively low, so for different initializations, the algorithm tends to converge toward similar results. Moreover, we tried a version of our model where we let the graph be an undirected graph, meaning that for each edge type the weight for the ingoing and the outgoing edge is the same. The results show that the directed graph outperforms its undirected version. For the educational social network, weights based on the user study result in worse performance, even lower than UB-KNN.

Table 4. Performance results (†mean/std). All results marked with * are significantly different with a confidence level of 0.05 from the genetically optimized graph (ANOVA [22] is used for MRR@k and McNemar's test [9] is used for HR@k). The best results are in bold.

Dataset	Educational social network				Insurance	
Measure	MRR@5	MRR@10	HR@5	HR@10	MRR@3	HR@3
Most Popular	0.0797*	0.0901*	0.1978*	0.2729*	0.4982*	0.6791*
SVD	0.3639*	0.3767*	0.5275*	0.6209*	0.5787*	0.7399*
NeuMF	0.3956*	0.4110*	0.5604*	0.6740*	0.5937*	0.7448*
UB-KNN	0.4304*	0.4456*	0.6172*	0.7271*	0.6238*	0.7569*
NGCF	0.4471	0.4592	0.6245*	0.7143*	**0.6517**	**0.8043***
Uniform Graph	0.4600	0.4735	0.6300*	0.7289*	0.6263*	0.7730*
User Study Graph	0.4162*	0.4330*	0.5952*	0.7179*	-	-
Genetically Undirected Graph	0.4809	0.4957	0.6410	0.7509	0.6339	0.7760*
Genetically Directed Graph†	**0.4907/** 0.0039	**0.5045/** 0.0037	**0.6505/** 0.0052	**0.7520/** 0.0055	0.6435/ 0.0029	0.7875/ 0.0044

Overall, scores are higher on the insurance data. This might happen because: (1) data from the educational social network is sparser than insurance data (see Sect. 4.1); (2) the insurance data has a considerably larger training set; (3) there are fewer items to recommend in the insurance domain (16 vs. 388).

Figure 2 shows MRR at varying cutoffs k. We have similar results for HR, which are omitted due to space limitations. It appears that the results are consistent for varying thresholds. Only on the insurance dataset, we see that the UB-KNN is slightly better than the uniform graph-based model for smaller thresholds.

Inspecting the optimal weights in Table 3, we see that for the educational social network all the interaction types associated with courses (following course, creating resource, creating comment, creating and liking posts) are highly weighted. This is reasonable since courses are the recommended items. Moreover, a higher weight is assigned when a user follows a user compared to when a user is followed by a user. This reasonably suggests that users are interested in courses attended by the users they follow, rather than the courses of their followers. For the insurance dataset, we observe that the greatest weights are given when a user clicks on items in the information and personal account section, when items are purchased by a user, and when items and services are clicked in the e-commerce section, which are all closely related to the process of purchasing items.

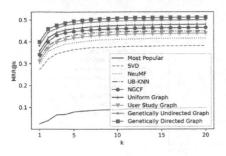
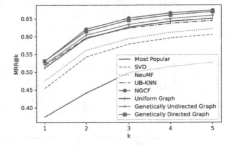

(a) Educational social network dataset.   (b) Insurance dataset.

Fig. 2. MRR@k for varying choices of the cutoff threshold k.

Table 5. Results on smaller samples of the insurance dataset. The notation is as in Table 4.

Percentage of insurance dataset	10%		25%		50%		100%	
Measure	MRR@3	HR@3	MRR@3	HR@3	MRR@3	HR@3	MRR@3	HR@3
Most popular	0.5003*	0.6856*	0.4981*	0.6789*	0.5003*	0.6856*	0.4982*	0.6791*
SVD	0.5785*	0.7336*	0.5794*	0.7395*	0.5758*	0.7379*	0.5787*	0.7399*
NeuMF	0.5792*	0.7326*	0.5759*	0.7291*	0.5849*	0.7466*	0.5937	0.7448
UB-KNN	0.6183	0.7661*	0.6180*	0.7494*	0.6243	0.7524*	0.6238*	0.7569*
NGCF	0.5937*	0.7199*	0.6030*	0.7405*	**0.6397**	0.7894	**0.6517**	**0.8043***
Uniform Graph	0.6196	0.7687*	0.6206*	0.7687*	0.6253	0.7746	0.6263*	0.7730*
Genetically Undirected Graph	0.6238	0.7672*	0.6224	0.7601*	0.6286	0.7741	0.6339	0.7760
Genetically Directed Graph[†]	**0.6267/** 0.0052	**0.7784/** 0.0084	**0.6353/** 0.0039	**0.7845/** 0.0073	**0.6397/** 0.0045	**0.7903/** 0.0071	0.6435/ 0.0029	0.7875/ 0.0044

We further evaluate the performance of our models, when trained on smaller samples of the insurance dataset. We randomly sample 10%, 25% and 50% from the training data, which is then split into training and validation set as described in Sect. 4.3. In order to account for randomness of the genetic algorithm, we sample 5 times for each sample size and report the mean and standard deviation of the results. We evaluate the models on the original test set for comparable results. The results are presented in Table 5. While the genetically optimized graph-based model only partially outperforms the NGCF model on large sample sizes (50% and 100%) it outperforms all the baselines on small sample sizes (10% and 25%). This shows that our genetically optimized model is more robust for small data problems than a neural graph-based model. In addition, it is still able to compete with the neural graph-based model when larger datasets are available.

In Fig. 3 we inspect how the outgoing edge weights evolve when optimized on different sizes of the insurance dataset. We have similar results for the ingoing edge weights, which are omitted due to space limitation. It appears that the optimized weights only change a little when more data is added to the training

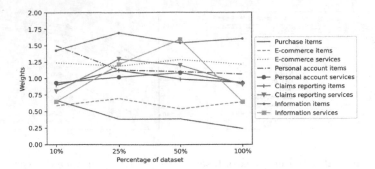

Fig. 3. Plot of how the outgoing edge weights evolve for different sizes of the insurance dataset.

set, and the relative importance of the interaction types remains stable across the different sizes of the dataset. Only the interaction type "information services" has large variations across the different dataset sizes, and in general, the biggest development of the weights happens when the dataset is increased from 10% to 25%. It shows that once the genetic algorithm has found the optimal weights in offline mode, the weights can be held fixed while the RS is deployed online, and the weights only need to be retrained (offline) once in a while, reducing the need for a fast optimization algorithm.

5   Conclusions and Future Work

We have introduced a novel recommender approach able to cope with very low data scenarios. This is a highly relevant problem for SMEs that might not have access to large amounts of data. We use a heterogeneous graph with users, content and their interactions to generate recommendations. We assign different weights to edges depending on the interaction type and use a genetic algorithm to find the optimal weights. Experimental results on two different use cases show that our model outperforms state-of-the-art baselines for two real-world small data scenarios. We make our code and datasets publicly available.

As future work we will consider possible extensions of the graph structure, for example, we can include contextual and demographic information as additional layers, similarly to what is done in [53]. Moreover, we can account for the temporal dimension, by encoding the recency of the actions in the edge weight, as done in [52]. We will further experiment with the more recent particle swarm [21] and ant colony optimization algorithms [10] instead of the genetic algorithm to find the optimal weights. Finally, we will investigate how to incorporate the edge weights into an explainability model, so that we can provide explanations to end users in principled ways as done in [4].

Acknowledgements. This paper was partially supported by the European Union's Horizon 2020 research and innovation programme under the Marie Sklodowska-Curie grant agreement No. 893667 and the Industriens Fond, AI Denmark project.

References

1. Aggarwal, C.C.: Context-sensitive recommender systems. In: Recommender Systems, pp. 255–281. Springer, Cham (2016). https://doi.org/10.1007/978-3-319-29659-3_8
2. Aggarwal, K., Yadav, P., Keerthi, S.S.: Domain adaptation in display advertising: an application for partner cold-start. In: Bogers, T., Said, A., Brusilovsky, P., Tikk, D. (eds.) Proceedings of the 13th ACM Conference on Recommender Systems, (RecSys 2019), pp. 178–186. ACM (2019). https://doi.org/10.1145/3298689.3347004
3. Anelli, V.W., Noia, T.D., Sciascio, E.D., Ferrara, A., Mancino, A.C.M.: Sparse feature factorization for recommender systems with knowledge graphs. In: Pampín, H.J.C., et al. (eds.) Proceedings of the 15th ACM Conference on Recommender Systems, (RecSys 2021), pp. 154–165. ACM (2021). https://doi.org/10.1145/3460231.3474243
4. Atanasova, P., Simonsen, J.G., Lioma, C., Augenstein, I.: Diagnostics-guided explanation generation. Proceed. AAAI Conf. Artif. Intell. $36(10)$, 10445–10453 (2022). https://doi.org/10.1609/aaai.v36i10.21287. https://ojs.aaai.org/index.php/AAAI/article/view/21287
5. Barkan, O., Koenigstein, N., Yogev, E., Katz, O.: CB2CF: a neural multiview content-to-collaborative filtering model for completely cold item recommendations. In: Bogers, T., Said, A., Brusilovsky, P., Tikk, D. (eds.) Proceedings of the 13th ACM Conference on Recommender Systems, (RecSys 2019), pp. 228–236. ACM (2019). https://doi.org/10.1145/3298689.3347038
6. Biasini, M.: Design and implementation of gamification in a social e-learning platform for increasing learner engagement, Master's thesis, Danmarks Tekniske Universitet and Università degli Studi di Padova (2020)
7. Bruun, S.B., Maistro, M., Lioma, C.: Learning recommendations from user actions in the item-poor insurance domain. In: Golbeck, J. (eds.) Proceedings of the 16th ACM Conference on Recommender Systems, (RecSys 2022), pp. 113–123. ACM (2022). https://doi.org/10.1145/3523227.3546775
8. Cremonesi, P., Koren, Y., Turrin, R.: Performance of recommender algorithms on top-n recommendation tasks. In: Amatriain, X., Torrens, M., Resnick, P., Zanker, M. (eds.) Proceedings of the 4th ACM Conference on Recommender Systems, (RecSys 2010), pp. 39–46. ACM (2010). https://doi.org/10.1145/1864708.1864721
9. Dietterich, T.G.: Approximate statistical tests for comparing supervised classification learning algorithms. Neural Comput. $10(7)$, 1895–1923 (1998). https://doi.org/10.1162/089976698300017197
10. Dorigo, M., Birattari, M., Stutzle, T.: Ant colony optimization. IEEE Comput. Intell. Mag. $1(4)$, 28–39 (2006). https://doi.org/10.1109/MCI.2006.329691
11. Group, T.W.B.: Small and Medium Enterprises (SMEs) Finance (2022). https://www.worldbank.org/en/topic/smefinance. Accessed 04 Oct 2022
12. Hansen, C., Hansen, C., Simonsen, J.G., Alstrup, S., Lioma, C.: Content-aware neural hashing for cold-start recommendation. In: Proceedings of the 43rd International ACM SIGIR Conference on Research and Development in Information Retrieval, pp. 971–980. SIGIR 2020, Association for Computing Machinery, New York, NY, USA (2020). https://doi.org/10.1145/3397271.3401060
13. Hansen, C., Hansen, C., Hjuler, N., Alstrup, S., Lioma, C.: Sequence modelling for analysing student interaction with educational systems. In: Hu, X., Barnes, T., Hershkovitz, A., Paquette, L. (eds.) EDM. International Educational Data Mining

Society (IEDMS) (2017). http://dblp.uni-trier.de/db/conf/edm/edm2017.html#HansenHHAL17

14. Hansen, C., Hansen, C., Simonsen, J.G., Lioma, C.: Projected hamming dissimilarity for bit-level importance coding in collaborative filtering. In: Proceedings of the Web Conference 2021, pp. 261–269. WWW 2021, Association for Computing Machinery, New York, NY, USA (2021). https://doi.org/10.1145/3442381.3450011

15. Harper, F.M., Konstan, J.A.: The MovieLens datasets: history and context. ACM Trans. Interact. Intell. Syst. 5(4), 1–19 (2016). https://doi.org/10.1145/2827872

16. He, X., Liao, L., Zhang, H., Nie, L., Hu, X., Chua, T.: Neural collaborative filtering. In: Barrett, R., Cummings, R., Agichtein, E., Gabrilovich, E. (eds.) Proceedings of the 26th International Conference on World Wide Web, (WWW 2017), pp. 173–182. ACM (2017). https://doi.org/10.1145/3038912.3052569

17. Holland, J.H.: Genetic algorithms. Sci. Am. 267(1), 66–73 (1992). http://www.jstor.org/stable/24939139

18. Inozemtseva, L., Holmes, R., Walker, R.J.: Recommendation systems in-the-small. In: Robillard, M.P., Maalej, W., Walker, R.J., Zimmermann, T. (eds.) Recommendation Systems in Software Engineering, pp. 77–92. Springer, Heidelberg (2014). https://doi.org/10.1007/978-3-642-45135-5_4

19. Kaminskas, M., Bridge, D., Foping, F., Roche, D.: Product recommendation for small-scale retailers. In: Stuckenschmidt, H., Jannach, D. (eds.) EC-Web 2015. LNBIP, vol. 239, pp. 17–29. Springer, Cham (2015). https://doi.org/10.1007/978-3-319-27729-5_2

20. Kaminskas, M., Bridge, D., Foping, F., Roche, D.: Product-seeded and basket-seeded recommendations for small-scale retailers. J. Data Semantics 6(1), 3–14 (2016). https://doi.org/10.1007/s13740-016-0058-3

21. Kennedy, J., Eberhart, R.: Particle swarm optimization. In: Proceedings of ICNN1995 - International Conference on Neural Networks, vol. 4, pp. 1942–1948 (1995). https://doi.org/10.1109/ICNN.1995.488968

22. Kutner, M., Nachtsheim, C.J., Neter, J., Li, W., et al.: Applied Linear Statistical Models. McGraw-Hill, Irwin (2005)

23. Kużelewska, U.: Effect of dataset size on efficiency of collaborative filtering recommender systems with multi-clustering as a neighbourhood identification strategy. In: Krzhizhanovskaya, V.V., et al. (eds.) ICCS 2020. LNCS, vol. 12139, pp. 342–354. Springer, Cham (2020). https://doi.org/10.1007/978-3-030-50420-5_25

24. Latifi, S., Mauro, N., Jannach, D.: Session-aware recommendation: a surprising quest for the state-of-the-art. Inf. Sci. 573, 291–315 (2021). https://doi.org/10.1016/j.ins.2021.05.048

25. Lee, D., Kang, S., Ju, H., Park, C., Yu, H.: bootstrapping user and item representations for one-class collaborative filtering. In: Diaz, F., Shah, C., Suel, T., Castells, P., Jones, R., Sakai, T. (eds.) Proceedings of the 44th International ACM Conference on Research and Development in Information Retrieval, (SIGIR 2021), pp. 1513–1522. ACM (2021). https://doi.org/10.1145/3404835.3462935

26. Lee, S., Park, S., Kahng, M., Lee, S.: PathRank: ranking nodes on a heterogeneous graph for flexible hybrid recommender systems. Expert Syst. Appl. 40(2), 684–697 (2013). https://doi.org/10.1016/j.eswa.2012.08.004

27. Lee, Y., Cheng, T., Lan, C., Wei, C., Hu, P.J.: Overcoming small-size training set problem in content-based recommendation: a collaboration-based training set expansion approach. In: Chau, P.Y.K., Lyytinen, K., Wei, C., Yang, C.C., Lin, F. (eds.) Proceedings of the 11th International Conference on Electronic Commerce, (ICEC 2009), pp. 99–106. ACM (2009). https://doi.org/10.1145/1593254.1593268

28. Ludewig, M., Mauro, N., Latifi, S., Jannach, D.: Empirical analysis of session-based recommendation algorithms. User Model. User-Adap. Inter. **31**(1), 149–181 (2020). https://doi.org/10.1007/s11257-020-09277-1

29. Ng, A.Y.T.: Why AI Projects Fail, Part 4: Small Data (2019). https://www.deeplearning.ai/the-batch/why-ai-projects-fail-part-4-small-data/. Accessed 04 Oct 2022

30. Odili, J.: The dawn of metaheuristic algorithms. Int. J. Softw. Eng. Comput. Syst. **4**, 49–61 (2018). https://doi.org/10.15282/ijsecs.4.2.2018.4.0048

31. Page, L., Brin, S., Motwani, R., Winograd, T.: The pagerank citation ranking : bringing order to the web. In: WWW 1999 (1999)

32. Pan, X., et al.: MetaCVR: conversion rate prediction via meta learning in small-scale recommendation scenarios. In: Amigó, E., Castells, P., Gonzalo, J., Carterette, B., Culpepper, J.S., Kazai, G. (eds.) Proceedings of the 45th International ACM SIGIR Conference on Research and Development in Information Retrieval, (SIGIR 2022), pp. 2110–2114. ACM (2022). https://doi.org/10.1145/3477495.3531733

33. Raziperchikolaei, R., Liang, G., Chung, Y.: Shared neural item representations for completely cold start problem. In: Pampín, H.J.C., (eds.) Proceedings of the 15th ACM Conference on Recommender Systems, (RecSys 2021), pp. 422–431. ACM (2021). https://doi.org/10.1145/3460231.3474228

34. Salha-Galvan, G., Hennequin, R., Chapus, B., Tran, V., Vazirgiannis, M.: Cold start similar artists ranking with gravity-inspired graph autoencoders. In: Pampín, H.J.C., et al. (eds.) Proceedings of the 15th ACM Conference on Recommender Systems, (RecSys 2021), pp. 443–452. ACM (2021). https://doi.org/10.1145/3460231.3474252

35. Sankar, A., Wang, J., Krishnan, A., Sundaram, H.: ProtoCF: prototypical collaborative filtering for few-shot recommendation. In: Pampín, H.J.C., et al. (eds.) Proceedings of the 15th ACM Conference on Recommender Systems, (RecSys 2021), pp. 166–175. ACM (2021). https://doi.org/10.1145/3460231.3474268

36. Schnabel, T., Bennett, P.N.: Debiasing item-to-item recommendations with small annotated datasets. In: Santos, R.L.T., et al. (eds.) Proceedings of the 14th ACM Conference on Recommender Systems, (RecSys 2020), pp. 73–81. ACM (2020). https://doi.org/10.1145/3383313.3412265

37. Shuai, J., ct al.: Proceedings of the 45th international ACM SIGIR conference on research and development in information retrieval, (SIGIR 2022), pp. 1283–1293. ACM (2022). https://doi.org/10.1145/3477495.3531927

38. Strickroth, S., Pinkwart, N.: High quality recommendations for small communities: the case of a regional parent network. In: Cunningham, P., Hurley, N.J., Guy, I., Anand, S.S. (eds.) Proceedings of the 6th ACM Conference on Recommender Systems, (RecSys 2012), pp. 107–114. ACM (2012). https://doi.org/10.1145/2365952.2365976

39. Sun, H., Xu, J., Zheng, K., Zhao, P., Chao, P., Zhou, X.: MFNP: a meta-optimized model for few-shot next POI recommendation. In: Zhou, Z. (ed.) Proceedings of the 30th International Joint Conference on Artificial Intelligence, (IJCAI 2021), pp. 3017–3023. ijcai.org (2021). https://doi.org/10.24963/ijcai.2021/415

40. Sun, X., Shi, T., Gao, X., Kang, Y., Chen, G.: FORM: follow the online regularized meta-leader for cold-start recommendation. In: Diaz, F., Shah, C., Suel, T., Castells, P., Jones, R., Sakai, T. (eds.) Proceedings of the 44th International ACM Conference on Research and Development in Information Retrieval, (SIGIR 2021), pp. 1177–1186. ACM (2021). https://doi.org/10.1145/3404835.3462831

41. Sun, Y., Han, J.: Mining heterogeneous information networks: principles and methodologies. Synthesis Lectures on Data Mining and Knowledge Discovery, Morgan & Claypool Publishers (2012). https://doi.org/10.2200/S00433ED1V01Y201207DMK005

42. Sun, Z., Yu, D., Fang, H., Yang, J., Qu, X., Zhang, J., Geng, C.: Are we evaluating rigorously? benchmarking recommendation for reproducible evaluation and fair comparison. In: Santos, R.L.T., Marinho, L.B., Daly, E.M., Chen, L., Falk, K., Koenigstein, N., de Moura, E.S. (eds.) Proceedings of the 14th ACM Conference on Recommender Systems, (RecSys 2020), pp. 23–32. ACM (2020). https://doi.org/10.1145/3383313.3412489

43. Volkovs, M., Yu, G.W., Poutanen, T.: DropoutNet: Addressing Cold Start in Recommender Systems. In: Guyon, I., et al. (eds.) Proceedings of the 30th Annual Conference on Neural Information Processing Systems (NeurIPS 2017), pp. 4957–4966 (2017). https://proceedings.neurips.cc/paper/2017/hash/dbd22ba3bd0df8f385bdac3e9f8be207-Abstract.html

44. Wang, Q., Yin, H., Wang, H., Nguyen, Q.V.H., Huang, Z., Cui, L.: enhancing collaborative filtering with generative augmentation. In: Teredesai, A., Kumar, V., Li, Y., Rosales, R., Terzi, E., Karypis, G. (eds.) Proceedings of the 25th ACM International Conference on Knowledge Discovery and Data Mining, (SIGKDD 2019), pp. 548–556. ACM (2019). https://doi.org/10.1145/3292500.3330873

45. Wang, S., et al.: Graph learning based recommender systems: a review. In: Zhou, Z. (ed.) Proceedings of the 30th International Joint Conference on Artificial Intelligence, (IJCAI 2021), pp. 4644–4652. ijcai.org (2021). https://doi.org/10.24963/ijcai.2021/630

46. Wang, S., Zhang, K., Wu, L., Ma, H., Hong, R., Wang, M.: Privileged graph distillation for cold start recommendation. In: Diaz, F., Shah, C., Suel, T., Castells, P., Jones, R., Sakai, T. (eds.) Proceedings of the 44th International ACM Conference on Research and Development in Information Retrieval, (SIGIR 2021), pp. 1187–1196. ACM (2021). https://doi.org/10.1145/3404835.3462929

47. Wang, X., He, X., Cao, Y., Liu, M., Chua, T.: KGAT: knowledge graph attention network for recommendation. In: Teredesai, A., Kumar, V., Li, Y., Rosales, R., Terzi, E., Karypis, G. (eds.) Proceedings of the 25th ACM International Conference on Knowledge Discovery & Data Mining, (SIGKDD 2017), pp. 950–958. ACM (2019). https://doi.org/10.1145/3292500.3330989

48. Wang, X., He, X., Wang, M., Feng, F., Chua, T.S.: Neural graph collaborative filtering. In: Proceedings of the 42nd International ACM SIGIR Conference on Research and Development in Information Retrieval. p. 165–174. SIGIR2019, Association for Computing Machinery, New York, NY, USA (2019). https://doi.org/10.1145/3331184.3331267

49. Wu, J., Wang, X., Feng, F., He, X., Chen, L., Lian, J., Xie, X.: sn. In: Diaz, F., Shah, C., Suel, T., Castells, P., Jones, R., Sakai, T. (eds.) Proceedings of the 44th International ACM Conference on Research and Development in Information Retrieval, (SIGIR 2021), pp. 726–735. ACM (2021). https://doi.org/10.1145/3404835.3462862

50. Wu, J., Xie, Z., Yu, T., Zhao, H., Zhang, R., Li, S.: Dynamics-aware adaptation for reinforcement learning based cross-domain interactive recommendation. In: Amigó, E., Castells, P., Gonzalo, J., Carterette, B., Culpepper, J.S., Kazai, G. (eds.) Proceedings of the 45th International ACM SIGIR Conference on Research and Development in Information Retrieval, (SIGIR 2022), pp. 290–300. ACM (2022). https://doi.org/10.1145/3477495.3531969

51. Xia, L., Huang, C., Xu, Y., Zhao, J., Yin, D., Huang, J.X.: Hypergraph contrastive collaborative filtering. In: Amigó, E., Castells, P., Gonzalo, J., Carterette, B., Culpepper, J.S., Kazai, G. (eds.) Proceedings of the 45th International ACM SIGIR Conference on Research and Development in Information Retrieval, (SIGIR 2022), pp. 70–79. ACM (2022). https://doi.org/10.1145/3477495.3532058
52. Xiang, L., et al.: Temporal Recommendation on Graphs via Long- and Short-term Preference Fusion. In: Rao, B., Krishnapuram, B., Tomkins, A., Yang, Q. (eds.) Proceedings of the 16th ACM International Conference on Knowledge Discovery & Data Mining, (SIGKDD 2010), pp. 723–732. ACM (2010). https://doi.org/10.1145/1835804.1835896
53. Yao, W., He, J., Huang, G., Cao, J., Zhang, Y.: Personalized recommendation on multi-layer context graph. In: Lin, X., Manolopoulos, Y., Srivastava, D., Huang, G. (eds.) WISE 2013. LNCS, vol. 8180, pp. 135–148. Springer, Heidelberg (2013). https://doi.org/10.1007/978-3-642-41230-1_12
54. Yu, X., et al.: Personalized entity recommendation: a heterogeneous information network approach. In: Carterette, B., Diaz, F., Castillo, C., Metzler, D. (eds.) Proceedings of the 7th ACM International Conference on Web Search and Data Mining, (WSDM 2014), pp. 283–292. ACM (2014). https://doi.org/10.1145/2556195.2556259

It's Just a Matter of Time: Detecting Depression with Time-Enriched Multimodal Transformers

Ana-Maria Bucur[1,2]($\boxtimes$)(iD), Adrian Cosma[3](iD), Paolo Rosso[2](iD),
and Liviu P. Dinu[4](iD)

[1] Interdisciplinary School of Doctoral Studies, University of Bucharest,
Bucharest, Romania
ana-maria.bucur@drd.unibuc.ro
[2] PRHLT Research Center, Universitat Politècnica de València, Valencia, Spain
prosso@dsic.upv.es
[3] Politehnica University of Bucharest, Bucharest, Romania
ioan_adrian.cosma@upb.ro
[4] Faculty of Mathematics and Computer Science, University of Bucharest,
Bucharest, Romania
ldinu@fmi.unibuc.ro

Abstract. Depression detection from user-generated content on the internet has been a long-lasting topic of interest in the research community, providing valuable screening tools for psychologists. The ubiquitous use of social media platforms lays out the perfect avenue for exploring mental health manifestations in posts and interactions with other users. Current methods for depression detection from social media mainly focus on text processing, and only a few also utilize images posted by users. In this work, we propose a flexible time-enriched multimodal transformer architecture for detecting depression from social media posts, using pre-trained models for extracting image and text embeddings. Our model operates directly at the user-level, and we enrich it with the relative time between posts by using time2vec positional embeddings. Moreover, we propose another model variant, which can operate on randomly sampled and unordered sets of posts to be more robust to dataset noise. We show that our method, using EmoBERTa and CLIP embeddings, surpasses other methods on two multimodal datasets, obtaining state-of-the-art results of 0.931 F1 score on a popular multimodal Twitter dataset, and 0.902 F1 score on the only multimodal Reddit dataset.

Keywords: Depression detection · Mental health · Social media · Multimodal learning · Transformer · Cross-attention · Time2vec

1 Introduction

More than half of the global population uses social media[1]. People use platforms such as Twitter and Reddit to disclose and discuss their mental health problems

---

[1] https://datareportal.com/reports/digital-2022-july-global-statshot.

J. Kamps et al. (Eds.): ECIR 2023, LNCS 13980, pp. 200–215, 2023.
https://doi.org/10.1007/978-3-031-28244-7_13

online. On Twitter, users feel a sense of community, it is a safe place for expression, and they use it to raise awareness and combat the stigma around mental illness or as a coping mechanism [5]. On Reddit, more so than on Twitter, users are pseudo-anonymous, and they can choose to use "throw-away" accounts to be completely anonymous, subsequently encouraging users to disclose their mental health problems. Users talk about symptoms and their daily struggles, treatment, and therapy [13] on dedicated subreddits such as *r/depression, r/mentalhealth*. To date, there have been many methods that aim to estimate signs of mental disorders (i.e., depression, eating disorders) [30,58] from the social media content of users. The primary focus has been on processing the posts' text, fuelled partly by the widespread availability and good performance of pretrained language models (e.g., BERT) [1,25,56]. Recently, however, both textual and visual information has been used for multimodal depression detection from social media data, on datasets collected from Twitter [20,39,41], Instagram [10,32], Reddit [51], Flickr [57], and Sina Weibo [54]. These methods obtain good performance, but nevertheless assume that social media posts are synchronous and uploaded at regular intervals.

We propose a time-enriched multimodal transformer for user-level depression detection from social media posts (i.e., posts with text and images). Instead of operating at the token-level in a low-level manner, we utilize the cross- and self-attention mechanism across posts to learn the high-level posting patterns of a particular user. The attention layers process semantic embeddings obtained by pretrained state-of-the-art language and image processing models. As opposed to current time-aware methods for mental disorders detection [4,9], our method does not require major architectural modifications and can easily accommodate temporal information by simply manipulating the transformer positional encodings (e.g., using time-enriched encodings such as time2vec [24]). We propose two viable ways to train our architecture: a time-aware regime using time2vec and a set-based training regime, in which we do not employ positional encodings and regard the user posts as a set. The second approach is motivated by the work of Dufter et al. [16], which observed that positional encodings are not universally necessary to obtain good downstream performance. We train and evaluate our method on two social media multimodal datasets, each with its own particularities in user posting behavior, and obtain state-of-the-art results in depression detection. We make our code publicly available on github[2].

This work makes the following contributions:

1. We propose a time-enriched multimodal transformer for user-level depression detection from social media posts. Using EmoBERTa and CLIP embeddings, and time2vec positional embeddings, our method achieves 0.931 F1 on a popular multimodal Twitter dataset [20], surpassing current methods by a margin of 2.3%. Moreover, using no positional embeddings, we achieve 0.902 F1 score on *multiRedditDep* [51], the only multimodal Reddit dataset to date.
2. We perform extensive ablation studies and evaluate different types of image and text encoders, window sizes, and positional encodings. We show that a

---

[2] https://github.com/cosmaadrian/time-enriched-multimodal-depression-detection.

time-aware approach is suitable when posting frequency is high, while a set-based approach is robust to dataset noise (i.e., many uninformative posts).
3. We perform a qualitative error analysis using Integrated Gradients [46], which proves that our model is interpretable and allows for the selection of the most informative posts in a user's social media timeline.

2   Related Work

Although research in depression detection was focused on analyzing language cues uncovered from psychology literature (e.g., self-focused language reflected in the greater use of the pronoun "I" [37,38], dichotomous thinking expressed in absolute words (e.g., "always", "never") [18]), studies also began to investigate images posted on social media. Reece et al. [35] showed that the images posted online by people with depression were more likely to be sadder and less happy, and to have bluer, darker and grayer tones than those from healthy individuals. Users with depression posted more images with faces of people, but they had fewer faces per image, indicating reduced social interactivity and an increased self-focus [21,35]. Guntuku et al. [21] and Uban et al. [51] revealed that users diagnosed with depression have more posts with animal-related images.

Deep learning methods such as CNNs [34,58], LSTMs [43,49] and transformer-based architectures [1,7,56] achieved good results on depression detection using only the textual information from users' posts. Further, multimodal methods incorporating visual features achieve even greater results [32,39,57]. Shen et al. [41] collected the first user-level multimodal dataset from social media for identifying depression with textual, behavioral, and visual information from Twitter users and proposed a multimodal depressive dictionary learning method. The same Twitter dataset was later explored by Gui et al. [20], who used a cooperative multi-agent reinforcement learning method with two agents for selecting only the relevant textual and visual information for classification. An et al. [2] proposed a multimodal topic-enriched auxiliary learning approach, in which the performance of the primary task on depression detection is improved by auxiliary tasks on visual and textual topic modeling. By not taking the time component into account, the above methods assume that social media posts are synchronous, and are sampled at regular time intervals. Realistically, posts from online platforms are asynchronous and previous studies have shown differences in social media activity, partly due to the worsening of depression symptoms at night [14,31,45]. Motivated by this, methods that use time-adapted weights [10] or Time-Aware LSTMs [9,40] to include the time component of data for mental health problems detection report higher performance.

3   Method

The problem of depression detection from social media data is usually formulated as follows: given a user with an ordered, asynchronous, sequence of multimodal social media posts containing text and images, determine whether or

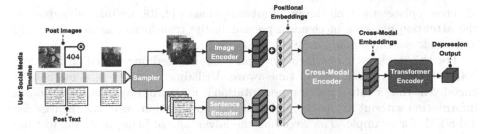

Fig. 1. The overall architecture of our proposed method. From a user's social media timeline, we sample a number of posts containing text and images. Each image and text is encoded with pretrained image and text encoders. The sequence of encoded images and texts is then processed using a cross-modal encoder followed by a transformer encoder, together with the relative posting time encoded into the positional embeddings. After mean pooling, we perform classification for the particular user.

not the user has symptoms of depression. This formulation corresponds to user-level binary classification. The main difficulty in this area is modeling a large number of user posts (i.e., tens of thousands), in which not all posts contain relevant information for depression detection. This aspect makes the classification inherently noisy. Formally, we consider a user i to have multiple social media posts U_i, each post containing the posting date Δ, a text T and an accompanying image I. A post-sequence P_i is defined as K posts sampled from U_i: $P_i = \{(T^j, I^j, \Delta^j) \sim U_i, j \in (1 \ldots K)\}$. During training, we used an input batch defined by the concatenation of n such post-sequences: $B = \{S_{b_1}, S_{b_2}, \ldots S_{b_n}\}$. Since the users' posts are asynchronous (i.e., are not regularly spaced in time), time-aware approaches based on T-LSTM [9,40] have become the norm in modeling users' posts alongside with the relative time between them. However, in T-LSTM [4], including a relative time component involves the addition of new gates and hand-crafted feature engineering of the time. Moreover, T-LSTM networks are slow, cannot be parallelized and do not allow for transfer learning.

To address this problem, we propose a transformer architecture that can perform user-level multimodal classification, as shown in Fig. 1. To process the multimodal data from posts, we first encode the visual and textual information using pretrained models (e.g., CLIP [33] for images and EmoBERTa [26] for text). The embeddings are linearly projected to a fixed size using a learnable feed-forward layer, are augmented with a variant of positional encodings (expanded below) and are further processed with a cross-modal encoder based on LXMERT [48]. Finally, self-attention is used to process the cross-modal embeddings further, and classification is performed after mean pooling. The network is trained using the standard binary cross-entropy loss. In this setting, the transformer attention does not operate on low-level tokens such as word pieces or image patches. Rather, the cross- and self-attention operates across posts, allowing the network to learn high-level posting patterns of the user, and be more robust to individual uninformative posts. As opposed to other related works in which a vector of zeros replaces the embeddings of missing images [2,20], we take advantage of the attention masking mechanism present in the transformer architecture [52], and mask out the missing images.

For this work, we experiment with positional embeddings based on *time2vec* [24] to make the architecture time-aware. Utilizing *time2vec* as a method to encode the time τ into a vector representation is a natural way to inject temporal information without any major architectural modification, as it was the case for T-LSTM, for example. *Time2vec* has the advantage of being invariant to time rescaling, avoiding hand-crafted time features, it is periodic, and simple to define and consume by various models. It is a vector of $k + 1$ elements, defined as:

$$\mathbf{t2v}(\tau)[i] = \begin{cases} \omega_i g(\tau) + \phi_i & i = 0 \\ \mathcal{F}(\omega_i g(\tau) + \phi_i) & 1 \leq i \leq k \end{cases} \qquad (1)$$

where $\mathbf{t2v}(\tau)[\mathrm{i}]$ is the ith element of $\mathbf{t2v}(\tau)$, $\mathcal{F}$ is a periodic activation function (in our case it is $sin(x)$), and ω_is and ϕ_is are learnable parameters. To avoid arbitrarily large τ values, we transform τ with $g(\tau) = \frac{1}{(\tau+\epsilon)}$, with $\epsilon = 1$. For processing user's posts, we use sub-sequence sampling and sample K *consecutive* posts from a user's timeline (Fig. 2). We name the model utilizing *time2vec* embeddings *Time2VecTransformer*.

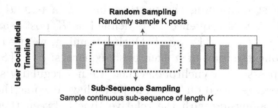

Fig. 2. The two sampling methods used in this work. Methods that use positional embeddings (either *time2vec* or *learned*) employ sub-sequence sampling, and the *SetTransformer*, with *zero* positional embeddings, uses random sampling of posts. K refers to the number of posts in the post window.

However, processing user posts sequentially is often not desirable, as clusters of posts might provide irrelevant information for depression detection. For instance, sub-sequence sampling of posts from users with mental health problems might end up in an interval of positive affect, corresponding to a sudden shift in mood [50]. Somewhat orthogonal to the time-aware school of thought, we also propose a *SetTransformer* for processing sets of user posts for depression detection, to alleviate the issues mentioned above. Our proposed *SetTransformer* randomly samples texts from a user and assumes no order between them by omitting the positional encoding, essentially making the transformer permutation invariant [52]. For this method, K posts in a post-sequence are randomly sampled (Fig. 2). For *SetTransformer*, we treat the user timeline as a "bag-of-posts" motivated by the work of Dufter et al. [16], in which the authors show that treating sentences as "bag-of-words" (i.e., by not utilizing any positional embeddings) results in a marginal loss in performance for transformer architectures.

4 Experiments

4.1 Datasets

To benchmark our method, we use in our experiments *multiRedditDep* [51] and the Twitter multimodal depression dataset from Gui et al. [20][3]. Reddit and Twitter are two of the most popular social media platforms where users choose to disclose their mental health problems [5,13]. Moreover, the data coming from these two platforms have different particularities: both social media platforms support images, but the textual information is richer on Reddit, with posts having up to 40,000 characters, as opposed to Twitter, where the limit is 280 characters. In some subreddits from Reddit, the image cannot be accompanied by text; the post is composed only of image(s) and a title with a maximum length of 300 characters. On Twitter, posts with images have the same character limit as regular posts. For both datasets, the users from the depression class were annotated by retrieving their mention of diagnosis, while users from the control group did not have any indication of depression diagnosis.

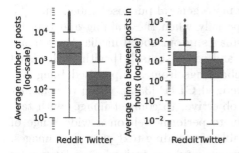

Fig. 3. *Left* - Distribution of posts per user on both datasets. Users from Reddit have significantly (Mann Whitney U Test, p < 0.001) more posts than users from Twitter. *Right* - Distribution of average time duration in hours between posts at the user level. Users from Twitter post significantly (p < 0.001) more frequently than users from Reddit.

Table 1. Statistics for the Reddit [51] and Twitter [20] datasets. #T represents the number of posts with only text, #(T+I) represents the number of posts with both text (title, in the case of Reddit) and images. $\overline{\#T}$ and $\overline{\#(T+I)}$ represent the average number of posts with only text and text + image per user, respectively.

Dataset	Class	#T	#(T+I)	$\overline{\#T}$	$\overline{\#(T+I)}$
Reddit	Depr	6,6M	46.9k	4.7k	33.33
	Non-Depr	8,1M	73.4k	3.5k	31.51
Twitter	Depr	213k	19.3k	152.43	13.81
	Non-Depr	828k	50.6k	590.82	36.16

The Reddit dataset contains 1,419 users from the depression class and 2,344 control users. The authors provided the train, validation and test splits, with

---

[3] We also attempted to perform our experiments on a multimodal dataset gathered from Instagram [9,10], but the authors did not respond to our request.

2633, 379 and 751 users, respectively. The dataset from Twitter contains 1,402 users diagnosed with depression and 1,402 control users. Shen et al. [41] collected only tweets in the span of one month for users from both classes. Due to the lack of access to the benchmark train/test splits from An et al. [2], we are running the experiments following the same experimental settings as Gui et al. [20], and perform five-fold cross-validation. In Table 1, we showcase the statistics for both datasets. There is a greater total number of posts from the non-depressed group, but depressed users have more posts, on average, than non-depressed users, in the case of Reddit. For Twitter, the opposite is true, with non-depressed users having more posts on average. It is important to note that not all posts have images, only a small amount of social media posts contain both text and image data. Figure 3 (left) shows that Reddit users have a greater average number of posts than Twitter users. Regarding posting frequency, Twitter users post more frequently than Reddit users, as shown in Fig. 3 (right), in which the average time between posts is 26.9 h for Reddit and 11.2 h for Twitter.

4.2   Experimental Settings

Image Representation Methods. For encoding the images in users' posts, we opted to use two different encoders trained in a self-supervised manner. Many approaches for transfer learning employ a supervised network, usually pretrained on ImageNet [15]. However, transfer learning with self-supervised trained networks is considered to have more general-purpose embeddings [8], which aid downstream performance, especially when the training and transfer domains qualitatively differ [17]. Since images in our social media datasets are very diverse, including internet memes and screenshots [51] with both visual and textual information, specialized embeddings from a supervised ImageNet model are not appropriate. Therefore, we used **CLIP** [33], a vision transformer trained with a cross-modal contrastive objective, to connect images with textual descriptions. CLIP has been shown to perform well on a wide range of zero-shot classification tasks, and is capable of encoding text present in images. CLIP embeddings are general, multi-purpose, and are trained on a large-scale internet-scraped dataset. Additionally, we used **DINO** [8], a vision transformer pretrained in a self-supervised way on ImageNet, achieving notable downstream performance. Moreover, compared to a supervised counterpart, DINO automatically learns class-specific features without explicitly being trained to do so.

Text Representation Methods. We explored three pretrained transformers models to extract contextual embeddings from users' texts. **RoBERTa** [29], pretrained in a self-supervised fashion on a large corpus of English text, was chosen given its state-of-the-art performance on downstream tasks. Emotion-informed embeddings from **EmoBERTa** [26] that incorporate both linguistic and emotional information from users' posts were also used. Emotions expressed in texts are a core feature used for identifying depression [3,28], users with depression showing greater negative emotions [14]. EmoBERTa is based on a

RoBERTa model trained on two conversational datasets and adapted for identifying the emotions found in users' utterances. **Multilingual MiniLM** [53], the distilled version of the multilingual language model XLM-RoBERTa [12], was used for encoding text because both Twitter and Reddit datasets contain posts in various languages besides English, such as Spanish, German, Norwegian, Japanese, Romanian and others. Moreover, MiniLM is a smaller model, providing 384-dimensional embeddings, as opposed to 768-dimensional embeddings from RoBERTa and EmoBERTa.

Positional Encodings. Since the network is processing sequences of posts, we explored three different methods for encoding the relative order between posts. Firstly, we used a standard **learned** positional encoding, used in many pretrained transformer models, such as BERT [25], RoBERTa [29] and GPT [6]. Wang and Chen [55] showed that for transformer encoder models, learned positional embeddings encode local position information, which is effective in masked language modeling. However, this type of positional encoding assumes that users' posts are equally spaced in time. Second, we used **time2vec** [24] positional encoding, which allows the network to learn a vectorized representation of the relative time between posts. Lastly, we omit to use any positional encoding, and treat the sequence of user posts as a set. We refer to this type of positional encodings as **zero** encodings. For both *learned* and *time2vec* we used a sub-sequence sampling of posts, while for *zero*, we used a random sample of user posts (Fig. 2).

4.3 Comparison Models

We evaluate our proposed method's performance against existing multimodal and text-only state-of-the-art models on the two multimodal datasets from Twitter and Reddit, each with its different particularities in user posting behavior. For *multiRedditDep*, since it is a new dataset, there are no public benchmarks besides the results of Uban et al. [51]. We report Accuracy, Precision, Recall, F1, and AUC as performance measures. The baselines are as follows. **Time-Aware LSTM (T-LSTM)** [4] - we implement as text-only baseline a widely used [9,40] T-LSTM-based neural network architecture that integrates the time irregularities of sequential data in the memory unit of the LSTM. **EmoBERTa Transformer** - we train a text-only transformer baseline on user posts. Both text-only models are based on EmoBERTa embeddings. **LSTM + RL** and **CNN + RL** [19] - two text-only state-of-the-art models which use a reinforcement learning component for selecting the posts indicative of depression. **Multimodal Topic-Enriched Auxiliary Learning (MTAL)** [2] - a model capturing the multimodal topic information, in which two auxiliary tasks accompany the primary task of depression detection on visual and textual topic modeling. **Multimodal Time-Aware Attention Networks (MTAN)** [9] - a multimodal model that uses as input BERT [25] textual features, InceptionResNetV2 [47] visual features, posting time features and incorporates T-LSTM for taking into account the time intervals between posts and self-attention. **GRU + VGG-Net + COMMA** [20] - in which a reinforcement learning component is used for selecting posts with text and images which are indicative of depression and are classified with an MLP. For extracting the textual and visual features, GRU [11] and VGGNet [42] were used. **BERT + word2vec embeddings** - baseline proposed by Uban et al. [51], which consists of a neural network architecture that uses as input BERT features from posts' titles and word2vec embeddings for textual information found in images (i.e., ImageNet labels and text extracted from images). **VanillaTransformer** - the multimodal transformer proposed in this work, with standard *learned* positional encodings. **SetTransformer** - the set-based multimodal transformer proposed in this work employing *zero* positional encoding, alongside a random sampling of user posts. **Time2VecTransformer** - the time-aware multimodal transformer proposed in this work using time-enriched positional embeddings (i.e., *time2vec* [24]) and sub-sequence sampling.

4.4   Training and Evaluation Details

We train all models using Adam [27] optimizer with a base learning rate of 0.00001. The learning rate is modified using Cyclical Learning Rate schedule [44], which linearly varies the learning rate from 0.00001 to 0.0001 and back across 10 epochs. The model has 4 cross-encoder layers with 8 heads each and an embedding size of 128. The self-attention transformer has 2 layers of 8 heads each and the same embedding size.

At test time, since it is unfeasible to use all users' posts for evaluation, with some users having more than 50k posts, we make 10 random samples of post-sequences for a user, and the final decision is taken through majority voting on the decisions for each post-sequence. In this way, the final classification is more robust to dataset noise and uninformative posts.

5   Results

5.1   Performance Comparison with Prior Works

In Table 2, we present the results for models trained on the Twitter dataset. For our models and proposed baselines, each model was evaluated 10 times and we report mean and standard deviation. Our architecture, *Time2VecTransformer* achieves state-of-the-art performance in multimodal depression detection, obtaining a 0.931 F1 score. The model uses as input textual embeddings extracted from EmoBERTa and visual embeddings extracted from CLIP. The best model uses sequential posts as input from a 512 window size, and the time component is modeled by *time2vec* positional embeddings. Our time-aware multimodal architecture surpasses other time-aware models such as T-LSTM [4] and MTAN [9]. Moreover, our method surpasses text-only methods such as T-LSTM and EmoBERTa Transformer - a unimodal variant of our model using only self-attention on text embeddings.

As opposed to the previous result on Twitter data, which include the time component, for Reddit, we achieved good performance by regarding the users' posts as a set. In Table 3, we showcase our model's performance on the Reddit dataset, compared to Uban et al. [51], our trained text-only T-LSTM and text-only EmoBERTa Transformer. Our *SetTransformer* model (with *zero* positional encodings), using as input EmoBERTa and CLIP, obtains the best performance, a 0.902 F1 score. While counter-intuitive, treating posts as a "bag-of-posts" outperforms *Time2VecTransformer* in the case of Reddit. However,

Table 2. Results for multimodal depression detection on Twitter dataset [20]. Our models use EmoBERTa and CLIP for extracting embeddings. *Vanilla Transformer* and *SetTransformer* were trained on 128 sampled posts, while *Time2VecTransformer* was trained on a window size of 512 posts. Denoted with **bold** are the best results for each column. *An et al. [2] report the performance on a private train/dev/test split, not on five-fold cross-validation. ** Cheng et al. [9] are not explicit in their experimental settings for the Twitter data. † indicates that the result is a statistically significant improvement over SetTransformer ($p < 0.005$, using Wilcoxon signed-rank test). ‡ indicates that there is a statistically significant improvement over Time2VecTransformer ($p < 0.05$, using Wilcoxon signed-rank test).

Method	Modality	F1	Prec.	Recall	Acc.
T-LSTM [4]	T	$0.848\pm$8e-3	$0.896\pm$2e-2	$0.804\pm$1e-2	$0.855\pm$5e-3
EmoBERTa Transformer	T	$0.864\pm$1e-2	$0.843\pm$1e-2	$0.887\pm$3e-2	$0.861\pm$1e-2
LSTM + RL [19]	T	0.871	0.872	0.870	0.870
CNN + RL [19]	T	0.871	0.871	0.871	0.871
MTAL [2]*	T+I	0.842	0.842	0.842	0.842
GRU + VGG-Net + COMMA [20]	T+I	0.900	0.900	0.901	0.900
MTAN [9]**	T+I	0.908	0.885	0.931	-
Vanilla Transformer (**ours**)	T+I	$0.886\pm$1e-2	$0.868\pm$2e-2	$0.905\pm$2e-2	$0.883+$5e-3
SetTransformer (**ours**)	T+I	$0.927\pm$8e-3	$0.921\pm$1e-2	**$0.934\pm$2e-2‡**	$0.926\pm$8e-3
Time2VecTransformer (**ours**)	T+I	**$0.931\pm$4e-3†**	**$0.931\pm$2e-2†**	$0.931\pm$1e-2	**$0.931\pm$4e-3†**

Table 3. Results for multimodal depression detection on *multiRedditDep*. Our models use EmoBERTa and CLIP for extracting embeddings. *Vanilla Transformer* was trained on 128 sampled posts, while *Time2VecTransformer* and *SetTransformer* were trained on a window size of 512 posts. We denote with **bold** the best results for each column. *Uban et al. [51] conducted experiments using the visual and textual features from images and titles of the posts. † indicates that the result is a statistically significant improvement over Time2VecTransformer ($p < 0.005$, using Wilcoxon signed-rank test).

Method	Modality	F1	Prec.	Recall	Acc.	AUC
T-LSTM [4]	T	$0.831\pm$1e-2	$0.825\pm$8e-3	$0.837\pm$1e-2	$0.872\pm$7e-3	$0.946\pm$2e-3
EmoBERTa Transformer	T	$0.843\pm$6e-3	$0.828\pm$3e-3	$0.858\pm$1e-2	$0.879\pm$4e-3	$0.952\pm$2e-3
Uban et al. [51]*	T+I	-	-	-	0.663	0.693
Vanilla Transformer (**ours**)	T+I	$0.837\pm$8e-3	$0.827\pm$1e-2	$0.848\pm$1e-2	$0.876\pm$6e-3	$0.956\pm$3e-3
SetTransformer (**ours**)	T+I	**$0.902\pm$7e-3†**	**$0.878\pm$6e-3†**	**$0.928\pm$1e-2†**	**$0.924\pm$5e-3†**	**$0.976\pm$1e-3†**
Time2VecTransformer (**ours**)	T+I	$0.869\pm$7e-3	$0.869\pm$7e-3	$0.869\pm$8e-3	$0.901\pm$5e-3	$0.967\pm$1e-3

the Reddit dataset contains a considerable amount of noise and uninformative posts (links, short comments, etc.), which dilutes discriminative information for depression detection. A random sampling of posts seems to alleviate this problem to some degree.

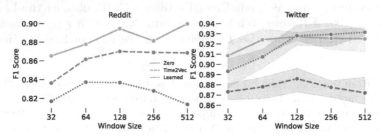

Fig. 4. Comparison among different window sizes used in our experiments, using EmoBERTa for text embeddings and CLIP for image encodings. For Twitter, the results were averaged across the 5 folds.

5.2 Ablation Study

To gauge the effect of the sampling window size, we performed experiments using CLIP as an image encoder and EmoBERTa as a text encoder, in which we varied the window size in $K = \{32, 64, 128, 256, 512\}$, as presented in Fig. 4. For Reddit, the 128 window size is the best suited for all three kinds of positional embeddings, as evidenced by the high F1 score. Even if Reddit users have an average of over 3,000 posts (Table 1), 128 posts contain enough information to make a correct decision. On Twitter, for *learned* and *zero* positional embeddings, the model with 128 posts window size performs best, while for *time2vec*, 512 has the best F1 score. This may be because the 512 window size covers the average number of posts from users in the Twitter dataset (see Table 1). Given the short time span of one month for the posts coming from Twitter, we can hypothesize that for datasets in which the time between posts is very small (average of 11.2 h for Twitter), the time component modeled by *time2vec* positional encodings may be more informative than other positional embedding methods. For Reddit, many of the users' posts are, in fact, comments or links, which are usually not informative to the model decision. Nevertheless, we achieve a performance comparable to the previous state-of-the-art, even in low-resource settings, by processing only 32 posts. Including the time component modeled by *time2vec* has a more important contribution when the time between posts is shorter (as in the data from Twitter) as opposed to larger periods between the posts (as is the case of Reddit).

In Table 4, we showcase different combinations of text encoders (RoBERTa/EmoBERTa/MiniLM) and image encoders (CLIP/DINO). The difference between image encoders is marginal, due to the small number of images in the users' timeline (Table 1). Interestingly, RoBERTa embeddings are more appropriate for Twitter, while EmoBERTa is better suited for modeling posts from Reddit. We hypothesize that this is due to the pseudo-anonymity offered through Reddit, encouraging users to post more intimate and emotional texts [13]. Using RoBERTa text embeddings and CLIP image embeddings, we obtain an F1 score of 0.943, which is even higher than the state-of-the-art. Further, the performance using MiniLM for text embeddings is lacking behind other encoders.

Table 4. Model comparison with different text and image encoding methods using the Reddit and Twitter datasets, with *time2vec* positional embeddings and 128 window size. The best results are with **bold**, and with <u>underline</u> the second best results.

Text+Image Enc.	Reddit			Twitter		
	F1	Prec.	Recall	F1	Prec.	Recall
MiniLM+CLIP	0.789±6e-3	0.686±7e-3	**0.929±9e-3**	0.827±8e-3	0.803±3e-2	0.854±4e-2
MiniLM+DINO	0.799±9e-3	0.745±8e-3	<u>0.862±1e-2</u>	0.792±8e-3	0.782±3e-2	0.806±4e-2
RoBERTa+CLIP	0.845±8e-3	0.829±8e-3	<u>0.862±1e-2</u>	**0.943±6e-3**	**0.951±1e-2**	**0.936±2e-2**
RoBERTa+DINO	0.840±9e-3	0.820±7e-3	0.861±1e-2	<u>0.936±1e-2</u>	0.946±2e-2	<u>0.926±2e-2</u>
EmoBERTa+CLIP	**0.871±7e-3**	**0.883±8e-3**	0.858±8e-3	0.928±1e-2	0.933±1e-2	0.924±2e-2
EmoBERTa+DINO	0.863±6e-3	<u>0.865±4e-3</u>	<u>0.862±1e-2</u>	0.915±1e-2	0.918±2e-2	0.913±2e-2

5.3 Error Analysis

We perform an error analysis on the predictions of *Time2VecTransformer* on the Twitter data. We use Integrated Gradients [46] to extract posts' attributions scores for predictions. In Fig. 5 (*Left*), the posts with depression cues have the strongest attribution scores, and the user is correctly labeled by the model. Given the way in which the mental health datasets are annotated by users' mention of diagnosis, some users from the non-depressed class may also have depression, but without mentioning it on social media. This may be the case of the user

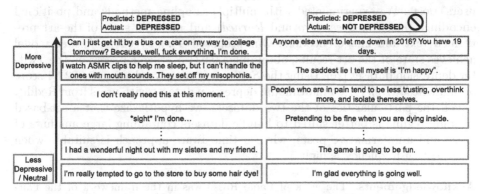

Fig. 5. Error analysis on two predictions, one correct (*Left*), and another incorrect (*Right*). The posts are sorted by their attribution scores given by Integrated Gradients [46]. The top posts have strong attribution for a positive (depressed) prediction, the bottom texts have a weak attribution to a positive prediction. *All examples were paraphrased, and only the texts are shown to maintain anonymity.*

from Fig. 5 (*Right*), who is showing definite signs of sadness and was incorrectly predicted by the model as having depression. Since our model operates across posts, using a feature attribution method such as Integrated Gradients naturally enables the automatic selection of the most relevant posts from a user, similar to [20,36], but without relying on a specialized procedure to do so.

5.4 Limitations and Ethical Considerations

The method proposed in this paper is trained on data with demographic bias [23] from Reddit and Twitter, two social media platforms with a demographic skew towards young males from the United States; thus our method may not succeed in identifying depression from other demographic groups. The aim of our system is to help in detecting cues of depression found in social media, and not to diagnose depression, as the diagnosis should only be made by a health professional following suitable procedures. Further, the dataset annotations rely on the users' self-reports, but without knowing the exact time of diagnosis. Harrigian et al. [22] studied the online content of users with a self-report of depression diagnosis and observed a decrease in linguistic evidence of depression over time that may be due to the users receiving treatment or other factors.

6   Conclusion

In this work, we showcased our time-enriched multimodal transformer architecture for depression detection from social media posts. Our model is designed to operate directly at the user-level: the attention mechanisms (both cross-modal and self-attention) attend to text and images across posts, and not to individual tokens. We provided two viable ways to train our method: a time-aware approach, in which we encode the relative time between posts through *time2vec* positional embeddings, and a set-based approach, in which no order is assumed between users' posts. We experimented with multiple sampling methods and positional encodings (i.e., *time2vec*, *zero* and *learned*) and multiple state-of-the-art pretrained text and image encoders. Our proposed *Time2VecTransformer* model achieves state-of-the-art results, obtaining a 0.931 average F1 score on the Twitter depression dataset [20]. Using the *SetTransformer*, we obtain a 0.902 F1 score on *multiRedditDep* [51], a multimodal depression dataset with users from Reddit. Given the particularities of the two datasets, we hypothesize that a set-based training regime is better suited to handle datasets containing large amounts of noise and uninformative posts, while a time-aware approach is suitable when user posting frequency is high.

Acknowledgements. The work of Paolo Rosso was in the framework of the Fair-TransNLP research project funded by MCIN, Spain (PID2021-124361OB-C31). Liviu P. Dinu was partially supported by a grant on Machine Reading Comprehension from Accenture Labs.

References

1. Alhuzali, H., Zhang, T., Ananiadou, S.: sign of depression via using frozen pre-trained models and random Predicting forest classifier. In: CLEF (Working Notes), pp. 888–896 (2021)
2. An, M., Wang, J., Li, S., Zhou, G.: Multimodal topic-enriched auxiliary learning for depression detection. In: Proc. of COLING, pp. 1078–1089 (2020)
3. Aragon, M.E., Lopez-Monroy, A.P., Gonzalez-Gurrola, L.C.G., Montes, M.: Detecting mental disorders in social media through emotional patterns-the case of anorexia and depression. In: IEEE Trans. Affect. Comput. (2021)
4. Baytas, I.M., Xiao, C., Zhang, X., Wang, F., Jain, A.K., Zhou, J.: Patient subtyping via time-aware lSTM networks. In: Proc. of SIGKDD. KDD '17, Assoc. Comput. Mach, pp. 65–74 (2017)
5. Berry, N., Lobban, F., Belousov, M., Emsley, R., Nenadic, G., Bucci, S., et al.: # whywetweetmh: understanding why people use twitter to discuss mental health problems. JMIR **19**(4), e6173 (2017)
6. Brown, T., et al.: Language models are few-shot learners. Adv. Neural. Inf. Process. Syst. **33**, 1877–1901 (2020)
7. Bucur, A.M., Cosma, A., Dinu, L.P.: Early risk detection of pathological gambling, self-harm and depression using bert. In: CLEF (Working Notes) (2021)
8. Caron, M., et al.: Emerging properties in self-supervised vision transformers. In: Proc. of ICCV, pp. 9650–9660 (2021)
9. Cheng, J.C., Chen, A.L.: Multimodal time-aware attention networks for depression detection. J. Intell. Inf. Syst, pp. 1–21 (2022)
10. Chiu, C.Y., Lane, H.Y., Koh, J.L., Chen, A.L.: Multimodal depression detection on instagram considering time interval of posts. J. Intell. Inf. Syst. **56**(1), 25–47 (2021)
11. Chung, J., Gulcehre, C., Cho, K., Bengio, Y.: Empirical evaluation of gated recurrent neural networks on sequence modeling. In: Proc. of NIPS 2014 Workshop on Deep Learning (2014)
12. Conneau, A., et al.: Unsupervised cross-lingual representation learning at scale. In: Proc. of ACL. Association for Computational Linguistics, pp. 8440–8451 (2020)
13. De Choudhury, M., De, S.: Mental health discourse on reddit: Self-disclosure, social support, and anonymity. In: Proc. of ICWSM 8(1), 71–80 (2014)
14. De Choudhury, M., Gamon, M., Counts, S., Horvitz, E.: Predicting depression via social media. In: Proc. of ICWSM, pp. 128–137 (2013
15. Deng, J., Dong, W., Socher, R., Li, L.J., Li, K., Fei-Fei, L.: ImageNet: a large-scale hierarchical image database. In: Proc. of CVPR, pp. 248–255. IEEE (2009)
16. Dufter, P., Schmitt, M., Schütze, H.: Position information in transformers: an overview. Computational Linguistics, pp. 1–31 (2021)
17. Ericsson, L., Gouk, H., Hospedales, T.M.: How well do self-supervised models transfer? In: Proc. of CVPR, pp. 5414–5423 (2021)
18. Fekete, S.: The Internet - A New Source of Data on Suicide, Depression and Anxiety: A Preliminary Study. Arch. Suicide Res. **6**(4), 351–361 (2002). https://doi.org/10.1080/13811110214533
19. Gui, T., Zhang, Q., Zhu, L., Zhou, X., Peng, M., Huang, X.: Depression Detection on Social Media with Reinforcement Learning. In: Sun, M., Huang, X., Ji, H., Liu, Z., Liu, Y. (eds.) Chinese Computational Linguistics: 18th China National Conference, CCL 2019, Kunming, China, October 18–20, 2019, Proceedings, pp. 613–624. Springer International Publishing, Cham (2019). https://doi.org/10.1007/978-3-030-32381-3_49

20. Gui, T., Zhu, L., Zhang, Q., Peng, M., Zhou, X., Ding, K., Chen, Z.: Cooperative Multimodal Approach to Depression Detection in Twitter. Proc. AAAI Conf. Artif. Intell. **33**(01), 110–117 (2019). https://doi.org/10.1609/aaai.v33i01.3301110
21. Guntuku, S.C., Preotiuc-Pietro, D., Eichstaedt, J.C., Ungar, L.H.: What twitter profile and posted images reveal about depression and anxiety. In: Proc. of ICWSM. **13**, pp. 236–246 (2019)
22. Harrigian, K., Dredze, M.: Then and now: Quantifying the longitudinal validity of self-disclosed depression diagnoses. In: Proc. of CLPsych Workshop. Association for Computational Linguistics, pp. 59–75 (2022)
23. Hovy, D., Spruit, S.L.: The social impact of natural language processing. In: Proc. of ACL, pp. 591–598 (2016)
24. Kazemi, S.M., et al.: Time2vec: Learning a vector representation of time. arXiv preprint arXiv:1907.05321 (2019)
25. Kenton, J.D.M.W.C., Toutanova, L.K.: Bert: Pre-training of deep bidirectional transformers for language understanding. In: Proc. of NAACL, pp. 4171–4186 (2019)
26. Kim, T., Vossen, P.: Emoberta: Speaker-aware emotion recognition in conversation with roberta. arXiv preprint arXiv:2108.12009 (2021)
27. Kingma, D.P., Ba, J.: Adam: A method for stochastic optimization. In: Proc. of ICLR (2015)
28. Lara, J.S., Aragón, M.E., González, F.A., Montes-y Gómez, M.: Deep bag-of-sub-emotions for depression detection in social media. In: Proc. of TSD. pp. 60–72. (2021) https://doi.org/10.1007/978-3-030-83527-9_5
29. Liu, Y., et al.: Roberta: a robustly optimized bert pretraining approach. arXiv preprint arXiv:1907.11692 (2019)
30. Losada, D., Crestani, F., Parapar, J.: Overview of eRisk 2019 Early Risk Prediction on the Internet. In: Crestani, F., et al. (eds.) CLEF 2019. LNCS, vol. 11696, pp. 340–357. Springer, Cham (2019). https://doi.org/10.1007/978-3-030-28577-7_27
31. Lustberg, L., Reynolds, C.F.: Depression and insomnia: questions of cause and effect. Sleep Med. Rev. **4**(3), 253–262 (2000)
32. Mann, P., Paes, A., Matsushima, E.H.: See and read: detecting depression symptoms in higher education students using multimodal social media data. In: Proc. of ICWSM. vol. 14, pp. 440–451 (2020)
33. Radford, A., et al.: Learning transferable visual models from natural language supervision. In: Proc. of ICML, pp. 8748–8763. PMLR (2021)
34. Rao, G., Zhang, Y., Zhang, L., Cong, Q., Feng, Z.: MGL-CNN: a hierarchical posts representations model for identifying depressed individuals in online forums. IEEE Access **8**, 32395–32403 (2020)
35. Reece, A.G., Danforth, C.M.: Instagram photos reveal predictive markers of depression. EPJ Data Sci. **6**(1), 15 (2017)
36. Ríssola, E.A., Bahrainian, S.A., Crestani, F.: A dataset for research on depression in social media. In: Proc. of UMAP, pp. 338–342 (2020)
37. Ríssola, E., Aliannejadi, M., Crestani, F.: Beyond Modelling: Understanding Mental Disorders in Online Social Media. In: Jose, J., et al. (eds.) ECIR 2020. LNCS, vol. 12035, pp. 296–310. Springer, Cham (2020). https://doi.org/10.1007/978-3-030-45439-5_20
38. Rude, S., Gortner, E.M., Pennebaker, J.: Language use of depressed and depression-vulnerable college students. Cognition Emotion **18**(8), 1121–1133 (2004)
39. Safa, R., Bayat, P., Moghtader, L.: Automatic detection of depression symptoms in twitter using multimodal analysis. J. Supercomputing 78(4) (2022)

40. Sawhney, R., Joshi, H., Gandhi, S., Shah, R.: A time-aware transformer based model for suicide ideation detection on social media. In: Proc. of EMNLP, pp. 7685–7697 (2020)

41. Shen, G., et al.: Depression detection via harvesting social media: a multimodal dictionary learning solution. In: Proc. of IJCAI, pp. 3838–3844 (2017)

42. Simonyan, K., Zisserman, A.: Very deep convolutional networks for large-scale image recognition. In: Bengio, Y., LeCun, Y. (eds.) Proc. of ICLR (2015)

43. Skaik, R., Inkpen, D.: Using twitter social media for depression detection in the canadian population. In: Proc. of AICCC, pp. 109–114 (2020)

44. Smith, L.N.: No more pesky learning rate guessing games. arXiv preprint arXiv:1206.1106 (2015)

45. Stankevich, M., Isakov, V., Devyatkin, D., Smirnov, I.V.: Feature engineering for depression detection in social media. In: Proc. of ICPRAM, pp. 426–431 (2018)

46. Sundararajan, M., Taly, A., Yan, Q.: Axiomatic attribution for deep networks. In: International conference on machine learning, pp. 3319–3328. PMLR (2017)

47. Szegedy, C., Ioffe, S., Vanhoucke, V., Alemi, A.A.: Inception-v4, inception-resnet and the impact of residual connections on learning. In: Proc. of AAAI, pp. 4278–4284. AAAI Press (2017)

48. Tan, H., Bansal, M.: Lxmert: Learning cross-modality encoder representations from transformers. In: Proc. of EMNLP-IJCNLP, pp. 5100–5111 (2019)

49. Trotzek, M., Koitka, S., Friedrich, C.M.: Utilizing neural networks and linguistic metadata for early detection of depression indications in text sequences. IEEE Trans. Knowl. Data Eng. **32**(3), 588–601 (2018)

50. Tsakalidis, A., Nanni, F., Hills, A., Chim, J., Song, J., Liakata, M.: Identifying moments of change from longitudinal user text. In: Proc. of ACL, pp. 4647–4660 (2022)

51. Uban, A.S., Chulvi, B., Rosso, P.: Explainability of depression detection on social media: From deep learning models to psychological interpretations and multimodality. In: Early Detection of Mental Health Disorders by Social Media Monitoring, pp. 289–320. Springer (2022). https://doi.org/10.1007/978-3-031-04431-1_13

52. Vaswani, A., et al.: Attention is all you need. Adv. Neural. Inf. Process. Syst. 30 (2017)

53. Wang, W., Wei, F., Dong, L., Bao, H., Yang, N., Zhou, M.: Minilm: Deep self-attention distillation for task-agnostic compression of pre-trained transformers. Adv. Neural. Inf. Process. Syst. **33**, 5776–5788 (2020)

54. Wang, Y., Wang, Z., Li, C., Zhang, Y., Wang, H.: A multimodal feature fusion-based method for individual depression detection on sina weibo. In: Proc. of IPCCC, pp. 1–8. IEEE (2020)

55. Wang, Y.A., Chen, Y.N.: What do position embeddings learn? an empirical study of pre-trained language model positional encoding. In: Proc. of EMNLP, pp. 6840–6849. Association for Computational Linguistics, (2020)

56. Wu, S.H., Qiu, Z.J.: A roberta-based model on measuring the severity of the signs of depression. In: CLEF (Working Notes), pp. 1071–1080 (2021)

57. Xu, Z., Pérez-Rosas, V., Mihalcea, R.: Inferring social media users' mental health status from multimodal information. In: Proc. of LREC, pp. 6292–6299 (2020)

58. Yates, A., Cohan, A., Goharian, N.: Depression and self-harm risk assessment in online forums. In: Proc. of EMNLP, pp. 2968–2978 (2017)

Recommendation Algorithm Based on Deep Light Graph Convolution Network in Knowledge Graph

Xiaobin Chen and Nanfeng Xiao[✉]

School of Computer Science and Engineering, South China University of Technology, GuangZhou, China
cssmall44992@mail.scut.edu.cn, xiaonf@scut.edu.cn

Abstract. Recently, recommendation algorithms based on Graph Convolution Network (GCN) have achieved many surprising results thanks to the ability of GCN to learn more efficient node embeddings. However, although GCN shows powerful feature extraction capability in user-item bipartite graphs, the GCN-based methods appear powerless for knowledge graph (KG) with complex structures and rich information. In addition, all of the existing GCN-based recommendation systems suffer from the over-smoothing problem, which results in the models not being able to utilize higher-order neighborhood information, and thus these models always achieve their best performance at shallower layers. In this paper, we propose a Deep Light Graph Convolution Network for Knowledge Graph (KDL-GCN) to alleviate the above limitations. Firstly, the User-Entity Bipartite Graph approach (UE-BP) is proposed to simplify knowledge graph, which leverages entity information by constructing multiple interaction graphs. Secondly, a Deep Light Graph Convolution Network (DLGCN) is designed to make full use of higher-order neighborhood information. Finally, experiments on three real-world datasets show that the KDL-GCN proposed in this paper achieves substantial improvement compared to the state-of-the-art methods.

Keywords: Recommendation · Graph convolution network · Knowledge graph

1 Introduction

Information overload has now become a nuisance for people, and it is difficult for information consumers to find their demands from the cluttered data, but the emergence of the recommendation systems (RS) have alleviated the problems, because the RS can recommend items of interest to target users based on their historical behavior or similar relationships among the users. A highly effective recommendation system has great benefits for information consumers, information providers and platforms. Collaborative filtering (CF) [22]

© The Author(s), under exclusive license to Springer Nature Switzerland AG 2023
J. Kamps et al. (Eds.): ECIR 2023, LNCS 13980, pp. 216–231, 2023.
https://doi.org/10.1007/978-3-031-28244-7_14

enables effective personalized recommendation services which focus on historical user-item interactions and learn user and item representations by assuming
that users with similar behaviors have the same preferences. Many CF-based
results [2,18,19,34,35,43] have been achieved based on different proposals.

In recent years, Graph Convolution Network [25] has shown unprecedented
feature extraction capabilities on graph data structures. Coincidentally, the data
in the recommendation systems is converted into bipartite graphs after processing. Therefore, many researchers [1,3,17,37,42,43,49] apply GCN in the recommendation systems, and various models are built to solve specific problems.
Although the GCN-based models have achieved great success, they face two
dilemmas.

Firstly, GCN shows excellent feature extraction ability in user-item bipartite
graph, but is powerless in the face of knowledge graph. KG contains multiple
types of entities, and there are complex interactions between different entities.
However, it is difficult for GCN to distinguish entity types in knowledge graphs,
and the interaction between different classes of entities generates noise interference. The UE-BP proposed in this paper first fuses KG and user-item bipartite
graphs, and then decomposes the fused graphs to obtain multiple user-entity
bipartite graphs, which enables GCN to make full use of entity information
while eliminating the noise interference between different entities.

Secondly, most current GCN-based models [1,3,17,42,43] achieve their best
performance at a shallow layer without utilizing higher-order neighborhood information. For instance, GCMC [1] uses one layer of GCN, NGCF [42] and DGCF
[43] are both 3-layer models, and LR-GCCF [3] and LightGCN [17] simplify GCN
to alleviate the over-smoothing problem, but still can only exploit information
of 3 to 4 steps. The deeper GCN-based model is supposed to have better performance because it receives more information about the higher-order neighborhoods. However, the over-smoothing problem makes the performance of deeper
networks degrade dramatically.

The over-smoothing problem states that as the GCN deepens, the embeddings of the graph nodes become similar. As the aggregation path lengthens, the
central node will aggregate an extremely large number of nodes, where the number of higher-order nodes will be much larger than the lower-order nodes, which
leads to the neglect of the lower-order information. However, the lower-order
nodes are closer to the central node and have more influence on the representation of the central node. The DLGCN is designed to alleviate the over-smoothing
problem, which assigns higher weights to the low-order nodes and allows the low-
order neighborhood information to propagate farther.

In summary, the KDL-GCN model is proposed to alleviate two challenges
faced by GCN-based recommendation algorithms. Firstly, the UE-BP method is
adopted to simplify KG and obtain multiple user-entity bipartite graphs, providing an idea of GCN for KG. Secondly, a neglected factor is noticed: deeper networks
aggregate too many higher-order nodes, resulting in lower-order features being
neglected. Therefore, DLGCN is designed to alleviate the over-smoothing problem.

Finally, the node embeddings extracted from the KG are fused for the recommendation task.

2   Related Work

2.1   Problem Formulation

In this section, we introduce the basic definition and symbols of recommendation systems. In a typical recommendation scenario, let $\mathcal{U} = \{u_1, u_2, \ldots, u_N\}$ be a user set of N users, and $\mathcal{I} = \{i_1, i_2, \ldots, i_M\}$ denote the item collection with M items. The task of the RS is to find more suitable user and item embedding. Therefore, let $e_u \in \mathbb{R}^{1 \times d}$ be final embedding of user u and $e_i \in \mathbb{R}^{1 \times d}$ be final embedding of item i, where d is embedding size.

2.2   Recent Work

The recommendation systems have received a lot of attention since its emergence and they are capable of making personalized recommendations for users even when they do not have explicit needs. Among them, CF-based models are fruitful as the most popular technique in the modern recommendation systems [5,6,17, 19,21,27,32,41,42].

CF usually takes the rating matrix as the input to the algorithm and retrieves similar users or items as the basis for rating prediction. For example, matrix decomposition (MF) [27] project user or item IDs as hidden embeddings and use inner products to predict interactions. Fast MF [20] proposes a method to weight missing data based on item popularity, and designs a new learning algorithm based on the element-wise Alternating Least Squares (eALS) technique, which is effective and efficient in terms of both improved MF models.

Subsequently, there has been a boom in neural recommendation models [4,8,11,19]. For instance, deep collaborative filtering model [8] applied deep learning to the field of recommendation algorithms and proved its excellent learning ability. NCF [19] replaces the interaction function in MF with non-linear neural networks, and these improvements and optimizations allow better learning of user and item representations.

Additionally, to alleviate data sparsity, some researchers further suggest borrowing users' historical behaviors for better user representation modeling, e.g., FISM [23] aggregates users' historical interaction items embedding as user representation vectors. SVD++ [26] additionally utilizes user ID embedding on the basis of FISM. Recent studies have found that users' historical interaction behavior should be given different importance. On account of this reason, attention mechanisms are applied to recommendation algorithms, such as ACF [2] and NAIS [18], which automatically assign different weighting factors to each historical interaction.

Another research direction is the combination of the Graph Neural Networks (GNNs) and the Recommendation System [1,17,30,31,37,41,42,44,45,48–51,53]. GNNs fit well with RS, because most data in RS is essentially graph structure, and on the other hand, GNNs are better able to capture the relationships between nodes and have strong capabilities in representation learning of graph data [47]. The Graph Convolution Network is one of the various GNNs frameworks, which have shown powerful feature extraction capabilities in recent years, and as a result, many GCN-based recommendation models [1,3,17,30,31,37,42,44] have emerged.

The basic idea of GCN is to aggregate neighborhood information to update the representation of the central node. GCMC [1] considers the matrix completion problem of RS from the perspective of link prediction of graphs, which employs user-item bipartite graphs to represent interaction data, and proposes a framework for autoencoder based on bipartite interaction graphs with promising performance. However, GCMC applies only one layer of graph convolution operations and ignores the higher order information in the data. Neural Graph Collaborative Filtering (NGCF) [42] proposes a new approach to encode nodes based on the development of Graph Neural Networks. More specifically, multiple GCN layers are leveraged to propagate node embedding, and in this way, information about higher-order neighborhoods is obtained, which is then injected into the embedding in an efficient manner. LightGCN [17] removes the nonlinear activation function and the feature transformation matrix in the GCN, conducts extensive experiments to prove the correctness of its model, and has made great progress.

IMP-GCN [30] argues that indiscriminate use of higher-order nodes introduces negative information and leads to performance degradation when stacking more layers. In view of that, it groups the users and constructs subgraphs of the users with similar hobbies, which reduces the propagation of negative information among higher-order neighbors, alleviates the over-smoothing problem, and deepens the Graph Convolutional Network. LGC-ACF [31] proposes that most GCN-based methods focus on the user-item bipartite graph, but ignore the information carried by the items themselves, introducing multi-faceted information of the items and thus improves the model. However, we argue that LGC-ACF simply mean-pooling the item information embedding and item ID embedding produces noise effects, and conducted derivations and extensive experiments to show that these noises affect the performance of the model.

In real life, KG is more extensive and there are many KG-based recommendation system models [9,28,33,36,39–41,46,52]. IntentGC [52] reconstructs user-user relationship and item-item relationship based on multi-entity knowledge graph, turning the multi-relationship graph into two homogeneous graphs. Unfortunately, the simplified graph structure will lose some information. AKGE [33] automatically extracts higher-order subgraphs that link user-item pairs with rich semantics based on the shortest path algorithm. KGNN-LS [39] is designed to learn user-specific item embeddings by identifying important knowledge graph relationships for a given user, and ablation studies demonstrate the efficacy of constraining neighboring nodes with the same label. KGAT [41] considers the user node as an entity in the knowledge graph and the interaction between the user and the item as a relationship. Consequently, the user-item bipartite graph is incorporated into the KG, which unifies the propagation operations while introducing noise interference.

3 Method

In this section, we introduce KDL-GCN as an example of a user-item bipartite graph and a knowledge graph with two types of entity information. Firstly, the KDL-GCN model framework is briefly described. Then each component of the model is presented in detail. Finally, a description of the loss function is given.

3.1 Overall Structure of KDL-GCN

Figure 1 shows the overall framework of KDL-GCN, which takes user-item bipartite graph and KG as input, and the output is the probability that user u selects item i. Firstly, the UE-BP method is proposed to decompose the relationship between user-item-entity and construct multiple user-entity bipartite graphs. Secondly, DLGCN is designed to extract the features of user-entity bipartite graphs and represent node ID as node embedding. It is worth mentioning that DLGCN alleviates the over-smoothing problem and makes full use of higher-order neighborhood information. Finally, the node embeddings obtained from the training are fused to obtain the final representation of the nodes, and then the inner product operation is performed to simulate the recommendation task.

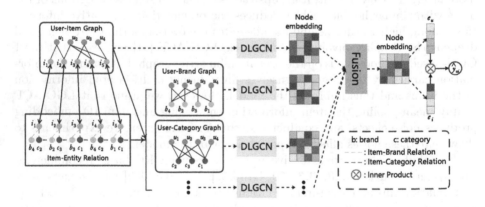

Fig. 1. An overview of KDL-GCN with two item-entity relations as illustration.

3.2 User-Entity Bipartite Graph Method

The knowledge graph contains rich entity information, for example, the movie KG includes movies, actors, directors, and genres; the commodity KG is composed of products, brands, and categories. However, the complex structure of the KG is not compatible with GCN, and the interaction between entities in KG generates noise interference. Therefore, the application of entity information in GCN becomes a difficult problem and there are fewer related studies. UE-BP method takes the item node as the medium to link the user-item bipartite graph with the KG, and then generates the user-entity bipartite graph based on the class of entities in the KG.

As shown in Fig. 2, using the example of users purchasing items to illustrate, KG contains brand and category information. There are four users, five items, three brands, and two categories. Using u for user, i for item, b for brand, and c for category. In Fig. 2 (a), u_1 purchased i_1, i_3. In the knowledge graph, the entity information of i_1 is c_1 and b_1, and i_3 corresponds to c_1, b_2. Thus, in the user-brand interaction graph, u_1 interacts with b_1, b_2; in the user-category bipartite graph, u_1 is associated with c_1. And so on, decompose all the user-item-entity relationships and construct user-entity bipartite graphs, namely $\{G_1, G_2, G_3\}$, as shown in Fig. 2 (b). Please note that in this paper, the user-item bipartite graph is treated as a special type of user-entity bipartite graph, which is because the item node is a kind of entity information.

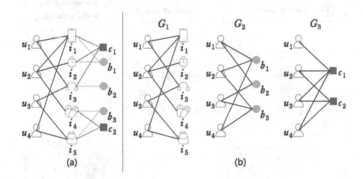

Fig. 2. (a) user-item-entity relations. (b) user-entity bipartite graphs.

3.3   Deep Light Graph Convolution Network

The GCN-based recommendation models update the node embedding at the $(l + 1)^{th}$ layer by aggregating the representations of its neighbor nodes at the lth layer. It is undeniable that the GCN-based algorithms have obtained great success, but the over-smoothing problem causes the models to achieve only suboptimal performance.

Figure 3 depicts the overall architecture of DLGCN, combining the Residual Network [14] and LightGCN [17]. The addition of the residual network gives higher weights to the lower-order information, thus ensuring that nodes embedding do not assimilate by aggregating too many higher-order features. Based on the GCN [25], the updated formulation of DLGCN on graph G for node embeddings between layers is specified as follows:

$$e_{G,u}^{(l+1)} = \sum_{v \in N_{G,u}} \frac{1}{\sqrt{\|N_{G,u}\|\|N_{G,v}\|}} e_{G,v}^{(l)} + \gamma e_{G,u}^{(l-1)} \tag{1}$$

$$e_{G,v}^{(l+1)} = \sum_{u \in N_{G,v}} \frac{1}{\sqrt{\|N_{G,v}\|\|N_{G,u}\|}} e_{G,u}^{(l)} + \gamma e_{G,v}^{(l-1)} \tag{2}$$

where, $e_{G,u}^{(l)}$ and $e_{G,v}^{(l)}$ denote respectively the embedding of user u and entity v after l layers aggregation on graph G, $N_{G,u}$ indicates the set of entities that interact with user u on G, $N_{G,v}$ represents the set of users that interact with entity v on G, γ is the proportion of feature information in the residual network passed from shallow to deep layers. The interval of the residual network is 2. This is considering that even if the GCN is deepened, its depth is still shallow. Besides, we adopt the symmetric normalization $\frac{1}{\sqrt{\|N_{G,u}\|\|N_{G,v}\|}}$ to prevent the increase in embedding scale with graph convolution operations.

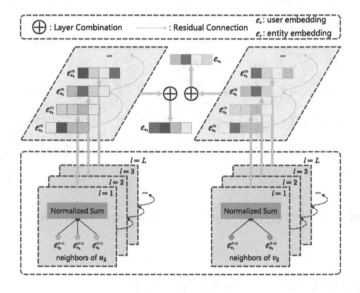

Fig. 3. An illustration of the DLGCN model architecture.

Layer-level Combination. When the DLGCN model with L-layers is adopted to a user-entity bipartite graph G, each node has $(L + 1)$ embeddings, namely $\{e_{G,u}^{(0)}, e_{G,u}^{(1)}, \ldots, e_{G,u}^{(L)}\}$ and $\{e_{G,v}^{(0)}, e_{G,v}^{(1)}, \ldots, e_{G,v}^{(L)}\}$. A fair strategy is applied to assign equal weights to the embeddings at each layer. The embeddings formula for the nodes can be obtained as follows [17]:

$$e_{G,u} = \frac{1}{L+1} \sum_{l=0}^{L} e_{G,u}^{(l)}, \qquad e_{G,v} = \frac{1}{L+1} \sum_{l=0}^{L} e_{G,v}^{(l)} \tag{3}$$

where, L represents the depths of the DLGCN, $e_{G,u}^{(l)}$ and $e_{G,v}^{(l)}$ indicate the embedding of user u and item v on graph G, respectively, which aggregate the node information within l steps.

3.4 Node Embedding Fusion and Rating Prediction

The concatenation is chosen to fuse the node embeddings learned from different user-entity interaction graphs because it eliminates the noise generated by the interaction between entities. The inner product of the final embeddings of users and items is used to predict users' preferences, calculated as follows:

$$\hat{y}_{ui} = e_u e_i^\top \tag{4}$$

where, $\hat{y}_{ui}$ is the prediction result of the model, which indicates the preference of user u for item i, and e_u, e_i is the final embedding of user u and item i.

Analysis. In this section, the advantages of concatenation are illustrated by comparing the concatenation with the sum operation and the mean-pooling method. For illustration, two user-entity bipartite graphs are used, then there are embedding of user u as $e_{G_1,u}, e_{G_2,u}$, and embedding of item i as $e_{G_1,v}, e_{G_2,v}$.

Adopting the concatenation [12], $e_u = (e_{G_1,u}, e_{G_2,u}), e_i = (e_{G_1,v}, e_{G_2,v})$, are obtained, and then $\hat{y}_{ui}$ is calculated as follows:

$$\hat{y}_{ui} = e_{G_1,u} e_{G_1,v}^\top + e_{G_2,u} e_{G_2,v}^\top \tag{5}$$

with the sum operation [38], getting $e_u = (e_{G_1,u} + e_{G_2,u}), e_i = (e_{G_1,v} + e_{G_2,v})$, applying the model prediction, $\hat{y}_{ui}$ is represented as follows:

$$\hat{y}_{ui} = e_{G_1,u} e_{G_1,v}^\top + e_{G_2,u} e_{G_2,v}^\top + \underbrace{e_{G_1,u} e_{G_2,v}^\top + e_{G_2,u} e_{G_1,v}^\top}_{Noise} \tag{6}$$

After performing the mean-pooling operation [29], $e_u = \frac{1}{2}(e_{G_1,u} + e_{G_2,u})$, $e_i = \frac{1}{2}(e_{G_1,v} + e_{G_2,v})$, and $\hat{y}_{ui}$ is depicted as follows via model prediction:

$$\hat{y}_{ui} = \frac{1}{4} \left(e_{G_1,u} e_{G_1,v}^\top + e_{G_2,u} e_{G_2,v}^\top + \underbrace{e_{G_1,u} e_{G_2,v}^\top + e_{G_2,u} e_{G_1,v}^\top}_{Noise} \right) \tag{7}$$

Comparing the three methods, it can be found that the sum operation and the mean-pooling method lead to the interplay of node embeddings learned on different graphs, which generates noise interference. However, the purpose of the UE-BP method is to separate the user-item-entity relations. Sum operations and mean-pooling approach complicate the relationship and run counter to our original intent. In Sect. 4.3, a comparative test of the three methods is performed to prove the correctness of our theory.

3.5 Optimization

The parameters trained in KDL-GCN are the 0th layer embedding in each user-entity bipartite graph, i.e., $\Theta = \{E_{G_1}^{(0)}, E_{G_2}^{(0)}, \ldots, E_{G_s}^{(0)}\}$. In the training process of the model, we use Bayesian Personalized Ranking (BPR) loss [32], which is based on the idea of making the difference between the scores of positive and negative samples as large as possible. To implement the optimization process, the set of triples (u, i^+, i^-) is constructed, where interactions are observed between u and i^+, while no interactions are observed between u and i^-. Therefore the BPR loss function equation is as follows [32]:

$$\mathcal{L} = \sum_{(u,i^+,i^-)\in O} -\ln \sigma \left(\hat{y}_{ui^+} - \hat{y}_{ui^-} \right) + \lambda \|\Theta\|_2^2 \tag{8}$$

where, $O = \{(u, i^+, i^-)|(u, i^+) \in R^+, (u, i^-) \in R^-\}$ represents the training data, R^+ denotes the positive samples, which imply that there is an observable interaction between u and i^+, and R^- indicates the sampled unobserved interactions. In addition, L_2 regularization is used to prevent overfitting, λ is the regularization weight and Θ represents the parameters in the model. The Adam [24] optimizer in a mini-batch manner is applied in KDL-GCN, which calculates the adaptive learning rate for each parameter and works well in practical applications.

3.6 Time Complexity Analysis

Cluster-GCN [7] states that for the $l - th$ propagation layer of GCN [25], the computational complexity is $O\left(\|A\|_0 d_{l-1} + N d_l d_{l-1}\right)$, where $\|A\|_0$ is number of nonzeros in the adjacency matrix, n is the number of nodes, d_l and d_{l-1} are the embedding size of the current layer and the previous layer, respectively.

The nonlinear activation functions and feature transformation matrices are removed from LightGCN [17] and only the aggregation operations are saved. Its time complexity is $O\left(L\|A\|_0 d\right)$, where L denotes the number of network layers and d is the embedding size, which is fixed in each layer.

In contrast to LightGCN, the residual network is added to DLGCN and the residual interval is 2, so the computational complexity is $O\left(L\|A\|_0 d + \frac{LN}{2} d\right)$.

KDL-GCN fuses features of multiple user-entity bipartite graphs, therefore the time complexity of KDL-GCN is $O\left(L \sum_{G=G_1}^{G_s} \left(\|A_G\|_0 + \frac{N}{2}\right) d\right)$, where G is the user-entity bipartite graph.

4 Experiments

4.1 Experiment Setup

Datasets. To evaluate the effectiveness of KDL-GCN, extensive experiments are conducted on three benchmark datasets: Movielens [13], Amazon-Electronic [15], TaoBao [54], which are widely used for training and evaluation of the recommendation models. These benchmark datasets are open sourced real-world data with

Table 1. The statistics of the knowledge graph.

Datasets	Users	Items	Interactions	Density	Attribute	Counts
Movielens	610	9606	100551	0.01716	Genres	21
					directors	3942
					actors	4034
Amazon	13201	14094	390581	0.00210	Brands	2771
					categories	15
TaoBao	69166	39406	918431	0.00034	Categories	1671

various domains, sizes, and density. The demographic data of the datasets are shown in Table 1.

Movielens: Movielens latest small is chosen. Regarding the entity information, three relations were used: genre, director and actor. Finally, the explicit ratings were converted to implicit data.

Amazon: Amazon-Electronic from the public Amazon review dataset. We removed items with less than 5 interactions and ensured that each user interacted with at least 20 items. For KG, two relationships are selected: brand and category, and finally the explicit ratings are converted to implicit data.

TaoBao: TaoBao dataset is a publicly e-commerce dataset. The users with less than 10 interactions were removed, and also treat 'buy' and 'add to cart' as positive samples. In terms of knowledge graph, we choose the category.

Baseline Models. Comparing KDL-GCN with the following methods: (1) MF [27] adopts the user-item direct interactions, which is a classical model based on matrix factorization. (2) GCMC [1] applies the one layer GCN to utilize the connections between users and items. (3) NGCF [42] employs 3-layer graph convolution that captures both direct and indirect information between users and items. (4) LR-GCCF [3] removes the non-linear activation function to overcome the over-smoothing problem. (5) LightGCN [17] simplifies the GCN by removing the nonlinear activation and the feature transformation matrices. (6) IMP-GCN [30] mitigates the over-smoothing problem and deepens the Light-GCN by using a user grouping strategy, i.e., separating users that are not similar. (7) LGC-ACF [31] follows the structure of LightGCN and further exploits the multi-aspect user-item interaction information.

Evaluation Metrics. Two mainstream metrics are used to evaluate the KDL-GCN, Recall and Normalized Discounted Cumulative Gain (NDCG) [16]. The top-20 results for both assessment metrics were calculated.

Hyper-parameters Setting. The embedding size is fixed to 64 for all models and embedding parameters are initialized with Xavier method [10]. The embedding size may be set larger, such as 128 or 256, which can further improve the accuracy of the model, but the computational efficiency will be reduced. And the embedding size of each comparison model is the same, which does not affect the fairness of the experiment. Adam [24] with a default learning rate of 0.001 is used to optimize KDL-GCN. Meanwhile, the mini-batch size is set to 2048 to accelerate the training of the model. The L_2 regularization coefficient λ is search from $1e^{-6}, 1e^{-5}, \ldots, 1e^{-2}$, and in most cases the optimal value is $1e^{-4}$. In addition, setting the propagation ratio γ in residual network to 0.1, and L is tested in the range of 1 to 8.

The triplet (u, i^+, i^-) is constructed by random sampling and recombination, where the seed of the random number is set to 2022. Specifically, there are two types of relationships between users and items in the training set, i.e., POSITIVE and NEGATIVE. POSITIVE indicates interaction with the user and NEGATIVE means no interaction with the user, so each user has independent POSITIVE and NEGATIVE samples. In each epoch, all users perform a random sampling of their own POSITIVE and NEGATIVE to get the i^+ and i^- corresponding to user u. Finally, the data obtained by random sampling are combined into a triple (u, i^+, i^-).

4.2   Performance Comparison

Table 2 shows the results of model performance comparison. From the experimental results, we have the following observations:

Table 2. The comparison of the overall performance of Recall@20 and NDCG@20. The best performance is highlighted in bold, and the second best is underlined. Note that these values are in the form of percentages with the '%' omitted.

Dataset	Movielens		Amazon		TaoBao	
Method	Recall	NDCG	Recall	NDCG	Recall	NDCG
MF	21.87	27.67	4.35	3.04	3.23	1.72
GCMC	22.65	29.98	4.68	3.38	3.39	1.82
NGCF	23.68	30.85	4.98	3.74	3.59	1.98
LR-GCCF	24.42	33.23	6.17	4.55	3.73	2.05
LightGCN	24.59	33.11	5.79	4.39	4.88	2.82
IMP-GCN	24.65	33.58	6.72	5.22	5.66	3.23
LGC-ACF	24.91	34.76	6.21	4.70	6.76	3.80
KDL-GCN	**25.86**	**35.53**	**7.52**	**5.77**	**7.61**	**4.30**
Improve(%)	3.81	2.22	11.90	10.54	12.57	13.16

– The KDL-GCN achieves the best performance on all datasets, indicating that UE-BP method and DLGCN significantly improve the performance of the recommendation system.

- Of all the methods, MF performs the worst because it does not use the higher-order connection between the user and the item. GCMC also does not achieve good performance because it uses only one layer of GCN. However, GCMC performs better than MF, which indicates that the GCN-based method is stronger in feature extraction.
- LGC-ACF utilizes the mean-pooling operation to introduce the item information, and KDL-GCN outperforms it on every dataset, which indicates that mean-pooling generates noise interference and concatenation is a more appropriate fusion method.
- In contrast to LGC-ACF, the improvement of KDL-GCN is more obvious on large datasets, which indicates that there are more high-order nodes on the big datasets and the adoption of high-order neighborhood information will improve the performance of the model.

4.3 Ablation Analysis

Impact of Network Depth. To investigate the influence of the number of layers L on KDL-GCN, varying L in the range of $\{1,\dots,8\}$ and summarizing the experimental results as shown in Fig. 4.

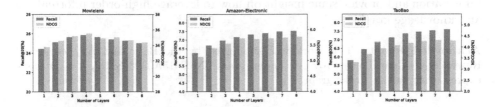

Fig. 4. The impact of network depth on KDL-GCN.

The best results of KDL-GCN are achieved at $L = 4$ for the Movielens dataset, and then the model performance gradually decreases as the number of layers increases. This indicates that small datasets contain few higher-order nodes. Focusing on Amazon and Taobao, the performance of the model increases gradually as L increases from 1 to 8, but the metric increases at a slower rate and eventually levels off. This denotes that higher-order neighborhood information plays a significant role in the improvement of model performance.

Impact of Fusion Methods. In Sect. 3.4, a theoretical comparison of the three approaches to fusion embeddings is presented. In this section, three fusion methods are tested on three datasets based on DLGCN. Table 3 shows the experimental results, which are consistent with the theoretical analysis that the sum operation and mean-pooling method produce noise interference and decreases the performance of the model.

Table 3. Performance comparison of different fusion methods. Note that these values are in the form of percentages with the '%' omitted.

Dataset	Movielens		Amazon		TaoBao	
Metrics	Recall	NDCG	Recall	NDCG	Recall	NDCG
Concatenation	**25.86**	**35.53**	**7.52**	**5.77**	**7.61**	**4.30**
Sum operation	24.45	31.15	6.20	4.59	7.12	3.99
Mean-pooling	24.55	34.37	6.24	4.65	6.47	3.56

5   Conclusion

In this paper, a new GCN-based recommendation model, KDL-GCN, is proposed, which unites the user-item bipartite graph and the knowledge graph. KDL-GCN, firstly, constructs the user-entity interaction graph, then extracts the features of each user-entity bipartite graph using the DLGCN module, and finally the features on different graphs are fused to get the final embedding of the nodes. The experiments on the real datasets demonstrate the effectiveness of the KDL-GCN model and achieve the state-of-the-art performance. We believe that this study can demonstrate the importance of higher-order nodes and entity information and provide some insight on how to leverage high-order information and knowledge graphs.

Acknowledgements. This work was supported by the Basic and Applied Basic Research of Guangdong Province under grant [No.2015A030308018], the authors express their thanks to the grant.

References

1. Berg, R.V.d., Kipf, T.N., Welling, M.: Graph convolutional matrix completion. arXiv preprint arXiv:1706.02263 (2017)
2. Chen, J., Zhang, H., He, X., Nie, L., Liu, W., Chua, T.S.: Attentive collaborative filtering: multimedia recommendation with item-and component-level attention. In: Proceedings of the 40th International ACM SIGIR Conference on Research and Development in Information Retrieval, pp. 335–344 (2017)
3. Chen, L., Wu, L., Hong, R., Zhang, K., Wang, M.: Revisiting graph based collaborative filtering: a linear residual graph convolutional network approach. In: Proceedings of the AAAI Conference on Artificial Intelligence, pp. 27–34 (2020)
4. Cheng, H.T., et al.: Wide & deep learning for recommender systems. In: Proceedings of the 1st Workshop on Deep Learning for Recommender Systems, pp. 7–10 (2016)
5. Cheng, Z., Chang, X., Zhu, L., Kanjirathinkal, R.C., Kankanhalli, M.: MMALFM: explainable recommendation by leveraging reviews and images. ACM Trans. Inf. Syst. (TOIS) **37**(2), 1–28 (2019)
6. Cheng, Z., Ding, Y., Zhu, L., Kankanhalli, M.: Aspect-aware latent factor model: rating prediction with ratings and reviews. In: Proceedings of the 2018 World Wide Web Conference, pp. 639–648 (2018)

7. Chiang, W.L., Liu, X., Si, S., Li, Y., Bengio, S., Hsieh, C.J.: Cluster-GCN: an efficient algorithm for training deep and large graph convolutional networks. In: Proceedings of the 25th ACM SIGKDD international conference on knowledge discovery & data mining, pp. 257–266 (2019)
8. Covington, P., Adams, J., Sargin, E.: Deep neural networks for youtube recommendations. In: Proceedings of the 10th ACM Conference on Recommender Systems, pp. 191–198 (2016)
9. Fan, W., et al.: Graph neural networks for social recommendation. In: The World Wide Web Conference, pp. 417–426 (2019)
10. Glorot, X., Bengio, Y.: Understanding the difficulty of training deep feedforward neural networks. In: Proceedings of the Thirteenth International Conference on Artificial Intelligence and Statistics, pp. 249–256. JMLR Workshop and Conference Proceedings (2010)
11. Guo, H., Tang, R., Ye, Y., Li, Z., He, X.: DeepFM: a factorization-machine based neural network for CTR prediction. arXiv preprint arXiv:1703.04247 (2017)
12. Hamilton, W.L., Ying, R., Leskovec, J.: Inductive representation learning on large graphs. In: Proceedings of the 31st International Conference on Neural Information Processing Systems, pp. 1025–1035 (2017)
13. Harper, F.M., Konstan, J.A.: The movielens datasets: history and context. ACM Trans. Interact. Intell. Syst. (TIIS) 5(4), 1–19 (2015)
14. He, K., Zhang, X., Ren, S., Sun, J.: Deep residual learning for image recognition. In: Proceedings of the IEEE Conference on Computer Vision and Pattern Recognition, pp. 770–778 (2016)
15. He, R., McAuley, J.: Ups and downs: modeling the visual evolution of fashion trends with one-class collaborative filtering. In: Proceedings of the 25th International Conference on World Wide Web, pp. 507–517 (2016)
16. He, X., Chen, T., Kan, M.Y., Chen, X.: Trirank: review-aware explainable recommendation by modeling aspects. In: Proceedings of the 24th ACM International on Conference on Information and Knowledge Management, pp. 1661–1670 (2015)
17. He, X., Deng, K., Wang, X., Li, Y., Zhang, Y., Wang, M.: LightGCN: simplifying and powering graph convolution network for recommendation. In: Proceedings of the 43rd International ACM SIGIR Conference on Research and Development in Information Retrieval, pp. 639–648 (2020)
18. He, X., He, Z., Song, J., Liu, Z., Jiang, Y.G., Chua, T.S.: Nais: neural attentive item similarity model for recommendation. IEEE Trans. Knowl. Data Eng. 30(12), 2354–2366 (2018)
19. He, X., Liao, L., Zhang, H., Nie, L., Hu, X., Chua, T.S.: Neural collaborative filtering. In: Proceedings of the 26th International Conference on World Wide Web, pp. 173–182 (2017)
20. He, X., Zhang, H., Kan, M.Y., Chua, T.S.: Fast matrix factorization for online recommendation with implicit feedback. In: Proceedings of the 39th International ACM SIGIR Conference on Research and Development in Information Retrieval, pp. 549–558 (2016)
21. Hsieh, C.K., Yang, L., Cui, Y., Lin, T.Y., Belongie, S., Estrin, D.: Collaborative metric learning. In: Proceedings of the 26th International Conference on World Wide Web, pp. 193–201 (2017)
22. Hu, Y., Koren, Y., Volinsky, C.: Collaborative filtering for implicit feedback datasets. In: 2008 Eighth IEEE International Conference on Data Mining, pp. 263–272. IEEE (2008)

23. Kabbur, S., Ning, X., Karypis, G.: FISM: factored item similarity models for top-n recommender systems. In: Proceedings of the 19th ACM SIGKDD International Conference on Knowledge Discovery and Data Mining, pp. 659–667 (2013)

24. Kingma, D.P., Ba, J.: Adam: a method for stochastic optimization. arXiv preprint arXiv:1412.6980 (2014)

25. Kipf, T.N., Welling, M.: Semi-supervised classification with graph convolutional networks. arXiv preprint arXiv:1609.02907 (2016)

26. Koren, Y.: Factorization meets the neighborhood: a multifaceted collaborative filtering model. In: Proceedings of the 14th ACM SIGKDD International Conference on Knowledge Discovery and Data Mining, pp. 426–434 (2008)

27. Koren, Y., Bell, R., Volinsky, C.: Matrix factorization techniques for recommender systems. Computer **42**(8), 30–37 (2009)

28. Li, M., Gan, T., Liu, M., Cheng, Z., Yin, J., Nie, L.: Long-tail hashtag recommendation for micro-videos with graph convolutional network. In: Proceedings of the 28th ACM International Conference on Information and Knowledge Management, pp. 509–518 (2019)

29. Li, Y., Tarlow, D., Brockschmidt, M., Zemel, R.: Gated graph sequence neural networks. arXiv preprint arXiv:1511.05493 (2015)

30. Liu, F., Cheng, Z., Zhu, L., Gao, Z., Nie, L.: Interest-aware message-passing GCN for recommendation. In: Proceedings of the Web Conference 2021, pp. 1296–1305 (2021)

31. Mei, D., Huang, N., Li, X.: Light graph convolutional collaborative filtering with multi-aspect information. IEEE Access **9**, 34433–34441 (2021)

32. Rendle, S., Freudenthaler, C., Gantner, Z., Schmidt-Thieme, L.: BPR: Bayesian personalized ranking from implicit feedback. arXiv preprint arXiv:1205.2618 (2012)

33. Sha, X., Sun, Z., Zhang, J.: Attentive knowledge graph embedding for personalized recommendation. arXiv preprint arXiv:1910.08288 (2019)

34. Shi, C., et al.: Deep collaborative filtering with multi-aspect information in heterogeneous networks. IEEE Transactions on Knowledge and Data Engineering (2019)

35. Shi, Y., Larson, M., Hanjalic, A.: List-wise learning to rank with matrix factorization for collaborative filtering. In: Proceedings of the fourth ACM Conference on Recommender Systems, pp. 269–272 (2010)

36. Song, W., Xiao, Z., Wang, Y., Charlin, L., Zhang, M., Tang, J.: Session-based social recommendation via dynamic graph attention networks. In: Proceedings of the Twelfth ACM International Conference on Web Search and Data Mining, pp. 555–563 (2019)

37. Sun, J., et al.: Multi-graph convolution collaborative filtering. In: 2019 IEEE International Conference on Data Mining (ICDM), pp. 1306–1311. IEEE (2019)

38. Veličković, P., Cucurull, G., Casanova, A., Romero, A., Lio, P., Bengio, Y.: Graph attention networks. arXiv preprint arXiv:1710.10903 (2017)

39. Wang, H., et al.: Knowledge-aware graph neural networks with label smoothness regularization for recommender systems. In: Proceedings of the 25th ACM SIGKDD International Conference on Knowledge Discovery & Data Mining, pp. 968–977 (2019)

40. Wang, H., Zhao, M., Xie, X., Li, W., Guo, M.: Knowledge graph convolutional networks for recommender systems. In: The World Wide Web Conference, pp. 3307–3313 (2019)

41. Wang, X., He, X., Cao, Y., Liu, M., Chua, T.S.: KGAT: knowledge graph attention network for recommendation. In: Proceedings of the 25th ACM SIGKDD International Conference on Knowledge Discovery & Data Mining, pp. 950–958 (2019)

42. Wang, X., He, X., Wang, M., Feng, F., Chua, T.S.: Neural graph collaborative filtering. In: Proceedings of the 42nd International ACM SIGIR Conference on Research and Development in Information Retrieval, pp. 165–174 (2019)
43. Wang, X., Jin, H., Zhang, A., He, X., Xu, T., Chua, T.S.: Disentangled graph collaborative filtering. In: Proceedings of the 43rd International ACM SIGIR Conference on Research and Development in Information Retrieval, pp. 1001–1010 (2020)
44. Wang, X., Wang, R., Shi, C., Song, G., Li, Q.: Multi-component graph convolutional collaborative filtering. In: Proceedings of the AAAI Conference on Artificial Intelligence, pp. 6267–6274 (2020)
45. Wu, L., Li, J., Sun, P., Hong, R., Ge, Y., Wang, M.: Diffnet++: a neural influence and interest diffusion network for social recommendation. IEEE Trans. Knowl. Data Eng. (2020)
46. Wu, Q., et al.: Dual graph attention networks for deep latent representation of multifaceted social effects in recommender systems. In: The World Wide Web Conference, pp. 2091–2102 (2019)
47. Wu, S., Sun, F., Zhang, W., Cui, B.: Graph neural networks in recommender systems: a survey. arXiv preprint arXiv:2011.02260 (2020)
48. Wu, S., Zhang, M., Jiang, X., Ke, X., Wang, L.: Personalizing graph neural networks with attention mechanism for session-based recommendation. arXiv preprint arXiv:1910.08887 (2019)
49. Ying, R., He, R., Chen, K., Eksombatchai, P., Hamilton, W.L., Leskovec, J.: Graph convolutional neural networks for web-scale recommender systems. In: Proceedings of the 24th ACM SIGKDD International Conference on Knowledge Discovery & Data Mining, pp. 974–983 (2018)
50. Zhang, J., Shi, X., Zhao, S., King, I.: Star-GCN: stacked and reconstructed graph convolutional networks for recommender systems. arXiv preprint arXiv:1905.13129 (2019)
51. Zhang, M., Chen, Y.: Inductive matrix completion based on graph neural networks. arXiv preprint arXiv:1904.12058 (2019)
52. Zhao, J., et al.: IntentGC: a scalable graph convolution framework fusing heterogeneous information for recommendation. In: Proceedings of the 25th ACM SIGKDD International Conference on Knowledge Discovery & Data Mining, pp. 2347–2357 (2019)
53. Zheng, L., Lu, C.T., Jiang, F., Zhang, J., Yu, P.S.: Spectral collaborative filtering. In: Proceedings of the 12th ACM Conference on Recommender Systems, pp. 311–319 (2018)
54. Zhu, H., et al.: Learning tree-based deep model for recommender systems. In: Proceedings of the 24th ACM SIGKDD International Conference on Knowledge Discovery & Data Mining, pp. 1079–1088 (2018)

Query Performance Prediction for Neural IR: Are We There Yet?

Guglielmo Faggioli[1]([✉]), Thibault Formal[2,3], Stefano Marchesin[1],
Stéphane Clinchant[2], Nicola Ferro[1], and Benjamin Piwowarski[3,4]

[1] University of Padova, Padova, Italy
guglielmo.faggioli@unipd.it
[2] Naver Labs Europe, Meylan, France
[3] Sorbonne Université, ISIR, Paris, France
[4] CNRS, Paris, France

Abstract. Evaluation in Information Retrieval (IR) relies on post-hoc empirical procedures, which are time-consuming and expensive operations. To alleviate this, Query Performance Prediction (QPP) models have been developed to estimate the performance of a system without the need for human-made relevance judgements. Such models, usually relying on lexical features from queries and corpora, have been applied to traditional sparse IR methods – with various degrees of success. With the advent of neural IR and large Pre-trained Language Models, the retrieval paradigm has significantly shifted towards more semantic signals. In this work, we study and analyze to what extent current QPP models can predict the performance of such systems. Our experiments consider seven traditional bag-of-words and seven BERT-based IR approaches, as well as nineteen state-of-the-art QPPs evaluated on two collections, Deep Learning '19 and Robust '04. Our findings show that QPPs perform statistically significantly worse on neural IR systems. In settings where semantic signals are prominent (e.g., passage retrieval), their performance on neural models drops by as much as 10% compared to bag-of-words approaches. On top of that, in lexical-oriented scenarios, QPPs fail to predict performance for neural IR systems on those queries where they differ from traditional approaches the most.

1   Introduction

The advent of Neural IR (NIR) and Pre-trained Language Models (PLM) induced considerable changes in several central IR research and application areas, with implications that are yet to be fully tamed by the research community. Query Performance Prediction (QPP) is defined as the prediction of the performance of an IR system without human-crafted relevance judgements and is one of the areas the most interested by advancements in NIR and PLM domains. In fact, *i)* PLM can help developing better QPP models, and *ii)* it is not fully clear yet whether current QPP techniques can be successfully applied to NIR. With this paper, we aim to explore the connection between PLM-based first-stage retrieval techniques and the available QPP models. We are interested in investigating to what extent QPP techniques can be applied to such IR systems, given

J. Kamps et al. (Eds.): ECIR 2023, LNCS 13980, pp. 232–248, 2023.
https://doi.org/10.1007/978-3-031-28244-7_15

i) their fundamentally different underpinnings compared to traditional lexical IR approaches, *ii)* that they hold the promise to replace – or at least complement – them in multi-stage ranking pipelines. In return, QPP advantages are multi-fold: it can be used to select the best-performing system for a given query, help users in reformulating their needs, or identify pathological queries that require manual intervention from the system administrators. Said otherwise, the need for QPP still holds for NIR methods. Among the plethora of available QPP methods, most of them rely on lexical aspects of the query and the collection. Such approaches have been devised, tested, and evaluated in predicting the performance of lexical bag-of-words IR systems – from now on referred to as Traditional IR (TIR) – with various degrees of success. Recent advances in Natural Language Processing (NLP) led to the advent of PLM-based IR systems, which shifted the retrieval paradigm from traditional approaches based on lexical matching to exploiting contextualized semantic signals – thus alleviating the semantic gap problem. To ease the readability throughout the rest of the manuscript, with an abuse of notation, we use the more general term NIR to explicitly refer to first-stage IR systems based on BERT [13].

At the current time, no large-scale work has been devoted to assessing whether traditional QPP models can be used for NIR systems – which is the goal of this study. We compare the performance of nineteen QPP methods applied to seven traditional TIR systems, with those achieved on seven state-of-the-art first-stage NIR approaches based on PLM. We consider both pre- and post-retrieval QPPs, and include in our analyses post-retrieval QPP models that exploit lexical or semantic signals to compute their predictions. To instantiate our analyses on different scenarios we consider two widely adopted experimental collections: Robust '04 and Deep Learning '19. Our contributions are as follows:

– we apply and evaluate several state-of-the-art QPP approaches to multiple NIR retrievers based on BERT, on Robust '04 and Deep Learning '19;
– we observe a correlation between QPPs performance and how different NIR architectures perform lexical match;
– we show that currently availableQPPs perform reasonably well when applied to TIR systems, while they fail to properly predict the performance for NIR systems, even on NIR oriented collections;
– we highlight how such decrease in QPP performance is particularly prominent on queries where TIR and NIR performances differ the most – which are those queries where QPPs would be most beneficial.

The remainder of this paper is organized as follows: Sect. 2 outlines the main related endeavours. Section 3 details our methodology, while Sect. 4 contains the experimental setting. Empirical results are reported in Sect. 5. Section 6 summarizes the main conclusions and future research directions.

2   Related Work

The rise of large PLM like BERT [13] has given birth to a new generation of NIR systems. Initially employed as re-rankers in a standard learning-to-rank framework [35], a real paradigm shift occurred when the first PLM-based retrievers outperformed standard TIR models as candidate generators in a multi-stage ranking setting. For such a task, dense representations, based on a simple pooling of contextualized embeddings, combined with approximate nearest neighbors algorithms, have proven to be both highly effective and efficient [22,28–30,37,49]. ColBERT [31,41] avoids this pooling mechanism, and directly models semantic matching at the token level – allowing it to capture finer-grained relevance signals. In the meantime, another research branch brought lexical models up to date, by taking advantage of BERT and the proven efficiency of inverted indices in various manners. Such sparse approaches for instance learn contextualized term weights [10,33,34,55], query or document expansion [36], or both mechanisms jointly [20,21]. This new wave of NIR systems, which substantially differ from lexical ones – and from each other – demonstrate state-of-the-art results on several datasets, from MS MARCO [3] on which models are usually trained, to zero-shot settings such as the BEIR [46] or LoTTE [41] benchmarks.

A well-known problem linked to IR evaluation is the variation in performance achieved by different IR systems, even on a single query [4,9]. To partially account for it, a large body of work has focused on predicting the performance that a system would achieve for a given query, using QPP models. Such models are typically divided into pre- and post-retrieval predictors. Traditional pre-retrieval QPPs leverage statistics on the query terms occurrences [26]. For example, SCQ [53], VAR [53] and IDF [8,42] combine query tokens' occurrence indicators, such as Collection Frequency (CF) and Inverse Document Frequency (IDF), to compute their performance prediction score. Post-retrieval QPPs exploit the results of IR models for the given query [4]. Among them, Clarity [7] compares the language model of the first k retrieved documents with the one of the entire corpus. NQC [43], WIG [54] and SMV [45] exploit the retrieval scores distribution for the top-ranked documents to compute their predictive score. Finally, Utility Estimation Framework (UEF) [44] serves as a general framework that can be instantiated with many of the mentioned predictors, pre-retrieval ones included. Post-retrieval predictors are based on lexical signals – SMV, NQC and WIG rely on the Language Model scores estimated from top-retrieved documents, while Clarity and UEF exploit the language models of the top-k documents.

We further divide QPP models into traditional and neural approaches. Among neural predictors, one of the first approaches is NeuralQPP [50] which computes its predictions by combining semantic and lexical signals using a feedforward neural network. Notice that NeuralQPP is explicitly designed for TIR and is hence not expected to work better with NIR [50]. A similar approach for Question Answering is NQA-QPP [24], which also relies on three neural components but, unlike NeuralQPP, exploits BERT [13] to embed tokens semantics. Similarly, BERT-QPP [2] encodes semantics via BERT, but directly *fine-tunes* it to predict query performance based on the first retrieved document.

Subsequent approaches extend BERT-QPP by employing a groupwise predictor to jointly learn from multiple queries and documents [5] or by transforming its pointwise regression into a classification task [12]. Since we did not consider multiple formulations, we did not experiment with such approach in our empirical evaluation.

Although traditional QPP methods have been widely used over the years, only few works have been done to apply them on NIR models. Similarly, neural QPP methods – which model the semantic interactions between query and document terms – have been mostly *designed for* and *evaluated on* TIR models. Two noteworthy exceptions concerning the tested IR models are [24] who evaluate the devised QPP on pre-BERT approaches for Question Answering (QA), while [11] assess the performance of their approach on DRMM [23] (pre-BERT) and ColBERT [31] (BERT-based) as NIR models. Hence, there is an urgent need to deepen the evaluation of QPP on state-of-the-art NIR models to understand where we are, what are the challenges, and which directions are more promising.

A third category that can be considered a hybrid between the groups of predictors mentioned above is passage retrieval QPP [38]. In [38], authors exploit lexical signals obtained from passages' language models to devise a predictor meant to better deal with passage retrieval prediction.

3   Methodology

Evaluating Query Performance Predictors. QPP models compute a score for each query, that is expected to correlate with the quality of the retrieval for such query. Traditional evaluation of QPP models relies on measuring the correlation between the predicted QPP scores and the observed performance measured with a traditional IR measure. Typical correlation coefficients include Kendall's τ, Spearman's ρ and the Pearson's r. This evaluation procedure has the drawback of summarizing, through the correlation score, the performance of a QPP model into a single observation for each system and collection [15,16]. Therefore, Faggioli et al. [15] propose a novel evaluation approach based on the scaled Absolute Rank Error (sARE) measure that, given a query q, is defined as $\text{sARE}(q) = \frac{|R_q^e - R_q^p|}{|Q|}$, where R_q^e and R_q^p are the ranks of the query q induced by the IR measure and the QPP score respectively, over the entire set of queries of size $|Q|$. With "rank" we refer to the ordinal position of the query if we sort all the queries of the collection either by IR performance or prediction score. By switching from a single-point estimation to a distribution of performance, sARE has the advantage of allowing conducting more powerful statistical analyses and carrying out failure analyses on queries where the predictors are particularly bad. To be comparable with previous literature, we report in Sect. 5.1 the performance of the analyzed predictors using the traditional Pearson's r correlation-based evaluation. On the other hand, we use sARE as the evaluation measure for the statistical analyses, to exploit its additional advantages. Such analyses, whose results are reported in Sect. 5.2, are described in the remainder of this section.

ANOVA. To assess the effect induced by NIR systems on QPP performance, we employ the following ANalysis Of VAriance (ANOVA) models. The first model, dubbed MD1, aims at explaining the sARE performance given the predictor, the type of IR model and the collection. Therefore, we define it as follows:

$$sARE_{ijpqr} = \mu + \pi_p + \eta_i + \chi_j + (\eta\chi)_{ij} + \epsilon_{ijpqr}, \qquad \text{(MD1)}$$

where μ is the grand mean, π_p is the effect of the p-th predictor, η_i represents the type of IR model (either TIR or NIR), χ_j stands for the effect of the j-th collection on QPP's performance, and $(\eta\chi)_{ij}$ describes how much the type of run and the collection interact and ϵ is the associated error.

Secondly, since we are interested in determining the effect of different predictors in interaction with each query, we define a second model, dubbed MD2, that also includes the interaction factor and is formulated as follows:

$$sARE_{ipqr} = \mu + \pi_p + \tau_q + \eta_i + (\pi\tau)_{qp} + (\pi\eta)_{pi} + (\tau\eta)_{iq} + \epsilon_{ipqr}, \qquad \text{(MD2)}$$

Differently from MD1, we apply MD2 to each collection separately. Therefore, having a single collection, we replace the effect of the collection with τ_q, the effect for the q-th topic. Furthermore, the model includes also all the first-order interactions.

The Strength of Association (SOA) [39] is assessed using ω^2 measure computed as:

$$\omega^2_{<fact>} = \frac{\text{df}_{<fact>} * F_{<fact>}}{\text{df}_{<fact>} * (F_{<fact>} - 1) * N},$$

where N is the number of experimental data-points, $\text{df}_{<fact>}$ is the factor's number of Degrees of Freedom (DF), and $F_{<fact>}$ it the F statistics computed by ANOVA. As a rule-of-thumb, $\omega^2 < 6\%$ indicates a small SOA, $6\% \leq \omega^2 < 14\%$ is a medium-sized effect, while $\omega^2 \geq 14\%$ represent a large-sized effect.

ANOVA Models have been fitted using **anovan** function from the stats MAT-LAB package. In terms of sample size, depending on the model and collection at hand, we considered 19 predictors, 249 topics in the case of Robust '04 and 43 for Deep Learning '19 and 14 different IR systems for a total of 66234 and 11438 observations for Robust '04 and Deep Learning '19 respectively.

4   Experimental Setup

Our analyses focus on two distinct collections: Robust '04 [47], and TREC Deep Learning 2019 Track (Deep Learning '19) [6]. The collections have respectively 249 and 43 topics each and are based on TIPSTER and MS MARCO passages corpora. Robust '04 is one of the most used collections to test lexical approaches, while providing a reliable benchmark for NIR models [48] – even though they struggle to perform well on this collection, especially when evaluated in a zero-shot setting [46]. Deep Learning '19 concerns passage retrieval from natural questions – the formulation of queries and the nature of the documents (passages) make the retrieval harder for TIR approaches, while NIR systems tend to have an edge in retrieving relevant documents.

Our main objective is to assess whether existing QPPs are effective in predicting the performance of different state-of-the-art NIR models. As reference points, we consider seven TIR methods: Language Model with Dirichlet (LMD) and Jelinek-Mercer (LMJM) smoothing [52], BM25, vector space model [40] (TFIDF), InExpB2 [1] (InEB2), Axiomatic F1-EXP [17] (AxF1e), and Divergence From Independence (DFI) [32]. TIR runs have been computed using Lucene. For the NIR methods, we focus on BERT-based first-stage models. We consider state-of-the-art models from the three main families of NIR models, which exhibit different behavior, and thus might respond to QPPs differently. We consider *dense* models, *i)* a "standard" bi-encoder (bi) trained with negative log-likelihood, *ii)* TAS-B [28] (bi-tasb) whose training relies on topic-sampling and knowledge distillation *iii)* and finally CoCondenser [22] (bi-cc) and Contriever [29] (bi-ct) which are based on contrastive pre-training. We also consider two models from the *sparse* family: SPLADE [21] (sp) with default training strategy, and its improved version SPLADE++ [19,20] (sp++) based on distillation, hard-negative mining and pre-training. We finally consider the *late-interaction* ColBERTv2 [41] (colb2). Models are fine-tuned on the MS MARCO passage dataset; given the absence of training queries in Robust '04, they are evaluated in a zero-shot manner, similarly to previous settings [41,46]. Besides the bi-encoder we trained on our own, we rely on open-source weights available for every model. The advantage of considering multiple TIR and NIR models is that *i)* we achieve more generalizable results: different models, either TIR or NIR perform the best in different scenarios and therefore our conclusions should be as generalizable as possible; *ii)* it allows to achieve more statistical power in the experimental evaluation. We focus our analyses on Normalized Discounted Cumulated Gain (nDCG) with cutoff 10, as it is employed across NIR benchmarks consistently. This is not the typical setting for evaluating traditional QPP – which usually considers Average Precision (AP) @1000. Nevertheless, given our objective – determining how QPP performs on settings where NIR models can be used successfully – we are also interested in selecting the most appropriate measure.

Concerning QPP models, we select the most popular state-of-the-art approaches. In details, we consider 9 pre-retrieval models: Simplified query Clarity Score (SCS) [27], Similarity Collection-Query (SCQ) [53], VAR [53], IDF and Inverse Collection Term Frequency (ICTF) [8,42]. For SCS, we use the sum aggregation, while for others we use max and mean, which empirically produce the best results. In terms of post-retrieval QPP models, our experiments are based on Clarity [7], Normalized Query Commitment (NQC) [43], Score Magnitude and Variance (SMV) [45], Weighted Information Gain (WIG) [54] and their UEF [44] counterparts. Among post-retrieval predictors, we also include a supervised approach, BERT-QPP [2], using both bi-encoder (*bi*) and cross-encoder (*ce*) formulations. We train BERT-QPP[1] for each IR system on the MS

---

[1] We use the implementation provided at https://github.com/Narabzad/BERTQPP.

Table 1. nDCG@10 for the selected TIR and NIR systems. NIR outperform traditional approaches on Deep Learning '19, and have comparable performance on Robust '04.

	axF1e	BM25	LMD	LMJM	TFIDF	DFI	InEB2	bi	bi-tasb	bi-cc	bi-ct	sp	sp++	colbv2
Deep Learning '19	0.45	0.48	0.45	0.48	0.37	0.47	0.49	0.64	0.72	0.72	0.67	0.71	0.73	0.75
Robust '04	0.39	0.44	0.43	0.40	0.31	0.44	0.44	0.23	0.45	0.30	0.46	0.39	0.45	0.47

MARCO training set, as proposed in [2]. Similarly to what is done for NIR models, we apply BERT-QPP models on Robust '04 queries in a zero-shot manner.

5   Experimental Results

5.1   QPP Models Performance

Table 1 reports the absolute nDCG@10 performance for the selected TIR and NIR models. Figures 1a and 1b refer, respectively, to Robust '04 and Deep Learning '19 collections and report the Pearson's r correlation between the scores predicted by the chosen predictors and the nDCG@10, for both TIR and NIR runs[2]. The presence of negative values indicates that some predictors fail in specific contexts and has been observed before in the QPP setting [25].

For Robust '04, we notice that – following previous literature – pre-retrieval (top) predictors (mean correlation: 15.9%) tend to perform 52.3% worse than post-retrieval ones (bottom) (mean correlation: 30.2%). Pre-retrieval results are in line with previous literature [51]. The phenomenon is more evident (darker colors) for NIR runs (right) than TIR ones (left). Pre-retrieval predictors fail in predicting the performance of NIR systems (mean correlation 6.2% vs 25.6% for TIR), while in general, to our surprise, we notice that post-retrieval predictors tend to perform similarly on TIR and NIR (34.5% vs 32.3%) – with some exceptions. For instance, for bi, post-retrieval predictors either perform extremely well or completely fail. This happens particularly on Clarity, NQC, and their UEF counterparts. Note that bi is the worst performing approach on Robust '04, with 23% of nDCG@10 – the second worst is bi-cc which achieves 30% nDCG@10.

The patterns observed for Robust '04 hold only partially on Deep Learning '19. For example, we notice again that pre-retrieval predictors (mean correlation: 14.7%) perform 58.3% worse than post-retrieval ones (mean correlation: 35.3%). On the contrary, the difference in performance is far more evident between NIR and TIR. On TIR runs, almost all predictors perform particularly well (mean correlation: 38.1%) – even better than on Robust '04 collection. The only three exceptions are SCQ (both in avg and max formulations) and VAR using max formulation. Conversely, on NIR the performance is overall lower (13.1%) and relatively more uniform between pre- (5.4%) and post-retrieval (19.9%) models. In absolute value, maximum correlation achieved by pre-retrieval predictors for

---

[2] Additional IR measures and correlations, as well as full ANOVA tables are available at: https://github.com/guglielmof/ECIR2023-QPP.

	axF1e	BM25	LMD	bi	bi-cc	bi-ct	bi-tasb	sp	sp++	colbv2
SCSsum	0.27	0.27	0.34	-0.1	0.01	0.02	0.03	0.06	0.08	0.02
SCQavg	0.17	0.19	0.25	-0.00	0.08	0.08	0.13	0.12	0.10	0.09
SCQmax	0.2	0.21	0.29	0.1	0.13	0.10	0.14	0.12	0.10	0.1
ICTFavg	0.28	0.27	0.34	-0.06	0.02	0.02	0.02	0.06	0.08	0.03
ICTFmax	0.27	0.26	0.34	0.02	0.03	0.04	0.03	0.06	0.09	0.03
IDFavg	0.29	0.28	0.35	-0.04	0.02	0.04	0.03	0.08	0.09	0.04
IDFmax	0.28	0.27	0.36	0.04	0.04	0.05	0.05	0.07	0.10	0.04
VARavg	0.15	0.13	0.18	0.04	0.05	0.07	0.06	0.11	0.11	0.06
VARmax	0.2	0.21	0.29	0.1	0.13	0.10	0.14	0.12	0.10	0.1
Clarity	0.24	0.29	0.43	-0.32	0.22	0.1	0.24	0.3	0.25	0.41
NQC	0.29	0.43	0.46	-0.25	0.27	0.28	0.32	0.41	0.37	0.31
SMV	0.24	0.45	0.46	0.6	0.32	0.33	0.36	0.47	0.44	0.41
WIG	0.35	0.34	0.44	0.63	0.39	0.26	0.34	0.42	0.33	0.51
UEFClarity	0.32	0.41	0.48	0.02	0.43	0.24	0.39	0.44	0.4	0.51
UEFNQC	0.23	0.42	0.48	-0.23	0.38	0.3	0.4	0.5	0.39	0.44
UEFSMV	0.19	0.4	0.49	0.61	0.4	0.32	0.41	0.5	0.44	0.5
UEFWIG	0.37	0.43	0.5	0.63	0.48	0.28	0.42	0.48	0.42	0.52
BERTQPPce	0.22	0.24	0.25	0.51	0.28	0.22	0.15	0.36	0.08	0.34
BERTQPPbi	0.19	0.13	0.13	0.43	0.02	0.03	0.08	0.16	0.01	0.23

(a) Robust '04

	axF1e	BM25	LMD	bi	bi-cc	bi-ct	bi-tasb	sp	sp++	colb2
SCSsum	0.44	0.42	0.46	-0.07	0.09	0.16	0.08	0.2	-0.08	-0.04
SCQavg	-0.03	-0.11	-0.05	0.07	-0.18	-0.07	-0.17	0.02	-0.15	-0.18
SCQmax	-0.18	-0.29	-0.24	0.12	-0.23	-0.13	-0.15	-0.01	-0.10	-0.12
ICTFavg	0.49	0.45	0.49	-0.00	0.1	0.22	0.13	0.24	-0.03	0.03
ICTFmax	0.47	0.45	0.46	0.02	0.17	0.32	0.26	0.26	0.1	0.18
IDFavg	0.5	0.45	0.48	0.02	0.11	0.23	0.14	0.25	-0.02	0.04
IDFmax	0.49	0.46	0.46	0.03	0.18	0.32	0.26	0.26	0.1	0.18
VARavg	0.58	0.51	0.49	0.09	0.16	0.17	0.16	0.22	-0.00	0.05
VARmax	-0.18	-0.29	-0.24	0.12	-0.23	-0.13	-0.15	-0.01	-0.10	-0.12
Clarity	0.45	0.31	0.23	0.31	0.04	0.05	0.2	0.03	-0.00	0.06
NQC	0.54	0.62	0.58	0.18	0.21	0.32	0.12	0.24	0.11	0.03
SMV	0.51	0.56	0.56	0.28	0.34	0.31	0.18	0.26	0.12	0.04
WIG	0.49	0.48	0.53	0.52	0.12	0.32	0.25	0.31	0.11	0.3
UEFClarity	0.52	0.48	0.58	0.55	0.36	0.24	0.3	0.01	0.31	0.25
UEFNQC	0.5	0.61	0.57	0.3	0.32	0.27	0.16	0.18	0.23	0.14
UEFSMV	0.47	0.53	0.55	0.4	0.4	0.33	0.22	0.22	0.23	0.12
UEFWIG	0.56	0.55	0.64	0.6	0.34	0.31	0.32	0.24	0.27	0.37
BERTQPPce	0.47	0.38	0.58	0.31	0.23	0.15	0.05	-0.05	-0.04	-0.08
BERTQPPbi	0.36	0.35	0.53	0.12	0.19	0.052	-0.04	-0.18	-0.09	0.01

(b) Deep Learning '19

Fig. 1. Pearson's r correlation observed for different pre (top) and post (bottom) retrieval predictors on lexical (left) and neural (right) runs. To avoid cluttering, we report the results for the 3 main TIR models, other models achieve highly similar results.

NIR on Deep Learning '19 is much higher than the one achieved on Robust '04, especially for bi-ct, sp, and bi-tasb runs. On the other hand, post-retrieval predictors, perform worse than on the Robust '04. The only exception to this pattern is again represented by bi, on which some post-retrieval predictors, namely WIG, UEFWIG, and UEFClarity work surprisingly well. The supervised BERT-QPP shows a trend similar to other post-retrieval predictors on Deep Learning '19 (42.3% mean correlation against 52.9% respectively) for what concerns TIR, with performance in line with the one reported in [2]. This is exactly the setting where BERT-QPP has been devised and tested. If we focus on Deep Learning

Table 2. Pearson's r QPP performance for three versions of sp++ applied on Robust '04, with varying degree of sparsity (sp++$_2$ ≻ sp++$_1$ ≻ sp++$_0$ in terms of sparsity). The more "lexical" are the models, the better QPP performs. d_l and q_l represent respectively the average document/query sizes (i.e. non-zero dimensions in SPLADE) on Robust '04.

d_l/q_l		Clarity	NQC	SMV	WIG	UEFClarity	UEFNQC	UEFSMV	UEFWIG
sp++$_2$	55/22	0.26	0.31	0.46	0.42	0.44	0.4	0.48	0.5
sp++$_1$	79/29	0.2	0.34	0.47	0.35	0.38	0.4	0.46	0.43
sp++$_0$	204/45	0.25	0.37	0.44	0.33	0.4	0.39	0.44	0.42

'19 and NIR systems, its performance (mean correlation: 4.5%) is far lower than those of other post-retrieval predictors (mean correlation without BERT-QPP: 23.8%). Finally, its performance on Robust '04 – applied in zero-shot – is considerably lower compared to other post-retrieval approaches.

Interestingly, on Robust '04, post-retrieval QPPs achieve, on average, top performance on the late interaction model (colb2), followed by sparse approaches (sp and sp++). Finally, excluding bi, where predictors achieve extremely inconsistent performance, dense approaches are those where QPP perform the worst. In this sense, the performance that QPP methods achieve on NIR systems seems to correlate with the importance these systems give to lexical signals. In this regard, Formal et al. [20] observed how late-interaction and sparse architectures tend to rely more on lexical signals, compared to dense ones.

To further corroborate this observation, we apply the predictors to three versions of SPLADE++ with various levels of sparsit as controlled by the regularization hyperparameter. Increasing the sparsity of representations leads to models that cannot rely as much on expansion – emphasizing the importance given to lexical signals in defining the document ranking. Therefore, as a first approximation, we can deem sparser methods to be also more lexical. Given the low performance achieved by pre-retrieval QPPs, we focus this analysis on post-retrieval methods only. Table 2 shows the Pearson's r for the considered predictors and different SPLADE++ versions. Interestingly, in the majority of the cases, QPPs perform the best for the sparser version (sp++$_2$), followed by sp++$_1$ and sp++$_0$ – which is the one used in Fig. 1. There are a few switches, often associated with very close correlation values (SMV and UEFClarity). Only one predictor, NQC, completely reverses the order. This goes in favour of our hypothesis that indeed QPP performance tends to correlate with the degree of lexicality of the NIR approaches. Although not directly comparable, following this line of thought, sp, being handled better by QPPs (cfr. Fig. 1a), is more lexical than all the sp++ versions considered: this is reasonable, given the different training methodology. Finally, colb2, being the method where QPPs achieve the best performance, might be the one that, at least for what concerns the Robust '04 collection, gives the highest importance to lexical signals – in line with what was observed in [21].

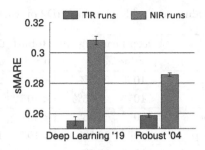

Fig. 2. Comparison between the mean sARE (sMARE) achieved over TIR or NIR when changing the corpus. Observe the large distance between results on NIR – especially for Deep Learning '19 – compared to the one on TIR runs.

5.2 ANOVA Analysis

To further statistically quantify the phenomena observed in the previous subsection, we apply MD1 to our data, considering both collections at once. From a quantitative standpoint, we notice that all the factors included in the model are statistically significant (p-value $< 10^{-4}$). In terms of SOA, the collection factor has a small effect (0.02%). The run type, on the other hand, impacts for $\omega^2 = 0.48\%$. Finally, the interaction between the collection and run type, although statistically significant, has a small impact on the performance ($\omega^2 = 0.05\%$): in both collections QPPs perform better on TIR models. All factors are significant but have small-size effects. This is in contrast with what was observed for the performance of IR systems [9,18], where most of the SOA range between medium to large. Nevertheless, it is in line with what was observed by Faggioli et al. [15] for the performance QPP methods, who showed that all the factors besides the topic are small to medium. A second observation is that it is likely that the small SOAs are due to a model unable to accrue for all the aspects of the problem – more factors should be considered. Model MD2, introducing also the topic effect, allows for further investigation of this hypothesis.

We are now interested in breaking down the performance of the predictors according to the collection and type of run. Figure 2 reports the average performance (measured with sMARE, the lower the better) for QPPs applied on NIR or TIR runs over different collections, with their confidence intervals as computed using ANOVA. Interestingly, regardless of the type of collection, the performance achieved by predictors on NIR models will *on average* be worse than those achieved on TIR runs. QPP models perform better on TIR than NIR on both collections: this explains the small interaction effect between collections and run types. Secondly, there is no statistical difference QPPs applied to TIR models when considering Deep Learning '19 and Robust '04– the confidence intervals are overlapping. This goes in contrast with what happens on Robust '04 and Deep Learning '19 when considering NIR models: QPPs approaches applied on the latter dataset perform by far worse than on the former.

Table 3. p-values and ω^2 SOA using MD2 on each collection

	Deep Learning '19		Robust '04	
	p-value	ω^2	p-value	ω^2
Topic	$< 10^{-4}$	22.5%	$< 10^{-4}$	24.0%
qpp	$< 10^{-4}$	1.65%	$< 10^{-4}$	2.21%
Run type	$< 10^{-4}$	4.35%	$< 10^{-4}$	0.11%
Topic*qpp	$< 10^{-4}$	22.7%	$< 10^{-4}$	17.2%
Topic*run type	$< 10^{-4}$	15.2%	$< 10^{-4}$	10.0%
qpp*run type	0.0012	0.23%	$< 10^{-4}$	0.30%

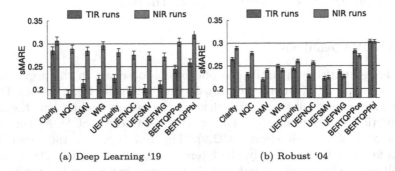

(a) Deep Learning '19   (b) Robust '04

Fig. 3. sMARE observed for different predictors on Deep Learning '19 (left) and Robust '04 (right). On Deep Learning '19, predictors behave differently on TIR and NIR runs, while they are more uniform on Robust '04.

While *on average* we will be less satisfied by QPP predictors applied to NIR regardless of the type of collection, there might be some noticeable exceptions of good performing predictors also for NIR systems. To verify this hypothesis, we apply MD2 to each collection separately, and measure what happens to each predictor individually[3]. Table 3 reports the p-values and ω^2 SOA for the factors included in MD2, while Fig. 3 depicts the phenomena visually. We observe that, concerning Deep Learning '19, the run type (TIR or NIR) is significant, while the interaction between the predictor and the run type is small: indeed predictors always perform better on TIR runs than on NIR ones. The only model that behaves slightly differently is Clarity, with far closer performance for both classes of runs – this can be explained by the fact that Clarity is overall the worst-performing predictor. Notice that, the best predictor on TIR runs – NQC – performs almost 10% worse on NIR ones. Finally, we notice a large-size interaction between topics and QPP models – even bigger than the topic or QPP themselves. This indicates that whether a model will be better than another strongly depends on the topic considered. An almost identical pattern was observed also

---

[3] To avoid cluttering, we report the subsequent analyses only for post-retrieval predictors – similar observations hold for pre-retrieval ones.

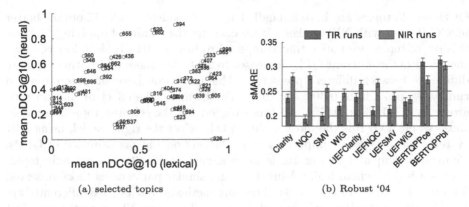

(a) selected topics

(b) Robust '04

Fig. 4. left: topics selected to maximize the difference between lexical and neural models; right: results of MD2 applied on Robust '04 considering only the selected topics.

in [15]. Therefore, to improve QPP's generalizability, it is important not only to address challenges caused by differences in NIR and TIR but also to take into consideration the large variance introduced by topics. We analyze more in detail this variance later, where we consider only "semantically defined" queries.

If we consider Robust '04, the behaviour changes deeply: Fig. 3 shows that predictors performances are much more similar for TIR and NIR runs compared to Deep Learning '19. This is further highlighted by the far smaller ω^2 for run type on Robust '04 in Table 3 – 4.35% against 0.11%. The widely different pattern between Deep Learning '19 and Robust '04 suggests that current QPPs are doomed to fail when used to predict the performance of IR approaches that learned the semantics of a collection – which is the case for Deep Learning '19 that was used to fine-tune the models. Current QPPs evaluate better IR approaches that rely on lexical clues. Such approaches include both TIR models and NIR models applied in a zero-shot fashion, as it is the case for Robust '04. Thus, QPP models are expected to fail where NIR models behave differently from the TIR ones. This poses at stake one of the major opportunities provided by QPP: if we fail in predicting the performance of NIR models where they behave differently from TIR ones, then a QPP cannot be safely used to carry out model selection. To further investigate this aspect, we carry out the following analysis: we select from Robust '04 25% of the queries that are mostly "semantically defined" and rerun MD2 on the new set of topics. We call "semantically defined" those queries where NIR behave, on average, oppositely w.r.t. the TIR, either failing or succeeding at retrieving documents. In other terms, we select queries in the top quartile for the absolute difference in performance (nDCG), averaged over all TIR or NIR models.

Figure 4a shows the performance of topics that maximize the difference between TIR and NIR and can be considered as more "semantically defined" [14]. There are 62 topics selected (25% of the 249 topics available on Robust '04).

Of these, 35 topics are better handled by TIR models, while 27 obtain better nDCG if dealt with NIR rankers. If we consider the results of applying MD2 on this set of topics, we notice that compared to Robust '04 (Table 3, last column) the effect of the different QPPs increases to 2.29%: on these topics, there is more difference between different predictors. The interaction between predictors and run types grows from 0.30% to 0.91%. Furthermore, the effect of the run type grows from 0.11% to 0.67% – 6 times bigger. On the selected topics, arguably those where a QPP is the most useful to help select the right model, using NIR systems has a negative impact (6 times bigger) on the performance of QPPs. Figure 4b, compared to Fig. 3b, is more similar to Fig. 3a – using only topics that are highly semantically defined, we get similar patterns as those observed for Deep Learning '19 on Fig. 3a. The only methods that behave differently are BERT-QPP approaches, whose performance is better on NIR runs than on TIR ones, but are the worst approaches in terms of predictive capabilities for both run types. In this sense, even though the contribution of the semantic signals appears to highly important to define new models with improved performance in the NIR setting, it does not suffice to compensate for current QPPs limitations.

6   Conclusion and Future Work

With this work, we assessed to what extent current QPPs are applicable to the recent family of first-stage NIR models based on PLM. To verify that, we evaluated 19 diverse QPP models, used on seven traditional bag-of-words lexical models (TIR) and seven first-stage NIR methods based on BERT, applied to the Robust '04 and Deep Learning '19 collections. We observed that if we consider a collection where NIR systems had the chance to learn the semantics – i.e., Deep Learning '19 – QPPs are effective in predicting TIR systems performance, but fail in dealing with NIR ones. Secondly, we considered Robust '04. In this collection, NIR models were applied in a zero-shot fashion, and thus behave similarly to TIR models. In this case, we observed that QPPs tend to work better on NIR models than in the previous scenario, but they fail on those topics where NIR and TIR models differ the most. This, in turn, impairs the possibility of using QPP models to choose between NIR and TIR approaches where it is most needed. On the other hand, semantic QPP approaches such as BERT-QPP do not solve the problem: being devised and tested on lexical IR systems, they work properly on such category of approaches but fail on neural systems. These results highlight the need for QPPs specifically tailored to Neural IR.

As future work, we plan to extend our analysis by considering other factors, such as the query variations to understand the impact that changing how a topic is formulated has on QPP. Furthermore, we plan to devise QPP methods explicitly designed to synergise with NIR models, but that also take into consideration the large variance introduced by topics.

Acknowledgements. The work was partially supported by University of Padova Strategic Research Infrastructure Grant 2017: "CAPRI: Calcolo ad Alte Pre-stazioni per la Ricerca e l'Innovazione", ExaMode project, as part of the EU H2020 program under Grant Agreement no. 825292.

References

1. Amati, G., van Rijsbergen, C.J.: Probabilistic models of information retrieval based on measuring the divergence from randomness. ACM Trans. Inf. Syst **20**(4), 357–389 (2002)
2. Arabzadeh, N., Khodabakhsh, M., Bagheri, E.: BERT-QPP: contextualized pre-trained transformers for query performance prediction. In: CIKM '21: The 30th ACM International Conference on Information and Knowledge Management, Virtual Event, Queensland, Australia, November 1–5, 2021, pp. 2857–2861 (2021)
3. Bajaj, P., et al.: MS MARCO: a human generated machine reading comprehension dataset (2016)
4. Carmel, D., Yom-Tov, E.: Estimating the Query Difficulty for Information Retrieval. Morgan & Claypool Publishers, San Rafael (2010)
5. Chen, X., He, B., Sun, L.: Groupwise query performance prediction with BERT. In: Advances in Information Retrieval - 44th European Conference on IR Research, ECIR 2022, Stavanger, Norway, April 10–14, 2022, Proceedings, Part II. Lecture Notes in Computer Science, vol. 13186, pp. 64–74 (2022)
6. Craswell, N., Mitra, B., Yilmaz, E., Campos, D., Voorhees, E.M., Soboroff, I.: TREC Deep Learning Track: reusable test collections in the large data regime. In: SIGIR '21: The 44th International ACM SIGIR Conference on Research and Development in Information Retrieval, Virtual Event, Canada, July 11–15, 2021, pp. 2369–2375 (2021)
7. Cronen-Townsend, S., Zhou, Y., Croft, W.B.: Predicting query performance. In: SIGIR 2002: Proceedings of the 25th Annual International ACM SIGIR Conference on Research and Development in Information Retrieval, August 11–15, 2002, Tampere, Finland, pp. 299–306 (2002)
8. Cronen-Townsend, S., Zhou, Y., Croft, W.B.: A language modeling framework for selective query expansion. Tech. rep, CIIR, UMass (2004)
9. Culpepper, J.S., Faggioli, G., Ferro, N., Kurland, O.: Topic Difficulty: collection and query formulation effects. ACM Trans. Inf. Syst. 40(1), 19:1–19:36 (2022)
10. Dai, Z., Callan, J.: Context-aware term weighting for first stage passage retrieval. In: Proceedings of the 43rd International ACM SIGIR Conference on Research and Development in Information Retrieval, SIGIR 2020, Virtual Event, China, July 25–30, 2020, pp. 1533–1536 (2020)
11. Datta, S., Ganguly, D., Mitra, M., Greene, D.: A relative information gain-based query performance prediction framework with generated query variants. ACM Trans. Inf. Syst., pp. 1–31 (2022)
12. Datta, S., MacAvaney, S., Ganguly, D., Greene, D.: A 'Pointwise-Query, Listwise-Document' based query performance prediction approach. In: SIGIR 2022: The 45th International ACM SIGIR Conference on Research and Development in Information Retrieval, Madrid, Spain, July 11–15, 2022, pp. 2148–2153 (2022)
13. Devlin, J., Chang, M., Lee, K., Toutanova, K.: BERT: pre-training of deep bidirectional transformers for language understanding. In: Proceedings of the 2019 Conference of the North American Chapter of the Association for Computational Linguistics: Human Language Technologies, NAACL-HLT 2019, Minneapolis, MN, USA, June 2–7, 2019, Volume 1 (Long and Short Papers), pp. 4171–4186 (2019)

14. Faggioli, G., Marchesin, S.: What makes a query semantically hard? In: Proceedings of the Second International Conference on Design of Experimental Search & Information REtrieval Systems, Padova, Italy, September 15–18, 2021. CEUR Workshop Proceedings, vol. 2950, pp. 61–69. CEUR-WS.org (2021), http://ceur-ws.org/Vol-2950/paper-06.pdf

15. Faggioli, G., Zendel, O., Culpepper, J.S., Ferro, N., Scholer, F.: An Enhanced Evaluation Framework for Query Performance Prediction. In: Advances in Information Retrieval - 43rd European Conference on IR Research, ECIR 2021, Virtual Event, March 28 - April 1, 2021, Proceedings, Part I. vol. 12656, pp. 115–129 (2021)

16. Faggioli, G., Zendel, O., Culpepper, J.S., Ferro, N., Scholer, F.: sMARE: a new paradigm to evaluate and understand query performance prediction methods. Inf. Retr. J. **25**(2), 94–122 (2022)

17. Fang, H., Zhai, C.: An exploration of axiomatic approaches to information retrieval. In: SIGIR 2005: Proceedings of the 28th Annual International ACM SIGIR Conference on Research and Development in Information Retrieval, Salvador, Brazil, August 15–19, 2005. pp. 480–487 (2005)

18. Ferro, N., Silvello, G.: Toward an anatomy of IR system component performances. J. Assoc. Inf. Sci. Technol. **69**(2), 187–200 (2018)

19. Formal, T., Lassance, C., Piwowarski, B., Clinchant, S.: SPLADE v2: sparse lexical and expansion model for information retrieval. CoRR abs/2109.10086 (2021)

20. Formal, T., Lassance, C., Piwowarski, B., Clinchant, S.: From distillation to hard negative sampling: making sparse neural IR models more effective. In: SIGIR 2022: The 45th International ACM SIGIR Conference on Research and Development in Information Retrieval, Madrid, Spain, July 11–15, 2022, pp. 2353–2359 (2022)

21. Formal, T., Piwowarski, B., Clinchant, S.: SPLADE: sparse lexical and expansion model for first stage ranking. In: SIGIR 2021: The 44th International ACM SIGIR Conference on Research and Development in Information Retrieval, Virtual Event, Canada, July 11–15, 2021, pp. 2288–2292 (2021)

22. Gao, L., Callan, J.: Unsupervised corpus aware language model pre-training for dense passage retrieval. In: Proceedings of the 60th Annual Meeting of the Association for Computational Linguistics (Volume 1: Long Papers), ACL 2022, Dublin, Ireland, May 22–27, 2022, pp. 2843–2853 (2022)

23. Guo, J., Fan, Y., Ai, Q., Croft, W.B.: A deep relevance matching model for ad-hoc retrieval. In: Proceedings of the 25th ACM International Conference on Information and Knowledge Management, CIKM 2016, Indianapolis, IN, USA, October 24–28, 2016, pp. 55–64 (2016)

24. Hashemi, H., Zamani, H., Croft, W.B.: Performance prediction for non-factoid question answering. In: Proceedings of the 2019 ACM SIGIR International Conference on Theory of Information Retrieval, ICTIR 2019, Santa Clara, CA, USA, October 2–5, 2019, pp. 55–58 (2019)

25. Hauff, C.: Predicting the effectiveness of queries and retrieval systems. SIGIR Forum **44**(1), 88 (2010)

26. Hauff, C., Hiemstra, D., de Jong, F.: A survey of pre-retrieval query performance predictors. In: Proceedings of the 17th ACM Conference on Information and Knowledge Management, CIKM 2008, Napa Valley, California, USA, October 26–30, 2008, pp. 1419–1420 (2008)

27. He, J., Larson, M.A., de Rijke, M.: Using coherence-based measures to predict query difficulty. In: Advances in Information Retrieval, 30th European Conference on IR Research, ECIR 2008, Glasgow, UK, March 30-April 3, 2008. Proceedings. vol. 4956, pp. 689–694 (2008)

28. Hofstätter, S., Lin, S., Yang, J., Lin, J., Hanbury, A.: Efficiently teaching an effective dense retriever with balanced topic aware sampling. In: SIGIR 2021: The 44th International ACM SIGIR Conference on Research and Development in Information Retrieval, Virtual Event, Canada, July 11–15, 2021, pp. 113–122 (2021)

29. Izacard, G., et al.: Towards unsupervised dense information retrieval with contrastive learning. CoRR abs/2112.09118 (2021)

30. Karpukhin, V., et al.: Dense passage retrieval for open-domain question answering. In: Proceedings of the 2020 Conference on Empirical Methods in Natural Language Processing (EMNLP), pp. 6769–6781 (2020)

31. Khattab, O., Zaharia, M.: ColBERT: efficient and effective passage search via contextualized late interaction over BERT. In: Proceedings of the 43rd International ACM SIGIR Conference on Research and Development in Information Retrieval, SIGIR 2020, Virtual Event, China, July 25–30, 2020, pp. 39–48. ACM (2020)

32. Kocabas, I., Dinçer, B.T., Karaoglan, B.: A nonparametric term weighting method for information retrieval based on measuring the divergence from independence. Inf. Retr. **17**(2), 153–176 (2014)

33. Lin, J., Ma, X.: A few brief notes on DeepImpact, COIL, and a conceptual framework for information retrieval techniques. CoRR abs/2106.14807 (2021)

34. Mallia, A., Khattab, O., Suel, T., Tonellotto, N.: Learning passage impacts for inverted indexes. In: SIGIR 2021: The 44th International ACM SIGIR Conference on Research and Development in Information Retrieval, Virtual Event, Canada, July 11–15, 2021, pp. 1723–1727 (2021)

35. Nogueira, R.F., Cho, K.: Passage re-ranking with BERT. CoRR abs/1901.04085 (2019)

36. Nogueira, R.F., Yang, W., Lin, J., Cho, K.: Document expansion by query prediction. CoRR abs/1904.08375 (2019)

37. Reimers, N., Gurevych, I.: Sentence-BERT: sentence embeddings using siamese BERT-networks. In: Proceedings of the 2019 Conference on Empirical Methods in Natural Language Processing and the 9th International Joint Conference on Natural Language Processing, EMNLP-IJCNLP 2019, Hong Kong, China, November 3–7, 2019, pp. 3980–3990 (2019)

38. Roitman, H.: An extended query performance prediction framework utilizing passage-level information. In: Song, D., et al. (eds.) Proceedings of the 2018 ACM SIGIR International Conference on Theory of Information Retrieval, ICTIR 2018, Tianjin, China, September 14–17, 2018, pp. 35–42. ACM (2018). https://doi.org/10.1145/3234944.3234946

39. Rutherford, A.: ANOVA and ANCOVA: a GLM approach. John Wiley & Sons (2011)

40. Salton, G., Buckley, C.: Term-weighting approaches in automatic text retrieval. Inf. Process. Manag. **24**(5), 513–523 (1988)

41. Santhanam, K., Khattab, O., Saad-Falcon, J., Potts, C., Zaharia, M.: ColBERTv2: effective and efficient retrieval via lightweight late interaction. In: Proceedings of the 2022 Conference of the North American Chapter of the Association for Computational Linguistics: Human Language Technologies, NAACL 2022, Seattle, WA, United States, July 10–15, 2022, pp. 3715–3734 (2022)

42. Scholer, F., Williams, H.E., Turpin, A.: Query association surrogates for web search. J. Assoc. Inf. Sci. Technol. **55**(7), 637–650 (2004)

43. Shtok, A., Kurland, O., Carmel, D.: Predicting query performance by query-drift estimation. In: Advances in Information Retrieval Theory, Second International Conference on the Theory of Information Retrieval, ICTIR 2009, Cambridge, UK, September 10–12, 2009, Proceedings. vol. 5766, pp. 305–312 (2009)

44. Shtok, A., Kurland, O., Carmel, D.: Using statistical decision theory and relevance models for query-performance prediction. In: Proceeding of the 33rd International ACM SIGIR Conference on Research and Development in Information Retrieval, SIGIR 2010, Geneva, Switzerland, July 19–23, 2010, pp. 259–266 (2010)
45. Tao, Y., Wu, S.: Query performance prediction by considering score magnitude and variance together. In: Proceedings of the 23rd ACM International Conference on Conference on Information and Knowledge Management, CIKM 2014, Shanghai, China, November 3–7, 2014. pp. 1891–1894 (2014)
46. Thakur, N., Reimers, N., Rücklé, A., Srivastava, A., Gurevych, I.: BEIR: a heterogeneous benchmark for zero-shot evaluation of information retrieval models. In: Proceedings of the Neural Information Processing Systems Track on Datasets and Benchmarks 1, NeurIPS Datasets and Benchmarks 2021, December 2021, virtual (2021)
47. Voorhees, E.M.: The TREC robust retrieval track. SIGIR Forum **39**(1), 11–20 (2005)
48. Voorhees, E.M., Soboroff, I., Lin, J.: Can Old TREC collections reliably evaluate modern neural retrieval models? CoRR abs/2201.11086 (2022)
49. Xiong, L., et al.: Approximate nearest neighbor negative contrastive learning for dense text retrieval. In: 9th International Conference on Learning Representations, ICLR 2021, Virtual Event, Austria, May 3–7, 2021 (2021)
50. Zamani, H., Croft, W.B., Culpepper, J.S.: Neural query performance prediction using weak supervision from multiple signals. In: The 41st International ACM SIGIR Conference on Research & Development in Information Retrieval, SIGIR 2018, Ann Arbor, MI, USA, July 08–12, 2018, pp. 105–114 (2018)
51. Zendel, O., Shtok, A., Raiber, F., Kurland, O., Culpepper, J.S.: Information needs, queries, and query performance prediction. In: Piwowarski, B., Chevalier, M., Gaussier, É., Maarek, Y., Nie, J., Scholer, F. (eds.) Proceedings of the 42nd International ACM SIGIR Conference on Research and Development in Information Retrieval, SIGIR 2019, Paris, France, July 21–25, 2019, pp. 395–404. ACM (2019). https://doi.org/10.1145/3331184.3331253,https://doi.org/10.1145/3331184.3331253
52. Zhai, C.: Statistical language models for information retrieval: a critical review. Found. Trends Inf. Retr. **2**(3), 137–213 (2008)
53. Zhao, Y., Scholer, F., Tsegay, Y.: Effective pre-retrieval query performance prediction using similarity and variability evidence. In: Advances in Information Retrieval, 30th European Conference on IR Research, ECIR 2008, Glasgow, UK, March 30-April 3, 2008. Proceedings. vol. 4956, pp. 52–64 (2008)
54. Zhou, Y., Croft, W.B.: Query performance prediction in web search environments. In: SIGIR 2007: Proceedings of the 30th Annual International ACM SIGIR Conference on Research and Development in Information Retrieval, Amsterdam, The Netherlands, July 23–27, 2007, pp. 543–550 (2007)
55. Zhuang, S., Zuccon, G.: TILDE: Term independent likelihood moDEl for passage re-ranking. In: SIGIR 2021: The 44th International ACM SIGIR Conference on Research and Development in Information Retrieval, Virtual Event, Canada, July 11–15, 2021, pp. 1483–1492 (2021)

Item Graph Convolution Collaborative Filtering for Inductive Recommendations

Edoardo D'Amico[✉][iD], Khalil Muhammad[iD], Elias Tragos[iD], Barry Smyth[iD], Neil Hurley[iD], and Aonghus Lawlor[iD]

Insight Centre for Data Analytics, Dublin, Ireland
{edoardo.damico,khalil.muhammad,elias.tragos,barry.smyth,
neil.hurley,aonghus.lawlor}@insight-centre.org

Abstract. Graph Convolutional Networks (GCN) have been recently employed as core component in the construction of recommender system algorithms, interpreting user-item interactions as the edges of a bipartite graph. However, in the absence of *side information*, the majority of existing models adopt an approach of randomly initialising the user embeddings and optimising them throughout the training process. This strategy makes these algorithms inherently *transductive*, curtailing their ability to generate predictions for users that were unseen at training time. To address this issue, we propose a convolution-based algorithm, which is *inductive* from the user perspective, while at the same time, depending only on implicit user-item interaction data. We propose the construction of an item-item graph through a weighted projection of the bipartite interaction network and to employ convolution to inject higher order associations into item embeddings, while constructing user representations as weighted sums of the items with which they have interacted. Despite not training individual embeddings for each user our approach achieves state-of-the-art recommendation performance with respect to *transductive* baselines on four real-world datasets, showing at the same time robust inductive performance.

Keywords: Recommender systems · Inductive recommendations · Graph convolution · Collaborative filtering

1 Introduction

Recent years have witnessed the success of Graph Convolutional Networks based algorithm in many domains, such as social networks [3,15], natural language processing [29] and computer vision [25]. The core component of Graph Convolutional Networks algorithms is the iterative process of aggregating information mined from node neighborhoods, with the intent of capturing high-order associations between nodes in a graph. GCNs have opened a new perspective for recommender systems in light of the fact that user-item interactions can be interpreted as the edges of a bipartite graph [4,10,24]. Real-world recommender system scenarios must contend with the issue that user-item graphs change dynamically

© The Author(s), under exclusive license to Springer Nature Switzerland AG 2023
J. Kamps et al. (Eds.): ECIR 2023, LNCS 13980, pp. 249–263, 2023.
https://doi.org/10.1007/978-3-031-28244-7_16

over time. New users join the system on a daily basis, and existing users can produce additional knowledge by engaging with new products (introducing new edges in the user-item interaction graph). The capacity to accommodate new users to the system - those who were not present during training - and fast leverage novel user-item interactions is a highly desirable characteristic for recommender systems meant to used in real-world context. Delivering high quality recommendations under these circumstances poses a severe problem for many existing *transductive* recommender system algorithms. Models such as [4,10,24] need to be completely re-trained to produce the embedding for a new user that joins the system post-training and the same happens when new user-item interactions must be considered; this limitation restricts their use in real-world circumstances. [28].

One solution present in literature, is to leverage side information (user and item metadata) beyond the pure user-item interactions in order to learn a mapping function from user and item features to embeddings [8,12,23,30]. However, it can be difficult to obtain this additional side information in many real-world scenarios, as it may be hard to extract, unreliable, or simply unavailable. For example, when new users join a system, there may be very little or no information available about them, making it difficult or impossible to generate their embeddings. Even when it is possible to gather some information about these users, it may not be useful in inferring their preferences. Another way to account for new users and rapidly create embeddings which exploit new user-item interactions is to resort to *item-based* models [5,13]. In this setting only the item representations are learnt and then exploited to build the user embeddings. Anyway these category of models do not directly exploit the extra source of information present in the user-item interaction graph, which have been shown to benefit the performance of the final model. Furthermore the application of standard Graph Convolution methods recently presented for the collaborative filtering problem have not been extended to work in a setting where only the item representations are learnt.

In this paper we propose a novel item-based model named Item Graph Convolutional Collaborative Filtering (IGCCF), capable of handling dynamic graphs while also leveraging the information contained in the user-item graph through graph convolution. It is designed to learn rich item embeddings capturing the higher-order relationships existing among them. To extract information from the user-item graph we propose the construction of an item-item graph through a weighted projection of the bipartite network associated to the user-item interactions with the intent of mining high-order associations between items. We then construct the user representations as a weighted combination of the item embeddings with which they have previously interacted, in this way we remove the necessity for the model to learn static one-hot embeddings for users, reducing the space complexity of previously introduced GCN-based models and, at the same time, unlocking the ability to handle dynamic graphs, making straightforward the creation of the embeddings for new users that join the system post training as well as the ability of updating them when new user-item interactions have been gathered, all of that without the need of an expensive retraining procedure.

2   Preliminaries and Related Work

In this paper we consider the extreme setting for inductive recommendation in which user preferences are estimated by leveraging only past user-item interactions without any additional source of information. We focus on implicit user feedback [21], with the understanding that explicit interactions are becoming increasingly scarce in real-world contexts. More formally, denoting with $\mathcal{U}$ and $\mathcal{I}$ the sets of users and items, and with $U = |\mathcal{U}|$ and $I = |\mathcal{I}|$ their respective cardinalities, we define the user-item interaction matrix $\mathbf{R}_{U \times I}$, where cell $r_{ui} = 1$ if user u has interacted with item i, and 0 otherwise, as the only source of information.

2.1   GCN-Based Recommender

GCN-based models have recently been applied to recommender system models, by virtue of the fact that historical user-item interactions can be interpreted as the edges of a graph. It is possible to define the adjacency matrix $\mathbf{A}$, associated with an undirected bipartite graph, exploiting the user-item interaction matrix $\mathbf{R}_{U \times I}$, as:

$$\mathbf{A} = \begin{bmatrix} \mathbf{0}_{U \times I} & \mathbf{R} \\ \mathbf{R}^T & \mathbf{0}_{I \times U} \end{bmatrix}$$

The set of the graph's nodes is $\mathcal{V} = \mathcal{U} \bigcup \mathcal{I}$ and there exists an edge between a user u and an item i if the corresponding cell of the interaction matrix $r_{ui} = 1$. He et al. [24], first applied graph convolution in a setting where no side information was available, and proposed to initialise the node representations with free parameters. This formulation is a variant of the one proposed in [15] but includes information about the affinity of two nodes, computed as the dot product between embeddings. Subsequently, Chen et al. [4] have shown how the affinity information as well as the non-linearities tend to complicate the training process as well as degrade the overall performance. Finally, He et al. [10], confirmed the results of [26] by showing how the benefits of graph convolution derive from smoothing the embeddings and that better performance can be achieved by removing all the intermediary weight matrices. In this formulation, the embeddings of users and items at depth k can be simply computed as the linear combination of the embeddings of the previous step with weights assigned from a suitably chosen propagation matrix $\mathbf{P}$.

2.2   Item-Based Recommender

Item-based models aim to learn item embeddings which are subsequently used to infer user representations. As a result, this model category is capable of providing recommendations to new users who join the system post training. Cremonesi et al.[5], proposed PureSVD which uses singular value decomposition to retrieve item representations from the user-item interaction matrix, and subsequently

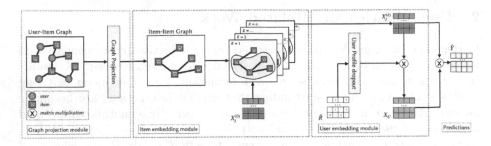

Fig. 1. Model architecture.

compute the user embeddings as a weighted combination of item representations. Later, Kabbur *et al.*in [13] also propose to compute users as a weighted combination of items, but instead of computing them after the creation of the item embeddings, they are jointly used together with the item representation as part of an optimisation process.

Our proposed *IGCCF* model inherits from the item-based model the core idea of inferring user embeddings from items, but it is also capable of leveraging the information contained in the graph-structure during the item representation learning phase through graph convolution.

3   Methodology

In this section we present details of the proposed model. *IGCCF* comprises three different elements: (1) a graph projection module, which is used to transform a user-item bipartite graph into a homogeneous item-item graph; (2) an item embedding module, which is used to learn item embeddings starting from the item-item graph; (3) a user embedding module, which is used to build user embeddings given the user-item interaction matrix and the items embeddings. The overall architecture is presented in Fig. 1.

3.1   Graph Projection Module

The graph convolution module operates over item embeddings which are optimised during training while the explicit representation and optimisation of separate user embeddings is not required. This gives the model the flexibility to easily make recommendations for unseen users. To fully capture the item relationships we construct an item-item relational graph from which extract knowledge regarding item associations during the representation learning process. The purpose of the graph projection module is to transform the bipartite user-item graph into a homogeneous item-item graph. The simplest means of achieving this is to use a one-mode projection onto the set of item nodes $\mathcal{I}$, creating an unweighted graph with exactly I nodes where two item nodes share an edge when they have at least one common neighbour in $\mathcal{U}$ [19]. This technique ignores the frequency with which two nodes share neighbors, resulting in information loss. To account for this, we build the projected item-item graph by weighting the edges based on the cosine similarity of item profiles. The edge between nodes i and j has weight $w_{ij} = \frac{\mathbf{r}_i \cdot \mathbf{r}_j}{||\mathbf{r}_i|| \cdot ||\mathbf{r}_j||}$ where $i, j \in \mathcal{I}$ and indicating with $\mathbf{r}_i$ the ith column of the matrix $\mathbf{R}$. In this way we are able to retain information about the frequency with which two items share neighbors. The model can easily adapt to different bipartite graph projection methodologies such as hyperbolic weighting [20] that takes into account the saturation effect; or weighting based on resource allocation [31], which doesn't assume symmetric weights between pairs of nodes.

Top-K Pruning. Previous works on GCNs have highlighted how the size of the neighbourhood included in the convolution operation, as well as the convolution depth, can lead to an *oversmoothing* of the embeddings. The oversmoothing leads to a loss of embedding uniqueness, and results in the degradation of recommendation performance [4,16,27]. To address this problem we apply a top-K pruning preprocessing step on the edges of the item-item graph, keeping only the K edges associated to the highest similarity score, for each item node. In this way only the most important neighbours are included in every convolution operation reducing the effect of the smoothing phenomenon. In Sect. 4.5 we show how the top-K pruning is beneficial to both training time and recommendation performance of the presented algorithm.

3.2   Item Embedding Module

The item embedding module uses information from the item-item graph to generate refined item embeddings. The primary difference between this module and previously described graph convolution modules [4,10,24] is that we use the item-item similarity matrix as propagation matrix, allowing us to directly leverage the information provided by the weighted projection used to construct the homogeneous item graph.

At the first iteration, $k = 0$, the item embedding matrix $\mathbf{X}^{(0)}$ is randomly initialised. At each subsequent iteration k, the item embedding matrix is a weighted combination of the embedding matrix at the previous layer $k-1$ with the propagation matrix, formed from the cosine similarity measure:

$$\mathbf{X}^{(k)} = \mathbf{P}\mathbf{X}^{(k-1)} = \mathbf{P}(\mathbf{P}\mathbf{X}^{(k-2)}) = \mathbf{P}^k\mathbf{X}^0 \tag{1}$$

The representation of an item i at convolution depth k can be written explicitly as:

$$\mathbf{x}_i^{(k)} = \sum_{j \in \mathcal{N}_i} w_{ij}\mathbf{x}_j^{(k-1)}$$

where $\mathcal{N}_i$ represents the 1-hop neighbourhood of item i.

The embedding at depth k can be directly computed using the power of the propagation matrix as shown in Eq. 1, which demonstrates that, at depth k, the embedding can be seen as the linear combination of neighbourhoods representations up to k-hop distance with weights given by the kth power of the cosine similarity matrix $\mathbf{P}^k$.

3.3 User Embedding Module

As there are no separate user embeddings, a method to map users into the item embedding space is required. We propose to map a user inside the item latent space as a weighted combination of the items in their profile. Given the item embeddings, a user embedding is created as:

$$\mathbf{x}_u = \sum_{i \in \mathcal{I}} \lambda_{ui} r_{ui} \mathbf{x}_i \tag{2}$$

where λ_{ui} is a scalar weighting the contribution of item i to the embedding of user u and $\mathbf{x}_i$ represents the embedding of item i. We can compute the user embeddings in matrix form as follows:

$$\mathbf{U} = (\mathbf{R} \odot \Lambda)\mathbf{X} = \tilde{\mathbf{R}}\mathbf{X}$$

where $\odot$ indicates the Hadamard product, $\tilde{\mathbf{R}}$ represents a weighted version of the interaction matrix and $\mathbf{X}$ is the item embedding matrix. In the proposed work, we assign uniform weights to all user interactions and leave the investigation of different weighting mechanisms as future work.

We want to emphasize the key advantages of modeling a user as a weighted sum of item embeddings in their profiles over having a static one-hot representation for each of them. First, it makes the model inductive from the user perspective and endows IGCCF with the ability to perform real-time updates of the user-profile as it is possible to create the embedding of a new user as soon as they start interacting with items in the system using Eq. 2. Second, it improves the model's space complexity from $\mathcal{O}(I + U)$ to $\mathcal{O}(I)$ when compared to transductive models. Finally, different importance scores may be assigned to user-item interactions when generating the user embeddings, this might be beneficial in situations where recent interactions are more significant than older ones.

3.4 Model Training

To learn the model parameters, we adopt the *Bayesian Personalised Ranking* (BPR) loss [21]:

$$L_{BPR} = \sum_{(u,i^+,i^-) \in \mathcal{O}} -\ln \sigma(\hat{y}_{ui^+} - \hat{y}_{ui^-}) + \lambda \|\Theta\|_2^2$$

where $\mathcal{O} = \{(u, i^+, i^-) | (u, i^+) \in \mathcal{R}^+, (u, i^-) \in \mathcal{R}^-\}$ denotes the pairwise training data, $\mathcal{R}^+$ indicates the observed interactions, and $\mathcal{R}^-$ the unobserved interactions; $\sigma(\cdot)$ represents the sigmoid activation function; Θ are the parameters of the model which correspond to the item embeddings.

We use the Glorot initialisation for the item embeddings [6] and mini-batch stochastic gradient descent with Adam as optimiser [14]. The preference of a user for an item is modelled through the standard dot product of their embeddings $\hat{y}(u, i) = \mathbf{x}_u^T \cdot \mathbf{x}_i$

User-Profile Dropout. It is well-known that machine learning models can suffer from overfitting. Following previously presented works on GCNs [1, 22, 24], we design a new dropout mechanism called *user-profile* dropout. Before applying Eq. 2 to form the user embeddings, we randomly drop entries of the weighted user interaction matrix $\tilde{R}$ with probability $p \in [0, 1]$. The proposed regularisation mechanism is designed to encourage the model to rely on strong patterns that exist across items rather than allowing it to focus on a single item during the construction of user embeddings.

4   Experiments

We perform experiments on four real-world datasets to evaluate the proposed model. We answer to the following research questions. [**RQ1**]: How does IGCCF perform against transductive graph convolutional algorithms? [**RQ2**]: How well does IGCCF generalise to unseen users? [**RQ3**]: How do the hyperparameters of the algorithm affect its performance?

4.1   Datasets

To evaluate the performance of the proposed methodology we perform experiments on four real world datasets gathered in different domains. **LastFM**: Implicit interactions from the Last.fm music website. In particular, the user *listened* artist relation expressed as listening counts [2]. We consider a positive interaction as one where the user has listened to an artist. **Movielens1M**: User ratings of movies from the MovieLens website [7]. Rating values range from 1 to 5, we consider ratings ≥ 3 as positive interactions. **Amazon Electronics**: User ratings of electronic products from the Amazon platform [9, 18]. The rating values also range from 1 to 5, so we consider ratings ≥ 3 as positive interactions. **Gowalla** User *check-ins* in key locations from Gowalla [17]. Here, we consider a positive interaction between a user and a location, if the user has checked-in at least once. To ensure the integrity of the datasets, following [10, 24], we perform a k-core preprocessing step setting $k_{core} = 10$, meaning we discard all users and items with less than ten interactions.

4.2   Baselines

To demonstrate the benefit of our approach we compare it against the following baselines: **BPRMF** [21] Matrix factorisation optimised by the BPR loss function. **iALS** [11] matrix factorization learned by implicit alternating least squares.

Table 1. Transductive performance comparison. Bold and underline indicate the first and second best performing algorithm respectively.

user/item/int	LastFM 1,797/1,507/6,376				Ml1M 6,033/3,123/834,449			
Model	NDCG		Recall		NDCG		Recall	
	@5	@20	@5	@20	@5	@20	@5	@20
BPR-MF	0.2162	0.3027	0.2133	0.4206	0.1883	0.3173	0.1136	0.2723
iALS	0.2232	0.3085	0.2173	0.4227	<u>0.2057</u>	<u>0.3410</u>	<u>0.1253</u>	<u>0.2893</u>
PureSVD	0.1754	0.2498	0.1685	0.3438	0.2024	0.3369	0.1243	0.2883
FISM	0.2143	0.2978	0.2145	0.4139	0.1929	0.3188	0.1203	0.2805
NGCF	0.2216	0.3085	0.2185	0.4299	0.1996	0.3309	0.1206	0.2821
LightGCN	<u>0.2293</u>	<u>0.3157</u>	<u>0.2287</u>	<u>0.4379</u>	0.1993	0.3319	0.1218	0.2864
IGCCF (Ours)	**0.2363**	**0.3207**	**0.2372**	**0.4405**	**0.2070**	**0.3456**	**0.1249**	**0.2954**
user/item/int	Amazon 13,455/8,360/234,521				Gowalla 29,858/40,988/1,027,464			
Model	NDCG		Recall		NDCG		Recall	
	@5	@20	@5	@20	@5	@20	@5	@20
BPR-MF	0.0247	0.0419	0.0336	0.0888	0.0751	0.1125	0.0838	0.1833
iALS	0.0273	0.0432	0.0373	0.0876	0.0672	0.1013	0.0763	0.1667
PureSVD	0.0172	0.0294	0.0244	0.0631	0.0795	0.1032	0.0875	0.1861
FISM	0.0264	0.0424	0.0353	0.0865	0.0812	0.1191	0.0915	0.1925
NGCF	0.0256	0.0436	0.0346	0.0926	0.0771	0.1156	0.0867	0.1896
LightGCN	<u>0.0263</u>	<u>0.0455</u>	<u>0.0358</u>	<u>0.0978</u>	<u>0.0874</u>	<u>0.1279</u>	<u>0.0975</u>	<u>0.2049</u>
IGCCF (Ours)	**0.0336**	**0.0527**	**0.0459**	**0.1072**	**0.0938**	**0.1373**	**0.1049**	**0.2203**

PureSVD [5]Compute item embeddings through a singular value decomposition of the user-item interaction matrix, which will be then used to infer user representations. **FISM** [13] Learn item embeddings through optimisation process creating user representations as a weighted combination of items in their profile. Additional user and item biases as well as an agreement term are considered in the score estimation. **NGCF** [24] Work that introduces graph convolution to the collaborative filtering scenario, it uses dense layer and inner product to enrich the knowledge injected in the user item embeddings during the convolution process. **LightGCN** [10] Simplified version of graph convolution applied to collaborative filtering directly smooth user and item embeddings onto the user-item bipartite graph. We follow the original paper [10] and use $a_k = 1/(k+1)$.

For each baseline, an exhaustive grid-search has been carried out to ensure optimal performance. Following [10], for all adopted algorithms the batch size has been set to 1024 and embedding size to 64. Further details on the ranges of the hyperparameter search as well as the data used for the experiments are available in the code repository[1].

4.3   Transductive Performance

In this section we evaluate the performance of IGCCF against the proposed baselines in a transductive setting, meaning considering only users present at training time. To evaluate every model, following [10, 24], for each user, we randomly sample 80% of his interactions to constitute the training set, 10% to be the test set, while the remaining 10% are used as a validation set to tune the algorithm hyper-parameters. Subsequently, validation and training data are merged together and used to retrain the model, which is then evaluated on the test set. In order to asses the quality of the recommendations produced by our system, we follow the approach outlined in [4, 24, 26]. For each user in the test data, we generate a ranking of items and calculate the average *Recall@N* and *NDCG@N* scores across all users, considering two different cutoff values $N = 5$ and $N = 20$. The final results of this analysis are presented in Table 1.

Based on the results obtained, we can establish that IGCCF outperforms NGCF and LightGCN on all four datasets examined for each metric and cutoff. This confirms that explicitly parametrizing the user embeddings is not necessary to get the optimum performance; on the contrary, it might result in an increase in the number of parameters of the model, which is detrimental to both training time and spatial complexity of the model. Furthermore, IGCCF shows superior performance with respect to the item-based baseline models. This demonstrates that interpreting user-item interaction as graph-structured data introduces relevant knowledge into the algorithm learning process, leading to improved model performance.

4.4   Inductive Performance

A key feature of the proposed IGCCF algorithm, is the ability to create embeddings and consequently retrieve recommendations for *unseen* users who are not present at training time. IGCCF does not require an additional learning phase to create the embeddings. As soon as a new user begins interacting with the items in the catalogue, we may construct its embedding employing Eq. 2.

To assess the inductive performance of the algorithm we hold out 10% of the users, using the remaining 90% as training data. For every unseen user we use 90% of their profile interactions to create their embedding (Eq. 2) and we evaluate the performance on the remaining 10% of interactions. We compare the performance of our model against the inductive baselines corresponding to the item-based models (PureSVD and FISM) since the transductive models are not able to make predictions for users who are not present at training time without an additional learning phase. Recommendation performance is evaluated using the same metrics and cutoffs reported in Subsect. 4.3. The overall results are reported

---

[1] https://github.com/damicoedoardo/IGCCF.

Table 2. Inductive performance on *unseen* users. Bold indicates the performance of the best ranking algorithm.

Model	LastFM				MllM			
	NDCG		Recall		NDCG		Recall	
	@5	@20	@5	@20	@5	@20	@5	@20
PureSVD	0.1640	0.2279	0.1610	0.3124	0.2064	0.3418	0.1165	0.2759
FISM	0.1993	0.2921	0.1927	0.4165	0.1974	0.3221	0.1105	0.2638
IGCCF (Ours)	**0.2374**	**0.3227**	**0.2355**	**0.4395**	**0.2089**	**0.3474**	**0.1177**	**0.2817**
Model	Amazon				Gowalla			
	NDCG		Recall		NDCG		Recall	
	@5	@20	@5	@20	@5	@20	@5	@20
PureSVD	0.0221	0.0345	0.0320	0.0721	0.0815	0.1213	0.0862	0.1910
FISM	0.0330	0.0468	0.0424	0.0891	0.0754	0.1102	0.0829	0.1763
IGCCF (Ours)	**0.0356**	**0.0513**	**0.0477**	**0.0978**	**0.0910**	**0.1341**	**0.1009**	**0.2172**

in Table 2. IGCCF outperforms the item-based baselines on all the datasets. These results strongly confirm our insight that the knowledge extracted from the constructed item-item graph is beneficial to the item-embedding learning phase, even when making predictions for unseen users.

Robustness of Inductive Performance. We are interested in the extent to which IGCCF can maintain comparable recommendation performance between *seen* and *unseen* users as we train the model with less data. For this experiment, we increasingly reduce the percentage of seen users which are used to train the model and consequently increase the percentage of unseen users which are presented for inductive inference. We train the model 5 times on each different

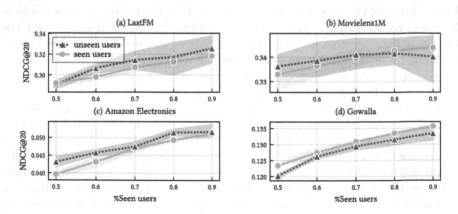

Fig. 2. For each dataset we vary the percentage of users in the training data, and evaluate the performance of IGCCF on both *seen* and *unseen* users.

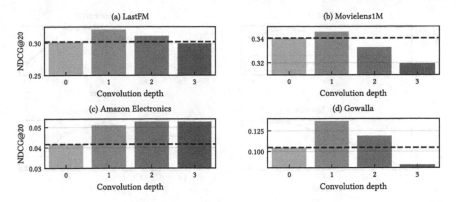

Fig. 3. Ablation study: Effect of the Convolution depth parameter on the algorithm performance.

split used and we report the average performance (NDCG@20). From the results in Fig. 2 we can observe: IGCCF exhibits comparable performance on both seen and unseen user groups for all the splits analysed, showing how the inductive performance of IGCCF is robust with respect to the amount of training data available. As expected, reducing the amount of available training data results in a lower $NDCG$@20, anyway is interesting to notice how the drop in performance is minimal even when the model is trained with half of the data available.

4.5 Ablation Study

Convolution Depth. The convolution operation applied during the learning phase of the item embeddings, is beneficial in all the studied datasets, the results are reported in Fig. 3. It is interesting to consider the relationship between the dataset density and the effect of the convolution operation. We can see that the largest improvement of 31% is found on Gowalla, which is the least dense dataset (0.08%). As the density increases, the benefit introduced by the convolution operation decreases. We have an improvement of 26% and 6% on Amazon Electronics (0.21%) and LastFM (2.30%) respectively while there is a very small increase of 1.5% on Movielens1M (4.43%). The results obtained suggest an inverse correlation between the dataset density and the benefit introduced by the convolution operation.

User-Profile Dropout. From the analysis reported in Fig. 4, it is clearly visible that user profile dropout regularisation have a strong impact on the performance of the proposed method. In all four datasets, the utilisation of the suggested regularisation technique enhance the quality of the recommendation performance, resulting in a gain over the $NDCG$@20 metric of 4.4%, 3.0%, 10.5%, 1.5% for LastFM, Movielens1M, Amazon Electronics and Gowalla respectively. Dropping a portion of the user profiles during the embeddings creation phase, force the algorithm to not heavily rely on information coming from specific items.

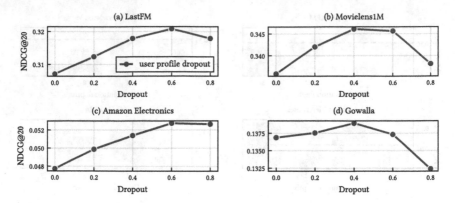

Fig. 4. Ablation study: Effect of the dropout mechanisms on the algorithm performance.

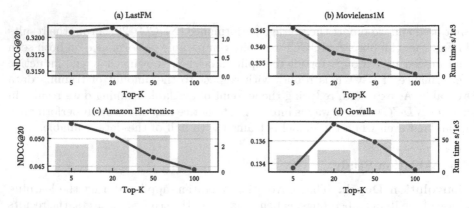

Fig. 5. Effect of the top-K preprocessing step on the algorithm performance and training time.

Top-K Pruning. To prevent the well-known oversmoothing issue caused by graph convolution, we trim the edges of the item-item graph to maintain only the most strong connections between items. Figure 5 illustrates the results of the ablation study. In all of the datasets investigated, utilising at most 20 neighbours for each item node yields the highest performance; this demonstrates how retaining edges associated with weak item links can worsen model performance while also increasing the algorithm training time.

5   Conclusion and Future Work

In this work we presented IGCCF, an item-based model that employs graph convolution to learn refined item embeddings. We build upon the previously presented graph convolution models by removing the explicit parameterisation of users. The benefits of that are threefold: first, it reduces model complexity;

second, it allows real-time user embeddings updates as soon as new interactions are gathered; and third, it enables inductive recommendations for new users who join the system post-training without the need for a new expensive training procedure. To do this, we devised a novel procedure that first constructs an item-item graph from the user-item bipartite network. A top-K pruning procedure is then employed to refine it, retaining only the most informative edges. Finally, during the representation learning phase, we mine item associations using graph convolution, building user embeddings as a weighted combination of items with which they have interacted. In the future, we will extend the provided methodology to operate in settings where item side-information are available.

Acknowledgement. This research was supported by Science Foundation Ireland (SFI) under Grant Number SFI/12/RC/2289_P2.

References

1. van der Berg, R., Kipf, T.N., Welling, M.: Graph convolutional matrix completion. arXiv preprint arXiv:1706.02263 (2017)
2. Cantador, I., Brusilovsky, P., Kuflik, T.: Second workshop on information heterogeneity and fusion in recommender systems (hetrec2011). In: Proceedings of the fifth ACM Conference on Recommender systems, pp. 387–388. ACM, New York, NY, USA (2011)
3. Chen, J., Ma, T., Xiao, C.: FastGCN: fast learning with graph convolutional networks via importance sampling. In: 6th International Conference on Learning Representations, ICLR 2018, Vancouver, BC, Canada, 30 Apr - 3 May 2018, Conference Track Proceedings (2018)
4. Chen, L., Wu, L., Hong, R., Zhang, K., Wang, M.: Revisiting graph based collaborative filtering: A linear residual graph convolutional network approach. In: The Thirty-Fourth AAAI Conference on Artificial Intelligence, AAAI 2020, New York, NY, USA, 7–12 Feb 2020, pp. 27–34 (2020)
5. Cremonesi, P., Koren, Y., Turrin, R.: Performance of recommender algorithms on top-n recommendation tasks. In: Proceedings of the Fourth ACM Conference on Recommender Systems, pp. 39–46. RecSys 2010, Association for Computing Machinery, New York, NY, USA (2010)
6. Glorot, X., Bengio, Y.: Understanding the difficulty of training deep feedforward neural networks. In: Teh, Y.W., Titterington, D.M. (eds.) Proceedings of the Thirteenth International Conference on Artificial Intelligence and Statistics, AISTATS 2010, Chia Laguna Resort, Sardinia, Italy, 13–15 May 2010. JMLR Proceedings, vol. 9, pp. 249–256 (2010)
7. Harper, F.M., Konstan, J.A.: The movielens datasets: History and context. ACM Trans. Interact. Intell. Syst. **4**, 1–19 (2015)
8. Hartford, J., Graham, D., Leyton-Brown, K., Ravanbakhsh, S.: Deep models of interactions across sets. In: Dy, J., Krause, A. (eds.) Proceedings of the 35th International Conference on Machine Learning. Proceedings of Machine Learning Research, vol. 80, pp. 1909–1918. PMLR (10–15 Jul 2018)

9. He, R., McAuley, J.J.: Ups and downs: modeling the visual evolution of fashion trends with one-class collaborative filtering. In: Proceedings of the 25th International Conference on World Wide Web, WWW 2016, Montreal, Canada, 11–15 April 2016, pp. 507–517 (2016)

10. He, X., Deng, K., Wang, X., Li, Y., Zhang, Y., Wang, M.: LightGCN: simplifying and powering graph convolution network for recommendation. In: Proceedings of the 43rd International ACM SIGIR conference on research and development in Information Retrieval, SIGIR 2020, Virtual Event, China, 25–30 July 2020, pp. 639–648 (2020)

11. Hu, Y., Koren, Y., Volinsky, C.: Collaborative filtering for implicit feedback datasets. In: 2008 Eighth IEEE International Conference on Data Mining, pp. 263–272. IEEE (2008)

12. Jain, P., Dhillon, I.S.: Provable inductive matrix completion (2013)

13. Kabbur, S., Ning, X., Karypis, G.: FISM: factored item similarity models for top-n recommender systems. In: Proceedings of the 19th ACM SIGKDD International Conference on Knowledge Discovery and Data Mining, pp. 659–667 (2013)

14. Kingma, D.P., Ba, J.: Adam: a method for stochastic optimization. In: 3rd International Conference on Learning Representations, ICLR 2015, San Diego, CA, USA, 7–9 May 2015, Conference Track Proceedings (2015)

15. Kipf, T.N., Welling, M.: Semi-supervised classification with graph convolutional networks. In: 5th International Conference on Learning Representations, ICLR 2017, Toulon, France, 24–26 Apr 2017, Conference Track Proceedings (2017)

16. Li, Q., Han, Z., Wu, X.: Deeper insights into graph convolutional networks for semi-supervised learning. In: Proceedings of the Thirty-Second AAAI Conference on Artificial Intelligence, (AAAI-18), the 30th innovative Applications of Artificial Intelligence (IAAI-18), and the 8th AAAI Symposium on Educational Advances in Artificial Intelligence (EAAI-18), New Orleans, Louisiana, USA, 2–7 Feb 2018, pp. 3538–3545 (2018)

17. Liang, D., Charlin, L., McInerney, J., Blei, D.M.: Modeling user exposure in recommendation. In: Proceedings of the 25th International Conference on World Wide Web, WWW 2016, Montreal, Canada, 11–15 April 2016, pp. 951–961 (2016)

18. McAuley, J.J., Targett, C., Shi, Q., van den Hengel, A.: Image-based recommendations on styles and substitutes. In: Proceedings of the 38th International ACM SIGIR Conference on Research and Development in Information Retrieval, Santiago, Chile, 9–13 August 2015, pp. 43–52. ACM (2015)

19. Newman, M.E.J.: The structure of scientific collaboration networks. Proc. Natl. Acad. Sci. **98**(2), 404–409 (2001)

20. Newman, M.E.: Scientific collaboration networks. ii. shortest paths, weighted networks, and centrality. Phys. Rev. E Stat. Nonlin. Soft. Matter Phys. **64**(1), 016132 (2001)

21. Rendle, S., Freudenthaler, C., Gantner, Z., Schmidt-Thieme, L.: BPR: Bayesian personalized ranking from implicit feedback. In: Bilmes, J.A., Ng, A.Y. (eds.) UAI 2009, Proceedings of the Twenty-Fifth Conference on Uncertainty in Artificial Intelligence, Montreal, QC, Canada, 18–21 June 2009, pp. 452–461. AUAI Press (2009)

22. Rong, Y., Huang, W., Xu, T., Huang, J.: DropEdge: towards deep graph convolutional networks on node classification. In: 8th International Conference on Learning Representations, ICLR 2020, Addis Ababa, Ethiopia, 26–30 April 2020 (2020)

23. Volkovs, M., Yu, G.W., Poutanen, T.: Dropoutnet: Addressing cold start in recommender systems. In: Guyon, I., et al. (eds.) Advances in Neural Information Processing Systems 30: Annual Conference on Neural Information Processing Systems 2017, 4–9 Dec 2017, Long Beach, CA, USA, pp. 4957–4966 (2017)
24. Wang, X., He, X., Wang, M., Feng, F., Chua, T.S.: Neural graph collaborative filtering. In: Proceedings of the 42nd International ACM SIGIR Conference on Research and Development In Information Retrieval, pp. 165–174 (2019)
25. Wang, X., Ye, Y., Gupta, A.: Zero-shot recognition via semantic embeddings and knowledge graphs. In: Proceedings of the IEEE Conference on Computer Vision and Pattern Recognition, pp. 6857–6866 (2018)
26. Wu, F., Jr., A.H.S., Zhang, T., Fifty, C., Yu, T., Weinberger, K.Q.: Simplifying graph convolutional networks. In: Proceedings of the 36th International Conference on Machine Learning, ICML 2019, 9–15 June 2019, Long Beach, California, USA. vol. 97, pp. 6861–6871 (2019)
27. Xu, K., Li, C., Tian, Y., Sonobe, T., Kawarabayashi, K., Jegelka, S.: Representation learning on graphs with jumping knowledge networks. In: Proceedings of the 35th International Conference on Machine Learning, ICML 2018, Stockholmsmässan, Stockholm, Sweden, 10–15 July 2018. Proceedings of Machine Learning Research, vol. 80, pp. 5449–5458. PMLR (2018)
28. Yang, L., Schnabel, T., Bennett, P.N., Dumais, S.: Local factor models for large-scale inductive recommendation. In: Fifteenth ACM Conference on Recommender Systems, pp. 252–262. RecSys 2021, Association for Computing Machinery, New York, NY, USA (2021)
29. Yao, L., Mao, C., Luo, Y.: Graph convolutional networks for text classification. In: Proceedings of the AAAI Conference on Artificial Intelligence, vol. 33, pp. 7370–7377 (2019)
30. Zhang, M., Chen, Y.: Inductive matrix completion based on graph neural networks. In: 8th International Conference on Learning Representations, ICLR 2020, Addis Ababa, Ethiopia, 26–30 April 2020. OpenReview.net (2020)
31. Zhou, T., Ren, J., Medo, M., Zhang, Y.C.: Bipartite network projection and personal recommendation. Phys. Rev. E $76(4)$, 046115 (2007)

CoLISA: Inner Interaction via Contrastive Learning for Multi-choice Reading Comprehension

Mengxing Dong[1], Bowei Zou[2], Yanling Li[1], and Yu Hong[1(✉)]

[1] Computer Science and Technology, Soochow University, Suzhou, China
tianxianer@gmail.com
[2] Institute for Infocomm Research, Singapore, Singapore
zou_bowei@i2r.a-star.edu.sg

Abstract. Multi-choice reading comprehension (MC-RC) is supposed to select the most appropriate answer from multiple candidate options by reading and comprehending a given passage and a question. Recent studies dedicate to catching the relationships within the triplet of passage, question, and option. Nevertheless, one limitation in current approaches relates to the fact that confusing distractors are often mistakenly judged as correct, due to the fact that models do not emphasize the differences between the answer alternatives. Motivated by the way humans deal with multi-choice questions by comparing given options, we propose CoLISA (Contrastive Learning and In-Sample Attention), a novel model to prudently exclude the confusing distractors. In particular, CoLISA acquires option-aware representations via contrastive learning on multiple options. Besides, in-sample attention mechanisms are applied across multiple options so that they can interact with each other. The experimental results on QuALITY and RACE demonstrate that our proposed CoLISA pays more attention to the relation between correct and distractive options, and recognizes the discrepancy between them. Meanwhile, CoLISA also reaches the state-of-the-art performance on QuALITY (Our code is available at https://github.com/Walle1493/CoLISA..).

Keywords: Machine reading comprehension · Multi-choice question answering · Contrastive learning

1 Introduction

Machine Reading Comprehension (MRC) requires models to answer questions through reasoning over given documents. Multi-choice reading comprehension (MC-RC) [17], as one of the variants of MRC tasks, aims at choosing the most appropriate answer from multiple options to respond to the question for a given passage. It requires models to identify the validity of each candidate option by reading and comprehending the referential passage.

Existing studies on MC-RC usually focus on solving the gap between the passage and a single option for a given question [16,26]. The models encode

© The Author(s), under exclusive license to Springer Nature Switzerland AG 2023
J. Kamps et al. (Eds.): ECIR 2023, LNCS 13980, pp. 264–278, 2023.
https://doi.org/10.1007/978-3-031-28244-7_17

each option independently. In this way, each option corresponding to a certain question cannot intuitively interact with each other, which limits the inference capability of models. Besides, there exists a more troublesome condition in some of the real cases. For a certain question, some options are literally and even semantically similar to the gold answer, in other words, they appear to be plausible when identifying the authenticity of only a single option. Existing methods fail to deal with such kinds of cases. We believe that an extra and more elaborate operation is supposed to be applied between those so-called confusing distractors and the correct answer. Table 1 depicts an example of an indistinguishable distractor from QuALITY[1] [15]. Reading through the entire passage, we consider that both the correct answer O_2 (bold and underlined) and the confusing distractor O_1 (underlined) can be regarded as precisely correct. To select the most appropriate option given a question, a model needs to find the discrepancy between the representations of gold answers and the distractors.

Table 1. An example from QuALITY.

Passage:
(...) I'm sure that 'justifiable yearnings for territorial self-realization' would be more appropriate to the situation (...) (Over 4,000 words)
Question:
According to Retief what would happen if the Corps did not get involved in the dispute between the Boyars and the Aga Kagans?
Options:
O_1: The Aga Kagans would enslave the Boyars
O_2: **The Boyars and the Aga Kagans would go to war**
O_3: The Aga Kagans would leave Flamme to find a better planet
O_4: The Boyars would create a treaty with the Aga Kagans without the Corps' approval

Humans usually exclude plausible distractors by carefully comparing between options to answer MC-RC questions [4]. Motivated by such a procedure, we come up with a framework with **C**ontrastive **L**earning and **In-S**ample **A**ttention (CoLISA) including two main characteristics. First of all, with two different hidden dropout masks applied, we acquire two slightly different representations including two correct answers and multiple distractors from the same input. Our proposed CoLISA aims to pull two correct answers together and push the answer-distractor pairs away by means of contrastive learning, therefore, the model is expected to learn a more effective representation. In addition, self-attention mechanisms [21] are applied across multiple options within a specific

---

[1] The issue that input passages exceed the length constraint exists in QuALITY, we seriously consider it in our work as well.

sample to allow information to flow through them. As a result, the model learns to incline to the correct answer through self-attention interaction across multiple candidate options. We conduct extensive experiments on two MC-RC datasets, QuALITY [15] and RACE [12]. Experimental results demonstrate that CoLISA significantly outperforms the existing methods.

Our contributions are summarized as follows.

- We introduce the contrastive learning method into multi-choice reading comprehension, which is capable of distinguishing the correct answer from distractors. Our approach pays sufficient attention to distractive candidates by distributing more weight to them.
- We apply in-sample attention mechanisms across multiple options within a specific sample to have them interact with each other.
- Our proposed model reaches the state-of-the-art performance on QuALITY and achieves considerable improvements compared with solid baselines on RACE.

2   Task Formulation

Given a referential passage, a target question, and several candidate options, the multi-choice reading comprehension (MC-RC) task aims to predict one option as the final answer. Formally, we define the passage as $p = [s_1^p, s_2^p, ..., s_n^p]$, where $s_i^p = [w_1^s, w_2^s, ..., w_l^s]$ denotes the i-th sentence in p and w_j^s denotes the j-th word of s_i^p. The question is defined as $q = [w_1^q, w_2^q, ..., w_t^q]$, where w_i^q denotes the i-th word of q. The option set is defined as $O = [o_1, o_2, ..., o_r]$, where $o_i = [w_1^o, w_2^o, ..., w_k^o]$ denotes the i-th option and w_k^o denotes the k-th word of o_i. The target of MC-RC is to maximize the probability of the predicted option:

$$a = \underset{i}{\mathrm{argmax}}(\mathcal{P}(o_i|p, q)). \tag{1}$$

When the length of p exceeds the maximum input length of encoders, we compress it to a shorter context by retrieving relevant sentences. The shorter context is represented as $c = [s_1^c, s_2^c, ..., s_m^c]$, where $s_i^c = [w_1^s, w_2^s, ..., w_l^s]$ denotes the i-th sentence of c, and w_j^s denotes the i-th sentence of s_i^c.

3   Methodology

As illustrated in Fig. 1, we come up with a novel MC-RC framework with contrastive learning and in-sample attention (CoLISA), which is composed of 1) *DPR-based retriever* that selects relevant sentences from a long passage according to the given question and its multiple options, to construct a new context with their original order in the passage (Sect. 3.1), and 2) *CoLISA reader* that predicts the final answer from several candidate options according to the given question and context. In particular, the CoLISA reader consists of two modules. 1) *In-Sample Attention (ISA)* mechanism, whereby we introduce a long-sequence

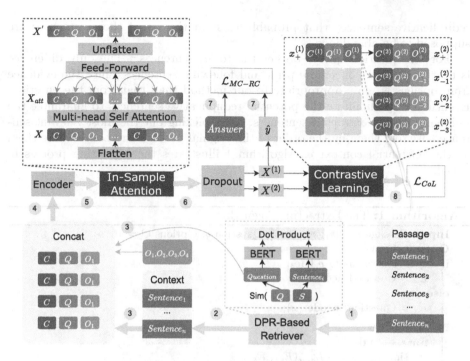

Fig. 1. Architecture of CoLISA.

network with a multi-head self-attention operation to enhance the interaction across multiple options within one sample (Sect. 3.2). 2) *Contrastive Learning (CoL)* with a distractive factor (DiF) to represent the sequences composed of the context, the question, and the options (Sect. 3.3).

3.1 DPR-Based Retriever

To select relevant sentences from a long passage, we employ a Dense Passage Retrieval (DPR)-based sentence retriever, which is a dense retriever for latent semantic encoding [11]. Note that two encoders from DPR are already pre-trained for extracting diverse sorts of sentences, we correspondingly utilize them to ensure the diversity of the retrieved sentences. A context encoder E_S encodes all sentences s of the referential p to d-dimensional vectors. Similarly, a query encoder E_R encodes the question q and the option set O to d-dimensional vectors, as two sorts of retrieval queries r. The global representations of the $[CLS]$ tokens of s and r are fetched to calculate their negative Euclidean (L^2) distance:

$$- L_{dist}^2(r, s) = -||E_R(r) - E_S(s)||^2. \tag{2}$$

For the option query, we select top-k sentences in descending order of the relevance distance between s and r. Meanwhile, prior and following sentences of these k sentences are selected for the sake of semantic coherence. Furthermore, reduplicative sentences that probably exist in the set of these sentences are supposed to be abandoned.

For the question query, we get the top-n sentences.[2] The only difference is that we do not collect the prior and following sentences, since the evidence from options is more appropriate to our method than that from the question. Eventually, we eliminate the potential reduplicative sentences to guarantee the uniqueness of all extracted sentences. After picking out the most relevant sentences, we sort them by the order of the original passage and concatenate them as the referential context c. Algorithm 1 illustrates the extracting process in detail.

Algorithm 1: The Extracting Algorithm

Input: Passage $p = [s_1, s_2, ..., s_n]$, Question q, Options $O = [o_1, o_2, ..., o_m]$
Output: Context
if *Input (x) belongs to p* **then**
$\quad\mid\quad E_x = \text{ContextEncoder}(x)$;
else
$\quad\mid\quad E_x = \text{QuestionEncoder}(x)$;
for o_i *in* O **do**
$\quad$**for** s_j *in* p **do**
$\qquad\mid\quad \text{sim}(o_i, s_j) = -L^2_{dist}(E_{o_i}, E_{s_j})$
$\quad$Select: Top-k relevant sentences $s_{i_1}, s_{i_2}, ..., s_{i_k}$;
$\quad$Select: Their previous and next sentences $s^{prev}_{i_1}, s^{next}_{i_1}, ..., s^{prev}_{i_k}, s^{next}_{i_k}$;
$\quad$Context $\leftarrow$ Selected sentences;
for s_j *in* p **do**
$\quad\mid\quad \text{sim}(q, s_j) = -L^2_{dist}(E_q, E_{s_j})$
Select: Top-n relevant sentences $s_{j_1}, s_{j_2}, ..., s_{j_n}$;
Context $\leftarrow$ Selected sentences;
Context.unique().sort().to_str();

3.2   In-Sample Attention Mechanism

When encoding q and o_i, different options for the same question are encoded independently of each other, which causes a lack of attention among these options. To address this problem, we borrow from human behaviors when answering multi-choice questions. Generally, one prefers to eliminate distractors by simultaneously comparing multiple indistinguishable options to ultimately decide which option to choose. Inspired by such a way, we come up with an in-sample attention (ISA) mechanism to enhance the interaction between representations of different options.

---

[2] Considering an appropriate value of the extracted context length, we simply define k as 2 and n as 1 in our experiments.

In general, each candidate option o_i is concatenated with its corresponding context c and question q to form a triplet $x_i = [c; q; o_i]$. We get series of complete sequential representations by feeding each x_i into a pre-trained encoder one by one. To model the interaction across multiple o_i, our proposed ISA module first collects all of x_i that are corresponding to the same question q, with an operation of concatenation to construct a single sequence $X = [x_1; x_2; ...; x_n]$. Then, we calculate the self-attention representation of X to learn the distant dependency across multiple options. Specifically, we reuse the architecture of vanilla self-attention mechanisms and simultaneously pack the sequence X into three matrices query Q, key K, and value V. The self-attention matrix of outputs is calculated as:

$$SA(Q, K, V) = softmax(\frac{QK^T}{\sqrt{d_k}})V, \tag{3}$$

where d_k is a scaling factor denoting the dimension of K against causing extremely small gradients [21]. Furthermore, a multi-head mechanism is introduced to comprehensively represent X at various vector dimensions. The process of multi-head self-attention is defined as:

$$head_i = SA(QW_i^Q, KW_i^K, VW_i^V), \tag{4}$$

$$H = Concat(head_1, ..., head_h), \tag{5}$$

$$MSA(Q, K, V) = HW^O, \tag{6}$$

where h denotes the number of parallel attention heads, and W_i^Q, W_i^K, W_i^V, W^O are parameter matrices. We note the vector representation of multi-head self-attention as X_{att}. As illustrated in the ISA mechanism of Fig. 1, the self-attention mechanisms over the concatenated sequence X achieve the interaction across multiple options, by calculating its crossing-option representations X_{att}.

In addition, to avoid the collapse of multi-head self-attention output [19,21], we add a fully-connected feed-forward layer after the multi-head self-attention layer:

$$X_{ffn} = FFN(X_{att}). \tag{7}$$

We employ GeLU [9] as the activation function. Finally, X_{ffn} is supposed to be unpacked to multiple representations corresponding to multiple input triplets x_i. We denote the set of output triplets as X'.

3.3 Contrastive Learning for Inner Interaction

To encourage CoLISA to explicitly distinguish the representation of the answer from other distractors, we introduce a contrastive learning (CoL) module. Inspired by the common contrastive learning framework [7], our CoL module promotes two positive triplets to get closer to each other while pushes all representations within a sample well-distributed in a specific vector space. Following the ISA module, the input vector with interacted attention X' is passed to the dropout layer twice, to produce two slightly different representations for each input, $X^{(1)}$ and $X^{(2)}$, respectively.

We first calculate the cross-entropy loss of MC-RC between the target label y and the output $X^{(1)}$. Specifically, we apply a linear layer to convert $X^{(1)}$ to the prediction $\hat{y}$ that is the same size as y. The loss function is defined as:

$$\mathcal{L}_{MC-RC} = -\frac{1}{N}\Sigma_{i=1}^{N}(y^{(i)}\log\hat{y}^{(i)} + (1 - y^{(i)})\log(1 - \hat{y}^{(i)})). \tag{8}$$

Note that both $X^{(1)}$ and $X^{(2)}$ from each sample consistently consist of the exact amount of triplets x_i: one gold triplet and the rest distractive triplets. Therefore, the contrastive loss is defined as the average of negative log-likelihood losses for the two outputs over each sample. More specifically, we keep the gold triplet as the anchor and remove the distractive triplets from $X^{(1)}$. While for $X^{(2)}$, all triplets are reserved: the gold one is regarded as the positive instance and the rest as negative instances. Each loss term discriminates the positive instance from the negative ones. The loss function of CoL is defined as:

$$\mathcal{L}_{CoL} = -\frac{1}{N}\Sigma_{j=1}^{N}\log\frac{e^{sim(x_+^{(1)},x_+^{(2)})/\tau}}{\Sigma_{i=1}^{S}e^{sim(x_+^{(1)},x_i^{(2)})/\tau}}, \tag{9}$$

where $x_+^{(1)}$ is the representation of the gold triplet from $X^{(1)}$, $x_i^{(2)}$ is each vector representation of the triplets of $X^{(2)}$, in which $x_+^{(2)}$ is the positive sample, and τ is the configurable temperature hyper-parameter. $sim(\cdot)$ is a similarity metric (we typically use cosine similarity in all our experiments), S indicates the number of triplets for each sample, and N denotes the batch size. The aggregated contrastive loss $\mathcal{L}_{CoL}$ is preliminarily obtained by averaging over all the samples within a batch.

Distractive Factor. Finding out that diverse distractors within a sample variously interfere the inference capability of the model, we introduce a distractive factor (DiF) into CoLISA by combining it with the process of contrastive learning. Annotators from QuALITY treat candidate options that mislead their judgment worst as the distractor items for each question. Therefore, we construct a list of confidence factors to present the contribution of each option to $\mathcal{L}_{CoL}$, according to their corresponding annotation scores. Specifically, we enumerate the annotated votes for each option to construct the DiF $\Theta = [\theta_1, \theta_2, ..., \theta_n]$, where n is the number of options. Then a softmax function helps scale the Θ to differentiate each θ_i more obviously. Each θ_i is revised as below:

$$\theta_i = \frac{e^{\theta_i}}{\Sigma_i^n e^{\theta_i}}. \tag{10}$$

The exponential operation distinctly discriminates n coefficients θ_i. When calculating the contrastive loss, θ_i is multiplied by the similarity value of its corresponding option, to weight the correct options and the confusing distractors. The more prominent the θ_i is, the more the corresponding option contributes to the loss. We improve the contrastive loss function as below:

$$\mathcal{L}_{CoL} = -\frac{1}{N}\Sigma_{j=1}^{N}log\frac{\theta_{+}e^{sim(x_{+}^{(1)},x_{+}^{(2)})/\tau}}{\Sigma_{i=1}^{S}\theta_{i}e^{sim(x_{+}^{(1)},x_{i}^{(2)})/\tau}}, \tag{11}$$

where θ_{+} and θ_{i} correspond to the positive sample and the i-th sample from Θ, respectively. The final loss function is formulated as:

$$\mathcal{L} = \alpha \cdot \mathcal{L}_{MC-RC} + (1-\alpha) \cdot \mathcal{L}_{CoL}, \tag{12}$$

where $\alpha \in [0,1]$ is the balance coefficient.

4 Experimentation

We describe our experimental settings and provide facts about the experiments that we execute in this section. Then we further depict the comparison with existing models and conduct some ablation studies.

4.1 Experimental Settings

We employ the benchmark datasets QuALITY and RACE to validate the effectiveness of CoLISA. We mainly conduct experiments on QuALITY[3], and simultaneously report experimental results on RACE.

- QuALITY [15]. The multi-choice reading comprehension (MC-RC) dataset with an average length of about 5,000 tokens for context passages. The most noteworthy feature of the dataset is that some distractors influence the cognition of the model badly. Skimming and simple search are not adequate to perform consistently and reasonably without relying on abstracts or excerpts. Articles are mainly sourced from science fiction and magazines, which have been read and validated by contributors.
- RACE [12]. We embrace another gigantic MC-RC dataset to verify the model performance since QuALITY comprises only 6,737 source data that cannot thoroughly estimate the results. RACE is collected from English entrance exams for Chinese middle and high school students. Most of the questions also need reasoning. What's more, the domains of passages are diversified, ranging from news and story to advertisements, making the dataset challengeable.

We mainly adopt *accuracy* (*acc*) in QuALITY [15] and RACE [12] as our evaluation criterion to assess the percentage of questions answered correctly. Following the divided manner of the two datasets, we also employ *acc* in the

---

[3] DPR-based retriever is available for long input, which is the characteristic owned by QuALITY. Besides, the best distractors are only annotated in QuALITY, therefore, we execute experiments mainly on QuALITY.

full/hard subset of QuALITY and in the middle/high subset of RACE[4] in our main experiments.

For the DPR-based retriever, we employ the identical component as QuALITY does to alleviate the impact of retrieval disparity for a fair comparison. Expressly, we utilize a question encoder to encode a query and a context encoder for the passage.

For the CoLISA reader, we predominantly use DeBERTaV3-large as our pretrained language model. For all the experiments, we train our models on QuALITY and RACE with a learning rate of 1e-5 and a warm-up proportion of 0.1. We maintain a default dropout rate of 0.1 in all dropout layers and use the GeLU activation function [9]. Our implementation adopts a batch size of 16 with a maximum length of 512 tokens. We fine-tune our model for 20 epochs on QuALITY and 3 epochs on RACE.

For contrastive learning, we adjust the temperature τ on the DeBERTaV3-base and RoBERTa-base models due to the experimental cost. The optimal $\tau = 0.1$ on base models is then directly applied to large models. Cross-entropy loss and contrastive loss are 50-50 split contributed to the synthetic loss. In addition, the distractive factor (DiF) is only adopted in experiments of QuALITY, for the reason that RACE does not annotate such labels for each option.

All our experiments are executed on 1 T V100-32GB GPU for training. FP16 from Apex is adopted for accelerating the training process.

Note that manifold approaches are applied on QuALITY, we employ two typical baselines.

- Longformer [1]. The pre-trained model combines sliding-window local attention and global attention to encode longer sequences. Longformer supports up to an utmost of 4,096 tokens. We pick up Longformer as one of our baselines because it contains most of the context needed to respond to the questions for samples in QuALITY.
- DPR & DeBERTaV3 [8,11]. The pipeline framework consists of a retriever and a reader. The DPR-based retriever is adopted to extract the most relevant context corresponding to the question from an article. The selected context, as part of inputs, is then fed to the following module. Finally, a standard DeBERTaV3 model for MC-RC is supposed to determine the correct option.

4.2   Main Results

We compare our CoLISA model with two solid baselines on QuALITY and three pre-trained language models on RACE. As the overall results are displayed in Table 2, our CoLISA surpasses all other models. The deficient performance of Longformer suggests that Longformer blunders to locate pivotal hints from a

---

[4] Source data in QuALITY is divided into full and hard subsets according to the question difficulty, while in RACE, middle and high subsets represent two levels of school entrance exams.

Table 2. *Acc* results on the development and test set over QuALITY (full/hard) and RACE (middle/high). Models on QuALITY are intermediately trained on RACE and then fine-tuned on QuALITY. It is worth mentioning that the results of DPR & DeBERTaV3 architecture baseline on the development set are our reimplementations (marked with *) and appreciably higher than records from QuALITY (53.6/47.4). And so are the results of DeBERTaV3-large. While other baseline results are from either related works or relevant leaderboards. Experiments on RACE do not demand a DPR-based retriever and the distractive factor.

Model	QuALITY		RACE	
	Dev	Test	Dev	Test
Longformer-base [1]	38.1/32.8	39.5/35.3	–	–
DPR & DeBERTaV3-large [15]	56.7*/48.6*	55.4/46.1	–	–
DeBERTaV3-base [8]	–	–	81.1 (85.2/79.4)	79.7 (82.8/78.4)
XLNet-large [24]	–	–	80.1 (–/–)	81.8 (85.5/80.2)
RoBERTa-large [14]	–	–	– (–/–)	83.2 (86.5/81.8)
DeBERTaV3-large [8]	–	–	88.3* (91.4*/87.0*)	87.5* (90.5*/86.8*)
CoL (DeBERTaV3-base)	–/–	–/–	82.9 (87.3/81.0)	81.6 (85.3/80.1)
CoLISA (DeBERTaV3-base)	–/–	–/–	83.2 (86.4/81.9)	81.6 (84.6/80.4)
CoL (DeBERTaV3-large)	60.1/52.6	62.1/54.3	88.6 (**91.6**/87.3)	**87.9** (**90.8**/86.9)
CoLISA (DeBERTaV3-large)	**61.7/53.6**	**62.3/54.7**	**88.8** (91.1/**87.8**)	87.8 (90.0/**87.0**)

more extensive range of source documents. We speculate that long-text pre-trained language models require more training data. Besides, the sequence length in QuALITY still exceeds the encoding limit of Longformer, which probably causes the lack of key information. Comparatively, the DPR & DeBERTaV3 architecture appears to be more predominant than Longformer due to their extraction strategy. Compared with the two baselines mentioned above, our proposed CoLISA learns more effective contextual representations and identifies the discrepancy between multiple options with contrastive learning (CoL) introduced. Meanwhile, adding an in-sample attention (ISA) mechanism after the DeBERTaV3 encoder consistently enhances the performance. This effort implies that inner interaction across multiple options in an ISA-style manner further captures the discrepant relationship across all candidates.

4.3   Analysis

Contrastive Learning. In order to assess how our CoL module influences the entire model performance, we complete an ablation study as illustrated in the first column in Table 3. At first, we discard the ISA module to only evaluate how contrastive learning works on our architecture. The experimental result validates our assumption that the CoL component alone significantly improves the performance compared to the baseline model, where the baseline model indicates the DPR & DeBERTaV3-large in the first three columns.

For the DPR-based retriever, CoL employs both the question and multiple options as queries to extract context from referential passages. In contrast, we find out merely using a question as the query exhibits a sharp performance degradation. The explanation is that our CoL method, which aims at distinguishing an accurate answer from other options, can better extract the question/option-related evidence.

Applying the standard dropout twice, we acquire two different representations. And each representation includes a positive instance, respectively, for contrastive learning. While collecting negative instances, previous works take the rest samples in the same batch as negative instances, usually named in-batch negatives [7]. We also take a crack at in-batch negatives. Following this method, the experimental result reveals that changing the way we construct negative instances from in-sample to in-batch leads to an inferior performance[5]. We conjecture that various samples have nothing to do with each other within a batch, which violates the target of pushing the gold answer and distractors away.

Table 3. Ablation study on contrastive learning module, distractive factor, and in-sample attention mechanism on the development set of QuALITY.

Model	Acc.	Model	Acc.	Model	Acc.	Model	Acc.
Baseline	56.7	Baseline	56.7	Baseline	56.7	Baseline	39.6
CoL	**60.1**	CoL	60.1	CoL	60.1	CoL	**40.8**
Question query	58.9	w/ DiF	**60.9**	w/ self-attention	60.4	In-batch negatives	37.1
In-batch negatives	59.4	KL Loss	57.0	Context masked	60.4		
				w/ transformer	**61.0**		
				w/ transformer*2	60.8		
				w/ modified transformer	58.9		

Distractive Factor. As mentioned before, we specifically design a distractive factor (DiF) for QuALITY to purposely emphasize confusing distractors. The second column in Table 3 displays our implementation results about DiF, where we multiply a DiF Θ with the similarity of corresponding instances to weight the contribution of confusing distractors. It can be observed that the DiF action significantly boosts the performance of our CoL module. A straightforward explanation is that the DiF forces confusing distractors to contribute more to the contrastive loss, making the model more inclined to learn how to recognize distractive options.

Out of a hypothesis that the likelihood of each candidate option being the definitive answer follows a specific probability distribution, we intuitively substitute KL divergence loss [10] for the cross-entropy loss. The fiercely degraded

---

[5] The performance on base models is actually far lower than listed, we transfer identical experiments from large models to base models, which display worse results. Results are listed in the last column in Table 3. The performance drops fiercely from 40.8 to 37.1. Our baseline here is Roberta-base. We have to deploy a small batch size on large models due to device limitations. Hence, both manners of in-batch and in-sample on large models do not show such an enormous dissimilarity.

result validates that KL divergence appears to function similarly to the DiF. Meanwhile, the cross-entropy loss is indispensable for identifying the authenticity of each triplet concatenated with context, question, and option.

In-Sample Attention. We exploit variants of ISA mechanisms to verify their effectiveness, as illustrated in the third column in Table 3. We observe that, as expected, when our ISA mechanism is self-attention, the performance rises due to interaction across multiple triplets. Then, we further attempt to mask context tokens, which means only the question and option attend to each other without context information. The performance remains the same, indicating the interaction across the question and multiple options is truly the crucial operation.

The ablation study further demonstrates that the transformer architecture is superior to a single self-attention layer. It is because the feed-forward network inside the transformer maintains considerable parameters to assure the propagation of attention output. An intriguing phenomenon is that assembling an additional transformer layer leads to a tiny decline in model performance. We believe that taking over the checkpoint successively fine-tuned on RACE and QuALITY expands the amount of randomly initialized parameters, which burdens the training process.

In addition, we modify the inner structure of the fundamental encoder by transplanting attention mechanisms inside the encoder. The whole pre-trained encoder consists of n layers[6], where each layer shares completely the identical structure: a multi-head self-attention sub-layer and a feed-forward sub-layer. Consequently, we add an extra attention layer between such two sub-layers for multi-choice interaction. Note that lower layers principally present shallow semantics, while higher layers are for deep semantics, we only replenish attention mechanisms at top-4 layers in practice. Such modified transformer intuitively models both inner-sequence relationship in lower layers and multi-sequence interaction in higher layers. Experimental results indicate that modifying the encoder structure is not so unexceptionable as the best method due to incomplete exploitation from pre-trained outcomes.

5 Related Work

Multi-choice Reading Comprehension. The multi-choice reading comprehension (MC-RC) task aims at selecting the most appropriate answer from given multiple options. Research in this domain primarily concerns tasks that involve complex passages or distractive options, including datasets like QuALITY [15], RACE [12], and DREAM [20]. QuALITY is an MC-RC dataset with source passages whose average lengths are around 5,000 tokens. Each question is written by professional authors and requires consolidating reasoning or sophisticated comparison over multiple options. RACE covers English entrance exams for Chinese

---

[6] The parameter n is 12 for base models or 24 for large models.

middle and high school students and demands miscellaneous reasoning strategies. DREAM is a dialogue-based dataset that considers unspoken world knowledge and information across multi-turn involves multi-party. To solve MC-RC problems, recent works usually focus on the interaction between the passage and an option for a given question. DUMA [28] captures relationships among them with a pattern of dual attention for the whole combined sequence. RekNet [27] extracts the critical information from passages and quotes relevant explicit knowledge from such information. Besides, an alternative solution for multi-choice question answering is to retrieve knowledge from external KBs and select valid features [13].

Long-Text Comprehension. Long text is infeasible to be processed since the sequence length of text usually exceeds the constraint of pre-trained language models (PLM) [5]. Two practical approaches are proposed to address this limitation: extractive strategies and sparse attention for PLMs. DPR [11], a dense passage retrieval trained for open-domain QA [22], outperforms other sparse retrievers for extracting critical spans from the long text. Longformer [1] and other sparse-attention PLMs [25] can process longer sequences by applying the combination of local sliding window attention and global attention mechanism. In general, a two-stage framework [6] is usually adopted to solve the long-text MRC task.

Contrastive Learning. Contrastive learning, which has been recently applied to various PLMs, is supposed to learn precise representations by pulling semantically similar instances together and pushing all instances apart as possible. For example, SimCSE [7] proposes two contrastive learning frameworks: an unsupervised approach of using standard dropout twice, and a supervised way by collecting positive and negative instances in a mini-batch on natural language inference (NLI) [3] tasks. Besides, [2] utilizes contrastive learning to explicitly identify supporting evidence spans from a long document by maximizing the similarity of the question and evidence. Moreover, xMoCo [23] proposes a contrastive learning method to learn a dual-encoder for query-passage matching. In contrast, our proposed model CoLISA is devoted to explicitly distinguishing the representation among multiple options corresponding to the same question.

Self-attention. The attention mechanism is proposed originally for interaction between two input sequences. BiDAF [18] utilizes a bi-directional attention flow to couple query and context vectors and produce a set of query-aware representations for each word within a sequence. Transformer [21] employs a self-attention mechanism to represent relations of specific tokens in a single sequence, which has been proven to be extraordinarily effective in all kinds of tasks.

6   Conclusion

This paper concentrates on dealing with confusing distractors in multi-choice reading comprehension by a contrastive learning method and an in-sample attention mechanism. The proposed model, CoLISA, achieves the interaction across multiple options to solve hard cases that urgently demand a rigorous comparison between the correct answer and distractive options. The experiments demonstrate a consistent improvement of CoLISA over two different benchmarks. For future work, we would like to explore more variants of contrastive learning methods.

Acknowledgements. The research is supported by National Key R&D Program of China (2020YFB1313601) and National Science Foundation of China (62076174, 62076175).

References

1. Beltagy, I., Peters, M.E., Cohan, A.: Longformer: the long-document transformer. arXiv preprint arXiv:2004.05150 (2020)
2. Caciularu, A., Dagan, I., Goldberger, J., Cohan, A.: Utilizing evidence spans via sequence-level contrastive learning for long-context question answering. arXiv preprint arXiv:2112.08777 (2021)
3. Conneau, A., Kiela, D., Schwenk, H., Barrault, L., Bordes, A.: Supervised learning of universal sentence representations from natural language inference data. arXiv preprint arXiv:1705.02364 (2017)
4. Daniel, K.: Thinking, fast and slow (2017)
5. Devlin, J., Chang, M.W., Lee, K., Toutanova, K.: BERT: pre-training of deep bidirectional transformers for language understanding. arXiv preprint arXiv:1810.04805 (2018)
6. Dong, M., Zou, B., Qian, J., Huang, R., Hong, Y.: ThinkTwice: a two-stage method for long-text machine reading comprehension. In: Wang, L., Feng, Y., Hong, Y., He, R. (eds.) Natural Language Processing and Chinese Computing, pp. 427–438. Springer International Publishing, Cham (2021). https://doi.org/10.1007/978-3-030-88480-2_34
7. Gao, T., Yao, X., Chen, D.: Simcse: simple contrastive learning of sentence embeddings. arXiv preprint arXiv:2104.08821 (2021)
8. He, P., Gao, J., Chen, W.: DeBERTaV 3: improving deberta using electra-style pre-training with gradient-disentangled embedding sharing. arXiv preprint arXiv:2111.09543 (2021)
9. Hendrycks, D., Gimpel, K.: Gaussian error linear units (GELUs). arXiv preprint arXiv:1606.08415 (2016)
10. Hershey, J.R., Olsen, P.A.: Approximating the kullback leibler divergence between gaussian mixture models. In: 2007 IEEE International Conference on Acoustics, Speech and Signal Processing-ICASSP'07, vol. 4, pp. IV-317. IEEE (2007)
11. Karpukhin, V., et al.: Dense passage retrieval for open-domain question answering. arXiv preprint arXiv:2004.04906 (2020)
12. Lai, G., Xie, Q., Liu, H., Yang, Y., Hovy, E.: Race: Large-scale reading comprehension dataset from examinations. arXiv preprint arXiv:1704.04683 (2017)

13. Li, Y., Zou, B., Li, Z., Aw, A.T., Hong, Y., Zhu, Q.: Winnowing knowledge for multi-choice question answering. In: Findings of the Association for Computational Linguistics: EMNLP 2021, pp. 1157–1165. Association for Computational Linguistics (2021)
14. Liu, Y., et al.: RoBERTa: a robustly optimized BERT pretraining approach. ArXiv abs/1907.11692 (2019)
15. Pang, R.Y., et al.: Quality: question answering with long input texts, yes! arXiv preprint arXiv:2112.08608 (2021)
16. Ran, Q., Li, P., Hu, W., Zhou, J.: Option comparison network for multiple-choice reading comprehension. arXiv preprint arXiv:1903.03033 (2019)
17. Richardson, M., Burges, C.J., Renshaw, E.: MCTest: a challenge dataset for the open-domain machine comprehension of text. In: Proceedings of the 2013 Conference on Empirical Methods in Natural Language Processing, pp. 193–203 (2013)
18. Seo, M., Kembhavi, A., Farhadi, A., Hajishirzi, H.: Bidirectional attention flow for machine comprehension. arXiv preprint arXiv:1611.01603 (2016)
19. Sukhbaatar, S., Grave, E., Lample, G., Jegou, H., Joulin, A.: Augmenting self-attention with persistent memory. arXiv preprint arXiv:1907.01470 (2019)
20. Sun, K., Yu, D., Chen, J., Yu, D., Choi, Y., Cardie, C.: DREAM: a challenge data set and models for dialogue-based reading comprehension. Trans. Assoc. Comput. Linguist. **7**, 217–231 (2019)
21. Vaswani, A., et al.: Attention is all you need. In: Advances in Neural Information Processing Systems, vol. 30 (2017)
22. Voorhees, E.M., et al.: The TREC-8 question answering track report. In: TREC, vol. 99, pp. 77–82 (1999)
23. Yang, N., Wei, F., Jiao, B., Jiang, D., Yang, L.: xMoCo: cross momentum contrastive learning for open-domain question answering. In: Proceedings of the 59th Annual Meeting of the Association for Computational Linguistics and the 11th International Joint Conference on Natural Language Processing (Volume 1: Long Papers), pp. 6120–6129 (2021)
24. Yang, Z., Dai, Z., Yang, Y., Carbonell, J.G., Salakhutdinov, R., Le, Q.V.: XLNet: generalized autoregressive pretraining for language understanding. In: Neural Information Processing Systems (2019)
25. Zaheer, M., et al.: Big Bird: transformers for longer sequences. Adv. Neural. Inf. Process. Syst. **33**, 17283–17297 (2020)
26. Zhang, S., Zhao, H., Wu, Y., Zhang, Z., Zhou, X., Zhou, X.: DCMN+: dual co-matching network for multi-choice reading comprehension. In: Proceedings of the AAAI Conference on Artificial Intelligence, vol. 34, pp. 9563–9570 (2020)
27. Zhao, Y., Zhang, Z., Zhao, H.: Reference knowledgeable network for machine reading comprehension. IEEE/ACM Trans. Audio Speech Lang. Process. **30**, 1461–1473 (2022)
28. Zhu, P., Zhang, Z., Zhao, H., Li, X.: DUMA: reading comprehension with transposition thinking. IEEE/ACM Trans. Audio Speech Lang. Process. **30**, 269–279 (2021)

Viewpoint Diversity in Search Results

Tim Draws[1][✉][iD], Nirmal Roy[1][iD], Oana Inel[2][iD], Alisa Rieger[1][iD],
Rishav Hada[3][iD], Mehmet Orcun Yalcin[4][iD], Benjamin Timmermans[5][iD],
and Nava Tintarev[6][iD]

[1] Delft University of Technology, Delft, Netherlands
{t.a.draws,n.roy,a.rieger}@tudelft.nl
[2] University of Zurich, Zurich, Switzerland
inel@ifi.uzh.ch
[3] Microsoft Research, Bangalore, India
[4] Independent Researcher, Istanbul, Türkiye
[5] IBM, Amsterdam, Netherlands
b.timmermans@nl.ibm.com
[6] Maastricht University, Maastricht, Netherlands
n.tintarev@maastrichtuniversity.nl

Abstract. Adverse phenomena such as the *search engine manipulation effect* (SEME), where web search users change their attitude on a topic following whatever most highly-ranked search results promote, represent crucial challenges for research and industry. However, the current lack of automatic methods to comprehensively measure or increase viewpoint diversity in search results complicates the understanding and mitigation of such effects. This paper proposes a viewpoint bias metric that evaluates the divergence from a pre-defined scenario of ideal viewpoint diversity considering two essential viewpoint dimensions (i.e., *stance* and *logic of evaluation*). In a case study, we apply this metric to actual search results and find considerable viewpoint bias in search results across queries, topics, and search engines that could lead to adverse effects such as SEME. We subsequently demonstrate that viewpoint diversity in search results can be dramatically increased using existing diversification algorithms. The methods proposed in this paper can assist researchers and practitioners in evaluating and improving viewpoint diversity in search results.

Keywords: Viewpoint diversity · Metric · Evaluation · Bias · Search results

1 Introduction

Web search is increasingly used to inform important personal decisions [16,31,45] and users commonly believe that web search results are accurate, trustworthy, and unbiased [53]. However, especially for search results related to debated topics, this perception may often be false [30,54,65,66]. Recent research has demonstrated that a lack of viewpoint diversity in search results can lead to undesired outcomes such as the *search engine manipulation effect* (SEME), which occurs when users change their attitude on a topic following whichever viewpoint

J. Kamps et al. (Eds.): ECIR 2023, LNCS 13980, pp. 279–297, 2023.
https://doi.org/10.1007/978-3-031-28244-7_18

happens to be predominant in highly-ranked search results [5,6,24,27,52]. For instance, SEME can lead users to judge medical treatments as (in-)effective [52] or prefer a particular political candidate over another [27]. To mitigate potential large-scale negative consequences of SEME for individuals, businesses, and society, it is essential to evaluate and foster viewpoint diversity in search results.

Measuring and increasing the diversity of search results has been studied extensively in recent years, e.g., to satisfy pluralities of search intents [2,19,58] or ensure fairness towards protected classes [9,70,72,73]. First attempts in specifically evaluating [23,43] and fostering [47,61] *viewpoint diversity* in ranked outputs have also been made. However, two essential aspects have not been sufficiently addressed yet: (1) current methods only allow for limited viewpoint representations (i.e., one-dimensional, often binary) and (2) there is no clear conceptualization of viewpoint diversity or what constitutes viewpoint bias in search results. Current methods often assume that any top k portion of a ranked list should represent all available (viewpoint) categories proportionally to their overall distribution, i.e., analogous to the notion of *statistical parity* [23,42], without considering other notions of diversity [64]. This impedes efforts to meaningfully assess viewpoint bias in search results or measure improvements made by diversification algorithms. We thus focus on three research questions:

RQ1. What metric can thoroughly measure viewpoint diversity in search results?
RQ2. What is the degree of viewpoint diversity in actual search results?
RQ3. What method can foster viewpoint diversity in search results?

We address **RQ1** by proposing a metric that evaluates viewpoint bias (i.e., deviation from viewpoint diversity) in ranked lists using a two-dimensional viewpoint representation developed for human information interaction (Sect. 3). We show that this metric assesses viewpoint diversity in a more comprehensive fashion than current methods and apply it in a case study of search results from two popular search engines (**RQ2**; Sect. 4). We find notable differences in search result viewpoint diversity between queries, topics, and search engines and show that applying existing diversification methods can starkly increase viewpoint diversity (**RQ3**; Sect. 4.3). All code and data are available at https://osf.io/kz3je/.

2   Related Work

Viewpoint Representations. *Viewpoints*, sometimes called *arguments* [3,26] or *stances* [41], are positions or opinions concerning debated topics or claims [20]. To *represent* viewpoints in ranked lists of search results, each document needs to receive a label capturing the viewpoint(s) it expresses. Previous work has predominantly assigned binary (e.g., *con/pro*) or ternary (e.g., *against/neutral/in favor*) viewpoint labels [32,52,71]. However, these labels ignore the viewpoint's degree and reason behind opposing or supporting a given topic [20], e.g., two statements *in favor* of school uniforms could express entirely different viewpoints in strongly supporting school uniforms for productivity reasons and only somewhat supporting them for popularity reasons. To overcome these limitations, earlier work has represented viewpoints on ordinal scales [23,24,43,57], continuous scales [43], as multi-categorical perspectives [3,18,21], or computed the viewpoint *distance* between documents [47]. A recently proposed, more comprehensive viewpoint label [20], based on work in the communication sciences [7,8,11], consists of two dimensions: *stance* (i.e., an ordinal scale ranging from "strongly opposing" to "strongly supporting") and *logics of evaluation* (i.e., underlying reasons –sometimes called *perspectives* [18,21], *premises* [13,26] or *frames* [3,47]).

Viewpoint Diversity in Ranked Outputs. Previous research has shown that search results across topics and domains (e.g., politics [54], health [65,66]) may not always be viewpoint-diverse and that highly-ranked search results are often unbalanced concerning query subtopics [30,50]. Limited diversity, or bias, can root in the overall search result index but become amplified by biased queries and rankings [30,56,66]. Extensive research further shows that viewpoint-biased (i.e., *unbalanced*) search results can lead to undesired consequences for individuals, businesses, and society (e.g., SEME) [5,10,24,27,52,67]. That is why many studies now focus on understanding and mitigating cognitive user biases in this context [6,24,28,33,44,51,57,68,69,71]. However, because adverse effects in web search are typically an interplay of content and user biases [67], it is essential to also develop methods to evaluate and foster viewpoint diversity in search results.

Building on work that measured diversity or fairness in search results concerning more general subtopics [2,9,19,70,72,73], recent research has begun to evaluate *viewpoint diversity* in ranked outputs. Various metrics have been adapted from existing information retrieval (IR) practices to quantitatively evaluate democratic notions of diversity [37,63,64], though only few [63] crucially incorporate users' attention drop over the ranks [6,27,39,49]. *Ranking fairness metrics* such as *normalized discounted difference* (rND) [70] can assess viewpoint diversity by measuring the degree to which documents of a pre-defined protected viewpoint category are ranked lower than others [23]. The recently proposed *ranking bias* (RB) metric considers the full range of a continuous viewpoint dimension and evaluates viewpoint balance [43]. Existing metrics such as rND and RB, however, have a key limitation when measuring viewpoint diversity: they cannot accommodate comprehensive, multi-dimensional viewpoint representations. Incorporating such more comprehensive viewpoint labels is crucial because stances and the reasons behind them can otherwise not be considered simultaneously [20].

Search Result Diversification. To improve viewpoint diversity, we build on earlier work on diversifying search results concerning *user intents* [1,2,25,40, 59]. *xQuAD* [59] and *HxQuAD* [38] are two such models that re-rank search results with the aim of fulfilling diverse ranges of information needs at high ranks. Whereas xQuAD diversifies for single dimensions of (multi-categorical) subtopics, HxQuAD adapts xQuAD to accommodate multiple dimensions of subtopics and diversifies in a multi-level hierarchical fashion. For example, for the query *java*, two first-level subtopics may be *java island* and *java programming*.

For the former, queries such as *java island restaurant* and *java island beach* may then be second-level subtopics. To the best of our knowledge, such methods have so far not been used to foster viewpoint diversity in ranked lists.

3   Evaluating Viewpoint Diversity in Search Results

This section introduces a novel metric for assessing viewpoint diversity in ranked lists such as search results. To comprehensively capture documents' viewpoints, we adopt the two-dimensional viewpoint representation recently introduced by Draws et al. [20] (see Sect. 2). Each document thus receives a single *stance* label on a seven-point ordinal scale from strongly opposing (-3) to strongly supporting (3) a topic and anywhere from no to seven *logic of evaluation* labels that reflect the underlying reason(s) behind the stance (i.e., *inspired, popular, moral, civic, economic, functional, ecological*). Although other viewpoint diversity representations could be modeled, this 2D representation supports more nuanced viewpoint diversity analyses than current approaches, and it is still computationally tractable (i.e., only seven topic-independent categories per dimension).

We consider a set of documents retrieved in response to a query (e.g., "school uniforms well-being") related to a particular debated topic (e.g., mandatory school uniforms). R is a ranked list of N retrieved documents (i.e., by the search engine), $R_{1...k}$ is the top-k portion of R, and R_k refers to the kth-ranked document. We refer to the sets of stance and logic labels of the documents in R as S and $\mathcal{L}$, respectively, and use $\mathcal{S}_k$ or $\mathcal{L}_k$ to refer to the labels of the particular document at rank k. For instance, a document at rank k may receive the label $[\mathcal{S}_k = 2; \mathcal{L}_k = (popular, functional)]$ if the article **supports** (stance) school uniforms because they supposedly are popular among students (i.e., **popular** logic) and lead to better grades (i.e., **functional** logic). S and L, respectively, are the (multinomial) stance and logic distributions of the documents in R.

Defining Viewpoint Diversity. Undesired effects such as SEME typically occur when search result lists are one-sided and unbalanced in terms of viewpoints [6,27,52]. To overcome this, we follow the normative values of a *deliberative democracy* [37], and counteract these problems through viewpoint plurality and balance. We put these notions into practice by following three intuitions:

1. *Neutrality.* A set of documents should feature both sides of a debate equally and not take any particular side when aggregated. We consider a search result list as neutral if averaging its stance labels results in 0 (a neutral stance score).
2. *Stance Diversity.* A set of documents should have a balanced stance distribution so that different stance strengths (e.g., 1, 2, and 3) are covered. For example, we consider a search result list as stance-diverse if it contains equal proportions of all seven different stance categories, but not if it contains only the stance categories -3 and 3 (albeit satisfying *neutrality* here).
3. *Logic Diversity.* A set of documents should include a plurality of reasons for different stances (i.e., balanced logic distribution *within* each stance category). For example, a search result list may not satisfy *logic diversity* if documents containing few reasons (here, logics) are over-represented.

Our metric *normalized discounted viewpoint bias* (nDVB) measures the degree to which a ranked list *diverges* from a pre-defined scenario of ideal viewpoint diversity. It combines the three sub-metrics *normalized discounted polarity bias* (nDPN), *normalized discounted stance bias* (nDSB), and *normalized discounted logic bias* (nDLB), which respectively assess the three characteristics of a viewpoint-diverse search result list (i.e., *neutrality*, *stance diversity*, and *logic diversity*).

3.1   Measuring Polarity, Stance, and Logic Bias

We propose three sub-metrics that contribute to nDVB by considering different document aspects. They all ignore irrelevant during their computation and – like other IR evaluation metrics [55] – apply a discount factor for rank-awareness.

Normalized Discounted Polarity Bias (nDPB). Polarity bias considers the mean stance label balance. *Neutrality*, the first trait in our viewpoint diversity notion, posits that the stance labels for documents in any top k portion should balance each other out (mean stance = 0). We assess how much a top k search result list *diverges* from this ideal scenario (i.e., *polarity bias*; PB; see Eq. 1) by computing the average normalized stance label. Here, $S_{1...k}$ is the set of stance labels for all documents in the top k portion of the ranking. PB normalizes all stance labels S_i in the top k to a score between -1 and 1 (by dividing it by its absolute maximum, i.e., 3) and takes their average. To evaluate the neutrality of an entire search result list τ with N documents, we compute PB iteratively for the top $1, 2, \ldots, N$ ranking portions, aggregate the results in a discounted fashion, and apply min-max normalization to produce nDPB (see Eq. 2). Here, Z is a normalizer equal to the highest possible value for the aggregated and discounted absolute PB values and I is an indicator variable equal to -1 if $\sum_{k=1}^{N} \frac{PB(S,k)}{log_2(k+1)} < 0$ and 1 otherwise. nDPB quantifies a search result list's bias toward opposing or supporting a topic and ranges from -1 to 1 (more extreme values indicate greater bias, values closer to 0 indicate neutrality).

$$\text{PB}(\mathcal{S}, k) = \frac{\sum_{i=1}^{k} \frac{S_i}{3}}{|\mathcal{S}_{1...k}|} \quad (1) \qquad \text{nDPB}(\tau) = \frac{1}{Z} I \sum_{k=1}^{N} \frac{|\text{PB}(\mathcal{S}, k)|}{\log_2(k+1)} \quad (2)$$

Normalized Discounted Stance Bias (nDSB). Stance bias evaluates how much the stance distribution diverges from the viewpoint-diverse scenario. *Stance diversity*, the second trait of our viewpoint diversity notion, suggests that all stance categories are equally covered in any top k ranked list portion. We capture this ideal scenario of a balanced stance distribution in the uniform target distribution $T = \left(\frac{1}{7}, \frac{1}{7}, \frac{1}{7}, \frac{1}{7}, \frac{1}{7}, \frac{1}{7}, \frac{1}{7} \right)$. The stance distribution of the top k-ranked documents is given by $S_{1...k} = \left(\frac{|S_{1...k}^{-3}|}{k}, \ldots, \frac{|S_{1...k}^{3}|}{k} \right)$, where each numerator refers to the number of top-k search results in a stance category. We assess how much $S_{1...k}$ diverges from T by computing their *Jensen-Shannon divergence* (JSD), a symmetric distance metric for discrete probability distributions [29]. This approach is inspired by work suggesting divergence metrics to measure viewpoint diversity [23,63,64]. We then normalize JSD between $S_{1...k}$ and T by dividing it by the maximal divergence, i.e., $JSD(U\|T)$ where $U = (1,0,0,0,0,0,0)$ and call the result *stance bias* (SB; see Eq. 3). SB ranges from 0 (desired scenario of stance diversity) to 1 (maximal stance bias). Notably, SB will deliberately *always* return high values for the very top portions (e.g., top one or two) of any search result list, as it is impossible to get a balanced distribution of the seven stance categories in just a few documents. We evaluate an entire search result list using nDSB (see Eq. 4), by computing SB iteratively for the top $1, 2, \ldots, N$ ranking portions, aggregating the results in a discounted fashion, and normalizing.

$$\text{SB}(S, k) = \frac{JSD(S_{1\ldots k}\|T)}{JSD(U\|T)} \quad (3) \quad \text{nDSB}(\tau) = \frac{1}{Z} \sum_{k=1}^{N} \frac{\text{SB}(S, k)}{\log_2(k + 1)} \quad (4)$$

Normalized Discounted Logic Bias (nDLB). Logic bias measures how balanced documents in each stance category are in terms of logics. *Logic diversity* suggests that all logics are equally covered in each document group when splitting documents by stance category. Thus, when a search result list contains documents, e.g., with stances -1, 0, and 1, the logic distributions of each of those three groups should be balanced. The logic distribution of all top k results belonging to a particular stance category s is given by $L_{1\ldots k}^s = \left(\frac{|\mathcal{L}_{1\ldots k}^{s,l_1}|}{|\mathcal{L}_{1\ldots k}^s|}, \ldots, \frac{|\mathcal{L}_{1\ldots k}^{s,l_7}|}{|\mathcal{L}_{1\ldots k}^s|}\right)$, where each numerator $|\mathcal{L}_{1\ldots k}^{s,l}|$ refers to the number of times logic l (e.g., *inspired*) appears in the top k documents with stance category s. Each denominator $|\mathcal{L}_{1\ldots k}^s|$ is the total number of logics that appear in the top k documents with stance category s. $L_{1\ldots k}^s$ reflects the relative frequency of each logic in the top k documents in a specific stance category. Similar to SB, we evaluate the degree to which $L_{1\ldots k}^s$ diverges from T by computing the normalized JSD for the logic distributions of each available stance category and then produce *logic bias* (LB) by averaging the results (Eq. 5). Here, $\mathcal{S}_k^*$ is the set of unique stance categories among the top k-ranked documents. LB thus quantifies, on a scale from 0 to 1, the average degree to which the logic distributions diverge from the ideal, viewpoint-diverse scenario where all logics are equally present within each stance category. We produce nDLB by computing LB iteratively for the top $1, 2, \ldots, N$ documents and applying our discounted aggregation and normalization procedures (Eq. 6).

$$\text{LB}(\mathcal{S}, L, k) = \frac{1}{|\mathcal{S}_k^*|} \sum_{s \in \mathcal{S}_k^*} \frac{JSD(L_{1\ldots k}^s\|T)}{JSD(U\|T)} \quad (5) \quad \text{nDLB}(\tau) = \frac{1}{Z} \sum_{k=1}^{N} \frac{\text{LB}(\mathcal{S}, L, k)}{\log_2(k + 1)} \quad (6)$$

3.2   Normalized Discounted Viewpoint Bias

To evaluate overall viewpoint diversity, we combine nDPB, nDSB, and nDLB into a single metric, called *normalized discounted viewpoint bias* (nDVB):

$$\text{nDVB}(\tau) = I \frac{\alpha|\text{nDPB}(\tau)| + \beta\text{nDSB}(\tau) + \gamma\text{nDLB}(\tau)}{\alpha + \beta + \gamma}.$$

Here, I is an indicator variable that equals -1 when $\text{nDPB}(\tau) < 0$ and 1 otherwise. The parameters α, β, and γ are weights that control the relative importance of the three sub-metrics. Thus, nDVB measures the degree to which a ranked list of documents diverges from an ideal, viewpoint-diverse scenario. It ranges from -1 to 1, indicating the direction and severity with which such a ranked list (e.g., search results) is biased (values closer to 0 imply greater viewpoint diversity).

Our proposed metric nDVB allows for a more comprehensive assessment of viewpoint diversity in search results compared to metrics such as rND or RB. It does so by allowing for comprehensive viewpoint representations of search results, simultaneously considering *neutrality*, *stance diversity*, and *logic diversity*.

4 Case Study: Evaluating, Fostering Viewpoint Diversity

This section presents a case study in which we show how to practically apply the viewpoint bias metric we propose (nDVB; see Sect. 3.2) and examine the viewpoint diversity of real search results from commonly used search engines, using relevant queries for currently debated topics (i.e., *atheism, school uniforms*, and *intellectual property*). Finally, we demonstrate how viewpoint diversity in search results can be enhanced using existing diversification algorithms. More details on the materials and results (incl. figures) are available in our repository.

4.1 Materials

Topics. We aimed to include in our case study three topics that (1) are not scientifically answerable (i.e., with legitimate arguments in both the opposing and supporting directions) and (2) cover a broad range of search outcomes (i.e., consequences for the individual user, a business, or society). To find such topics, we considered the *IBM-ArgQ-Rank-30kArgs* data set [35], which contains arguments on controversial issues. The three topics we (manually) selected from this data set were *atheism* (where attitude change may primarily affect the user themselves, e.g., they become an atheist), *intellectual property rights* (where attitude change may affect a business, e.g., the user decides to capitalize on intellectual property they own), and *school uniforms* (where attitude change may affect society, e.g., the user votes to abolish school uniforms in their municipality).

Queries. We conducted a user study (approved by a research ethics committee) to find, per topic, five different queries that users might enter into a web search engine if they were wondering whether one should be an atheist (individual use case), intellectual property rights should exist (business use case), or students should have to wear school uniforms (societal use case). In a survey, we asked participants to imagine the three search scenarios and select, for each, three "neutral" and four "biased" queries from a pre-defined list. The neutral queries did not specify a particular debate side (e.g., `school uniforms opinions`), while the biased queries prompted opposing (e.g., `school uniforms disadvantages`) or supporting results (e.g., `school uniforms pros`).

We recruited 100 participants from *Prolific* (https://prolific.co) who completed our survey for a reward of $0.75 (i.e., $8.09 per hour). All participants were fluent English speakers older than 18. For our analysis, we excluded data from two participants who had failed at least one of two attention checks. The remaining 98 participants were gender-balanced (49% female, 50% male, 1% non-binary) and rather young (50% were between 18 and 24). We selected five queries per topic: the three most commonly selected neutral queries and the single most commonly selected opposing- and supporting-biased queries (see Table 1).[1]

Table 1. Viewpoint diversity evaluation for all 30 search result lists from Engine 1 and 2: rND, RB, and nDVB (incl. its sub-metrics DPB, DSB, and DLB). Queries were designed to retrieve neutral (neu), opposing (opp), or supporting (sup) results (↔).

Query	↔	Engine 1						Engine 2					
		rND	RB	nDPB	nDSB	nDLB	nDVB	rND	RB	nDPB	nDSB	nDLB	nDVB
why people become atheists or theists	neu	.70	.27	.32	.33	.38	.34	.69	.14	.21	.36	.33	.30
should I be atheist or theist	neu	.68	.13	.24	.39	.44	.35	.80	.04	.05	.51	.40	.32
atheism vs theism	neu	.58	−.06	−.07	.52	.37	−.32	.77	.01	.03	.53	.39	.32
why theism is better than atheism	opp	.47	.19	.22	.28	.35	.29	.53	−.04	−.15	.45	.30	−.30
why atheism is better than theism	sup	.35	.05	.15	.23	.43	.27	.68	.10	.15	.45	.34	.31
why companies maintain or give away IPRs	neu	.77	.46	.49	.41	.45	.45	.97	.61	.60	.48	.51	.53
should we have IPRs or not	neu	.80	.34	.34	.35	.33	.34	.93	.47	.44	.42	.41	.43
IPRs vs open source	neu	.80	.10	.09	.45	.43	.32	.92	.18	.19	.57	.53	.43
why IPRs don't work	opp	.69	.30	.33	.42	.40	.38	.54	.18	.19	.40	.35	.31
should we respect IPRs	sup	.90	.48	.49	.41	.36	.42	.95	.60	.59	.50	.35	.48
why countries adopt or ban school unif.	neu	.59	−.01	.14	.37	.25	.26	.54	−.10	−.11	.37	.20	−.23
should students wear school unif. or not	neu	.62	−.10	−.10	.45	.20	−.25	.85	.14	.15	.42	.19	.26
school unif. well-being	neu	.55	.07	.09	.28	.25	.21	.54	.13	.23	.31	.35	.30
why school unif. don't work	opp	.30	−.22	−.31	.33	.18	−.27	.59	−.01	−.03	.37	.21	−.20
why school unif. work	sup	.89	.43	.49	.38	.27	.38	.92	.45	.03	.50	.39	.36
Overall mean absolute bias		.65	.21	.26	.37	.34	.32	.75	.21	.24	.44	.34	.34

Note. In contrast to the actual queries, we here abbreviate *intellectual property rights* (IPRs) and *uniforms* (unif.).

Search Results. We retrieved the top 50-ranked search results for each of the $3 \times 5 = 15$ queries listed in Table 1 from two of the most commonly used search engines, through web crawling or an API.[2] This resulted in a data set of $15 \times 2 \times 50 = 1500$ search results, 25 of which (mostly the last one or two results) were not successfully retrieved. The remaining 1475 (i.e., 973 unique) search results were recorded, including their query, URL, title, and snippet.

Viewpoint Annotations. To assign each search result the 2D (stance, logic) viewpoint label (see Sect. 3), we employed six experts, familiar with the three topics, the annotation task, and the viewpoint labels. This is more than the one to three annotators typically employed for IR annotation practices [34,62]. The viewpoint label consists of *stance* (i.e., position on the debated topic on an ordinal scale ranging from −3; strongly opposing; to 3; strongly supporting)

---

[1] Due to error, we used the 2nd most common supporting query for the *IPR* topic.

[2] The retrieval took place on December 12th, 2021 in the Netherlands.

and *logics of evaluation* (i.e., motivations behind the stance).[3] First, the experts discussed annotation guidelines and examples before individually annotating the same set of 30 search results (i.e., two results randomly chosen per query). Then, they discussed their disagreements, created an improved, more consistent set of annotation guidelines, and revised their annotations. Following discussions, their overall agreement increased to satisfactory levels for stance (Krippendorff's $\alpha = .90$) and the seven logics ($\alpha = \{.79, .66, .73, .86, .77, .36, .57\}$). Such agreement values represent common ground in the communication sciences, where, e.g., two trained annotators got $\alpha = \{.21, .58\}$ when annotating *morality* and *economical* frames in news [15]. Each expert finally annotated an equal and topic-balanced share of the remaining 943 unique search results.

4.2 Viewpoint Diversity Evaluation Results

We conducted viewpoint diversity analyses per topic, search engine, and query. Specifically, we examined the overall viewpoint distributions and then measured viewpoint bias in each of the ($15 \times 2 =$) 30 different top 50 search result lists retrieved from the two search engines, by computing the existing metrics rND and RB (see Sect. 2) and our proposed metric incl. its sub-metrics (see Sect. 3).

Overall Viewpoint Distributions. Among the 973 unique URLs in our search results data set, 306, 334, and 263 respectively related to the topics *atheism*, *intellectual property rights* (IPRs), and *school uniforms*. A total of 70 unique search results were judged irrelevant to their topic and excluded from the analysis. Search Engine 1 (SE_1) provided a somewhat greater proportion of unique results for the 15 queries (77%) than Search Engine 2 (SE_2, 69%). For all three topics, supporting stances were more common. Regarding logics, the *school uniforms* topic was overall considerably more balanced than the others. Atheism-related documents often focused on *inspired, moral,* and *functional* logics (e.g., religious people have higher moral standards, atheism explains the world better). Documents related to IPRs often referred to *civic, economic,* and *functional* logics (e.g., IPRs are an important legal concept, IPRs harm the economy).

Viewpoint Diversity per Query, Topic, and Search Engine. We analyzed the viewpoint diversity of search results using the existing metrics rND, RB, and our proposed (combined) metric nDVB. We slightly adapted rND and RB to make their outcomes better comparable; aggregating both in steps of one and measuring viewpoint *imbalance* (or bias) rather than ranking fairness. Our rND implementation considered all documents with negative stance labels as *protected*, all documents with positive stance labels as *non-protected*, and ignored neutral documents. Computing RB required standardizing all stance labels to scores ranging from -1 to 1. To compute nDVB, we set the parameters to $\alpha = \beta = \gamma = 1$, i.e., giving all sub-metrics equal weights. Table 1 shows the evaluation

---

[3] Note that viewpoint labels do not refer to specific web search queries, but always to the topic (or claim) at hand. For example, a search result supporting the idea that students should have to wear school uniforms always receives a positive stance label (i.e., 1, 2, or 3), no matter what query was used to retrieve it.

results for all metrics across the 30 different search result lists from the two search engines. Scores closer to 0 suggest greater diversity (i.e., less distance to the ideal scenario), whereas scores further away from 0 suggest greater bias.

Neutrality. As we note in Sect. 3, viewpoint-diverse search result lists should feature both sides of debates equally. While rND does not indicate whether a search result list is biased against or in favor of the protected group [23], the RB and nDPB outcomes suggest that most of the search result lists we analyzed are biased towards *supporting* viewpoints. We observed that results on IPRs tended to be more biased than results on the other topics but, interestingly, we did not observe clear differences between query types. Moreover, except for the *school uniforms* topic, supposedly neutral queries generally returned results that were just as biased as queries targeted specifically at opposing or supporting results.

Stance Diversity. Another trait of viewpoint-diverse search result lists is a balanced stance distribution. Since rND, RB, and nDPB cannot clarify whether all stances (i.e., all categories ranging from -3 to 3) are uniformly represented, we here only inspect the nDSB outcomes. While we did not observe a noteworthy difference between topics or queries, we found that SE_2 returned somewhat more biased results than SE_1. Closer examination of queries where the two engines differed most in terms of nDSB (e.g., *why theism is better than atheism*) revealed that SE_2 was biased in the sense that it often returned fewer opinionated (and more neutral) results than SE_1. Regarding their balance between mildly and extremely opinionated results, both engines behaved similarly.

Logic Diversity. The final characteristic of viewpoint-diverse search result lists concerns their distribution of logics, i.e., the diversity of reasons brought forward to oppose or support topics. When inspecting the nDLB outcomes, we found that logic distributions in the search result lists were overall more balanced than stance distributions (see nDSB results) and similar across search engines and queries. However, we did observe that nDLB on the *school uniforms* topic tended to be lower than for other topics, suggesting that greater diversities of reasons opposing or supporting school uniforms were brought forward.

Overall Viewpoint Diversity. To evaluate overall viewpoint diversity in the search result lists, we examined nDVB, the only metric that simultaneously evaluates divergence from neutrality, stance diversity, and logic diversity. Bias *magnitude* per nDVB ranged from .20 to .53 across results from search engines, with only four out of 30 search result lists being biased against the topic. Regarding topics, search results for neutral queries were somewhat less biased on *school uniforms* compared to *atheism* or *intellectual property rights*.

Interestingly, search results for neutral queries on all topics were often just as viewpoint-biased as those from directed queries. Some queries returned search results with different bias magnitudes (e.g., *school uniforms well-being*) or bias directions (e.g., *atheism vs theism*) depending on the search engine. Moreover, whereas search results for supporting-biased queries were indeed always biased in the supporting direction (i.e., positive nDVB score), results for opposing-biased queries were often also biased towards supporting viewpoints. Figure 1 shows, per topic and search engine, how the absolute nDVB developed on average when evaluated at each rank. It illustrates that nDVB tended to decrease over the ranks across engines, topics, and queries but highlights that the top, say 10, search results that users typically examine are often much more viewpoint-biased than even the top 30 (i.e., more search results could offer more viewpoints).

4.3 Viewpoint Diversification

We implemented four diversification algorithms to foster viewpoint diversity in search results by (1) re-ranking and (2) creating viewpoint-diverse top 50 search result lists using all unique results from each topic. Specifically, we performed *ternary stance diversification, seven-point stance diversification, logic diversification* (all based on xQuAD; i.e., diversifying search results according to stance labels in the common ternary format, the seven-point ordinal format, or logic labels, respectively), and *hierarchical viewpoint diversification* (based on HxQuAD; i.e., diversifying search results hierarchically: first for seven-point ordinal stance labels and then, within each stance category, for logic labels; giving both dimensions equal weights). We evaluated the resulting search result lists using nDVB.

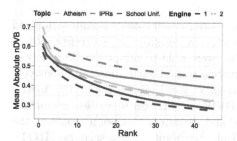 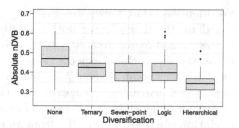

Fig. 1. Development of mean absolute nDVB@*k* across search result ranks, split by topic and search engine.

Fig. 2. Mean absolute viewpoint diversity (nDVB@10) per diversification algorithm across the 30 search result lists.

Re-ranked Top 50 Search Result Lists. Fig. 2 compares absolute nDVB between the original top 50 search result lists and the four diversification strategies. All strategies improved the viewpoint diversity of our lists. Whereas the ternary stance diversification only showed marginal improvements (mean abs. nDVB@10 = .42, nDVB@50 = .35) compared to the original search result lists (mean abs. nDVB@10 = .47, nDVB@50 = .33), the hierarchical viewpoint diversification based on stances and logics was the most effective in fostering viewpoint diversity (mean abs. nDVB@10 = .35, nDVB@50 = .27) . Viewpoint diversity for the seven-point stance diversification (mean abs. nDVB@10 = .39, nDVB@50 = .29) and logic diversification (mean abs. nDVB@10 = .42, nDVB@50 = .31) were comparable, and in between the ternary stance and hierarchical diversification.

"Best-case" Comparison. Despite the promising re-ranking results, diversification methods can only work with the specific sets of documents they are given. To show a "best-case" scenario for comparison, we employed our diversification algorithms to create, per topic, one maximally viewpoint-diverse search result list using all topic-relevant search results (i.e., from across queries and search engines). We found that all four diversification algorithms yielded search result lists with much less bias when given more documents compared to when they only re-ranked top 50 search results lists. Here, the hierarchical diversification was again most effective (mean abs. nDVB@10 = .29, nDVB@50 = .20); improving by a magnitude of .07 on average over the re-ranked top 50 search result lists. Compared to the average search result list we had retrieved from the two search engines, the "best-case" hierarchical diversification improved viewpoint diversity by margins of .17 (nDVB@10) and .13 (nDVB@50), reflecting a mean improvement of 39%. The other diversification algorithms showed similar improvements, albeit not as impactful as the hierarchical method (i.e., mean abs. nDVB@10 was .37, .37, .34 and mean abs. nDVB@50 was .31, .24, .24 for the ternary stance, seven-point stance, and logic diversifications, respectively).

5 Discussion

We identified that viewpoint diversity in search results can be conceptualized based on the deliberative notion of diversity by looking at *neutrality*, *stance diversity*, and *logics diversity*. Although we were able to adapt existing metrics to partly assess these aspects, a novel metric was needed to comprehensively measure viewpoint diversity in search results. We thus proposed the metric *normalized discounted viewpoint bias* (nDVB), which considers two important viewpoint dimensions (*stances* and *logics*) and measures viewpoint bias, i.e., the deviation of a search result list from an ideal, viewpoint-diverse scenario (**RQ1**). Findings from our case study suggest that nDVB is sensitive to expected data properties, such as aligning with the query polarity and bias decreasing for larger lists of search results. Although further refinement and investigation of the metric are required (e.g., to find the most practical and suitable balance between the three notions of diversity or outline interpretation guidelines), our results indicate that the metric is a good foundation for measuring viewpoint diversity.

The degree of viewpoint diversity across search engines in our case study was comparable: neither engine was consistently more biased than the other (**RQ2**). However, we found notable differences in bias magnitude and even bias direction between search engines *regarding the same query* and queries related to the same topic. This lends credibility to the idea that nDVB indeed measures viewpoint diversity, and is able to detect different kinds of biases. Further work is required to compare different metrics and types of biases. Similar to previous research [65], we found that search results were mostly biased in the *supporting* direction. This suggests that actual search results on debated topics may often not reflect a satisfactory degree of viewpoint diversity and instead be systemically biased in terms of viewpoints. More worryingly, depending on where (which search engine) or how (which query) users search for information, they may not only be exposed to different viewpoints, but ones representing a different bias than their peers. We also found that neutrally formulated queries often returned similarly biased search results as queries calling for specific viewpoints. In light of findings surrounding SEME and similar effects, this could have serious ramifications for individual users' well-being, business decision-making, and societal polarization.

Our case study further showed that diversification approaches based on xQuAD and HxQuAD can improve the viewpoint diversity in search results. Here, the hierarchical viewpoint diversification (based on HxQuAD, and able to consider both documents' *stances* and *logics of evaluation*) was most effective (**RQ3**).

Limitations and Future Work. Although our case study covered debated topics with consequences for individuals, businesses, and society, it is important to note that our results may not generalize to all search engines and controversial issues. We carefully selected the deliberative notion of diversity to guide our work as we believe it suits many debated topics, especially those with legitimate arguments on all sides of the viewpoint spectrum. However, we note that some scenarios may require applying other diversity notions and that presenting search results according to the deliberative notion of diversity (i.e., representing all viewpoints equally) may even cause harm to individual users or help spread fake news (e.g., considering health-related topics where only one viewpoint represents the scientifically correct answer [5,10,52,67]). Future work could measure search result viewpoint bias for larger ranges of topics, explore whether different diversity notions apply when debated topics have clear scientific answers [14,48,63], and capture user perceptions of diversity [36,46,57].

Another limitation of our work is that, despite providing a diverse range of queries to choose from, queries may not have represented all users adequately. Future work could collect topics and queries via open text fields [67]. Furthermore, our proposed metric nDVB is still limited in several ways, e.g., it does not yet incorporate document relevance, other viewpoint diversity notions, or the personal preferences and beliefs of users. We encourage researchers and practitioners to build on our work to help improve the measurement of viewpoint diversity in search results. Finally, annotating viewpoints is a difficult, time-consuming task even for expert annotators [15,20]. Recent work has already applied automatic stance detection methods to search results [22] but did so far not attempt to identify logics of evaluation. However, once such automatic systems have become more comprehensive, researchers and practitioners could easily combine them with existing methods for extracting arguments [12,60] and visualize viewpoints [4,17] in search results.

6 Conclusion

We proposed a metric for evaluating viewpoint diversity in search results, measuring the divergence from an ideal scenario of equal viewpoint representation. In a case study evaluating search results on three different debated topics from two popular search engines, we found that search results may often not be viewpoint-diverse, even if queries are formulated neutrally. We also saw notable differences between search engines concerning bias *magnitude* and *direction*. Our hierarchical viewpoint diversification method, based on HxQuAD, consistently improved the viewpoint diversity of search results. In sum, our results suggest that, while viewpoint bias in search results is not pervasive, users may unknowingly be exposed to high levels of viewpoint bias, depending on the query, topic, or search engine. These factors may influence (especially vulnerable and undecided) users' attitudes by means of recently demonstrated search engine manipulation effects and thereby affect individuals, businesses, and society.

Acknowledgements. This activity is financed by IBM and the Allowance for Top Consortia for Knowledge and Innovation (TKI's) of the Dutch ministry of economic affairs.

References

1. Abid, A., et al.: A survey on search results diversification techniques. Neural Comput. Appl. **27**(5), 1207–1229 (2015). https://doi.org/10.1007/s00521-015-1945-5
2. Agrawal, R., Gollapudi, S., Halverson, A., Ieong, S.: Diversifying search results. In: Proceedings of the Second ACM International Conference on Web Search and Data Mining - WSDM 2009, p. 5. ACM Press, Barcelona, Spain (2009). https://doi.org/10.1145/1498759.1498766, http://portal.acm.org/citation.cfm?doid=1498759.1498766
3. Ajjour, Y., Alshomary, M., Wachsmuth, H., Stein, B.: Modeling frames in argumentation. In: Proceedings of the 2019 Conference on Empirical Methods in Natural Language Processing and the 9th International Joint Conference on Natural Language Processing (EMNLP-IJCNLP), pp. 2922–2932. Association for Computational Linguistics, Hong Kong, China (Nov 2019). https://doi.org/10.18653/v1/D19-1290, https://aclanthology.org/D19-1290
4. Ajjour, Y., et al.: Visualization of the topic space of argument search results in args. me. In: Proceedings of the 2018 Conference on Empirical Methods in Natural Language Processing: System Demonstrations, pp. 60–65 (2018)
5. Allam, A., Schulz, P.J., Nakamoto, K.: The impact of search engine selection and sorting criteria on vaccination beliefs and attitudes: two experiments manipulating google output. J. Med. Internet Res. 16(4), e100 (Apr 2014). https://doi.org/10.2196/jmir.2642, http://www.jmir.org/2014/4/e100/
6. Azzopardi, L.: Cognitive Biases in Search: a review and reflection of cognitive biases in information retrieval. In: Proceedings of the 2021 Conference on Human Information Interaction and Retrieval, pp. 27–37. ACM, Canberra ACT Australia (Mar 2021). https://doi.org/10.1145/3406522.3446023, https://dl.acm.org/doi/10.1145/3406522.3446023
7. Baden, C., Springer, N.: Com(ple)menting the news on the financial crisis: the contribution of news users' commentary to the diversity of viewpoints in the public debate. Euro. J. Commun. 29(5), 529–548 (Oct 2014). https://doi.org/10.1177/0267323114538724, http://journals.sagepub.com/doi/10.1177/0267323114538724
8. Baden, C., Springer, N.: Conceptualizing viewpoint diversity in news discourse. Journalism 18(2), 176–194 (Feb 2017). https://doi.org/10.1177/1464884915605028, http://journals.sagepub.com/doi/10.1177/1464884915605028

9. Biega, A.J., Gummadi, K.P., Weikum, G.: Equity of Attention: amortizing individual fairness in rankings. In: The 41st International ACM SIGIR Conference on Research & Development in Information Retrieval, pp. 405–414. ACM, Ann Arbor MI USA (Jun 2018). https://doi.org/10.1145/3209978.3210063, https://dl.acm.org/doi/10.1145/3209978.3210063

10. Bink, M., Zimmerman, S., Elsweiler, D.: Featured snippets and their influence on users' credibility judgements. In: ACM SIGIR Conference on Human Information Interaction and Retrieval, pp. 113–122. CHIIR '22, Association for Computing Machinery, New York, NY, USA (2022). https://doi.org/10.1145/3498366.3505766, https://doi.org/10.1145/3498366.3505766

11. Boltanski, L., Thévenot, L.: On justification: economies of worth, vol. 27. Princeton University Press (2006)

12. Bondarenko, A., Ajjour, Y., Dittmar, V., Homann, N., Braslavski, P., Hagen, M.: Towards understanding and answering comparative questions. In: Proceedings of the Fifteenth ACM International Conference on Web Search and Data Mining, pp. 66–74 (2022)

13. Bondarenko, A., et al.: Overview of touché 2021: argument retrieval. In: International Conference of the Cross-Language Evaluation Forum for European Languages, pp. 450–467. Springer (2021)

14. Boykoff, M.T., Boykoff, J.M.: Balance as bias: global warming and the us prestige press. Glob. Environ. Chang. **14**(2), 125–136 (2004)

15. Burscher, B., Odijk, D., Vliegenthart, R., De Rijke, M., De Vreese, C.H.: Teaching the computer to code frames in news: comparing two supervised machine learning approaches to frame analysis. Commun. Methods Measures **8**(3), 190–206 (2014)

16. Carroll, N.: In Search We Trust: exploring how search engines are shaping society. Int. J. Knowl. Soc. Res. 5(1), 12–27 (Jan 2014). https://doi.org/10.4018/ijksr.2014010102, http://services.igi-global.com/resolvedoi/resolve.aspx?doi=10.4018/ijksr.2014010102

17. Chamberlain, J., Kruschwitz, U., Hoeber, O.: Scalable visualisation of sentiment and stance. In: Proceedings of the Eleventh International Conference on Language Resources and Evaluation (LREC 2018). European Language Resources Association (ELRA), Miyazaki, Japan (May 2018), https://aclanthology.org/L18-1660

18. Chen, S., Khashabi, D., Yin, W., Callison-Burch, C., Roth, D.: Seeing things from a different angle: discovering diverse perspectives about claims. In: Proceedings of NAACL-HLT, pp. 542–557 (2019)

19. Clarke, C.L., et al.: Novelty and diversity in information retrieval evaluation. In: Proceedings of the 31st annual international ACM SIGIR conference on Research and development in information retrieval - SIGIR 2008, p. 659. ACM Press, Singapore, Singapore (2008). https://doi.org/10.1145/1390334.1390446, http://portal.acm.org/citation.cfm?doid=1390334.1390446

20. Draws, T., Inel, O., Tintarev, N., Baden, C., Timmermans, B.: Comprehensive viewpoint representations for a deeper understanding of user interactions with debated topics. In: ACM SIGIR Conference on Human Information Interaction and Retrieval, pp. 135–145 (2022)

21. Draws, T., Liu, J., Tintarev, N.: Helping users discover perspectives: enhancing opinion mining with joint topic models. In: 2020 International Conference on Data Mining Workshops (ICDMW), pp. 23–30. IEEE, Sorrento, Italy (Nov 2020). https://doi.org/10.1109/ICDMW51313.2020.00013, https://ieeexplore.ieee.org/document/9346407/

22. Draws, T., et al.: Explainable cross-topic stance detection for search results. In: CHIIR 2023: ACM SIGIR Conference on Human Information Interaction and Retrieval. CHIIR 2023, ACM SIGIR Conference on Human Information Interaction and Retrieval (2023)

23. Draws, T., Tintarev, N., Gadiraju, U., Bozzon, A., Timmermans, B.: Assessing viewpoint diversity in search results using ranking fairness metrics. ACM SIGKDD Explorations Newsletter 23(1), 50–58 (May 2021). https://doi.org/10.1145/3468507.3468515, https://dl.acm.org/doi/10.1145/3468507.3468515

24. Draws, T., Tintarev, N., Gadiraju, U., Bozzon, A., Timmermans, B.: This is not what we ordered: exploring why biased search result rankings affect user attitudes on debated topics. In: Proceedings of the 44th International ACM SIGIR Conference on Research and Development in Information Retrieval, pp. 295–305. ACM, Virtual Event Canada (Jul 2021). https://doi.org/10.1145/3404835.3462851, https://dl.acm.org/doi/10.1145/3404835.3462851

25. Drosou, M., Pitoura, E.: Search result diversification. SIGMOD Record 39(1), 7 (2010)

26. Dumani, L., Neumann, P.J., Schenkel, R.: A framework for argument retrieval. In: Jose, J.M., et al. (eds.) ECIR 2020. LNCS, vol. 12035, pp. 431–445. Springer, Cham (2020). https://doi.org/10.1007/978-3-030-45439-5_29

27. Epstein, R., Robertson, R.E.: The search engine manipulation effect (SEME) and its possible impact on the outcomes of elections. In: Proceedings of the National Academy of Sciences 112(33), E4512–E4521 (Aug 2015). https://doi.org/10.1073/pnas.1419828112, http://www.pnas.org/lookup/doi/10.1073/pnas.1419828112

28. Epstein, R., Robertson, R.E., Lazer, D., Wilson, C.: Suppressing the search engine manipulation effect (SEME). In: Proceedings of the ACM on Human-Computer Interaction 1(CSCW), 1–22 (Dec 2017). https://doi.org/10.1145/3134677, https://dl.acm.org/doi/10.1145/3134677

29. Fuglede, B., Topsoe, F.: Jensen-shannon divergence and hilbert space embedding. In: International Symposium on Information Theory, 2004. ISIT 2004. Proceedings, p. 31. IEEE (2004)

30. Gao, R., Shah, C.: Toward creating a fairer ranking in search engine results. Inf. Process. Manag. 57(1), 102138 (Jan 2020). https://doi.org/10.1016/j.ipm.2019.102138, https://linkinghub.elsevier.com/retrieve/pii/S0306457319304121

31. Gevelber, L.: It's all about 'me'-how people are taking search personally. Tech. rep. (2018). https://www.thinkwithgoogle.com/marketing-strategies/search/personal-needs-search-trends/

32. Gezici, G., Lipani, A., Saygin, Y., Yilmaz, E.: Evaluation metrics for measuring bias in search engine results. Inf. Retrieval J. 24(2), 85–113 (Apr 2021). https://doi.org/10.1007/s10791-020-09386-w, http://link.springer.com/10.1007/s10791-020-09386-w

33. Ghenai, A., Smucker, M.D., Clarke, C.L.: A think-aloud study to understand factors affecting online health search. In: Proceedings of the 2020 Conference on Human Information Interaction and Retrieval, pp. 273–282. ACM, Vancouver BC Canada (Mar 2020). https://doi.org/10.1145/3343413.3377961, https://dl.acm.org/doi/10.1145/3343413.3377961

34. Grady, C., Lease, M.: Crowdsourcing document relevance assessment with mechanical turk. In: NAACL HLT Workshop on Creating Speech and Language Data with Amazon's Mechanical Turk, pp. 172–179 (2010)

35. Gretz, S., et al.: A large-scale dataset for argument quality ranking: construction and analysis. Proc. AAAI Conf. Artif. Intell. 34, 7805–7813 (2020). https://doi.org/10.1609/aaai.v34i05.6285

36. Han, B., Shah, C., Saelid, D.: Users' perception of search-engine biases and satisfaction. In: Boratto, L., Faralli, S., Marras, M., Stilo, G. (eds.) BIAS 2021. CCIS, vol. 1418, pp. 14–24. Springer, Cham (2021). https://doi.org/10.1007/978-3-030-78818-6_3

37. Helberger, N.: On the democratic role of news recommenders. Digital Journalism 7(8), 993–1012 (Sep 2019). https://doi.org/10.1080/21670811.2019.1623700, https://www.tandfonline.com/doi/full/10.1080/21670811.2019.1623700

38. Hu, S., Dou, Z., Wang, X., Sakai, T., Wen, J.R.: Search result diversification based on hierarchical intents. In: Proceedings of the 24th ACM International on Conference on Information and Knowledge Management, pp. 63–72. ACM, Melbourne Australia (Oct 2015). https://doi.org/10.1145/2806416.2806455, https://dl.acm.org/doi/10.1145/2806416.2806455

39. Joachims, T., Granka, L., Pan, B., Hembrooke, H., Gay, G.: Accurately interpreting clickthrough data as implicit feedback. ACM SIGIR Forum 51(1), 8 (2016)

40. Kaya, M., Bridge, D.: Subprofile-aware diversification of recommendations. User Modeling and User-Adapted Interaction 29(3), 661–700 (Jul 2019). https://doi.org/10.1007/s11257-019-09235-6, http://link.springer.com/10.1007/s11257-019-09235-6

41. Küçük, D., Can, F.: Stance detection: a survey. ACM Comput. Surv. (CSUR) 53(1), 1–37 (2020)

42. Kulshrestha, J., et al.: Quantifying search bias: investigating sources of bias for political searches in social media. In: Proceedings of the 2017 ACM Conference on Computer Supported Cooperative Work and Social Computing, pp. 417–432. ACM, Portland Oregon USA (Feb 2017). https://doi.org/10.1145/2998181.2998321, https://dl.acm.org/doi/10.1145/2998181.2998321

43. Kulshrestha, J., et al.: Search bias quantification: investigating political bias in social media and web search. Information Retrieval Journal 22(1–2), 188–227 (Apr 2019). https://doi.org/10.1007/s10791-018-9341-2, http://link.springer.com/10.1007/s10791-018-9341-2

44. Ludolph, R., Allam, A., Schulz, P.J.: Manipulating google's knowledge graph box to counter biased information processing during an online search on vaccination: application of a technological debiasing strategy. J. Med. Internet Res. 18(6), e137 (Jun 2016). https://doi.org/10.2196/jmir.5430, http://www.jmir.org/2016/6/e137/

45. McKay, D., et al.: We are the change that we seek: information interactions during a change of viewpoint. In: Proceedings of the 2020 Conference on Human Information Interaction and Retrieval, pp. 173–182 (2020)

46. McKay, D., Owyong, K., Makri, S., Gutierrez Lopez, M.: Turn and face the strange: investigating filter bubble bursting information interactions. In: ACM SIGIR Conference on Human Information Interaction and Retrieval, pp. 233–242. CHIIR '22, Association for Computing Machinery, New York, NY, USA (2022). https://doi.org/10.1145/3498366.3505822

47. Mulder, M., Inel, O., Oosterman, J., Tintarev, N.: Operationalizing framing to support multiperspective recommendations of opinion pieces. In: Proceedings of the 2021 ACM Conference on Fairness, Accountability, and Transparency, pp. 478–488. ACM, Virtual Event Canada (Mar 2021). https://doi.org/10.1145/3442188.3445911, https://dl.acm.org/doi/10.1145/3442188.3445911

48. Munson, S.A., Resnick, P.: Presenting diverse political opinions: how and how much. In: Proceedings of the SIGCHI Conference on Human Factors in Computing Systems, pp. 1457–1466. CHI 2010, Association for Computing Machinery, New York, NY, USA (2010). https://doi.org/10.1145/1753326.1753543

49. Pan, B., Hembrooke, H., Joachims, T., Lorigo, L., Gay, G., Granka, L.. In Google We Trust: Users' Decisions on Rank, Position, and Relevance. J. Comput.-Mediated Commun. 12(3), 801–823 (Apr 2007). https://doi.org/10.1111/j.1083-6101.2007.00351.x, https://academic.oup.com/jcmc/article/12/3/801-823/4582975

50. Pathiyan Cherumanal, S., Spina, D., Scholer, F., Croft, W.B.: Evaluating fairness in argument retrieval. In: Proceedings of the 30th ACM International Conference on Information & Knowledge Management, pp. 3363–3367 (2021)

51. Pennycook, G., Rand, D.G.: Lazy, not biased: susceptibility to partisan fake news is better explained by lack of reasoning than by motivated reasoning. Cognition 188, 39–50 (Jul 2019). https://doi.org/10.1016/j.cognition.2018.06.011, https://linkinghub.elsevier.com/retrieve/pii/S001002771830163X

52. Pogacar, F.A., Ghenai, A., Smucker, M.D., Clarke, C.L.: The positive and negative influence of search results on people's decisions about the efficacy of medical treatments. In: Proceedings of the ACM SIGIR International Conference on Theory of Information Retrieval, pp. 209–216. ACM, Amsterdam The Netherlands (Oct 2017). https://doi.org/10.1145/3121050.3121074, https://dl.acm.org/doi/10.1145/3121050.3121074

53. Purcell, K., Rainie, L., Brenner, J.: Search engine use 2012 (2012)

54. Puschmann, C.: Beyond the bubble: assessing the diversity of political search results. Digital Journalism 7(6), 824–843 (Jul 2019). https://doi.org/10.1080/21670811.2018.1539626, https://www.tandfonline.com/doi/full/10.1080/21670811.2018.1539626

55. Radlinski, F., Craswell, N.: Comparing the sensitivity of information retrieval metrics. In: Proceedings of the 33rd International ACM SIGIR Conference on Research and Development in Information Retrieval, pp. 667–674 (2010)

56. Reimer, J.H., Huck, J., Bondarenko, A.: Grimjack at touché 2022: axiomatic reranking and query reformulation. Working Notes Papers of the CLEF (2022)

57. Rieger, A., Draws, T., Theune, M., Tintarev, N.: This item might reinforce your opinion: obfuscation and labeling of search results to mitigate confirmation bias. In: Proceedings of the 32nd ACM Conference on Hypertext and Social Media, pp. 189–199 (2021)

58. Sakai, T., Craswell, N., Song, R., Robertson, S., Dou, Z., Lin, C.Y.: Simple evaluation metrics for diversified search results, p. 9 (2010)

59. Santos, R.L., Macdonald, C., Ounis, I.: Exploiting query reformulations for web search result diversification. In: Proceedings of the 19th international conference on World wide web, pp. 881–890 (2010)

60. Stab, C., et al.: ArgumenText: searching for arguments in heterogeneous sources. In: Proceedings of the 2018 Conference of the North American Chapter of the Association for Computational Linguistics: Demonstrations, pp. 21–25 (2018)

61. Tintarev, N., Sullivan, E., Guldin, D., Qiu, S., Odjik, D.: Same, same, but different: algorithmic diversification of viewpoints in news. In: Adjunct Publication of the 26th Conference on User Modeling, Adaptation and Personalization, pp. 7–13. ACM, Singapore Singapore (Jul 2018). https://doi.org/10.1145/3213586.3226203, https://dl.acm.org/doi/10.1145/3213586.3226203

62. Voorhees, E.M.: Variations in relevance judgments and the measurement of retrieval effectiveness. Inf. Process. Manag. 36(5), 697–716 (2000)

63. Vrijenhoek, S., Bénédict, G., Gutierrez Granada, M., Odijk, D., De Rijke, M.: Radio-rank-aware divergence metrics to measure normative diversity in news recommendations. In: Proceedings of the 16th ACM Conference on Recommender Systems. pp. 208–219 (2022)

64. Vrijenhoek, S., Kaya, M., Metoui, N., Möller, J., Odijk, D., Helberger, N.: Recommenders with a mission: Assessing diversity in news recommendations. In: Proceedings of the 2021 Conference on Human Information Interaction and Retrieval, pp. 173–183. CHIIR 2021, Association for Computing Machinery, New York, NY, USA (2021). https://doi.org/10.1145/3406522.3446019

65. White, R.: Beliefs and biases in web search. In: Proceedings of the 36th International ACM SIGIR Conference on Research and Development in Information Retrieval, pp. 3–12. ACM, Dublin Ireland (Jul 2013). https://doi.org/10.1145/2484028.2484053, https://dl.acm.org/doi/10.1145/2484028.2484053

66. White, R.W., Hassan, A.: Content bias in online health search. ACM Transactions on the Web 8(4), 1–33 (Nov 2014). https://doi.org/10.1145/2663355, https://dl.acm.org/doi/10.1145/2663355

67. White, R.W., Horvitz, E.: Belief dynamics and biases in web search. ACM Trans. Inf. Syst. 33(4), 1–46 (May 2015). https://doi.org/10.1145/2746229, https://dl.acm.org/doi/10.1145/2746229

68. Xu, L., Zhuang, M., Gadiraju, U.: How do user opinions influence their interaction with web search results?, pp. 240–244. Association for Computing Machinery, New York, NY, USA (2021), https://doi.org/10.1145/3450613.3456824

69. Yamamoto, Y., Shimada, S.: Can disputed topic suggestion enhance user consideration of information credibility in web search? In: Proceedings of the 27th ACM Conference on Hypertext and Social Media, pp. 169–177. ACM, Halifax Nova Scotia Canada (Jul 2016). https://doi.org/10.1145/2914586.2914592, https://dl.acm.org/doi/10.1145/2914586.2914592

70. Yang, K., Stoyanovich, J.: Measuring fairness in ranked outputs. In: Proceedings of the 29th International Conference on Scientific and Statistical Database Management, pp. 1–6. ACM, Chicago IL USA (Jun 2017). https://doi.org/10.1145/3085504.3085526, https://dl.acm.org/doi/10.1145/3085504.3085526

71. Yom-Tov, E., Dumais, S., Guo, Q.: Promoting civil discourse through search engine diversity. Soc. Sci. Comput. Rev. 32(2), 145–154 (Apr 2014). https://doi.org/10.1177/0894439313506838, http://journals.sagepub.com/doi/10.1177/0894439313506838

72. Zehlike, M., Bonchi, F., Castillo, C., Hajian, S., Megahed, M., Baeza-Yates, R.: FA*IR: A Fair Top-k ranking algorithm. In: Proceedings of the 2017 ACM on Conference on Information and Knowledge Management, pp. 1569–1578. ACM, Singapore Singapore (Nov 2017). https://doi.org/10.1145/3132847.3132938, https://dl.acm.org/doi/10.1145/3132847.3132938

73. Zehlike, M., Yang, K., Stoyanovich, J.: Fairness in ranking, part i: score-based ranking. ACM Comput. Surv. 55(6), 1–36 (2022)

COILcr: Efficient Semantic Matching in Contextualized Exact Match Retrieval

Zhen Fan[(✉)], Luyu Gao, Rohan Jha, and Jamie Callan[iD]

Carnegie Mellon University, Pittsburgh, PA 15213, USA
{zhenfan,luyug,rjha,callan}@cs.cmu.edu

Abstract. Lexical exact match systems that use inverted lists are a fundamental text retrieval architecture. A recent advance in neural IR, COIL, extends this approach with *contextualized inverted lists* from a deep language model backbone and performs retrieval by comparing contextualized query-document term representation, which is effective but computationally expensive. This paper explores the effectiveness-efficiency tradeoff in COIL-style systems, aiming to reduce the computational complexity of retrieval while preserving term semantics. It proposes COILcr, which explicitly factorizes COIL into intra-context term importance weights and cross-context semantic representations. At indexing time, COILcr further maps term semantic representations to a smaller set of canonical representations. Experiments demonstrate that canonical representations can efficiently preserve term semantics, reducing the storage and computational cost of COIL-based retrieval while maintaining model performance. The paper also discusses and compares multiple heuristics for canonical representation selection and looks into its performance in different retrieval settings.

Keywords: First-stage retrieval · Lexical exact match · Deep language models · Contextualized inverted lists · Approximation

1 Introduction

Lexical exact matching [21] has been a fundamental component of classic information retrieval (IR). In the new era of neural IR and deep language models (LM) [2,24], lexical retrievers are still used in large-scale settings and initial stages of ranking pipelines due to their simplicity and efficiency: lexical exact match signals are captured at the token (word) level, and the matching process can be highly accelerated with *inverted indexes* built during offline preprocessing. Such simplicity, however, is accompanied by the natural gap between explicit lexical form and the implicit semantics of a concept. Lexical retrievers have long suffered from the *vocabulary mismatch* (different lexical forms for the same concept) and *semantic mismatch* (different semantic meanings for the same lexical form) between the two spaces.

The introduction of deep LMs has led to large improvements in search accuracy [18]. Deep LM-augmented lexical retrieval systems fine-tune pretrained

© The Author(s), under exclusive license to Springer Nature Switzerland AG 2023
J. Kamps et al. (Eds.): ECIR 2023, LNCS 13980, pp. 298–312, 2023.
https://doi.org/10.1007/978-3-031-28244-7_19

language models on different retrieval-specific tasks to aid classic non-neural systems or directly perform retrieval. Particularly, to address semantic mismatch, recent work uses deep LMs to generate contextualized term representations and calculates term scores by vector similarity instead of scalar weight product.

Gao et al. proposed COIL (contextualized inverted lists) [8], a framework for incorporating semantic matching in lexical exact match systems. COIL augments traditional inverted lists with contextualized representations of each term occurrence. At search time, it keeps the constraint of lexical exact match, but calculates term scores based on the similarity of the representations. Compared to previous systems using term weights to model *in-sequence term importance*, COIL uses term representation vectors to additionally measure *cross-sequence term similarity* between query and document terms. This leads to improved retrieval accuracy but also increased storage and computational costs. Every term is represented by a d-dimensional vector during indexing, and each term score is calculated by a d-dimensional vector dot product at retrieval time.

This paper builds on the COIL framework, focusing on its semantic matching capability. It investigates whether it is it possible to effectively recognize semantic mismatch at a lower cost. COIL utilizes a dense vector to directly model a term's importance as well as the fine-grained semantics of its unique context. However, for a vocabulary term, the number of its important meanings or base senses across the corpus is usually much smaller than its actual collection term frequency. While modeling fine-grained term semantic match requires precise vector similarity comparison, modeling the mismatch of coarse term senses may not require such high representation capacity, and can be performed more efficiently by approximation of term representations.

Following these ideas, we propose COILcr, **CO**ntextualized **I**nverted **L**ists with **C**anonical **R**epresentations, to efficiently model coarse term semantics in COIL-based lexical exact match systems. We first factorize term representations and decouple term importance weight and semantic representation. We proceed to build a set of canonical representations (CR) for term semantic representations via spherical k-means clustering [3], and map all individual term occurrences to the set of base CRs. This approximation reduces the inverted index storage size and number of similarity calculations at retrieval time. We demonstrate through multiple experiments that COILcr's approximation is almost as effective as precise lexical COIL systems, but at much lower storage and computational cost.

The next section discusses related work, and provides a detailed description of the COIL framework. Section 3 describes the proposed canonical representation-based approach to the recognition of semantic mismatch. Section 4 discusses our experimental methodology, and Sect. 5 discusses experiment results and findings. Section 6 concludes.

2   Related Work

The introduction of deep language models [2] has led to a new era in neural IR. Cross-encoder models [18] first demonstrated that deep LMs can be tuned to understand context and handle the gap between explicit text and implicit semantics, and achieve state-of-the-art performance in neural reranking tasks.

Large-scale neural ranking systems have also benefited from the finetuned LMs' capability to generate semantic representations both at the text sequence level and at the lexical token level. *Dense retrieval* systems [14] directly encode text segments into a dense semantic representation, and score query-document pairs with some vector similarity metric.

$$\mathcal{S}(q,d) = \sigma(\mathbf{v}_q, \mathbf{v}_d)$$

where $\mathbf{v}_q$ and $\mathbf{v}_d$ are dense representations of the query and document, usually the [CLS] outout of the language model, and σ is a similarity function such as dot product or cosine similarity. Recent work investigates various training techniques to improve the quality of representations, including hard negative mining [11,25], pretraining task design [6,7,11] and knowledge distillation [9,10,20].

Lexical match systems, on the other hand, perform encoding and matching at the lexical token level, and score query-document pairs by aggregating term match scores.

$$\mathcal{S}(q,d) = \sum_{t \in \mathcal{V}_q \cap \mathcal{V}_d} s_t(q,d)$$

where $\mathcal{S}(q,d)$ is the overall document score, $s_t(q,d)$ is term matching score of term t, and $\mathcal{V}_q$ and $\mathcal{V}_d$ are the sets of terms in the query and document respectively. For non-neural lexical exact match retrievers such as BM25 [21], a document term is represented by a scalar weight $w_{t,d}$ that represents its *importance* and is stored in an inverted list. Term scoring is modeled as $s_t(q,d) = w_{t,q} w_{t,d}$, a product of query and document term importance weights. Finetuned language models were first introduced to improve existing non-neural weighting-based systems by performing term reweighting [1] and explicit vocabulary expansion [19].

In such lexical exact match systems, storing scalar weights ensures efficient storage and computational cost, but does not preserve extra semantic information or context. At retrieval time, the system can not distinguish the actual semantic agreement between query and document terms, thereby suffering from semantic mismatch. To tackle this problem, researchers explored using contextualized representations in lexical retrieval [15,22,29] under soft match settings. Gao et al. further proposed COIL[1] [8], which introduces contextualized term

---

[1] The full COIL retrieval model is a hybrid model combining dense document scoring and sparse token scoring. In this paper we mainly focus on the lexical exact match retrieval setting, and mainly refer to COIL as the basic concept of contextualized term representation and inverted index. We compare our system to the lexical-only model form of the COIL retriever, referred to as COIL-*tok* in the original work.

representations under lexical exact match settings, and expands the term weight product into a vector similarity calculation to further model the semantic similarity between query and document terms.

$$\mathbf{v}_{q_i} = \phi(LM(q, i))$$
$$\mathbf{v}_{d_j} = \phi(LM(d, j))$$
$$s_t(q_i, d_j) = \mathbf{v}_{q_i}^{\mathsf{T}} \mathbf{v}_{d_j}$$

where $q_i = d_j$ are the i-th query term and j-th document term with vector representation $\mathbf{v}_{q_i}$ and $\mathbf{v}_{d_j}$ respectively. ϕ denotes a linear transformation layer that maps the LM output to token representations of a lower dimension.

COIL's lexical-only model variant COIL-tok is a natural extension of traditional lexical retrieval systems. The vector representations of document terms $\mathbf{v}_{d_j}$ are precomputed and indexed in inverted lists, and the overall query-document score is the aggregation of exact match term scores. Replacing term weights with vectors leads to clear performance gain in accuracy and recall but also increases storage and computational cost. Lin and Ma further proposed UNICOIL [16], a followup study to COIL in which the generated representation dimension is lowered to $d_v = 1$ and the COIL language model directly predicts scalar term weight, and demonstrates that the model achieves decent accuracy with much lower cost under term weighting-only settings.

In this paper, we look into the necessity and methodology of preserving term semantics in COIL-style systems and balancing its effectiveness-efficiency tradeoff. Index compression and retrieval efficiency has gained growing research interest with the development of neural IR systems. Recent systems investigate multiple methods such as dimension reduction [12], hashing [26], product quantization [27,28] and residual compression [22].

3 COIL$_{CR}$: Contextualized Inverted Lists with Canonical Representations

COILCR is based on two key ideas: i) factorizing COIL token representations into intra- and cross-context components, and ii) approximating the cross-context component with canonical representations.

3.1 Term Score Factorization

COIL-tok implicitly models two distinct aspects of term match scoring: *intra-context importance*, which measures the importance of a term (q_i or d_j) to its own text (q or d), and *cross-context similarity*, which measures whether matching terms q_i and d_j are used in a similar context and require actual interaction at retrieval time. As shown in previous term-weighting systems and COIL model variants (e.g., UNICOIL [16]), the term importance component can be effectively represented with a scalar. A more critical question lies in the capacity and cost of representing term semantics and calculating query-document similarity.

COILCR explicitly factorizes COIL's contextualized token representations into a scalar *term weight* value and a *term semantic representation* vector for each term, using separate linear projection layers:

$$w_{d_j} = \phi_w(LM(d, j))$$
$$\mathbf{v}_{d_j} = \phi_v(LM(d, j))$$
$$\hat{\mathbf{v}}_{d_j} = \frac{\mathbf{v}_{d_j}}{||\mathbf{v}_{d_j}||}$$

where w_{d_j} is a non-negative value denoting term weight, and $\hat{\mathbf{v}}_{d_j}$ is a normalized vector denoting term semantics.

COILCR uses the same language model to encode query and document terms. The factorized contextualized exact match score between overlapping terms $q_i = d_j$ is defined as:

$$s(q_i, d_j) = w_{q_i} w_{d_j} cos(\mathbf{v}_{q_i}, \mathbf{v}_{d_j})$$
$$= w_{q_i} w_{d_j} \hat{\mathbf{v}}_{q_i}^{\mathsf{T}} \hat{\mathbf{v}}_{d_j} \tag{1}$$

where w, v and s represents the weighting component, semantic similarity component and overall token matching score respectively. Equation 1 can be viewed as a factorization of COIL's dot-product scoring function to an equivalent cosine similarity form. It explicitly decouples term weighting to enable more direct analysis and approximation of term semantics.

The exact overall score between a query q and document d is the sum of all lexical matching scores of overlapping terms. Following COIL, we train COILCR with an NLL loss defined on query q, document set $\{d^+, d_1^-, d_2^-, ..., d_{n-1}^-\}$, and the scoring function.

$$\mathcal{S}_e(q, d) = \sum_{q_i \in \mathcal{V}_q \cap \mathcal{V}_d} \max_{d_j = q_i} s(q_i, d_j)$$
$$= \sum_{q_i \in \mathcal{V}_q \cap \mathcal{V}_d} \max_{d_j = q_i} w_{q_i} w_{d_j} \hat{\mathbf{v}}_{q_i}^{\mathsf{T}} \hat{\mathbf{v}}_{d_j}$$
$$\mathcal{L}_{NLL} = -\log \frac{\exp(\mathcal{S}_e(q, d^+))}{\exp(\mathcal{S}_e(q, d^+)) + \sum_{i=1}^{n-1} \exp(\mathcal{S}_e(q, d_i^-))}$$

3.2   Approximate Term Semantic Interaction

The main additional cost of COIL compared to other lexical exact-match systems lies in storing a unique vector for each document term occurrence during indexing, and having to compute vector products $\mathbf{v}_{q_i}^{\mathsf{T}} \mathbf{v}_{d_j}$ or $\hat{\mathbf{v}}_{q_i}^{\mathsf{T}} \hat{\mathbf{v}}_{d_j}$ for each document term occurrence at retrieval time. COILCR mainly focuses on approximating the vector product by reducing the space of vocabulary representations. For a term t, instead of using unique vectors for every term occurrence, COILCR selects a fixed set of semantic *canonical representations* C_t after the encoding stage, and maps each term semantic representation to its closest vector in C_t.

$$c_{d_j} = \arg\max_{c \in C_t} \cos(\hat{\mathbf{v}}_{d_j}, c)$$

where c_{d_j} can be viewed as an approximate representation of the original term $\mathbf{v}_{d_j}$. At retrieval time, c_{d_j} is used to calculate an approximated term matching score and the final document score.

$$s_c(q_i, d_j) = w_{q_i} w_{d_j} \hat{\mathbf{v}}_{q_i}^{\mathsf{T}} c_{d_j}$$
$$\mathcal{S}_c(q, d) = \sum_{q_i \in q \cap d} \max_{d_j = q_i} w_q w_d \, \hat{\mathbf{v}}_{q_i}^{\mathsf{T}} c_{d_j}$$

Mapping unique term occurrence representations to canonical term occurrence representations reduces the storage cost of each individual term occurrence from a unique $|d|$-dim vector to just its term weight w and the index of its canonical representation. At retrieval time, instead of calculating $\hat{\mathbf{v}}_{q_i}^{\mathsf{T}} \hat{\mathbf{v}}_{d_j}$ for each term occurrence d_j, COILcr only needs to calculate $\hat{\mathbf{v}}_{q_i}^{\mathsf{T}} c$ for each CR $c \in C_t$. The actual term representation scoring is reduced to a lookup operation of $\hat{\mathbf{v}}_{q_i}^{\mathsf{T}} c_{d_j}$ from the set of candidate scores.

Canonical semantic representations C_t can be generated in varied ways. COILcr generates them using weighted spherical k-means clustering [3]. For each term, it iterates to optimize

$$\mathbf{F}_t = \sum_{d_j = t} w_{d_j} \cos(\hat{\mathbf{v}}_{d_j}, c_{d_j})$$

where $\mathbf{F}_t$ is a weighted sum of cosine similarity between $\hat{\mathbf{v}}_{d_j}$ and c_{d_j}. This is aligned with the scoring function of COILcr.

The number of canonical representations $|C_t|$, or the number of clusters, directly determines the granularity of term semantics and the degree of approximation. In this work we experiment with three cluster selection methods.

– Constant: A fixed number of clusters $|C|$ is generated for all terms.
– Dynamic: The cluster size is determined dynamically based on a *clustering error threshold*. Given an error threshold ϵ and a set of candidate cluster sizes $\{\mathbf{K}^d\}$, for each term the minimum cluster size $k_t^d \in \{\mathbf{K}^d\}$ is selected such that the clustering error $E_t = 1 - \frac{1}{|d_t|}\mathbf{F}_t$ falls below ϵ.
– Universal: Following previous work [22], we include a separate experiment where all terms share a fixed set of universal canonical representations. The centroids are generated by randomly sampling term representations in the entire corpus and performing clustering.

We perform detailed analysis of the effect of cluster size selection in Sect. 5.2. In the sections below, we refer to COILcr variants that perform clustering as COILcr-t,k where t is the type of clustering strategy (c/d/u for constant, dynamic, universal) and k is the respective parameter (cluster size $|C|$ for constant and universal, error threshold ϵ for dynamic). When no clustering approximation is performed, COILcr is equivalent to COIL-tok with factorized term scoring. We refer to this model variant as COILcr-∞ (infinite clusters).

4   Experimental Methodology

Implementation. COILCR mostly follows COIL's implementation[2] and training settings, using a default token representation with $d_v = 32$ dimensions. We analyze the effect of token representation dimension in Sect. 5. All COILCR variants are trained for 5 epochs with a batch size of 8 queries and 8 documents (1 positive, 7 negative) per query. At indexing time, we randomly sample representations and perform spherical k-means clustering with Faiss [13]. We experiment with $k \in \{1, 4, 16, 64, 256, 1024\}$ clusters for constant and dynamic cluster generation, error thresholds $\epsilon \in \{0.05, 0.1, 0.15, 0.2, 0.25\}$ for dynamic cluster generation, and $k \in \{256, 1024, 4096\}$ for universal cluster generation.

Extensions. COILCR does not perform expansion or remapping of original terms, but can be used with document expansion systems such as DocT5Query [19]. We also experiment with model initialization using coCondenser [7], a deep LM trained on retrieval-related tasks, as this has been effective in prior work [4, 22].

Experiments. We train our model on the MSMARCO passage dataset [17] and report model performance on MSMARCO dev queries and TREC DL 2019 manual queries. We report MRR@10 and recall@1000 for MSMARCO dev evaluation, and report NDCG@10 and recall@1000 for TREC DL queries. We mainly focus our comparison to previous COIL-based lexical exact match retrieval systems COIL-tok and its term-weighting variant UNICOIL. We also train and report results for UNICOIL and COIL-tok baselines with coCondenser initialization and DocT5Query augmentation. We additionally report the performance of two related retrieval systems: (1) COIL-full, a hybrid retriever that additionally utilizes a dense scoring component, and (2) SPLADE [5], an end-to-end term weighting-based lexical retrieval system with vocabulary expansion.

5   Experiments

In this section, we discuss the retrieval performance of COILCR and the effect of its components. We first separately analyze the effectiveness of explicit score factorization and post-hoc CR-based approximation. We further perform a quantitative analysis on the two main factors of COILCR's effectiveness-efficiency tradeoff: the vector representation dimension and the post-hoc approximation. We finally look into the semantic information of canonical representations through domain-transfer experiments and analysis.

5.1   Passage Retrieval Effectiveness

We first report the passage retrieval performance of COILCR-∞ variants on MSMARCO in Table 1. Under the same training settings, COILCR-∞ achieves

---

Table 1. Passage retrieval performance for COILcr on MSMARCO. Baselines labeled with * were retrained. We perform significance testing for COILcr variants with coCondenser initialization. Under the same training settings, [†] denotes equal or better performance compared to COIL-tok and [‡] denotes better performance compared to UniCOIL.

Model		MSMARCO dev		Trec DL 2019	
Retriever	Init	MRR@10	R@1000	NDCG@10	R@1000
Lexical retrievers w/o implicit vocabulary expansion					
UniCOIL	BERT	0.320	0.922	0.652	–
UniCOIL + DocT5Q	BERT	0.351	-	0.693	–
COIL-tok	BERT	0.341	0.949	0.660	–
UniCOIL*	coCondenser	0.328	0.929	0.646	0.778
UniCOIL + DocT5Q*	coCondenser	0.357	0.961	0.702	0.823
COIL-tok*	coCondenser	0.353	0.949	0.692	0.801
COIL-tok + DocT5Q*	coCondenser	0.365	0.967	0.707	0.833
Hybrid systems or lexical retrievers with implicit expansion					
COIL-full	BERT	0.355	0.963	0.704	–
COIL-full	coCondenser	0.374	0.981	–	–
SPLADE	BERT	0.322	0.955	0.665	0.813
COILcr-∞	BERT	0.341	0.944	0.673	0.787
COILcr-∞ + DocT5Q	BERT	0.358	0.964	0.692	0.830
COILcr-∞	coCondenser	$0.355^{†‡}$	$0.951^{†‡}$	0.717	0.794
COILcr-∞ + DocT5Q	coCondenser	$0.370^{†‡}$	$0.968^{†‡}$	0.711	0.832
COILcr-c256	BERT	0.331	0.941	0.676	0.784
COILcr-c256 + DocT5Q	BERT	0.352	0.963	0.698	0.831
COILcr-c256	coCondenser	$0.346^{‡}$	$0.948^{†‡}$	0.704	0.797
COILcr-c256 + DocT5Q	coCondenser	$0.362^{‡}$	$0.966^{†‡}$	0.714	0.836

[†] TOST testing with $\alpha = 5\%$ and equivalence bound of ± 0.005.
[‡] Paired t-test with $\alpha = 5\%$.

very similar accuracy and Recall compared to COIL-tok. This demonstrates the extra capacity of modeling term semantics, and that COILcr's score factorization step does not limit such model capacity by itself.

Performing coCondenser initialization and DocT5Query augmentation improves the retrieval performance of all COILcr variants with different effects, as expected. Initialization with coCondenser, a system pretrained on the MSMARCO dataset and on a dense retrieval-related task, also helps lexical-only retrieval systems COIL-tok and COILcr learn higher quality term representations and more accurate matching signals, leading to improvement in model accuracy. On the other hand, COIL-based models do not implicitly resolve the vocabulary mismatch between the query and document. Under comparable training setups, COILcr-∞ and COIL-tok systems outperform SPLADE on accuracy at top positions (MRR@10 and NDCG@10 on respective datasets) but underperform on recall@1000. The addition of DocT5Query augmentation introduces explicit document expansion which leads to better overall performance, especially for Recall@1000 ($0.95 \rightarrow 0.97$ for MSMARCO dev, $0.79 \rightarrow 0.83$ for Trec DL 2019). Specifically, on the Trec DL 2019 queries, UniCOIL achieves close performance to COILCR and COIL-tok with DocT5Query augmentation.

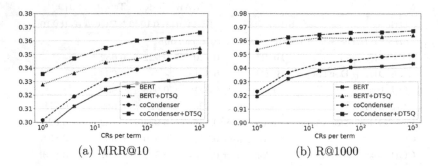

(a) MRR@10   (b) R@1000

Fig. 1. Passage retrieval performance on MSMARCO-dev for COILCR model variants with different degrees of approximation and different training setup.

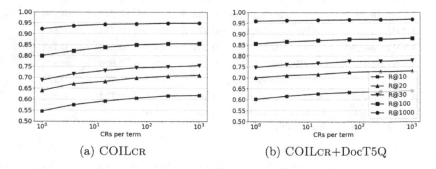

(a) COILCR   (b) COILCR+DocT5Q

Fig. 2. Recall at different depths (Recall@k) for COILCR model variants with different degree of approximation ($|C_t|$). Models are initialized with coCondenser.

After performing clustering with $|C_t| = 256$ CRs per term, we observe only a slight drop in MRR@10 and Recall. To further explore how post-hoc approximation affects COILCR's retrieval performance, we report the MRR@10 and Recall@1000 of COILCR on MSMARCO dev queries with different CR size $|C_t|$ in Fig. 1, and the change in Recall at different depths with different $|C_t|$ in Fig. 2. Under all training settings, the degree of approximation mainly affects the *precision* of lexical exact match signals and documents at the top of the ranking. It has particularly little impact on recall at lower positions, where the more critical bottleneck is vocabulary mismatch and sufficient lexical exact match signals do not exist.

Table 2. Passage retrieval accuracy and storage cost of COILCR with varying numbers of representation dimensions and CRs per term. Models initialized with coCondenser. † and ‡ respectively denotes equal or better performance compared to COIL-tok, and better performance compared to UNICOIL.

Model		MSMARCO dev		Storage (GB)		
		MRR@10	R@1000	CR Index	Inv. Index	Total
COIL-tok		0.353	0.949	n/a	45	45
UniCOIL		0.328	0.929	n/a	4.8	4.8
COILCR:						
32	∞	$0.355^{\dagger\ddagger}$	$0.951^{\dagger\ddagger}$	n/a	55	55
16	∞	$0.350^{\ddagger}$	$0.950^{\dagger\ddagger}$	n/a	34	34
8	∞	$0.350^{\ddagger}$	$0.946^{\dagger\ddagger}$	n/a	21	21
4	∞	$0.345^{\dagger}$	$0.941^{\ddagger}$	n/a	14	14
32	c256	$0.346^{\ddagger}$	$0.948^{\dagger\ddagger}$	0.7	5.5	6.2
16	c256	$0.343^{\ddagger}$	$0.947^{\dagger\ddagger}$	0.4	5.4	5.8
8	c256	$0.349^{\ddagger}$	$0.945^{\ddagger}$	0.2	5.5	5.7
4	c256	$0.343^{\ddagger}$	$0.941^{\ddagger}$	0.1	5.4	5.4
32	c256	$0.346^{\ddagger}$	$0.948^{\ddagger}$	0.7	5.5	6.2
32	c64	$0.340^{\ddagger}$	$0.946^{\ddagger}$	0.2	5.4	5.6
32	c16	0.331	$0.943^{\ddagger}$	0.1	5.2	5.3
32	c4	0.320	$0.938^{\ddagger}$	0.02	5.1	5.1
32	c1	0.302	0.923	<0.01	4.9	4.9

5.2   Balancing Model Efficiency

Next, we examine the effectiveness-efficiency tradeoff of COILCR and its two main factors, the number of term representation dimensions and the degree of approximation from original term representations to canonical representations.

Table 2 shows the model accuracy and storage cost of COILCR on the MSMARCO passage dataset with varying representation sizes d_v and CRs per term $|C_t|$. By reducing each inverted index entry to a term weight and a CR index, COILCR significantly lowers the storage cost of the COIL index. The content and storage cost of the inverted index entry remains the same regardless of representation dimension changes.

All COILCR-∞ systems outperform the UNICOIL baseline where $d_v=1$. We further report the performance of COILCR variants with different representation dimensions d_v in Fig. 3. Higher dimension representations lead to a higher ceiling in model accuracy, but require more CRs per term to reach such performance. On the other hand, the overall difference in Recall@1000 for different d_v and different CR size becomes relatively small after very coarse term semantic modeling ($|C_t| > 16$). This may be beneficial in Recall-oriented settings such as first-stage ranking in a reranking pipeline, when a lower d_v and $|C_t|$ reduces run time and storage cost while not affecting the overall performance of the system.

Table 3. Passage retrieval accuracy and retrieval cost of COILCR with different CR generation strategies. c/d/u respectively denotes the constant, dynamic and universal clustering approaches, as discussed in Sect. 3.2. Models initialized with coCondenser.

Model	Model Performance		Run Cost	
	MRR@10	R@1000	CR Storage	Avg. Ops
c1024	0.351	0.949	2.5	1024
c256	0.346	0.948	0.77	256
c64	0.340	0.946	0.2	64
c16	0.331	0.942	0.06	16
c4	0.320	0.938	0.02	4
c1	0.302	0.923	<0.01	1
d0.05	0.349	0.949	2.3	997.4
d0.1	0.349	0.948	1.1	591.5
d0.15	0.344	0.947	0.27	152.5
d0.2	0.337	0.945	0.1	41.34
d0.25	0.330	0.942	0.05	14.76
u4096	0.346	0.946	<0.01	2740
u1024	0.339	0.945	<0.01	823
u256	0.336	0.943	<0.01	233
Ctok-c1024	0.339	0.946	–	–
Ctok-c256	0.329	0.944	–	–
Ctok-c64	0.309	0.939	–	–

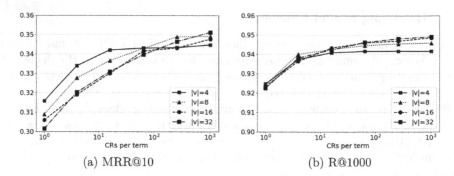

(a) MRR@10   (b) R@1000

Fig. 3. Passage retrieval performance on MSMARCO-dev for COILCR model variants with different representation dimensions.

5.3   Canonical Representation Analysis

In this section, we take a deeper look into the construction process and properties of canonical representations in COILCR. We first compare different CR selection strategies discussed in Sect. 3.2, and report model performance on MSMARCO in Table 3. As discussed in Sect. 3.2, for a query term at retrieval

Table 4. Zero-shot retrieval accuracy (nDCG@10) on the BEIR benchmark. CoCondenser initialization and DocT5Q augmentation were applied for all models. Best performance of each dataset is underlined.

Corpus	UNICOIL	COILCR-∞	COILCR-c256	COILCR-c256-tr
ArguAna	0.365	0.342	0.339	0.341
C-FEVER	0.178	0.186	0.188	0.188
DBPedia	0.360	0.378	0.380	0.376
FEVER	0.778	0.782	0.793	0.797
FiQA	0.293	0.310	0.303	0.297
HotpotQA	0.662	0.683	0.679	0.675
NFCorpus	0.336	0.338	0.338	0.336
NQ	0.446	0.485	0.483	0.477
Quora	0.732	0.773	0.762	0.750
SCIDOCS	0.150	0.153	0.154	0.151
SciFact	0.696	0.698	0.699	0.697
T-COVID	0.739	0.735	0.739	0.745
Touche2020	0.279	0.287	0.289	0.292

time, COILCR only performs vector product with its canonical representations instead of the representation of every term occurrence. In addition to CR index storage cost, we report the average number of retrieval-time *vector product operations*, or the average number of canonical representations a query term matches, to compare the computational cost between COILCR variants. Compared to term-specific CR selection, universal CR selection introduces much less storage cost, but naturally requires a larger set of CRs to preserve the semantics of *all terms*, and leads to extra operations at retrieval time. The two term-specific CR selection approaches have similar performance trends, as they require similar storage and operation costs to achieve the same level of performance.

To investigate the effect of factorizing term weight and term semantics, we additionally perform a side experiment where we directly generate canonical representations from COIL-tok term representations via k-means clustering. We denote this retrieval approach as Ctok-ck and report performance in Table 3. Compared to COILCR, the canonical representations generated from COIL-tok need to preserve extra information of the representation *norm*, which affects distance and loss calculation and leads to inefficient K-means clustering. Thus, this approach naturally requires much more CRs per term to reach the same retrieval performance as COILCR.

Additionally, to investigate the robustness of the COILCR system and the CR approximation approach, we take COILCR trained on MSMARCO and perform a *zero-shot* retrieval experiment on the BEIR [23] benchmark, which consists of datasets covering a wide range of different domains. We also introduce an extra COILCR variant, denoted as COILCR-tr, where we also directly transfer the CRs generated from MSMARCO representations, instead of generating from the new dataset. We report performance results on 13 datasets in the BEIR benchmark in Table 4. We observe that COILCR-∞ maintains its extra model capacity over UNICOIL, with larger than 3% gains on 7 of 13 datasets. the only dataset where COILCR clearly underperforms UNICOIL is ArguAna, which involves retrieval of counterarguments given a query argument, and is very different from classic web search settings. Moreover, across all datasets, the model accuracy of COILCR and COILCR-tr remains similar and close to the performance of COILCR-∞. This demonstrates the robustness of the CR approximation approach with sufficient clusters and suggests that the main bottleneck for COILCR in the zero-shot retrieval setting lies in the language model base, at the step of term representation generation.

6   Conclusion and Future Work

This paper investigates the model capacity and runtime cost of COIL-style lexical retrievers. We present COILCR, an extension to COIL which factorizes term representations into weighting and semantic components. At indexing time, COILCR constructs semantic canonical representations to approximate term semantics and precise matching between query and document terms, leading to reduced index storage and retrieval runtime cost.

Without approximation, COILCR-∞ maintains the model capacity of COIL-tok and consistently outperforms UNICOIL. Performing CR-based approximation for COILCR only slightly affects model accuracy, but drastically reduces the inverted index storage cost by 90% while also transforming most run-time vector product operations to a simple lookup operation.

Our experiments examine the effectiveness-efficiency balance of COILCR, and discuss the effects of different term representation sizes and clustering heuristics on model performance. We find that model accuracy is more prone to error from approximation, while consistent Recall performance can be achieved with very coarse term semantics. Experiments under different approximation and retrieval settings further demonstrate the robustness of the CR approximation approach.

Throughout this work, we observe the necessity of modeling term semantics in lexical exact match retrieval, as well as the potential of very efficiently doing so. In this paper, we utilize spherical clustering as a simple post-hoc approach for CR generation and note the possibility of finding improved methods to build canonical representation sets which reflect term senses. We hope this is an encouraging step towards building both effective and efficient lexical retrieval models and indexes in the future.

References

1. Dai, Z., Callan, J.: Context-aware document term weighting for Ad-hoc search. In: Proceedings of The Web Conference 2020, pp. 1897–1907 (2020)
2. Devlin, J., Chang, M.W., Lee, K., Toutanova, K.: BERT: pre-training of deep bidirectional transformers for language understanding. arXiv preprint arXiv:1810.04805 (2018)

3. Dhillon, I.S., Modha, D.S.: Concept decompositions for large sparse text data using clustering. Mach. Learn. **42**(1), 143–175 (2001)
4. Formal, T., Lassance, C., Piwowarski, B., Clinchant, S.: SPLADE v2: sparse lexical and expansion model for information retrieval. arXiv preprint arXiv:2109.10086 (2021)
5. Formal, T., Piwowarski, B., Clinchant, S.: SPLADE: sparse lexical and expansion model for first stage ranking. In: Proceedings of the 44th International ACM SIGIR Conference on Research and Development in Information Retrieval, pp. 2288–2292 (2021)
6. Gao, L., Callan, J.: Condenser: a pre-training architecture for dense retrieval. arXiv preprint arXiv:2104.08253 (2021)
7. Gao, L., Callan, J.: Unsupervised corpus aware language model pre-training for dense passage retrieval. arXiv preprint arXiv:2108.05540 (2021)
8. Gao, L., Dai, Z., Callan, J.: COIL: revisit exact lexical match in information retrieval with contextualized inverted list. In: Proceedings of the 2021 Conference of the North American Chapter of the Association for Computational Linguistics: Human Language Technologies, NAACL-HLT 2021, Online, 6–11 June 2021. pp. 3030–3042. Association for Computational Linguistics (2021). https://doi.org/10.18653/v1/2021.naacl-main.241
9. Hofstätter, S., Althammer, S., Schröder, M., Sertkan, M., Hanbury, A.: Improving efficient neural ranking models with cross-architecture knowledge distillation. arXiv preprint arXiv:2010.02666 (2020)
10. Hofstätter, S., Lin, S.C., Yang, J.H., Lin, J., Hanbury, A.: Efficiently teaching an effective dense retriever with balanced topic aware sampling. In: Proceedings of the 44th International ACM SIGIR Conference on Research and Development in Information Retrieval, pp. 113–122 (2021)
11. Izacard, G., et al.: Towards unsupervised dense information retrieval with contrastive learning. arXiv preprint arXiv:2112.09118 (2021)
12. Izacard, G., Petroni, F., Hosseini, L., De Cao, N., Riedel, S., Grave, E.: A memory efficient baseline for open domain question answering. arXiv preprint arXiv:2012.15156 (2020)
13. Johnson, J., Douze, M., Jégou, H.: Billion-scale similarity search with GPUs. IEEE Trans. Big Data **7**(3), 535–547 (2019)
14. Karpukhin, V., et al.: Dense passage retrieval for open-domain question answering. In: Proceedings of the 2020 Conference on Empirical Methods in Natural Language Processing (EMNLP), pp. 6769–6781 (2020)
15. Khattab, O., Zaharia, M.: ColBERT: efficient and effective passage search via contextualized late interaction over bert. In: Proceedings of the 43rd International ACM SIGIR Conference on Research and Development in Information Retrieval, pp. 39–48 (2020)
16. Lin, J., Ma, X.: A few brief notes on deepimpact, coil, and a conceptual framework for information retrieval techniques. CoRR abs/2106.14807 (2021). https://arxiv.org/abs/2106.14807
17. Nguyen, T., et al.: MS MARCO: a human generated machine reading comprehension dataset. In: CoCo@ NIPS (2016)
18. Nogueira, R., Cho, K.: Passage re-ranking with BERT. arXiv preprint arXiv:1901.04085 (2019)
19. Nogueira, R., Lin, J., Epistemic, A.: From doc2query to docTTTTTquery. Online preprint 6 (2019)
20. Qu, Y., et al.: RocketQA: an optimized training approach to dense passage retrieval for open-domain question answering. arXiv preprint arXiv:2010.08191 (2020)

21. Robertson, S., Zaragoza, H., et al.: The probabilistic relevance framework: Bm25 and beyond. Found. Trends® Inf. Retriev. **3**(4), 333–389 (2009)
22. Santhanam, K., Khattab, O., Saad-Falcon, J., Potts, C., Zaharia, M.: ColBERTv2: effective and efficient retrieval via lightweight late interaction. arXiv preprint arXiv:2112.01488 (2021)
23. Thakur, N., Reimers, N., Rücklé, A., Srivastava, A., Gurevych, I.: BEIR: a heterogenous benchmark for zero-shot evaluation of information retrieval models. arXiv preprint arXiv:2104.08663 (2021)
24. Vaswani, A., et al.: Attention is all you need. In: Advances in Neural Information Processing Systems 30 (2017)
25. Xiong, L., et al.: Approximate nearest neighbor negative contrastive learning for dense text retrieval. arXiv preprint arXiv:2007.00808 (2020)
26. Yamada, I., Asai, A., Hajishirzi, H.: Efficient passage retrieval with hashing for open-domain question answering. arXiv preprint arXiv:2106.00882 (2021)
27. Zhan, J., Mao, J., Liu, Y., Guo, J., Zhang, M., Ma, S.: Jointly optimizing query encoder and product quantization to improve retrieval performance. In: Proceedings of the 30th ACM International Conference on Information & Knowledge Management, pp. 2487–2496 (2021)
28. Zhan, J., Mao, J., Liu, Y., Guo, J., Zhang, M., Ma, S.: Learning discrete representations via constrained clustering for effective and efficient dense retrieval. In: Proceedings of the Fifteenth ACM International Conference on Web Search and Data Mining, pp. 1328–1336. WSDM 2022, Association for Computing Machinery, New York, NY, USA (2022). https://doi.org/10.1145/3488560.3498443
29. Zhao, T., Lu, X., Lee, K.: SPARTA: efficient open-domain question answering via sparse transformer matching retrieval. In: Proceedings of the 2021 Conference of the North American Chapter of the Association for Computational Linguistics: Human Language Technologies, pp. 565–575. Association for Computational Linguistics, Online (2021). https://doi.org/10.18653/v1/2021.naacl-main.47. https://aclanthology.org/2021.naacl-main.47

Bootstrapped nDCG Estimation
in the Presence of Unjudged Documents

Maik Fröbe[1(✉)], Lukas Gienapp[2], Martin Potthast[2], and Matthias Hagen[1]

[1] Friedrich-Schiller-Universität Jena, Jena, Germany
maik.froebe@uni-jena.de
[2] Leipzig University and ScaDS.AI, Leipzig, Germany

Abstract. Retrieval studies often reuse TREC collections after the corresponding tracks have passed. Yet, a fair evaluation of new systems that retrieve documents outside the original judgment pool is not straightforward. Two common ways of dealing with unjudged documents are to remove them from a ranking (condensed lists), or to treat them as non- or highly relevant (naïve lower and upper bounds). However, condensed list-based measures often overestimate the effectiveness of a system, and naïve bounds are often very "loose"—especially for nDCG when some top-ranked documents are unjudged. As a new alternative, we employ bootstrapping to generate a distribution of nDCG scores by sampling judgments for the unjudged documents using run-based and/or pool-based priors. Our evaluation on four TREC collections with real and simulated cases of unjudged documents shows that bootstrapped nDCG scores yield more accurate predictions than condensed lists, and that they are able to strongly tighten upper bounds at a negligible loss of accuracy.

1 Introduction

The Cranfield experiments [12,13] were conducted on a collection of 1,400 documents and complete relevance judgments for 225 topics. Since collection sizes grew substantially, complete judgments became infeasible almost immediately thereafter. The current best practice at shared tasks in IR is to create per-topic pools of the submitted systems' top-ranked documents and then judge each topic's pool [40]. Systems that did not contribute to the pools may then later retrieve some unjudged documents. Thakur et al. [36] recently observed this for TREC-COVID [41], where dense retrieval models in post-hoc experiments retrieved many unjudged documents that turned out to be relevant. Typical reasons for "incomplete" judgments are lacking run diversity or time constraints—which was the case for TREC-COVID as per Roberts et al. [29]. When reusing shared task data, one thus often has to deal with unjudged documents.

Unjudged documents can be judged post hoc, but this can be costly and inconsistent with the original judging process. Typically, post-hoc evaluations either remove unjudged documents (condensing the results lists of a new system to the included judged documents in their relative order) [31], or the unjudged documents are assumed to either all being non- or highly relevant (naïve lower/upper bounds) [25]. Both ideas have drawbacks: Condensed lists

© The Author(s), under exclusive license to Springer Nature Switzerland AG 2023
J. Kamps et al. (Eds.): ECIR 2023, LNCS 13980, pp. 313–329, 2023.
https://doi.org/10.1007/978-3-031-28244-7_20

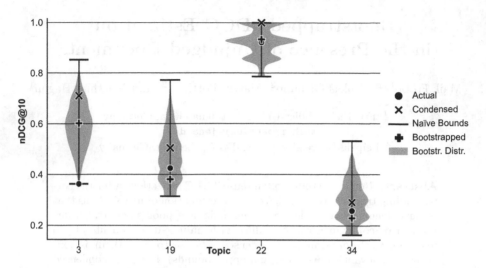

Fig. 1. Actual (obtained via post-judging) and estimated nDCG@10 of the dense retrieval model ANCE for selected TREC-COVID topics with unjudged documents.

often overestimate effectiveness [33], and the difference between naïve lower and upper bounds can be very large [25]—especially for a recall-oriented measure such as nDCG [23], one of the most reported measures for many retrieval tasks [11,15,17,36]. We further show that lower/upper bounds on nDCG are potentially incomparable to results reported based on complete judgments on the same data (Sect. 3.3).

To address the outlined problems, we propose a new bootstrapping approach to estimate nDCG in the presence of unjudged documents (Sect. 3). By repeatedly sampling judgments for unjudged documents using run- and/or pool-based priors, we derive a distribution of possible nDCG scores for a retrieval system on a topic. Figure 1 compares such distributions with the estimates of condensed lists and the naïve lower/upper bounds on selected TREC-COVID topics for the dense retrieval model ANCE [43] (which retrieved many unjudged documents deemed relevant [36]). The distributions help to identify topics with an extremely unlikely naïve upper bound (Topics 3, 19, 34), or where only a few nDCG scores between the bounds are very likely (Topic 22). In an evaluation on the Robust04, ClueWeb09, ClueWeb12, and TREC-COVID collections with real and simulated unjudged documents, we show the mode of the bootstrapped nDCG score distribution to be a more accurate estimate than those obtained from condensed lists and the, often default, naïve lower bound (Sect. 4). Moreover, bootstrapped nDCG bounds can be configured to be a lot tighter than the naïve upper bound at a negligible loss of accuracy. For future nDCG evaluations with unjudged documents, we share our data and code compatible with TrecTools [28].

2   Background and Related Work

We briefly review the nDCG evaluation measure, methods for dealing with unjudged documents, and previous applications of bootstrapping in IR.

Normalized Discounted Cumulative Gain (nDCG). The nDCG [23] is one of the most widely used IR evaluation measures (e.g., in the TREC Web and Deep Learning tracks [8,17] or in the BEIR benchmark [36]). It is a normalized version of the discounted cumulative gain (DCG) that combines result ranks and graded relevance so that lower-ranked results contribute less "gain". The DCG is usually defined as

$$\mathrm{DCG@}k = \sum_{i=1}^{k} \frac{2^{rel(d_i, q)} - 1}{\log_2(1 + i)} ,$$

where k is the maximum rank to consider, $rel(d_i, q)$ is the graded relevance judgment of the document returned at rank i for the query q, the logarithm ensures smooth reduction, and $2^{rel(d_i, q)}$ emphasizes highly relevant documents. The nDCG@k normalizes a system's DCG@k score by dividing by the DCG*@k score of the "ideal" top-k ranking of the pool (i.e., the ranking of the judged documents by relevance). Note that the ideal ranking may easily include documents that some systems do not return in their results.

Methods to Deal with Unjudged Documents. Only a few "specialized" retrieval effectiveness measures specifically target situations with unjudged documents (e.g., bpref [4] or RBP [27]). Yet, these measures are used in only a few scenarios like the TREC 2009 Web track [8] that aimed for minimal judgment pools [6]. Most retrieval studies instead usually report measures that assume all documents in the evaluated part of a ranking to have relevance judgments (e.g., nDCG). When evaluating a new retrieval system in the scenario of such a study, retrieved documents that were not in the original judgment pool cause problems [4,46].

Typical methods [25] to deal with unjudged documents are: (1) assuming non-relevance, (2) predicting relevance, (3) condensing result lists, or (4) computing naïve bounds. Assuming non-relevance for unjudged documents is the standard in trec_eval, but only yields good results for "essentially" complete judgments [42] and favors systems that retrieve many (relevant) judged documents [35]. Since systems that retrieve unjudged but relevant documents might be severely underestimated [46], there have been attempts to automatically predict relevance [1,2,5,7] (e.g., based on document content). However, such predictions can be problematic given that even experienced human assessors can struggle [38]. Also inferred measures like infAP [44] and infNDCG [45] could be viewed as prediction approaches. They exploit the probabilities with which documents were sampled for incomplete judgment pools with reduced overall effort [39]. But inference does not really work for post-hoc evaluation of systems that did not contribute to the original pool sampling since the sampling probabilities for newly retrieved high-ranked documents then can be undefined. Still, the general idea of sampling inspired our approach.

In the condensed list approach, all unjudged documents are removed from a ranked list before calculating effectiveness. The conceptual simplicity and the experimental evidence [31] that condensed lists give better results than the specially designed bpref helped condensed lists to become widely used—also in Trec-Tools [28] or PyTerrier [26]. But like relevance prediction, compressed lists also have the disadvantage of hiding the potential uncertainty created by unjudged documents. This motivates approaches that make this uncertainty "visible," such as calculating (naïve) lower or upper effectiveness bounds [25, 27].

Naïve bounds contrast the worst case with the best case by calculating the score a system would achieve if all unjudged documents were non-relevant or highly relevant. In the context of utility-based (based only on ranking) and recall-based (normalized by a "best possible" ranking) evaluation measures, the naïve bounds are designed for the former [25]. For utility-based measures, any actual effectiveness score of a system is guaranteed to be within the naïve bounds. However, for recall-oriented measures like nDCG, we show that the actual effectiveness of a system may lie outside the naïve bounds (cf. Sect. 3.3) and that expanding them often leads to meaningless 0.0 (lower) and 1.0 (upper) bounds.

Our new bootstrapping approach addresses the outlined shortcomings of the existing ideas for dealing with unjudged documents when using nDCG. By deriving a distribution of possible nDCG scores, we allow tighter bounds and more informed point estimates. Both improvements are based on the same underlying distribution of possible nDCG values, which also simplifies uncertainty assessment and interpretation.

Bootstrapping in Information Retrieval. Bootstrapping is a statistical technique in which repeated samples are drawn from data to obtain a distribution for subsequent statistical analyses [18]. It has been applied to various statistical problems in information retrieval, either as topic bootstrapping or corpus bootstrapping. Topic bootstrapping was probably the first use of bootstrapping in IR [34]. It refers to the repeated sampling of queries for some statistical analyses and has been used in significance tests [34, 35] or to assess the discriminatory power of effectiveness measures [30, 32, 47]. However, topic bootstrapping is not intended to assess the uncertainty created by unjudged documents.

In corpus bootstrapping, documents are sampled from a corpus to simulate different corpora [47]. Previous use cases of corpus bootstrapping include assessing the transferability of system comparisons between different corpora [16] or the robustness of evaluation measures [47] and significance tests [19]. The assumption underlying corpus bootstrapping is that observations should be stable between (slightly) different corpora. This inspired our idea of applying bootstrapping to evaluations with unjudged documents in the sense that an unjudged document should "behave" similarly to the judged documents in a run and/or pool. Bootstrapping has not yet been applied to the evaluation of unjudged documents, although the research reviewed above shows that bootstrapping enables similar applications. By making our code publicly available, we try to support Sakai's call for bootstrapping to get more attention in IR [30].

3   Bootstrapping nDCG Scores

After preparatory theoretical considerations, we propose a bootstrapping approach to generate nDCG score distributions by repeatedly sampling judgments
for unjudged documents. Based on the lessons learned, we then reconsider current methods for estimating lower and upper bounds and propose improvements.

3.1   Preparatory Theoretical Considerations

As briefly discussed in Sect. 2, nDCG requires judgments to be complete up to
the desired scoring depth k. Unjudged documents in the top-k results of a system
must therefore either be post-judged, or be estimated otherwise based on some
strategy. Post-judgments are costly and may lead to inconsistencies with prior
judgments. This often leaves automatically estimating unjudged documents as
the most feasible practical option.

A first idea could be to simply randomly sample relevance labels for unjudged
documents. But without any further corrections, this approach can lead to
invalid results. For instance, consider an evaluation setting with three relevance
grades $\{0; 1; 2\}$ and a fictional judgment pool that contains nine highly relevant
documents (grade 2), one relevant document (grade 1), and arbitrarily many
non-relevant documents (grade 0) for some topic. Assume that a to-be-evaluated
system A returns in its top-10 results the nine highly relevant documents from
the pool and one unjudged document not part of the pool. Suppose that relevance grade 2 is randomly sampled for the unjudged document. Adding this
sampled highly relevant document to the pool then improves the ideal ranking:

$$\text{DCG}^*_{\text{original pool}}@10 \quad < \quad \text{DCG}^*_{\text{pool with sample}}@10 \, .$$

If $\text{DCG}^*_{\text{pool with sample}}@10$ is used as the normalization denominator for computing the nDCG@10 of system A, the resulting scores are thus not directly
comparable to nDCG scores of other systems calculated based on complete
judgments for the original pool and $\text{DCG}^*_{\text{original pool}}@10$. Comparability could
be reestablished by recalculating the nDCG scores of the other systems using
$\text{DCG}^*_{\text{pool with sample}}@10$. Yet, recalculating scores might be biased towards the
newly added system: in case the randomly sampled score is higher than the
unjudged document's true relevance, recomputing diminishes the original systems' nDCG scores below their true value, yet increases the newly added systems' nDCG beyond its true value.

Conversely, also using $\text{DCG}^*_{\text{original pool}}@10$ as the denominator to maintain
comparability is not valid. In the example case of system A, this would cause

$$\text{DCG}_{\text{system } A}@10 \quad > \quad \text{DCG}^*_{\text{original pool}}@10 \quad \leadsto \quad \frac{\text{DCG}_{\text{system } A}@10}{\text{DCG}^*_{\text{original pool}}@10} \quad > \quad 1 \, ,$$

which exceeds the range of nDCG expected from normalization.

It follows that theoretically sound and empirically viable nDCG estimation
approaches to handle unjudged documents *must not* change the pool's initial
number of judgments per relevance grade in order to preserve the $\text{DCG}^*@k$.

Algorithm 1.   Bootstrapping nDCG Scores

Input:	R	top-k ranking for query q that contains unjudged documents
	J	pool of pairs $(d, rel(d, q))$ (i.e., documents with relevance judgments)
	b	number of desired bootstrapped nDCG scores
	prior	pool-, run-, or pool+run-based sampling probability
Output:	*Scores*	multiset of b bootstrapped nDCG scores for R based on J and *prior*

1:   $Scores \leftarrow \emptyset$
2:   **repeat** b-times   ▷ following Sakai [30], we usually set $b = 1,000$
3:      $J' \leftarrow J, \quad S' \leftarrow \emptyset$ ▷ buffers for pool and judgm. sample of unjudg. documents
4:      **for all** unjudged documents $d \in R$ **do**   ▷ try to sample *prior*-based judgment
5:          select target relevance label r for d based on *prior*
6:          **if** J' contains a document $d' \notin R$ with $rel(d', q) = r$ **then**
7:              $J' = J' \setminus \{(d', r)\}$
8:              $S' = S' \cup \{(d, r)\}$ ▷ desired judgment can be sampled from pool
9:          **else if** J' contains a document $d' \notin R$ with $0 \le rel(d', q) < r$ **then**
10:            let $d' \notin R$ be a document in J' with highest $rel(d', q) < r$
11:            $J' = J' \setminus \{(d', rel(d', q))\}$
12:            $S' = S' \cup \{(d, rel(d', q))\}$ ▷ otherwise, sample best possible lower judgm.
13:          **else**
14:            $S' = S' \cup \{(d, 0)\}$ ▷ fallback: standard assumption of non-relevance
15:      $Scores \leftarrow Scores \cup \left\{ \frac{\text{DCG@}k \text{ of } R \text{ based on } J' \cup S'}{\text{DCG*@}k \text{ of } J} \right\}$

3.2   Our Bootstrapped nDCG Estimation Approach

Algorithm 1 shows our approach. It meets the constraint of preserving the number of judgments per relevance grade in the pool by restricting the random sampling of relevance degrees to a *prior*. In each of the b bootstrap iterations, a relevance grade r is sampled for an unjudged document in the top-k ranking R from the judgment pool J according to one of three sampling *priors*:

$$\text{pool-based} \qquad P(rel = r \mid J) \quad = \quad \frac{|\{d \in J \, : \, rel(d, q) = r\}|}{|J|},$$

$$\text{run-based} \qquad P(rel = r \mid R) \quad = \quad \frac{|\{d \in R \, : \, rel(d, q) = r\}|}{|\{d \in R \, : \, d \text{ is judged}\}|}, \text{ and}$$

$$\text{pool+run-based} \quad P(rel = r \mid J, R) \quad = \quad \frac{P(rel = r \mid J) + P(rel = r \mid R)}{2}.$$

During sampling, our approach checks in each iteration whether the desired relevance grade r is still present in the pool. If not, the highest possible judgment that is below the desired grade is selected, with grade 0 as the default fallback option. This sampling strategy guarantees that the ideal ranking of the original pool J and the ideal ranking of the final "sampled" judgments $J' \cup S'$ have the same DCG*@k. The bootstrapped nDCG scores for R are thus directly comparable to nDCG scores of other rankings derived from the same pool J (e.g., to completely judged runs with nDCG scores computed on the initial pool).

Table 1. Examples with incorrect RBP-inspired/naïve nDCG@2 bounds or with very broad guaranteed nDCG bounds; relevance labels from 0 (not rel.) to 3 (highly rel.).

Bound	Input	Truth	Estimated nDCG Bounds vs. Actual Score		
	? = unjudged		Lower Bound $\leq$	Actual	$\leq$ Upper Bound
RBP-insp.	[1, ?]	[1, **2**]	$\frac{DCG([1,0])}{DCG([1,0])} = 1.00 \nleq$	$\frac{DCG([1,2])}{DCG([2,1])} = 0.80 \nleq$	$\frac{DCG([1,3])}{DCG([3,1])} = 0.71$
Naïve	[?, 1]	[**2**, 1]	$\frac{DCG([0,1])}{DCG([1,0])} = 0.63 \leq$	$\frac{DCG([2,1])}{DCG([2,1])} = 1.00 \nleq$	$\frac{DCG([0,1])}{DCG([1,0])} = 0.63$
Guarant.	[1, ?]	[1, **2**]	$\frac{DCG([1,0])}{DCG([\mathbf{3,3}])} = 0.09 \leq$	$\frac{DCG([1,2])}{DCG([2,1])} = 0.80 \leq$	$\frac{DCG([1,0])}{DCG([1,0])} = 1.00$

Efficient Implementation. Our bootstrapping approach computes nDCG scores in each iteration. To ensure efficiency, we precompute and tabulate the possible discounted gain values for each relevance grade at each of the top-k ranks, the DCG*@k of the ideal ranking of the given pool J, and the sum of the discounted gain values of the judged documents in R—all of these values do not change during bootstrapping. The nDCG score computation can then look up the sampled discounted gain values for unjudged documents, add them to the precomputed intermediate DCG of the judged part of R, and divide by the precomputed DCG*@k of J. On an AMD Epyc 1.8 GHz CPU, a TrecTools-based tabulated implementation of our approach takes an average of 2.84 s per topic (stddev: 0.01 s) to bootstrap nDCG@10 scores for the four runs that have the most unjudged documents in TREC-COVID (9–32% unjudged documents) as per Thakur et al. [36]—without tabulation: 17.62 s (stddev: 0.91 s). The fast run time shows that bootstrapping is practically applicable, especially since further massive parallelization is possible.

3.3 Conceptual Comparison

Our preparatory considerations from Sect. 3.1 also apply to the derivation of lower/upper bounds for nDCG. Bounds for nDCG inspired by RBP [25,27] can be incomparable, too. Naïve bounds can easily be made comparable but we show that they and RBP-inspired bounds are not guaranteed to be correct. We thus devise guaranteed bounds, but show that they then "necessarily" are very broad.

Error Bounds for nDCG. Inspired by the error bounds proposed for the utility-based measure RBP [25,27], lower/upper bounds for nDCG may be derived by either assigning a relevance grade of 0 or the highest relevance grade to all unjudged documents. But since the latter changes the ideal ranking, such an upper bound can lead to incomparable nDCG scores. Therefore, in order to yield comparable scores, we propose that an RBP-inspired "naïve" upper bound for nDCG should iteratively greedily assign the highest still available relevance judgment from the pool to the highest ranked unjudged document. If the pool's available non-zero grades are exhausted, 0 is assigned. This naïve bounding does not change the DCG*@k and thus yields scores comparable to other rankings on

Table 2. Characteristics of methods to deal with unjudged documents in nDCG scoring. Some are deterministic, some not (Det.), and they use different strategies with pool-and/or run-based priors. All are "comparable" (i.e., do not change the ideal DCG*@k).

Approach	Det.	Selection/Sampling Strategy	Prior		Comp.
			Run	Pool	
Condensed lists [31]	✓	Remove unjudged documents.	✓	✗	✓
Naïve low. b. [25, 27]	✓	Unj. = Non-relevant.	✗	✗	✓
Naïve upper bound	✓	Unj. = Highest remaining judgm.	✗	✓	✓
Pool-based bootstr.	✗	$P(rel = r \mid J) = \frac{|\{d \in J : rel(d,q)=r\}|}{|J|}$	✗	✓	✓
Run-based bootstr.	✗	$P(rel = r \mid R) = \frac{|\{d \in R : rel(d,q)=r\}|}{|\{d \in R : d \text{ is judged}\}|}$	✓	✗	✓
Pool+run-based bs.	✗	$P(rel = r \mid J, R) = \frac{P(r \mid J) + P(r \mid R)}{2}$	✓	✓	✓

the pool. However, the examples in Table 1 show that both the RBP-inspired and the naïve bounds can be incorrect. The RBP-inspired lower bound (and thus also the equivalent naïve lower bound) can be too high (first row; the actual grade of 2 for the unjudged document increases DCG*@k more than DCG@k). Similarly, also the upper RBP-inspired and naïve bounds can be incorrect (first and second row). For a guaranteed correct lower bound, a hypothetical ideal ranking needs to be assumed that consists of only documents with the highest relevance grade, and all unjudged documents get a grade of 0. Computing a guaranteed correct upper bound is more complicated but in the end usually uses a different ideal ranking which makes the guaranteed bounds incomparable.

Discussion. Table 2 summarizes characteristics of methods that deal with unjudged documents but that preserve the ideal ranking. The methods rely on different priors (none, pool-, run-, or pool+run-based)—some only implicitly, like the upper bound method, which uses the pools highest remaining judgments. Our bootstrapping idea incorporates priors from both run and pool, and indicates the uncertainty introduced by unjudged documents through a probability distribution. Condensed lists and naïve bounds only generate point scores.

4 Evaluation

We experimentally compare our bootstrapping approach to naïve bounds and condensed lists on real and simulated scenarios with unjudged documents on the Robust04, ClueWeb09, ClueWeb12, and TREC-COVID collections. In the comparison, we assess the ability to predict actual nDCG scores, their effects on system rankings, and the tightness of potential bounds. For score prediction and the creation of subsequent system rankings, our approach uses the most likely nDCG score from the bootstrapped distribution, for tighter bounds, our approach uses fixed percentiles in the bootstrapped distribution. All experiments use nDCG@10, since it is predominant in shared tasks and the highest cut-off for which the four collections have complete judgments for the submitted runs.

Table 3. The prevalence of each relevance label in the judgment pool and the unjudged documents, respectively. For Robust04, ClueWeb09, and ClueWeb12, we show the simulated incompleteness averaged over groups; TREC-COVID is real incompleteness.

Corpus	Judgement Pool					Unjudged Documents				
	0	1	2	3	4	0	1	2	3	4
ClueWeb09	0.74	0.17	0.07	0.01	0.01	0.80	0.15	0.03	0.01	0.01
ClueWeb12	0.64	0.25	0.09	0.02	0.02	0.67	0.24	0.08	0.02	0.01
Robust04	0.80	0.18	0.02	0.00	0.00	0.96	0.04	0.00	0.00	0.00
TREC-COVID	0.63	0.16	0.21	0.00	0.00	0.75	0.02	0.23	0.00	0.00

4.1 Experimental Setup

We compare a run with unjudged documents in two setups against (1) runs without unjudged documents (measuring the accuracy of lower and upper bounds), and (2) other runs without unjudged documents (measuring correlations in system rankings). Score ties in a run are solved via alphanumeric ordering by document ID (following a recommendation by Lin and Yang [24]). To reduce the impact of low-performing systems, only the 75% of runs with the highest nDCG@10 are included (following a similar setup by Bernstein and Zobel [3]). The ClueWeb corpora have a high number of near-duplicates [20] that might invalidate subsequent evaluations [3,21,22]. We use pre-calculated lists [20] to deduplicate the run and qrel files. We follow trec_eval and replace negative relevance judgments with 0. All experiments use TrecTool's nDCG@10 implementation with default parameters, and we report statistical significance where applicable according to the Students' t-test with Bonferroni correction at $p = 0.05$.

Test Collections. Our evaluation is based on four collections: (1) Robust04 [37] (528,155 documents, 249 topics, 311,410 relevance judgments, pool: 111 runs by 14 groups), (2) ClueWeb09 (1 billion web pages, 200 topics, 58,414 judgments from TREC Web tracks [8–11], pools: 32–71 runs by 12–23 groups), (3) ClueWeb12 (0.7 billion web pages, 100 topics, 23,233 judgments from TREC Web tracks [14,15], pools: 34 + 30 runs by 14 + 12 groups), (4) TREC-COVID [41] (171,332 documents, 50 topics, 66,336 judgments).

Establishing Incompleteness. TREC-COVID allows a real case study on incompleteness. In post-hoc experiments [36], three models retrieved 17% to 41% unjudged documents in their top-10 that were post-judged [36]. For Robust04, ClueWeb09, and ClueWeb12, we simulate incomplete pools with the "leave one group out" method [38], adjusting the pool by removing documents solely contributed by the group submitting a run (i.e., only their runs have the document in the top-10 results), simulating that the group did not participate. This yields one incomplete pool per group, where runs of other groups remain fully judged.

Table 3 provides an overview of the ratios of relevance degrees in the pools and the unjudged documents. For simulated incompleteness, we report averages over

Table 4. Overview of nDCG score prediction assessed by the actual RMSE, and the lower and upper bound RMSE (ignoring under/overestimations) on Robust04 (R04), ClueWeb09 (CW09), and ClueWeb12 (CW12). We report statistical significance according to Student's t-test with Bonferroni correction at p=0.05 to the naïve lower (†) and upper bound (‡), respectively condensed lists (∗).

Approach	RMSE on R04			RMSE on CW09			RMSE on CW12		
	Lower	Actual	Upper	Lower	Actual	Upper	Lower	Actual	Upper
Naïve (L)	**.004***‡	.058*‡	.058*‡	**.009***‡	.076*‡	.076*‡	**.007***‡	.113*‡	.113*‡
Conden.	.062†‡	.068†‡	.027†‡	.081†‡	.087†‡	.034†‡	.081†‡	.092†‡	.043†‡
Naïve (U.)	.210†*	.210†*	**.002**†*	.338†*	.338†*	**.000**†*	.307†*	.307†*	**.001**†*
Bootstr.$_P$	.078†*‡	.083†*‡	.027†‡	.086†‡	.097†*‡	.046†*‡	.093†*‡	.105*‡	.048†‡
Bootstr.$_R$	.007†*‡	.058*‡	.058*‡	.021†*‡	.077*‡	.075*‡	.059†*‡	.108*‡	.091†*‡
Bootstr.$_{P+R}$	.037†*‡	**.056***‡	.041†*‡	.046†*‡	**.074***‡	.058†*‡	.058†*‡	**.083**†‡	.060†*‡

all groups. None of the collections are complete, as all have relevant documents among the unjudged ones. However, for Robust04, the high number of submitted runs and deep pooling ensured that the pools are "essentially complete", even for simulated incompleteness (4% of the unjudged documents are relevant). The remaining collections have 20% to 33% relevant documents among the unjudged ones, providing a good range of (in)completeness for our experiments.

4.2 Evaluation Results

For nDCG prediction experiments, accuracy is reported as root-mean-square error (RMSE), contrasted by two RMSE variants that assess lower and upper bounds. Furthermore, we measure the correlation of system rankings obtained by predicted nDCG scores to the ground truth rankings as Kendall's τ and Spearman's ρ. For experiments on tightening naïve bounds, we measure precision and recall in reconstructing per-topic system rankings. Evaluation is first conducted on simulated incompleteness and concludes with the TREC-COVID case study.

nDCG Score Predicion. Table 4 reports the nDCG@10 prediction accuracy of all tested approaches. We report the actual RMSE, a lower-bound RMSE (ignoring underestimations), and an upper-bound RMSE (ignoring overestimations). Cases with incorrect naïve bounds occur in practice but are rare. The naïve lower bound is slightly more inaccurate than the naïve upper bound (maximum violations of 0.009 on ClueWeb09 for the lower bound vs. 0.002 for the upper bound on Robust04). Similar to the incompleteness degrees of the collections (Table 3), the actual RMSE is rather small on Robust04, larger on ClueWeb09, and the highest on ClueWeb12. Consequently, the naïve lower bound that assumes unjudged documents are non-relevant has high accuracy on both collections, but is outperformed by condensed lists on ClueWeb12 (RMSE 0.113 vs. 0.92).

Our three bootstrapping variants with a prior from the pool (Bootstr.$_P$), the run (Bootstr.$_R$), or both (Bootstr.$_{P+R}$) show that priors from the run yield more accurate results than from the pool, and combining both yields the highest accuracy in all cases, significantly improving upon the naïve lower and upper bound,

Table 5. Overview of the correlation between system rankings obtained via predicted nDCG@10 scores on incompletely judged runs to those runs with complete judgments. We report Kendall's τ and Spearman's ρ on Robust04, ClueWeb09, ClueWeb12, and the mean over those three corpora.

Approach	Robust04		ClueWeb09		ClueWeb12		Mean	
	τ	ρ	τ	ρ	τ	ρ	τ	ρ
Naïve (L)	.936	**.997**	**.821**	**.959**	.646	.837	.801	.931
Conden.	.924	.978	.610	.744	.786	.889	.773	.870
Naïve (U.)	.189	-.268	-.411	-.656	-.097	-.250	-.106	-.391
Bootstr.$_P$	.911	.975	.644	.824	.781	.909	.779	.903
Bootstr.$_R$	.943	**.997**	.721	.878	.764	.908	.810	.927
Bootstr.$_{P+R}$	**.966**	.996	.716	.885	**.814**	**.924**	**.832**	**.935**

and condensed lists. This result is reasonable, as the combination of run priors and pool priors allows the bootstrapping approach to account for relationships between the topic and the run. The results show that bootstrapped nDCG scores from run and pool priors are highly applicable in practice as they yield the most accurate nDCG predictions in all our experiments. Additionally, by comparing the lower- and upper-bound RMSE of condensed lists with those of pool/run-based bootstrapping, we observe that condensed lists are inclined to overestimate on all corpora. In contrast, bootstrapped predictions are more balanced with a tendency for underestimations, which is preferable in practice [35].

System Ranking Reconstruction Against Incompletely Judged Runs. We contrast our experiments on the accuracy of predicted nDCG@10 scores by measuring the correlation of system rankings obtained via predicted scores on incompletely judged runs to the ground truth system ranking obtained via fully judged runs. Therefore, we predict the nDCG@10 sores of each run using the incomplete judgments for the run obtained via the "leave one group out" method [38]. Table 5 reports the correlation of the system rankings obtained on the incomplete judgments with the ground-truth system ranking measured as Kendall's τ and Spearman's ρ. Again, we observe that the judgment pool for Robust04 is, even with simulated incompleteness, highly reusable as all approaches (besides the naïve upper bound) achieve high correlations (pool/run- based bootstrapping having the highest Kendall's τ of 0.966). Our pool/run-based bootstrapping substantially outperforms condensed lists in all cases, and also achieves the highest correlation on average over all three corpora (Kendall's τ of 0.832).

System Ranking Reconstruction Against Fully Judged Runs. To assess pool/run-based bootstrapping for tightening naïve bounds, we compare different methods for score prediction w.r.t. their ability to reconstruct the topic-level ground-truth ranking of systems. Given a run with unjudged documents, we first calculate point estimates: the naïve lower bound, condensed list, and the most likely score according to pool/run-based bootstrapping. Then, score ranges are established,

Table 6. Precision, recall, and F1 in reconstructing topic-level system rankings with unjudged documents. We report significance (Student's t-test with Bonferroni correction at p=0.05) to the point estimate of list condensation (∗) and score ranges starting at the lower bound, ending at the naïve upper bound (†), resp. list condensation (‡).

Approach		Reconstr. on R04			Reconstr. on CW09			Reconstr. on CW12		
		Prec.	Rec.	F1	Prec.	Rec.	F1	Prec.	Rec.	F1
Point	Naïve (L.)	.954†*	.954†*‡	.954†*	.921†*‡	.921†*‡	.921†*‡	.866†‡	.866†‡	.866†
	Conden.	.931†‡	.931†‡	.931†	.886†‡	.886†‡	.886†	.891†‡	.891†‡	.891†
	BS$_{R/P}$	.946†*‡	.946†*‡	.946†*	.916†*‡	.916†*‡	.916†*‡	.903†‡	.903†‡	.903†‡
Range	Naïve (U.)	.987*	.775*‡	.865*‡	.995*‡	.606*‡	.741*‡	.998*‡	.547*‡	.693*‡
	Cond.	.973*	.906†*	.936†	.969†*	.833†*	.892†	.957†*	.791†*	.862†
	BS$_{P+R@75}$	.977*	.868†*‡	.917†	.972†*	.822†*	.888†	.971†*	.758†*	.847†*
	BS$_{P+R@90}$	.985*	.831†*‡	.898†*‡	.985†*‡	.766†*‡	.857†*‡	.986†*‡	.707†*‡	.817†*‡
	BS$_{P+R@95}$	.986*	.815†*‡	.890†*‡	.988*‡	.739†*‡	.840†*‡	.990*‡	.673†*‡	.793†*‡

starting at the naïve lower bound and ending at different high points: the naïve upper bound, the score of condensed lists, and the upper 75%, 90%, or 95% percentiles of the bootstrapped distributions. Score ranges and point estimates for each run are compared against the scores of all other runs that contributed to the respective pool, emitting corresponding system preferences if the range/estimate is strictly below or above the exact score of another system.

Table 6 reports the reconstruction effectiveness as precision, recall, and F1 score. In recall-oriented settings, where score ranges are unsuitable, the naïve lower bound (recall of 0.954 on Robust04), or the bootstrapped prediction (recall of 0.903 on the ClueWeb12) should be used. In precision-oriented scenarios, naïve bounds achieve the highest precision at a high cost in recall (only 0.547 on the ClueWeb12). The pool/run-based bootstrapping at the 95% percentile provides significantly tighter naïve bounds (recall is always significantly better) at a negligible loss in precision (not significant in all cases). Hence, nDCG bounds can be substantially tightened without loss in accuracy using bootstrapping.

Real Incompleteness on TREC-COVID. As a final case study, we apply naïve bounds, condensed lists, and our pool/run-based bootstrapping to estimate the nDCG@10 of three dense retrieval models on the original TREC-COVID collection, for which the unjudged documents were post-judged [36]. The three dense retrieval systems operated in a zero-shot setting. Thus we compare them against the best run submitted to the first round of TREC-COVID, as those systems also had no access to training data.

Table 7 shows the results on the original (incomplete) TREC-COVID qrels and the post-hoc (complete) qrels for three selections of topics: (1) moderate levels of incompleteness (between 25% to 50% unjudged documents), (2) high incompleteness (more than 50% unjudged documents), and (3) all topics (only nDCG@10 scores in the setup with all topics are comparable between different systems). The original run files were not stored in the BEIR experiments [36], so we reproduced them (only minor differences for ANCE, TAS-B, and ColBERT,

Table 7. The nDCG@10 on the original qrels (unjudged documents) from TREC-COVID and the expanded qrels (all documents judged) for topics with 25% to 50% unjudged documents (.25 to .5), topics with more than 50% unjudged documents (.5 to 1), and all topics. We report the proportion of unjudged documents (U@10), and predictions of the lower bound (Default), condensed lists (Cond.), pool/run-based bootstrapping (BS_{P+R}), and naïve and tightened upper bounds ($BS_{P+R@95}$).

Model		Original Qrels					Ex. Qrels
		nDCG@10			Upper Bound		nDCG@10
	U@10	Default	Cond.	BS_{P+R}	Naïve	$BS_{P+R@95}$	
.25 to .5 ANCE	35.6%	$0.489_{-0.161}$	$0.683_{+0.033}$	$0.660_{+0.010}$	$0.838_{+0.188}$	$0.795_{+0.145}$	0.650
ColBERT	33.3%	$0.485_{-0.141}$	$0.641_{+0.015}$	$0.614_{-0.012}$	$0.770_{+0.144}$	$0.741_{+0.115}$	0.626
TAS-B	32.5%	$0.597_{\pm0.000}$	$0.875_{+0.278}$	$0.847_{+0.250}$	$0.902_{+0.305}$	$0.894_{+0.297}$	0.597
.5 to 1 ANCE	65.6%	$0.207_{-0.150}$	$0.547_{+0.190}$	$0.385_{-0.028}$	$0.769_{-0.412}$	$0.542_{+0.185}$	0.357
ColBERT	62.9%	$0.337_{-0.110}$	$0.679_{+0.232}$	$0.517_{+0.070}$	$0.881_{+0.434}$	$0.645_{+0.198}$	0.447
TAS-B	73.8%	$0.211_{-0.119}$	$0.584_{+0.254}$	$0.459_{+0.129}$	$0.918_{+0.588}$	$0.623_{+0.293}$	0.330
All Topics ANCE	22.4%	$0.652_{-0.083}$	$0.772_{+0.037}$	$0.747_{-0.012}$	$0.853_{-0.118}$	$0.804_{+0.069}$	0.735
ColBERT	17.2%	$0.680_{-0.054}$	$0.770_{+0.036}$	$0.741_{+0.007}$	$0.826_{+0.092}$	$0.789_{+0.055}$	0.734
TAS-B	41.0%	$0.481_{-0.074}$	$0.705_{-0.150}$	$0.633_{+0.078}$	$0.871_{+0.316}$	$0.729_{-0.174}$	0.555
1st@TREC	0.0%	$0.679_{\pm0.000}$	$0.679_{\pm0.000}$	$0.679_{\pm0.000}$	$0.679_{\pm0.000}$	$0.679_{\pm0.000}$	0.679

but for DPR, we scores were substantially different and still had unjudged documents, so we exclude DPR). The default behaviour of assuming that unjudged documents are non-relevant (i.e., the naïve lower bound) underestimates the effectiveness for all dense retrieval models. At the same time, condensed lists substantially overestimate the effectiveness (e.g., for TAS-B by 0.150). Our proposed pool/run-based bootstrapping produces the best estimates in all cases. Tightening upper bounds with bootstrapping is very valuable, as the 95% percentile of bootstrapped nDCG scores is much tighter as the naïve upper bound.

5 Conclusion

Our new bootstrapping method to account for unjudged documents in post-hoc nDCG evaluations is efficient in practice and more effective than previous methods that derive a point estimate or bounds for a system's true nDCG. Packaged as a TrecTools-compatible software that is publicly available, bootstrapped estimation is directly applicable to retrieval studies.

As interesting directions for future work, we want to expand our bootstrapping approach to more evaluation measures (e.g., Q-Measure, MAP, or RBP) and combine it with approaches that predict the relevance of unjudged documents based on their content. This combination could lead to more informed bootstrap priors and might also tighten the resulting bootstrapped score distributions.

Acknowledgments. This work has received funding from the European Union's Horizon Europe research and innovation programme under grant agreement No 101070014 (OpenWebSearch.EU, https://doi.org/10.3030/101070014).

References

1. Aslam, J.A., Pavlu, V., Yilmaz, E.: A statistical method for system evaluation using incomplete judgments. In: Efthimiadis, E.N., Dumais, S.T., Hawking, D., Järvelin, K. (eds.) SIGIR 2006: Proceedings of the 29th Annual International ACM SIGIR Conference on Research and Development in Information Retrieval, Seattle, Washington, USA, 6–11 August 2006, pp. 541–548. ACM (2006)
2. Aslam, J.A., Yilmaz, E.: Inferring document relevance from incomplete information. In: Silva, M.J., Laender, A.H.F., Baeza-Yates, R.A., McGuinness, D.L., Olstad, B., Olsen, Ø.H., Falcão, A.O. (eds.) Proceedings of the Sixteenth ACM Conference on Information and Knowledge Management, CIKM 2007, Lisbon, Portugal, 6–10 November 2007, pp. 633–642. ACM (2007)
3. Bernstein, Y., Zobel, J.: Redundant documents and search effectiveness. In: Herzog, O., Schek, H., Fuhr, N., Chowdhury, A., Teiken, W. (eds.) Proceedings of the 2005 ACM CIKM International Conference on Information and Knowledge Management, Bremen, Germany, October 31–November 5, 2005, pp. 736–743. ACM (2005)
4. Buckley, C., Voorhees, E.M.: Retrieval evaluation with incomplete information. In: Sanderson, M., Järvelin, K., Allan, J., Bruza, P. (eds.) SIGIR 2004: Proceedings of the 27th Annual International ACM SIGIR Conference on Research and Development in Information Retrieval, Sheffield, UK, 25–29 July 2004, pp. 25–32. ACM (2004)
5. Büttcher, S., Clarke, C.L.A., Yeung, P.C.K., Soboroff, I.: Reliable information retrieval evaluation with incomplete and biased judgements. In: Kraaij, W., de Vries, A.P., Clarke, C.L.A., Fuhr, N., Kando, N. (eds.) SIGIR 2007: Proceedings of the 30th Annual International ACM SIGIR Conference on Research and Development in Information Retrieval, Amsterdam, The Netherlands, 23–27 July 2007, pp. 63–70. ACM (2007)
6. Carterette, B., Allan, J., Sitaraman, R.K.: Minimal test collections for retrieval evaluation. In: Efthimiadis, E.N., Dumais, S.T., Hawking, D., Järvelin, K. (eds.) SIGIR 2006: Proceedings of the 29th Annual International ACM SIGIR Conference on Research and Development in Information Retrieval, Seattle, Washington, USA, 6–11 August 2006, pp. 268–275. ACM (2006)
7. Carterette, B., Jones, R.: Evaluating search engines by modeling the relationship between relevance and clicks. In: Platt, J.C., Koller, D., Singer, Y., Roweis, S.T. (eds.) Advances in Neural Information Processing Systems 20, Proceedings of the Twenty-First Annual Conference on Neural Information Processing Systems, Vancouver, British Columbia, Canada, 3–6 December 2007, pp. 217–224. Curran Associates, Inc. (2007)
8. Clarke, C.L.A., Craswell, N., Soboroff, I.: Overview of the TREC 2009 Web track. In: Voorhees, E.M., Buckland, L.P. (eds.) Proceedings of The Eighteenth Text REtrieval Conference, TREC 2009, Gaithersburg, Maryland, USA, 17–20 November 2009, NIST Special Publication, vol. 500–278. National Institute of Standards and Technology (NIST) (2009)
9. Clarke, C.L.A., Craswell, N., Soboroff, I., Cormack, G.V.: Overview of the TREC 2010 Web track. In: Proceedings of The Nineteenth Text REtrieval Conference, TREC 2010, Gaithersburg, Maryland, USA, 16–19 November 2010 (2010)
10. Clarke, C.L.A., Craswell, N., Soboroff, I., Voorhees, E.M.: Overview of the TREC 2011 web track. In: Proceedings of The Twentieth Text REtrieval Conference, TREC 2011, Gaithersburg, Maryland, USA, 15–18 November 2011 (2011)

11. Clarke, C.L.A., Craswell, N., Voorhees, E.M.: Overview of the TREC 2012 Web track. In: Voorhees, E.M., Buckland, L.P. (eds.) Proceedings of The Twenty-First Text REtrieval Conference, TREC 2012, Gaithersburg, Maryland, USA, 6–9 November 2012, NIST Special Publication, vol. 500–298. National Institute of Standards and Technology (NIST) (2012)

12. Cleverdon, C.: The Cranfield tests on index language devices. In: ASLIB Proceedings, pp. 173–192, MCB UP Ltd. (Reprinted in Readings in Information Retrieval, Karen Sparck-Jones and Peter Willett, editors, Morgan Kaufmann, 1997) (1967)

13. Cleverdon, C.W.: The significance of the Cranfield tests on index languages. In: Bookstein, A., Chiaramella, Y., Salton, G., Raghavan, V.V. (eds.) Proceedings of the 14th Annual International ACM SIGIR Conference on Research and Development in Information Retrieval. Chicago, Illinois, USA, 13–16 October 1991 (Special Issue of the SIGIR Forum), pp. 3–12. ACM (1991)

14. Collins-Thompson, K., Bennett, P.N., Diaz, F., Clarke, C., Voorhees, E.M.: TREC 2013 Web track overview. In: Proceedings of The Twenty-Second Text REtrieval Conference, TREC 2013, Gaithersburg, Maryland, USA, 19–22 November 2013 (2013)

15. Collins-Thompson, K., Macdonald, C., Bennett, P.N., Diaz, F., Voorhees, E.M.: TREC 2014 Web track overview. In: Proceedings of the Twenty-Third Text REtrieval Conference, TREC 2014, Gaithersburg, Maryland, USA, 19–21 November 2014 (2014)

16. Cormack, G.V., Lynam, T.R.: Statistical precision of information retrieval evaluation. In: Efthimiadis, E.N., Dumais, S.T., Hawking, D., Järvelin, K. (eds.) SIGIR 2006: Proceedings of the 29th Annual International ACM SIGIR Conference on Research and Development in Information Retrieval, Seattle, Washington, USA, 6–11 August 2006, pp. 533–540. ACM (2006)

17. Craswell, N., Mitra, B., Yilmaz, E., Campos, D., Voorhees, E.M.: Overview of the TREC 2019 Deep Learning track. In: Voorhees, E., Ellis, A. (eds.) 28th International Text Retrieval Conference, TREC 2019. Maryland, USA, NIST Special Publication, National Institute of Standards and Technology (NIST) (Nov, Gaithersburg (2019)

18. Efron, B., Tibshirani, R.: An introduction to the bootstrap. CRC Press (1994)

19. Ferro, N., Sanderson, M.: How do you test a test?: A multifaceted examination of significance tests. In: Candan, K.S., Liu, H., Akoglu, L., Dong, X.L., Tang, J. (eds.) WSDM '22: The Fifteenth ACM International Conference on Web Search and Data Mining, Virtual Event/Tempe, AZ, USA, 21–25 February, 2022, pp. 280–288. ACM (2022)

20. Fröbe, M., Bevendorff, J., Gienapp, L., Völske, M., Stein, B., Potthast, M., Hagen, M.: CopyCat: Near-duplicates within and between the ClueWeb and the Common Crawl. In: Diaz, F., Shah, C., Suel, T., Castells, P., Jones, R., Sakai, T. (eds.) 44th International ACM Conference on Research and Development in Information Retrieval (SIGIR 2021), pp. 2398–2404. ACM, July 2021

21. Fröbe, M., Bevendorff, J., Reimer, J., Potthast, M., Hagen, M.: Sampling bias due to near-duplicates in learning to rank. In: 43rd International ACM Conference on Research and Development in Information Retrieval (SIGIR 2020), pp. 1997–2000. ACM, July 2020

22. Fröbe, M., Bittner, J.P., Potthast, M., Hagen, M.: The effect of content-equivalent near-duplicates on the evaluation of search engines. In: Jose, J.M., Yilmaz, E., Magalhães, J., Castells, P., Ferro, N., Silva, M.J., Martins, F. (eds.) ECIR 2020. LNCS, vol. 12036, pp. 12–19. Springer, Cham (2020). https://doi.org/10.1007/978-3-030-45442-5_2

23. Järvelin, K., Kekäläinen, J.: Cumulated gain-based evaluation of IR techniques. ACM Trans. Inf. Syst. **20**(4), 422–446 (2002)
24. Lin, J., Yang, P.: The impact of score ties on repeatability in document ranking. In: Piwowarski, B., Chevalier, M., Gaussier, É., Maarek, Y., Nie, J., Scholer, F. (eds.) Proceedings of the 42nd International ACM SIGIR Conference on Research and Development in Information Retrieval, SIGIR 2019, Paris, France, 21–25 July 2019, pp. 1125–1128. ACM (2019)
25. Lu, X., Moffat, A., Culpepper, J.S.: The effect of pooling and evaluation depth on IR metrics. Inf. Retrieval J. **19**(4), 416–445 (2016). https://doi.org/10.1007/s10791-016-9282-6
26. Macdonald, C., Tonellotto, N.: Declarative experimentation in information retrieval using PyTerrier. In: Balog, K., Setty, V., Lioma, C., Liu, Y., Zhang, M., Berberich, K. (eds.) ICTIR '20: The 2020 ACM SIGIR International Conference on the Theory of Information Retrieval, Virtual Event, Norway, 14–17 September 2020, pp. 161–168. ACM (2020)
27. Moffat, A., Zobel, J.: Rank-biased precision for measurement of retrieval effectiveness. ACM Trans. Inf. Syst. **27**(1), 2:1–2:27 (2008)
28. Palotti, J.R.M., Scells, H., Zuccon, G.: TrecTools: an open-source Python library for information retrieval practitioners involved in TREC-like campaigns. In: Piwowarski, B., Chevalier, M., Gaussier, É., Maarek, Y., Nie, J., Scholer, F. (eds.) Proceedings of the 42nd International ACM SIGIR Conference on Research and Development in Information Retrieval, SIGIR 2019, Paris, France, 21–25 July 2019, pp. 1325–1328, ACM (2019)
29. Roberts, K., Alam, T., Bedrick, S., Demner-Fushman, D., Lo, K., Soboroff, I., Voorhees, E.M., Wang, L.L., Hersh, W.R.: TREC-COVID: rationale and structure of an information retrieval shared task for COVID-19. J. Am. Medical Informatics Assoc. **27**(9), 1431–1436 (2020)
30. Sakai, T.: Evaluating evaluation metrics based on the bootstrap. In: Efthimiadis, E.N., Dumais, S.T., Hawking, D., Järvelin, K. (eds.) SIGIR 2006: Proceedings of the 29th Annual International ACM SIGIR Conference on Research and Development in Information Retrieval, Seattle, Washington, USA, 6–11 August 2006, pp. 525–532. ACM (2006)
31. Sakai, T.: Alternatives to bpref. In: Kraaij, W., de Vries, A.P., Clarke, C.L.A., Fuhr, N., Kando, N. (eds.) SIGIR 2007: Proceedings of the 30th Annual International ACM SIGIR Conference on Research and Development in Information Retrieval, Amsterdam, The Netherlands, 23–27 July 2007, pp. 71–78. ACM (2007)
32. Sakai, T.: On the reliability of information retrieval metrics based on graded relevance. Inf. Process. Manag. **43**(2), 531–548 (2007)
33. Sakai, T.: Comparing metrics across TREC and NTCIR: The robustness to system bias. In: Shanahan, J.G., Amer-Yahia, S., Manolescu, I., Zhang, Y., Evans, D.A., Kolcz, A., Choi, K., Chowdhury, A. (eds.) Proceedings of the 17th ACM Conference on Information and Knowledge Management, CIKM 2008, Napa Valley, California, USA, October 26–30, 2008, pp. 581–590. ACM (2008)
34. Savoy, J.: Statistical inference in retrieval effectiveness evaluation. Inf. Process. Manag. **33**(4), 495–512 (1997)
35. Smucker, M.D., Allan, J., Carterette, B.: A comparison of statistical significance tests for information retrieval evaluation. In: Silva, M.J., Laender, A.H.F., Baeza-Yates, R.A., McGuinness, D.L., Olstad, B., Olsen, Ø.H., Falcão, A.O. (eds.) Proceedings of the Sixteenth ACM Conference on Information and Knowledge Management, CIKM 2007, Lisbon, Portugal, 6–10 November 2007, pp. 623–632. ACM (2007)

36. Thakur, N., Reimers, N., Rücklé, A., Srivastava, A., Gurevych, I.: BEIR: A heterogeneous benchmark for zero-shot evaluation of information retrieval models. In: Vanschoren, J., Yeung, S. (eds.) Proceedings of the Neural Information Processing Systems Track on Datasets and Benchmarks 1, NeurIPS Datasets and Benchmarks 2021, December 2021, virtual (2021)
37. Voorhees, E.: The TREC robust retrieval track. SIGIR Forum **39**(1), 11–20 (2005)
38. Voorhees, E.M.: The philosophy of information retrieval evaluation. In: Peters, C., Braschler, M., Gonzalo, J., Kluck, M. (eds.) CLEF 2001. LNCS, vol. 2406, pp. 355–370. Springer, Heidelberg (2002). https://doi.org/10.1007/3-540-45691-0_34
39. Voorhees, E.M.: The effect of sampling strategy on inferred measures. In: Geva, S., Trotman, A., Bruza, P., Clarke, C.L.A., Järvelin, K. (eds.) The 37th International ACM SIGIR Conference on Research and Development in Information Retrieval, SIGIR '14, Gold Coast, QLD, Australia–July 06–11, 2014, pp. 1119–1122. ACM (2014)
40. Voorhees, E.M.: The evolution of cranfield. In: Information Retrieval Evaluation in a Changing World. TIRS, vol. 41, pp. 45–69. Springer, Cham (2019). https://doi.org/10.1007/978-3-030-22948-1_2
41. Voorhees, E.M., et al.: TREC-COVID: constructing a pandemic information retrieval test collection. SIGIR Forum **54**(1), 1:1–1:12 (2020)
42. Voorhees, E.M., Soboroff, I., Lin, J.: Can old TREC collections reliably evaluate modern neural retrieval models? CoRR abs/2201.11086 (2022)
43. Xiong, L., Xiong, C., Li, Y., Tang, K., Liu, J., Bennett, P.N., Ahmed, J., Overwijk, A.: Approximate nearest neighbor negative contrastive learning for dense text retrieval. In: 9th International Conference on Learning Representations, ICLR 2021, Virtual Event, Austria, 3–7 May 2021, OpenReview.net (2021)
44. Yilmaz, E., Aslam, J.A.: Estimating average precision with incomplete and imperfect judgments. In: Yu, P.S., Tsotras, V.J., Fox, E.A., Liu, B. (eds.) Proceedings of the 2006 ACM CIKM International Conference on Information and Knowledge Management, Arlington, Virginia, USA, 6–11 November 2006, pp. 102–111. ACM (2006)
45. Yilmaz, E., Kanoulas, E., Aslam, J.A.: A simple and efficient sampling method for estimating AP and NDCG. In: Myaeng, S., Oard, D.W., Sebastiani, F., Chua, T., Leong, M. (eds.) Proceedings of the 31st Annual International ACM SIGIR Conference on Research and Development in Information Retrieval, SIGIR 2008, Singapore, 20–24 July 2008, pp. 603–610. ACM (2008)
46. Zobel, J.: How reliable are the results of large-scale information retrieval experiments? In: Croft, W.B., Moffat, A., van Rijsbergen, C.J., Wilkinson, R., Zobel, J. (eds.) SIGIR '98: Proceedings of the 21st Annual International ACM SIGIR Conference on Research and Development in Information Retrieval, 24–28 August 1998, Melbourne, Australia, pp. 307–314. ACM (1998)
47. Zobel, J., Rashidi, L.: Corpus bootstrapping for assessment of the properties of effectiveness measures. In: d'Aquin, M., Dietze, S., Hauff, C., Curry, E., Cudré-Mauroux, P. (eds.) CIKM '20: The 29th ACM International Conference on Information and Knowledge Management, Virtual Event, Ireland, 19–23 October 2020, pp. 1933–1952. ACM (2020)

Predicting the Listening Contexts of Music Playlists Using Knowledge Graphs

Giovanni Gabbolini[✉][iD] and Derek Bridge[iD]

Insight Centre for Data Analytics, School of Computer Science & IT,
University College Cork, Cork, Ireland
giovanni.gabbolini@insight-centre.org, d.bridge@cs.ucc.ie

Abstract. Playlists are a major way of interacting with music, as evidenced by the fact that streaming services currently host billions of playlists. In this content overload scenario, it is crucial to automatically characterise playlists, so that music can be effectively organised, accessed and retrieved. One way to characterise playlists is by their listening context. For example, one listening context is "workout", which characterises playlists suited to be listened to by users while working out. Recent work attempts to predict the listening contexts of playlists, formulating the problem as multi-label classification. However, current classifiers for listening context prediction are limited in the input data modalities that they handle, and on how they leverage the inputs for classification. As a result, they achieve only modest performance. In this work, we propose to use knowledge graphs to handle multi-modal inputs, and to effectively leverage such inputs for classification. We formulate four novel classifiers which yield approximately 10% higher performance than the state-of-the-art. Our work is a step forward in predicting the listening contexts of playlists, which could power important real-world applications, such as context-aware music recommender systems and playlist retrieval systems.

Keywords: Music playlists · Context-awareness · Recommender systems

1 Introduction

Music is commonly organised in some form of a playlist. According to a standard definition, a playlist is a sequence of music songs [5]. Playlists are a popular feature of music streaming services. Users consume playlists for 31% of their total listening time [31]; and 55% of users create their own playlists [27]. Playlists are also created for users by professional editors and by algorithms. For instance, the popular music streaming service Spotify was hosting more than four billion playlists in 2021.[1] In this content overload scenario, it is crucial to automatically characterise playlists, so that music can be effectively organised, accessed and

---

[1] https://backlinko.com/spotify-users.

© The Author(s), under exclusive license to Springer Nature Switzerland AG 2023
J. Kamps et al. (Eds.): ECIR 2023, LNCS 13980, pp. 330–345, 2023.
https://doi.org/10.1007/978-3-031-28244-7_21

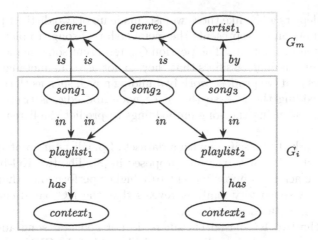

Fig. 1. A knowledge graph representing two playlists and three songs in total. The bottom and top boxes indicate two portions: G_i, which contains song, playlist and listening context nodes, and G_m, which contains metadata nodes, such as musical genres.

retrieved [9]. A common approach is playlist tagging, which is the task of assigning to a playlist one or more tags, drawn from a fixed vocabulary of tags. For example, [14] proposes a dataset of playlists annotated with a variety of different tags, like musical genres or decades. Similarly, [8] proposes a dataset of playlists annotated with listening context tags. Examples of listening context tags are "workout" and "party", which characterise playlists suited to be listened to by users while working out, and while having a party.

Listening context tags are interesting because they are user-centered, rather than music-centered [30]. For example, musical genre and decade tags refer to music. On the other hand, listening context tags refer to how people listen to music. As such, the accurate prediction of listening contexts can allow advances at the intersection of music information retrieval (MIR) and human-computer interaction (HCI), such as context-aware music recommendation [33]. In fact, recommending the right playlist at the right time is only possible if the listening context suited to listening to the playlist is known.

To the best of our knowledge, there exists only one attempt to predict the listening context of music playlists: [8]. The authors of [8] set up a multi-label classification problem, in which playlists are classified for their listening contexts, and they propose four classifiers: two matrix factorisation (MF)-based classifiers, that work by counting how many times a song is associated with each playlist listening context, and two convolutional neural network (CNN)-based classifiers, that work with song audio. However, these classifiers are limited in that they do not incorporate song metadata, such as musical genres.

In this paper, we formulate two novel knowledge graph (KG)-based classifiers. KGs are a powerful data model, suitable for storing heterogeneous information

[34]. Figure 1 depicts a KG like those we use, made up of two distinct portions: G_i and G_m. The portion G_i represents the membership of songs to playlists, and of playlists to listening contexts. The portion G_m represent song metadata, solving the limitation of existing classifiers that they do not use song metadata. The KG-based classifiers that we propose work by building a KG, such as the one depicted in Fig. 1, embedding the KG, so that each node and edge is transformed to a feature vector, and using the song embeddings to predict the listening contexts of playlists.

We benchmark the classifiers with a dataset of playlists annotated with their listening contexts, similar to the one proposed in [8]. The two KG-based classifiers we propose achieve approximately 10% higher performance than the existing predictors. A sensitivity analysis reveals that the KG-based classifiers can incorporate song metadata effectively.

However, the two KG-based classifiers do not consider song audio. So, we formulate another two novel predictors, as the hybrid of the CNN-based and KG-based classifiers. As expected, the hybrid classifiers outperform MF-based, KG-based and CNN-based predictors, setting the new state-of-the-art performance in the task.

We release the source code and the dataset that supports our work here, so as to allow reproducibility and foster new research on the subject.[2]

In summary, our contributions are:

1. The first two KG-based listening context predictors of music playlists that incorporate song metadata;
2. Another two novel predictors that incorporate KGs and song audio;
3. A comparison of the predictors reporting approximately 10% higher performance than the state-of-the-art, and showing the impact of song metadata on performance.

The rest of the paper is organised as follows: in Sect. 2, we review related work on music listening contexts, and especially work that looks into how music consumption changes in different listening contexts. We also review related work in music tagging. In Sect. 3, we describe our four novel classifiers for predicting the listening context of music playlists. In Sect. 3, we present extensive experiments that compare the novel classifiers to existing classifiers, and validate the design of the novel classifiers with a sensitivity analysis. Section 5 concludes the paper and outline future work.

2   Related Work

The task of tagging can be defined as marking content with descriptive terms, also called keywords or tags, drawn from a fixed vocabulary [16]. Content can refer to different objects, such as text, audio, images or video. For example, [7] propose an approach for tagging an image with its objects, such as: "fish",

---

[2] https://github.com/GiovanniGabbolini/playlist-context-prediction.

"plane" or "shoe". And, [25] survey tagging systems in the text, image and music domains.

In this work, we focus on the music domain, as tagging is a major topic in music information retrieval (MIR). Music tagging is the task of classifying music in one or more tag classes. As such, the vocabulary of tags is typically assumed to be fixed. One common setup is song tagging, where single songs are classified. [36], for example, offers a comparison of recent Convolutional Neural Network (CNN)-based classifiers: a CNN extracts learned features from the audio of a song, and leverages these features to output appropriate tags. Similarly, the state-of-the-art classifiers proposed in [10,29,35] are CNN-based. Progress in song tagging is enabled by the availability of large scale datasets, such as the Million Songs Dataset [3], the MagnaTagATune Dataset [24] and the MG-Jamendo Dataset [4]. These datasets contain songs annotated with tags of several categories: genre tags (*e.g.* "jazz"), instrumentation tags (*e.g.* "guitar"), decade tags (*e.g.* "80s"), mood tags (*e.g.* "happy") and listening context tags (*e.g.* "party"). A related (but different) task to song tagging is playlist tagging, where a list of songs is tagged, instead of a single song. [14] proposes a dataset of playlists annotated with a variety of different tags, like genre tags or decade tags. Classifiers for song tagging can be extended to do playlist tagging. For example, [8] proposes a CNN-based playlist classifier, with an architecture similar to the CNN-based song classifiers.

Previous work shows that music listening behaviour depends on the listening context [11,17]. For example, users listen to one type of music while having a party, to another type of music while spending time alone, and to another type while working. Context-aware music recommender systems [33] address the user's need to access the right music in the right context. Applications include: context-aware song and playlist recommendation, and context-aware playlist continuation [31]. Predicting the listening context that suits some music is a first step towards context-awareness. Hence, some of the recent work on music tagging focuses on listening context tags only. For example, [19,20] propose a dataset of songs annotated with listening context tags, and a baseline CNN-based classifier. And, [8] proposes a dataset of playlists annotated with listening context tags, such as "workout" and "party", and four baseline classifiers: two CNN-based and another two MF-based classifiers.

Our work here is on playlist tagging, as we focus on predicting the listening contexts of playlists. We build on [8], as we propose four novel classifiers, which outperform the four classifiers that they propose, setting the new state-of-the-art performance in the task.

3   Method

Predicting the listening contexts of playlists is framed by the authors of [8] as a multi-label classification problem. The same authors propose four such classifiers (MF-AVG, MF-SEQ, CNN-AVG and CNN-SEQ). Here, we propose another four such classifiers (KG-AVG, KG-SEQ, HYBRID-AVG, HYBRID-SEQ). As we will explain below, six of the classifiers that we consider follow the schema depicted in Fig. 2. The two hybrid classifiers follow the schema depicted in Fig. 3. In the rest of this section, we summarise the four classifiers that were proposed in [8], and we describe the four classifiers that we propose.

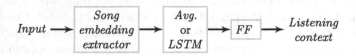

Fig. 2. Schematic architecture of MF-AVG, MF-SEQ, CNN-AVG, CNN-SEQ, KG-AVG and KG-SEQ.

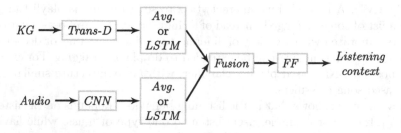

Fig. 3. Schematic architecture of HYBRID-AVG and HYBRID-SEQ.

3.1 Matrix Factorisation-Based

The two matrix factorisation (MF)-based classifiers (MF-AVG and MF-SEQ), originally proposed in [8], take as input a matrix $\mathbf{X} \in \mathbb{R}^{N,M}$ where N is the number of songs and M is the number of listening contexts. The element at row n and column m of $\mathbf{X}$ is equal to the number of times the nth song appears in playlists that have the mth context. The matrix $\mathbf{X}$ is factorised into two matrices, $\mathbf{S} \in \mathbb{R}^{N,H}$ and $\mathbf{C} \in \mathbb{R}^{H,M}$, using WR-MF, which is the MF procedure described in [18], so that $\mathbf{SC} \approx \mathbf{X}$. H is the embedding dimension, which is a hyper-parameter of WR-MF. The rows of $\mathbf{S}$ and the columns of $\mathbf{C}$ contain, respectively, song and listening context embeddings. Then, the song embedding vectors for the songs in a given playlist (a subset of the embeddings contained in $\mathbf{S}$) are either averaged song-wise (in MF-AVG) or input to a single-layered LSTM network (in MF-SEQ), to get a playlist embedding vector, which is fed into a single-layered feed-forward (FF) network that outputs a score for each listening context.

The architecture of MF-AVG and MF-SEQ fit into the schema of Fig. 2 as the matrix $\mathbf{X}$ is the *input*, and WR-MF is the *song embedding extractor*. Notice that MF-AVG and MF-SEQ work in two steps, that is the *song embedding extractor* is trained separately from the rest.

3.2 Convolutional Neural Network-Based

The two convolutional neural network (CNN)-based classifiers (CNN-AVG and CNN-SEQ), originally proposed in [8], extend the state-of-the-art in song tagging to playlist tagging. Given a song, they consider the full audio, and compute mel-spectrograms for every contiguous 3-seconds of audio. The mel-spectrogram is a hand-crafted feature extracted from audio, commonly used in many music information retrieval (MIR) tasks, such as song tagging, e.g. [10,29,35]. The mel-spectrograms are input to a Convolutional Neural Network (CNN) with five *1D*-convolutional layers, which outputs an embedding vector for every 3-seconds of audio. Such embeddings are averaged point-wise, to get one song embedding vector. Given a playlist, the song embedding vectors are computed as above, and either averaged song-wise (in CNN-AVG) or input to a single-layered LSTM network (in CNN-SEQ), to get a playlist embedding vector, which is fed into a single-layered (FF) network that outputs a score for each listening context.

The architecture of CNN-AVG and CNN-SEQ fit into the schema of Fig. 2 as the mel-spectrograms are the *input*, and the CNN is the *song embedding extractor*. Notice, however, that CNN-AVG and CNN-SEQ are end-to-end, that is the *song embedding extractor* is trained jointly with the rest.

3.3 Knowledge Graph-Based

A knowledge graph (KG) is a set of triples $G = \{(e, r, e') \mid e, e' \in E, r \in R\}$, where E and R denote, respectively, the sets of entities (nodes) and relationships (edges). KGs are suitable for representing heterogeneous information [34]. For example, [28] builds a KG representing users, their interactions with songs, and acoustical metadata, such as what musical instruments are played in the songs.

The information we want to represent is: songs; playlists; listening contexts; and song metadata. So, we build a KG composed of two portions. (1) G_i: the portion containing song nodes, playlist nodes and listening context nodes. These nodes are connected by edges according to membership: a song node is connected to the playlist nodes the song belongs to, and a playlist node is connected to its listening context node. (2) G_m: the portion containing song metadata, *i.e.* the record label associated with the song, the musical genres associated with the song, the year and the month when the song was released, the artist of the song, the city and the country where the artist is currently based, and where they were born. We selected these items of metadata empirically, through informal experimentation, and by taking inspiration from previous work; for example, [21] finds that the release year of a song can be a predictor for the listening context. In future work, G_m can be readily expanded to include more song metadata, such as information extracted from song lyrics. For each piece of song metadata, there is a node in G_m. Song nodes are connected by edges to their metadata nodes. Song metadata may be missing, *e.g.* we may not know the record label for a particular song. We obtain metadata from the crowd-sourced database MusicBrainz.[3]

Figure 1 depicts a KG, like those that we build.

We embed the KG using the TRANS-D algorithm, which is a state-of-the-art KG embedding algorithm [12]. TRANS-D produces an embedding vector for every node and edge in the KG, in such a way that the topology of the KG is preserved. In particular, given a KG G, and given a triple $(e, r, e') \in G$, TRANS-D produces three embedding vectors v_e, v_r and $v_{e'}$ that satisfy a relationship similar to $v_e + v_r \approx v_{e'}$, for every triple in G. The embedding vectors of the song nodes in the KG for the songs in a playlist are either averaged song-wise (in KG-AVG) or input to a single-layered LSTM network (in KG-SEQ), to get a playlist embedding vector, which is fed into a (FF) network, that outputs a score for each listening context.

The architecture of KG-AVG and KG-SEQ fit into the schema of Fig. 2 as the KG we build is the *input*, and TRANS-D is the *song embedding extractor*. Notice that KG-AVG and KG-SEQ work in two steps, that is the *song embedding extractor* is trained separately from the rest.

The MF-based and KG-based algorithms both leverage information about listening contexts when computing song embeddings. However, KG-based algorithms exploit that information more effectively. For example, let us consider the scenario depicted by the portion G_i of the KG in Fig. 1 where there are two playlists, $playlist_1$ and $playlist_2$, whose listening contexts are respectively $context_1$ and $context_2$, and which contain respectively the songs $song_1$ & $song_2$ and $song_2$ & $song_3$. MF song embeddings are aligned with their listening contexts, as explained in Sect. 3.1. In the example above, the MF embedding of $song_1$ is aligned with $context_1$, the MF embedding of $song_3$ is aligned with $context_2$, and the embedding of $song_2$ is aligned with both $context_1$ and $context_2$. However, $song_1$ and $song_2$ are in the same playlist ($playlist_1$). As such, we expect the embedding of $song_1$ to be aligned, to some extent, also with $context_2$, and not only with $context_1$; similarly for the embedding of $song_3$. That is, MF-based algorithms 'short-circuit' the representation of playlists by modelling the association of songs to playlist listening contexts directly. KG embeddings preserve the topology of the KG, and so can overcome the short-circuiting problem of the MF-algorithms. In the example above, the songs in G_i are all connected with each other, via the explicit representation of the playlists as well as the listening contexts. That is, the embeddings of $song_1$, $song_2$ and $song_3$ are all aligned, to some extent with $context_1$, and to some other extent with $context_2$. The short-circuiting problem undermines the performance of the MF-based classifiers, as we empirically prove in Sect. 4.3. In a similar vein, [26] propagates tags among songs in the same playlists, and measure an increase in performance.

---

[3] https://musicbrainz.org.

3.4   Hybrid

The CNN-based classifiers and the KG-based classifiers differ on their input data, as the CNN-based classifiers rely on song audio, while the KG-based classifiers rely on a KG representation of songs, playlists, listening contexts, and song metadata. The audio and the KG differ in modality, as well as availability. For example, while song audio is available for every song in the catalogue, a KG such as the one we use may represent the most famous songs well, but it may fail to represent properly more niche songs, which is a manifestation of the long-tail problem [22], and it may also fail to represent newly-released songs. To address this limitation, we complement the KG-based classifiers with the CNN-based classifiers, by formulating two hybrid classifiers.

The hybrids work by jointly running a KG-based classifier (KG-AVG or KG-SEQ) and a CNN-based classifier (CNN-AVG or CNN-SEQ), and by fusing the two playlist embedding vectors that they compute, before they are passed to a single-layered FF network that outputs a score for each listening context. We refer to HYBRID-AVG as the hybrid of KG-AVG & CNN-AVG and to HYBRID-SEQ as the hybrid of KG-SEQ & CNN-SEQ. The architecture of the two hybrids follow the schema of Fig. 3.

For the embedding fusion, both the audio and KG-based playlist embedding vectors are input to two separate linear layers, two separate non-linearities, and then summed point-wise, as suggested by [2]. We did experiment with other simple fusion strategies, e.g. concatenation, but they achieved lower performance.

3.5   Implementation Details

Our implementation of CNN-AVG and CNN-SEQ is a little different from the original paper [8] as we make two simplifications. First, we use Spotify's 30-second audio previews of the songs instead of their full audio. These audio previews are freely available, unlike the full audio, which is expensive to access due to copyright restrictions. Moreover, the usage of audio previews make our work reproducible. Second, we average the 3-second mel-spectrograms of a song point-wise in input to the CNN. As such, the CNN receives only one spectrogram, and outputs the song embedding directly. This second simplification saves computing resources. In Sect. 4, we show that our implementation of the CNN-based models outperforms the MF-based models, which is consistent with the original paper. More specifically, our implementation of CNN-SEQ achieves 7% higher performance than MF-SEQ, which is consistent with the original paper; similarly for CNN-AVG and MF-AVG. Given those results, we are confident that our implementations of the CNN-based models, although simplified, are as valid as the original implementations presented in [8].

We compute the mel-spectrograms for the CNN-AVG and CNN-SEQ classifiers with 22,050 Hz sampling rate, 1,024 FFT size, 512 hop size, and 128 mel bins. We set hyper-parameters of the MF and CNN-based classifiers as in the original paper [8]. That is, we set the song embedding dimension to 50, and we use *ReLU* as the non-linearity. We do the same in the KG-based and hybrid classifiers. We train the classifiers with early-stopping, monitoring *FH@1* on the validate set, with patience equal to ten. We tune other hyper-parameters of the eight classifiers (learning rate, weight decay and batch size) using bayesian optimisation [32]. We fix the number of trials of the bayesian optimiser to 20. For the WR-MF and Trans-D embedding procedures, we use the default parameters and we set the number of epochs to ensure convergence of the loss function.

For other implementation details, we refer the reader to the source code that supports our work here.

4   Experiments

We compare the classifiers described in Sect. 3, and variants of those, on their performance in predicting the listening context of music playlists.

4.1   Dataset

We use a dataset of playlists annotated with their listening contexts. The dataset was annotated by the authors of [8], starting from user playlists contained in the Spotify Million Playlist Dataset (MPD) [6], and retaining only the portion of playlists that have a listening context as title.[4] Examples of listening contexts present in the dataset are: driving, studying and summertime. For other examples, we refer the reader to the dataset that supports our work here. Also, we refer the reader to [8] for more information on the annotation procedure. Each playlist is annotated with one listening context. We split the dataset randomly into train, validate and test sets, accounting respectively for 60%, 20% and 20% of the total playlists. Similar to [8], we filter out songs that occur in the validate and test sets but not in the train set, as some classifiers cannot handle at testing time songs not seen at training time. The classifiers that have this limitation are MF-AVG, MF-SEQ, KG-AVG and KG-SEQ. They work by training a *song embedding extractor* model in a first step, separately from the classifier that outputs the listening context, see Sects. 3.1 and 3.3. As a result, embeddings for songs not present at training time are not available at test time. In a real world scenario, where new releases are frequently added to the songs catalogue, it would be necessary to incrementally train the models so that the training set covers all songs in the catalogue. An alternative is to use CNN-AVG and CNN-SEQ, as they rely on the audio signal, which is available for songs not seen at training time.

Table 1 contains statistics of the dataset that we use (train, validate and test splits together).

---

[4] The dataset we use is not the one used in [8], which is proprietary, but it was supplied to us by the authors of [8] as a dataset annotated with the same procedure, and in which similar results can be obtained.

Table 1. Dataset statistics.

Statistic	Value
Number of playlists	114,689
Average playlist length	62.6
Number of unique songs	418,767
Number of unique listening contexts	102

4.2 Metrics

We call D the test set, and we call p a playlist in the test set, that is $p \in D$. We call $|D|$ the number of playlists in the test set. The classifiers described in Sect. 3 predict a score for each listening context. As such, given a playlist, a classifier predicts a ranking of listening contexts, by decreasing score. Given a ranking of listening contexts for a playlist p, we call $rank_p$ the position of the correct listening context in the ranking. For example, if a classifier assigns the highest score to the correct listening context, then $rank_p = 1$. Instead, if the classifier assigns the lowest score to the correct listening context, then $rank_p = 102$ (see Table 1).

We compare the classifiers for their performance in predicting the listening contexts of the playlists in D. We measure performance with four metrics, as in [8]:

Flat hits ($FH@1$, $FH@5$). Flat hits is the percentage of playlists D such that $rank_p \leq k$. In our case, since the goal is classification rather than retrieval, we consider only $k = 1$ and $k = 5$ and no higher values for k. In formulas:

$$FH@k = \frac{1}{|D|} \sum_{p \in D} \mathbb{1}(rank_p \leq k)$$

where $\mathbb{1}(rank_p \leq k)$ is the indicator function. That is, $\mathbb{1}(rank_p \leq k) = 1$ if $rank_p \leq k$ and 0 otherwise. In other words, $FH@1$ is the percentage of playlists for which the classifier predicts the listening context correctly. And, $FH@5$ is the percentage of playlists for which the classifier predicts the correct listening context among the first five predictions.

Mean reciprocal rank (MRR). The reciprocal rank is the reciprocal of $rank_p$. The MRR is the average of those reciprocals ranks. In formulas:

$$MRR = \frac{1}{|D|} \sum_{p \in D} \frac{1}{rank_p}.$$

Mean average precision ($MAP@5$). MAP is equivalent to MRR, except that we set the reciprocal rank to 0 when $rank_p > k$[5]. That is, if $rank_p > k$ for

---

[5] Our formulation of MAP is different from others, which allow for multiple relevant items. In our case, there is only one relevant item: the correct listening context.

Table 2. Performance of the classifiers.

	FH@1	*FH@5*	*MRR*	*MAP@5*
MF-AVG	0.299	0.536	0.416	0.386
MF-SEQ	0.327	0.595	0.452	0.423
CNN-AVG	0.291	0.583	0.425	0.395
CNN-SEQ	0.352	0.639	0.484	0.456
KG-AVG	0.388	0.678	0.521	0.495
KG-SEQ	0.389	0.678	0.520	0.494
HYBRID-AVG	**0.395**	**0.687**	**0.528**	**0.503**
HYBRID-SEQ	0.389	0.678	0.520	0.495

every $p \in D$, then $MAP@k = 0$. In our case, we consider $k = 5$. In formulas:

$$MAP@k = \frac{1}{|D|} \sum_{p \in D} \frac{1}{rank_p} \times \mathbb{1}(rank_p \leq k).$$

On the one hand, *FH@k* disregards the actual position of the correct listening context in the ranking, but counts how frequently this position is lower than a threshold k. On the other hand, *MAP@k* and *MRR* do account for the actual position of the correct listening context in the ranking. Therefore, these metrics give a multi-sided view of the classifiers' performance.

We set up significance tests to check whether differences in performance are statistically significant or not. Following [13], we set up a t-test for *MRR* and *MAP@5*, and a paired bootstrap test for *FH@1* and *FH@5*. Similar to [23], we fix the number of bootstrap replicas to 1000.

4.3 Results

We conduct two experiments: a comparison with the state-of-the-art, and a sensitivity analysis.

Comparison with State-of-the-Art. We measure the performance of the classifiers that we propose (KG-AVG, KG-SEQ, HYBRID-AVG, HYBRID-SEQ), and the performance of the state-of-the-art baselines, *i.e.* the existing listening context classifiers (MF-AVG, MF-SEQ, CNN-AVG, CNN-SEQ). The results are in Table 2.

The classifiers that we propose outperform the baselines by a considerable amount. HYBRID-AVG scores highest performance, improving by approximately 10% over the baselines. The improvement in performance is statistically significant ($p < 10^{-4}$). In general, all the classifiers we propose improve performance over the baselines ($p < 10^{-4}$).

The improvement in performance has real world relevance. For example, HYBRID-AVG achieves 12% higher *FH@1* than the best baseline (0.395 *vs* 0.352),

Table 3. Performance of KG-based classifiers with (w) and without (wo) song metadata.

	FH@1	*FH@5*	*MRR*	*MAP@5*
KG-AVG wo metadata	0.375	0.665	0.507	0.481
KG-AVG w metadata	0.388	0.679	0.521	0.495
KG-SEQ wo metadata	0.382	0.668	0.513	0.487
KG-SEQ w metadata	0.388	0.679	0.520	0.495

which means than in a sample of 1000 playlists, our algorithm predicts the listening context correctly 395 *vs* 352 times, on average. Considering that the current databases contain millions of playlists, the 12% increase over the baselines is particularly 'tangible'.

We notice that the more complex SEQ variants of the algorithms are not always superior to their simpler AVG variant. MF-SEQ and CNN-SEQ have higher performance than, respectively, MF-AVG and CNN-AVG ($p < 10^{-4}$). But we do not find any statistically significant differences between the performance of KG-AVG and KG-SEQ, while HYBRID-SEQ has lower performance than HYBRID-AVG ($p < 10^{-4}$). Probably, the architecture of HYBRID-SEQ is too complex for the task at hand, and may overfit the training set, while the simpler HYBRID-AVG generalises better to new data. Moreover, the result corroborates previous work [8], where the SEQ variant is found to be sometimes superior and sometimes inferior to the AVG variant.

The hybrid classifiers are the combination of the (audio) CNN-based and KG-based classifiers. Accordingly, HYBRID-AVG has higher performance than CNN-AVG and KG-AVG. Though statistically significant ($p < 10^{-4}$), the increase in performance is only slight. We can understand the result by looking at the literature on the well-studied task of music similarity [1]. Flexer [15] shows that increasing the performance of similarity algorithms is particularly challenging after a certain threshold, as there exists an upper bound to performance, caused by the low agreement of different users in the perception of music similarity. Likewise, humans can have different perceptions of the right listening context for a given playlist. In the dataset we use, each song is associated with 17 different playlist listening contexts, on average. As such, we expect that increasing the performance of classifiers can become particularly challenging after a certain threshold. For example, HYBRID-SEQ has higher performance than CNN-SEQ ($p < 10^{-4}$), but not over KG-SEQ (no statistically significant difference).

Sensitivity Analysis. KG-based classifiers have as input a KG with songs, playlists, their listening contexts (portion G_i) and song metadata (portion G_m). We measure the performance of variants of the KG-based classifiers that have as input only the portion G_i of the full KG. The results are in Table 3, and show an increase in performance when using metadata ($p < 10^{-4}$). This indicates that the KG-based classifiers make effective use of song metadata for predicting listening contexts. However, the increase in performance is only slight, and again can be explained by the work of Flexer [15], as in Sect. 4.3.

The G_i portion of the KG contains the same information as the input to the MF-based classifiers, *i.e.* playlist listening contexts. However, as argued in Sect. 3.3, MF-based classifiers suffer from what we called the playlist short-circuiting problem, *i.e.* they model the association of songs to playlist listening contexts directly, while KG-based classifiers do not. A comparison of the results of the KG-based classifiers without metadata in Table 3 and the MF-based classifiers in Table 2 reveals the consequences of these two ways of modelling the information. The comparison shows that the KG-based algorithms exploit that information more effectively, since their results are significantly superior to those of the MF-based algorithms ($p < 10^{-4}$).

5   Conclusions and Future Work

We propose four novel systems for predicting the listening contexts of music playlists, which include, for the first time, song metadata in their models. In two of them, we represent songs, playlists, listening contexts and song metadata in a KG, that we embed, and we use the song embeddings to make predictions. In the other two, we combine the KG and song audio in a unique hybrid model. We benchmark the performance of the predictors we propose, reporting an increase in performance of approximately 10% over the state-of-the-art. We also show, through a sensitivity analysis, that the KG-based predictors can incorporate the song metadata effectively. We argued that the improvement in performance that we have achieved has real world relevance.

Our work can power a number of real applications that make use of listening contexts, such as context-aware recommender systems. More generally, our work introduces a way to use KGs for effective music classification, which is an under-explored direction.

Future work include the construction of a novel playlist extender *i.e.* one that recommends songs to add to a playlist but that ensures that the new songs are suited to the playlist listening context.

Acknowledgements. This publication has emanated from research conducted with the financial support of Science Foundation Ireland under Grant number 12/RC/2289-P2 which is co-funded under the European Regional Development Fund. For the purpose of Open Access, the author has applied a CC BY public copyright licence to any Author Accepted Manuscript version arising from this submission. We are grateful to Elena Epure & Romain Hennequin from Deezer, for sharing the dataset we used in the experiments, and for assisting us in the process of replicating the baselines. We are also grateful to Jeong Choi from NAVER, for assisting us in the process of replicating the baselines.

References

1. Aucouturier, J.J., Pachet, F., et al.: Music similarity measures: What's the use? In: 3rd International Society for Music Information Retrieval Conference, pp. 13–17 (2002)
2. Baltrušaitis, T., Ahuja, C., Morency, L.P.: Multimodal machine learning: a survey and taxonomy. IEEE Trans. Pattern Anal. Mach. Intell. **41**(2), 423–443 (2018)
3. Bertin-Mahieux, T., Ellis, D.P., LabROSA, E., Whitman, B., Lamere, P.: The million song dataset. In: Proceedings of the 12th International Conference on Music Information Retrieval (2011)
4. Bogdanov, D., Won, M., Tovstogan, P., Porter, A., Serra, X.: The MTG-Jamendo dataset for automatic music tagging. In: Machine Learning for Music Discovery Workshop, International Conference on Machine Learning (ICML 2019). Long Beach, CA, United States (2019). http://hdl.handle.net/10230/42015
5. Bonnin, G., Jannach, D.: Automated generation of music playlists: survey and experiments. ACM Comput. Surv. (CSUR) **47**(2), 1–35 (2014)
6. Chen, C.W., Lamere, P., Schedl, M., Zamani, H.: RecSys challenge 2018: Automatic music playlist continuation. In: Proceedings of the 12th ACM Conference on Recommender Systems, pp. 527–528. RecSys 2018, Association for Computing Machinery, New York, NY, USA (2018). https://doi.org/10.1145/3240323.3240342
7. Chen, M., Zheng, A., Weinberger, K.: Fast image tagging. In: International Conference on Machine Learning, pp. 1274–1282. PMLR (2013)
8. Choi, J., Khlif, A., Epure, E.: Prediction of user listening contexts for music playlists. In: Proceedings of the 1st Workshop on NLP for Music and Audio (NLP4MusA), pp. 23–27 (2020)
9. Choi, K., Fazekas, G., McFee, B., Cho, K., Sandler, M.: Towards music captioning: generating music playlist descriptions. arXiv preprint arXiv:1608.04868 (2016)
10. Choi, K., Fazekas, G., Sandler, M., Cho, K.: Convolutional recurrent neural networks for music classification. In: 2017 IEEE International Conference on Acoustics, Speech and Signal Processing (ICASSP), pp. 2392–2396. IEEE (2017)
11. Cunningham, S.J., Bainbridge, D., Falconer, A.: 'More of an art than a science': supporting the creation of playlists and mixes. In: 7th International Society for Music Information Retrieval Conference (2006)
12. Dai, Y., Wang, S., Xiong, N.N., Guo, W.: A survey on knowledge graph embedding: approaches, applications and benchmarks. Electronics **9**(5), 750 (2020)
13. Dror, R., Baumer, G., Shlomov, S., Reichart, R.: The hitchhiker's guide to testing statistical significance in natural language processing. In: Proceedings of the 56th Annual Meeting of the Association for Computational Linguistics (Volume 1: Long Papers), pp. 1383–1392. Association for Computational Linguistics, Melbourne, Australia (2018). https://doi.org/10.18653/v1/P18-1128. https://aclanthology.org/P18-1128
14. Ferraro, A., et al.: Melon playlist dataset: a public dataset for audio-based playlist generation and music tagging. In: ICASSP 2021–2021 IEEE International Conference on Acoustics, Speech and Signal Processing (ICASSP), pp. 536–540. IEEE (2021)
15. Flexer, A.: On inter-rater agreement in audio music similarity. In: Proceedings of the 15th International Conference on Music Information Retrieval, pp. 245–250 (2014)
16. Golder, S.A., Huberman, B.A.: The structure of collaborative tagging systems. J. Inf. Sci. **32**(2), 0508082 (2006)

17. Greasley, A.E., Lamont, A.: Exploring engagement with music in everyday life using experience sampling methodology. Music Sci. **15**(1), 45–71 (2011)
18. Hu, Y., Koren, Y., Volinsky, C.: Collaborative filtering for implicit feedback datasets. In: 2008 Eighth IEEE International Conference on Data Mining, pp. 263–272. IEEE (2008)
19. Ibrahim, K., Epure, E., Peeters, G., Richard, G.: Should we consider the users in contextual music auto-tagging models? In: 21st International Society for Music Information Retrieval Conference (2020)
20. Ibrahim, K.M., Royo-Letelier, J., Epure, E.V., Peeters, G., Richard, G.: Audio-based auto-tagging with contextual tags for music. In: ICASSP 2020–2020 IEEE International Conference on Acoustics, Speech and Signal Processing (ICASSP), pp. 16–20. IEEE (2020)
21. Kamehkhosh, I., Bonnin, G., Jannach, D.: Effects of recommendations on the playlist creation behavior of users. User Model. User-Adap. Inter. **30**(2), 285–322 (2020)
22. Knees, P., Schedl, M.: Contextual music meta-data: comparison and sources. In: Music Similarity and Retrieval. TIRS, vol. 36, pp. 107–132. Springer, Heidelberg (2016). https://doi.org/10.1007/978-3-662-49722-7_5
23. Koehn, P.: Statistical significance tests for machine translation evaluation. In: Proceedings of the 2004 Conference on Empirical Methods in Natural Language Processing, pp. 388–395. Association for Computational Linguistics, Barcelona, Spain (2004). https://aclanthology.org/W04-3250
24. Law, E., West, K., Mandel, M.I., Bay, M., Downie, J.S.: Evaluation of algorithms using games: The case of music tagging. In: 10th International Society for Music Information Retrieval Conference, pp. 387–392 (2009)
25. Lee, S., Masoud, M., Balaji, J., Belkasim, S., Sunderraman, R., Moon, S.J.: A survey of tag-based information retrieval. Int. J. Multimedia Inf. Retrieval **6**(2), 99–113 (2017)
26. Lin, Y.H., Chung, C.H., Chen, H.H.: Playlist-based tag propagation for improving music auto-tagging. In: 2018 26th European Signal Processing Conference (EUSIPCO), pp. 2270–2274. IEEE (2018)
27. Muligan, M.: Announcing MIDiA's state of the streaming nation 2 report. https://midiaresearch.com/blog/announcing-midias-state-of-the-streaming-nation-2-report (2017). Accessed 15 Mar 2022
28. Oramas, S., Ostuni, V.C., Noia, T.D., Serra, X., Sciascio, E.D.: Sound and music recommendation with knowledge graphs. ACM Trans. Intell. Syst. Technol. (TIST) **8**(2), 1–21 (2016)
29. Pons, J., Nieto, O., Prockup, M., Schmidt, E., Ehmann, A., Serra, X.: End-to-end learning for music audio tagging at scale. In: Proceedings of the 19th International Conference on Music Information Retrieval (2018)
30. Schedl, M., Flexer, A., Urbano, J.: The neglected user in music information retrieval research. J. Intell. Inf. Syst. **41**(3), 523–539 (2013)
31. Schedl, M., Zamani, H., Chen, C.-W., Deldjoo, Y., Elahi, M.: Current challenges and visions in music recommender systems research. Int. J. Multimedia Inf. Retrieval **7**(2), 95–116 (2018). https://doi.org/10.1007/s13735-018-0154-2
32. Victoria, A.H., Maragatham, G.: Automatic tuning of hyperparameters using Bayesian optimization. Evol. Syst. **12**(1), 217–223 (2021)
33. Wang, X., Rosenblum, D., Wang, Y.: Context-aware mobile music recommendation for daily activities. In: Proceedings of the 20th ACM International Conference on Multimedia, pp. 99–108 (2012)

34. Wilcke, X., Bloem, P., De Boer, V.: The knowledge graph as the default data model for learning on heterogeneous knowledge. Data Sci. **1**(1–2), 39–57 (2017)
35. Won, M., Chun, S., Nieto, O., Serra, X.: Data-driven harmonic filters for audio representation learning. In: Proceedings of International Conference on Acoustics, Speech and Signal Processing (ICASSP), pp. 536–540. IEEE (2020)
36. Won, M., Ferraro, A., Bogdanov, D., Serra, X.: Evaluation of CNN-based automatic music tagging models. arXiv preprint arXiv:2006.00751 (2020)

Keyword Embeddings for Query Suggestion

Jorge Gabín[1,2]([✉])[iD], M. Eduardo Ares[1][iD], and Javier Parapar[2][iD]

[1] Linknovate Science, Rúa das Flores 33, Roxos, Santiago de Compostela, A Coruña
15896, Spain
jorge.gabin@udc.es, eduardo@linknovate.com
[2] IRLab, CITIC, Computer Science Department, University of A Coruña, A Coruña
15071, Spain
javier.parapar@udc.es

Abstract. Nowadays, search engine users commonly rely on query suggestions to improve their initial inputs. Current systems are very good at recommending lexical adaptations or spelling corrections to users' queries. However, they often struggle to suggest semantically related keywords given a user's query. The construction of a detailed query is crucial in some tasks, such as legal retrieval or academic search. In these scenarios, keyword suggestion methods are critical to guide the user during the query formulation. This paper proposes two novel models for the keyword suggestion task trained on scientific literature. Our techniques adapt the architecture of `Word2Vec` and `FastText` to generate keyword embeddings by leveraging documents' keyword co-occurrence. Along with these models, we also present a specially tailored negative sampling approach that exploits how keywords appear in academic publications. We devise a ranking-based evaluation methodology following both known-item and ad-hoc search scenarios. Finally, we evaluate our proposals against the state-of-the-art word and sentence embedding models showing considerable improvements over the baselines for the tasks.

Keywords: Keyword suggestion · Keyword embeddings · Negative sampling · Academic search

1   Introduction

The use of word embeddings [6] has improved the results of many Natural Language Processing (NLP) tasks, such as name entity recognition, speech processing, part-of-speech tagging, semantic role labelling, chunking, and syntactic parsing, among others. These techniques represent words as dense real-valued vectors which preserve semantic and syntactic similarities between words. More recently, the so-called document and sentence embedding models allow the computing of embeddings for larger pieces of text directly (instead of doing it word by word), getting state-of-the-art results on different tasks.

Nowadays, search engines are outstanding in recommending lexical adaptations or corrections to users' queries [14,17]. However, there is still room

J. Kamps et al. (Eds.): ECIR 2023, LNCS 13980, pp. 346–360, 2023.
https://doi.org/10.1007/978-3-031-28244-7_22

for improvement in suggesting semantically similar phrases to users' inputs. The few systems that address this task do it by simply using existing word or sentence embedding models, which leads to poor keyword suggestions.

Keyword suggestion, which consists in recommending keywords similar to the user's input, is critical in search scenarios where the completeness of the query clauses may dramatically affect the recall, such as academic or legal search [13, 24]. An incomplete query may end up in a *null search session* [15], that is, the system presents an empty result list to the user. In tasks where use cases are recall-oriented, having an incomplete or even empty results set greatly diminishes the search experience.

Having good keyword suggestion models in a keyword-based search engine has many benefits. (i) It will help to identify the query intent [10], as users will be able to refine their search by adding new query clauses that may be clarifying. (ii) It will promote serendipity [1,5], as users may see a recommended keyword that they were not considering but perfectly fits their search interest. (iii) It will help prevent null sessions or incomplete results lists. (iv) Systems may use semantically similar keywords to the user input to perform query expansion [7, 22] without further user interaction.

In this paper, we leverage Word2Vec [19] and FastText [4] models' architecture to generate keyword embeddings instead of word embeddings, this meaning that embeddings represent sequences of words rather than a single word, similar to sentence embeddings. Unlike the base models, which use bag-of-words to build the representations, our approaches are based on bag-of-keywords. Thus, the proposed models exploit the annotated keywords' co-occurrence in the scientific literature for learning the dense keyword representations.

Our main aim for producing keyword embeddings is to represent concepts or topics instead of representing just words' semantics or even their contextual meaning like word embedding models do. The keywords' conceptual or topical semantics rely on an explicit human annotation process. First, the annotator will only select proper or commonly used descriptors for the presented documents. In this way, we may assume that those keywords are a good proxy when searching for the documents' topics. Second, authors use a limited number of keywords when annotating documents, so the strength of the semantic relationships among them is higher than the traditional approach of word co-occurrence in windows of free text.

Along with these models, we propose a new method to perform negative sampling, which leverages connected components to select the negative inputs used in the training phase. This method uses the keywords co-occurrence graph to extract the connected components and then select inputs that are not in the same connected component as negative samples.

To evaluate our models' performance, we compare them against a set of baselines composed of state-of-the-art word and sentence embedding models. We both trained the baselines in the keyword data and used the pre-trained models. In particular, we further trained Word2Vec [19] and Sentence-BERT [21] (SBERT) on the task data, and we used SBERT, SciBERT [2] and FastText [4] base models. We carry out the evaluation in Inspec [11] and KP20k [18], two classical keyword extraction datasets compound of scientific publications which count with a set of human-annotated keywords (for now on, we refer to keywords annotated by documents' authors or professional annotators as annotated keywords). The results show significant improvements for the task over state-of-the-art word and sentence embedding models.

The main contributions of our work can be summarized as follows:

- `Keywords2Vec`, a keyword embedding model based on `Word2Vec`.
- `FastKeywords`, a keyword embedding model based on `FastText`.
- A new method for negative sampling based on connected components.
- Baselines and evaluation methods for the similar keyword suggestion task.

2   Related Work

This section briefly overviews the existing word and sentence embedding models, particularly those used as baselines in the evaluation stage, along with some previous work related to the keyword suggestion task.

Dense vector representations, a.k.a. embeddings, have an appealing, intuitive interpretation and can be the subject of valuable operations (e.g. addition, subtraction, distance measures, etc.). Because of those features, embeddings have massively replaced traditional representations in most Machine Learning algorithms and strategies. Many word embedding models, such as `Word2Vec` [19], `GloVe` [20] or `FastText` [4], have been integrated into widely used toolkits, resulting in even more precise and faster word representations.

`Word2Vec` [19] was one of the first widely used neural network-based techniques for word embeddings. These representations preserve semantic links between words and their contexts by using the surrounding words to the target one. The authors proposed two methods for computing word embeddings [19]: skip-gram (SG), which predicts context words given a target word, and continuous bag-of-words (CBOW), which predicts a target word using a bag-of-words context.

`FastText` [4] is a `Word2Vec` add-on that treats each word as a collection of character n-grams. `FastText` can estimate unusual and out-of-vocabulary words thanks to the sub-word representation. In [12], authors employed `FastText` word representation in conjunction with strategies such as bag of n-gram characteristics and demonstrated that `FastText` outperformed deep learning approaches while being faster.

Sentence embeddings surged as a natural progression of the word embedding problem. Significant progress has been made in sentence embeddings in recent years, particularly in developing universal sentence encoders capable of producing good results in a wide range of downstream applications.

`Sentence-BERT` [21] is one of the most popular sentence embedding models and state-of-the-art on the sentence representation task. It is a modification of the `BERT` [9] network using siamese and triplet networks that can derive semantically meaningful sentence embeddings.

There are other sentence embedding models based on pre-trained language models. An example of those language models which achieves excellent results when working with scientific data is SciBERT. SciBERT [2] is an adaptation of BERT [9] to address the lack of high-quality, large-scale labelled scientific data. This model leverages BERT unsupervised pre-training capabilities to further train the model on a large multi-domain corpus of scientific publications, producing significant improvements in downstream scientific NLP tasks.

Even though word and sentence embeddings have been widely studied, the work on keyword embeddings is still limited. Researchers have employed them mainly on tasks like keyword extraction, phrase similarity or paraphrase identification. Several approaches for the keyword extraction task, like EmbedRank [3], directly rely on pre-trained sentence embedding models to rank the extracted keywords. However, other approaches such as Key2Vec [16] train their own model for the phrase embedding generation. In particular, the authors propose directly training multi-word phrase embeddings using FastText instead of a classic approach that learns a model for unigram words combining the words' dense vectors to build multi-word embeddings later.

Yin and Schütze [25] presented an embedding model for generalized phrases to address the paraphrase identification task. This approach aims to train the Word2Vec SG model without any modification to learn phrase embeddings. They pre-process the corpus by reformatting the sentences with the continuity information of phrases. The final collection contains two-word phrases whose parts may occur next to each other (continuous) or separated from each other (discontinuous).

The phrase semantic similarity task is akin to the one we address in this paper. Many sentence embedding models, including SBERT, are pre-trained in that downstream task. In [26], the authors present a composition model for building phrase embeddings with semantic similarity in mind. This model, named Feature-rich Compositional Transformation (FCT), learns transformations for composing phrase embeddings from the component words based on extracted features from a phrase.

Finally, the keyword suggestion task is also a sub-type of query suggestion where all the input and output queries are represented as keywords. Previous works in query term suggestion did not leverage the power of keyword co-occurrence to recommend new terms. Instead, existing query suggestion systems usually approach this task by suggesting terms extracted from the document's content without relying on the relations between these terms. For example, in [24], authors propose indexing a set of terms extracted from the corpus to rank them using a language model given an input query. The problem with these approaches is that they depend on the appearance of semantically related terms in the analyzed documents.

3   Proposal

In this section, we present two novel keyword embedding models that leverage `Word2Vec` [19] and `FastText` [4] architectures to produce keyword embeddings. We name these models `Keywords2Vec` and `FastKeywords`, respectively. Unlike `Word2Vec` and `FastText`, these models are not trained on rolling windows over the documents' full text; instead, we only use combinations from the documents' set of keywords as inputs.

The first of them, `Keywords2Vec`, modifies `Word2Vec` CBOW architecture to represent each keyword as one item. That is, we use keywords as token inputs instead of words. Additionally, we change how to perform the training of the models; we will explain it later for both proposals. We also evaluated the SG counterpart, but it performed considerably worse than the CBOW, so we do not report them here for brevity.

The second one, `FastKeywords`, adapts `FastText` CBOW variant in a more complex way. First, instead of working with words as the bigger information unit, it works with keywords. Second, it always selects each word of the keyword and the keyword itself as inputs, and then, during the n-grams selection process, it generates each word's n-grams.

Taking the keyword "*search engine*" and $n = 3$ as an example, it will be represented by the following n-grams:

the special sequences for words:

| search | engine |

and the special sequence for the whole keyword:

| search engine |

The model uses special sequences to capture the semantic meaning of both words and keywords. Finally, we implemented a weighting system to ignore "fill" n-grams used when a keyword does not have enough n-grams to fill the desired input size. The inclusion of this kind of n-grams is needed because the model requires the same input length on every iteration.

Another novel contribution of this paper is how we generate the training inputs. The model needs both positive and negative contexts for the target keyword. They are called positive and negative samples. For producing the positive samples, we select combinations of annotated keywords from the document to which the target keyword belongs. In the case of the negative samples, we represent the keywords' co-occurrences in the dataset as a graph. In this graph, each keyword is a node and edges are created when two keywords appear together in a document. We select the negative samples from connected components different from that of the target keyword.

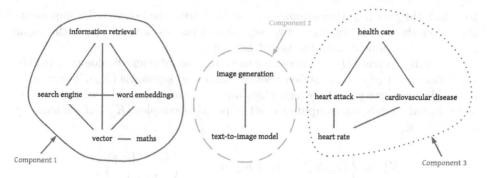

Fig. 1. Example of a keyword co-occurrence graph with its connected components.

Figure 1 shows an example of a keyword co-occurrence graph and its connected components. In that example, the positive samples for the keyword "*information retrieval*" would be combinations of "*search engine*", "*word embeddings*" and "*vector*", and its negative samples will always be extracted from connected components 2 or 3. Note that this means that the keyword "*maths*" will never be a negative sample for "*information retrieval*", even though it does not co-occur with it in any document.

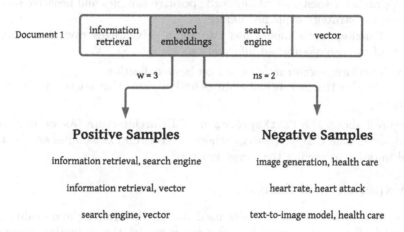

Fig. 2. Positive and negative samples generation example. (*The target keyword is highlighted*).

Figure 2 shows how the positive and negative sampling is performed for a document extracted from the graph shown in the Fig. 1. Having the keyword "*word embeddings*" as the target, we first select its positive samples. In this example, we select every combination of 2 ($w - 1 = 2$) keywords that belong to the same document as the target. Then, for each pair of keywords we have to select the two negative samples ($ns = 2$). Given the connected components shown in the Fig. 1, we select two keywords that do not belong to the same connected component than the target keyword.

Formally, to train these models, we use a dataset of scientific documents (D). Each document ($d \in D$) contains a set of annotated keywords (K_d). Then, given a combination size ($w - 1$), which plays an analogous role to the window size in the original models, we compute the set of positive samples (K_d^{ps}) of a document d as follows:

$$K_d^{ps} = \left\{ (k_i, K_d^{ps_j}) \mid k_i \in K_d, \ K_d^{ps_j} \in \binom{K_d - \{k_i\}}{w - 1} \right\},$$

that is, for each keyword (k_i) of the document's keywords set (K_d), we compute its positive samples (K_d^{ps}), which are the combinations of size $w - 1$ of the document's keywords set excluding the target keyword ($K_d - \{k_i\}$). Finally, for each positive samples set ($K_d^{ps_j}$) we obtain a pair ($k_i, K_d^{ps_j}$).

As for the negative samples for each document (K_d^{ns}) we have followed the subsequent novel approach. First, we build the aforementioned keywords co-occurrence graph for the collection and compute its connected components. Then, for each pair ($k_i, K_d^{ps_j}$) we select as negative samples ns keywords belonging to a different connected component than the target keyword (k_i).

To adapt the previous process to our FastKeywords model, we have to generate n-grams for each context keyword (positive samples and negative samples) following the strategy explained before.

The FastKeywords model has two main advantages over Keywords2Vec because of the use of subword information:

- It will perform significantly better on large collections.
- It will be able to generate embeddings for keywords that are not in the training corpus.

Figure 3 shows the FastKeywords model's architecture. As we may see, it follows the classic CBOW strategy where several context samples are fed to the model in order to predict the target keyword.

4   Experimental Setup

This section describes the datasets used during the training and evaluation of the models, the baselines used to compare our model, the evaluation process and metrics, and finally, the parameters and setup used to train our models.

Our objective is to suggest similar keywords to the user input in the search process. With this in mind, we designed two experiments that approach the keyword suggestion problem as a keyword ranking task. In particular, the evaluation considers if the model can find keywords that belong to the same document as the target keyword. The rationale for this evaluation is that keywords with which the authors annotate a document tend to be semantically related. Alternatively, we may perform user studies to evaluate the perceived quality of the suggestions. We leave that for future work.

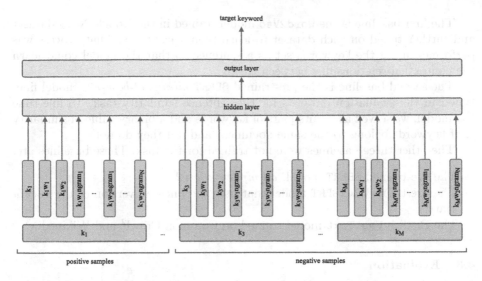

Fig. 3. FastKeywords model high-level architecture diagram.

4.1 Datasets

We selected two datasets commonly used on the classical keyword extraction task: the Inspec [11] and the KP20k [18] collections.

Inspec. This collection consists of 2,000 titles and abstracts from scientific journal papers. The annotators assigned two sets of keywords to each document in this collection: controlled keywords that occur in the Inspec thesaurus and uncontrolled keywords that can be any suitable terms. We will only use the uncontrolled keywords created by professional indexers in our experiments, following the standard approach in the keyword extraction task.

KP20k. This dataset is a keyword generation collection that includes 567,830 articles of the computer science domain obtained from various online digital libraries, including ACM Digital Library, ScienceDirect, Wiley and Web of Science. In this case, the authors made no differentiation during the annotation process, so we used all of the keywords in both training and testing.

4.2 Baselines

We include five baselines in our evaluation to compare our embedding models against other models that have already proven their effectiveness in capturing words or sentence syntax and semantic information.

The first baseline is the `Word2Vec` model trained in the Google News dataset and further tuned on each dataset to learn their vocabulary. That process was performed using the keywords set as a sentence so that the model could learn the contextual relationships between them.

The second baseline is the pre-trained `SBERT` *all-mpnet-base-v2*[1] model fine-tuned in the evaluation datasets for the sentence similarity task. To fine-tune the model, we served as inputs pairs of keywords and a binary value of similarity (1 if keywords belong to the same document and 0 if they do not).

The other three baselines were not trained for the task. These baselines are:

- The pre-trained `SBERT` model, *all-mpnet-base-v2*.
- The pre-trained `SciBERT` model that uses the uncased version of the vocabulary.
- The English `FastText` model trained on Common Crawl[2] and Wikipedia[3].

4.3   Evaluation

As we mentioned before, to assess our models' performance in the keyword suggestion task, we devise two evaluation strategies representing two specific cases in the academic search field.

The first task consists of retrieving all the keywords with which a document was annotated, given one of them. The keyword retrieval is done via cosine similarity between the query keyword and all the other keywords. Then, we compute each average precision and aggregate all the results to compute the MAP (Mean Average Precision). Specifically, we calculate MAP@20.

The second task we propose follows a masking problem. For each document, we mask one keyword from the keywords set and then use the remaining keywords to try to find the masked one. For each of the remaining keywords, we retrieve the most similar ones using cosine similarity. Then, combining the scores of all the non-masked keywords, we compute the final ranking. Finally, we compute each masked keyword's reciprocal rank to calculate the MRR (Mean Reciprocal Rank). Precisely, we compute MRR@100. Note that for this evaluation method, we only used 50 random documents from the test split of each dataset.

Regarding methodologies, we evaluated both tasks under all-items and test-items approaches. In the all-items fashion, the keywords to be ranked are the whole set of keywords in the dataset, while in the test-items indexing approach, only the keywords in the test subset are ranked.

We also perform statistical significance tests for all the evaluations. On the one hand, for comparing more than two systems, we use the Randomised Tukey HSD Test [8,23]. On the other hand, we follow the classic permutation test approach when comparing a pair of systems. These tests verify that the family-wise error does not exceed the confidence level α.

---

[1] https://huggingface.co/sentence-transformers/all-mpnet-base-v2.
[2] https://commoncrawl.org/.
[3] https://dumps.wikimedia.org/.

Table 1. Models' training parameters on the Inspec and KP20k datasets.

Model	Inspec					KP20k				
	Batch	Epochs	Dim	w	ns	Batch	Epochs	Dim	w	ns
Word2Vec	–	*early stopping*	300	6	–	–	20	300	6	–
SBERT	2^9	*early stopping*	768	–	4	2^9	20	768	–	4
Keywords2Vec	2^{17}	*early stopping*	300	3	4	2^{17}	20	300	3	4
FastKeywords	2^{15}	*early stopping*	300	3	4	2^{15}	20	300	3	4

4.4   Experimental Settings

This section will describe the parameters used to train the baselines and our proposed models. Table 1 shows a summary of the models' training parameters detailed hereunder.

To train the Word2Vec model, we followed an early-stopping strategy for the Inspec dataset, while for KP20k, we trained the model for 20 epochs. For both datasets, the selected embedding size was 300. Finally, we used a window size of 6 and 1 as the minimum count (lowest frequency of a word to be considered).

On the other hand, as we mentioned before, we fine-tuned the SBERT model in the sentence similarity task. For each positive sample (two keywords that co-occur in a document), we selected four negative samples (two keywords that never co-occur in a document). Again, we followed an early-stopping strategy for the Inspec dataset, while for KP20k, we trained the model for 20 epochs. For both datasets, we used a batch size of 2^9 and an embedding size of 768.

We trained both our models using an early-stopping strategy for the Inspec dataset. On the other hand, when using the KP20k dataset, we trained each model for 20 epochs. Regarding the batch size, the Keywords2Vec model was trained with a batch size of 2^{17} on both datasets, while for the FastKeywords model, we used a batch size of 2^{15} on Inspec and KP20k. Also, both were trained to generate embeddings of size 300. The combination size and the number of selected negative samples were the same for both models on both datasets, 2 ($w = 3$) and 4, respectively. Finally, regarding FastKeywords parameters, we used a minimum n-gram size of 3, a maximum n-gram size of 6 and a maximum number of n-grams of 20.

When trained, we used the whole set of document keywords from the dataset in the tuning process of the models[4]. When not trained, we used the default pre-trained models for the baselines. In the evaluation process, we selected a subset of documents on which we performed the aforementioned tasks.

---

[4] All experiments were run on an NVIDIA A100 GPU with 80GB of memory.

Table 2. MAP@20 for the document's keywords identification task. Statistically significant improvements according to the Randomized Tukey HSD test (*permutations* = 100, 000; $\alpha = 0.05$) are superscripted.

Model	Inspec		KP20K	
	All items	Test items	All items	Test items
SBERT (*no train*) (†)	0.0323	$0.0544^{ᵂ}$	0.0039	$0.0134^{ᵂ}$
SciBERT (*no train*) (‡)	0.0205	0.0345	0.0040	0.0110
FastText (*no train*) (ᵂ)	0.0147	0.0217	0.0039	0.0095
Word2Vec (∓)	$0.0463^{ᵂ}$	$0.0832^{†‡ᵂ}$	0.0048	$0.0130^{‡ᵂ}$
SBERT (±)	$0.1481^{†‡ᵂ∓}$	$0.5880^{†‡ᵂ∓}$	$0.0072^{†‡ᵂ∓}$	$0.0182^{†‡ᵂ∓}$
Keywords2Vec (⊗)	$0.8634^{†‡ᵂ∓±}$	$0.9090^{†‡ᵂ∓±}$	$0.0690^{†‡ᵂ∓±}$	$0.0918^{†‡ᵂ∓±}$
FastKeywords (⊙)	$\mathbf{0.8659}^{†‡ᵂ∓±}$	$\mathbf{0.9161}^{†‡ᵂ∓±}$	$\mathbf{0.0762}^{†‡ᵂ∓±⊗}$	$\mathbf{0.1060}^{†‡ᵂ∓±⊗}$

5 Results

This section reports how Keywords2Vec and FastKeywords perform against the selected baselines and how they perform against each other. As we mentioned in Sect. 4.3 we report results using two tasks and two evaluation techniques.

Table 2 shows the results for the document's keywords identification task. We can see that our models significantly outperform the established baselines in both datasets and evaluation approaches. This is especially remarkable in the case of the state-of-the-art sentence BERT model, which, even for the fine-tuned scenario and using embeddings with twice the dimensions, lies quite behind our proposals. As expected, using the test-items strategy produces better results since the keyword set where the search is performed is much smaller.

Moreover, we can see that the dataset size is a determinant factor in this task. A much larger set of keywords (this is the case of the KP20k dataset versus the Inspec dataset) greatly impacts the final score. This makes sense because increasing the keyword set size makes finding the ones we are looking for more challenging (something analogous happens to a lesser degree between the all-items and test-items results).

In terms of comparing FastKeywords against Keywords2Vec we can see that the former performs better and that the more considerable difference between them appears when the keyword set is the biggest (KP20k dataset). The main reason behind this relies on the capability of FastKeywords to leverage subword information to build the keywords embeddings.

Table 3 shows results for the masked keyword discovering task. These results confirm what we had already seen in the first one: our models significantly outperform all the established baselines in every experiment we performed. Again, we can see that increasing the keyword set length produces worse results as the task becomes more and more challenging. Again, when comparing the proposed models against each other on this task, we can see that FastKeywords performs better than Keywords2Vec when increasing the dataset size, getting statistically significant improvements over the Word2Vec-based method.

Table 4 shows the performance of the proposed negative sampling method, comparing it with a FastKeywords model that uses a random negative sampling

Table 3. MRR@100 for the masked keyword discovering task. Statistically significant improvements according to the Randomized Tukey HSD test (*permutations* = $1,000,000$; $\alpha = 0.05$) are superscripted.

Model	Inspec		KP20K	
	All items	Test items	All items	Test items
SBERT *no train* (†)	0.0342	0.0479	0.0018	0.0152
SciBERT *no train* (‡)	0.0220	0.0345	0.0056	0.0129
FastText *no train* (ʊ)	0.0068	0.0097	0.0023	0.0047
Word2Vec (∓)	0.0751	0.1187ʊ	0.0008	0.0079
SBERT (±)	0.1688†‡ʊ	0.5890†‡ʊ∓	0.0138	0.0448
Keywords2Vec (⊗)	0.8914†‡ʊ∓±	0.9100†‡ʊ∓±	0.0778†‡ʊ∓±	0.0828†‡ʊ∓
FastKeywords (⊙)	0.8988†‡ʊ∓±	0.9102†‡ʊ∓±	0.1402†‡ʊ∓±⊗	0.1467†‡ʊ∓±⊗

Table 4. Negative sampling methods comparison on the Inspec dataset. Statistically significant improvements according to the permutation test (*permutations* = $1,000,000$; $\alpha = 0.05$) are superscripted.

Model	MAP@20		MRR@100	
	All items	Test items	All items	Test items
FastKeywords *random negative sampling* (⊕)	0.8420	0.9093	0.8750	0.8890
FastKeywords (⊙)	0.8659⊕	0.9161⊕	0.8988	0.9102⊕

Table 5. Top 10 nearest neighbours for the keyword "*information retrieval*" on the KP20k collection (*test-items index*).

FastKeywords		Keywords2Vec	
keyword	score	keyword	score
ranking	0.9133	retrieval model	0.6045
query expansion	0.9104	collection selection	0.5587
text classification	0.9067	link topic detection	0.5472
relevance	0.9044	dempstershafer evidence theory	0.5418
text mining	0.9018	retrospective evidence event detection	0.5371
information extraction	0.9012	instance based learning	0.5361
relevance feedback	0.8980	query expansion	0.5282
knowledge discovery	0.8979	terminology extraction	0.5260
document clustering	0.8971	viral marketing	0.5253
text categorization	0.8970	query formulation	0.5248

strategy. The results ratify that the proposed method works significantly better than a naive random approach in all cases except on MRR@100 using the all-items indexing strategy. For this case, the p-value is 0.06.

Finally, for illustrative purposes, Table 5 shows the top 10 nearest neighbours for the keyword "*information retrieval*" on the KP20k collection retrieved by both proposed models. We also show each keyword's associated score, which represents the similarity with the query keyword.

6 Conclusions

This paper explored the potential of keyword embedding models in the keyword suggestion task. We also propose several baselines and new evaluation methods to assess the performance of the models, as not much previous work has been published for this task.

The proposed models adapt `Word2Vec` and `FastText` CBOW architectures to compute keyword embeddings instead of word embeddings. Along with these two keyword embedding models, we present a novel strategy for the negative sampling task, which leverages the potential of the keyword co-occurrence graph's connected components to perform a better selection of the negative samples.

Results show that our methods significantly outperform the selected baselines on both evaluation datasets. We also demonstrated the potential of sub-keyword and sub-word information to represent keywords as embeddings. In future work we aim to:

- Use the designed weights system to give more relevance to full keywords and words than to n-grams.
- Assess the models' performance using popularity-based negative sampling.
- Combine negative samples extracted from the target keyword connected component and from different connected components.
- Use special delimiters to differentiate if a word is a part of a keyword or a keyword itself or if a n-gram is a part of a word or a word itself.
- Train and test the models on non-scientific keyword-style annotated data.
- Study how the offline findings of this work align with live user testing.

Acknowledgements. This work was supported by projects PLEC2021-007662 (MCIN/AEI/10.13039/501100011033, Ministerio de Ciencia e Innovación, Agencia Estatal de Investigación, Plan de Recuperación, Transformación y Resiliencia, Unión Europea-Next Generation EU) and RTI2018-093336-B-C22 (Ministerio de Ciencia e Innovación, Agencia Estatal de Investigación). The first and third authors also thank the financial support supplied by the Consellería de Cultura, Educación e Universidade Consellería de Cultura, Educación, Formación Profesional e Universidades (accreditation 2019-2022 ED431G/01, ED431B 2022/33) and the European Regional Development Fund, which acknowledges the CITIC Research Center in ICT of the University of A Coruña as a Research Center of the Galician University System. The fist author also acknowledges the support of grant DIN2020-011582 financed by the MCIN/AEI/10.13039/501100011033.

References

1. André, P.A., Schraefel, M.C., Teevan, J., Dumais, S.T.: Discovery is never by chance: designing for (un)serendipity. In: Bryan-Kinns, N., Gross, M.D., Johnson, H., Ox, J., Wakkary, R. (eds.) Proceedings of the 7th Conference on Creativity & Cognition, Berkeley, California, USA, 26–30 October 2009, pp. 305–314. ACM (2009). doi: https://doi.org/10.1145/1640233.1640279

2. Beltagy, I., Lo, K., Cohan, A.: SciBERT: a pretrained language model for scientific text. In: Proceedings of the 2019 Conference on Empirical Methods in Natural Language Processing and the 9th International Joint Conference on Natural Language Processing (EMNLP-IJCNLP), pp. 3615–3620. Association for Computational Linguistics, Hong Kong, China (2019). https://doi.org/10.18653/v1/D19-1371

3. Bennani-Smires, K., Musat, C., Hossmann, A., Baeriswyl, M., Jaggi, M.: Simple unsupervised keyphrase extraction using sentence embeddings. In: Korhonen, A., Titov, I. (eds.) Proceedings of the 22nd Conference on Computational Natural Language Learning, CoNLL 2018, Brussels, Belgium, 31 Oct - 1 Nov 2018, pp. 221–229. Association for Computational Linguistics (2018). https://doi.org/10.18653/v1/k18-1022

4. Bojanowski, P., Grave, E., Joulin, A., Mikolov, T.: Enriching word vectors with subword information. Trans. Assoc. Comput. Linguist. **5**, 135–146 (2017)

5. Buchanan, S.A., Sauer, S., Quan-Haase, A., Agarwal, N.K., Erdelez, S.: Amplifying chance for positive action and serendipity by design. Proceed. Assoc. Inf. Sci. Technol. **57**(1), e288 (2020)

6. Camacho-Collados, J., Pilehvar, M.T.: From word to sense embeddings: a survey on vector representations of meaning. J. Artif. Intell. Res. **63**, 743–788 (2018)

7. Carpineto, C., Romano, G.: A survey of automatic query expansion in information retrieval. ACM Comput. Surv. (CSUR) **44**(1), 1–50 (2012)

8. Carterette, B.A.: Multiple testing in statistical analysis of systems-based information retrieval experiments. ACM Trans. Inf. Syst. **30**(1), 1–34 (2012). https://doi.org/10.1145/2094072.2094076

9. Devlin, J., Chang, M.W., Lee, K., Toutanova, K.: BERT: pre-training of deep bidirectional transformers for language understanding. In: Proceedings of the 2019 Conference of the North American Chapter of the Association for Computational Linguistics: Human Language Technologies, Volume 1 (Long and Short Papers), pp. 4171–4186. Association for Computational Linguistics, Minneapolis, Minnesota (2019). https://doi.org/10.18653/v1/N19-1423

10. Hu, J., Wang, G., Lochovsky, F., Sun, J.t., Chen, Z.: Understanding user's query intent with wikipedia. In: Proceedings of the 18th International Conference on World Wide Web, pp. 471–480. WWW 2009, Association for Computing Machinery, New York, NY, USA (2009). https://doi.org/10.1145/1526709.1526773

11. Hulth, A.: Improved automatic keyword extraction given more linguistic knowledge. In: Proceedings of the 2003 Conference on Empirical Methods in Natural Language Processing, pp. 216–223 (2003)

12. Joulin, A., Grave, E., Bojanowski, P., Mikolov, T.: Bag of tricks for efficient text classification. In: Lapata, M., Blunsom, P., Koller, A. (eds.) Proceedings of the 15th Conference of the European Chapter of the Association for Computational Linguistics, EACL 2017, Valencia, Spain, 3–7 April 2017, Volume 2: Short Papers, pp. 427–431. Association for Computational Linguistics (2017). https://doi.org/10.18653/v1/e17-2068

13. Kim, Y., Seo, J., Croft, W.B.: Automatic Boolean query suggestion for professional search. In: Proceedings of the 34th International ACM SIGIR Conference on Research and Development in Information Retrieval, pp. 825–834. SIGIR 2011, Association for Computing Machinery, New York, NY, USA (2011). https://doi.org/10.1145/2009916.2010026

14. Li, M., Zhu, M., Zhang, Y., Zhou, M.: Exploring distributional similarity based models for query spelling correction. In: Proceedings of the 21st International Conference on Computational Linguistics and 44th Annual Meeting of the Association for Computational Linguistics, pp. 1025–1032 (2006)

15. Li, X., Schijvenaars, B.J.A., de Rijke, M.: Investigating queries and search failures in academic search. Inf. Process. Manag. **53**(3), 666–683 (2017). https://doi.org/10.1016/j.ipm.2017.01.005

16. Mahata, D., Kuriakose, J., Shah, R., Zimmermann, R.: Key2Vec: automatic ranked keyphrase extraction from scientific articles using phrase embeddings. In: Proceedings of the 2018 Conference of the North American Chapter of the Association for Computational Linguistics: Human Language Technologies, Volume 2 (Short Papers), pp. 634–639 (2018)

17. Martins, B., Silva, M.J.: Spelling correction for search engine queries. In: Vicedo, J.L., Martínez-Barco, P., Muñoz, R., Saiz Noeda, M. (eds.) EsTAL 2004. LNCS (LNAI), vol. 3230, pp. 372–383. Springer, Heidelberg (2004). https://doi.org/10.1007/978-3-540-30228-5_33

18. Meng, R., Zhao, S., Han, S., He, D., Brusilovsky, P., Chi, Y.: Deep keyphrase generation. In: Proceedings of the 55th Annual Meeting of the Association for Computational Linguistics (Volume 1: Long Papers), pp. 582–592. Association for Computational Linguistics, Vancouver, Canada (2017). https://doi.org/10.18653/v1/P17-1054

19. Mikolov, T., Chen, K., Corrado, G., Dean, J.: Efficient estimation of word representations in vector space. In: Bengio, Y., LeCun, Y. (eds.) 1st International Conference on Learning Representations, ICLR 2013, Scottsdale, Arizona, USA, 2–4 May 2013, Workshop Track Proceedings (2013). https://arxiv.org/abs/1301.3781

20. Pennington, J., Socher, R., Manning, C.D.: Glove: global vectors for word representation. In: Proceedings of the 2014 Conference on Empirical Methods in Natural Language Processing (EMNLP), pp. 1532–1543 (2014)

21. Reimers, N., Gurevych, I.: Sentence-BERT: sentence embeddings using Siamese bert-networks. In: Proceedings of the 2019 Conference on Empirical Methods in Natural Language Processing. Association for Computational Linguistics (2019). https://arxiv.org/abs/1908.10084

22. Russell-Rose, T., Gooch, P., Kruschwitz, U.: Interactive query expansion for professional search applications. Bus. Inf. Rev. **38**(3), 127–137 (2021)

23. Laboratory Experiments in Information Retrieval. TIRS, vol. 40. Springer, Singapore (2018). https://doi.org/10.1007/978-981-13-1199-4

24. Verberne, S., Sappelli, M., Kraaij, W.: Query term suggestion in academic search. In: de Rijke, M., et al. (eds.) ECIR 2014. LNCS, vol. 8416, pp. 560–566. Springer, Cham (2014). https://doi.org/10.1007/978-3-319-06028-6_57

25. Yin, W., Schütze, H.: An exploration of embeddings for generalized phrases. In: Proceedings of the 52nd Annual Meeting of the Association for Computational Linguistics, ACL 2014, 22–27 June 2014, Baltimore, MD, USA, Student Research Workshop, pp. 41–47. The Association for Computer Linguistics (2014). https://doi.org/10.3115/v1/p14-3006

26. Yu, M., Dredze, M.: Learning composition models for phrase embeddings. Trans. Assoc. Comput. Linguistics **3**, 227–242 (2015). https://doi.org/10.1162/tacl_a_00135

Domain-Driven and Discourse-Guided Scientific Summarisation

Tomas Goldsack[1] , Zhihao Zhang[2] , Chenghua Lin[1](✉) ,
and Carolina Scarton[1]

[1] University of Sheffield, Sheffield, UK
{tgoldsack1,c.lin,c.scarton}@sheffield.ac.uk
[2] Beihang University, Beijing, China
zhhzhang@buaa.edu.cn

Abstract. Scientific articles tend to follow a standardised discourse that enables a reader to quickly identify and extract useful or important information. We hypothesise that such structural conventions are strongly influenced by the scientific domain (e.g., Computer Science, Chemistry, etc.) and explore this through a novel extractive algorithm that utilises domain-specific discourse information for the task of abstract generation. In addition to being both simple and lightweight, the proposed algorithm constructs summaries in a structured and interpretable manner. In spite of these factors, we show that our approach outperforms strong baselines on the arXiv scientific summarisation dataset in both automatic and human evaluations, confirming that a scientific article's domain strongly influences its discourse structure and can be leveraged to effectively improve its summarisation. Our code can be found at: https://github.com/TGoldsack1/DodoRank.

Keywords: Summarisation · Scientific documents · Scientific discourse

1 Introduction

Scientific abstracts are used by researchers in determining whether a given article is relevant to their own work. Therefore, a well-written scientific abstract should concisely describe the essential content of an article from the author's perspective [18], whilst following some standardised discourse structure [15,26]. Several structural classification schemes exist that attempt to model the sentence-level discourse of scientific articles within a particular scientific domain (e.g., Computer Science, Chemistry, etc.), with a focus on categorising sentences according to factors such as rhetorical status [3,20,27]. These schemes have proven utility in the automatic summarisation of scientific articles [6,21,28].

T. Goldsack and Z. Zhang—Equal contribution

J. Kamps et al. (Eds.): ECIR 2023, LNCS 13980, pp. 361–376, 2023.
https://doi.org/10.1007/978-3-031-28244-7_23

Neural network-based approaches have gained increasing popularity for abstractive summarisation in recent years [11]. Although these models are capable of producing coherent summaries, they are prone to generating hallucinations (i.e., text that is unfaithful to the source document) which can result in factual inconsistencies in their output [2,23]. When it comes to scientific content, this is especially problematic as it may result in the dissemination of false or misleading information relating to important research findings [24]. On the other hand, extractive summarisation systems typically ensure the basic syntactic and semantic correctness of their output and thus remain widely used [11].

Recent extractive works have attempted to leverage coarse *section-level* discourse structure of scientific articles, assuming that this provides a strong signal in determining the most informative content of the input. Dong et al. [10] build on the unsupervised graph-based approach of Zheng and Lapata [31], introducing a hierarchical document representation that accounts for both intra- and intersection connections and exploits positional cues to determine sentence importance. Similarly, Zhu et al. [33] exploit discourse structural information, proposing a supervised approach that extracts article sections based on predicted salience before ranking their sentences via a hierarchical graph-based summariser. While the aforementioned works have explored modelling the discourse of articles to some extent, few have addressed how it can be used to directly influence the structure and content of their generated output. Furthermore, to our knowledge, no prior work has studied how the *scientific domain* of an article impacts the discourse structure of its abstract.

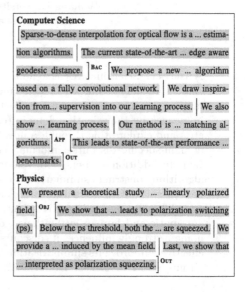

Fig. 1. Examples of the rhetorical abstract structures typical of different scientific domains (Computer Science and Physics). Highlighted text donates rhetorical labels: BACKGROUND, OBJECTIVE, APPROACH, and OUTCOME.

In this work, we tackle the problem of generating scientific abstracts, proposing a novel summarisation method that explicitly models and exploits *domain-specific* discourse, a valuable source of information that has not been explored in the prior literature. We hypothesise that the scientific domain of an article has a strong influence over the rhetorical structure of its abstract (exemplified in Fig. 1), and thus can be leveraged in the context of abstract generation to improve both the content and structure of the generated output. As such, we conduct the first study on the influence of scientific domain over the discourse structure of abstracts, using it to inform our summarisation method, DODORANK (**Do**main and **D**iscourse-oriented **Rank**ing model). DODORANK consists of two primary components: (i) *a discourse extraction module* that, for a given dataset, determines which sections contain the most salient content for abstract generation and computes domain-specific, sentence-level rhetorical abstract structures for governing the generation process; and (ii) an *unsupervised extractive summariser*, which produces a scientific abstract based on the extracted domain-specific discourse information. Specifically, discourse information is used to *both* reduce the input to the most salient sections and impose a rhetorical structure upon our generated output that conforms to the conventions of the scientific domain. The sentences of salient sections are ranked and extracted in an unsupervised fashion, using sentence *centrality* as a domain-independent measure of importance [10,12,31]. Consequently, DODORANK constitutes a lightweight and interpretable approach to summarisation that, despite its simplicity, gives better or comparable performance to strong supervised baselines on the multi-domain arXiv dataset [4], whilst also reducing the size of the input by an average of 66.24%. We further illustrate the effectiveness our approach via human evaluation, achieving superior performance to the state-of-the-art centrality-based model. Finally, we provide a domain-specific breakdown of results on arXiv, conclusively demonstrating that consideration of a scientific article's domain is beneficial for its summarisation.

2   Related Work

Scientific Discourse. The formalised nature of scientific writing has led to the creation of several classification schemes which categorise sentences according to their role within the larger discourse structure. By capturing the different types of information scientific articles contain at a fine-grained level, these schemes have the ability to support both the manual study and automatic analysis of scientific literature and its inherent structural discourse [14,16,28]. The two most prevalent of these schemes are Argumentative Zoning [27] and Core Scientific Concepts (CoreSC) [20], both of which categorise sentences according to their rhetorical status and are provided alongside manually annotated corpora with scientific articles from the Computational Linguistics and Chemistry domains, respectively. Subsequent works have since introduced corpora annotated with similar schemes, typically focusing on a single scientific domain [3,16,17,29]. Several of these datasets are used within this work (see Table 1).

Table 1. Scientific datasets used in this work. Here, † denotes that sentences are manually annotated with rhetorical labels and * denotes the official data splits.

Dataset	Domain	# Train / Val / Test
ART CORPUS† (2010)	Chemistry	169 / 28 / 28
CSABSTRUCT† * (2019)	CS	1668 / 295 / 226
AZ ABSTRACT† (2010)	Biomedical	750 / 150 / 100
AZ ARTICLE† (2013)	Biomedical	37 / 7 / 6
ARXIV* (2018)	Multi-domain	203K / 6.4K / 6.4K

Extractive Summarisation of Scientific Articles. Extractive summarisation approaches aim to build a summary using text spans extracted from the source document. These approaches remain attractive as they prevent factual hallucinations in the generated output, resulting in more reliable and usable summaries.

Prior to the rise of deep neural models, the use of traditional supervised algorithms proved popular, commonly used in combination with statistical text features such as sentence length and location, TFIDF scores, and frequency of citations [5,6,21,28]. Rhetorical classification schemes, such as those previously described, have also been shown to have value as features to these algorithms [6,28]. We make use of rhetorical classes in a style similar to that of Liakata et al. [21] who use the distribution of CoreSC classes within Chemistry articles to create a rhetorical plan for their generated output. In contrast to this work, we derive rhetorical structures directly from abstracts themselves and deploy them on *multiple scientific domains*. These earlier works also fail to address the section-based structure of scientific documents, which has since been shown to have influence over the distribution of summary-worthy content [10,33].

More recently, Xiao and Carenini [30] incorporate section information within a bi-directional RNN document encoder, before outputting confidence scores for each sentence. As covered in §1, both Dong et al. [10] and Zhu et al. [33] model coarse-level structural discourse information within hierarchical document graphs. Additionally, Dong et al. [10] determine the importance of a sentence calculating its centrality within a group of sentences [12,31]. In this work, we also make use of centrality to compute sentence importance. However, where Dong et al. [10] group sentences based on the origin section and place emphasis on the position of sentences within sections, we group sentences based on rhetorical status and place emphasis on structuring output in a way that conforms with the conventions of the specific scientific domain.

3   Method

Our summarisation framework consists of two key components: A) a discourse extraction component and B) a summarisation component, as illustrated in Fig. 2. The discourse extraction component operates offline during an initial 'learning phase'. It is responsible for (i) determining the most salient article sections for abstract generation (Sect. 3.1), and (ii) capturing domain specific, sentence-level rhetorical abstract structures (Sect. 3.2). The summarisation component then employs the extracted discourse information in an online setting for guided abstract generation based on the centrality ranking of sentences (Sect. 3.3).

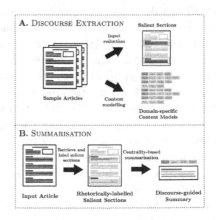

Fig. 2. DODORANK model overview.

3.1 Salient Section Determination

A scientific abstract is a concise summary of a research paper, typically high-lighting a few important components such as the research motivation, problem, approach, and key results and/or findings. We hypothesise that some of the article sections serve a more similar communicative goal to the abstract than others, and hence contribute more significantly to its content.

To deduce the most salient article sections for a given dataset, we propose to assess their individual contribution to the content of the abstract. We conduct our analysis and discourse information extraction based on a *sample set*, derived from the arXiv dataset, that contains articles from all eight scientific domains.[1] We balance sampling in terms of the scientific domain to ensure that the discourse information extracted is representative of all domains in the dataset. Further details on the sample set are given in Sect. 4.

For each abstract within the sample, we calculate the similarity between each sentence it contains and every sentence in the main body of the respective article. Due to the scientific nature of the text, similarity is measured using cosine similarity between SciBERT embeddings [1]. Subsequently, the sentence with the greatest similarity to each abstract sentence (referred to as the *oracle sentence*) and the heading of the section in which it is located are retrieved.

Table 2. Matching terms used for each conflated heading.

Conflated heading	Matching terms
Introduction	"introduction"
Conclusion	"conclu", "summary"
Discussion	"discussion"
Result/analysis	"result", "analys", "ablat"
Background/motivation	"background", "motivation"
Method	"implement", "method"
Model	"architec", "system", "model"
Future work	"direction", "future"
References	"referenc"
Acknowledgements	"acknowledg"
Related work	"related"

In verbatim form, the retrieved section headings are noisy, with much variation in phrasing and formatting when referring to semantically similar concepts (e.g., the concluding section of an article can be titled "conclusion", "summary", "concluding remarks", etc., all of which are semantically identical). Therefore, following Ermakova et al. [13], verbatim headings are conflated into a standardised form by matching selected words and sub-words against them using regular expressions. Matching terms for all conflated headings are derived empirically based on the sample set and given in Table 2.

The most important sections are regarded as those which the oracle sentences most frequently originate from, and we use only sentences from these sections as input to our summarisation component. Specifically, we select the minimum amount of sections that cumulatively contribute to at least 50% of all oracle sentences (see Table 5), to ensure sufficient coverage of salient content. Our analysis based on the arXiv dataset shows that the Introduction and Conclusion sections are the most salient contributors for the abstracts across *all* tested domains. Please refer to Sect. 5.1 for detailed analysis.

---

[1] This sample set is also used within §3.2, as indicated in Fig. 2.

3.2 Rhetorical Content Modelling

To govern the generation process of our summariser, we aim to extract a rhetorical structure which is representative of a typical abstract for each specific scientific domain. We refer to these structures as *content models*. To this end, we adopt a sentence-level classification scheme similar to that of CSAbstruct [3], a dataset consisting of rhetorically labelled Computer Science abstracts. Specifically, our scheme contains the rhetorical labels: {BACKGROUND, OBJECTIVE, APPROACH, OUTCOME, OTHER}, where each label represents a high-level rhetorical role that may be assigned to any sentence in a given article, regardless of scientific domain.

Applying this classification scheme requires us to obtain the rhetorical labels for the unlabelled scientific abstracts. Therefore, we train a SciBERT-based sequential classifier [1,3,9] on a combination of four datasets, all of which contain scientific articles and/or abstracts manually annotated with similar sentence-level rhetorical schemes by their creators. Specifically, we convert the labels of CSAbstruct [3], the ART corpus [20], AZ Abstract [16], and AZ Article [17] to our given

Table 3. Mapping of the rhetorical labels of other datasets to our classification scheme.

Our label	Verbatim label			
	ART Corpus	CSAbstruct	AZ Article	AZ Abstract
BACKGROUND	Background	Background	Background Connection Difference	Background Related Work
OBJECTIVE	Motivation Goal Hypothesis Object	Objective	Problem Future Work	Objective Future Work
APPROACH	Experiment Model Method	Method	Method	Method
OUTCOME	Observation Result Conclusion	Result	Result Conclusion	Result Conclusion
OTHER	-	Other	-	-

scheme via a simple label mapping procedure, illustrated in Table 3. In combining these datasets and exposing the classifier to instances of our rhetorical classes from different scientific domains (see Table 1), we aim to make it more robust to unseen domains. We validate the reliability of this mapping by evaluating the trained classifier on the test set of CSAbstruct (Sect. 5.1), by far the largest of the contributing datasets.

Following label prediction on the abstracts of the sample set, we employ a frequency-based approach to extract the domain-specific content models. The core idea of this approach is to find the most common pattern of rhetorical labels observed in the abstracts of a given domain. A content model M corresponds to a sequence of K rhetorical sentence labels, where K is a hyperparameter to our model and determined based on overall model performance on the validation split (e.g., for a value of $K = 3$, an example M could be [BACKGROUND, APPROACH, OUTCOME]).

To extract our domain-specific content models, one intuitive solution would be to sequentially compute the rhetorical label that occurs most frequently within its sample set abstracts for each sentence position from 0 to K. However, abstracts vary in length, and therefore a simple frequency-based method without taking this into account will be sub-optimal in capturing the rhetorical label distributions of the sample set. To tackle this challenge, we propose to normalise

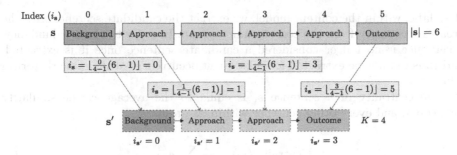

Fig. 3. Example showing how a sample abstract s is compressed to s' using (1).

a sample abstract's rhetorical label sequence s by performing position-aware up/down sampling, thus producing a normalised label sequence s' of length K that approximates the original rhetorical label distribution. This normalisation process (i.e., construction of s') is formally defined below:

$$i_s = \lfloor \frac{i_{s'}}{K-1}(|s|-1) \rfloor, \tag{1}$$

where $|s|$ is the number of sentence labels (equivalent to the number of sentences) in the sample abstract. Equation (1) essentially samples an index i_s in s from which we retrieve a label value for sentence position $i_{s'}$ ($0 < i_{s'} < K$) of the normalised sequence s'. More specifically, if $|s|$ is larger than the content model length K, Eq. (1) is used to perform position-aware down-sampling of s, retrieving a subset of its labels to form a condensed version of its label distribution; likewise, if $|s|$ is smaller than K, it will be used to perform position-aware up-sampling of s to form an expanded version of its label distribution. Figure 3 exemplifies how Eq. (1) samples the labels from s to form a representation s' with K elements.[2] After deriving an s' for each s within a given domain, we construct a content model M by calculating the most frequent label observed at each normalised sentence position. By following this position-aware approach rather than truncating/padding sample abstracts to K labels, we ensure that our content models better reflect the true rhetorical distributions of the sample abstracts.

3.3 Centrality-based Summariser

As per Sect. 3.1, the summariser receives the previously identified salient sections as input (i.e., Introduction and Conclusion). Prior to summarisation, the sentences of these sections are assigned a rhetorical label using the classifier described in Sect. 3.2. To generate an extractive summary guided by a content model, we first group input sentences by their assigned rhetorical class. For each

---

[2] Note that we also experimented with both rounding up and rounding to the nearest integer value for Eq. (1), but found that rounding down gave the best performance.

class label within the content model, we extract the candidate sentence with the greatest centrality from the corresponding group. In order to avoid redundancy, a sentence is no longer considered a candidate sentence once it is extracted, and thus can not be extracted for subsequent occurrences of the same rhetorical label.

The centrality of a sentence s_i is equal to the average cosine similarity between s_i and every other sentence within its group:

$$\text{Centrality}(s_i) = \frac{1}{N} \sum_{j=1}^{N} e_{i,j}, \tag{2}$$

$$e_{i,j} = \text{Cosine}(v_i, v_j) \tag{3}$$

Here v_i and v_j are the SciBERT embeddings of sentences s_i and s_j, and N is the number of sentences of the same rhetorical class.

4   Experimental Setup

Datasets. For the task of abstract generation, we experiment on the test split of the popular arXiv scientific summarisation dataset [4], for which document-summary pairs consist of full research articles (taken from the arXiv online repository) and their author-written abstracts. We assume a scenario where the domain of the input article is known by the user (as such information is typically available or easily predictable). We retrieve the scientific domain of each article within arXiv using the unique article IDs.[3] As shown in Table 4, arXiv contains articles from eight scientific domains,

Table 4. The frequencies of different scientific domains within the different data splits of arXiv (EESS = Electrical Engineering and Systems Science).

Domain	Frequency		
	Train	Valid	Test
Physics	169,827	5,666	5,715
Mathematics	20,141	360	322
Computer Science	9,041	280	258
Quantitative Biology	1,842	59	71
Statistics	1,399	47	46
Quantitative Finance	701	24	28
EESS	80	0	0
Economics	6	0	0
Total	203,037	6,436	6,440

with Physics being the most frequent. For discourse extraction (i.e., the derivation of salient sections and domain-specific content models), we sample 5,000 instances from the train split of arXiv ($\approx$ 840 samples from each domain, or all training instances when less than this is available), allowing for meaningful statistics to be calculated for each domains whilst remaining relatively computationally inexpensive. Also note that we exclude EESS and Economics domains for content modelling and the main experiment due to their limited size and no valid/test sets.

---

[3] The domain names retrieved are equal to highest-level categories as defined in the arXiv category taxonomy: https://arxiv.org/category_taxonomy.

Implementation Details. DODORANK contains only one hyperparameter, the length of the content model K. We found that a value of $K = 6$ gave the best performance on the validation set of arXiv, meaning our output summaries exclusively consist of 6 sentences.[4] Moreover, this happens to be the median number of sentences for abstracts within arXiv [19].

Baselines. We compare our model with traditional baselines ORACLE, LEAD, LEXRANK [12], and LSA [25]. We include supervised baselines DISCOURSE-AWARE [4], SEQ2SEQ-LOC&GLOB [30], MATCH-SUM [32], TOPIC-GRAPHSUM [8] and SSN-DM [7]. For unsupervised models, we include the results of PAC-SUM [31] and HIPORANK [10], the latter of which achieves state-of-the-art performance for an unsupervised model.

Evaluation. We perform automatic evaluation using standard ROUGE metrics [22].[5] Specifically, we give the average F1-scores for ROUGE-1, ROUGE-2 and ROUGE-L. Following Dong et al. [10], we also perform human evaluation on a subset of abstracts within the arXiv test set, allowing for a more comprehensive comparison of model performance.

5 Experimental Results

5.1 Structural Discourse Analyses

Section Contribution to Abstract Content. Table 5 gives the contribution of different sections to the content of the abstracts within the *sample set*, highlighting the selected sections. As stated in §3.1, this is measured as the percentage of oracle sentences that originate from a given section. We find that over 50% of abstract content is derived from the Introduction and Conclusion sections. The provided compression ratio indicates that these sections constitute 33.76% of an article on average. To validate the input reduction process, we also calculate the average ROUGE F-scores obtained by two oracle-based summaries (cf. §3.1) when compared to the reference abstracts: one containing oracle sentences extracted from the full text, and the other containing oracle sentences extracted from only the salient sections (Table 6).

Table 5. The average section contribution to abstract content within the arXiv sample set (% of oracle sentences originating from each section). Underlined text denotes selected input sections.

Conflated heading	Contribution (%)
Introduction	**35.98**
Conclusion	14.94
Results/Analysis	8.06
Discussion	5.13
Model	4.15
Method	2.68
Background/motivation	0.86
Compression ratio	**33.76**

---

[4] Increasing or decreasing K (which directly influences the number of sentences in the summaries produced by DODORANK) invariably led to a worse average performance, as measured by the metrics described in this Section.

[5] All ROUGE calculations are performed using the `rouge-score` Python package.

The results given in Table 6 support this, showing only a small disparity between the oracle ROUGE scores obtained using only these sections and those obtained using the full text (average difference < 1). As further validation, we carry out significance testing on oracle ROUGE scores

Table 6. Sample set oracle ROUGE scores using the full text and extracted sections (I = Introduction, C = Conclusion).

Oracle type	R-1	R-2	R-L
Full text	47.45	21.53	41.99
Extracted (I + C)	46.35	20.77	40.67

(t-test), the results of which indicate that differences between performance using the full text and the extracted sections are is statistically insignificant for all variants (p = 0.26, 0.48, and 0.19 for ROUGE-1, 2, and L, respectively).

Rhetorical Label Prediction. Table 7 provides statistics on the performance of the SciBERT-based classifier for rhetorical label prediction on the Computer Science abstracts of the CSAbstruct test set. Although the classifier exhibits strong performance when using only the training data of CSAbstruct, the additional out-of-

Table 7. Classifier performance for 5-way rhetorical label classification on the CSAbstruct test set. Results are average scores of 3 runs with different random seeds.

Training data	F-score	Acc.
CSABSTRUCT	0.794	0.819
Combined datasets	0.811	0.824

domain data provided by our combined datasets further improves performance in terms of both accuracy and F-score, attesting to the domain-independence of our classification scheme (described in §3.2). This suggests that we can rely on predicted rhetorical labels for subsequent experiments.

Rhetorical Structure of Abstracts. Fig. 4 provides a visualisation of how the frequency of each rhetorical class changes according to the sentence position within the abstracts of different scientific domains. For each sub-graph, by observing the pattern of most frequent labels (tallest bars) across all positions, we can identify the dominant rhetorical structure for a given domain. It is evident that these structures differ significantly depending on the scientific

Table 8. The Jensen-Shannon divergence between the distributions given in Fig. 4, averaged across all rhetorical labels (CSA = CSAbstruct, aX = arXiv, P = Physics, M = Mathematics, CS = Computer Science, QB = Quantitative Biology, S = Statistics).

	aX[P]	aX[M]	aX[CS]	aX[QB]	aX[S]	CSA
aX[P]	–	0.215	0.250	0.098	0.217	0.307
aX[M]	0.215	–	0.140	0.213	0.137	0.204
aX[CS]	0.250	0.140	–	0.201	0.094	0.138
aX[QB]	0.098	0.213	0.201	–	0.169	0.246
aX[S]	0.217	0.137	0.094	0.169	–	0.160
CSA	0.307	0.204	0.138	0.246	0.160	–

domain. To quantify this difference, we compute the average Jensen Shannon divergence (JSD) between the distributions (Table 8).

We show the five domains most prevalent within the arXiv train set: Physics (83.6%), Mathematics (9.8%), Computer Science (4.5%), Quantitative Biology (0.9%), and Statistics (0.7%). As an additional point of comparison with our predicted label distributions, we also include the distribution of the manually labelled CSAbstruct. We find that the most similar rhetor-

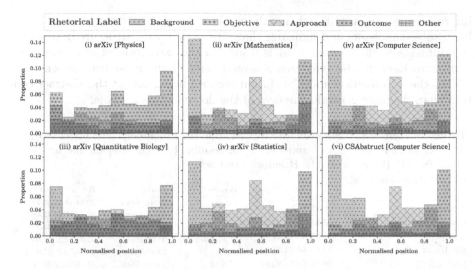

Fig. 4. Visualisation of the rhetorical distributions of abstracts for different domains.

ical structures are shared by the Statistics and Computer Science instances within arXiv (0.094 JSD). Both domains, in addition to Mathematics which also attains a low JSD score with each, follow the rhetorical flow of BACK-GROUND→APPROACH→OUTCOME. Similar patterns can also be observed in the extracted content models for all of these domains. Additionally, we find that predicted distributions for arXiv's Computer Science instances are similar to that CSAbstract (0.138 JSD), further supporting the reliability of the predicted labels. Interestingly, the abstracts of both the Physics and Quantitative Biology domain instances within arXiv exhibit a rhetorical structure that differs significantly from the other presented domains, placing a much greater emphasis on the OUTCOME class. Furthermore, their abstract structures are judged to be very similar by way of JSD score (0.098).

5.2 Abstract Generation

Automatic Evaluation. Table 9 presents the performance of DODORANK and selected baselines on the arXiv test split. We include the results of two ablated versions of DODORANK, DODORANK$_{no\_ss}$ and DODORANK$_{no\_cm}$. Here, DODORANK$_{no\_ss}$ denotes omission of section selection (full article text is used) and DODORANK$_{no\_cm}$, the omission of the sentence-level content models (i.e., K most central sentences are selected, regardless of rhetorical class).

DODORANK achieves strong performance, exceeding that of supervised models DISCOURSE-AWARE (in all metrics) and MATCH-SUM (in R-2 and R-L, whilst achieving comparable performance in R-1), despite employing an unsupervised extractive summariser using centrality. DODORANK also outperforms both closely-related centrality-based unsupervised baselines in all metrics. Furthermore, the use of both discourse-based sub-components results in a large improvement compared to when only one is used. This attests to the utility of both sub-components and indicates that: 1) the rhetorical structure imposed by the domain-specific content models results in the extraction of sentences more similar to those in the reference abstracts, and 2) the most central sentences within the Introduction and Conclusion sections better reflect the abstract content than the most central sentences of the document as a whole.

Table 9. Test set results on arXiv (ROUGE F1). Results with [†] and [*] are taken from [7,10], respectively. Results with [‡] are reproduced.

Model	R-1	R-2	R-L
ORACLE[*]	53.88	23.05	34.90
LEAD[*]	33.66	8.94	22.19
LEXRANK (2004)[*]	33.85	10.73	28.99
LSA (2004)[*]	29.91	7.42	25.67
Supervised models			
DISCOURSE-AWARE (2018)[†]	35.80	11.05	31.80
SEQ2SEQ-LOC&GLOB (2019)[†]	43.62	17.36	29.14
MATCH-SUM (2020)[†]	40.59	12.98	32.64
TOPIC-GRAPHSUM (2020)[†]	44.03	18.52	32.41
SSN-DM+DISCOURSE (2021)[†]	44.90	19.06	32.77
Unsupervised models			
PACSUM (2019)[*]	38.57	10.93	34.33
HIPORANK (2021)[*]	39.34	12.56	34.89
DODORANK (ours)	40.11	14.20	35.31
DODORANK$_{no\_ss}$ (ours)	36.66	10.87	31.12
DODORANK$_{no\_cm}$ (ours)	36.51	12.15	31.99

Table 10. Domain-specific breakdown of results (ROUGE F1) on the arXiv test set.

Domain	Model	R-1	R-2	R-L
Physics	DODORANK	40.34	14.47	35.49
	DODORANK$_{no\_ss}$	37.07	11.24	32.48
	DODORANK$_{no\_cm}$	36.40	12.19	31.87
Mathematics	DODORANK	34.65	10.21	30.53
	DODORANK$_{no\_ss}$	30.94	7.12	27.03
	DODORANK$_{no\_cm}$	33.76	10.17	29.77
Computer Science	DODORANK	41.13	14.01	36.27
	DODORANK$_{no\_ss}$	34.51	8.38	30.27
	DODORANK$_{no\_cm}$	39.79	13.46	35.13
Quantitive Biology	DODORANK	38.69	10.79	33.90
	DODORANK$_{no\_ss}$	34.86	7.99	30.59
	DODORANK$_{no\_cm}$	37.99	10.54	33.41
Statistics	DODORANK	39.06	10.93	34.24
	DODORANK$_{no\_ss}$	35.19	7.28	31.88
	DODORANK$_{no\_cm}$	38.64	11.20	33.89
Quantitive Finance	DODORANK	39.84	13.18	35.59
	DODORANK$_{no\_ss}$	36.72	10.06	31.88
	DODORANK$_{no\_cm}$	38.34	13.25	33.81

Table 10 provides a domain-specific breakdown of ROUGE scores on the arXiv test set for standard and ablated versions of DODORANK. Again, we observe a universal improvement in performance across all domains when both discourse-based analyses are included, providing further indication that our analyses capture the nuances of each scientific domain and, in doing so, improves the summarisation of their articles.

Human Evaluation. Results for human evaluation are given in Table 11. We chose to replicate the procedure of Dong et al. [10] to facilitate a direct comparison between our model and the state-of-the-art for centrality-based extractive summarisation model DODORANK. For this evaluation, two human judges[6] are repeatedly pre-

Table 11. Human evaluation results on 20 sampled reference summaries with 307 system summary sentences from arXiv. Results for both criteria are statistically significant with Mann-Whitney U tests.

Model	Content-coverage	Importance
HIPORANK	12.17	37.21
DODORANK	**22.03**	**59.04**

---

[6] Judges are native English speakers holding a bachelor's degree in scientific disciplines.

sented with a reference abstract and a sentence extracted by either DODORANK or HIPORANK model in a random anonymised order. Each sentence is evaluated on whether it contains content from the reference abstract (*content coverage*) and whether it contains content that would be important for a goal-oriented reader, regardless of whether it is covered in the reference abstract (*importance*). Each reference summary-sentences pair is annotated by two annotators (i.e., a score 1 will be assigned if a sentence is deemed content relevant or important; and 0 otherwise). Scores are aggregated and then averaged across the sentences from the tested models. The average annotator agreement of 73.31%, attesting to their reliability.

DODORANK significantly outperforms the HIPORANK model for both content coverage and importance, in line with automatic evaluation results. Overall, these results provide further indication that summarising only the salient sections of an article and arranging output according to a domain-specific rhetorical structure improves the overall quality of the summaries produced, particularly in the selection of important content.

6 Conclusions

In this paper we proposed an extractive, discourse-guided approach to the generation of scientific abstracts which adapts to the scientific domain. For a given domain, the model extracts rich discourse information which is used to both reduce the input and guide the output of a simple centrality-based summariser. Our approach exhibits impressive performance on the multi-domain arXiv dataset, exceeding that of strong baselines, both supervised and unsupervised. This demonstrates that the scientific domain of an article can effectively be leveraged in a summarisation context and supports our original hypothesis, that the domain of an article has a strong influence over the structural discourse of its abstract.

Acknowledgements. This work was supported by the Centre for Doctoral Training in Speech and Language Technologies (SLT) and their Applications funded by UK Research and Innovation [grant number EP/S023062/1].

References

1. Beltagy, I., Lo, K., Cohan, A.: SciBERT: a pretrained language model for scientific text. In: Proceedings of the 2019 Conference on Empirical Methods in Natural Language Processing and the 9th International Joint Conference on Natural Language Processing (EMNLP-IJCNLP), pp. 3615–3620. Association for Computational Linguistics, Hong Kong, China (2019). https://doi.org/10.18653/v1/D19-1371. https://aclanthology.org/D19-1371
2. Cheng, Y., et al.: Guiding the growth: difficulty-controllable question generation through step-by-step rewriting. In: Proceedings of the 59th Annual Meeting of the Association for Computational Linguistics and the 11th International Joint Conference on Natural Language Processing (ACL-IJCNLP), pp. 5968–5978. Online (2021)

3. Cohan, A., Beltagy, I., King, D., Dalvi, B., Weld, D.: Pretrained language models for sequential sentence classification. In: Proceedings of the 2019 Conference on Empirical Methods in Natural Language Processing and the 9th International Joint Conference on Natural Language Processing (EMNLP-IJCNLP), pp. 3693–3699. Association for Computational Linguistics, Hong Kong, China (2019)

4. Cohan, A., et al.: A discourse-aware attention model for abstractive summarization of long documents. In: Proceedings of the 2018 Conference of the North American Chapter of the Association for Computational Linguistics: Human Language Technologies, Volume 2 (Short Papers), pp. 615–621. Association for Computational Linguistics, New Orleans, Louisiana (2018). https://doi.org/10.18653/v1/N18-2097. https://aclanthology.org/N18-2097

5. Collins, E., Augenstein, I., Riedel, S.: A supervised approach to extractive summarisation of scientific papers. In: Proceedings of the 21st Conference on Computational Natural Language Learning (CoNLL 2017), pp. 195–205. Association for Computational Linguistics, Vancouver, Canada (2017)

6. Contractor, D., Guo, Y., Korhonen, A.: Using argumentative zones for extractive summarization of scientific articles. In: Proceedings of COLING 2012, pp. 663–678. The COLING 2012 Organizing Committee, Mumbai, India (2012). https://aclanthology.org/C12-1041

7. Cui, P., Hu, L.: Sliding selector network with dynamic memory for extractive summarization of long documents. In: Proceedings of the 2021 Conference of the North American Chapter of the Association for Computational Linguistics: Human Language Technologies, pp. 5881–5891 (2021)

8. Cui, P., Hu, L., Liu, Y.: Enhancing extractive text summarization with topic-aware graph neural networks. In: Proceedings of the 28th International Conference on Computational Linguistics, pp. 5360–5371 (2020)

9. Devlin, J., Chang, M.W., Lee, K., Toutanova, K.: BERT: pre-training of deep bidirectional transformers for language understanding. In: Proceedings of the 2019 Conference of the North American Chapter of the Association for Computational Linguistics: Human Language Technologies, Volume 1 (Long and Short Papers), pp. 4171–4186. Association for Computational Linguistics, Minneapolis, Minnesota (2019). https://doi.org/10.18653/v1/N19-1423. https://aclanthology.org/N19-1423

10. Dong, Y., Mircea, A., Cheung, J.C.K.: Discourse-Aware unsupervised summarization for long scientific documents. In: Proceedings of the 16th Conference of the European Chapter of the Association for Computational Linguistics: Main Volume, pp. 1089–1102. Association for Computational Linguistics, Online (2021)

11. El-Kassas, W., Salama, C., Rafea, A., Mohamed, H.: Automatic text summarization: a comprehensive survey. Expert Syst. Appl. **165**, 113679 (2020). https://doi.org/10.1016/j.eswa.2020.113679

12. Erkan, G., Radev, D.R.: LexRank: graph-based lexical centrality as salience in text summarization. J. Artif. Intell. Res. **22**, 457–479 (2004)

13. Ermakova, L., Bordignon, F., Turenne, N., Noel, M.: Is the abstract a mere teaser? Evaluating generosity of article abstracts in the environmental sciences. Front. Res. Metrics Anal. **3**, 16 (2018). https://doi.org/10.3389/frma.2018.00016. https://www.frontiersin.org/articles/10.3389/frma.2018.00016

14. Goldsack, T., Zhang, Z., Lin, C., Scarton, C.: Making science simple: corpora for the lay summarisation of scientific literature. In: Proceedings of the 2022 Conference on Empirical Methods in Natural Language Processing, pp. 10589–10604. Association for Computational Linguistics, Abu Dhabi (2022)

15. Graetz, N.: Teaching EFL students to extract structural information from abstracts. International Symposium on Language for Special Purposes (1982)
16. Guo, Y., Korhonen, A., Liakata, M., Silins, I., Sun, L., Stenius, U.: Identifying the information structure of scientific abstracts: an investigation of three different schemes. In: Proceedings of the 2010 Workshop on Biomedical Natural Language Processing, pp. 99–107. Association for Computational Linguistics, Uppsala, Sweden (2010)
17. Guo, Y., Silins, I., Stenius, U., Korhonen, A.: Active learning-based information structure analysis of full scientific articles and two applications for biomedical literature review. Bioinformatics **29**(11), 1440–1447 (2013)
18. Johnson, F.: Automatic abstracting research. Libr. Rev. **44**(8), 28–36 (1995). https://www.proquest.com/scholarly-journals/automatic-abstracting-research/docview/218330298/se-2?accountid=13828
19. Ju, J., Liu, M., Koh, H.Y., Jin, Y., Du, L., Pan, S.: Leveraging information bottleneck for scientific document summarization. In: Findings of the Association for Computational Linguistics: EMNLP 2021, pp. 4091–4098. Association for Computational Linguistics, Punta Cana, Dominican Republic (2021). https://doi.org/10.18653/v1/2021.findings-emnlp.345. https://aclanthology.org/2021.findings-emnlp.345
20. Liakata, M.: Zones of conceptualisation in scientific papers: a window to negative and speculative statements. In: Proceedings of the Workshop on Negation and Speculation in Natural Language Processing, pp. 1–4. University of Antwerp, Uppsala, Sweden (2010)
21. Liakata, M., Dobnik, S., Saha, S., Batchelor, C., Rebholz-Schuhmann, D.: A discourse-driven content model for summarising scientific articles evaluated in a complex question answering task. In: Proceedings of the 2013 Conference on Empirical Methods in Natural Language Processing, pp. 747–757. Association for Computational Linguistics, Seattle, Washington, USA (2013)
22. Lin, C.Y.: ROUGE: a package for automatic evaluation of summaries. In: Text Summarization Branches Out, pp. 74–81. Association for Computational Linguistics, Barcelona, Spain (2004). https://aclanthology.org/W04-1013
23. Maynez, J., Narayan, S., Bohnet, B., McDonald, R.: On faithfulness and factuality in abstractive summarization. In: Proceedings of the 58th Annual Meeting of the Association for Computational Linguistics. pp. 1906–1919. Association for Computational Linguistics, Online (2020). https://doi.org/10.18653/v1/2020.acl-main.173. https://aclanthology.org/2020.acl-main.173
24. Peng, K., Yin, C., Rong, W., Lin, C., Zhou, D., Xiong, Z.: Named entity aware transfer learning for biomedical factoid question answering. IEEE/ACM Trans. Comput. Biol. Bioinform. **19**(4), 2365–2376 (2021)
25. Steinberger, J., Jezek, K.: Using latent semantic analysis in text summarization and summary evaluation. In: Proceedings of the 7th International Conference ISIM (2004)
26. Swales, J.: Genre analysis: Eenglish in academic and research settings. Cambridge University Press (1990)
27. Teufel, S.: Argumentative zoning: information extraction from scientific text, Ph. D. thesis, University of Edinburgh (1999)
28. Teufel, S., Moens, M.: Articles summarizing scientific articles: experiments with relevance and rhetorical status. Comput. Linguist. **28**(4), 409–445 (2002)

29. Teufel, S., Siddharthan, A., Batchelor, C.: Towards discipline-independent argu-mentative zoning: evidence from chemistry and computational linguistics. In: Pro-ceedings of the 2009 Conference on Empirical Methods in Natural Language Pro-cessing, vol. 3, pp. 1493–1502. EMNLP 2009, Association for Computational Lin-guistics, USA (2009)

30. Xiao, W., Carenini, G.: Extractive summarization of long documents by combin-ing global and local context. In: Proceedings of the 2019 Conference on Empirical Methods in Natural Language Processing and the 9th International Joint Confer-ence on Natural Language Processing (EMNLP-IJCNLP), pp. 3011–3021. Associ-ation for Computational Linguistics, Hong Kong, China (2019). https://doi.org/10.18653/v1/D19-1298. https://aclanthology.org/D19-1298

31. Zheng, H., Lapata, M.: Sentence centrality revisited for unsupervised summariza-tion. In: Proceedings of the 57th Annual Meeting of the Association for Computa-tional Linguistics, pp. 6236–6247 (2019)

32. Zhong, M., Liu, P., Chen, Y., Wang, D., Qiu, X., Huang, X.J.: Extractive sum-marization as text matching. In: Proceedings of the 58th Annual Meeting of the Association for Computational Linguistics, pp. 6197–6208 (2020)

33. Zhu, T., Hua, W., Qu, J., Zhou, X.: summarizing long-form document with rich dis-course information. In: Proceedings of the 30th ACM International Conference on Information & Knowledge Management, pp. 2770–2779. CIKM 2021, Association for Computing Machinery, New York, NY, USA (2021)

Injecting Temporal-Aware Knowledge in Historical Named Entity Recognition

Carlos-Emiliano González-Gallardo[1(✉)] [iD], Emanuela Boros[1] [iD],
Edward Giamphy[1,2] [iD], Ahmed Hamdi[1] [iD], José G. Moreno[1,3] [iD],
and Antoine Doucet[1] [iD]

[1] University of La Rochelle, L3i, 17000 La Rochelle, France
{carlos.gonzalez_gallardo,emanuela.boros,ahmed.hamdi,
antoine.doucet}@univ-lr.fr
[2] Preligens, 75009 Paris, France
edward.giamphy@preligens.com
[3] University of Toulouse, IRIT, 31000 Toulouse, France
jose.moreno@irit.fr

Abstract. In this paper, we address the detection of named entities in multilingual historical collections. We argue that, besides the multiple challenges that depend on the quality of digitization (e.g., misspellings and linguistic errors), historical documents can pose another challenge due to the fact that such collections are distributed over a long enough period of time to be affected by changes and evolution of natural language. Thus, we consider that detecting entities in historical collections is time-sensitive, and explore the inclusion of temporality in the named entity recognition (NER) task by exploiting temporal knowledge graphs. More precisely, we retrieve semantically-relevant additional contexts by exploring the time information provided by historical data collections and include them as mean-pooled representations in a Transformer-based NER model. We experiment with two recent multilingual historical collections in English, French, and German, consisting of historical newspapers (19C-20C) and classical commentaries (19C). The results are promising and show the effectiveness of injecting temporal-aware knowledge into the different datasets, languages, and diverse entity types.

Keywords: Named entity recognition · Temporal information extraction · Digital humanities

1 Introduction

Recent years have seen the delivery of an increasing amount of textual corpora for the Humanities and Social Sciences. Representative examples are offered by the digitization of the gigantic *Gallica* collection by the National Library of France[1] and the *Trove* online Australian library[2], database aggregator and

---

[1] https://gallica.bnf.fr/.
[2] https://trove.nla.gov.au/.

© The Author(s), under exclusive license to Springer Nature Switzerland AG 2023
J. Kamps et al. (Eds.): ECIR 2023, LNCS 13980, pp. 377–393, 2023.
https://doi.org/10.1007/978-3-031-28244-7_24

service of full-text documents, digital images and data storage of digitized documents. Access to this massive data offers new perspectives to a growing number of disciplines, going from socio-political and cultural history to economic history, and linguistics to philology. Billions of images from historical documents including digitized manuscript documents, medieval registers and digitized old press are captured and their content is transcribed, manually through dedicated interfaces, or automatically using optical character recognition (OCR) or handwritten text recognition (HTR). The mass digitization process, initiated in the 1980s s with small-scale internal projects, led to the "rise of digitization", which grew to reach a certain maturity in the early 2000s s with large-scale digitization campaigns across the industry [12,16]. As this process of mass digitization continues, increasingly advanced techniques from the field of natural language processing (NLP) are dedicated to historical documents, offering new ways to access full-text semantically enriched archives [33], such as NER [4,10,19], entity linking (EL) [26] and event detection [5,32].

However, for developing such techniques, historical collections present multiple challenges that depend either on the quality of digitization, the need to handle documents deteriorated by the effect of time, the poor quality printing materials or inaccurate scanning processes, which are common issues in historical documents [20]. Moreover, historical collections can pose another challenge due to the fact that documents are distributed over a long enough period of time to be affected by language change and evolution. This is especially true in the case of Western European languages, which only acquired their modern spelling standards roughly around the 18th or 19th centuries [29]. With existing collections [12,15,16] providing such metadata as the year of publication, we propose to take advantage of the temporal context of historical documents in order to increase the quality of their semantic enrichment. When this metadata is not available, due to the age of the documents, the year has often been estimated and a new NLP task recently emerged, aiming to predict a document's year of publication [36].

NER corresponds to the identification of entities of interest in texts, generally of the type person, organization, and location. Such entities act as referential anchors that underlie the semantics of texts and guide their interpretation. For example, in Europe, by the medieval period, most people were identified simply by a mononym or a single proper name. Family names or surnames began to be expected in the 13th century but in some regions or social classes much later (17th century for the Welsh). Many people shared the same name and the spelling was diverse across vernacular and Latin languages, and also within one language (e.g., Guillelmus, Guillaume, Willelmus, William, Wilhelm). Locations may have disappeared or changed completely, for those that survived well into the 21st century from prehistory (e.g., Scotland, Wales, Spain), they are very ambiguous and also have very different spellings, making it very difficult to identify them [6]. In this article, we focus on exploring temporality in entity detection from historical collections. Thus, we propose a novel technique for injecting additional temporal-aware knowledge by relying on Wikipedia and Wikidata to provide related context information. More exactly, we retrieve semantically-relevant additional contexts by exploring the time information provided by the historical data collections and include them as mean-pooled representations in our Transformer-based NER model. We consider that adding grammatically correct contexts could improve the error-prone texts due to digitization errors while adding temporality could further be beneficial to handle changes in language or entity names.

The paper is structured as follows: we present the related work and datasets in Sect. 2 and 3 respectively. Our methodology for retrieving additional context through temporal knowledge graphs and how contexts are included within the proposed model is described in Sect. 4. We, then, perform several experiments in regards to the relativity of the time span when selecting additional context and present our findings in Sect. 5. Finally, conclusions and future work are drawn in Sect. 6[3].

2 Related Work

Named Entity Recognition in Historical Data. Due to the multiple challenges posed by the quality of digitization or the historical variations of a language, NER in historical and digitized documents is less noticeable in terms of high performance than in modern documents [47,52]. Recent evaluation campaigns such as the one organized by the *Identifying Historical People, Places, and other Entities* (HIPE) lab at CLEF 2020[4] [16] and 2022[5] [17] proposed tasks of NER and EL in ca. 200 years of historical newspapers written in multiple languages (English, French, German, Finnish and Swedish) and successfully showed that these tasks benefit from the progress in neural-based NLP (specifically driven by the latest advances in Transformer-based pre-trained language models approaches) as a considerable improvement in performance was observed on the historical collections, especially for NER [24,42,44].

The authors of [10] present an extensive survey on NER over historical datasets and highlight the challenges that state-of-the-art NER methods applied to historical and noisy inputs need to address. For overcoming the impact of the OCR errors, contextualized embeddings at the character level were utilized to find better representations of out-of-vocabulary words (OOVs) [2]. The contextualized embeddings are learned using language models and allow predicting the next character of strings given previous characters. Moreover, further research showed that the fine-tuning of several Transformer encoders on historical collections could alleviate digitization errors [4]. To deal with the lack of historical resources, [40] proposed to use transfer learning in order to learn models on large contemporary resources and then adapt them to a few corpora of historical nature. Finally, in order to address the spelling variations, some works developed transformation rules to model the diachronic evolution of words and generate a

---

[3] The code is available at https://github.com/EmanuelaBoros/clef-hipe-2022-13i.
[4] https://impresso.github.io/CLEF-HIPE-2020/.
[5] https://hipe-eval.github.io/HIPE-2022/.

normalized version processable by existing NER systems [8,23]. While most of these approaches rely generally on the local textual context for detecting entities in such documents, temporal information has generally been disconsidered. To the best of our knowledge, several approaches have been proposed for named entity disambiguation by utilizing temporal signatures for entities to reflect the importance of different years [1], and entity linking, such as the usage of time-based filters [26], but not for historical NER.

Named Entity Recognition with Knowledge Bases. Considering the complementary behaviors of knowledge-based and neural-based approaches for NER, several studies have explored knowledge-based approaches including different types of symbolic representations (e.g., knowledge bases, static knowledge graphs, gazetteers) and noticed significant improvements in token representations and the detection of entities over modern datasets (e.g., CoNLL [43], OntoNotes 5.0 [35]) [27,43]. Gazetteer knowledge has been integrated into NER models alongside word-level representations through gating mechanisms [31] and Wikipedia has mostly been utilized to increase the semantic representations of possible entities by fine-tuning recent pre-trained language models on the fill-in-the-blank (cloze) task [39,52].

When well-formed text is replaced with short texts containing long-tail entities, symbolic knowledge has also been utilized to increase the contextual information around possible entities [31]. Introducing external contexts into NER systems has been shown to have a positive impact on the entities' identification performance, even with these complications. [48] constructed a knowledge base system based on a local instance of Wikipedia to retrieve relevant documents given a query sentence. The retrieved documents and query sentences, after concatenation, were fed to the NER system. Our proposed methodology could be considered inspired by their work, however, we include the additional contexts at the model level by generating a mean-pooled representation for each context instead of concatenating the contexts with the initial sentence. We consider that having pooled representations for each additional context can reduce the noise that could be created by other entities found in these texts.

Temporality in Knowledge Graphs. Recent advances have shown a growing interest in learning representations of entities and relations including time information [7]. Other work [50] proposed a temporal knowledge graph (TKG) embedding model for representing facts involving time intervals by designing the temporal evolution of entity embeddings as rotation in a complex vector space. The entities and the relations were represented as single or dual complex embeddings and temporal changes were the rotations of the entity embeddings in the complex vector space. Since the knowledge graphs change over time in evolving data (e.g., the fact *The President of the United States is Barack Obama* is valid only from 2009 to 2017), A temporal-aware knowledge graph embedding approach [49] was also proposed by moving beyond the complex-valued representations and introducing multivector embeddings from geometric algebras to

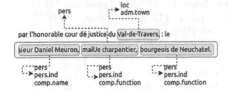

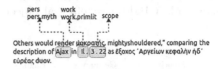

Fig. 1. An example from the hipe-2020 dataset.

Fig. 2. An example from the ajmc dataset.

model entities, relations, and timestamps for TKGs. Further research [51] presented a graph neural network (GNN) model treating timestamp information as an inherent property of the graph structure with a self-attention mechanism to associate appropriate weights to nodes according to their relevant relations and neighborhood timestamps. Therefore, timestamps are considered properties of links between entities.

TKGs, however, show many inconsistencies and a lack of data quality across various dimensions, including factual accuracy, completeness, and timeliness. In consequence, other research [9] further explores TKGs by targeting the completion of knowledge with accurate but missing information. Moreover, since such TKGs often suffer from incompleteness, the authors of [53] introduced a temporal-aware representation learning model that helps to infer the missing temporal facts by taking interest in facts occurring recurrently and leverage a copy mechanism to identify facts with repetition. The aforementioned methods demonstrate that the usage of TKGs is considered an emerging domain that is being explored, in particular in the field of NLP. The availability of information about the temporal evolution of entities, not only could be a promising solution for improving their semantic knowledge representations but also could provide additional contextual information for efficient NER. To the best of our knowledge, our work is the first attempt to leverage time information provided by TKGs to improve NER.

3   Datasets

In this study, we utilize two collections composed of historical newspapers and classical commentaries covering circa 200 years. Recently proposed by the CLEF-HIPE-2022 evaluation campaign [14], we experiment with the hipe-2020 and the Ajax Multi-Commentary (ajmc) datasets.

hipe-2020 includes newspaper articles from Swiss, Luxembourgish, and American newspapers in French, German, and English (19C-20C) and contains 19,848 linked entities as part of the training sets [12,15,16]. For each language, the corpus is divided into train, development, and test, with the only exception of English for which only development and test sets were produced [13]. In this case,

Table 1. Overview of the `hipe-2020` and `ajmc` datasets. LOC = Location, ORG = Organization, PERS = Person, PROD = Product, TIME = Time, WORK = human work, OBJECT = physical object, and SCOPE = specific portion of work.

Type	hipe-2020 French			German			English			ajmc French			German			English		
	train	dev	test	train	dev	test	train	dev	test	train	dev	test	train	dev	test	train	dev	test
LOC	3,089	774	854	1,740	588	595	–	384	181	15	0	9	31	10	2	39	3	3
ORG	836	159	130	358	164	130	–	118	76	–	–	–	–	–	–	–	–	–
PERS	2,525	679	502	1,166	372	311	–	402	156	577	123	139	620	162	128	618	130	96
PROD	200	49	61	112	49	62	–	33	19	–	–	–	–	–	–	–	–	–
TIME	276	68	53	118	69	49	–	29	17	2	0	3	2	0	0	12	5	3
WORK	–	–	–	–	–	–	–	–	–	378	99	80	321	70	74	467	116	95
OBJECT	–	–	–	–	–	–	–	–	–	10	0	0	6	4	2	3	0	0
SCOPE	–	–	–	–	–	–	–	–	–	639	169	129	758	157	176	684	162	151

we utilized the French and German datasets for training the proposed models in our experimental setup. An example from the French dataset is presented in Fig. 1.

`ajmc` is composed of classical commentaries from the Ajax Multi-Commentary project that includes digitized 19C commentaries published in French, German, and English [41] annotated with both universal and domain-specific named entities (NEs). An example in English is presented in Fig. 2.

These two collections pose several important challenges: the multilingualism (both containing three languages: English, French and German), the code-mixed documents (e.g., commentaries, where Greek is mixed with the language of the commentator), the granularity of annotations and the richness of the texts characterized by a high density of NEs. Both datasets provide different document metadata with different granularity (e.g., language, document type, original source, date) and have different entity tag sets that were built according to different annotation guidelines. Table 1 presents the statistics regarding the number and type of entities in the aforementioned datasets divided according to the training, development, and test sets.

4   Temporal Knowledge-based Contexts for Named Entity Recognition

The OCR output contains errors that produce noisy text and complications, similar to those studied by [30]. It has long been observed that adapting NER systems to deal with the OCR noise is more appropriate than adapting NER corpora [11]. Furthermore, [22] showed that applying post-OCR correction algorithms before running NER systems does not often have a positive impact on NER results since post-OCR may degrade clean words during the correction of the noisy ones. To deal with OCR errors, we introduce external grammatically correct contexts into NER systems which have a positive impact on the entity identification performance even in spite of these challenges [48]. Moreover, the inclusion of such contexts by taking into consideration temporality could further improve the detection of time-sensitive entities. Thus, we propose several settings for including additional context based on Wikidata5m[6] [46], a knowledge graph with five million Wikidata[7] entities which contain entities in the general domain (e.g., celebrities, events, concepts, things) and are aligned to a description that corresponds to the first paragraph of the matching Wikipedia page.

4.1   Temporal Information Integration

A TKG contains time information and facts associated with an entity that provides information about spontaneous changes or smooth temporal transformations of the entity while informing about the relations with other entities. We aggregate temporality into Wikidata5m including the TKG created by [25] and tuned by [18][8]. This TKG contains over 11 thousand entities, 150 thousand facts, and a temporal scope between the years 508 and 2017. For a given entity, it provides a set of time-related facts describing the interactions of the entity in time. It is thus necessary to combine these facts into a singular element through an aggregation operator over their temporal elements.

We perform a transformation on the temporal information of every fact of an entity in order to combine them into only one piece of temporal information. Let e be an entity described by the facts:

$$\{F_e\}_{i=1}^n = \{(e, r_1, e_1, t_1)(e, r_2, e_2, t_2), \ldots (e, r_i, e_i, t_i), \ldots (e, r_n, e_n, t_n)\},$$

where a fact (e, r_i, e_i, t_i) is composed of two entities e and e_i that are connected by the relation r_i and the timestamp t_i. A timestamp is a discrete point in time which corresponds to a year in this work. The aggregation operator is the function $AGG \rightarrow t_e$ that takes as input the time information from F_e and outputs the time information that is associated with e. Several aggregation operators are possible. Among them, natural options are mean, median, minimum, and maximum operations. The minimum of a set of facts is defined as the oldest fact, and the maximum is the most recent fact. If an entity is associated with four facts spanning over years 1891, 1997, 2006, and 2011, the minimum aggregation operator consists in keeping the oldest, resulting in the year 1891 the time information of the entity. Given that our datasets correspond to documents between 19C and 20C, the minimum operation is more likely to create an appropriate temporal context for the entities. Therefore it is a convenient choice to highlight entities matching the corresponding time period by accentuating older facts. At the end of the aggregation operation 8,176 entities of Wikidata5m are associated with a year comprised between 508 and 2001, filtering out most of the facts occurring during 21C.

---

[6] https://deepgraphlearning.github.io/project/wikidata5m.
[7] https://www.wikidata.org/.
[8] https://github.com/mniepert/mmkb/tree/master/TemporalKGs/wikidata.

4.2   Context Retrieval

Our knowledge base system relies on a local ElasticSearch[9] instance and follows a multilingual semantic similarity matching, which presents an advantage on multilingual querying and is achieved with dense vector field indexes. Thus given a query vector, a k-nearest neighbor search API retrieves the k closest vectors returning the corresponding documents as search hits. For each Wikidata5m entity, we create an ElasticSearch entry including an identifier field, a description field and a description embedding field which we obtain with a pre-trained multilingual Sentence-BERT model [37,38]. We build one index on the entity identifier and a dense vector index on the description embedding. We propose two different settings for context retrieval:

- `non-temporal`: This setting uses no temporal information. Given an input sentence during context retrieval, we first obtain the corresponding dense vector representation with the same Sentence-BERT model used during the indexing phase. Then, we query the knowledge base to retrieve the top-k semantically similar entities based on a k-nearest neighbors algorithm (k-NN) cosine similarity search over the description embedding dense vector index. The context C is finally composed of k entity descriptions.
- `temporal-`δ: This setting integrates the temporal information. For each semantically similar entity that is retrieved following `non-temporal`, we apply a filtering operation to keep or discard the entity as part of the context. Given the year t_{input} linked to the input sentence's metadata during context retrieval, the entity is kept if its associated year t_e is inside the interval $t_{input} - \delta \le t_e \le t_{input} + \delta$, where δ is the year interval threshold, otherwise it is rejected. As a result of AGG, t_e results to be the oldest year in the set of facts of entity e in the TKG. If t_e is nonexistent, e is also kept. This operation is repeated until $|C| = k$.

4.3   Named Entity Recognition Architecture

Base Model Our model consists of a hierarchical, multitask learning approach, with a fine-tuned encoder based on BERT. This model includes an encoder with two Transformer [45] layers with adapter modules [21,34] on top of the BERT pre-trained model. The adapters are added to each Transformer layer after the projection following multi-headed attention and they adapt not only to the task but also to the noisy input which proved to increase the performance of NER in such special conditions [4]. Finally, the prediction layer consists of a conditional random field (CRF) layer.

In detail, let $\{x_i\}_{i=1}^l$ be a token input sequence consisting of l words, denoted as $\{x_i\}_{i=1}^l = \{x_1, x_2, \ldots x_i, \ldots x_l\}$, where x_i refers to the i-th token in the sequence of length l. We first apply a pre-trained language model as *encoder* for further fine-tuning. The output is $\{h_i\}_{i=1}^l, H_{[CLS]} = encoder(\{x_i\}_{i=0}^l)$ where

---

[9] https://www.elastic.co/guide/en/elasticsearch/reference/8.1/release-highlights.html.

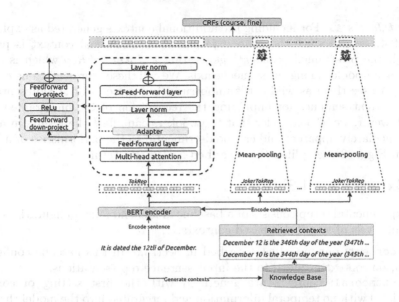

Fig. 3. NER model architecture with temporal-aware context 🃏 s (*context jokers*).

$\{h_i\}_{i=1}^l = [h_1, h_2, \ldots h_i, \ldots h_l]$ is the representation for each i-th position in x token sequence and $h_{[CLS]}$ is the final hidden state vector of $[CLS]$ as the representation of the whole sequence x. From now on, we refer to the *Token Representation* as $TokRep = \{x_i\}_{i=1}^l$ that is the token input sequence consisting of l words. The additional Transformer encoder contains a number of Transformer layers that takes as input the matrix $H = \{h_i\}_{i=1}^l \in R_{l \times d}$ where d is the input dimension (encoder output dimension). A Transformer layer includes a multi-head self-attention $Head(h)$: $Q^{(h)}, K^{(h)}, V^{(h)} = HW_q^{(h)}, HW_k^{(h)}, HW_v^{(h)}$ and $MultiHead(H) = [Head^{(1)}, \ldots, Head^{(n)}]W_O$[10] where n is the number of heads and the superscript h represents the head index. Q_t is the query vector of the t-th token, j is the token the t-th token attends. K_j is the key vector representation of the j-th token. The *Attn* softmax is along the last dimension. $MultiHead(H)$ is the concatenation on the last dimension of size $R^{l \times d}$ where d_k is the scaling factor $d_k \times n = d$. W_O is a learnable parameter of size $R^d \times d$.

By combining the position-wise feed-forward sub-layer and multi-head attention, we obtain a feed-forward layer $FFN(f(H)) = max(0, f(H)W_1)W_2$ where W_1, W_2 are learnable parameters and max is the *ReLU* activation. $W_1 \in R^{d \times d_{FF}}$, $W_2 \in R^{d_{FF} \times d}$ are trained projection matrices, and d_{FF} is a hyperparameter. The task adapter is applied at this level on $TokRep$ at each layer and consists of a down-projection $D \in R^{h \times d}$ where h is the hidden size of the Transformer model and d is the dimension of the adapter, also followed by a *ReLU* activation and an up-projection $U \in R^{d \times h}$.

---

[10] We leave out the details that can be consulted in [45].

Context Jokers 🃏 For including the additional contexts generated as explained in Sect. 4, we introduce the *context jokers*. Each additional context is passed through the pre-trained *encoder*[11] generating a *JokerTokRep* which is afterwards mean-pooled along the sequence axis. We call these representations *context jokers*. We see them as wild cards unobtrusively inserted in the representation of the current sentence for improving the recognition of the fine-grained entities. However, we also consider that these jokers can affect the results in a way not immediately apparent and can be detrimental to the performance of a NER system. Figure 3 exemplifies the described NER architecture.

5 Experimental Setup

Our experimental setup consists of a baseline model and four configurations with different levels of knowledge-based contexts:

- `no-context`: our model as described in Sect. 4.3. In this baseline configuration, no context is added to the input sentence representations.
- `non-temporal`: contexts are generated with the first setting of context retrieval with no temporal information and integrated into the model through *context jokers*.
- `temporal-(50|25|10)`: contexts are generated with the second setting of context retrieval with $\delta \in \{50, 25, 10\}$ (where δ is the time span or year interval threshold) and integrated into the model through *context jokers*.

Hyperparameters. In order to have a uniform experimental setting, we chose a BERT-based cased multilingual pre-trained model[12]. We denote the number of layers (i.e., adapter-based Transformer blocks) as L, the hidden size as H, and the number of self-attention heads as A. BERT has L=12, H=768 and A=12. We added two layers with H=128, A=12, and the adapters have 128×12 size. The adapters are trained on the task during training. For all context-retrieval configurations, the context size $|C|$ of an input sentence was set to $k = 10$. For indexing the documents in ElasticSearch, we utilized the multilingual pre-trained Sentence-BERT model[13].

Evaluation. The evaluation is performed over coarse-grained NER in terms of precision (P), recall (R), and F-measure (F1) at micro level [12,28] (i.e., consideration of all true positives, false positives, true negatives and false negatives over all samples) in a strict (exact boundary matching) and a fuzzy boundary matching setting[14]. Coarse-grained NER refers to the identification and categorization of entity mentions according to the high-level entity types listed in Table 1. We refer to these metrics as coarse-strict (*CS*) and coarse-fuzzy (*CF*).

---

[11] We do not utilize in this case the additional Transformer layers with adapters, since these were specifically proposed for noisy/non-standard text and they do not bring any increase in performance on standard text [4].

[12] https://huggingface.co/bert-base-multilingual-cased.

[13] https://huggingface.co/sentence-transformers/paraphrase-multilingual-MiniLM-L12-v2.

[14] We utilized the HIPE-scorer https://github.com/hipe-eval/HIPE-scorer.

Table 2. Results on French, German and English, for the `hipe-2020` and `ajmc` datasets.

	French						German						English					
	hipe-2020			ajmc			hipe-2020			ajmc			hipe-2020			ajmc		
	P	R	F1	P	R	F1	P	R	F1	P	R	F1	P	R	F1	P	R	F1
no-context																		
CS	0.755	0.757	0.756	0.829	0.806	0.817	0.754	0.730	0.742	0.910	0.877	0.893	0.604	0.563	0.583	0.789	0.859	0.823
CF	0.857	0.859	0.858	0.883	0.858	0.870	**0.853**	0.826	0.839	0.935	0.901	0.917	0.778	0.726	0.751	0.855	0.931	0.891
non-temporal																		
CS	0.762	**0.767**	**0.765**	0.829	0.783	0.806	0.759	**0.767**	**0.763**	**0.930**	0.898	0.913	0.565	0.601	0.583	0.828	0.871	0.849
CF	0.862	**0.869**	0.866	**0.906**	0.856	0.880	0.847	0.856	0.852	**0.949**	0.916	**0.932**	0.741	0.788	0.764	0.885	0.931	0.908
temporal-50																		
CS	**0.765**	0.765	**0.765**	0.839	0.822	0.830	0.748	0.756	0.752	0.921	**0.911**	**0.916**	**0.643**	0.617	**0.630**	0.855	0.882	0.868
CF	**0.867**	0.867	**0.867**	0.901	0.883	0.892	0.833	0.842	0.838	0.937	**0.927**	**0.932**	**0.794**	0.762	0.777	0.916	**0.945**	0.931
temporal-25																		
CS	0.759	0.757	0.757	**0.848**	**0.839**	**0.844**	0.757	0.743	0.750	0.925	0.903	0.914	0.621	0.630	0.625	0.833	0.876	0.854
CF	0.863	0.859	0.861	0.902	**0.892**	**0.897**	0.852	0.835	0.843	0.938	0.916	0.927	0.787	0.800	**0.793**	0.893	0.940	0.916
temporal-10																		
CS	0.762	0.764	0.763	**0.848**	**0.839**	**0.844**	**0.760**	0.765	0.762	0.917	0.898	0.907	0.605	**0.646**	0.625	**0.866**	**0.888**	**0.877**
CF	0.863	0.866	0.865	0.902	**0.892**	**0.897**	0.852	**0.857**	**0.854**	0.936	0.916	0.926	0.760	**0.811**	0.784	**0.922**	**0.945**	**0.933**
L3i@HIPE-2022																		
CS	0.782	0.827	0.804	0.810	0.842	0.826	0.780	0.787	0.784	0.946	0.921	0.934	0.624	0.617	0.620	0.824	0.876	0.850
CF	0.883	0.933	0.907	0.856	0.889	0.872	0.870	0.878	0.874	0.965	0.940	0.952	0.793	0.784	0.788	0.868	0.922	0.894

5.1 Results

Table 2 presents our results in all three languages and datasets (best results in bold). It can be seen that models with additional knowledge-based *context jokers* bring an improvement over the base model with no added contexts. Furthermore, including temporal information outperforms non-temporal contexts. `ajmc` scores show to be higher than `hipe-2020` independently of the language and contexts. We explain this behavior by the small diversity of some entity types of the `ajmc` dataset. For example, the ten most frequent entities from the "person" type represent the 55%, 51.5% and 62.5% from the train, development, and test sets respectively. It also exists an 80% top-10 intersection between train and test sets meaning that eight of the ten most frequent entities are shared between train and test sets. English `hipe-2020` presents the lowest scores compared to French and German independently from the contexts. We attribute this drop in performance to the utilization of the French and German sets during training given the absence of a specific English training set.

The last two rows of Table 2 show the results of our best system [3] during the HIPE-2022 evaluation campaign [15]. This system is similar to the one described in Sect. 4.3 but it stacks, for each language, a language-specific language model and does not include any temporal-aware knowledge. The additional language model motivates the slightly higher results[15]. For half of the datasets, this system outperforms the temporal-aware configurations (underlined values) but with the cost of being language dependent, a drawback that mainly impacts English `hipe-2020` dataset where no training data is available.

---

[15] We would expect higher results by utilising the temporal information, however, for this experimental setup, we were limited in terms of resources.

Table 3. Number of replaced contexts per time span.

	French		German		English	
	train	test	train	test	train	test
temporal-50/25/10						
hipe-2020	120/154/217	42/47/61	325/393/482	12/14/14	192/222/246	77/85/96
ajmc	10/12/12	0/0/0	71/71/73	20/20/20	2/2/2	0/0/0

5.2 Impact of Time Intervals

ajmc contains 19th-century commentaries to Greek texts [41] and was created in the context of the Ajax MultiCommentary project[16], and thus, the French, German and English dataset are about an Ancient Greek tragedy by Sophocles, the Ajax, from the early medieval period[17]. The German ajmc contains commentaries from two years (1853 and 1894), English ajmc, also two years (1881 and 1896), while French ajmc just one year (1886). Due to the size of the collection, hipe-2020 covers a larger range of years. In terms of span, French articles were collected from 1798 to 2018, German articles from 1798 to 1948, and English articles from 1790 to 1960. We, therefore, looked at the difference between the contexts retrieved by the non-temporal and the temporal configurations. Table 3 summarizes these differences for train and test sets and displays the number of contexts that had been filtered and replaced from non-temporal for each time span, i.e., $\delta \in \{50, 25, 10\}$. Overall, the smaller the interval of years, the greater the number of contexts that are replaced. It can be noticed that the number of replaced contexts is smaller for ajmc than for hipe-2020. This is explained by the restrained year span and the lack of entity diversity during these periods. When comparing with the results from Table 2, we can infer that, in general, it is beneficial to implement shorter time intervals such as $\delta = 10$. In fact, temporal-10 presents higher F1 scores for ajmc in almost all cases. However, this varies with the language and the year distribution of the dataset.

5.3 Impact of Digitization Errors

The ajmc commentaries on classical Greek literature present the typical difficulties of historical OCR. Having complex layouts, often with multiple columns and rows of text, the digitization quality of commentaries could severely impact NER and other downstream tasks like entity linking. Statistically, about 10% of NEs are affected by the OCR in the English and German ajmc datasets and 27.5% of NEs are contaminated in the French corpus. The models with additional context, especially the temporal approaches, contribute to recognizing NEs whether

---

[16] https://mromanello.github.io/ajax-multi-commentary/.

[17] Although the exact date of its first performance is unknown, most scholars date it to relatively early in Sophocles' career (possibly the earliest Sophoclean play still in existence), somewhere between 450 BCE to 430 BCE, possibly around 444 BCE.

contaminated or clean. This contribution is more significant on NEs with digitization errors. It has manifested in a better improvement in recognition of the contaminated NEs compared to the clean ones despite their dominance in the data. In the German corpus, for example, the gain is about 14% points using `temporal-50` compared to the baseline while only 2% points on the clean NEs. Additionally, three-quarters of NEs with 67% of character error rate are correctly recognized whereas the baseline recognized only one-quarter of them. Finally, all the models are completely harmed by error rates that exceed 70% on NEs.

5.4 Limitations

The system ideally requires metadata about the year when the datasets were written or at least a period interval. Otherwise, it will be necessary to use other systems for predicting the year of publication [36]. However, the errors of such systems will be propagated and may impact the NER results.

6 Conclusions & Future Work

In this paper, we explore a strategy to inject temporal information into the named entity recognition task on historical collections. In particular, we rely on using semantically-relevant contexts by exploring the time information provided in the collection's metadata and temporal knowledge graphs. Our proposed models include the contexts as mean-pooled representations in a Transformer-based model. We observed several trends regarding the importance of temporality for historical newspapers and classical commentaries, depending on the time intervals and the digitization error rate. First, our results show that a short time span works better for collections with restrained entity diversity and narrow year intervals, while a longer time span benefits wide year intervals. Second, we also show that our approach performs well in detecting entities affected by digitization errors even to a 67% of character error rate. Finally, we remark that the quality of the retrieved contexts is dependent on the affinity between the historical collection and the knowledge base, thus, in future work, it could be interesting to include temporality information by predicting the year spans of a large set of Wikipedia pages to be used as complementary contexts.

Acknowledgements. This work has been supported by the ANNA (2019-1R40226) and TERMITRAD (2020–2019-8510010) projects funded by the Nouvelle-Aquitaine Region, France.

References

1. Agarwal, P., Strötgen, J., del Corro, L., Hoffart, J., Weikum, G.: diaNED: Time-aware named entity disambiguation for diachronic corpora. In: Proceedings of the 56th Annual Meeting of the Association for Computational Linguistics (Short Papers). Assoc. Comput. Linguist. Melbourne, Australia. **2**, pp. 686–693(Jul 2018). https://doi.org/10.18653/v1/P18-2109. https://aclanthology.org/P18-2109

2. Bircher, S.: Toulouse and Cahors are French Cities, but Ti*louse and Caa. Qrs as well. Ph.D. thesis, University of Zurich (2019)
3. Boros, E., González-Gallardo, C.E., Giamphy, E., Hamdi, A., Moreno, J.G., Doucet, A.: Knowledge-based contexts for historical named entity recognition linking, pp. 1064–1078. http://ceur-ws.org/Vol-3180/#paper-84
4. Boroş, E., Hamdi, A., Pontes, E.L., Cabrera-Diego, L.A., Moreno, J.G., Sidere, N., Doucet, A.: Alleviating digitization errors in named entity recognition for historical documents. In: Proceedings of the 24th conference on computational natural language learning, pp. 431–441 (2020)
5. Boros, E., Nguyen, N.K., Lejeune, G., Doucet, A.: Assessing the impact of OCR noise on multilingual event detection over digitised documents. Int. J. Digital Libr, pp. 1–26 (2022)
6. Boroş, E., et al.: A comparison of sequential and combined approaches for named entity recognition in a corpus of handwritten medieval charters. In: 2020 17th International conference on frontiers in handwriting recognition (ICFHR), pp. 79–84. IEEE (2020)
7. Cai, B., Xiang, Y., Gao, L., Zhang, H., Li, Y., Li, J.: Temporal knowledge graph completion: a survey. arXiv preprint arXiv:2201.08236 (2022)
8. Díez Platas, M.L., Ros Munoz, S., González-Blanco, E., Ruiz Fabo, P., Alvarez Mellado, E.: Medieval spanish (12th-15th centuries) named entity recognition and attribute annotation system based on contextual information. J. Assoc. Inf. Sci. Technol. **72**(2), 224–238 (2021)
9. Dikeoulias, I., Amin, S., Neumann, G.: Temporal knowledge graph reasoning with low-rank and model-agnostic representations. arXiv preprint arXiv:2204.04783 (2022)
10. Ehrmann, M., Hamdi, A., Linhares Pontes, E., Romanello, M., Douvet, A.: A Survey of Named Entity Recognition and Classification in Historical Documents. ACM Comput. Surv. (2022). https://arxiv.org/abs/2109.11406
11. Ehrmann, M., Hamdi, A., Pontes, E.L., Romanello, M., Doucet, A.: Named entity recognition and classification on historical documents: a survey. arXiv preprint arXiv:2109.11406 (2021)
12. Ehrmann, M., Romanello, M., Bircher, S., Clematide, S.: Introducing the CLEF 2020 HIPE shared task: Named entity recognition and linking on historical newspapers. In: Jose, J.M., Yilmaz, E., Magalhães, J., Castells, P., Ferro, N., Silva, M.J., Martins, F. (eds.) Advances in information retrieval, pp. 524–532. Springer International Publishing, Cham (2020)
13. Ehrmann, M., Romanello, M., Clematide, S., Ströbel, P.B., Barman, R.: Language resources for historical newspapers. the impresso collection. In: Proceedings of The 12th Language Resources and Evaluation Conference, pp. 958–968 (2020)
14. Ehrmann, M., Romanello, M., Doucet, A., Clematide, S.: Introducing the HIPE 2022 shared task: Named entity recognition and linking in multilingual historical documents. In: European Conference on Information Retrieval, pp. 347–354 (2022). https://doi.org/10.1007/978-3-030-99739-7_44
15. Ehrmann, M., Romanello, M., Flückiger, A., Clematide, S.: Extended overview of clef HIPE 2020: named entity processing on historical newspapers. In: CEUR Workshop Proceedings. 2696, CEUR-WS (2020)
16. Ehrmann, M., Romanello, M., Flückiger, A., Clematide, S.: Overview of CLEF HIPE 2020: Named Entity Recognition and Linking on Historical Newspapers. In: Arampatzis, A., et al. (eds.) CLEF 2020. LNCS, vol. 12260, pp. 288–310. Springer, Cham (2020). https://doi.org/10.1007/978-3-030-58219-7_21

17. Bellot, P., et al. (eds.): CLEF 2018. LNCS, vol. 11018. Springer, Cham (2018). https://doi.org/10.1007/978-3-319-98932-7
18. García-Durán, A., Dumančić, S., Niepert, M.: Learning sequence encoders for temporal knowledge graph completion. arXiv preprint arXiv:1809.03202 (2018)
19. Hamdi, A., et al.: A multilingual dataset for named entity recognition, entity linking and stance detection in historical newspapers. In: Proceedings of the 44th International ACM SIGIR Conference on Research and Development in Information Retrieval, pp. 2328–2334 (2021)
20. Hamdi, A., Pontes, E.L., Sidere, N., Coustaty, M., Doucet, A.: In-depth analysis of the impact of OCR errors on named entity recognition and linking. Nat. Lang. Eng, pp. 1–24 (2022)
21. Houlsby, N., et al.: Parameter-efficient transfer learning for NLP. In: International Conference on Machine Learning, pp. 2790–2799. PMLR (2019)
22. Huynh, V., Hamdi, A., Doucet, A.: When to Use OCR Post-correction for Named Entity Recognition? In: Ishita, E., Pang, N., Zhou, L. (eds.) ICADL 2020. LNCS, vol. 12504, pp. 33–42. Springer, Cham (2020). https://doi.org/10.1007/978-3-030-64452-9_3
23. Kogkitsidou, E., Gambette, P.: Normalisation of 16th and 17th century texts in french and geographical named entity recognition. In: Proceedings of the 4th ACM SIGSPATIAL Workshop on Geospatial Humanities, pp. 28–34 (2020)
24. Kristanti, T., Romary, L.: Delft and entity-fishing: Tools for clef HIPE 2020 shared task. In: CLEF 2020-Conference and Labs of the Evaluation Forum. 2696. CEUR (2020)
25. Leblay, J., Chekol, M.W.: Deriving validity time in knowledge graph. In: Companion Proceedings of the The Web Conference 2018, pp. 1771–1776 (2018)
26. Linhares Pontes, E., Cabrera-Diego, L.A., Moreno, J.G., Boros, E., Hamdi, A., Doucet, A., Sidere, N., Coustaty, M.: Melhissa: a multilingual entity linking architecture for historical press articles. Int. J. Digit. Libr. **23**(2), 133–160 (2022)
27. Liu, T., Yao, J.G., Lin, C.Y.: Towards improving neural named entity recognition with gazetteers. In: Proceedings of the 57th Annual Meeting of the Association for Computational Linguistics, pp. 5301–5307 (2019)
28. Makhoul, J., Kubala, F., Schwartz, R., Weischedel, R., et al.: Performance measures for information extraction. In: Proceedings of DARPA broadcast news workshop, pp. 249–252. Herndon, VA (1999)
29. Manjavacas, E., Fonteyn, L.: Adapting vs pre-training language models for historical languages (2022)
30. Mayhew, S., Tsygankova, T., Roth, D.: ner and pos when nothing is capitalized. In: Proceedings of the 2019 Conference on Empirical Methods in Natural Language Processing and the 9th International Joint Conference on Natural Language Processing (EMNLP-IJCNLP) Association for Computational Linguistics, Hong Kong, China, pp. 6256–6261 (Nov 2019). https://doi.org/10.18653/v1/D19-1650. https://aclanthology.org/D19-1650
31. Meng, T., Fang, A., Rokhlenko, O., Malmasi, S.: Gemnet: Effective gated gazetteer representations for recognizing complex entities in low-context input. In: Proceedings of the 2021 Conference of the North American Chapter of the Association for Computational Linguistics. Human Language Technol, pp. 1499–1512 (2021)
32. Nguyen, N.K., Boros, E., Lejeune, G., Doucet, A.: Impact analysis of document digitization on event extraction. In: 4th workshop on natural language for artificial intelligence (NL4AI 2020) co-located with the 19th international conference of the Italian Association for artificial intelligence (AI* IA 2020). 2735, pp. 17–28 (2020)

33. Oberbichler, S., Boroş, E., Doucet, A., Marjanen, J., Pfanzelter, E., Rautiainen, J., Toivonen, H., Tolonen, M.: Integrated interdisciplinary workflows for research on historical newspapers: Perspectives from humanities scholars, computer scientists, and librarians. J. Assoc. Inf. Sci. Technol. **73**(2), 225–239 (2022)

34. Pfeiffer, J., Vulić, I., Gurevych, I., Ruder, S.: MAD-X: An Adapter-Based Framework for Multi-Task Cross-Lingual Transfer. In: Proceedings of the 2020 Conference on Empirical Methods in Natural Language Processing (EMNLP). Association for Computational Linguistics, pp. 7654–7673 (2020). https://doi.org/10.18653/v1/2020.emnlp-main.617. https://aclanthology.org/2020.emnlp-main.617

35. Pradhan, S., Moschitti, A., Xue, N., Ng, H.T., Björkelund, A., Uryupina, O., Zhang, Y., Zhong, Z.: Towards robust linguistic analysis using OntoNotes. In: Proceedings of the Seventeenth Conference on Computational Natural Language Learning. Assoc. Comput. Linguist. Sofia, Bulgaria, pp. 143–152. (2013). https://aclanthology.org/W13-3516

36. Rastas, I., et al.: Explainable publication year prediction of eighteenth century texts with the bert model. In: Proceedings of the 3rd Workshop on Computational Approaches to Historical Language Change, pp. 68–77 (2022)

37. Reimers, N., Gurevych, I.: Sentence-BERT: sentence embeddings using Siamese BERT-networks. In: Proceedings of the 2019 Conference on Empirical Methods in Natural Language Processing and the 9th International Joint Conference on Natural Language Processing (EMNLP-IJCNLP). Assoc. Comput. Linguist. Hong Kong, China, pp. 3982–3992 (2019). https://doi.org/10.18653/v1/D19-1410. https://aclanthology.org/D19-1410

38. Reimers, N., Gurevych, I.: Making monolingual sentence embeddings multilingual using knowledge distillation. In: Proceedings of the 2020 Conference on Empirical Methods in Natural Language Processing (EMNLP), pp. 4512–4525 (2020)

39. Ri, R., Yamada, I., Tsuruoka, Y.: mLUKE: The power of entity representations in multilingual pretrained language models. In: ACL 2022 (to appear) (2022)

40. Riedl, M., Padó, S.: A named entity recognition shootout for german. In: Proceedings of the 56th Annual Meeting of the Association for Computational Linguistics (Short Papers) **2**, pp. 120–125 (2018)

41. Romanello, M., Najem-Meyer, S., Robertson, B.: Optical character recognition of 19th century classical commentaries: the current state of affairs. In: The 6th International Workshop on Historical Document Imaging and Processing, pp. 1–6 (2021)

42. Suárez, P.J.O., Dupont, Y., Lejeune, G., Tian, T.: Sinner clef-hipe2020: sinful adaptation of SOTA models for named entity recognition in French and German. In: CLEF (Working Notes) (2020)

43. Tjong Kim Sang, E.F., De Meulder, F.: Introduction to the CoNLL-2003 shared task: Language-independent named entity recognition. In: Proceedings of the Seventh Conference on Natural Language Learning at HLT-NAACL 2003, pp. 142–147 (2003). https://www.aclweb.org/anthology/W03-0419

44. Todorov, K., Colavizza, G.: Transfer learning for named entity recognition in historical corpora. In: CLEF (Working Notes) (2020)

45. Vaswani, A., et al.: Attention is all you need. Adv. Neural Inf. Proc. Syst. **30** (2017)

46. Wang, X., Gao, T., Zhu, Z., Zhang, Z., Liu, Z., Li, J., Tang, J.: Kepler: A unified model for knowledge embedding and pre-trained language representation. Trans. Assoc. Comput. Linguist. **9**, 176–194 (2021)

47. Wang, X., et al.: Automated concatenation of embeddings for structured prediction. arXiv preprint arXiv:2010.05006 (2020)

48. Wang, X., et al.: Damo-nlp at semeval-2022 task 11: a knowledge-based system for multilingual named entity recognition. arXiv preprint arXiv:2203.00545 (2022)

49. Xu, C., Chen, Y.Y., Nayyeri, M., Lehmann, J.: Temporal knowledge graph completion using a linear temporal regularizer and multivector embeddings. In: Proceedings of the 2021 Conference of the North American Chapter of the Association for Computational Linguistics: Human Language Technologies. Assoc. Comput. Linguist, pp. 2569–2578 (2021). https://doi.org/10.18653/v1/2021.naacl-main.202. https://aclanthology.org/2021.naacl-main.202

50. Xu, C., Nayyeri, M., Alkhoury, F., Shariat Yazdi, H., Lehmann, J.: TeRo: A time-aware knowledge graph embedding via temporal rotation. In: Proceedings of the 28th International Conference on Computational Linguistics. Int. Committee Comput. Linguist. Barcelona, Spain, pp. 1583–1593 (2020). https://doi.org/10.18653/v1/2020.coling-main.139. https://aclanthology.org/2020.coling-main.139

51. Xu, C., Su, F., Lehmann, J.: Time-aware relational graph attention network for temporal knowledge graph embeddings (2021)

52. Yamada, I., Asai, A., Shindo, H., Takeda, H., Matsumoto, Y.: LUKE: Deep contextualized entity representations with entity-aware self-attention. In: Proceedings of the 2020 Conference on Empirical Methods in Natural Language Processing (EMNLP) Assoc. Comput. Linguist, pp. 6442–6454 (2020). https://doi.org/10.18653/v1/2020.emnlp-main.523, https://aclanthology.org/2020.emnlp-main.523

53. Zhu, C., Chen, M., Fan, C., Cheng, G., Zhang, Y.: Learning from history: modeling temporal knowledge graphs with sequential copy-generation networks. In: Proceedings of the AAAI Conference on Artificial Intelligence. **35**, pp. 4732–4740 (2021)

A Mask-based Logic Rules Dissemination Method for Sentiment Classifiers

Shashank Gupta[1]([envelope]) [iD], Mohamed Reda Bouadjenek[1] [iD],
and Antonio Robles-Kelly[2] [iD]

[1] School of Information Technology, Deakin University, Waurn Ponds Campus,
Geelong, VIC 3216, Australia
{guptashas,reda.bouadjenek}@deakin.edu.au
[2] Defence Science and Technology Group, Edinburg, SA 5111, Australia
antonio.robleskelly@defence.gov.au

Abstract. Disseminating and incorporating logic rules inspired by
domain knowledge in Deep Neural Networks (DNNs) is desirable to make
their output causally interpretable, reduce data dependence, and provide
some human supervision during training to prevent undesirable outputs.
Several methods have been proposed for that purpose but performing
end-to-end training while keeping the DNNs informed about logical constraints remains a challenging task. In this paper, we propose a novel
method to disseminate logic rules in DNNs for Sentence-level Binary
Sentiment Classification. In particular, we couple a Rule-Mask Mechanism with a DNN model which given an input sequence predicts a vector
containing binary values corresponding to each token that captures if
applicable a linguistically motivated logic rule on the input sequence.
We compare our method with a number of state-of-the-art baselines
and demonstrate its effectiveness. We also release a new Twitter-based
dataset specifically constructed to test logic rule dissemination methods
and propose a new heuristic approach to provide automatic high-quality
labels for the dataset.

Keywords: Logic rules · Sentiment classification · Explainable AI

1 Introduction

Deep Neural Networks (DNNs) provide a remarkable performance across a broad
spectrum of Natural Language Processing (NLP) tasks thanks to mainly their
Hierarchical Feature Representation ability [5], However, the complexity and
non-interpretability of the features extracted hinder their application in high-
stakes domains, where automated decision-making systems need to have a human
understanding of their internal process, and thus, require user trust in their
outputs [23]. Moreover, a huge amount of labeled training data is required to
construct these models, which is both expensive and time-consuming [2].

To fight against the above-mentioned drawbacks, it is desirable to make
DNNs inherently interpretable by augmenting them with domain-specific or

© The Author(s), under exclusive license to Springer Nature Switzerland AG 2023
J. Kamps et al. (Eds.): ECIR 2023, LNCS 13980, pp. 394–408, 2023.
https://doi.org/10.1007/978-3-031-28244-7_25

task-specific Expert Prior Knowledge [4]. This would complement the labeled training data [26], make their output causally interpretable [23] to answer the *why?* question and help the model learn real-world constraints to abstain from providing strange outputs, in particular for high-stakes domains. For example, for the binary sentiment classification task, given a sentence containing an *A-but-B* syntactic structure where A and B conjuncts have contrastive senses of sentiment (*A-but-B* contrastive discourse relation), we would like the model to base its decision on the B conjunct – following the *A-but-B* linguistically motivated logic rule [14]. However, in practice, such rules are difficult to learn directly from the data [10,13].

In this paper, we propose to model Expert Prior Knowledge as First Order Logic rules and disseminate them in a DNN model through our Rule-Mask mechanism. Specifically, we couple a many-to-many sequence layer with DNN to recognize contrastive discourse relations like *A-but-B* on input sequence and transfer that information to the DNN model via Feature Manipulation on input sequence features. The task of recognizing these relations is treated as binary token classification, where each token in the input sequence is classified as either 0 or 1 creating a rule-mask of either syntactic structure (e.g., $0 - 0 - 1$ or $1 - 0 - 0$), where only tokens corresponding to the rule-conjunct are classified as 1. This mask is then applied to the input sequence features via a dot product and the output is fed to the DNN model for the downstream task. Compared to existing methods, our method is jointly optimized with the DNN model and so it maintains the flexibility of end-to-end training, being straightforward and intuitive. Thus, the key contributions of this paper are summarized as follows:

1. We introduce a model agnostic Rule-Mask Mechanism that can be coupled with any DNN model to ensure that it will provide prediction following some logical constraints on the downstream task. We test this mechanism on the task of Sentence-level Binary Sentiment Classification where the DNN model is constrained to predict sentence sentiment as per linguistically motivated logic rules.
2. We release a dataset for the Sentence-level Binary Sentiment Classification task which contains an equal proportion of the sentences having various applicable logic rules as contrastive discourse relations. This dataset was constructed to test our method's ability to recognize the applicable logic rule in the input sentence and disseminate the information in the DNN model (i.e. help the DNN model to constrain its prediction as per the logic rules).
3. Instead of manual labeling of the dataset, we propose a new heuristic approach to automatically assign the labels based on Emoji Analysis and using a lexicon-based sentiment analysis tool called VADER [12]. We validate this approach by labeling a sample of tweets where we find high consistency between automatic labels and human labels.
4. We present a thorough experimental evaluation to demonstrate the empirically superior performance of our method on a metric specifically constructed to test logic rule dissemination performance and compare our results against a number of baselines.

2 Related Work

Even before the advent of modern Neural Networks, attempts to combine logic rules representing domain-specific or task-specific knowledge with hierarchical feature representation models have been studied in different contexts. For example, Towell and Shavlik [26] developed Knowledge-Based Artificial Neural Networks (KBANN) to combine symbolic domain knowledge abstracted as propositional logic rules with neural networks via a three-step pipelined framework. Garcez et al. [4] defined such systems as Neural-Symbolic Systems, which can be viewed as a hybrid model containing the representational capacity of a connectionist model like Neural Network and inherent interpretability of symbolic methods like Logical Reasoning. Our work is related to the broader field of Neural-Symbolic Systems, where we construct an end-to-end model, which embeds the representational capacity of a Neural Network and is aware of the logical rules when making inference decisions on the input. Thus, we review below both implicit and explicit methods to construct Neural-Symbolic Systems.

2.1 Implicit Methods to Construct Neural-Symbolic Systems

While not originally proposed to construct a Neural-Symbolic System, these works show that certain existing models can implicitly capture logical structures without any explicit modifications to their training procedure or architecture. For example, Krishna et al. [13] claimed that creating Contextualized Word Embeddings (CWE) from input sequence can inherently capture the syntactic logical rules when fine-tuned with the DNN model on downstream sentiment analysis task. They proposed to create these embeddings using a pre-trained language model called ELMo [19]. More recent state-of-the-art models like BERT [3] and GPT-2 [21] can also be used to create contextual representations of words in the input sequence.

However, as we show in our experimental results, such contextual representation of words alone is not sufficient to capture logical rules in the input sequence and pass the information to the DNN model. We instead show that implicit learning can be used to learn a rule-mask by a sequence model which then can be used to explicitly represent logic rule information on the input features to the downstream DNN model via Feature Manipulation.

2.2 Explicit Methods to Construct Neural-Symbolic Systems

These methods construct Neural-Symbolic systems by explicitly encoding logic rules information into the trainable weights of the neural network by modifying either its input training data, architecture, or its objective function.

Focusing on sentence-level sentiment classification, perhaps the most famous method is the Iterative Knowledge Distillation (IKD) [10], where first-order logic rules are incorporated with general off-the-shelf DNNs via soft-constrained optimization. An upgraded version of this method is proposed in [11] called Mutual Distillation, where some learnable parameters ϕ are introduced with logic rules when constructing the constrained posterior, which are learned from the input data. Instead of formulating constraints as regularization terms, Li and Srikumar [16] build Constrained Neural Layers, where logical constraints govern the forward computation operations in each neuron. Another work by Gu et al. [6] uses a task-guided pre-training step before fine-tuning the downstream task in which domain knowledge is injected into the pre-trained model via a selectively masked language modeling.

In contrast to these methods, our approach does not encode the rule information into the trainable parameters of the model but instead uses Feature Manipulation on the input through rule masking so as to disseminate the rule information into the downstream model. Thus, our method can incorporate logic rules without any such complicated ad-hoc changes to either input training data, architectures, or training procedures. Overall, the current literature lacks any method to construct a Neural-Symbolic model for sentiment classification which is straightforward, intuitive, end-to-end trainable jointly with the base neural network on training data and that can provide empirically superior performance.

3   Methodology

This section provides a detailed description of our method starting with the inception of Logic rules from domain knowledge to disseminating them with a DNN model.

3.1   Sources of Logic Rules

Previous work has shown that Contrastive Discourse Relations (CDRs) are hard to capture by general DNN models like CNNs or RNNs for sentence-level binary sentiment classification through purely data-driven learning [10,13]. Thus, Prasad et al. [20] define such relations as sentences containing A-$keyword$-B syntactic structure where two clauses A and B are connected through a discourse marker ($keyword$) and have contrastive polarities of sentiment. Sentences containing such relations can be further classified into (i) CDR_{Fol}, where the dominant clause is *following* and the rule conjunct is B (sentence sentiment is determined by B conjunct), or (ii) CDR_{Prev}, where the dominant clause is *preceding* and the rule conjunct is A. Mukherjee and Bhattacharyya [17] argue that these relations need to be learned by the model while determining the overall sentence sentiment. Hence, for our experiments, we identify these relations as expert prior knowledge, construct First Order Logic rules from them and incorporate these rules with the DNN model through our mask method. Table 1 lists all the logic rules we study in this paper.

3.2   Rule-Mask Mechanism to Disseminate Logical Information

Our task is to build an end-to-end system, which provides sentence-level sentiment predictions and bases its predictions on linguistically motivated logic rules.

Table 1. List of logic rules used in this analysis. Rule conjunct denotes the dominant clause during the sentiment determination and is italicized in examples.

Logic rule	Keyword	Rule conjunct	Example
$A - \textbf{but} - B$	*but*	B [17]	Yes there is an emergency called covid-19 **but** *victory is worth celebration*
$A - \textbf{yet} - B$	*yet*	B [17]	Even though we can't travel **yet** *we can enjoy each other and what we have*
$A - \textbf{though} - B$	*though*	A [17]	*You are having an amazing time* **though** we are having this awful pandemic
$A - \textbf{while} - B$	*while*	A [1]	*Stupid people are not social distancing* **while** there's a global pandemic

Specifically, given an input sentence S containing a rule-syntactic structure like $A - keyword - B$ where *keyword* indicates an applicable logic rule in Table 1 and A & B conjuncts have contrastive senses of sentiment, we would like the classifier to predict the sentiment of S as per the B conjunct if the rule conjunct is B, otherwise, to predict the sentence sentiment as per A if the rule conjunct is A.

A straightforward method to create such a system is to use Feature Extraction [8] on the input data, where features corresponding to the rule conjunct are extracted and fed as input to the classifier. Specifically, given the input sentence, S containing A-*keyword*-B syntactic structure, Gupta et al. [8] proposed to manually compute a rule mask M of the structure $0 - 0 - 1$ if the rule conjunct is B, otherwise, $1 - 0 - 0$ if the rule conjunct is A. Then, they propose to compute a post-processed instance $X_{conjunct} = X * M$ as the dot product between S and M, where $X_{conjunct}$ can be regarded as an explicit representation of the applicable logic rule. $X_{conjunct}$ is then passed as input to the sentiment classifier and hence, the classifier predicts the sentiment as per the rule conjunct. The mask M is applied during both the training and testing phases of the classifier.

Although the Feature Extraction method proposed in [8] is quite simple, intuitive, and can determine whether the sentence contains A-*keyword*-B structure, it lacks the adaptability to the more nuanced nature of language since it cannot determine whether the conjuncts have *contrastive* polarities of sentiment and hence, cannot determine whether the sentence has a CDR or not. Moreover, simply removing a part of the input sequence entirely often leads to a loss of sentiment-sensitive information which can affect the sentiment classification performance on sentences that contains rule-syntactic structure but no CDR. Besides, as pointed out in [11], human knowledge about a phenomenon is usually abstract, fuzzy, and built on high-level concepts (e.g., discourse relations, visual attributes) as opposed to low-level observations (e.g., word sequences, image pixels). Thus, logic rules constructed from human knowledge should have these traits in the context of the dataset under consideration.

This necessitates a mechanism based on predictive modeling for the rule mask, which can: (i) determine whether the input sentence has a CDR instead of just rule syntactic structure, (ii) be learned from the training data, (iii) coupled

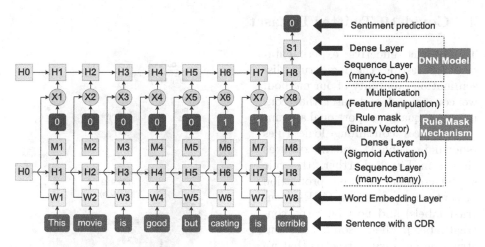

Fig. 1. Architecture of our rule-mask mechanism coupled with a DNN model. In mask block, the sequence layer predicts a rule mask M containing binary values corresponding to each token in input sentence S. Rule mask M is then multiplied with word embeddings W of S and the result is fed to the downstream DNN model.

with the classifier instead of being applied in a pipelined fashion, and (iv) jointly learned with the classifier on the training data to create a truly end-to-end system. Thus, we present a mechanism, in which given an input sentence S, it identifies whether it contains a logic rule structure like $A - keyword - B$ with A & B conjuncts having contrastive polarities of sentiment. If both conditions are met, it predicts a rule mask of a syntactic structure $0 - 0 - 1$ if the rule conjunct is B (mask values corresponding to tokens in A and $keyword$ parts are zero) or, otherwise, of structure $1 - 0 - 0$ if the rule conjunct is A (mask values corresponding to tokens in B and $keyword$ parts are zero). If there is no sentiment contrast between conjuncts or there is no rule-syntactic structure, it predicts a rule mask of a structure $1 - 1 - 1$. We optimize both the rule-mask mechanism and the DNN model jointly as:

$$\min_{\theta_1, \theta_2 \in \Theta} \quad L(y, p_{\theta_1}(y|x)) + \Sigma_{t=1}^{n} L(y_t, p_{\theta_2}(y_t|x_t)) \tag{1}$$

where $p_{\theta_1}(y|x)$ is the sentiment prediction of the DNN model and $p_{\theta_2}(y_t|x_t)$ is the mask value for t^{th} token in the input sequence $x = [x_1 \cdots x_n]$ and tackle the task of rule mask prediction by casting it as a token-level binary classification problem, where we predict either 0 or 1 tags corresponding to every token in the input sentence. We choose L as the Binary Cross-Entropy loss function.

Note that the proposed rule-mask mechanism can also be used with popular transformer-based DNN models BERT [3] where token embeddings can be first used to calculate the rule mask and then used to calculate the Masked Language Modeling (MLM) output.

4   Covid-19 Twitter Dataset

To conduct effective experimentation for testing the logic rule dissemination capability of our method,
we constructed a dataset that contains an equally proportional amount
of sentences containing logic rules
(shown in Table 1) and no rules as
shown in Fig. 2. Further, the rule
subset contains an equal proportion
of sentences containing CDRs (contrast labels) and no CDRs (no contrast labels). The reason behind constructing our own dataset is that we
wanted to get the specific distribution of sentences as shown in Fig. 2 to
test the logic rule dissemination performance of our method in an unbiased manner. Such distribution in
sufficient quantities is very difficult
to find in existing popular sentiment
classification datasets like SST2 [25],
MR [18], or CR [9].

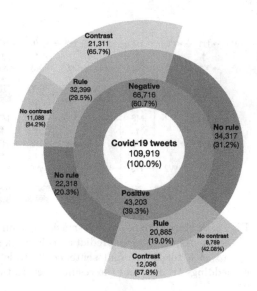

Fig. 2. Sector map of the constructed
dataset denoting the overall distribution of
tweets. The 1st layer denotes the proportion
of tweets containing negative and positive
sentiment polarities. In the 2nd layer, the
Rule sector denotes tweets having at-most
one of the logic rules applicable in Table 1
and the No-Rule sector denotes tweets with
no applicable logic rule. In the last layer,
the Contrast sector denotes tweets containing a CDR as defined in Sect. 3.1 and the
No-Contrast sector denotes tweets without
a CDR but contains a logic rule.

To get this distribution, we created a corpus of tweets from Twitter on the Covid-19 topic where the
tweet IDs were taken from the Covid-
19 Twitter dataset [15]. Raw tweets
were then pre-processed using a
tweet pre-processor[1], which removes
unwanted contents like hashtags,
URLs, @mentions, reserved keywords, and spaces. Each pre-processed tweet was
then passed through a series of steps as listed in Fig. 3 so as to obtain the
following: (1) Sentiment Label, which indicates the polarity of the sentence, (2)
Logic-Rule Label corresponding to either of the applicable rules listed in Table 1,
and (3) Contrast Label which determines if the sentence containing a logic rule
has a CDR or not (conjuncts A and B have a contrastive sense of sentiments).
In the following sub-sections, we provide more details on the definition of these
labels, why they need to be assigned, and how they were assigned to each tweet.

---

[1] Tweet pre-processing tool used here is accessible at https://pypi.org/project/tweet-
preprocessor/.

4.1 Sentiment Labels

Previous works [27] have shown that emojis indicate a strong correlation with associated sentence sentiment polarity and hence, we designed an Emoji Analysis method to assign sentiment labels to pre-processed tweets. Specifically, for each pre-processed tweet, we check whether it contains an emoji using an automatic emoji identification tool in texts[2], whether all emojis are present at the end of the tweet to make sure the tweet contains complete text[3] and whether at least one emoji is present in the EmoTag1200 table [24] which associates 8 types of positive and negative emotions scores with an emoji - anger, anticipation, disgust, fear, joy, sadness, surprise, and trust - and contains a score for each emotion assigned by human annotators. If the tweet passes the above checks, we then calculate the sum of all emotion scores for each emoji present and get an Aggregate Emotion Score for the tweet. This score is compared against emotion score thresholds for positive and negative polarities, which we found dynamically based on the dataset. These thresholds, 2.83 and -2.83, are such that they correspond to one standard deviation of aggregate emotion scores for a random sample of 1 million tweets. As a further consistency check, we used a lexicon-based sentiment analysis tool called VADER [12] and only kept those tweets in our dataset for which both VADER and emoji analysis assigns the same sentiment class.

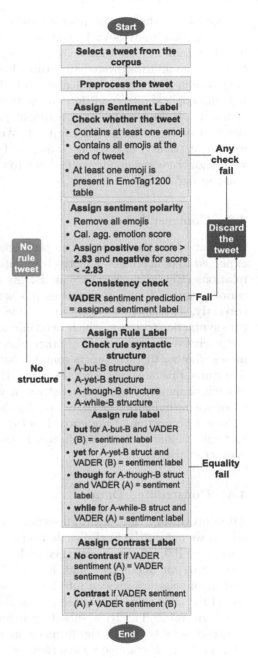

Fig. 3. Covid-19 tweets dataset construction flowchart.

---

[2] The emoji extraction tool is available at https://advertools.readthedocs.io/en/master/.

[3] This is so as to exclude tweets such as "I ♡NYC" as they are semantically incorrect.

4.2   Rule Labels

For each tweet that has been successfully assigned a sentiment label, we perform a conjunction analysis and identify if it contains any rule-syntactic structure listed in Table 1. Note that we only consider tweets that contain only one structure (i.e. no multiple nested structures like A-but-B-yet-C). The absence of any structure in the tweet is labeled as No-Rule otherwise, we check the corresponding rule applicability condition on the tweet which for example, for $A - but - B$ structure checks whether the sentiment polarity of the tweet is consistent with the sentiment polarity of B conjunct. We again use VADER to determine the sentiment polarity of the rule conjunct. If the rule applicability condition holds, we assign the corresponding rule label to the tweet, otherwise discard it to avoid noise in our dataset.

4.3   Contrast Labels

Contrast labels are important as performance on this subset (Rule-Contrast) is expected to indicate how effectively a method disseminates contrastive discourse relations (CDRs) in the DNN model. As mentioned in Sect. 3.1, general DNNs cannot capture CDRs in sentences and hence, cannot determine their sentiment correctly. Therefore, we need to provide another label to tweets containing a rule-syntactic structure, which determines whether they contain a CDR or not. For such tweets, we provide another binary label called Contrast, which determines whether their conjuncts contain contrastive senses of sentiments or not. To determine this label, we again use VADER and determine the sentiment polarity of each conjunct to compare whether they are similar indicating "No-Contrast" or opposite indicating a CDR and labeled as "Contrast". We again maintain an equal proportion of sentences labeled with "Contrast" and "No-Contrast" so to train classifiers that can effectively determine the CDR, not just the rule-syntactic structure.

4.4   Constructed Dataset

After processing the corpus (flowchart shown in Fig. 3) and assigning all the labels, we obtain the final distribution as shown in Fig. 2. The dataset contains a total of 109,919 tweets assigned either positive or negative sentiment labels accounting for about 60% and 40% of the dataset respectively. Further, each sentiment subset is divided into 2 subsets - Rule, which contains tweets having one of the logic rule labels listed in 1, and No-Rule tweets, which do not contain any logic rules. The Rule subsets are further divided into Contrast and No-Contrast subsets, where the former contains tweets containing logic rules and CDRs (A and B conjuncts have contrastive senses of sentiment), and the latter contains tweets having applicable logic rule but do not contain a CDR (A and B conjuncts do not have contrastive senses of sentiment). In Table 2, we show a small sample of tweets annotated manually for all the labels in our dataset as shown in Fig. 2 to validate our heuristic approach of dataset labeling.

Table 2. Sample of tweets labeled manually to validate the heuristic approach.

(a) No rule tweets labeled with Positive Sentiments.

finally have decent ppe in the care home.

love this idea we are living through history and this is a great way to capture it.

we went to crawley, it was well organised and we felt looked after so thanks indeed.

ederson still my best performing city player since lockdown.

u are well i hope you are staying safe much love from montreal canada.

ms dionne warwick you are giving me so much lockdown joy.

(b) No rule tweets labeled with Negative Sentiments.

the provincial governments are drastically failing its people.

this quarantine makes you to attend a funeral just to cry out.

duterte threatens to jail those who refuse covid vaccines.

my professor just sent us an email saying he got covid there will be no class.

got covid yesterday and today pumas lost what a shit weekend.

i told my mam i filled out my application for my vaccine and she called me a bitch.

(c) Rule tweets labeled with positive sentiment and contrast.

A lot has been said against our president **but** I think he is doing his best.

it's a covid 19 pandemic ravaged tennis season **yet** carlos alcaraz is still won 28 lost 3.

first game after lockdown started with a birdie **though** good scoring didnt last.

friends in brazil posting festivals **while** ive been in lockdown since march.

He's in quarantine **but** still looking good and handsome as always.

feku wrote the book on how to lie non stop **but** his supporters still believe him.

(d) Rule tweets labeled with positive sentiment and no contrast.

michael keaton is my favorite batman **but** lori lightfoot is my favorite beetlejuice.

best boy band and **yet** so down to earth and always down for fun bts best boy.

awww it's such a cute corona **though** i want to hug it.

happy birthday have all the fun **while** staying covid safe.

well said we always try to improve as human nature **but** corona teach us very well.

this research is funny **but** also might encourage some mask use.

(e) Rule tweets labeled with negative sentiment and contrast.

I want to get a massage **but** of course, that's not such a good idea during a pandemic.

kaaan it has been one freakin year **yet** people still dont take this pandemic seriously.

absolutely disgusting that fans would gather even **though** corona virus is a thing.

niggas having social events **while** its a pandemic out.

thats looks fun **but** covid 19 destroyed our habitat shame on that virus.

i got a plan for a trip **but** chuck it i know it's gonna get cancel.

Rule tweets labeled with negative sentiment and no contrast.

this is so sad i want churches to reopen too **but** i also dont want to see this happening.

stage 4 cancer **yet** its corona that killed him.

people are getting sick on the vaccine **though** i know people who have it very bad.

there is nothing safe about this **while** theres a pandemic still going on i mean wtf.

i may come off as rude **but** during the pandemic ive forgotten how to socialize sorry.

hes never stayed away from me **but** i know he misses them and i have to work.

5   Experimental Results

In this section, we discuss the performance results of our method and baselines under study for the task of sentence-level binary sentiment classification on our dataset[4].

5.1   Dataset Preparation

We divide the Covid-19 tweets dataset into the Train, Val, and Test splits containing 80%, 20%, and 20% proportion of sentences respectively. Each split contains similar distributions for various subsets - No-Rule Positive, No-Rule Negative, Rule Positive Contrast, Rule Negative Contrast, Rule Positive No-Contrast, and Rule Negative No-Contrast - as presented in the complete dataset Fig. 2. This ensures the classifiers are trained, tuned, and tested on splits containing proper distributions of every category of sentences.

5.2   Sentiment Classifiers

To conduct an exhaustive analysis, we train a range of DNN models as Base Classifiers - RNN, BiRNN, GRU, BiGRU, LSTM, and BiLSTM - to get the baseline measures of performances. Each model contains 1 hidden layer with 512 hidden units and does not have any mechanism to incorporate logic rules. We then train these models again coupled with a rule dissemination method proposed in Iterative Knowledge Distillation (IKD) [10], Contextualized Word Embeddings (CWE) [13] and our Rule-Mask Mechanism to construct Logic Rule Dissemination (LRD) Classifiers. For our method, we train a wide range of possible configurations to provide an exhaustive empirical analysis. These configurations are {RNN base classifier, BiRNN base classifier, GRU base classifier, BiGRU base classifier, LSTM base classifier, and BiLSTM base classifier} × {RNN mask layer, BiRNN mask layer, GRU mask layer, BiGRU mask layer, LSTM mask layer, and BiLSTM mask layer}, which totals up to 36 LRD classifiers to exhaustively test the empirical performance of our method. We want to compare the performance of our method with other dissemination methods and propose the best method for a particular base classifier.

5.3   Metrics

While Sentiment Accuracy is the obvious choice given the task is sentiment classification, it fails to assess whether the classifier based its decision on the applicable logic rule or not. For example, a classifier may correctly predict the sentiment of the sentence *"the casting was not bad but the movie was awful"* as negative but may base its decision as per the individual negative words like *not* in the A conjunct instead of using B conjunct. Hence, we decided to use an alternative metric called PERCY proposed in [7] which stands for *Post-hoc*

---

[4] Code and dataset are available at https://github.com/shashgpt/LRD-mask.git.

Explanation-based Rule ConsistencY. It assesses both the accuracy and logic rule consistency of a classifier for the sentiment classification task. Briefly, we compute this score as follows:

1. Given a sentence s which is an ordered sequence of terms $[t_1 t_2 \cdots t_n]$ and contains a logic rule structure like A-*keyword*-B, we use *LIME* explanation framework [22], which maps it to a vector $[\tilde{w}_1 \tilde{w}_2 \cdots \tilde{w}_n]$ with $\tilde{w}_n$ indicating how much the word t_n contributed to the final decision of the classifier.
2. Next we define the contexts $C(A) = [\tilde{w}_1 \cdots \tilde{w}_{i-1}]$ and $C(B) = [\tilde{w}_{i+1} \cdots \tilde{w}_n]$ as respectively the left and a right sub-sequences w.r.t the word *keyword* indexed by i.
3. Finally, we select top $k = 5$ tokens by their values from $C(A)$ as $C_k(A)$ and $C(B)$ as $C_k(B)$ and, propose that a classifier has based its decision on B conjunct if $\mathbb{E}_w[C_k(B)] > \mathbb{E}_w[C_k(A)]$ otherwise on A conjunct if $\mathbb{E}_w[C_k(A)] < \mathbb{E}_w[C_k(B)]$, where $\mathbb{E}$ is the expectation over conjunct weights. Hence, we define the PERCY score as the following:

$$PERCY(s) = (P(y|s) = y_{gt}) \wedge (\mathbb{E}_w[C_k(A)] \lessgtr \mathbb{E}_w[C_k(B)]) \qquad (2)$$

where the first condition $(P(y|s) = y_{gt})$ tests the classification accuracy $(P(y|s)$ denotes classifier prediction on sentence s and y_{gt} is the ground-truth sentiment) and the second condition $(\mathbb{E}_w[C_k(A)] < \mathbb{E}_w[C_k(B)]$ or $\mathbb{E}_w[C_k(A)] > \mathbb{E}_w[C_k(B)])$ checks whether the prediction was based as per the rule-conjunct (if the logic rule present is A-*but*-B or A-*yet*-B, the rule-conjunct is B whereas if the logic rule is A-*though*-B or A-*while*-B, the rule-conjunct is A).

5.4   Results

In this section, we analyze the PERCY scores for the classifiers as discussed in Sect. 5.2 obtained on **rule-contrast subset** of Covid-19 tweets test dataset (yellow color portion of the distribution as shown in Fig. 2), which contains sentences with Contrastive Discourse Relations as discussed in Sect. 3.1. Remember that the task of our method is to identify applicable CDRs in the sentences and disseminate the information in the downstream DNN model. Therefore, we show the results only on the rule-contrast subset.

Here, we find that our method outperforms all the base classifiers as well as the other logic rule dissemination methods proposed in [10] and [13]. This implies that the base classifiers cannot learn CDRs in sentences while determining their sentiments, and hence, they perform poorly. Further, we observe that the bidirectional mask models perform the best which implies that bidirectional models can identify the applicable CDRs and learn the rule mask better than unidirectional ones. It could be argued that the mask method uses the explicit representation of logic rules on input features instead of probabilistic modeling like other methods and, hence, is expected to provide the best empirical performance.

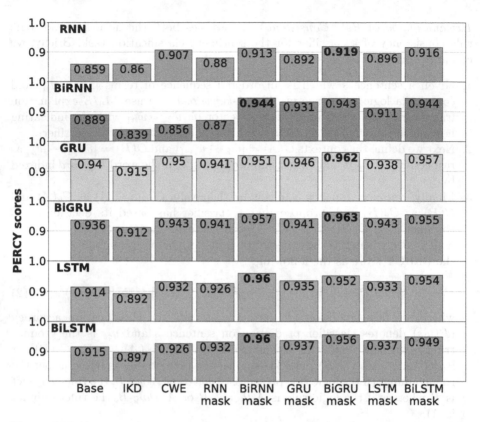

Fig. 4. PERCY scores for the classifiers obtained on **rule-contrast** subset of Covid-19 tweets test dataset. We show a total of 6 bar plots each corresponding to a base classifier (RNN, BiRNN, etc.) and each plot contains results for 9 classifiers as discussed in Sect. 5.2 with the best value highlighted in bold.

6   Conclusion

In this paper, we presented a novel method to disseminate contrastive discourse relations as logical information in a DNN for the sentiment classification task. This is done by coupling a rule mask mechanism with the DNN model, which identifies applicable CDR on the input sequence and transfers the information to the model via feature manipulation on the input sequence. Compared to existing methods, ours is end-to-end trainable jointly with the DNN model, does not require any ad-hoc changes to either training or, architecture, and is quite straightforward. We constructed our own dataset of tweets using a heuristic approach to conduct an unbiased analysis. We have shown results for various configurations of our method on different DNN models and compared it with existing dissemination methods. Our experimental results demonstrate that our method consistently outperforms all baselines on a both sentiment and rule consistency assessment metric (PERCY score) when applied to sentences with CDRs.

References

1. Agarwal, R., Prabhakar, T.V., Chakrabarty, S.: "I know what you feel": analyzing the role of conjunctions in automatic sentiment analysis. In: Nordström, B., Ranta, A. (eds.) GoTAL 2008. LNCS (LNAI), vol. 5221, pp. 28–39. Springer, Heidelberg (2008). https://doi.org/10.1007/978-3-540-85287-2_4
2. Bach, S.H., et al.: Snorkel DryBell: a case study in deploying weak supervision at industrial scale. In: Proceedings of the 2019 International Conference on Management of Data, pp. 362–375 (2019)
3. Devlin, J., Chang, M.W., Lee, K., Toutanova, K.: BERT: Pre-training of deep bidirectional transformers for language understanding. In: Proceedings of the 2019 Conference of the North American Chapter of the Association for Computational Linguistics: Human Language Technologies, Volume 1 (Long and Short Papers). pp. 4171–4186. Association for Computational Linguistics, Minneapolis, Minnesota (2019). https://doi.org/10.18653/v1/N19-1423
4. Garcez, A.S.D., Broda, K., Gabbay, D.M., et al.: Neural-symbolic learning systems: foundations and applications. Springer Science & Business Media (2002). https://doi.org/10.1007/978-1-4471-0211-3
5. Goodfellow, I., Bengio, Y., Courville, A.: Deep learning. MIT press (2016)
6. Gu, Y., Zhang, Z., Wang, X., Liu, Z., Sun, M.: Train no evil: selective masking for task-guided pre-training. In: EMNLP (2020)
7. Gupta, S., Bouadjenek, M.R., Robles-Kelly, A.: An analysis of logic rule dissemination in sentiment classifiers. In: Barrón-Cedeño, A., et al. (eds.) Experimental IR Meets Multilinguality, Multimodality, and Interaction. CLEF 2022. Lecture Notes in Computer Science, vol. 13390, pp. 118–124. Springer, Cham (2022). https://doi.org/10.1007/978-3-031-13643-6_9
8. Gupta, S., Robles-Kelly, A., Bouadjenek, M.R.: Feature extraction functions for neural logic rule learning. In: Torsello, A., Rossi, L., Pelillo, M., Biggio, B., Robles-Kelly, A. (eds.) S+SSPR 2021. LNCS, vol. 12644, pp. 98–107. Springer, Cham (2021). https://doi.org/10.1007/978-3-030-73973-7_10
9. Hu, M., Liu, B.: Mining and summarizing customer reviews. In: Proceedings of the Tenth ACM SIGKDD International Conference on Knowledge Discovery and Data Mining, pp. 168–177 (2004)
10. Hu, Z., Ma, X., Liu, Z., Hovy, E., Xing, E.: Harnessing deep neural networks with logic rules. In: Proceedings of the 54th Annual Meeting of the Association for Computational Linguistics (Volume 1: Long Papers). pp. 2410–2420. Association for Computational Linguistics, Berlin, Germany (2016). https://doi.org/10.18653/v1/P16-1228
11. Hu, Z., Yang, Z., Salakhutdinov, R., Xing, E.: Deep neural networks with massive learned knowledge. In: Proceedings of the 2016 Conference on Empirical Methods in Natural Language Processing, pp. 1670–1679. Association for Computational Linguistics (2016)
12. Hutto, C., Gilbert, E.: Vader: A parsimonious rule-based model for sentiment analysis of social media text. Proceed. Int. AAAI Conf. Web Soc. Media 8(1), 216–225 (2014). https://ojs.aaai.org/index.php/ICWSM/article/view/14550
13. Krishna, K., Jyothi, P., Iyyer, M.: Revisiting the importance of encoding logic rules in sentiment classification. In: Proceedings of the 2018 Conference on Empirical Methods in Natural Language Processing. pp. 4743–4751. Association for Computational Linguistics, Brussels, Belgium (2018). https://doi.org/10.18653/v1/D18-1505

14. Lakoff, R.: If's, and's and but's about conjunction. In: Fillmore, C.J., Langndoen, D.T. (eds.) Studies in Linguistic Semantics, pp. 3–114. Irvington (1971)

15. Lamsal, R.: Coronavirus (COVID-19) tweets dataset (2020). https://doi.org/10.21227/781w-ef42

16. Li, T., Srikumar, V.: Augmenting neural networks with first-order logic. In: Proceedings of the 57th Annual Meeting of the Association for Computational Linguistics, pp. 292–302. Association for Computational Linguistics, Florence, Italy (2019). https://doi.org/10.18653/v1/P19-1028

17. Mukherjee, S., Bhattacharyya, P.: Sentiment analysis in twitter with lightweight discourse analysis. In: COLING (2012)

18. Pang, B., Lee, L.: Seeing stars: Exploiting class relationships for sentiment categorization with respect to rating scales. In: Proceedings of the 43rd Annual Meeting of the Association for Computational Linguistics (ACL2005) (2005)

19. Peters, M.E., et al.: Deep contextualized word representations. In: Proceedings of the 2018 Conference of the North American Chapter of the Association for Computational Linguistics: Human Language Technologies, Volume 1 (Long Papers), pp. 2227–2237. Association for Computational Linguistics, New Orleans, Louisiana (2018). https://doi.org/10.18653/v1/N18-1202

20. Prasad, R., et al.: The Penn discourse TreeBank 2.0. In: Proceedings of the Sixth International Conference on Language Resources and Evaluation (LREC2008). European Language Resources Association (ELRA), Marrakech, Morocco (2008)

21. Radford, A., Narasimhan, K., Salimans, T., Sutskever, I.: Improving language understanding by generative pre-training. Tech. Rep, OpenAI (2018)

22. Ribeiro, M.T., Singh, S., Guestrin, C.: "why should i trust you?": explaining the predictions of any classifier. In: Proceedings of the 22nd ACM SIGKDD International Conference on Knowledge Discovery and Data Mining, pp. 1135–1144. KDD 2016, Association for Computing Machinery, New York, NY, USA (2016). https://doi.org/10.1145/2939672.2939778

23. Rudin, C.: Stop explaining black box machine learning models for high stakes decisions and use interpretable models instead. Nat. Mach. Intell. 1, 206–215 (2019)

24. Shoeb, A.A.M., de Melo, G.: EmoTag1200: understanding the association between emojis and emotions. In: Proceedings of the 2020 Conference on Empirical Methods in Natural Language Processing (EMNLP), pp. 8957–8967. Association for Computational Linguistics, Online (2020). https://doi.org/10.18653/v1/2020.emnlp-main.720

25. Socher, R., et al.: Recursive deep models for semantic compositionality over a sentiment treebank. In: Proceedings of the 2013 Conference on Empirical Methods in Natural Language Processing, pp. 1631–1642. Association for Computational Linguistics, Seattle, Washington, USA (2013). https://www.aclweb.org/anthology/D13-1170

26. Towell, G.G., Shavlik, J.W.: Knowledge-based artificial neural networks. Artif. Intell. 70(1), 119–165 (1994). https://doi.org/10.1016/0004-3702(94)90105-8

27. Yoo, B., Rayz, J.T.: Understanding emojis for sentiment analysis. The International FLAIRS Conference Proceedings 34 (2021). https://doi.org/10.32473/flairs.v34i1.128562

Contrasting Neural Click Models
and Pointwise IPS Rankers

Philipp Hager[1]([✉]) [iD], Maarten de Rijke[1] [iD], and Onno Zoeter[2] [iD]

[1] University of Amsterdam, Amsterdam, The Netherlands
{p.k.hager,m.derijke}@uva.nl
[2] Booking.com, Amsterdam, The Netherlands
onno.zoeter@booking.com

Abstract. Inverse-propensity scoring and neural click models are two popular methods for learning rankers from user clicks that are affected by position bias. Despite their prevalence, the two methodologies are rarely directly compared on equal footing. In this work, we focus on the pointwise learning setting to compare the theoretical differences of both approaches and present a thorough empirical comparison on the prevalent semi-synthetic evaluation setup in unbiased learning-to-rank. We show theoretically that neural click models, similarly to IPS rankers, optimize for the true document relevance when the position bias is known. However, our work also finds small but significant empirical differences between both approaches indicating that neural click models might be affected by position bias when learning from shared, sometimes conflicting, features instead of treating each document separately.

1 Introduction

Learning-to-rank a set of items based on their features is a crucial part of many real-world search [9,23,37,42] and recommender systems [15,20,55]. Traditional supervised learning-to-rank uses human expert annotations to learn the optimal order of items [8,9,31]. However, expert annotations are expensive to collect [9] and can be misaligned with actual user preference [41]. Instead, the field of unbiased learning-to-rank seeks to optimize ranking models from implicit user feedback, such as clicks [1,28,34,49,50]. One well-known problem when learning from click data is that the position at which an item is displayed affects how likely a user is to see and interact with it [16,27,28,47,50]. Click modeling [14,16,19,21,39] and inverse-propensity scoring (IPS) [1,25,28,35,45] are two popular methods for learning rankers from position-biased user feedback. IPS-based counterfactual learning-to-rank methods mitigate position bias by re-weighting clicks during training inversely to the probability of a user observing the clicked item [28,49]. In contrast, click models are generative models that represent position bias and item relevance as latent parameters to directly predict biased user behavior [14,16,19,21,39].

IPS approaches were introduced to improve over click models [28,49] by: (i) requiring less observations of the same query-document pair by representing

J. Kamps et al. (Eds.): ECIR 2023, LNCS 13980, pp. 409–425, 2023.
https://doi.org/10.1007/978-3-031-28244-7_26

items using features instead of inferring a separate relevance parameter for each document [1,28,49,50]; (ii) decoupling bias and relevance estimations into separate steps since the joint parameter inference in click models can fail [2,32,50]; and (iii) optimizing the order of documents through pairwise [25,28] and listwise loss [34] functions instead of inferring independent pointwise relevance estimations for each document [28,50].

At the same time, neural successors of click models have been introduced [7, 13,22,23,54,55] that can leverage feature inputs, similarly to IPS-based rankers. Moreover, pointwise IPS methods have been presented that address the same ranking setting as click models [5,40]. In this work, we ask if both approaches are two sides of the same coin when it comes to pointwise learning-to-rank?

To address this question, we first introduce both approaches (Sects. 2, 3) and show theoretically that both methods are equivalent when the position bias is known (Sect. 4). We then compare both approaches empirically on the prevalent semi-synthetic benchmarking setup in unbiased learning-to-rank (Sect. 5) and find small but significant differences in ranking performance (Sect. 6.1). We conclude by investigating the found differences by performing additional experiments (Sect. 6.2) and hypothesize that neural click models might be affected by position bias when learning from shared, sometimes conflicting, document features.

The main contributions of this work are:

(1) A theoretical analysis showing that a PBM click model optimizes for unbiased document relevance when the position bias is known.
(2) An empirical evaluation of both methods on three large semi-synthetic click datasets revealing small but significant differences in ranking performance.
(3) An analysis of the empirical differences that hint at neural click models being affected by position bias when generalizing over conflicting document features instead of treating each document separately.

2   Related Work

We provide an overview of probabilistic and neural click models, IPS-based counterfactual learning-to-rank, and comparisons between the two methodologies.

Click Models. Probabilistic click models emerged for predicting user interactions in web search [14,16,39]. Factors that impact a user's click decision, such as an item's probability to be seen or its relevance are explicitly modeled as random variables, which are jointly inferred using maximum likelihood estimation on large click logs [14]. An early but prevailing model is the position-based model (PBM), which assumes that a click on a given item only depends on its position and relevance [16,39]. Another prominent approach, the cascade model, assumes that users scan items from top to bottom and click on the first relevant item, not examining the documents below [16]. Follow-up work extends these approaches to more complex click behavior [11,19,21,48], more elaborate user interfaces [52,53], and feedback beyond clicks [18]. We refer to Chuklin et al. [14] for an overview.

Recent click models use complex neural architectures to model non-sequential browsing behavior [56] and user preference across sessions [12,30]. Additionally, exact identifiers of items are typically replaced by more expressive feature representations [22,54,56]. In contrast to ever more complicated click models, neural implementations of the classic PBM recently gained popularity in industry applications [23,54,55]. So-called two-tower models input bias and relevance-related features into two separate networks and combine the output to predict user clicks [22,54]. We use a neural PBM implementation similar to current two-tower models in this work and our findings on click model bias might be relevant to this community.

Counterfactual Learning-to-Rank. Joachims et al. introduced the concept of counterfactual learning-to-rank [28], relating to previous work by Wang et al. [49]. This line of work assumes a probabilistic model of user behavior, usually the PBM [25,28,34,40] or cascade click model [46], and uses inverse-propensity scoring to mitigate the estimated bias from click data. The first work by Joachims et al. [28] introduced an unbiased version of the pairwise RankSVM method, Hu et al. [25] introduced a modified pairwise LambdaMART, and Oosterhuis and de Rijke suggested an IPS-correction for the listwise LambdaLoss framework [35]. Given that click models are pointwise rankers [50], we use a pointwise IPS method introduced by Bekker et al. [5] and later Saito et al. [40].

Comparing Click Models and IPS. Lastly, we discuss related work comparing IPS and click models. To our knowledge, Wang et al. [50] conduct the only experiment that compares both approaches on a single proprietary dataset. Their RegressionEM approach extends a probabilistic PBM using logistic regression to predict document relevance from item features instead of inferring separate relevance parameters per document. While the main motivation behind their work is to obtain better position bias estimates to train a pairwise IPS model, the authors also report the ranking performance of the inferred logistic regression model which can be seen as a component of a single-layer neural click model. The authors find that the click model improves rankings over a baseline not correcting for position bias, but is outperformed by a pairwise IPS approach [50, Table 4]. The authors also include two pointwise IPS approximations which are less effective than the click model and also fail to outperform the biased baseline model. Therefore, it is unclear how current pointwise methods suggested by Bekker et al. [5] and Saito et al. [40] would compare. We compare a recent pointwise IPS method with a common neural PBM implementation and report experiments on three public LTR dataset unifying model architecture, hyperparameter tuning, and position bias estimation to avoid confounding factors.

Lastly, recent theoretical work by Oosterhuis [32] compares click models and IPS and their limits for unbiased learning-to-rank. Their work finds that IPS-based methods can only correct for biases that are an affine transformation of item relevance. For click models jointly inferring both relevance and bias parameters, they find no robust theoretical guarantees of unbiasedness and find settings in which even an infinite amount of clicks will not lead to inferring the true model parameters. We will discuss this work in more detail in Sect. 4 and extend their analysis to show that a click model only inferring item relevance should be in-fact unbiased.

3   Background

We introduce our assumptions on how position bias affects users, the neural click model, and IPS approach that we compare in this work.

A Model of Position Bias. We begin by assuming a model of how position bias affects the click behavior of users. For this work, we resort to the prevalent model in unbiased learning-to-rank, the position-based model (PBM) [16,39]. Let $P(Y = 1 \mid d, q)$ be the probability of a document d being relevant to a given search query q and $P(O = 1 \mid k)$ the probability of observing a document at rank $k \in K, K = \{1, 2, \ldots\}$; then we assume that clicks occur only on items that were observed and relevant:

$$P(C = 1 \mid d, q, k) = P(O = 1 \mid k) \cdot P(Y = 1 \mid d, q)$$
$$c_{d,k} = o_k \cdot y_d. \tag{1}$$

For brevity, we use the short notation above for the rest of the paper and drop the subscript q in all of our formulas assuming that the document relevance y_d is always conditioned on the current query context.

A Neural Position-Based Click Model. A neural click model directly mirrors the PBM user model introduced in the previous section in its architecture [7,13,22,54]. We use a neural network g to estimate document relevance $\hat{y}_d$ from features x_d and estimate position bias $\hat{o}_k$ using a single parameter per rank denoted by $f(k)$. We use sigmoid activations and multiply the resulting probabilities:

$$\hat{c}_{d,k} = \sigma(f(k)) \cdot \sigma(g(x_d))$$
$$\hat{c}_{d,k} = \hat{o}_k \cdot \hat{y}_d. \tag{2}$$

A common choice to fit neural click models is the binary cross-entropy loss between predicted and observed clicks in the dataset [22,23,54–56]:

$$\mathcal{L}_{\mathrm{pbm}}(\hat{y}, \hat{o}) = - \sum_{(d,k) \in D} c_{d,k} \cdot \log(\hat{y}_d \cdot \hat{o}_k) + (1 - c_{d,k}) \cdot \log(1 - \hat{y}_d \cdot \hat{o}_k). \tag{3}$$

A Pointwise IPS Model. Instead of predicting clicks, IPS directly predicts the document relevance $\hat{y}_d$ and assumes an estimation of the position bias $\hat{o}_k$ is given [28,40]. Thus, the IPS model we assume in this work only uses the relevance network g:

$$\hat{y}_d = g(x_d). \tag{4}$$

Bekker et al. [5] introduce a pointwise IPS loss that minimizes the binary cross-entropy between predicted and true document relevance. Note how the PBM assumption is used to recover the unbiased document relevance by dividing clicks by the estimated position bias $\hat{o}_k$:

$$\mathcal{L}_{\text{ips}}(\hat{y}, \hat{o}) = - \sum_{(d,k) \in D} \frac{c_{d,k}}{\hat{o}_k} \cdot \log(\hat{y}_d) + \left(1 - \frac{c_{d,k}}{\hat{o}_k}\right) \cdot \log(1 - \hat{y}_d). \tag{5}$$

4 Methods

4.1 Comparing Unbiasedness

In this section, we compare the ability of the neural click model and pointwise IPS ranker to recover the unbiased relevance of an item under position bias. We begin by noting that in the trivial case in which there is no position bias, i.e., clicks are an unbiased indicator of relevance, both approaches are identical.

Proposition 1. *When correctly assuming that no position bias exists, i.e., $\forall k \in K, o_k = \hat{o}_k = 1$, the click model and pointwise IPS method are equivalent:*

$$\mathbb{E}\left[\mathcal{L}_{ips}(\hat{y}, \hat{o})\right] = \mathbb{E}\left[\mathcal{L}_{pbm}(\hat{y}, \hat{o})\right] = - \sum_{(d,k) \in D} y_d \cdot \log(\hat{y}_d) + (1 - y_d) \cdot \log(1 - \hat{y}_d).$$

Second, both approaches also collapse to the same (biased) model in the case of not correcting for an existing position bias in the data.

Proposition 2. *When falsely assuming that no position bias exists, i.e., $\forall k \in K, \hat{o}_k = 1 \wedge o_k < 1$, the click model and pointwise IPS method are equivalently biased:*

$$\mathbb{E}\left[\mathcal{L}_{ips}(\hat{y}, \hat{o})\right] = \mathbb{E}\left[\mathcal{L}_{pbm}(\hat{y}, \hat{o})\right] = - \sum_{(d,k) \in D} y_d o_k \cdot \log(\hat{y}_d) + (1 - y_d o_k) \cdot \log(1 - \hat{y}_d).$$

However, how do both approaches compare when inferring the unbiased document relevance under an existing position bias? Saito et al. [40] show that $\mathcal{L}_{\text{ips}}(\hat{y})$ is unbiased if the position bias is correctly estimated, $\forall k \in K, \hat{o}_k = o_k$ and users actually behave according to the PBM [40, Proposition 4.3]. The notion of an unbiased estimator is harder to apply to neural click models, since relevance is a parameter to be inferred. Instead of unbiasedness, Oosterhuis [32] looks into consistency of click models and shows that click models jointly estimating both bias and relevance parameters are not consistent estimators of document relevance. This means that there are rankings in which even infinite click data will not lead to the true document relevance estimate.

But what happens if click models do not have to jointly estimate bias and relevance parameters, but only item relevance? Since IPS approaches often assume access to a correctly estimated position bias [1, 28, 34, 40, 45], we investigate this idealized setting for the click model and show that initializing the model parameters $\hat{o}_k$ with the true position bias leads to an unbiased relevance estimate.

Theorem 1. *The click model is an unbiased estimator of relevance when given access to the true position bias:*

$$\mathbb{E}\left[\hat{y}_d\right] = \frac{o_k y_d}{\hat{o}_k}, \forall k \in K, \hat{o}_k = o_k. \tag{6}$$

Proof. We begin by taking the partial derivative of $\mathcal{L}_{\mathrm{pbm}}$ with regard to the estimated document relevance $\hat{y}$ in our click model. Since the model factorizes, for ease of notation we will look at a single document and single observation:

$$
\begin{aligned}
\frac{\partial \mathcal{L}_{\mathrm{pbm}}}{\partial \hat{y}} &= -\left(c \cdot \frac{\partial}{\partial \hat{y}} \left[\log(\hat{o}\hat{y}) \right] + (1-c) \cdot \frac{\partial}{\partial \hat{y}} \left[\log(1-\hat{o}\hat{y}) \right] \right) \\
&= -\left(c \cdot \frac{\hat{o}}{\hat{o}\hat{y}} + (1-c) \cdot \frac{-\hat{o}}{1-\hat{o}\hat{y}} \right) \\
&= -\left(\frac{c}{\hat{y}} + \frac{-\hat{o} + \hat{o}c}{1-\hat{o}\hat{y}} \right) \\
&= -\frac{c - \hat{o}\hat{y}}{\hat{y}(1-\hat{o}\hat{y})}.
\end{aligned}
\tag{7}
$$

Next, we find the ideal model minimizing the loss by finding the roots of the derivative. We note that this function is convex and any extrema found will be a minimum:

$$
\begin{aligned}
\frac{\partial \mathcal{L}_{\mathrm{pbm}}}{\partial \hat{y}} &= 0 \\
-\frac{c - \hat{o}\hat{y}}{\hat{y}(1-\hat{o}\hat{y})} &= 0 \\
\hat{y} &= \frac{c}{\hat{o}}.
\end{aligned}
\tag{8}
$$

Lastly, in expectation we see that the obtained relevance estimate is the true document relevance when the estimated and true position bias are equal:

$$
\begin{aligned}
\mathbb{E}\left[\hat{y} \right] &= \frac{\mathbb{E}\left[c \right]}{\hat{o}} \\
\mathbb{E}\left[\hat{y} \right] &= \frac{oy}{\hat{o}}.
\end{aligned}
\tag{9}
$$

Thus, given the correct position bias, we find that both the click model and IPS objective optimize for the unbiased document relevance, suggesting a similar performance in an idealized benchmark setup. But before covering our empirical comparison, we want to note one additional difference of both loss functions.

4.2   A Difference in Loss Magnitude

We note one difference between the click model and IPS-based loss functions concerning their magnitude and relationship with position bias. While IPS-based loss functions are known to suffer from high variance due to dividing clicks by potentially small probabilities [44,51], the neural click model seems to suffer from the opposite problem since both $y_{d,k}$ and $\hat{y}_{d,k}$ (assuming our user model is correct) are multiplied by a potentially small examination probability. Thus, independent of document relevance, items at lower positions have a click probability closer to zero, impacting the magnitude of the loss (and gradient). Figure 1 visualizes the loss for a single item of relevance $y_d = 0.5$ under varying degrees of position bias. While the pointwise IPS loss in expectation of infinite clicks always converges to the same distribution, the click model's loss gets smaller in magnitude with an increase in position bias. While the magnitude differs, the minimum of the loss, as shown earlier in Sect. 4.1, is still correctly positioned at 0.5. We will explore if this difference in loss magnitude might negatively impact items at lower positions in our upcoming experiments.

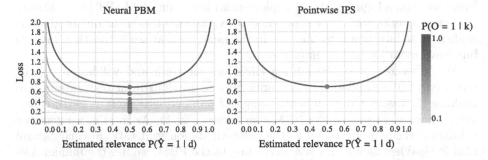

Fig. 1. Visualizing $\mathcal{L}_{\text{pbm}}$ on the left and $\mathcal{L}_{\text{ips}}$ on the right for a single document of relevance $y_d = 0.5$ under varying degrees of position bias.

Table 1. Overview of the LTR datasets used in this work.

Dataset	#Features	#Queries	%Train/val/test	#Documents per query				
				min	mean	med	p90	max
MSLR-WEB30K	136	31,531	60/20.0/20.0	1	120	109	201	1,251
Istella-S	220	33,018	58.3/19.9/21.8	3	103	120	147	182
Yahoo! Webscope	699	29,921	66.6/10.0/23.3	1	24	19	49	139

5   Experimental Setup

To compare click model and IPS-based approaches empirically, we use an evaluation setup that is prevalent in unbiased learning-to-rank [24,26,28,33,35,36, 45,47]. The main idea is to use real-world LTR datasets containing full expert annotations of item relevance to generate synthetic clicks according to our user model. Below, we describe the used datasets, the click generation procedure, as well as model implementation and training.

Datasets. We use three large-scale public LTR datasets to simulate synthetic user clicks: *MSLR-WEB30k* [37], *Istella-S* [17], and *Yahoo! Webscope* [9]. Each query-document pair is represented by a feature vector x_d and is accompanied by a score $s_d \in \{0, 1, 2, 3, 4\}$ indicating relevance as judged by a human annotator. Table 1 contains an overview of the dataset statistics. During preprocessing, we normalize the document feature vectors of *MSLR-WEB30k* and *Istella-S* using log1p(x_d) = log_e(1 + |x_d|) $\odot$ sign(x_d), as recently suggested by Qin et al. [38]. The features of *Yahoo! Webscope* come already normalized [9]. We use stratified sampling to limit each query to contain at most the 90th percentile number of documents (Table 1), improving computational speed while keeping the distribution of document relevance in the datasets almost identical.

Simulating User Behavior. Our click simulation setup closely follows [45, 47]. First, we train a LightGBM [29] implementation of LambdaMART [8] on 20 sampled train queries with fully supervised relevance annotations as our production ranker.[1] The intuition is to simulate initial rankings that are better than random but leave room for further improvement.

We generate up to 100 million clicks on our train and validation sets by repeatedly: (i) sampling a query uniformly at random from our dataset; (ii) ranking the associated documents using our production ranker; and (iii) generating clicks according to the PBM user model (Eq. 1). As in [45], we generate validation clicks proportional to the train/validation split ratio in each dataset (Table 1). When sampling clicks according to the PBM, we use the human relevance labels provided by the datasets as ground truth for the document relevance y_d. We use a graded notion of document relevance [3, 4, 10, 25] and add click noise of $\epsilon = 0.1$ to also sample clicks on documents of zero relevance:

$$y_d = \epsilon + (1 - \epsilon) \cdot \frac{2^{s_d} - 1}{2^4 - 1}. \tag{10}$$

We follow Joachims et al. [28] and simulate the position bias for a document at rank k after preranking as:

$$o_k = \left(\frac{1}{k}\right)^\eta \tag{11}$$

The parameter η controls the strength of position bias; $\eta = 0$ corresponds to no position bias. We use a default of $\eta = 1$. Lastly, we apply an optimization step from [34] and train on the average click-through-rate of each query-document pair instead of the actual sampled raw click data [34, Eq. 39]. This allows us to scale our simulation to millions of queries and multiple repetitions while keeping the computational load almost constant. Our experimental results hold up without this trick.

Model Implementation and Training. We estimate document relevance from features using the same network architecture $g(x_d)$ for both the click model and IPS-based ranker. Similar to [45, 46], we use a three layer feed-forward network with [512, 256, 128] neurons, ELU activations, and dropout 0.1 in the last two layers. We pick the best-performing optimizer[2] and learning rate[3] over five independent runs on the validation set for each model. In all experiments, we

---

[1] LightGBM Version 3.3.2, using 100 trees, 31 leafs, and learning rate 0.1.
[2] optimizer $\in \{Adam, Adagrad, SGD\}$.
[3] learning rate $\in \{0.1, 0.05, 0.01, 0.005, 0.001, 0.0005, 0.0001\}$.

train our models on the synthetic click datasets up to 200 epochs and stop early after five epochs of no improvement of the validation loss. We do not clip propensities in the IPS model to avoid introducing bias [1,28].

Experimental Runs. We follow related work and report the final evaluation metrics on the original annotation scores of the test set [1,28,34]. We test differences for significance using a two-tailed student's t-test [43], apply the Bonferroni correction [6] to account for multiple comparisons, and use a significance level of $\alpha = 0.0001$. All results reported in this work are evaluated over ten independent simulation runs with different random seeds. We compare five models:

IPS/PBM - Naive: A version of both models that does not compensate for position bias. In this case both models are equivalent (Proposition 2).

IPS-True bias: Pointwise IPS ranker with access to the true simulated position bias.

PBM-Estimated bias: Neural PBM jointly inferring position bias and document relevance during training.

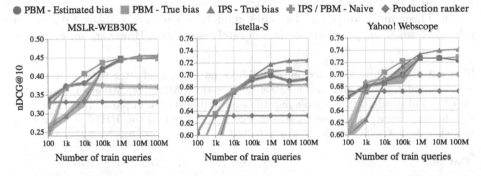

Fig. 2. Test performance after training on up to 100M simulated queries. All results are averaged over ten independent runs, and we display a bootstrapped 95% confidence interval.

PBM-True bias: Neural PBM initialized with the true position bias; the bias is fixed during training.

Production ranker: LambdaMART production ranker used to pre-rank queries during simulation.

6   Results and Analysis

We examine if the neural click model and pointwise IPS models are empirically equivalent in a semi-synthetic click simulation.

6.1 Main Findings

Fig. 2 displays the test performance of all model combinations when training up to 100M simulated queries; full tabular results are available in Table 2. Inspecting Fig. 2, we first note that all approaches improve over the initial rankings provided by the production ranker. The version of both models not correcting for position bias (*IPS/PBM - Naive*) converges to its final, suboptimal, performance after one million clicks. Significantly improving over the naive baseline on two out of three datasets (except *Istella-S*) is the neural click model jointly estimating position bias and relevance (*PBM - Estimated bias*).

Next, we see that providing the *PBM - True Bias* model with access to the correct position bias stabilizes and improves performance significantly over the naive baseline on all datasets. While having a lower variance, the improvements over *PBM - Estimated Bias* are not significant on any of the datasets. The *IPS - True bias* model is less effective than the neural click models for the first 100k clicks but ends up outperforming the click model significantly on two of the three LTR datasets (*Istella-S* and *Yahoo! Webscope*). These differences under idealized conditions between pointwise IPS and the click model are small, but significant. And to our surprise, the neural click model performs worse than the pointwise IPS model, even with access to the true position bias.

In Theorem 1, we prove that click models can recover unbiased document relevance when the position bias is accurately estimated. However, our empirical

Table 2. Ranking performance on the full-information test set after 100M train queries as measured in nDCG and Average Relevant Position (ARP) [28]. Results are averaged over ten independent runs, displaying the standard deviation in parentheses. We mark significantly higher ▲ or lower performance ▼ compared to the **PBM - True bias** model using a significance level of $\alpha = 0.0001$.

Dataset	Model	nDCG@5 ↑		nDCG@10 ↑		ARP ↓	
MSLR-WEB30K	Production	0.301	(0.027) ▼	0.330	(0.024) ▼	49.223	(0.693) ▲
	Naive	0.348	(0.022) ▼	0.370	(0.020) ▼	48.386	(0.538) ▲
	PBM - Est. Bias	0.429	(0.010)	0.449	(0.008)	44.835	(0.274)
	PBM - True Bias	0.428	(0.006)	0.447	(0.006)	44.965	(0.230)
	IPS - True Bias	0.432	(0.011)	0.454	(0.010)	44.418	(0.227)
Istella-S	Production	0.566	(0.012) ▼	0.632	(0.010) ▼	10.659	(0.207) ▲
	Naive	0.616	(0.005) ▼	0.683	(0.005) ▼	9.191	(0.154) ▲
	PBM - Est. Bias	0.629	(0.008)	0.692	(0.007)	10.605	(1.193)
	PBM - True Bias	0.638	(0.003)	0.703	(0.004)	8.911	(0.212)
	IPS - True Bias	0.656	(0.005) ▲	0.724	(0.004) ▲	8.274	(0.141) ▼
Yahoo! Webscope	Production	0.613	(0.012) ▼	0.671	(0.009) ▼	10.439	(0.095) ▲
	Naive	0.647	(0.006) ▼	0.699	(0.004) ▼	10.199	(0.052) ▲
	PBM - Est. Bias	0.673	(0.005)	0.722	(0.003)	9.848	(0.055)
	PBM - True Bias	0.680	(0.004)	0.728	(0.003)	9.812	(0.035)
	IPS - True Bias	0.695	(0.001) ▲	0.741	(0.001) ▲	9.658	(0.011) ▼

evaluation indicates a difference between click model and IPS-based approaches, even under the idealized conditions assumed in this setup: *unlike the IPS-based approach, the neural click model may suffer from bias.* Given this observed difference, we conduct further analyses by revisiting the effect of position bias on the magnitude of the click model's loss discussed earlier in Sect. 4.2.

6.2 Further Analyses

Our first hypothesis to explain the lower performance of the neural click model concerns hyperparameter tuning. Section 4.2 shows that the click model loss decreases with an increase in position bias. Through manual verification, we find that items at lower positions have smaller gradient updates, affecting the choice of learning rate and the number of training epochs. While this is a concern when using SGD, our extensive hyperparameter tuning and use of adaptive learning rate optimizers should mitigate this issue (Sect. 5). Hence, we reject this hypothesis.

Instead, we hypothesize that higher ranked items might overtake the gradient of lower ranked items, given their higher potential for loss reduction. This case might occur when encountering two documents with similar features but different relevance. The item at the higher position could bias the expected relevance towards its direction. This is indeed what we find when simulating a toy scenario with two documents in Fig. 3. There, we display one relevant but rarely observed document (red triangle) and one irrelevant but always observed item (orange square). Both click model and IPS approaches converge to the correct document relevance when computing the loss for each item separately, but when computing the combined average loss, the IPS approach converges to the mean relevance of both items while the click model is biased towards the item with the higher examination probability.

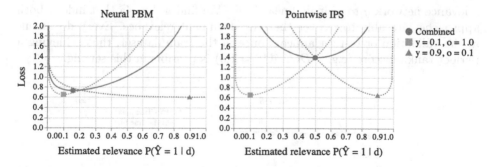

Fig. 3. Visualizing the loss and estimated document relevance of two documents when calculated separately (dotted lines) and combined (solid line).

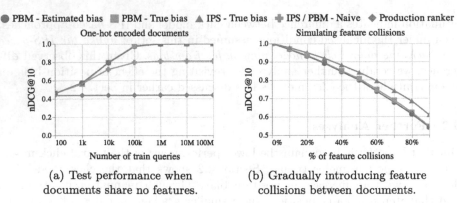

● PBM - Estimated bias   ■ PBM - True bias   ▲ IPS - True bias   ✚ IPS / PBM - Naive   ◆ Production ranker

(a) Test performance when documents share no features.

(b) Gradually introducing feature collisions between documents.

Fig. 4. Experiments on one-hot encoded documents. All results are averaged over ten independent runs. We display a bootstrapped 95% confidence interval.

One can frame this finding as an instance of *model misfit*. Theorem 1 demands a separate parameter $\hat{y}_d$ for each query-document pair, but by generalizing over features using the relevance network g, we might project multiple documents onto the same parameter $\hat{y}_d$, which might be problematic when features do not perfectly capture item relevance. We test our hypothesis that the click model's gradient updates are biased towards items with higher examination probabilities with three additional experiments.

No Shared Document Features. First, we should see an equivalent performance of both approaches in a setting in which documents share no features since the gradient magnitude should not matter in this setting. We create a fully synthetic dataset of 10,000 one-hot encoded vectors with uniform relevance scores between 0 and 4. To avoid feature interactions, we reduce the depth of the relevance network g to a single linear layer. We find in Fig. 4a that indeed both approaches are able to recover the true document relevance. Every document in the validation or test set appears once in the train dataset, thus achieving a perfect ranking score (e.g., $nDCG@10 = 1.0$) is possible in this setting.

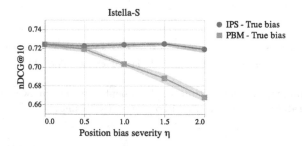

Fig. 5. Simulating an increasing (known) position bias. We report test performance after 100M clicks over 10 independent runs.

Feature Collisions. Second, gradually forcing documents to share features by introducing random feature collisions into our one-hot encoded dataset should lead to a stronger drop in performance for the click model. At the start of each simulation, we use a modulo operation to assign a share of documents based on their id on to the same one-hot encoded feature vectors. Figure 4b shows that both approaches perform equivalently when each document has its own feature vector. But when gradually introducing collisions, *PBM-Estimated bias* and *PBM-True bias* deteriorate faster in performance than *IPS-True bias*.

Mitigating Position Bias. A last interesting consequence is that this problem should get worse with an increase in (known) position bias. Simulating an increasing position bias and supplying the examination probabilities to both approaches on *Istella-S* shows that IPS can recover consistently from high position bias, while the click model deteriorates in performance with an increase in position bias (Fig. 5).

In summary, we found strong evidence that when encountering documents of different relevance but similar features, the neural click model biases its relevance estimate towards items with higher exposure.

7 Conclusion

We have considered whether recent neural click models and pointwise IPS rankers are equivalent for pointwise learning-to-rank from position-biased user clicks. We show theoretically and empirically that neural click models and pointwise IPS rankers achieve equal performance when the true position bias is known, and relevance is estimated for each item separately. However, we also find small but significant empirical differences, indicating that the neural click model may be affected by position bias when learning from shared and potentially conflicting document features.

Given the similarity of the neural PBM used in this work to current industry trends [22, 23, 54, 55], we urge practitioners to investigate if their model architecture is vulnerable to the described bias, especially when representing items using a small set of features or low dimensional latent embeddings. Potential diagnostic tools include simulating synthetic clicks or training a related pointwise IPS method to test for performance improvements.

We emphasize that our findings are specific to our neural PBM setup, and we make no claims about other architectures, such as additive two-tower models [54] or click models trained using expectation maximization [50]. We plan to further investigate connections and differences between IPS and click models, extending our evaluation beyond the pointwise setting to more sophisticated conditions such as mixtures of user behavior and bias misspecification. We share our code at https://github.com/philipphager/ultr-cm-vs-ips/

Acknowledgements. We thank our reviewers for their time and valuable feedback. For insightful discussions and their comments, we thank Shashank Gupta, Romain Deffayet, Kathrin Parchatka, and Harrie Oosterhuis.

This research was supported by the Mercury Machine Learning Lab, a collaboration between TU Delft, the University of Amsterdam, and Booking.com. Maarten de Rijke was supported by the Hybrid Intelligence Center, a 10-year program funded by the Dutch Ministry of Education, Culture and Science through the Netherlands Organisation for Scientific Research, https://hybrid-intelligence-centre.nl.

All content represents the opinion of the authors, which is not necessarily shared or endorsed by their respective employers and/or sponsors.

References

1. Agarwal, A., Takatsu, K., Zaitsev, I., Joachims, T.: A general framework for counterfactual learning-to-rank. In: International ACM SIGIR Conference on Research and Development in Information Retrieval (SIGIR) (2019)
2. Agarwal, A., Zaitsev, I., Wang, X., Li, C., Najork, M., Joachims, T.: Estimating position bias without intrusive interventions. In: International Conference on Web Search and Data Mining (WSDM) (2019)
3. Ai, Q., Bi, K., Luo, C., Guo, J., Croft, W.B.: Unbiased learning to rank with unbiased propensity estimation. In: International ACM SIGIR Conference on Research and Development in Information Retrieval (SIGIR) (2018)
4. Ai, Q., Yang, T., Wang, H., Mao, J.: Unbiased learning to rank: Online or offline? ACM Trans. Inform. Syst. (TOIS) **39**(2), 1–29 (2021)
5. Bekker, J., Robberechts, P., Davis, J.: Beyond the selected completely at random assumption for learning from positive and unlabeled data. In: Machine Learning and Knowledge Discovery in Databases: European Conference (ECML PKDD) (2019)
6. Bonferroni, C.: Teoria statistica delle classi e calcolo delle probabilita. Pubblicazioni del R. Istituto Superiore di Scienze Economiche e Commericiali di Firenze **8**, 3–62 (1936)
7. Borisov, A., Markov, I., de Rijke, M., Serdyukov, P.: A neural click model for web search. In: The World Wide Web Conference (WWW) (2016)
8. Burges, C.J.: From ranknet to lambdarank to lambdamart: An overview. Tech. Rep. MSR-TR-2010-82, Microsoft (2010)
9. Chapelle, O., Chang, Y.: Yahoo! learning to rank challenge overview. J. Mach. Learn. Res. (JMLR) **14**, 1–24 (2011)
10. Chapelle, O., Metlzer, D., Zhang, Y., Grinspan, P.: Expected reciprocal rank for graded relevance. In: International Conference on Information and Knowledge Management (CIKM) (2009)
11. Chapelle, O., Zhang, Y.: A dynamic bayesian network click model for web search ranking. In: The World Wide Web Conference (WWW) (2009)
12. Chen, J., Mao, J., Liu, Y., Zhang, M., Ma, S.: A context-aware click model for web search. In: International Conference on Web Search and Data Mining (WSDM) (2020)
13. Chu, W., Li, S., Chen, C., Xu, L., Cui, H., Liu, K.: A general framework for debiasing in ctr prediction (2021). https://doi.org/10.48550/arXiv.2112.02767
14. Chuklin, A., Markov, I., de Rijke, M.: Click Models for Web Search. Morgan & Claypool (2015), ISBN 9781627056489. https://doi.org/10.2200/S00654ED1V01Y201507ICR043
15. Covington, P., Adams, J., Sargin, E.: Deep neural networks for youtube recommendations. In: ACM Conference on Recommender Systems (RecSys) (2016)

16. Craswell, N., Zoeter, O., Taylor, M., Ramsey, B.: An experimental comparison of click position-bias models. In: International Conference on Web Search and Data Mining (WSDM) (2008)
17. Dato, D., et al.: Fast ranking with additive ensembles of oblivious and non-oblivious regression trees. ACM Trans. Inform. Syst. (TOIS) **35**(2), 1–31 (2016)
18. Diaz, F., White, R., Buscher, G., Liebling, D.: Robust models of mouse movement on dynamic web search results pages. In: International Conference on Information and Knowledge Management (CIKM) (2013)
19. Dupret, G.E., Piwowarski, B.: A user browsing model to predict search engine click data from past observations. In: International ACM SIGIR Conference on Research and Development in Information Retrieval (SIGIR) (2008)
20. Gomez-Uribe, C.A., Hunt, N.: The netflix recommender system: Algorithms, business value, and innovation. ACM Trans. Manage. Inform. Syst. (TMIS) **6**(4), 1–19 (2016)
21. Guo, F., et al.: Click chain model in web search. In: The World Wide Web Conference (WWW) (2009)
22. Guo, H., Yu, J., Liu, Q., Tang, R., Zhang, Y.: Pal: A position-bias aware learning framework for ctr prediction in live recommender systems. In: ACM Conference on Recommender Systems (RecSys) (2019)
23. Haldar, M., et al.: Improving deep learning for airbnb search. In: International Conference on Knowledge Discovery and Data Mining (SIGKDD) (2020)
24. Hofmann, K., Schuth, A., Whiteson, S., de Rijke, M.: Reusing historical interaction data for faster online learning to rank for ir. In: International Conference on Web Search and Data Mining (WSDM) (2013)
25. Hu, Z., Wang, Y., Peng, Q., Li, H.: Unbiased lambdamart: An unbiased pairwise learning-to-rank algorithm. In: The World Wide Web Conference (WWW) (2019)
26. Jagerman, R., Oosterhuis, H., de Rijke, M.: To model or to intervene: A comparison of counterfactual and online learning to rank from user interactions. In: International ACM SIGIR Conference on Research and Development in Information Retrieval (SIGIR) (2019)
27. Joachims, T., Granka, L., Pan, B., Hembrooke, H., Gay, G.: Accurately interpreting clickthrough data as implicit feedback. In: International ACM SIGIR Conference on Research and Development in Information Retrieval (SIGIR) (2005)
28. Joachims, T., Swaminathan, A., Schnabel, T.: Unbiased learning-to-rank with biased feedback. In: International Conference on Web Search and Data Mining (WSDM) (2017)
29. Ke, G., et al.: Lightgbm: A highly efficient gradient boosting decision tree. In: International Conference on Neural Information Processing Systems (NIPS) (2017)
30. Lin, J., et al.: A graph-enhanced click model for web search. In: International ACM SIGIR Conference on Research and Development in Information Retrieval (SIGIR) (2021)
31. Liu, T.Y., et al.: Learning to rank for information retrieval. Found. Trends Inf. Retr. **3**, 225–331 (2009)
32. Oosterhuis, H.: Reaching the end of unbiasedness: Uncovering implicit limitations of click-based learning to rank. In: International Conference on the Theory of Information Retrieval (ICTIR) (2022)
33. Oosterhuis, H., de Rijke, M.: Differentiable unbiased online learning to rank. In: International Conference on Information and Knowledge Management (CIKM) (2018)

34. Oosterhuis, H., de Rijke, M.: Policy-aware unbiased learning to rank for top-k rankings. In: International ACM SIGIR Conference on Research and Development in Information Retrieval (SIGIR) (2020)
35. Oosterhuis, H., de Rijke, M.: Unifying online and counterfactual learning to rank: A novel counterfactual estimator that effectively utilizes online interventions. In: International Conference on Web Search and Data Mining (WSDM) (2021)
36. Ovaisi, Z., Ahsan, R., Zhang, Y., Vasilaky, K., Zheleva, E.: Correcting for selection bias in learning-to-rank systems. In: The Web Conference (2020)
37. Qin, T., Liu, T.: Introducing letor 4.0 datasets (2013), 10.48550/arXiv. 1306.2597
38. Qin, Z., et al.: Are neural rankers still outperformed by gradient boosted decision trees? In: International Conference on Learning Representations (ICLR) (2021)
39. Richardson, M., Dominowska, E., Ragno, R.: Predicting clicks: Estimating the click-through rate for new ads. In: The World Wide Web Conference (WWW) (2007)
40. Saito, Y., Yaginuma, S., Nishino, Y., Sakata, H., Nakata, K.: Unbiased recommender learning from missing-not-at-random implicit feedback. In: International Conference on Web Search and Data Mining (WSDM) (2020)
41. Sanderson, M., et al.: Test collection based evaluation of information retrieval. Found. Trends Inf. Retr. 4, 247–375 (2010)
42. Sorokina, D., Cantu-Paz, E.: Amazon search: The joy of ranking products. In: International ACM SIGIR Conference on Research and Development in Information Retrieval (SIGIR) (2016)
43. Student: The probable error of a mean. Biometrika pp. 1–25 (1908)
44. Swaminathan, A., Joachims, T.: The self-normalized estimator for counterfactual learning. In: International Conference on Neural Information Processing Systems (NIPS) (2015)
45. Vardasbi, A., Oosterhuis, H., de Rijke, M.: When inverse propensity scoring does not work: Affine corrections for unbiased learning to rank. In: International Conference on Information and Knowledge Management (CIKM) (2020)
46. Vardasbi, A., de Rijke, M., Markov, I.: Cascade model-based propensity estimation for counterfactual learning to rank. In: International ACM SIGIR Conference on Research and Development in Information Retrieval (SIGIR) (2020)
47. Vardasbi, A., de Rijke, M., Markov, I.: Mixture-Based Correction for Position and Trust Bias in Counterfactual Learning to Rank (2021)
48. Wang, C., Liu, Y., Wang, M., Zhou, K., Nie, J.y., Ma, S.: Incorporating non-sequential behavior into click models. In: International ACM SIGIR Conference on Research and Development in Information Retrieval (SIGIR) (2015)
49. Wang, X., Bendersky, M., Metzler, D., Najork, M.: Learning to rank with selection bias in personal search. In: International ACM SIGIR Conference on Research and Development in Information Retrieval (SIGIR) (2016)
50. Wang, X., Golbandi, N., Bendersky, M., Metzler, D., Najork, M.: Position bias estimation for unbiased learning to rank in personal search. In: International Conference on Web Search and Data Mining (WSDM) (2018)
51. Wang, Y.X., Agarwal, A., Dudík, M.: Optimal and adaptive off-policy evaluation in contextual bandits. In: International Conference on Machine Learning (ICML) (2017)
52. Xie, X., et al.: Investigating examination behavior of image search users. In: International ACM SIGIR Conference on Research and Development in Information Retrieval (SIGIR) (2017)

53. Xie, X., Mao, J., de Rijke, M., Zhang, R., Zhang, M., Ma, S.: Constructing an interaction behavior model for web image search. In: International ACM SIGIR Conference on Research and Development in Information Retrieval (SIGIR) (2018)
54. Yan, L., Qin, Z., Zhuang, H., Wang, X., Bendersky, M., Najork, M.: Revisiting two-tower models for unbiased learning to rank. In: International ACM SIGIR Conference on Research and Development in Information Retrieval (SIGIR) (2022)
55. Zhao, Z., et al.: Recommending what video to watch next: A multitask ranking system. In: ACM Conference on Recommender Systems (RecSys) (2019)
56. Zhuang, H., et al.: Cross-positional attention for debiasing clicks. In: The Web Conference (2021)

Sentence Retrieval for Open-Ended Dialogue Using Dual Contextual Modeling

Itay Harel[1](✉), Hagai Taitelbaum[2], Idan Szpektor[2], and Oren Kurland[3]

[1] TSG IT Advanced Systems Ltd., Tel Aviv, Israel
itay.harel91@gmail.com
[2] Google Research, Tel Aviv, Israel
hagait@google.com, szpektor@google.com
[3] Technion—Israel Institute of Technology, Haifa, Israel
kurland@technion.ac.il

Abstract. We address the task of retrieving sentences for an open domain dialogue that contain information useful for generating the next turn. We propose several novel neural retrieval architectures based on dual contextual modeling: the dialogue context and the context of the sentence in its ambient document. The architectures utilize contextualized language models (BERT), fine-tuned on a large-scale dataset constructed from Reddit. We evaluate the models using a recently published dataset. The performance of our most effective model is substantially superior to that of strong baselines.

Keywords: Open domain dialogue · Dialogue retrieval · Sentence retrieval

1 Introduction

Throughout the last few years there has been a rapid increase in various tasks related to dialogue (conversational) systems [7,12,14,15,37,47]. Our work focuses on responses in an open-dialogue setup: two parties converse in turns on any number of topics with no restrictions to the topic shifts and type of discussion on each topic. In addition, the dialogue is not grounded to a specific document, in contrast to the setting used in some previous work (e.g., [28]). The task we pursue is to retrieve passages—specifically, sentences—from some document corpus that would be useful for generating the next response in a given dialogue; the response can be written either by humans or by conditional generative language models [12,17,34].

There has been much research effort on utilizing information induced from the context of the last turn in the dialogue—henceforth referred to as *dialogue context*—so as to retrieve a response from a corpus of available responses [4,20, 35,38,40,45,46,48]. However, these models address complete responses as the retrieved items. In our setting, the retrieved items are sentences from documents,

---

I. Harel—Work done while at the Technion.

© The Author(s), under exclusive license to Springer Nature Switzerland AG 2023
J. Kamps et al. (Eds.): ECIR 2023, LNCS 13980, pp. 426–440, 2023.
https://doi.org/10.1007/978-3-031-28244-7_27

which may aid in writing a complete response. Unlike full responses, sentences usually do not contain all the information needed to effectively estimate their relevance to an information need. Indeed, there is a line of work on ad hoc sentence retrieval [13,30] and question answering [18,24,41] that demonstrated the clear merits of using information induced from the document containing the sentence, henceforth referred to as *sentence context*.

To address sentence retrieval in a dialogue setting, we present a suite of novel approaches that employ dual contextual modeling: they utilize information not only from the dialogue context but also from the context of candidate sentences to be retrieved; specifically, from the documents that contain them. We are not aware of previous work on conversational search that utilizes the context of retrieved sentences. Using the context of the dialogue is important for modeling latent information needs that were explicitly or implicitly mentioned only in previous dialogue turns. Using the context of the sentence in its ambient document is important for inducing an enriched representation of the sentence which can help, for example, to fill in topical and referential information missing from the sentence.

Our sentence retrieval approaches employ the BERT [10] language model, fine-tuned for simultaneous modeling of the context of the last turn in the dialogue and the context of a candidate sentence for retrieval. We propose three different BERT-based architectures that differ in the way context in the dialogue and in the document are modeled and interact with each other. While our main architectural novelty lies in the study of the dialogue/sentence context interaction, some of the dialogue context modeling techniques we employ are also novel to this task.

We evaluated our models using a recently published dataset for sentence retrieval for open-domain dialogues [19]. The dataset was created from Reddit. It includes human generated relevance labels for sentences with respect to dialogues. As in [19], we used weakly supervised (pseudo) relevance labels for training our models.

We contrast the performance of our models with that of a wide suite of strong reference comparisons. The retrieval performance of our best performing approach is substantially better than that of the baselines. We also show that while using only the dialogue context results in performance superior to that of using only the sentence context, using them both is of clear merit. In addition, we study the performance effect of the type of document context used for sentences and the length of the context used from the dialogue.

To summarize, we address a research challenge that has not attracted much research attention thus far: retrieving sentences that contain information useful for generating the next turn in an open-domain dialogue. This is in contrast to retrieving responses and/or generating them, and to conversational retrieval where the information need is explicitly expressed via queries or questions. On the model side, our work is the first, to the best of our knowledge, to model both the dialogue context and the context of candidate sentences to be retrieved; specifically, using neural architectures that utilize a pre-trained language model.

2   Related Work

The two main lines of work on open domain dialogue-based systems [21] are response generation [12,17,34,49] and response selection [4,20,35,38,40,45,46, 48]; some of these methods are hybrid selection/generation approaches. Some recent generative dialogue systems include a retrieval component that generates a standalone query from the dialogue context against a fixed search engine (e.g. Bing or Google) [16,23,42]. Response selection is a retrieval task of ranking candidate full responses from a given pool. In contrast, our task is to retrieve from a document corpus sentences that could serve as a basis for generating a response, optimizing the retrieval model.

Related to our task is conversational search [7,15,36,47]. The goal of this task is to retrieve answers to questions posted in the conversation or to retrieve passages/documents that pertain to the information needs expressed in it. In our setting, we do not make any assumptions on the type of response to be generated from retrieved sentences. It could be an answer, an argument, or even a question. Consequently, the types of information needs our retrieval models have to satisfy are not necessarily explicitly mentioned in the dialogues (e.g., in the form of a question); they could be quite evolved, and should be inferred from the dialogue.

The last turn in a dialogue is often used as the basis for response selection or passage/answer retrieval. A large body of work utilized information induced from the dialogue *context* – the turns preceding the last one – to improve ranking. There are approaches that reformulate the last turn [26], expand it using terms selected from the context [27,43], expand it using entire previous turns [29,34,37], or use the context for cross referencing the last turn [40]. Yan et al. [45] proposed to create multiple queries from the context, assign a retrieval score to a candidate response w.r.t. each query, and fuse the retrieved scores. We demonstrate the merits of our joint representation approach w.r.t. a representative turn expansion method [43] and to Yan et al.'s [45] fusion approach.

Other models for dialogue-based retrieval include the dialogue context as part of a retrieval neural network. Several works [38,46,48] use a hierarchical architecture for propagating information from previous turns to the last turn representation. Qu et al. [35] embed selected previous answers in the conversation as an additional layer in BERT. In contrast, we focus on early cross-attention of all context in the input, simultaneously modeling the dialogue context and the context of the sentences to be retrieved.

Passage context from the ambient document was used for non-conversational passage retrieval [13,30] and question answering [18,24,41], but there was no dialogue context to utilize in contrast to our work. As already mentioned, there is much work on utilizing dialogue context for dialogue-based retrieval [26,27, 29,34,37,40,43], and we use some of these methods as reference comparisons, but the passage (response) context was not utilized in this line of work.

3 Retrieval Framework for Open Dialogues

Suppose that two parties are holding a dialogue g which is an open-ended conversation. Open domain means that there are no restrictions about the topics discussed in g and their shifts.

The parties converse in turns: $g \overset{def}{=} < t_1, \ldots, t_n >$; t_1 is the first turn and t_n is the current (last) turn. Herein, we refer to a turn and the text it contains interchangeably. Our goal is to retrieve *relevant* sentences from some document corpus C, where *relevance* means that the sentence contains information that can be used to generate the next turn, t_{n+1}. We address this sentence retrieval task via a two-stage ranking approach: an initial ranker (see Sect. 4 for details) followed by a more computationally-intensive ranker that reranks the top-k retrieved sentences and is our focus.

In open-ended dialogues there is often no explicit expression of an information need; e.g., a query or a question. We assume that the current turn, t_n, expresses to some extent the information need since t_{n+1} is a response to t_n. Because turns can be short, preceding turns in g are often used as the context of t_n in prior work on conversational search and dialogue-based retrieval [20,35,40, 45,46,48]. Accordingly, we define the sequence $CX(t_n) \overset{def}{=} t_{n-h}, \ldots, t_{n-1}$ to be the dialogue (historical) context, where h is a free parameter. We treat the sequence $t_{n-h}, \ldots, t_n$ as a *pseudo query*, Q, which is used to express the presumed information need[1].

Standard ad hoc sentence retrieval based on query-sentence similarities is prone to vocabulary mismatch since both the queries and sentences are relatively short. We use the BERT [10] language model to compare the pseudo query, Q, with a sentence, which should ameliorate the effects of token-level mismatch. We note that the pseudo query Q is not as short as queries in ad hoc retrieval. Nevertheless, we hypothesize that utilizing information from the document containing the sentence, which was found useful in ad hoc sentence retrieval [13,30] and question answering [18,24,41], will also be of merit in sentence retrieval for open dialogue.

Specifically, we define the context of sentence s in its ambient document as a sequence of m sentences selected from the document: $CX(s) \overset{def}{=} s_1, \ldots, s_m$; m is a free parameter. These sentences are ordered by their relative ordering in the document, but they need not be adjacent[2] to each other in the document. We treat the ordered sequence composed of s and its context as the *pseudo sentence*: $S \overset{def}{=} s_1, \ldots, s, \ldots, s_m$ (s may appear before s_1 or after s_m).

In what follows, we describe estimates for the relevance of sentence s with respect to dialogue g where relevance means, as noted above, inclusion of information useful for generating the next turn in the dialogue. The estimates are based on modeling the relation between the pseudo sentence S and the pseudo query Q.

---

[1] If $n - 1 < h$, we set $Q \overset{def}{=} t_1, \ldots, t_n$. If h=0, $Q \overset{def}{=} t_n$, $CX(t_n) \overset{def}{=} \{\}$.
[2] We evaluate a few approaches for selecting the sentences in $CX(s)$ in Sect. 5.

3.1 Sentence Retrieval Methods

We present three architectures for sentence retrieval in dialogue context that are based on matching between Q and S. These architectures utilize BERT for embedding texts, taking its pooler output vector as the single output, denoted by v^{BERT}.

From Dense Vectors to a Sentence Relevance Score. The architectures we present below utilize a generic neural component that induces a sentence relevance score from a set of k dense vectors. The vectors, denoted $v_1, \ldots, v_k$, are the outcome of representing parts of Q and S and/or their matches using the architectures.

Following work on estimating document relevance for ad hoc document retrieval using passage-level representations [25], we train an encoder-only Transformer [10] whose output is fed into a softmax function that induces a relevance score for sentence s with respect to pseudo query Q:

$$Score(s|v_1, \ldots, v_k) = Softmax(W_{score} \, v_{out} + b_{score}), \qquad (1)$$

$$v_{out} = TRANSFORMER_{enc}(v_0, \ldots, v_k)[0], \qquad (2)$$

where: a) W_{score} and b_{score} are learned parameters; b) v_0 is the word embedding of the [CLS] token in BERT, which is prepended to the vector sequence $v_1, \ldots, v_k$; and c) v_{out} is the contextual embedding of the [CLS] token v_0 in the topmost Transformer layer[3]. Figures 1, 2 and 3 depict three models with this high-level architecture.

Architectures. We next present three BERT-based architectures. The output of each architecture is a sequence of dense vectors which is the input to $Score(s|v_1, \ldots, v_k)$ in Eq. 1 to compute the final sentence score.

The Tower architecture (Fig. 1). Two instances of BERT with shared weights produce two output vectors to compute $Score(s|v_1^{BERT}, v_2^{BERT})$ in Eq. 1. The input to the first BERT is the pseudo query Q with separating tokens between the turns: "[CLS] t_{n-h} [SEP] ... [SEP] t_n [SEP]"[4]. The second BERT is fed with the pseudo sentence S to be scored: "[CLS] s_1 [SEP] ... s [SEP] ... s_m [SEP]". This architecture is similar to Dense Passage Retrieval (DPR) [23].

The Hierarchical architecture (Fig. 2). A potential drawback of the Tower architecture (cf., [33]) is that matching the pseudo query and the pseudo sentence is performed after their dense representations were independently induced. To address this, we present the Hierarchical architecture, which uses BERT to

---

[3] We tried replacing the Transformer-based embedding in Eq. 2 with a feed-forward network with the same number of parameters, but this resulted in inferior performance.

[4] We also tested a simpler scoring approach (without fine tuning), $Cosine(v_1, v_2)$, which performed significantly worse.

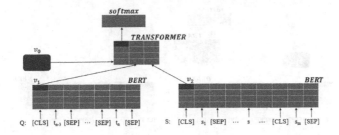

Fig. 1. The Tower model with dialog history $h = 3$.

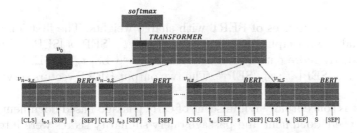

Fig. 2. The Hierarchical model with dialog history $h = 3$.

induce joint representations of each turn in the pseudo query with parts of the pseudo sentence.

This model uses $2*(h+1)$ instances of BERT with shared weights, constructed as follows. For each turn $i \in \{n - h, \ldots, n\}$ in Q, the model computes output $v_{i,s}^{BERT}$ by feeding BERT with turn i and the sentence s: "[CLS] t_i [SEP] s". Similarly, the output $v_{i,S}^{BERT}$ is computed by feeding BERT turn i and pseudo sentence S: "[CLS] t_i [SEP] s_1 [SEP] $\ldots s$ [SEP] $\ldots s_m$[SEP]". The output vectors $\{v_{n,s}^{BERT}, v_{n,S}^{BERT} \ldots v_{n-h,s}^{BERT}, v_{n-h,S}^{BERT}\}$ are fed as input to Eq. 1. This architecture is inspired by the PARADE ad hoc document retrieval model [25]. Unlike PARADE, we enrich all embeddings with positional encoding. The goal is to model sequential information, e.g., the order of turns in the dialogue.

The QJoint architecture (Fig. 3). The Hierarchical architecture enables early joint representations for each turn in the pseudo query Q and the pseudo sentence S. Still, turns are used independently to derive intermediate representations. We next present the QJoint architecture that represents jointly all turns in Q. The goal is to cross-attend the inter-relations between the turns in Q and their relations with S as early as possible.

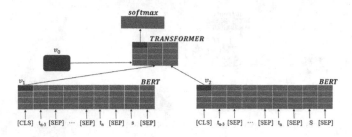

Fig. 3. The QJoint model with dialog history $h = 3$.

We use two instances of BERT with shared weights. The first jointly represents Q and s, with input "[CLS] t_{n-h} [SEP] ...t_n [SEP] s [SEP]". The second instance jointly represents Q and S, with input "[CLS] t_{n-h} [SEP] ...t_n [SEP] s_1 [SEP] ...s [SEP] ...s_m [SEP]". The two output BERT vectors serve as input to Eq. 2.

QJoint is conceptually reminiscent of early passage-based document retrieval methods where document and passage query similarity scores were interpolated [2]. Here, the pseudo sentence S is viewed as the document and the sentence s as the passage and their query matching induced using the BERT models is interpolated via a Transformer. The sentence context serves as a means to disambiguate, resolve references and offer global document topicality that may be missing from a single sentence. Yet, the BERT model that focuses on matching only the single sentence with the pseudo query offers a differentiator for ranking two consecutive sentences, which would share much of their pseudo sentence content.

Neural Reference Comparisons. As noted in Sect. 1, previous work on conversational search and dialogue-based retrieval did not utilize the context of candidate passages in contrast to our architectures. We use several such BERT-based sentence retrieval methods as reference comparisons.

RANK$_{BERT}$. This method, which was the best performing in [19] for sentence retrieval for open-domain dialogues, takes BERT with input "[CLS] q [SEP] s [SEP]" and uses its output as input to Eq. 1 (which includes a Transformer layer). In this method, q is set to be turn t_n and no context for the sentence s is utilized; see QuReTeC and CONCAT next for different q settings.

QuReTeC. The Query Resolution by Term Classification method [43] (QuReTeC in short) is a representative of methods (e.g., [27]) that use explicit term-based expansion, based on the dialog history, to enrich the current (last) turn t_n. Specifically, it applies a token-level classifier, utilizing turns in $CX(t_n)$ (with a [SEP] token separating between turns), to select a few tokens that will be added to t_n. The resultant text is provided as input q to RANK$_{BERT}$.

CONCAT. As an alternative to the term selection approach, represented by QuReTeC, we consider the CONCAT method [27] which uses all $CX(t_n)$ when constructing the input q to RANK$_{\text{BERT}}$: "[CLS] t_{n-h} [SEP] ... [SEP] t_n [SEP] s [SEP]" . We note that CONCAT is essentially a special case of QJoint (Sect. 3.1) where no sentence context is used.

External Fusion. The Hierarchical architecture (Sect. 3.1) fuses information induced from the turns in the pseudo query by aggregating turn/sentence representations using a Transformer. In contrast, QJoint, and its special case CONCAT, perform the fusion via a joint representation of all the turns in Q and the sentence.

We also consider an external fusion approach which fuses information induced from the turns in Q at the retrieval-score level. We employ RANK$_{\text{BERT}}$ to assign a score to each sentence s in an initially retrieved sentence list with respect to each turn in Q. Hence, each turn t_i induces a ranked list of sentences L_i. Let $rank_{L_i}(s)$ be s's rank in L_i; the highest rank is 1. We use reciprocal rank fusion (RRF) [6] to fuse the lists $\{L_i\}$: $ExtFuse(s) \stackrel{def}{=} \sum_{L_i} \mu_i \frac{1}{\nu + rank_{L_i}(s)}$, where ν is a free parameter and μ_i is a uniform weight; linear and exponential decay weights did not yield improvements. Fusion was also used in [45] at the retrieval score level for conversational search, but contextualized language models were not used.

4   Experimental Setting

Dataset and Evaluation Measures. We use a recent dataset of sentence retrieval for open-ended dialogues [19][5]. Dialogues were extracted from Reddit, and sentences were retrieved from Wikipedia. The test set contains 846 diagloues, each accompanied with an initially retrieved list of 50 sentences judged by crowd workers for relevance. The initial ranker, henceforth *Initial Ranker*, is based on unigram language models and utilizes the dialogue context [19]. All sentence retrieval methods that we evaluate re-rank the initial sentence list provided in the dataset. We use Harel et al.'s [19] 50 random equal-sized splits of the test dialogues to validation and test sets; the validation set is used for hyperparamter tuning. We report the average and standard deviation over the 50 test sets of mean average precision (MAP), NDCG of the 5 highest ranked sentences (NDCG@5) and mean reciprocal rank (MRR). The two tailed permutation (randomization) test with $10,000$ random permutations and $p \leq 0.05$ was used to determine statistical significance. Bonferroni correction is applied for multiple hypothesize testing.

We followed the guidelines on the weakly-supervised training data collection from [19], which showed much merit. Specifically, the sentences in the initial list retrieved for a dialogue were assigned pseudo relevance labels using a fusion

---

[5] https://github.com/SIGIR-2022/A-Dataset-for-Sentence-Retrieval-for-Open-Ended-Dialogues. git.

approach. Following these guidelines, we obtained ~73,000 dialogues with weakly annotated sentences, which were used to fine tune a basic BERT-based sentence retrieval model.

Other datasets are not a good fit for training and evaluating our models since (i) they do not include open-domain dialogues (e.g., TREC's CAsT datasets [7–9][6], CoQA [37], DoQA [3] and QuAC [4]) and/or (ii) they do not include the document context for training a dual context model.

Model and Training Settings. All neural models (Sect. 3.1) were fine-tuned end2end on the weakly supervised training set, unless stated otherwise. For a single text embedding in the Tower architecture, pre-trained BERT-Large [10] is used as a starting point. To embed a pair of texts, e.g., in $RANK_{BERT}$ and QJoint, as starting point we used a pre-trained BERT that was fine-tuned for ad hoc passage retrieval on the MS MARCO dataset [31]. We fine-tuned it using the $RANK_{BERT}$ architecture[7]; q and s were set to a query and a passage in MS MARCO, respectively, and trained with pointwise classification loss[8] [32].

We implemented and trained the QuReTeC model using the hyperparameter values detailed in [43] for 10 epochs. When generating the resolved queries, we applied the constraints mentioned below to the dialogue context; i.e., we set $h = 3$ with maximum of 70 tokens per turn. Then, $RANK_{BERT}$ was utilized for inference on the resolved queries. We tested all QuReTeC variants (different batch sizes, learning rates and number of epochs), each with the $RANK_{BERT}$ variant that was the best performing model in most of the validation splits.

Modeling the Pseudo Sentence Context. We tested three alternatives for modeling $CX(s)$, the context of sentence s in its ambient document: (i) *LocalSurround.* the sentence that precedes and the sentence that follows s, (ii) *LocalPrev.* the two sentences that precede s; and (iii) *Global.* the two sentences in the document whose TF-IDF vectors are most similar (via Cosine similarity) to the TF-IDF vector of s. Sentences in (i) and (ii) were limited to passage boundaries.

Unless stated otherwise, we used *LocalSurround* in our models, since it performed best in terms of MAP. We use only two sentences as context due to BERT's input-length limitation. If the input to BERT should still be truncated, first the sentences in the context are truncated, and only then the sentence s.

Bag-of-Terms Reference Comparisons. In addition to the neural reference comparisons described in Sect. 3.1, and the unigram language-model-based Initial Ranker, we applied **Okapi BM25** [39] on the last turn t_n as a reference comparison.

Hyperparameter Tuning. The values of the following hyperparameters were optimized for MAP over the validation sets. Okapi BM25's $k_1 \in \{1.2, 2, 4, 8, 12\}$

---

[6] In addition, these datasets include a too small number of dialogues which does not allow for effective training of the proposed architectures, even when used for weak supervision.

[7] Without the additional transformer in Eq. 2.

[8] Training with pairwise loss showed no improvement.

Table 1. Architectures which utilize both the dialogue and the sentence context. Statistical significance w.r.t. Tower and Hierarchical is marked with 't' and 'h', respectively.

	MAP	NDCG@5	MRR
Tower	$.212^{\pm.007}$	$.298^{\pm.015}$	$.291^{\pm.014}$
Hierarchical	$.451_t^{\pm.010}$	$.611_t^{\pm.012}$	$.588_t^{\pm.013}$
QJoint	$\mathbf{.477}_{th}^{\pm.010}$	$\mathbf{.644}_{th}^{\pm.012}$	$\mathbf{.609}_{th}^{\pm.013}$

and $b \in \{0.25, 0.5, 0.75, 1\}$. RRF's (external fusion) $\nu \in \{0, 10, 60, 100\}$. All BERT-based models were trained using the Adam optimizer with learning rate $\in \{3e-6, 3e-8\}$ and batch size $\in \{8, 16\}$. RANK$_{BERT}$ that is the starting point of all these models was fine tuned as in [32].

All models were trained for 10 epochs on Google's TPU[9] v3-8 and the best model snapshot was selected based on the validation set. The number of Transformer layers (Sect. 3.1) was set to 2 in all related architectures following [25]. The maximum sequence length for all BERT-based models is 512. The dialogue context length h is set to 3. (We analyze the effect of h in Sect. 5.)

5 Results

Main Result. Table 1 compares our architectures, which utilize both the dialogue context and the sentence context. We see that Hierarchical outperforms Tower. This shows that jointly modeling matched texts, in our case the pseudo query and the sentence, is superior to modeling the interaction between texts only at the top-most neural layer. This finding is aligned with previous reports on semantic matching [11, 22]. We also see that QJoint is the best performing model. This attests to the downside of "breaking up" the pseudo query at the lower representation levels of the network while early cross-representation between the pseudo query and the pseudo sentence results in higher-quality semantic retrieval modeling. One potential benefit of Hierarchical is increased input capacity, since concatenating both the query and its context and the sentence and its context in QJoint may exceed the input size limit and incur penalty due to truncation.

Table 2 compares our most effective architecture, QJoint, with the neural and bag-of-terms baselines. The main difference between QJoint and the other models is that QJoint utilizes both the dialogue context and the sentence context, while the other methods utilize only the dialogue context, with the exception of BM25, as is the case in all prior work as noted in Sect. 1.

We see in Table 2 that all trained neural methods significantly outperform the Initial Ranker and Okapi BM25. The superiority of ExtFuse (external fusion) to QuReTeC can potentially be attributed to the fact that it compares all the turns in the dialogue context with the sentence rather than "fuses" several selected

---
[9] https://cloud.google.com/tpu/.

Table 2. The best performing QJoint compared to reference models. Statistical significance w.r.t. Initial Ranker and QJoint is marked with 'i' and 'q' respectively.

	MAP	NDCG@5	MRR
Okapi BM25	$.185_{iq}^{\pm.006}$	$.259_{iq}^{\pm.010}$	$.258_{iq}^{\pm.009}$
Initial Ranker	$.238^{\pm.007}$	$.355^{\pm.012}$	$.353^{\pm.012}$
QuReTeC	$.375_{iq}^{\pm.009}$	$.543_{iq}^{\pm.014}$	$.517_{iq}^{\pm.013}$
ExtFuse	$.436_{iq}^{\pm.011}$	$.606_{iq}^{\pm.013}$	$.582_{iq}^{\pm.014}$
CONCAT	$.470_{iq}^{\pm.009}$	$.635_{iq}^{\pm.012}$	$.607_{i}^{\pm.012}$
QJoint	$\mathbf{.477^{\pm.010}}$	$\mathbf{.644^{\pm.012}}$	$\mathbf{.609^{\pm.013}}$

Table 3. QJoint with no context, only with sentence context, only with dialogue context, and both. The corresponding statistical significance marks are 'n', 's' and 'h', respectively.

Context Used	MAP	NDCG@5	MRR
None	$.354^{\pm.009}$	$.481^{\pm.014}$	$.468^{\pm.013}$
Sentence context only	$.351^{\pm.008}$	$.478^{\pm.014}$	$.463^{\pm.013}$
Dialogue context only	$.470_{ns}^{\pm.009}$	$.635_{ns}^{\pm.012}$	$.607_{ns}^{\pm.012}$
Both	$\mathbf{.477}_{nsh}^{\pm.010}$	$\mathbf{.644}_{nsh}^{\pm.012}$	$\mathbf{.609}_{ns}^{\pm.013}$

terms with the last turn to yield a single query compared with the sentence. It is also clear that CONCAT outperforms ExtFuse, which attests to the merit of using a joint representation for the entire pseudo query and the sentence compared to fusing retrieval scores attained from matching parts of the pseudo query with the sentence. We also point that CONCAT improves over Hierarchical (see Table 1), which does utilize the sentence context. This indicates that Hierarchical's use of the sentence context does not compensate for the performance drop due to breaking up the pseudo query in the model's lower layers. Finally, Table 2 shows that QJoint consistently and statistically significantly outperforms all other methods. The improvement over CONCAT shows the merit of utilizing the sentence context, since CONCAT is a special case of QJoint that does not use it.

Analysis of Retrieval Contexts. To further study the merit of using both the dialogue and sentence contexts, we trained the QJoint model (i) with no context, (ii) only with sentence context, (iii) only with dialogue context and (iv) with both (the full model).

Table 3 shows that using only the sentence context without the dialogue context yields performance that is statistically significantly indistiguishable from that of not using context at all, and statistically significantly inferior to using only the dialogue context. Yet, using both dialogue and sentence contexts yields statistically significant improvements over using just the dialogue context.

Table 4. Sentence context in QJoint. 's' and 'p': statistical significance w.r.t. *Local-Surround* and *LocalPrev*, respectively.

	MAP	NDCG@5	MRR
QJoint $_{LocalSurround}$	$\mathbf{.477}^{\pm.010}$	$\mathbf{.644}^{\pm.012}$	$.609^{\pm.013}$
QJoint $_{LocalPrev}$	$.476^{\pm.010}$	$.639^{\pm.013}$	$\mathbf{.612}^{\pm.015}$
QJoint $_{Global}$	$.467^{\pm.011}_{sp}$	$.623^{\pm.013}_{sp}$	$.601^{\pm.015}_{sp}$

Table 5. The effect of the number of past turns (h) used as dialogue context on QJoint's performance. Statistical significance w.r.t. $h = 0, 1, 2$ is marked with 0, 1 and 2, respectively.

	MAP	NDCG@5	MRR
QJoint ($h = 0$)	$.351^{\pm.008}$	$.478^{\pm.014}$	$.463^{\pm.013}$
QJoint ($h = 1$)	$.430^{\pm.009}_{0}$	$.589^{\pm.014}_{0}$	$.566^{\pm.014}_{0}$
QJoint ($h = 2$)	$.472^{\pm.010}_{01}$	$.638^{\pm.012}_{01}$	$.604^{\pm.012}_{01}$
QJoint ($h = 3$)	$\mathbf{.477}^{\pm.010}_{01}$	$\mathbf{.644}^{\pm.012}_{01}$	$\mathbf{.609}^{\pm.013}_{012}$

This result indicates that while the sentence context does not help by itself, it is beneficial when used together with the dialogue context.

Thusfar, the context of sentence s, $CX(s)$, was the two sentences that surround it in the document. Table 4 presents the performance of our best method, QJoint, with the alternative sentence contexts described in Sect. 4. We see that both *LocalSurround* and *LocalPrev* statistically significantly outperform *Global*, which is aligned with findings in some work on question answering [18]. This attests to the merits of using the "local context"; i.e., the sentences around the candidate sentence. There are no statistically significant differences between *LocalSurround* and *LocalPrev*, providing flexibility to choose local context based on other constraints; e.g., input size.

Heretofore, we used the $h = 3$ turns that precede the current (last) turn in the dialogue as the dialogue context. Table 5 shows that reducing the number of previous turns for dialogue context results in decreasing performance. The smaller difference between $h = 2$ and $h = 3$ is due to relatively few dialogues in the test set with history longer than 2 turns (about 15%). For these dialogues, the difference in performance between $h = 2$ and $h = 3$ is similar to that between $h = 1$ and $h = 2$.

6   Conclusions and Future Work

We addressed the task of retrieving sentences that contain information useful for generating the next turn in an open-ended dialogue. Our approaches utilize both the dialogue context and the context of candidate sentences in their ambient documents. Specifically, we presented architectures that utilize various hierarchies of the match between a sentence, its context and the dialogue context. Empirical evaluation demonstrated the merits of our best performing approach.

We intend to explore the use of transformers for long texts [1,5,44] to overcome the input size limitation. We also plan to ground generative language models with our retrieval models and study the conversations that emerge from such grounding.

Acknowledgements. We thank the reviewers for their comments. This work was supported in part by a grant from Google.

References

1. Beltagy, I., Peters, M.E., Cohan, A.: Longformer: The long-document transformer. CoRR abs/2004.05150 (2020), https://arxiv.org/abs/2004.05150
2. Callan, J.P.: Passage-level evidence in document retrieval. In: Proceedings of SIGIR, pp. 302–310 (1994)
3. Campos, J.A., Otegi, A., Soroa, A., Deriu, J., Cieliebak, M., Agirre, E.: DoQA - accessing domain-specific FAQs via conversational QA. In: Proceedings of the 58th Annual Meeting of the Association for Computational Linguistics, pp. 7302–7314 (2020)
4. Choi, E., et al.: QuAC: question answering in context. In: Proceedings of the 2018 Conference on Empirical Methods in Natural Language Processing, pp. 2174–2184 (2018)
5. Choromanski, K., et al.: Rethinking attention with performers. CoRR abs/2009.14794 (2020). https://arxiv.org/abs/2009.14794
6. Cormack, G.V., Clarke, C.L., Buettcher, S.: Reciprocal rank fusion outperforms condorcet and individual rank learning methods. In: Proceedings of the 32nd international ACM SIGIR Conference On Research and Development in Information Retrieval, pp. 758–759 (2009)
7. Dalton, J., Xiong, C., Callan, J.: TREC cast 2019: the conversational assistance track overview (2020)
8. Dalton, J., Xiong, C., Callan, J.: Cast 2020: the conversational assistance track overview. In: Proceedings of TREC (2021)
9. Dalton, J., Xiong, C., Callan, J.: TREC cast 2021: the conversational assistance track overview. In: Proceedings of TREC (2022)
10. Devlin, J., Chang, M., Lee, K., Toutanova, K.: BERT: pre-training of deep bidirectional transformers for language understanding. CoRR abs/1810.04805 (2018)
11. Devlin, J., Chang, M.W., Lee, K., Toutanova, K.: BERT: pre-training of deep bidirectional transformers for language understanding. In: Proceedings of the 2019 Conference of the North American Chapter of the Association for Computational Linguistics: Human Language Technologies, Volume 1 (Long and Short Papers), pp. 4171–4186 (2019)
12. Dinan, E., Roller, S., Shuster, K., Fan, A., Auli, M., Weston, J.: Wizard of wikipedia: knowledge-powered conversational agents. In: 7th International Conference on Learning Representations, ICLR 2019, New Orleans, LA, USA, 6–9 May 2019 (2019)
13. Fernández, R.T., Losada, D.E., Azzopardi, L.: Extending the language modeling framework for sentence retrieval to include local context. Inf. Retr. **14**(4), 355–389 (2011)

14. Gao, J., Galley, M., Li, L.: Neural approaches to conversational AI. CoRR (2018)
15. Gao, J., Xiong, C., Bennett, P., Craswell, N.: Neural approaches to conversational information retrieval. CoRR abs/2201.05176 (2022)
16. Glaese, A., et al.: Improving alignment of dialogue agents via targeted human judgements (2022)
17. Gopalakrishnan, K., et al.: Topical-Chat: towards Knowledge-Grounded Open-Domain Conversations. In: Proceedings Interspeech 2019, pp. 1891–1895 (2019). https://doi.org/10.21437/Interspeech.2019-3079
18. Han, R., Soldaini, L., Moschitti, A.: Modeling context in answer sentence selection systems on a latency budget. CoRR abs/2101.12093 (2021)
19. Harel, I., Taitelbaum, H., Szpektor, I., Kurland, O.: A dataset for sentence retrieval for open-ended dialogues. CoRR (2022)
20. Huang, H., Choi, E., Yih, W.: Flowqa: Grasping flow in history for conversational machine comprehension. In: Proceedings of ICLR (2019)
21. Huang, M., Zhu, X., Gao, J.: Challenges in building intelligent open-domain dialog systems. ACM Trans. Inf. Syst. **38**(3), 1–32 (2020)
22. Humeau, S., Shuster, K., Lachaux, M., Weston, J.: Poly-encoders: Architectures and pre-training strategies for fast and accurate multi-sentence scoring. In: 8th International Conference on Learning Representations, ICLR 2020, Addis Ababa, Ethiopia, 26–30 April 2020 (2020)
23. Komeili, M., Shuster, K., Weston, J.: Internet-augmented dialogue generation. In: Proceedings of the 60th Annual Meeting of the Association for Computational Linguistics (Volume 1: Long Papers), pp. 8460–8478 (2022)
24. Lauriola, I., Moschitti, A.: Answer sentence selection using local and global context in transformer models. In: Proceedings of ECIR, pp. 298–312 (2021)
25. Li, C., Yates, A., MacAvaney, S., He, B., Sun, Y.: Parade: Passage representation aggregation for document reranking. arXiv preprint arXiv:2008.09093 (2020)
26. Lin, S., Yang, J., Nogueira, R., Tsai, M., Wang, C., Lin, J.: Conversational question reformulation via sequence-to-sequence architectures and pretrained language models (2020)
27. Lin, S., Yang, J., Nogueira, R., Tsai, M., Wang, C., Lin, J.: Query reformulation using query history for passage retrieval in conversational search. CoRR (2020)
28. Ma, L., Zhang, W., Li, M., Liu, T.: A survey of document grounded dialogue systems (DGDS). CoRR abs/2004.13818 (2020)
29. Mehrotra, S., Yates, A.: MPII at TREC cast 2019: incoporating query context into a BERT re-ranker. In: Voorhees, E.M., Ellis, A. (eds.) Proceedings of TREC (2019)
30. Murdock, V., Croft, W.B.: A translation model for sentence retrieval. In: Proceedings of HLT/EMNLP, pp. 684–695 (2005)
31. Nguyen, T., Rosenberg, M., Song, X., Gao, J., Tiwary, S., Majumder, R., Deng, L.: MS MARCO: a human-generated machine reading comprehension dataset (2016)
32. Nogueira, R., Cho, K.: Passage re-ranking with BERT. arXiv preprint arXiv:1901.04085 (2019)
33. Qiao, Y., Xiong, C., Liu, Z., Liu, Z.: Understanding the behaviors of BERT in ranking. CoRR (2019)
34. Qin, L., et al.: Conversing by reading: Contentful neural conversation with on-demand machine reading. In: Proceedings of the 57th Annual Meeting of the Association for Computational Linguistics, pp. 5427–5436. Association for Computational Linguistics, Florence, Italy (2019). https://doi.org/10.18653/v1/P19-1539. https://www.aclweb.org/anthology/P19-1539

35. Qu, C., Yang, L., Qiu, M., Croft, W.B., Zhang, Y., Iyyer, M.: BERT with history answer embedding for conversational question answering. In: Proceedings SIGIR, pp. 1133–1136

36. Radlinski, F., Craswell, N.: A theoretical framework for conversational search. In: Proceedings of CHIIR, pp. 117–126 (2017)

37. Reddy, S., Chen, D., Manning, C.D.: CoQa: a conversational question answering challenge. Trans. Assoc. Comput. Linguistics 7, 249–266 (2019)

38. Ren, G., Ni, X., Malik, M., Ke, Q.: Conversational query understanding using sequence to sequence modeling. In: Proceedings of WWW, pp. 1715–1724 (2018)

39. Robertson, S., Walker, S., Jones, S., Hancock-Beaulieu, M., Gatford, M.: Okapi at trec-3. In: Proceedings of TREC-3, pp. 109–126 (1995)

40. Stamatis, V., Azzopardi, L., Wilson, A.: VES team at TREC conversational assistance track (cast) 2019. In: Proceedings of TREC (2019)

41. Tan, C., et al.: Context-aware answer sentence selection with hierarchical gated recurrent neural networks. IEEE ACM Trans. Audio Speech Lang. Process. 26(3), 540–549 (2018)

42. Thoppilan, R., et al.: LaMDA: language models for dialog applications. arXiv preprint arXiv:2201.08239 (2022)

43. Voskarides, N., Li, D., Ren, P., Kanoulas, E., de Rijke, M.: Query resolution for conversational search with limited supervision. In: Proceedings of SIGIR, pp. 921–930 (2020)

44. Wang, S., Li, B.Z., Khabsa, M., Fang, H., Ma, H.: Linformer: self-attention with linear complexity. CoRR abs/2006.04768 (2020). https://arxiv.org/abs/2006.04768

45. Yan, R., Song, Y., Wu, H.: Learning to respond with deep neural networks for retrieval-based human-computer conversation system. In: Proceedings of SIGIR, pp. 55–64 (2016)

46. Yan, R., Zhao, D.: Coupled context modeling for deep chit-chat: towards conversations between human and computer. In: Guo, Y., Farooq, F. (eds.) Proceedings of SIGKDD, pp. 2574–2583 (2018)

47. Zamani, H., Trippas, J.R., Dalton, J., Radlinski, F.: Conversational information seeking. CoRR (2022)

48. Zhang, Z., Li, J., Zhu, P., Zhao, H., Liu, G.: Modeling multi-turn conversation with deep utterance aggregation. In: Proceedings of COLING, pp. 3740–3752

49. Zhu, C., Zeng, M., Huang, X.: SDNet: contextualized attention-based deep network for conversational question answering. CoRR abs/1812.03593 (2018)

Temporal Natural Language Inference: Evidence-Based Evaluation of Temporal Text Validity

Taishi Hosokawa[1], Adam Jatowt[2(✉)], and Kazunari Sugiyama[3]

[1] Kyoto University, Kyoto, Japan
[2] University of Innsbruck, Innsbruck, Austria
adam.jatowt@uibk.ac.at
[3] Osaka Seikei University, Osaka, Japan

Abstract. It is important to learn whether text information remains valid or not for various applications including story comprehension, information retrieval, and user state tracking on microblogs and via chatbot conversations. It is also beneficial to deeply understand the story. However, this kind of inference is still difficult for computers as it requires temporal commonsense. We propose a novel task, *Temporal Natural Language Inference*, inspired by traditional natural language reasoning to determine the temporal validity of text content. The task requires inference and judgment whether an action expressed in a sentence is still ongoing or rather completed, hence, whether the sentence still remains valid, given its supplementary content. We first construct our own dataset for this task and train several machine learning models. Then we propose an effective method for learning information from an external knowledge base that gives hints on temporal commonsense knowledge. Using prepared dataset, we introduce a new machine learning model that incorporates the information from the knowledge base and demonstrate that our model outperforms state-of-the-art approaches in the proposed task.

1 Introduction

It is rather easy for humans to reason on the validity of sentences. Given a user's post: "I am taking a walk", and a subsequent post from the same user: "Ordering a cup of coffee to take away", we can guess that the person is very likely still taking a walk, and has just only stopped for a coffee during her walk. That is, the action stated in the former message is still ongoing, thus, the first sentence remains valid. On the other hand, if the subsequent post would be "I am preparing a dinner", it would be highly possible that the first message (the one about taking a walk) is no longer valid in view of this additional evidence. This kind of inference is usually smoothly done by the commonsense of humans.

Thanks to the emergence of pre-training models, computers have shown significant performance in the field of natural language understanding [59]. However, it is still a challenging task for computers to perform effective reasoning that requires commonsense knowledge [54]. As the amount of available text information is exploding

J. Kamps et al. (Eds.): ECIR 2023, LNCS 13980, pp. 441–458, 2023.
https://doi.org/10.1007/978-3-031-28244-7_28

nowadays, it is getting more and more important for machines to achieve much better understanding of natural language.

In this work, we propose a novel task called *Temporal Natural Language Inference* (TNLI), in which we evaluate the validity of text content using an additional related content used as evidence. Similarly to Natural Language Inference (NLI) task [55] an input is in the form of a sentence pair (we explain more on the differences of these two tasks in Sect. 2.5). We address the TNLI problem as a classification task using the following two input sentences: (1) hypothesis sentence and (2) premise sentence. The first one, the hypothesis sentence, is the one whose validity is to be judged. The second one, the premise sentence, is following the hypothesis sentence and is supposed to provide new information useful for classifying the hypothesis. The classification labels for the hypothesis sentence are as follows: SUPPORTED, INVALIDATED, and UNKNOWN. SUP-PORTED means that the hypothesis is still valid after seeing the premise, INVALIDATED means that the hypothesis ceased to be valid, and otherwise it is UNKNOWN.

Considering our earlier example, if we regard "I am taking a walk" as a hypothesis and "I am preparing a dinner" as a premise, the hypothesis becomes INVALIDATED since the user has clearly concluded her earlier action. If we consider "coffee for take away" as a premise instead, the hypothesis would be SUPPORTED.[1]

The potential applications of our proposed task are as follows:

Support for Story Understanding and Event Extraction: Effective methods trained for the proposed task can lead to better understanding of stories described in text and potentially also more effective event extraction [72]. Reading comprehension of stories would be improved if one incorporates a component that can reason about action completion given the evidence provided by the following sentences. We note that this kind of knowledge is often implicit in text.

Classification and Recommendation of Microblog Posts: Microblog posts can be valid for different lengths of time. In the era of information overload, users need to select valid messages among a large number of posts, as valid ones are typically the most relevant and important. This kind of information overload would be significantly alleviated when they could use an option to filter out invalid posts from their timelines, or the posts could be ranked by several factors including their estimated temporal validity.

User Tracking and Analysis: User tracking and analysis in social networks services (SNS) and chats [1,2,33] can be enhanced based on temporal processing of user's posts so that the user's current situation and action can be flagged. This could be useful for example for selecting suitable advertisements or in emergency situations like during the time of disasters to know the current state of users.

---

[1] Note that it is not always easy to determine the correct answer as the context or necessary details might be missing, and in such cases humans seem to rely on probabilistic reasoning besides the commonsense base.

Chatbots: Chatbots and AI assistants are becoming recently increasingly popular [43]. However, their use is still quite limited as they cannot understand well users' context such as their actions, goals and plans, and so on. To achieve much better communication with humans, it is necessary to develop user state-aware approach.

In addition to the proposal of a novel task, our second contribution is the construction of dedicated dataset for the proposed task. As we especially focus on sentences that describe actions and relatively dynamic states that can be relevant to the aforementioned applications, our dataset contains pairs of sentences describing concrete actions.

Finally, we also develop a new machine learning model incorporating information from a knowledge graph as we believe that successful model requires external knowledge about the world. Our proposed model combines the following two encoders: the first one incorporates commonsense knowledge via pre-training and the other one is purely based on text to jointly capture and reason with the commonsense knowledge. To develop these encoders, we construct a new model for learning the embeddings of concepts in knowledge bases. In particular, we employ the ATOMIC-2020 dataset [25] as an external knowledge base.

Our main contributions can be summarized as follows:

1. We propose a novel task called, Temporal Natural Language Inference (TNLI), requiring to identify the validity of a text content given an evidence in the form of another content. We formulate it as a text classification problem.
2. We construct a dedicated dataset[2] for our proposed task that contains over 10k sentence pairs, and analyze its relation to Natural Language Inference (NLI) task and NLI's corresponding datasets.
3. We design and develop an effective new machine learning model for the proposed task which utilizes information from commonsense knowledge bases.
4. Finally, we compare our model with some state-of-the-arts in natural language inference task and discuss the results. We also test if pre-training using standard NLI datasets is effective for TNLI.

2 Related Work

2.1 Temporal Information Retrieval and Processing

Temporal Information Retrieval is a subset of Information Retrieval domain that focuses on retrieving information considering their temporal characteristics. Many tasks and approaches have been proposed so far [1,9,27,28,42,45], including the understanding of story [23], temporal relation extraction [18,63], question answering [24,26], and so on. White and Awadallah [69] estimated the duration of tasks assigned by users in calendars. Takemura and Tajima [57] classified microblog posts to different lifetimes based on features specific to Twitter such as number of followers or presence of URLs. Almquist and Jatowt [3] examined the validity of sentences considering the time elapsed since their creation (more in Sect. 2.5).

---

[2] The dataset will be made freely available after paper publication.

2.2 Commonsense Reasoning

Implicit information that humans commonly know is addressed in, what is called, Commonsense Reasoning domain [60]. Winograd Schema Challenge [32] was one of the earliest challenges for machines in this regard, and many other challenges and approaches have also been proposed [21,34,36,38,44,49,58]. Temporal Commonsense is one of them, in which temporal challenges are addressed [78]. Zhou et al. [77] focused on comparing actions such as "going on a vacation" with others like "going for a walk" to assess which take longer, and constructed a dataset for question-answering including this kind of estimation. We further compare our task with other related ones in Sect. 2.5.

2.3 Natural Language Inference

Recently, Natural Language Understanding (NLU) by computers has attracted a lot of researchers' attention. Natural Language Inference (NLI) or Recognizing Textual Entailment is one of NLU domains, in which computers deal with input in the form of two sentences [55], similar to our proposed task. NLI problems require to determine that a premise sentence entails, contradicts, or is neutral to a hypothesis sentence (or in some settings, entails vs. not entails). In the early stages of NLI work, Dagan et al. [15] constructed a relatively small dataset. The first largely annotated dataset was *Stanford Natural Language Inference* (SNLI) dataset [6], which was annotated through crowdsourcing. After that, many NLI datasets [16,22], including *Multi-genre Natural Language Inference* (MNLI) [70] and Scitail [30], have been constructed. Notably, Vashishtha *et al.* [62] converted existing datasets for temporal reasoning into NLI format, pointing out that no NLI dataset focuses on temporal reasoning. Their task focuses on explicit temporal description while our task tackles implicit information. The emergence of these large scale datasets made it possible to train more complex models [7,55]. Remarkably, state-of-the-art large-scale pre-trained models such as BERT [17] and RoBERTa [37] demonstrated significant performance on NLI datasets, and are also used to train multi-task models [14].

2.4 Incorporation of Knowledge Bases

Generally, NLU works make use of Knowledge Graphs (KG) or Knowledge Bases (KB) to improve model performance [11,40,73]. Especially, Commonsense Reasoning works commonly incorporate knowledge from large KBs such as ConceptNet [35,53] and WikiData [64] in their architectures [48,75,76]. However, only a few works in NLI attempt to incorporate KGs into models [8,29]. Wang et al. [66], for example, improve performance on Scitail using knowledge in ConceptNet.

2.5 Comparison with Related Tasks

Similar to NLI, our work addresses a text classification problem, in which two sentences form an input. However, we focus on neither entailment nor contradiction but the validity of sentences (see Tables 3 and 4 for comparison).

The NLI dataset constructed by Vashishtha *et al.* [62] includes temporal phenomena. However, their task addresses explicit descriptions of temporal relations such as duration and order, while we focus on implicit temporal information that is latent in sentences. Temporal Commonsense task [77] includes implicit temporal information, too. The problem that the task deals with is reasoning about event duration, ordering, and frequency in a separate manner. However, our approach requires a more comprehensive understanding of temporal phenomena through a contrastive type inference. Also, their task is posed as a question-answering problem while ours is formalized as an NLI type problem. Almquist and Jatowt [3] also worked on the validity of sentences. Unlike their work, we use premises as the additional source instead of the information on the elapsed time from sentence creation as in [3], since, in practice, in many situations, additional text is available (e.g., sequences of tweets posted by the same user, or following sentences in a story or novel). Table 1 compares our task with the most related ones.

Table 1. Comparison our work with related tasks.

Task	Task Type	Temporal	Input	Output
McTaco [77]	Question Answering	✓	source, question, answer candidates	correct answers
NLI [12,20]	Classification		sentence pair	3 classes (entailment, contradiction, neutral)
Validity Period Estimation [3]	Classification	✓	sentence	5 classes (hours, days, weeks, months, years)
TNLI (Proposed Task)	Classification	✓	sentence pair	3 classes (SUPPORTED, INVALIDATED, UNKNOWN)

3 Task Definition

We first provide the definition of our task, in which a pair of sentences $p = (s_1, s_2)$ is given, where s_1 and s_2 are a hypothesis and a premise sentence, respectively.[3] The task is to assign one of the following three classes to s_1 based on the inference using the content of s_2:

$$c \in \{\text{SUPPORTED}, \text{INVALIDATED}, \text{UNKNOWN}\} \tag{1}$$

Here, the SUPPORTED class means that s_1 is still valid given the information in s_2. The INVALIDATED class, on the other hand, means that s_1 ceased to be valid in view of s_2. The third one, UNKNOWN class, indicates that the situation evidence is not conclusive or clear, and we cannot verify the validity of the hypothesis.

---

[3] Note that s_1 and s_2 may have temporal order: $t_{s_1} \leq t_{s_2}$, where $t_{s_{id}}$ $(id = 1, 2)$ is the creation time (or a reading order) of a sentence s_{id}. This may be for example in the case of receiving microblog posts issued by the user (or when reading next sentences of a story or a novel).

4   Proposed Method

We discuss next our proposed approach for content validity estimation task. We hypothesize that the task cannot be successfully solved without the incorporation of external knowledge given the inherent need of temporal commonsense reasoning. Therefore, we first attempt to find the useful knowledge base to provide knowledge of temporal properties. We then propose a new model by combining an encoder that incorporates information from this knowledge base and a text encoder that uses only text data. The output of the encoder using the knowledge base and the text encoder are combined and used as input to the softmax classifier. Figure 1 shows the model outline.

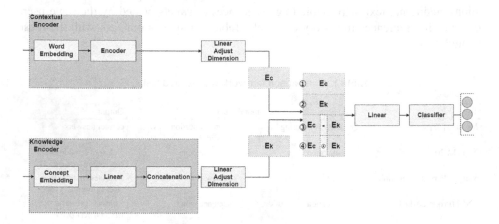

Fig. 1. Outline of our approach.

4.1   Encoding Knowledge

One of the key components of the proposed model is the knowledge encoder.

Knowledge Base. We have explored different knowledge bases (KBs) for our objective including FrameNet [19], WikiHow [31], Howto100m [39], and VerbNet [52]. We concluded that ATOMIC-2020 (*An ATlas Of MachIne Commonsense*) [25] is the most suitable KB to achieve our goal thanks to its temporal commonsense knowledge and relatively large scale (1.33 M commonsense knowledge tuples and 23 commonsense relations).

ATOMIC [51] is the predecessor KB of ATOMIC-2020 designed for commonsense reasoning that contains nine different if-then relations such as Cause, Effect, Intention, Reaction, and so on. Most of the entities in this KG are in the form of sentences or phrases. COMET [5] is a language model trained with ATOMIC and ConceptNet in order to generate entities that were not in the ATOMIC dataset. Then, ATOMIC-2020 adds new relations in comparison to ConceptNet and ATOMIC. The new relations include "IsAfter", "IsBefore", "HasSubevent", and so on, which represent the relations between events. For example, "PersonX pays PersonY a compliment" and "PersonX will want to chat with PersonY" are sentences belonging to the if-then relation in Atomic-2020, while "PersonX bakes bread" and "PersonX needed to buy ingredients" is an example of a pair of sentences connected by the "IsAfter" relation.

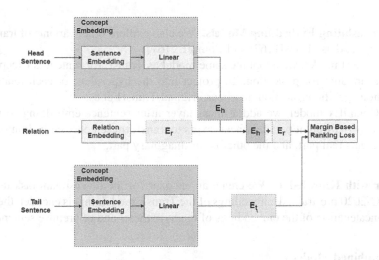

Fig. 2. TransE model for sentences.

TransE. We now briefly explain TransE [4] which we adapt for the purpose of KB relation embedding. TransE is a model for learning embeddings of KBs represented in the triple form of entities and relations <head entity, relation, tail entity>. Relations are considered to be translations in the embedding space. TransE learns embedding using the following loss function, which is an operation between entities and relations as in skip-gram [41], where head entity + relation = tail entity:

$$\mathcal{L} = \sum_{(\mathbf{h},\mathbf{l},\mathbf{t})\in S} \sum_{(\mathbf{h'},\mathbf{l},\mathbf{t'})\in S'_{(\mathbf{h},\mathbf{l},\mathbf{t})}} [\gamma + d(\mathbf{h}+\mathbf{l},\mathbf{t}) - d(\mathbf{h'}+\mathbf{l},\mathbf{t'})]_+, \qquad (2)$$

where $[x]_+$ denotes the positive part of x, γ is a margin parameter, d is the distance function, while $\mathbf{h}$, $\mathbf{l}$, and $\mathbf{t}$ are the embeddings of head entity, relation label, and tail entity, respectively. In addition, S is a set of positive examples, while S' is the set of negative examples.

TransE for Sentences. Since the entities in the ATOMIC-2020 dataset are mostly in the form of short sentences, it is difficult to train with original TransE as the number of potential entities is very large, and inference for sentences that are not in the training data is not possible. To solve this problem, we adapt the TransE model for sentences. First, we compute the sentence vector corresponding to each entity in the KG using Sentence-BERT (SBERT) [50]. We then train the weights W for the sentence vectors and the relation embedding E_r using Margin Based Ranking Loss as in TransE. Here, the weights of SBERT are fixed and not trained, so that if the data is not in the training set, the embeddings of similar sentences will remain similar. Figure 2 shows the model structure. Since information related to time is crucial in our work, we only use "IsAfter" and "IsBefore" ATOMIC-2020 relations.

Other Translating Embedding Models. We also explore other variants of translating embedding models: TransH [67] and ComplEx [61].

For TransH model, we adopt the same model as TransE for sentences except for an additional module for projection. To project into the hyperplane of each relation, we use relation-specific projection matrix as original TransH does.

For ComplEx model, we add a linear layer after sentence embedding so that the model has two different parallel linear layers to transform sentence embeddings, where one represents real part, and the other is for imaginary part.

Encoder with Knowledge. We create an encoder for the downstream task using the ATOMIC-2020 pre-trained embeddings of the TransE model. In this encoder, the output is the concatenation of the embeddings of the hypothesis and of premise sentence.

4.2   Combined Model

The entire model for the proposed downstream task consists of text encoder, knowledge encoder, and a classification layer on top of them. Since the dimensions of the pre-trained embeddings and the output of the text encoder are not the same, each output is linearly transformed to make the dimensions equal. The two vectors obtained in this way are compared and combined, and then linearly transformed. We use the concatenation, difference, and element-wise product for combination:

$$\mathbf{H} = Linear(\mathbf{H_t}; \mathbf{H_k}; \mathbf{H_t} - \mathbf{H_k}; \mathbf{H_t} \odot \mathbf{H_k}), \tag{3}$$

where $\mathbf{H_t}$ is the output of text encoder, $\mathbf{H_k}$ is the output of knowledge encoder, and $\odot$ denotes element-wise multiplication. The obtained output is linearly transformed, and then fed into a softmax classifier to decide the validity class.

5   Dataset

5.1   Dataset Construction

To create our dataset, we need hypotheses, premises, and ground truth labels. As mentioned before, we decided to focus on sentences similar to the typical setup of NLI task. As the hypothesis sentences we used randomly selected 5,000 premise sentences from SNLI dataset[4]. These sentences were originally taken from the captions of the Flickr30k corpus [74]. We conducted clustering over all the collected sentences and sampled equal number of sentences from each cluster to maintain high variation of

---

[4] SNLI dataset is licensed under CC-BY-SA 4.0.

Table 2. Average sentence length in our dataset.

	Average	Variance
Hypothesis	11.4	19.4
Premise	8.9	10.8
Invalidated	8.4	8.6
Supported	9.3	10.7
Unknown	8.9	12.7

sentence topics. For this, we first employed BERT [17] to vectorize each word, and then vectorized the sentences based on computing the arithmetic mean of word vectors. For clustering, we employed k-means to group sentence vectors into 100 clusters. We then extracted up to 50 sentences from each cluster with uniform probability and used them as the source of the hypotheses. Premise sentences and labels were collected using crowdsourcing with the Amazon Mechanical Turk[5]. For each hypothesis, we asked two crowdworkers to create a sentence corresponding to each label. To avoid sentences that are simply copied or modified with minimum effort, we accepted sentences only when 40% or more words were not overlapped with the corresponding hypothesis. Otherwise, crowdworkers could, for example, simply change one word in order to claim the added trivial sentence is of a given class (e.g., SUPPORTED class). Since the dataset should involve non-explicit temporal information and we wanted to make sure that the workers carefully consider it, we also asked for providing a description of the estimated time during which the hypotheses sentences could have been realistically valid, although this information was not included in the final dataset. In total, about 400 workers participated in the dataset creation.

Since we found out some spamming and dishonest activity, we later manually verified the validity of all the obtained data, corrected the grammar and words, as well as we manually filtered poor-quality, noisy, offensive, or too-personal sentences. For example, removed sentences included instances in which a single word was substituted with a different one that has the same or similar meaning or different case such as replacing "mother" with "MOM." We removed in total 19,341 pairs of sentences.

5.2 Dataset Statistics

The final dataset includes 10,659 sentence pairs. Since the previous research has pointed out that the number of words in sentences in some NLI datasets varies significantly depending on their labels [30], we examined the average number of words in our dataset. Table 2 shows the statistics indicating that the average number of words in our dataset does not change significantly for different labels. The variance tends to be higher, however, for the UNKNOWN class.

Note that the number of sentence pairs belonging to each class is the same (3,553). Table 3 shows some examples of the generated data, while, for contrast, we also show example sentences of NLI task in Table 4.

---

[5] https://www.mturk.com/.

6   Experiments

6.1   Experimental Settings

In our experiments, we perform 5-fold non-nested cross-validation for all the compared models. The batch size is 16, and the learning rate is determined by the performance on the validation fold chosen from the training folds among 0.005, 0.0005, 0.00005. For all the models, the optimal value was 0.00005. We evaluate our approach with accuracy, the percentage of correct answers, which is the most relevant metric and widely used for NLI task. We compare our proposed approach with the following four models: BERT (bert-base-uncased) [17], Siamese Network [10], SBERT [50] Embeddings with Feedforward Network, and Self-Explaining Model [56]. Except for ones including the self-explaining model, we train the models with cross-entropy loss.

Table 3. Example sentences of TNLI task in our dataset.

Hypothesis	Label	Premise
A woman in blue rain boots is eating a sandwich outside.	INVALIDATED	She takes off her boots in her house
A small Asian street band plays in a city park	SUPPORTED	Their performance pulls a large crowd as they used new tunes and songs today
A man jumping a rail on his skateboard.	UNKNOWN	His favorite food is pizza.

Table 4. Example sentences of NLI task from SNLI dataset (borrowed from SNLI website: https://nlp.stanford.edu/projects/snli/).

Hypothesis	Label	Premise
A man is driving down a lonely road.	Contradiction	A black race car starts up in front of a crowd of people
Some men are playing a sport.	Entailment	A soccer game with multiple males playing
A happy woman in a fairy costume holds umbrella.	Neutral	A smiling costumed woman is holding an umbrella

The architecture of the Siamese Network is similar to that of Bowman *et al.* [6] using the 8B version of GloVe [47] for word embedding and multiple layers of tanh. For SBERT with Feedforward Network, each layer has 500 dimensions and ReLU activation and dropout. The number of hidden layers on the top of SBERT equals to 3, and the output layer is a softmax classifier. The dropout rates are 75%.

The Self-Explaining [56] is a model with an attention-like Self-Explaining layer on top of a text encoder (RoBERTa-base [37]), and it achieves state-of-the-art results on SNLI. The Self-Explaining layer consists of three layers: Span Infor Collecting (SIC) layer, Interpretation layer, and an output layer.

For testing the proposed architecture, we experimented with the two types of contextual encoders: Siamese Network and Self-Explaining model. The dimensionality of the entity embedding was 256, and the combined embedding was linearly transformed to 128 to match the dimensionality of each encoder. We trained the entity embeddings of Bordes *et al.* [4] with a learning rate of 0.001.

We implemented the models using PyTorch [46] with HuggingFace's transformers [71] and we conducted our experiments on a machine equipped with GPU.

6.2   Experiments with NLI Pre-training

Generally, pre-training with NLI datasets improves accuracy in many downstream tasks [13]. As there is certain degree of relevance between our proposed task and NLI, we first experimented with pre-training selected models using the NLI datasets and then fine-tuning them on the proposed task. The datasets we used are the training sets of SNLI 1.0 and MNLI 0.9, which contain 550,152 and 392,702 examples, respectively. We mapped SUPPORTED to the entailment class of NLI, INVALIDATED to contradiction, and UNKNOWN to neutral.

Table 5 shows the results obtained by NLI pre-training, indicating that the NLI data has some relevance to our proposed task and can improve accuracy as it improves the results for Siamese network. However, when we use the Self-Explaining model, which is an already pre-trained model, we did not observe any improvement. This indicates that NLI datasets include effective information, yet it can likely be learned from general corpus, especially when using RoBERTa [37].

Table 5. NLI pre-training results in Siamese Network and Self-Explaining Model.

Model	Accuracy
Siamese	0.715
+SNLI	0.756
+MNLI	0.757
Self-Explaining	0.873
+SNLI	0.867
+MNLI	0.535

6.3   Incorporating Common-sense Knowledge

Table 6 shows the main experimental results. Among the compared models, the Self-Explaining model achieves the best accuracy. BERT, on the other hand, gives the worst results. We also observe that training of BERT is unstable as also pointed out in [17].

The incorporation of commonsense knowledge improves the accuracy in both the Siamese and Self-Explaining cases. While it can significantly improve the performance for the case of Siamese Network, it does not help much for Self-Explaining model. Self-explaining uses RoBERTa model which has been trained on larger data and has been carefully optimized, while other models use standard BERT. This might affect the results. In general, we believe that adding commonsense data through the architecture that we have proposed is a promising direction for TNLI task that calls for exploration of more sophisticated commonsense reasoning and more extensive datasets. This conclusion is also supported by the analysis of the confusion matrices shown in Fig. 3. Incorporating TransE to Siamese net helps to more correctly determine SUPPORTED and UNKNOWN classes (improvement by 28% and 6.5%, respectively), while only slightly confusing the INVALIDATED class (decrease of 1.4%).

6.4   Testing Different Knowledge Embedding Approaches

Finally, we explored different approaches of translating embedding models in addition to TransE. Table 7 shows the results of TransE variants combined with the Self-Explaining model. As it can be seen, the loss for pre-training does not go down in TransH [67] and ComplEx [61]. As the loss remains high, the accuracy with the proposed downstream task is also lower, indicating that the proposed architecture requires simpler way to construct the knowledge-based embeddings for TNLI because TransH and ComplEx are more complex models than TransE. Another possibility is that the other models were not supplied with sufficient knowledge to properly benefit from their more complex architectures.

Table 6. Results on TNLI task.

Model	Accuracy
Siamese	0.715
SBERT + FFN	0.806
BERT	0.441
Self-Explaining	0.873
Siamese+TransE	0.784
Self-Explaining+TransE	0.878

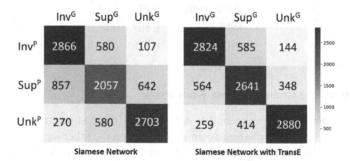

Fig. 3. Confusion matrices for TNLI prediction task of Siamese network (left) and Siamese network with TransE (right). The horizontal axis corresponds to the prediction (x^P) and vertical one to gold labels (x^G). The left (upper) blocks are INVALIDATED, middle ones are SUPPORTED, and right (bottom) ones are UNKNOWN.

Table 7. Results of TransE variants used with Self-Explaining model on TNLI task.

Model	Pre-Train Loss	Accuracy
TransE	0.19	0.878
TransH	0.48	0.868
ComplEx	1.24	0.856

7 Conclusion and Future Work

Computational processing of non-explicit temporal information in natural language still poses many challenges. In this work, we proposed a novel task for reasoning on the validity of sentences based on additional evidence and we trained a new model with an embedded knowledge base for this task. The motivation behind our idea is that humans can judge the temporal validity of a sentence using their commonsense, and our goal is to enable such reasoning ability for machines.

To achieve this goal, we first formally defined the task, constructed a dedicated dataset, and trained several baseline models. In addition, we proposed a new method of knowledge base embedding for sentences and a machine learning model that incorporates it. We believe that this work can contribute to our understanding of how to rely on knowledge bases that contain sentences as entities and of how to further improve the accuracy of TNLI task. We have also experimented with popular NLI datasets to answer a question on whether these can be useful for the proposed task.

Extending the dataset is one of our future goals. Our current work focused on sentences with relatively dynamic descriptions based on envisioned applications in microblogging. However, for more applications and training more robust models, it is necessary to construct datasets that also contain other forms of descriptions. More data, regardless of type, would be also necessary for larger-scale and less-biased training.

One way to achieve this would be to consider conversion methods from other datasets, as some NLI datasets such as Scitail and QNLI [16,65,68] have already employed them. Multi-modal datasets that include videos as well as their captions could be candidates for this. Another future direction is to extend the proposed task itself. More specifically, the timestamp of the premise sentences can be used as an additional signal to identify the validity of hypotheses sentences [3] in addition to judgments based on the content of premise sentences. This would lead to a more general task and a higher number of potential applications. It would be possible to address the cases where not only additional content is available as evidence for reasoning on the hypothesis's validity but also the time gap that elapsed from its statement is utilized (e.g., when using both the content and timestamps of user messages/posts or utterances). Therefore, the re-formulation of our task with added time information and the construction of a corresponding dataset are also in our plans. Finally, a further future work may focus on automatically generating premise sentences that would move their hypotheses into a required validity class to obtain the desired indication of action's completion or continuation.

References

1. Abe, S., Shirakawa, M., Nakamura, T., Hara, T., Ikeda, K., Hoashi, K.: Predicting the Occurrence of Life Events from User's Tweet History. In Proceedings of the 12th IEEE International Conference on Semantic Computing (ICSC 2018), pp. 219–226 (2018)
2. Abel, F., Gao, Q., Houben, G.-J., Tao, K.: Analyzing user modeling on twitter for personalized news recommendations. In Proceedings of the 19th International Conference on User Modeling, Adaptation, and Personalization (UMAP 2011), pp. 1–12 (2011)
3. Almquist, V., Jatowt, A.: Towards content expiry date determination: predicting validity periods of sentences. In Proceedings of the 41st European Conference on IR Research (ECIR 2019), pp. 86–101 (2019)
4. Bordes, A., Usunier, N., Garcia-Durán, A., Weston, J., Yakhnenko, O.: Translating Embeddings for Modeling Multi-relational Data. In Proceedings of the 27th International Conference on Neural Information Processing Systems (NIPS 2013), pp. 2787–2795 (2013)
5. Bosselut, A., Rashkin, H., Sap, M., Malaviya, C., Celikyilmaz, A., Choi. Y.: COMET: commonsense transformers for automatic knowledge graph construction. In Proceedings of the 57th Annual Meeting of the Association for Computational Linguistics, pp. 4762–4779, Florence, Italy, July (2019). Association for Computational Linguistics
6. Bowman, S.R., Angeli, G., Potts, C., Manning, C.D.: A large annotated corpus for learning natural language inference. In Proceedings of the 2015 Conference on Empirical Methods in Natural Language Processing, pp. 632–642, Lisbon, Portugal (2015). Association for Computational Linguistics
7. Chen, Q., Zhu, X., Ling, Z.-H., Wei, Z.-H., Jiang, H., Inkpen, D.: Enhanced LSTM for natural language inference. In Proceedings of the 55th Annual Meeting of the Association for Computational Linguistics (Volume 1: Long Papers), pp. 1657–1668, Vancouver, Canada (2017). Association for Computational Linguistics
8. Chen, Y., Huang, S., Wang, F., Cao, J., Sun, W., Wan, X.: Neural maximum subgraph parsing for cross-domain semantic dependency analysis. In Proceedings of the 22nd Conference on Computational Natural Language Learning, pp. 562–572, Brussels, Belgium (2018). Association for Computational Linguistics
9. Cheng, F., Miyao, Y.: Predicting event time by classifying sub-level temporal relations induced from a unified representation of time anchors. arXiv preprint arXiv:2008.06452 (2020)
10. Chicco, D.: Siamese Neural Networks: an Overview. Artificial Neural Networks - Third Edition, pp. 73–94 (2021)
11. Clark, P., Dalvi, B., Tandon, N.: What happened? leveraging VerbNet to predict the effects of actions in procedural text. arXiv preprint arXiv:1804.05435 (2018)
12. Condoravdi, C., Crouch, D., de Paiva, V., Stolle, R., Bobrow, D.G.: Entailment, intensionality and text understanding. In Proceedings of the HLT-NAACL 2003 Workshop on Text Meaning, pp. 38–45 (2003)
13. Conneau, A., Kiela, D., Schwenk, H., Barrault, L., Bordes, A.: Supervised learning of universal sentence representations from natural language inference data. In Proceedings of the 2017 Conference on Empirical Methods in Natural Language Processing, pp. 670–680, Copenhagen, Denmark (2017). Association for Computational Linguistics
14. Crawshaw, M.: Multi-task learning with deep neural networks: a survey. arXiv preprint arXiv:2009.09796 (2020)
15. Dagan, I., Glickman, O., Magnini, B.: The PASCAL recognising textual entailment challenge. In: Machine Learning Challenges Workshop (MLCW 2005), pp. 177–190 (2005)
16. Demszky, D., Guu, K., Liang, P.: Transforming question answering datasets into natural language inference datasets. arXiv preprint arXiv:1809.02922 (2018)

17. Devlin, J., Chang, M.-W., Lee, K., Toutanova, K.: BERT: pre-training of deep bidirectional transformers for language understanding. In Proceedings of the 2019 Conference of the North American Chapter of the Association for Computational Linguistics: Human Language Technologies, Volume 1 (Long and Short Papers), pp. 4171–4186, Minneapolis, Minnesota (2019). Association for Computational Linguistics

18. Dligach, D., Miller, T., Lin, C., Bethard, S., Savova, G.: Neural temporal relation extraction. In Proceedings of the 15th Conference of the European Chapter of the Association for Computational Linguistics: Volume 2, Short Papers, pp. 746–751, Valencia, Spain (2017). Association for Computational Linguistics

19. Fillmore, C.J., Baker, C.: A frames approach to semantic analysis. In: The Oxford Handbook of Linguistic Analysis. Oxford University Press (2010)

20. Fyodorov, Y., Winter, Y., Francez, N.: A natural logic inference system. In: Proceedings of the 2nd Workshop on Inference in Computational Semantics (ICoS-2) (2000)

21. Gao, Q., Yang, S., Chai, J., Vanderwende, L.: What action causes this? towards naive physical action-effect prediction. In Proceedings of the 56th Annual Meeting of the Association for Computational Linguistics (Volume 1: Long Papers), pp. 934–945, Melbourne, Australia (2018). Association for Computational Linguistics

22. Glockner, M., Shwartz, V., Goldberg, Y.: Breaking NLI systems with sentences that require simple lexical inferences. In: Proceedings of the 56th Annual Meeting of the Association for Computational Linguistics (Volume 2: Short Papers), pp. 650–655, Melbourne, Australia (2018). Association for Computational Linguistics

23. Han, R., Liang, M., Alhafni, B., Peng, N.: Contextualized word embeddings enhanced event temporal relation extraction for story understanding. arXiv preprint arXiv:1904.11942 (2019)

24. Harabagiu, S., Bejan, C.A.: Question answering based on temporal inference. In: Proceedings of the AAAI-2005 Workshop on Inference for Textual Question Answering, pp. 27–34 (2005)

25. Hwang, J.D., et al.: (COMET-) Atomic 2020: on symbolic and neural commonsense knowledge graphs. In: Proceedings of the 34th AAAI Conference on Artificial Intelligence (AAAI-21), pp. 6384–6392 (2021)

26. Jatowt, A.: Temporal question answering in news article collections. In: Companion of The Web Conference 2022, Virtual Event/Lyon, France, 25–29 April 2022, p. 895. ACM (2022)

27. Jatowt, A., Antoine, É., Kawai, Y., Akiyama, T.: Mapping temporal horizons: analysis of collective future and past related attention in Twitter. In: Proceedings Of The 24th International Conference on World Wide Web (WWW 2015), pp. 484–494 (2015)

28. Kanazawa, K., Jatowt, A., Tanaka, K.: Improving retrieval of future-related information in text collections. In: Proceedings of the 2011 IEEE/WIC/ACM International Conference on Web Intelligence (WI 2011), pp. 278–283 (2011)

29. Kapanipathi, P., et al.: Infusing knowledge into the textual entailment task using graph convolutional networks. In: Proceedings of the 34th AAAI Conference on Artificial Intelligence (AAAI-20), pp. 8074–8081 (2020)

30. Khot, T., Sabharwal, A., Clark, P.: SciTaiL: a textual entailment dataset from science question answering. In: Proceedings of the 32nd AAAI Conference on Artificial Intelligence (AAAI-18) (2018)

31. Koupaee, M., Wang, W.Y.: WikiHow: a large scale text summarization dataset. arXiv preprint arXiv:1810.09305 (2018)

32. Levesque, H., Davis, E., Morgenstern, L.: The winograd schema challenge. In: Proceedings of the 13th International Conference on the Principles of Knowledge Representation and Reasoning (KR 2012), pp. 552–561 (2012)

33. Li, P., Lu, H., Kanhabua, N., Zhao, S., Pan, G.: Location inference for non-geotagged tweets in user timelines. IEEE Trans. Knowl. Data Eng. (TKDE) 31(6), 1150–1165 (2018)

34. Lin, B.Y., Chen, X., Chen, J., Ren, X.: KagNet: knowledge-aware graph networks for commonsense reasoning. In Proceedings of the 2019 Conference on Empirical Methods in Natural Language Processing and the 9th International Joint Conference on Natural Language Processing (EMNLP-IJCNLP), pp. 2829–2839, Hong Kong, China (2019). Association for Computational Linguistics

35. Liu, H., Singh, P.: ConceptNet - a practical commonsense reasoning tool-kit. BT Technol. J. **22**(4), 211–226 (2004)

36. Liu, Q., Jiang, H., Ling, Z.-H., Zhu, X., Wei, S., Hu, Y.: Combing context and commonsense knowledge through neural networks for solving winograd schema problems. In: Proceedings of the AAAI 2017 Spring Symposium on Computational Context: Why It's Important, What It Means, and Can It Be Computed? pp. 315–321 (2017)

37. Liu, Y., et al.: RoBERTa: a robustly optimized BERt pretraining approach. arXiv preprint arXiv:1907.11692 (2019)

38. Luo, Z., Sha, Y., Zhu, K.Q., Hwang, S.-W., Wang, Z.: Commonsense causal reasoning between short texts. In: Proceedings of the 15th International Conference on the Principles of Knowledge Representation and Reasoning (KR 2016), pp. 421–431 (2016)

39. Miech, A., Zhukov, D., Alayrac, J.-B., Tapaswi, M., Laptev, I., Sivic, J.: HowTo100M: learning a text-video embedding by watching hundred million narrated video clips. In: Proceedings of the IEEE/CVF International Conference on Computer Vision (ICCV 2019), pp. 2630–2640 (2019)

40. Mihaylov, T., Frank, A.: Knowledgeable reader: enhancing cloze-style reading comprehension with external commonsense knowledge. In: Proceedings of the 56th Annual Meeting of the Association for Computational Linguistics (Volume 1: Long Papers), pp. 821–832, Melbourne, Australia (2018). Association for Computational Linguistics

41. Mikolov, T., Sutskever, I., Chen, K., Corrado, G.S., Dean, J.: Distributed representations of words and phrases and their compositionality. In: Proceedings of the 27th Annual Conference on Neural Information Processing Systems (NIPS 2013), pp. 3111–3119 (2013)

42. Minard, A.-L., et al.: SemEval-2015 task 4: timeline: cross-document event ordering. In: Proceedings of the 9th International Workshop on Semantic Evaluation (SemEval 2015), pp. 778–786, Denver, Colorado (2015). Association for Computational Linguistics

43. Mnasri, M.: Recent advances in conversational NLP: towards the standardization of chatbot building. arXiv preprint arXiv:1903.09025 (2019)

44. Mostafazadeh, N., et al.: A corpus and cloze evaluation for deeper understanding of commonsense stories. In: Proceedings of the 2016 Conference of the North American Chapter of the Association for Computational Linguistics: Human Language Technologies, pp. 839–849, San Diego, California (2016). Association for Computational Linguistics

45. Ning, Q., Wu, H., Roth, D.: A multi-axis annotation scheme for event temporal relations. In: Proceedings of the 56th Annual Meeting of the Association for Computational Linguistics (Volume 1: Long Papers), pp. 1318–1328, Melbourne, Australia (2018). Association for Computational Linguistics

46. Paszke, A., et al.: PyTorch: an imperative style, high-performance deep learning library. In: Proceedings of the 33rd Conference on Neural Information Processing Systems (NeurIPS 2019), pp. 8026–8037 (2019)

47. Pennington, J., Socher, R., Manning, C.: GloVe: global vectors for word representation. In: Proceedings of the 2014 Conference on Empirical Methods in Natural Language Processing (EMNLP), pp. 1532–1543, Doha, Qatar (2014). Association for Computational Linguistics

48. Peters, M.E., et al.: Knowledge enhanced contextual word representations. In Proceedings of the 2019 Conference on Empirical Methods in Natural Language Processing and the 9th International Joint Conference on Natural Language Processing (EMNLP-IJCNLP), pp. 43–54, Hong Kong, China (2019). Association for Computational Linguistics

49. Rashkin, H., Sap, M., Allaway, E., Smith, N.A., Choi, Y.: Event2Mind: commonsense inference on events, intents, and reactions. In: Proceedings of the 56th Annual Meeting of the Association for Computational Linguistics (Volume 1: Long Papers), pp. 463–473, Melbourne, Australia (2018). Association for Computational Linguistics

50. Reimers, N., Gurevych, I.: Sentence-BERT: sentence embeddings using Siamese BERT-networks. In: Proceedings of the 2019 Conference on Empirical Methods in Natural Language Processing and the 9th International Joint Conference on Natural Language Processing (EMNLP-IJCNLP), pp. 3982–3992, Hong Kong, China (2019). Association for Computational Linguistics

51. Sap, M., et al.: ATOMIC: an atlas of machine commonsense for if-then reasoning. In: Proceedings of the AAAI Conference on Artificial Intelligence (AAAI-19), pp. 3027–3035 (2019)

52. Schuler, K.: VerbNet: a broad-coverage, comprehensive verb lexicon, Ph. D. thesis, University of Pennsylvania (2005)

53. Speer, R., Chin, J., Havasi, C.: ConceptNet 5.5: an open multilingual graph of general knowledge. In: Proceedings of the 31st AAAI Conference on Artificial Intelligence (AAAI-17), pp. 4444–4451 (2017)

54. Storks, S., Gao, Q., Chai, J.Y.: Commonsense reasoning for natural language understanding: a survey of benchmarks, resources, and approaches. arXiv preprint arXiv:1904.01172, pp. 1–60 (2019)

55. Storks, S., Gao, Q., Chai, J.Y.: Recent advances in natural language inference: a survey of benchmarks, resources, and approaches. arXiv preprint arXiv:1904.01172 (2019)

56. Sun, Z., et al.: Self-explaining structures improve NLP models. arXiv preprint arXiv:2012.01786 (2020)

57. Takemura, H., Tajima, K.: Tweet classification based on their lifetime duration. In: Proceedings of the 21st ACM International Conference on Information and Knowledge Management (CIKM 2012), pp. 2367–2370 (2012)

58. Tamborrino, A., Pellicanò, N., Pannier, B., Voitot, P., Naudin, L.: Pre-training is (almost) all you need: an application to commonsense reasoning. In: Proceedings of the 58th Annual Meeting of the Association for Computational Linguistics, pp. 3878–3887, Online (2020). Association for Computational Linguistics

59. Torfi, A., Shirvani, R.A., Keneshloo, Y., Tavaf, N., Fox, E.A.: Natural language processing advancements by deep learning: a survey. arXiv preprint arXiv:2003.01200 (2020)

60. Trinh, T.H., Le, Q.V.: A simple method for commonsense reasoning. arXiv preprint arXiv:1806.02847 (2018)

61. Trouillon, T., Welbl, J., Riedel, S., Gaussier, É., Bouchard, G.: Complex Embeddings for Simple Link Prediction. In Proceedings of the 33nd International Conference on Machine Learning (ICML 2016), pp. 2071–2080 (2016)

62. Vashishtha, S., Poliak, A., Lal, Y.K., Van Durme, B., White, A.S.: Temporal reasoning in natural language inference. In: Findings of the Association for Computational Linguistics: EMNLP 2020, pp. 4070–4078, Online (2020). Association for Computational Linguistics

63. Vashishtha, S., Van Durme, B., White, A.S.: Fine-grained temporal relation extraction. In: Proceedings of the 57th Annual Meeting of the Association for Computational Linguistics, pp. 2906–2919, Florence, Italy (2019). Association for Computational Linguistics

64. Vrandečić, D., Krötzsch, M.: Wikidata: A Free Collaborative Knowledgebase. Commun. ACM 57(10), 78–85 (2014)

65. Wang, A., Singh, A., Michael, J., Hill, F., Levy, O., Bowman, S.: GLUE: a multi-task benchmark and analysis platform for natural language understanding. In: Proceedings of the 2018 EMNLP Workshop BlackboxNLP: Analyzing and Interpreting Neural Networks for NLP, pp. 353–355, Brussels, Belgium (2018). Association for Computational Linguistics

66. Wang, X., et al.: Improving natural language inference using external knowledge in the science questions domain. In: Proceedings of the 33rd AAAI Conference on Artificial Intelligence (AAAI-19), pp. 7208–7215 (2019)
67. Wang, Z., Zhang, J., Feng, J., Chen, Z.: Knowledge graph embedding by translating on hyperplanes. In: Proceedings of the 28th AAAI Conference on Artificial Intelligence (AAAI-14), pp. 1112–1119 (2014)
68. White, A.S., Rastogi, P., Duh, K., Van Durme, B.: Inference is everything: Recasting semantic resources into a unified evaluation framework. In: Proceedings of the Eighth International Joint Conference on Natural Language Processing (Volume 1: Long Papers), pp. 996–1005, Taipei, Taiwan (2017). Asian Federation of Natural Language Processing
69. White, R.W., Awadallah, A.H.: Task duration estimation. In: Proceedings of the 12th ACM International Conference on Web Search and Data Mining (WSDM 2019), pp. 636–644 (2019)
70. Williams, A., Nangia, N., Bowman, S.: A broad-coverage challenge corpus for sentence understanding through inference. In: Proceedings of the 2018 Conference of the North American Chapter of the Association for Computational Linguistics: Human Language Technologies, Volume 1 (Long Papers), pp. 1112–1122, New Orleans, Louisiana (2018). Association for Computational Linguistics
71. Wolf, T., et al.: HuggingFace's transformers: state-of-the-art natural language processing. arXiv preprint arXiv:1910.03771 (2019)
72. Xiang, W., Wang, B.: A survey of event extraction from text. IEEE Access 7, 173111–173137 (2019)
73. Yasunaga, M., Ren, H., Bosselut, A., Liang, P., Leskovec, J.: QA-GNN: reasoning with language models and knowledge graphs for question answering. In: Proceedings of the 2021 Conference of the North American Chapter of the Association for Computational Linguistics: Human Language Technologies, pp. 535–546, Online (2021). Association for Computational Linguistics
74. Young, P., Lai, A., Hodosh, M., Hockenmaier, J.: From image descriptions to visual denotations: new similarity metrics for semantic inference over event descriptions. Trans. Assoc. Comput. Linguist. 2, 67–78 (2014)
75. Zhang, T., et al.: HORNET: enriching pre-trained language representations with heterogeneous knowledge sources. In: Proceedings of the 30th ACM International Conference on Information and Knowledge Management (CIKM 2021), pp. 2608–2617 (2021)
76. Zhang, Z., Han, X., Liu, Z., Jiang, X., Sun, M., Liu, Q.: ERNIE: enhanced language representation with informative entities. In Proceedings of the 57th Annual Meeting of the Association for Computational Linguistics, pp. 1441–1451, Florence, Italy (2019). Association for Computational Linguistics
77. Zhou, B., Khashabi, D., Ning, Q., Roth, D.: "Going on a vacation" takes longer than "going for a walk": a study of temporal commonsense understanding. In: Proceedings of the 2019 Conference on Empirical Methods in Natural Language Processing and the 9th International Joint Conference on Natural Language Processing (EMNLP-IJCNLP), pp. 3363–3369, Hong Kong, China (2019). Association for Computational Linguistics
78. Zhou, B., Ning, Q., Khashabi, D., Roth, D. Temporal common sense acquisition with minimal supervision. In Proceedings of the 58th Annual Meeting of the Association for Computational Linguistics, pages 7579–7589, Online (2020). Association for Computational Linguistics

Theoretical Analysis on the Efficiency of Interleaved Comparisons

Kojiro Iizuka[1,2]([⊠]) [iD], Hajime Morita[1]([⊠]) [iD], and Makoto P. Kato[2]([⊠]) [iD]

[1] Gunosy Inc., Shibuya, Japan
iizuka.kojiro@gmail.com, hajime.morita@gunosy.com
[2] University of Tsukuba, Tsukuba, Japan
mpkato@acm.org

Abstract. This study presents a theoretical analysis on the efficiency of *interleaving*, an efficient online evaluation method for rankings. Although interleaving has already been applied to production systems, the source of its high efficiency has not been clarified in the literature. Therefore, this study presents a theoretical analysis on the efficiency of interleaving methods. We begin by designing a simple interleaving method similar to ordinary interleaving methods. Then, we explore a condition under which the interleaving method is more efficient than A/B testing and find that this is the case when users leave the ranking depending on the item's relevance, a typical assumption made in click models. Finally, we perform experiments based on numerical analysis and user simulation, demonstrating that the theoretical results are consistent with the empirical results.

Keywords: Interleaving · Online evaluation · A/B testing

1 Introduction

Online evaluation is one of the most important methods for information retrieval and recommender systems [8]. In particular, A/B tests are widely conducted in web services [10,18]. Although A/B testing is easy to implement, it may negatively impact service experiences if the test lasts a long time and the alternative system (B) is not as effective as the current system (A). In recent years, interleaving methods have been developed to mitigate the negative impacts of online evaluation, as interleaving is experimentally known to be much more efficient than A/B testing [3,24].

Although interleaving methods have already been applied to production systems, the source of their high efficiency has not been sufficiently studied. Hence, this study presents a theoretical analysis of the efficiency of interleaving methods. A precise understanding of what makes interleaving efficient will enable us to distinguish properly the conditions under which interleaving should and should not be employed. Furthermore, the theoretical analysis could lead to the further development of interleaving methods.

J. Kamps et al. (Eds.): ECIR 2023, LNCS 13980, pp. 459–473, 2023.
https://doi.org/10.1007/978-3-031-28244-7_29

The following example intuitively explains the efficiency of interleaving. Let us consider the case of evaluating the rankings A and B, where A has higher user satisfaction than B. Interleaving presents an interleaved ranking of A and B to the user and evaluates these rankings based on the number of clicks given to the items from each ranking. For example, A is considered superior to B if more items from A in the interleaved ranking are clicked than those from B. In the interleaved ranking, an item from A can be placed ahead of one from B and vice versa. If the user clicks on an item and is satisfied with the content, the user may leave the ranking and not click on the other items. Because we assume items from ranking A are more satisfactory than those from ranking B, items from B have a lesser chance of being clicked when ranking B alone is presented. Thus, interleaving enables a superior ranking to overtake implicitly the click opportunities of the other ranking, making it easier to observe the difference in the ranking effects on users' click behaviors.

Following this intuition, this study analyzes the efficiency of interleaving from a theoretical perspective. We begin by designing a simple interleaving method to generalize representative interleaving methods. Then, we probabilistically model our interleaving method by decomposing an item click into an examination probability and the item's relevance. By analyzing the model, we show that interleaving has a lower evaluation error rate than A/B testing in cases when users leave the ranking depending on the item's relevance.

We conduct experiments to validate the theoretical analysis, the results of which confirm that the efficiency of interleaving is superior to that of A/B testing when users leave the ranking based on the item's relevance. The results are consistent with those of the theoretical analysis. Therefore, this study theoretically and empirically verifies the source of the efficiency of interleaving.

The contributions of this study can be summarized as follows:

- We discuss the nature of the efficiency of interleaving, focusing on the theoretical aspects on which the literature is limited.
- We identify the condition under which interleaving is more efficient than A/B testing according to theoretical analysis.
- Our experiments confirm that the theoretical analysis results are consistent with the empirical results.

The structure of this paper is as follows. Section 2 describes the related research. Section 3 introduces the notations and other preparations. Section 4 analyzes the efficiency of interleaving. Sections 5 and 6 describe the numerical analysis and the user simulation experiments. Finally, Sect. 7 summarizes the study.

2   Related Works

2.1   User Click Behavior

Among the various types of user feedback, modeling of click behavior has been extensively studied. The user behavior model concerning clicks in the information retrieval area is called the *click model*, and it helps simulate user behaviors in the absence of real users or when experiments with real users would interfere with the user experience. The term click model was first used in the context of the cascade click model introduced by Craswell et al. [7]. Thereafter, basic click models [4, 9, 11] were developed and are still being improved in different ways. Our study uses *examination* and *relevance*, as defined in the click model, to model the interleaving method. The examination represents a user behavior in that the user examines an item on a ranking and relevance concerns how well an item meets the information needs of the user.

2.2   Online Evaluation

Online evaluation is a typical method used to perform ranking evaluations. Among the various methods of online evaluation, A/B testing is widely used and is easy to implement. It divides users into groups A and B and presents the ranking A and ranking B to each group of users. As an extension of A/B testing, improved methods exist that reduce the variance of the evaluation [8, 20, 21, 27].

Another method of online evaluation is interleaving [2, 12, 14–17, 24], which involves evaluating two rankings, whereas multileaving involves evaluating three or more rankings [2, 19]. In particular, Pairwise Preference Multileaving (PPM) [19] is a multileaving method that has a theoretical guarantee about *fidelity* [13] that is related to this study. Some studies suggest that interleaving methods are experimentally known to be 10 to 100 times more efficient than A/B testing [3, 24]. However, the discussion concerning from where this efficiency comes is limited. Related to the analysis in this study, Oosterhuis and de Rijke showed that some interleaving methods can cause systematic errors, including bias when compared with the ground truth [20]. In this study, we present the cases in which the examination probability of interleaved rankings affects the efficiency of the interleaving method.

3   Preliminary

In this section, we first introduce a simple interleaving method for analysis. We then define the notations to model A/B testing and the interleaving method. Finally, we define the efficiency.

3.1   Interleaving Method for Analysis (IMA)

Because theoretically analyzing existing interleaving methods is difficult, we introduce an interleaving method for analysis (IMA), which is a simplified version of existing interleaving methods. In the remainder of this paper, the term *interleaving* will refer to this IMA. Our method performs for each round l as follows. Items are probabilistically added according to the flip of a coin. If the coin is face up, the l-th item in ranking A is used as the l-th item in the interleaved ranking I. If the coin is face down, the l-th item in ranking B is used as the l-th item in the interleaved ranking I. A set of items selected from A (or B) is denoted by $TeamA$ (or $TeamB$), and this continues until the length of the interleaved ranking reaches the required length, which we assumed to be same for the input ranking length. In this study, we also assume that no duplication items exist in the ranking A and B to simplify the discussion.

We use almost the same scoring as that in Team Draft Interleaving (TDI) [24] to derive a preference between A and B from the observed clicking behavior in I. We let c_j denote the rank of the j-th click in the interleaved ranking $I = (i_1, i_2, \ldots)$. Then, we attribute the clicks to ranking A or B based on which team generated the clicked result. In particular, for the u-th impression, we obtain the score for IMA as follows:

$$score_{I,A}^u = |\{c_j : i_{c_j} \in TeamA_u\}|/|I| \tag{1}$$
$$\text{and } score_{I,B}^u = |\{c_j : i_{c_j} \in TeamB_u\}|/|I|,$$

where $TeamA_u$ and $TeamB_u$ denote the team attribution of ranking A and ranking B, respectively, at the u-th impressions and $|I|$ is the length of the interleaved ranking. Note that the scores $score_{I,A}^u$ and $score_{I,B}^u$ represent the expected values of the click per item in the interleaved ranking, as the scores are divided by the ranking length. At the end of the evaluation, we score IMA using the formulas $score_{I,A} = \sum_{u=1}^n score_{I,A}^u/n$ and $score_{I,B} = \sum_{u=1}^n score_{I,B}^u/n$, where n is the total number of impressions. We infer a preference for A if $score_{I,A} > score_{I,B}$, a preference B if $score_{I,A} < score_{I,B}$, and no preference if $score_{I,A} = score_{I,B}$.

An interleaved comparison method is *biased* if, under a random distribution of clicks, it prefers one ranker over another in expectation [12]. Some existing interleaving methods were designed so that the interleaved ranking would not be biased. For example, Probabilistic Interleaving (PI) [12], TDI [24], and Optimized Interleaving (OI) [23] were proven unbiased [12,19]. Our IMA is also unbiased, because every ranker is equally likely to add an item at each location in the ranking; in other words, the interleaved ranking is generated so that the examination probabilities for the input rankings will all be the same.

3.2   A/B Testing

In this paper, we re-define an A/B testing method for compatibility with the IMA. We use Eq. (1) for each u-th impression to score A/B testing, denoted as $score_{AB,A}^u$ and $score_{AB,B}^u$. At the end of the evaluation, we score A/B testing using $score_{AB,A} = \sum_{u=1}^n score_{AB,A}^u/n$ and $score_{AB,B} = \sum_{u=1}^n score_{AB,B}^u/n$, where n is the total number of impressions, the same as in IMA. The difference between A/B testing and the IMA is the policy for generating rankings. All items in the interleaved ranking are selected from either ranking A only or ranking B only in the A/B testing. In other words, the probability of selecting A or B is fixed to 1 when generating a single interleaved ranking, or either ranking A or B is presented to the user at an equal probability.

3.3   Notation

First, we introduce the notations of random variables. We denote random variables for clicks as $Y \in \{0,1\}$, the examination of the item on the ranking as $O \in \{0,1\}$, and the relevance of the item as $R \in \{0,1\}$. The value $O = 1$ means a user examines the item on the ranking. This study assumes $Y = O \cdot R$, meaning the user clicks on an item only if the item is examined and relevant. We denote the probability of some random variable T as $P(T)$, the expected value of T as $E(T)$, and the variance in T as $V(T)$. In addition, the expected value of and variance in the sample mean for T_i of each i-th impression over n times are denoted as $E(\bar{T}^n)$ and $V(\bar{T}^n)$, respectively.

Next, we introduce the notations of random variables for the rankings. In this study, we evaluate two rankings, A and B, and we do not distinguish between ranking and items to simplify the notations. We denote the random variable of clicks in $TeamA$ as Y_A and the random variable of relevance for ranking A as R_A. $O_{AB,A}$ and $O_{I,A}$ are defined as the random variables of the examination of A/B testing and of interleaving, respectively. Ranking A and ranking B are interchangeable in the above notations. We also use $\bullet$ to denote ranking A or B. For example, $O_{I,\bullet}$ refers to the random variable for the examination of interleaving when ranking A or B.

Probabilistic Models. This study assumes the probabilistic models of A/B testing and interleaving as follows:

- $Y_{AB,A} = S_{AB,A} \cdot O_{AB,A} \cdot R_A$ holds where a random variable for ranking assignment in A/B testing is denoted as $S_{AB,A} \in \{0,1\}$, where $S_{AB,A} = 1$ if ranking A is selected via A/B testing. We note that $E(S_{AB,A}) = 1/2$, as ranking A or B is randomly selected.
- $Y_{I,A} = S_{I,A} \cdot O_{I,A} \cdot R_A$ holds where a random variable for ranking assignment in interleaving is denoted as $S_{I,A} \in \{0,1\}$, where $S_{I,A} = 1$ if the item belongs to $TeamA$. We note that $E(S_{I,A}) = 1/2$, as the item is selected randomly at each position in I from ranking A or B.

We assume $Y_{AB,A}^i$, a random variable for clicks in the i-th impression, follows the Bernoulli distribution $B(p_{AB,A})$, where $p_{AB,A} = E(Y_{AB,A})$. Then, $\bar{Y}_{AB,A}^n$ is defined as $\bar{Y}_{AB,A}^n = \frac{1}{n} \sum_{i=1}^n Y_{AB,A}^i$. In this definition, $\bar{Y}_{AB,A}^n$ follows a binomial distribution, and $\bar{Y}_{AB,A}^n$ can be considered as a random variable that follows normal distribution when $n \to \infty$. We note that

$$E(\bar{Y}_{AB,A}^n) = E(\frac{1}{n} \sum_{i=1}^n Y_{AB,A}^i) = \frac{1}{n} \sum_{i=1}^n p_{AB,A} = E(Y_{AB,A}), \qquad (2)$$

$$V(\bar{Y}_{AB,A}^n) = V(\frac{1}{n} \sum_{i=1}^n Y_{AB,A}^i) = \frac{1}{n^2} \sum_{i=1}^n p_{AB,A}(1 - p_{AB,A}) = \frac{1}{n}V(Y_{AB,A}) \quad (3)$$

holds. The same holds for $\bar{Y}_{AB,B}^n$, $\bar{Y}_{I,A}^n$, and $\bar{Y}_{I,B}^n$. In addition,

$$E(Y_{AB,A}) = E(S_{AB,A})E(O_A \cdot R_A) = E(S_{AB,A})E(Y_A) \qquad (4)$$

holds, as $Y_{AB,A} = S_{AB,A} \cdot O_{AB,A} \cdot R_A = S_{AB,A} \cdot O_A \cdot R_A$, and $S_{AB,A}$ is independent of O_A and R_A. Finally, note that $E(\bar{Y}^n_{AB,A}) = score_{AB,A}$ holds from the definition. In the above equation, ranking A and ranking B are interchangeable.

3.4　Definition of Efficiency

In this study, efficiency is defined as the level of evaluation error probability, denoted as $P(Error)$. The efficiency reflects that the error probability is small, given the statistical parameters. We demonstrate below that interleaving is more efficient; $P(Error_I) < P(Error_{AB})$ holds with some conditions. More formally, the error probability for A/B testing can be defined as follows: if $E(Y_A) - E(Y_B) > 0$ and $\bar{Y}^n_{AB,\bullet} \sim \mathcal{N}(E(\bar{Y}^n_{AB,\bullet}), V(\bar{Y}^n_{AB,\bullet}))$ for n is sufficiently large, then:

$$
\begin{aligned}
&P(Error_{AB}) \\
&= P(\bar{Y}^n_{AB,A} - \bar{Y}^n_{AB,B} \leq 0) \\
&= \int_{-\infty}^{0} \mathcal{N}(x | E(\bar{Y}^n_{AB,A}) - E(\bar{Y}^n_{AB,B}), V(\bar{Y}^n_{AB,A}) + V(\bar{Y}^n_{AB,B})) dx \\
&= \int_{-\infty}^{-(E(\bar{Y}^n_{AB,A}) - E(\bar{Y}^n_{AB,B}))} \mathcal{N}(x | 0, V(\bar{Y}^n_{AB,A}) + V(\bar{Y}^n_{AB,B})) dx.
\end{aligned}
$$

The second line to the third line uses the reproductive property of a normal distribution. In the above equation, ranking A and ranking B are interchangeable, and the error probability of interleaving $P(Error_I)$ is also given in the same way.

4　Theoretical Analysis

This section discusses the theoretical efficiency of interleaving. From the definition of efficiency given in Sect. 3.4, we see that the A/B testing and interleaving error probability rates depend on the sum of the variances and the difference between the expected click values. In particular, the smaller the sum of the variances, the smaller the error probability, and the larger the difference between the expected click values, the smaller the error probability. We investigate the relationship between the variance and difference in the expected click values in the following two cases: when the examination probability is constant and when it is relevance-aware.

Case of Constant Examination. The case of constant examination means the examination probability is constant for the relevance; in other words, the examination probability does not depend on the relevance of a ranking. For example, a position-based click model [5] in which the examination probability depends only on the item's position in the ranking is a constant case. Another example is a perfect click model in the cascade click model used in Sect. 6.

We show that interleaving has the same efficiency as A/B testing for the case of constant examination based on two theorems under the following conditions.

Condition 1. $E(O_{AB,A}) = E(O_{I,A}) = E(O_{AB,B}) = E(O_{I,B}) = c$: *The expected value of the examination is constant and the same between A/B testing and interleaving.*

Condition 2. $O \perp R$: *The random variable of the examination O and the random variable of the relevance R are independent from each other.*

Theorem 1. *If conditions 1 and 2 are satisfied, $E(\bar{Y}_{AB,A}^n) - E(\bar{Y}_{AB,B}^n) = E(\bar{Y}_{I,A}^n) - E(\bar{Y}_{I,B}^n)$ holds.*

Proof. When $E(O_{AB,A}) = E(O_{I,A})$ from condition 1,

$$E(Y_{AB,A}) = E(Y_{I,A}) \tag{5}$$

holds   because $E(Y_{AB,A})$ $=$ $E(S_{AB,A} \cdot O_{AB,A} \cdot R_A)$ $=$ $E(S_{AB,A})E(O_{AB,A})E(R_A) = E(S_{I,A})E(O_{I,A})E(R_A) = E(Y_{I,A})$ from condition 2 and $E(S_{AB,A}) = E(S_{I,A}) = \frac{1}{2}$. The same holds for ranking B, as $E(Y_{AB,B}) = E(Y_{I,B})$. Thus, $E(Y_{AB,A}) - E(Y_{AB,B}) = E(Y_{I,A}) - E(Y_{I,B})$. Therefore, $E(\bar{Y}_{AB,A}^n) - E(\bar{Y}_{AB,B}^n) = E(\bar{Y}_{I,A}^n) - E(\bar{Y}_{I,B}^n)$ holds, as $E(\bar{Y}_{AB,A}^n) = E(Y_{AB,A})$, $E(\bar{Y}_{AB,B}^n) = E(Y_{AB,B})$, $E(\bar{Y}_{I,A}^n) = E(Y_{I,A})$, and $E(\bar{Y}_{I,B}^n) = E(Y_{I,B})$ from Eq. (2).

Theorem 2. *If conditions 1 and 2 are satisfied, $V(\bar{Y}_{AB,A}^n) + V(\bar{Y}_{AB,B}^n) = V(\bar{Y}_{I,A}^n) + V(\bar{Y}_{I,B}^n)$ holds.*

Proof. From Eqs. (3) and (5), $V(\bar{Y}_{AB,A}^n) = \frac{1}{n}E(Y_{AB,A})(1 - E(Y_{AB,A})) = \frac{1}{n}E(Y_{I,A})(1 - E(Y_{I,A})) = V(\bar{Y}_{I,A}^n)$. Similarly, $V(\bar{Y}_{AB,B}^n) = V(\bar{Y}_{I,B}^n)$. Thus, $V(\bar{Y}_{AB,A}^n) + V(\bar{Y}_{AB,B}^n) = V(\bar{Y}_{I,A}^n) + V(\bar{Y}_{I,B}^n)$ holds.

From Theorems 1 and 2, both the difference and variance in the expected click values are the same for interleaving and A/B testing. Thus, interleaving has the same efficiency as A/B testing when the examination is constant.

Case of Relevance-Aware Examination. In the case of relevance-aware examination, the examination probability depends on the relevance of a ranking. For example, the navigational click model [5] is a relevance-aware case in which the click action depends on the relevance, and the click action affects the examination.

We consider the error rate when the examination is relevance-aware under the following conditions.

Condition 3. $E(R_A) > E(R_B) \wedge E(Y_A) > E(Y_B)$: *The expected value of the relevance and the click of ranking A is greater than the expected value of the relevance and the click of ranking B.*

Condition 4. $E(O_{I,\bullet} \cdot R_\bullet) \simeq f(max[E(R_A), E(R_B)])E(R_\bullet)$, where f is a monotonically decreasing function.

The second condition in function f reflects that the user is more likely to leave the ranking after clicking on an item with high relevance. In particular, a decreasing condition in f means the user is leaving the ranking, and max in f means the leaving behavior is affected more by higher relevance items. For example, users leave the ranking after interacting with the high relevance items in the navigational click model [5]. We show below that interleaving is more efficient than A/B testing when these two conditions hold, based on the following two theorems.

Theorem 3. If conditions 3 and 4 are satisfied, then $E(\bar{Y}^n_{I,A}) - E(\bar{Y}^n_{I,B}) > E(\bar{Y}^n_{AB,A}) - E(\bar{Y}^n_{AB,B})$. In other words, the difference in the expected click value of the interleaved comparison is greater than the A/B testing value.

Proof. From conditions 3 and 4, $E(O_{I,A} \cdot R_A) = f(max[E(R_A), E(R_B)]) E(R_A) = f(E(R_A))E(R_A)$. By interpreting the A/B test as mixing the same input rankings together, $E(O_{AB,A} \cdot R_A) = f(max[E(R_A), E(R_A)])E(R_A) = f(E(R_A))E(R_A)$.

Thus,

$$E(Y_{I,A}) = E(Y_{AB,A}), \tag{6}$$

as $E(O_{I,A} \cdot R_A) = E(O_{AB,A} \cdot R_A)$ holds. Similarly, $E(O_{AB,B} \cdot R_B) = f(E(R_B))E(R_B)$ holds by interpreting the A/B test as mixing the same input rankings. In addition, $E(O_{I,B} \cdot R_B) = f(max[E(R_A), E(R_B)])E(R_B) = f(E(R_A))E(R_B)$. Then, $E(O_{AB,B} \cdot R_B) > E(O_{I,B} \cdot R_B)$ holds because f is a monotonically decreasing function and $E(R_A) > E(R_B)$. Therefore,

$$E(Y_{AB,B}) > E(Y_{I,B}). \tag{7}$$

Furthermore,

$$E(Y_{AB,A}) > E(Y_{AB,B}), \tag{8}$$

as $E(Y_{AB,A}) = 2E(Y_A) > 2E(Y_B) = E(Y_{AB,B})$ holds from Eq. (4) and condition 4.

From Eqs. (6), (7), and (8), $E(Y_{I,A}) = E(Y_{AB,A}) > E(Y_{AB,B}) > E(Y_{I,B})$ holds. Using this relationship, we get $E(\bar{Y}^n_{I,A}) - E(\bar{Y}^n_{I,B}) > E(\bar{Y}^n_{AB,A}) - E(\bar{Y}^n_{AB,B})$, as $E(\bar{Y}^n_{AB,A}) = E(Y_{AB,A})$, $E(\bar{Y}^n_{AB,B}) = E(Y_{AB,B})$, $E(\bar{Y}^n_{I,A}) = E(Y_{I,A})$ and $E(\bar{Y}^n_{I,B}) = E(Y_{I,B})$ from Eq. (2). $\qquad$

Next, we show that the sum of the variances of interleaving is less than that of A/B testing.

Theorem 4. If conditions 3 and 4 hold, $V(\bar{Y}^n_{AB,A}) + V(\bar{Y}^n_{AB,B}) > V(\bar{Y}^n_{I,A}) + V(\bar{Y}^n_{I,B})$.

Proof. Recall that $V(\bar{Y}^n_{AB,\bullet}) = \frac{1}{n}E(Y_{AB,\bullet})(1 - E(Y_{AB,\bullet}))$ from Eq. (3). We note that $V(\bar{Y}^n_{AB,\bullet})$ is monotonically increasing according to $E(Y_{AB,\bullet})$ if the value of $E(Y_{AB,\bullet})$ is less than or equal to $\frac{1}{2}$. In fact, $E(Y_{AB,\bullet})$ is less than or equal to 0.5 because $E(Y_{AB,\bullet}) = E(S_{AB,\bullet} \cdot O_{AB,\bullet} \cdot R_{\bullet}) = E(S_{AB,\bullet})E(O_{AB,\bullet} \cdot R_{\bullet})$, where $E(S_{AB,\bullet}) = \frac{1}{2}$ and $E(O_{AB,\bullet} \cdot R_{\bullet}) \leq 1$. The same holds for $E(Y_{I,\bullet})$. From equation $E(Y_{I,A}) = E(Y_{AB,A})$ in (6), we get $V(\bar{Y}_{AB,A}^n) = V(\bar{Y}_{I,A}^n)$. Furthermore, from equation $E(Y_{AB,B}) > E(Y_{I,B})$ in (7), we get $V(\bar{Y}_{AB,B}^n) > V(\bar{Y}_{I,B}^n)$. Thus,

$$V(\bar{Y}_{AB,A}^n) + V(\bar{Y}_{AB,B}^n) > V(\bar{Y}_{I,A}^n) + V(\bar{Y}_{I,B}^n).$$

Based on Theorems 3 and 4, the difference in the expected click values is greater and the variance is lesser in interleaving than in A/B testing. Thus, interleaving is more efficient than A/B testing when the examination probability depends on the relevance and when the user leaves the ranking according to the relevance.

5 Numerical Analysis

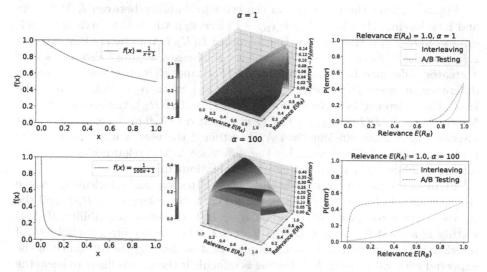

Fig. 1. Shape of the examination probability f

Fig. 2. Difference in the error probability between A/B testing and interleaving

Fig. 3. Error probability of A/B testing and interleaving

This section investigates how the examination probability function f affects the error probability in A/B testing and interleaving. We set $f(x) = \frac{1}{\alpha x+1}$, where $\alpha \in \{1, 100\}$ controls how likely a user is to leave the ranking based on the item's relevance. This function has the following properties that represent actual user behavior:

Table 1. Click Models

	$P(click = 1|R)$			$P(stop = 1|R)$		
R	0	1	2	0	1	2
Perfect	0.0	0.5	1.0	0.0	0.0	0.0
Navigational	0.0	0.5	1.0	0.0	0.5	1.0

- $f(0) = 1$; users must examine the item on the ranking at position $k + 1$ when the top-k items are irrelevant;
- f is a monotonically decreasing function; users leave the ranking based on its relevance.

Figure 1 shows the shape of function f, where the x-axis is the level of relevance and the y-axis is the examination probability. The bottom figure with $\alpha = 100$ represents cases when users are most likely to leave the ranking; the top figure with $\alpha = 1$ represents cases when they are least likely to do so.

Figure 2 shows the difference in the error probability between A/B testing and interleaving, that is, $P(Error_{AB}) - P(Error_I)$, where the x-axis is the relevance $E(R_A)$ and the y-axis is the relevance $E(R_B)$. We observe that when α is larger, the $P(Error_{AB}) - P(Error_I)$ level increases. Figure 2 also shows that a greater difference between relevances $E(R_A)$ and $E(R_B)$ indicates a greater difference between $P(Error_{AB})$ and $P(Error_I)$ when $\alpha = 100$, whereas the lesser the difference between relevance $E(R_A)$ and $E(R_B)$, the greater the difference between $P(Error_{AB})$ and $P(Error_I)$ at $\alpha = 1$. This result implies that interleaving is more efficient than A/B testing if the user is likely to leave the ranking based on relevance and if the difference in the relevance is large. We further validate these results using user simulations in Sect. 6.

Figure 3 shows the error probability of A/B testing and interleaving, where the relevance $E(R_A)$ is fixed at 1.0, the x-axis is the relevance $E(R_B)$, and the y-axis is the error probability. When $\alpha = 100$, the error probability of A/B testing is around 0.5 even if $E(R_B)$ is at 0.2, whereas the error probability of interleaving is around 0.0. Figure 3 implies that evaluating the difference in the expected click value using A/B testing is difficult if the user is likely to leave the ranking based on a small relevance level. In contrast, the interleaving method more stably evaluates the difference in the expected click value.

6   User Simulation

In this section, we present the results of our user simulations to answer the following research questions (RQs):

- **RQ1:** How does the user click model affect the efficiency of A/B testing and interleaving?
- **RQ2:** How does the variability in the relevance of the input rankings affect the error rate?

6.1 Datasets

We use multiple datasets previously adopted in interleaving research [19]. Most of the datasets are TREC web tracks from 2003 to 2008 [6,22,26]. HP2003, HP2004, NP2003, NP2004, TD2003, and TD2004 each have 50–150 queries and 1,000 items. Meanwhile, the OHSUMED dataset is based on the query logs of the search engine MEDLINE, an online medical information database, and contains 106 queries. MQ2007 and MQ2008 are TREC's million query track [1] datasets, which consist of 1,700 and 800 queries, respectively. The relevance labels are divided into three levels: irrelevant (0), relevant (1), and highly relevant (2).

We generate the input rankings A and B by sorting the items with the features used in past interleaving experiments [19,25]. We use the BM25, TF.IDF, TF, IDF, and LMIR.JM features for MQ2007. For the other datasets, we use the BM25, TF.IDF, LMIR.JM, and hyperlink features. Note that each feature's value is included in the dataset. We then generate a pair of input rankings A and B with $|I| = 5$ for all pairs of features. The source code of the ranking generations and user simulations are available in a public GitHub repository.[1]

6.2 User Behavior

User behavior is simulated in three steps. First, the ranking is displayed after the user issues a query. Next, the user decides whether to click on the items in the ranking. If the user clicks on an item, they will leave the ranking according to its relevance label. The details of this user behavior are as follows.

Ranking Impressions. First, the user issues a pseudo-query by uniformly sampling queries from the dataset. The interleaving model generates an interleaved ranking and displays the ranking to the user. The IMA was used as the interleaving method in this experiment. In each ranking impression, up to five items are shown to the user. After this ranking impression, the user simulates a click using the click model.

Click Model. The cascade click model is used to simulate clicks. Table 1 presents the two types of the cascade click model, where $P(click = 1|R)$ represents how the user clicks the item according to the relevance R, and $P(stop = 1|R)$ represents how the user leaves the ranking according to the relevance after the click. The perfect click model examines all items in the ranking and clicks on all relevant items. The navigational click model simulates a user looking for a single relevant item.

6.3 Results

RQ1: How Does the User Click Model Affect the Efficiency of A/B Testing and Interleaving? Figure 4 presents the error rate over impressions, that is, the efficiency of each click model. The x-axis represents the number

---

[1] https://github.com/mpkato/interleaving.

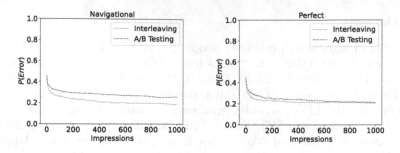

Fig. 4. Efficiency over impressions.

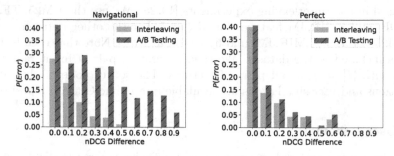

Fig. 5. Error rate for the nDCG difference between rankings.

of ranking impressions, and the y-axis represents the error rate. The values in Fig. 4 show the average values. The maximum number of impressions is 1,000, and each impression corresponds to a randomly selected query, i.e., one query has one impression. This evaluation procedure is repeated 10 times for each dataset.

The results show that interleaving is more efficient and has fewer evaluation errors than A/B testing. The results are consistent with those of the efficiency analysis, as explained in the numerical experiment showing that interleaving is more efficient than A/B testing if the user is likely to leave the ranking after clicking on the item with the highest relevance. Thus, the answer to RQ1 is that interleaving is efficient for the navigational click model, where users leave a ranking according to relevance.

RQ2: How Does the Variability in the Relevance of the Input Rankings Affect the Error Rate? Figure 5 presents the error rate for the nDCG difference between each pair of ranking A and B. The x-axis represents the nDCG difference, and the y-axis represents the error rate. The values in Fig. 5 show the average values for all queries. We randomly select 1,000 queries that allow query duplication for each dataset. The number of impressions is 1,000 for each query, i.e., one query has 1,000 impressions, and the evaluation error is calculated after 1,000 impressions for each query. This evaluation procedure is repeated 10 times for each dataset.

The navigational click model shows a lower error rate for interleaving than for A/B testing. The greater difference between the relevances of ranking A and B corresponds to the greater difference in error rates between interleaving and A/B testing. These results are consistent with those of our previously illustrated numerical experiment, which shows that interleaving has a lower error rate than A/B testing when the difference in the relevance is large and the user is likely to leave the ranking with relevance, whereas the perfect click model shows a small or equal difference in the error rate. From the above analysis, the answer to RQ2 is that based on the navigational click model, interleaving has a low error rate when the ranking pair has a large difference in relevance.

7   Conclusion

This study presented a theoretical analysis on the efficiency of interleaving, an online evaluation method, in the information retrieval field. We first designed a simple interleaving method similar to other interleaving methods, our analysis of which showed that interleaving is more efficient than A/B testing when users leave a ranking according to the item's relevance. The experiments verified the efficiency of interleaving, and the results according to the theoretical analysis were consistent with the empirical results.

Our theoretical analyses are limited in that they assume no duplication items exist in the A and B rankings. Item duplication might further contribute to efficiency in some cases beyond our analysis. However, we believe our basic analysis could be a first step toward discussing the efficiency of more general interleaving settings. The next challenge includes investigating what other examination probability functions satisfy the efficiency of interleaved or multileaved comparisons.

References

1. Allan, J., Carterette, B., Aslam, J.A., Pavlu, V., Dachev, B., Kanoulas, E.: Million query track 2007 overview. Tech. Rep., Massachusetts University Amherst Department of Computer Science (2007)
2. Brost, B., Cox, I.J., Seldin, Y., Lioma, C.: An improved multileaving algorithm for online ranker evaluation. In: Proceedings of the 39th International ACM SIGIR conference on Research and Development in Information Retrieval, pp. 745–748 (2016)
3. Chapelle, O., Joachims, T., Radlinski, F., Yue, Y.: Large-scale validation and analysis of interleaved search evaluation. ACM Trans. Inf. Syst. (TOIS) **30**(1), 1–41 (2012)
4. Chapelle, O., Zhang, Y.: A dynamic Bayesian network click model for web search ranking. In: Proceedings of the 18th International Conference on World Wide Web, pp. 1–10 (2009)
5. Chuklin, A., Markov, I., De Rijke, M.: Click models for web search. Morgan & Claypool Publishers (2015)
6. Clarke, C.L., Craswell, N., Soboroff, I.: Overview of the TREC 2009 web track. Tech. rep., Waterloo University (Ontario) (2009)

7. Craswell, N., Zoeter, O., Taylor, M., Ramsey, B.: An experimental comparison of click position-bias models. In: Proceedings of the 1st ACM International Conference on Web Search and Data Mining, pp. 87–94 (2008)
8. Deng, A., Xu, Y., Kohavi, R., Walker, T.: Improving the sensitivity of online controlled experiments by utilizing pre-experiment data. In: Proceedings of the 6th ACM International Conference on Web Search and Data Mining, pp. 123–132 (2013)
9. Dupret, G.E., Piwowarski, B.: A user browsing model to predict search engine click data from past observations. In: Proceedings of the 31st Annual International ACM SIGIR Conference on Research and Development in Information Retrieval, pp. 331–338 (2008)
10. Grbovic, M., Cheng, H.: Real-time personalization using embeddings for search ranking at airbnb. In: Proceedings of the 24th ACM SIGKDD International Conference on Knowledge Discovery and Data Mining, pp. 311–320. KDD 2018, Association for Computing Machinery (2018)
11. Guo, F., Liu, C., Wang, Y.M.: Efficient multiple-click models in web search. In: Proceedings of the Second ACM International Conference on Web Search and Data Mining, pp. 124–131 (2009)
12. Hofmann, K., Whiteson, S., De Rijke, M.: A probabilistic method for inferring preferences from clicks. In: Proceedings of the 20th ACM International on Conference on Information and Knowledge Management, pp. 249–258 (2011)
13. Hofmann, K., Whiteson, S., Rijke, M.D.: Fidelity, soundness, and efficiency of interleaved comparison methods. ACM Trans. Inf. Syst. (TOIS) 31(4), 1–43 (2013)
14. Iizuka, K., Seki, Y., Kato, M.P.: Decomposition and interleaving for variance reduction of post-click metrics. In: Proceedings of the 2021 ACM SIGIR International Conference on Theory of Information Retrieval, pp. 221–230 (2021)
15. Joachims, T.: Optimizing search engines using clickthrough data. In: Proceedings of the Eighth ACM SIGKDD International Conference on Knowledge Discovery and Data Mining, pp. 133–142 (2002)
16. Kharitonov, E., Macdonald, C., Serdyukov, P., Ounis, I.: Using historical click data to increase interleaving sensitivity. In: Proceedings of the 22nd ACM International Conference on Information & Knowledge Management, pp. 679–688 (2013)
17. Kharitonov, E., Macdonald, C., Serdyukov, P., Ounis, I.: Generalized team draft interleaving. In: Proceedings of the 24th ACM International on Conference on Information and Knowledge Management, pp. 773–782 (2015)
18. Okura, S., Tagami, Y., Ono, S., Tajima, A.: Embedding-based news recommendation for millions of users. In: Proceedings of the 23rd ACM SIGKDD International Conference on Knowledge Discovery and Data Mining, pp. 1933–1942. KDD 2017, Association for Computing Machinery (2017)
19. Oosterhuis, H., de Rijke, M.: Sensitive and scalable online evaluation with theoretical guarantees. In: Proceedings of the 2017 ACM on Conference on Information and Knowledge Management, pp. 77–86. CIKM 2017, Association for Computing Machinery (2017)
20. Oosterhuis, H., de Rijke, M.: Taking the counterfactual online: efficient and unbiased online evaluation for ranking. In: Proceedings of the 2020 ACM SIGIR on International Conference on Theory of Information Retrieval, pp. 137–144 (2020)
21. Poyarkov, A., Drutsa, A., Khalyavin, A., Gusev, G., Serdyukov, P.: Boosted decision tree regression adjustment for variance reduction in online controlled experiments. In: Proceedings of the 22nd ACM SIGKDD International Conference on Knowledge Discovery and Data Mining, pp. 235–244 (2016)

22. Qin, T., Liu, T.Y., Xu, J., Li, H.: LETOR: a benchmark collection for research on learning to rank for information retrieval. Inf. Retrieval **13**(4), 346–374 (2010)
23. Radlinski, F., Craswell, N.: Optimized interleaving for online retrieval evaluation. In: Proceedings of the 6th ACM International Conference on Web Search and Data Mining, pp. 245–254 (2013)
24. Radlinski, F., Kurup, M., Joachims, T.: How does clickthrough data reflect retrieval quality? In: Proceedings of the 17th ACM conference on Information and Knowledge Management, pp. 43–52 (2008)
25. Schuth, A., Sietsma, F., Whiteson, S., Lefortier, D., de Rijke, M.: Multileaved comparisons for fast online evaluation. In: Proceedings of the 23rd ACM International Conference on Information and Knowledge Management, pp. 71–80 (2014)
26. Voorhees, E.M., Harman, D.: Overview of TREC 2003. In: TREC, pp. 1–13 (2003)
27. Xie, H., Aurisset, J.: Improving the sensitivity of online controlled experiments: case studies at Netflix. In: Proceedings of the 22nd ACM SIGKDD International Conference on Knowledge Discovery and Data Mining, pp. 645–654 (2016)

Intention-Aware Neural Networks for Question Paraphrase Identification

Zhiling Jin, Yu Hong[✉], Rui Peng, Jianmin Yao, and Guodong Zhou

School of Computer Science and Technology, Soochow University, Suzhou, China
tianxianer@gmail.com, {jyao,guodongzhou}@suda.edu.cn

Abstract. We tackle Question Paraphrasing Identification (QPI), a task of determining whether a pair of interrogative sentences (i.e., questions) are paraphrases of each other, which is widely applied in information retrieval and question answering. It is challenging to identify the distinctive instances which are similar in semantics though holding different intentions. In this paper, we propose an intention-aware neural model for QPI. Question words (e.g., "*when*") and blocks (e.g., "*what time*") are extracted as features for revealing intentions. They are utilized to regulate pairwise question encoding explicitly and implicitly, within Conditional Variational AutoEncoder (CVAE) and multi-task VAE frameworks, respectively. We conduct experiments on the benchmark corpora QQP, LCQMC and BQ, towards both English and Chinese QPI tasks. Experimental results show that our method yields generally significant improvements compared to a variety of PLM-based baselines (BERT, RoBERTa and ERNIE), and it outperforms the state-of-the-art QPI models. It is also proven that our method doesn't severely reduce the overall efficiency, which merely extends the training time by 12.5% on a RTX3090. All the models and source codes will be made publicly available to support reproducible research.

Keywords: Information retrieval · Natural language processing · Paraphrase identification · Deep learning

1 Introduction

QPI aims to determine whether two interrogative sentences induce the same answer [25]. It serves as a crucial and practical technique in the field of factoid question answering, dialogue systems and information retrieval [1,13,23,24].

QPI can be boiled down to a binary classification task that determines either mutually-paraphrased cases or non-paraphrased. Different from the naive paraphrase identification, however, QPI is required to verify not only semantic equivalence of contexts but the consistency of intentions. For example, the two questions in 1) imply different intentions (i.e., *Person* and *Definition*), although their contexts are closely similar in semantics, where solely detecting semantic equivalence is not reliable enough for determination.

J. Kamps et al. (Eds.): ECIR 2023, LNCS 13980, pp. 474–488, 2023.
https://doi.org/10.1007/978-3-031-28244-7_30

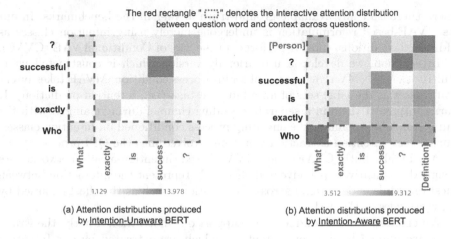

The red rectangle " ⸬ " denotes the interactive attention distribution
between question word and context across questions.

(a) Attention distributions produced
by Intention-Unaware BERT

(b) Attention distributions produced
by Intention-Aware BERT

Fig. 1. Attention heat-maps obtained by different BERT-based QPI models. A lighter color indicates lower attention.

1) *Q1: Who exactly is successful?*   [Intention: **Person**]
   Q2: What exactly is success?   [Intention: **Definition**]

A variety of sophisticated neural QPI approaches have been proposed [4, 8, 15, 18, 27, 31]. Recently, the large Pre-trained Language Models (abbr., PLMs) are leveraged as crucial supportive encoders for neural QPI [6, 7, 20, 21, 26, 32]. All the aforementioned arts achieve significant improvements, which increase QPI performance up to the accuracy rates of 91.6%, 88.3% and 85.3% on the benchmark corpora QQP, LCQMC and BQ.

We suggest that the existing PLM-based QPI models can be further enhanced by appropriately highlighting the interaction information of question words. This is motivated by the findings as below:

- Error analysis shows that there are at least 32.5% QPI errors occurred due to the failure to perceive intentions.
- Lower perception of intentions is ascribed to the negligible interactive attention between question words and contexts, such as that occurred in the interaction representation produced by the widely-used BERT-based QPI baselines [3, 21, 32]. Figure 1-(a) offers an example of underestimated interactive attention which is marked with red rectangles.

In this study, we propose an intention-aware approach to enhance PLM-based QPI models. An heuristic intention extractor is designed to determine intention classes in terms of question words or blocks. On the basis, the semantic representations of questions produced by PLMs will be reformulated with the intervention of intention classes, where Variational AutoEncoder (VAE) is used. It has been proven effective in answer-**aware** QPI [30] due to the resultant adaptation to vagueness. It is noteworthy that we tackle answer-**unaware** QPI, where there isn't any available ground-truth answer in the benchmarks. In our case, VAE-based reformulation is undergone merely using intention classes as additional conditions, thus, it performs in the way of Conditional VAE (CVAE).

In addition, we develop a user-friendly version which is constructed using Multi-task VAE (MVAE). Instead of an in-process condition, MVAE takes intention classes as the label sets of an additional task (i.e., intention prediction). It learns to predict them in a separate decoding channel during training, with the aim to implicitly regulate the encoding process conditioned on intention classes. During test, intention extraction won't be involved into the QPI process.

All in all, both CVAE and MVAE-based intention-aware networks are designed to sensitively perceive and effectively represent the interaction between question words and contexts across the questions, such as the effects marked by red rectangles in Fig. 1-(b).

An error analysis is conducted to support our motivation. Firstly, the wrong cases are filtered by different intentions which are extracted by our Intention Extraction module. After that, we manually annotate the cases with similar content but different intentions like case 1) did. Experiments on the benchmarks of English QQP as well as Chinese LCQMC and BQ show that our approach improves a broad set of PLM models like BERT [7], RoBERTa [20], ERNIE [26] and their variants [6]. Besides, it outperforms the state-of-the-art QPI models without using any external data. Our contributions can be summarized as below:

- The proposed intention-aware network improves the current PLMs-based QPI models.
- We provide a user-friendly version, i.e., the MVAE-based network. It performs slightly worse than the CVAE-based version, though it dispenses with intention extraction and in-process intention encoding during test.
- Both CVAE and MVAE-based networks are vest-pocket (Sect. 4). Each of them extends the training time by no more than 3 h on a RTX3090.

The rest of the paper is organized as follows. Section 2 overviews the related work. We briefly introduce the fundamental of VAE as well as our heuristic intention extractor in Sect. 3. We present our intent-aware networks and QPI models in Sect. 4. Section 5 provides the experimental results and analysis. We conclude the paper in Sect. 6.

2 Related Work

QPI has been widely studied due to its widespread application in downstream tasks. The existing approaches can be divided into two categories: representation-based approaches and interaction-based approaches.

The representation-based approaches generally leverages a Siamese architecture to separately encode two questions into different vectors of high-level features. On this basis, the semantic-level similarity is computed between their feature vectors. He et al. [9] propose a siamese Convolutional Neural Network (CNN) for encoding, and they verify the effectiveness of cosine similarity,

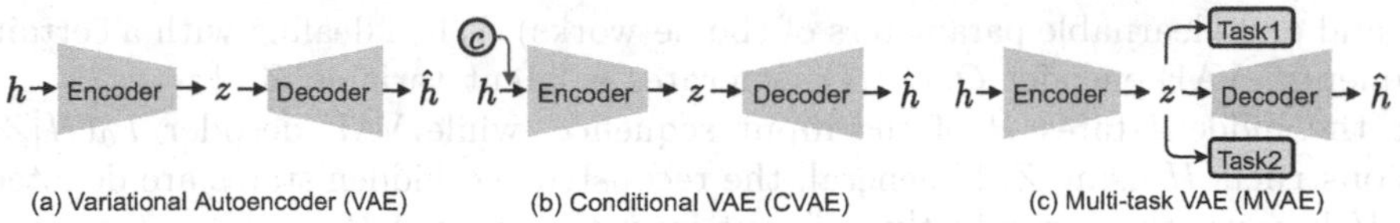

Fig. 2. Architectures of VAE, CVAE and MVAE.

Euclidean distance and element-wise difference. Wang et al. [28] decompose two questions to the sets of similar and different tokens. They use CNN to extract shareable and distinguishable features respectively from such token sets, so as to improve the reliability of feature vectors for similarity measurement. Lai et al. [18] intend to resolve the ambiguity caused by Chinese word segmentation and propose a lattice based CNN model (LCNs). LCNs successfully leverages multi-granularity information inherently implied in the word lattice.

Basically, an interaction-based QPI model is designed to perceive interaction information between questions, and incorporate such information into the representation learning process. Specifically, Wang et al. [27] propose a BiMPM model under the matching-aggregation framework. A variety of one-to-many token-level matching operations are conducted bidirectionally to obtain cross-sentence interaction information for every encoding time step. Chen's ESIM [4] model performs interactive inference based on LSTM representations, where element-wise dot product and subtraction are used to sharpen the interaction information. Inspired by ResNet [10], Kim et al. [15] propose a densely-connected recurrent and co-attentive Network (DRCN). It combines RNN with attention interaction and residual operation. Nowadays, PLMs are successfully leveraged in QPI. Fundamentally, they appear as a kind of distinctive interaction-based models due to their interactive attention computation among tokens. Zhang et al. [32] propose a Relation Learning Network (R^2-Net) based on BERT, i.e., a prototypical transformer-based PLM that contains multi-head interactive attention mechanism. R^2-Net is characterized by interactive relationship perception of multi-granularity linguistic units (tokens and sentences). Lyu's LET-BERT [21] model perceives multi-granularity interaction based on words lattice graph, where graph attention networks and external knowledge of *HowNet* are used.

3   Preliminary

To facilitate the understanding of our approach, we briefly introduce the family of VAE models (VAE, CVAE and MVAE) and the method of intention extraction.

3.1   VAE, CVAE and MVAE

VAE [16] is developed as encoder-decoder networks (as shown in Fig. 2-a), which possesses an encoder network $Q_\phi(Z|H)$ and a decoder network $P_\theta(H|Z)$

(ϕ and θ are learnable parameters of the networks). When dealing with a certain sequence, VAE encoder $Q_\phi(Z|H)$ generates a latent variable $Z=\{z_1, z_2, ..., z_t\}$ for the hidden states H of the input sequence, while VAE decoder $P_\theta(H|Z)$ reconstructs H using Z. In general, the reconstructed hidden states are denoted as $\hat{H}$ because they are slightly different from the original H.

During a self-supervised training process, VAE encoder learns to produce Z by random sampling from the variable distributions, the ones qualified by probability density function $F(H)$, which are not only akin to H, but approximate the contrastive distribution $P(Z)$ (e.g., Gaussian distribution). This enables VAE encoder to produce various homogeneous representations (i.e., variables Zs) for the same input (due to the uncertainty of random sampling). Therefore, VAE helps to improve training effects when training data is sparse or the training set is full of redundant and duplicate samples. More importantly, the variations of representations enable the decoder to adapt to vagueness.

CVAE [17] is a variant of VAE. As shown in Fig. 2.(b), CVAE utilizes an extra condition C as the additional input to regulate the encoding and decoding processes: $Q_\phi(Z|H, C)$ and $P_\theta(H|Z, C)$. The distribution of resultant latent variable Z of CVAE is inclined to the information of C, more or less.

MVAE involves VAE into a multi-task learning architecture. The latent variable Z is shareable by VAE decoder and other task-specific decoders, as shown in Fig. 2 (c). During training, the reconstruction loss (same as VAE) and task-oriented losses are combined to collaboratively optimize VAE encoder.

In our experiments, both encoders and decoders of the aforementioned VAE, CVAE and MVAE are constructed with BiGRU [5].

3.2   Heuristic Intention Extraction

Our approach relies on the intention extraction which classifies questions into different intention classes. In our experiments, the intention of an English question is extracted in terms of question word or block (e.g., "*When*", "*Why*", "*How*", etc.) and fixed patterns. We consider eight intention classes, including *Approach*, *Reason*, *Location*, *Time*, *Definition*, *Choice*, *Person* and *Normal*. For example, the questions that begins with "*What's a good way*" and "*How*" are classified into the "*Approach*"-type intention. A series of predefined regular expressions are used to extract the intentions, such as the case in 2).

2) "*What .* way .* ?*"   [Regular expression]
"***What*** would be the best ***way*** to control anger?" [Note: the underlined text span matches the expression]

To be specific, *Normal* is mainly kind of questions with *Yes or No*. We group questions outside the other seven types as *Normal*. However, it is challenging to extract intentions from Chinese questions due to the omission of explicit and canonical question words. Far more serious than time consuming, manually designing regular expressions for Chinese questions is not generally applicable. To address the issue, we utilize an off-the-shelf Machine Translation (MT) system

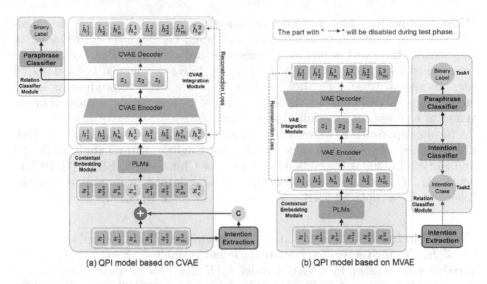

(a) QPI model based on CVAE   (b) QPI model based on MVAE

Fig. 3. CVAE-based intention-aware QPI model. Note that all components are considered during training, though CVAE decoder is disabled in the test phase.

to translate Chinese questions into English. On the basis, we use English regular expressions to extract intentions. The intention class will be used in the CVAE and MVAE-based networks (see next section).

4 Approach

Let us first define QPI task in an accurate way. Given a pair of questions $X_1=\{x_1^1, x_2^1, ...x_n^1\}$ and $X_2=\{x_1^2, x_2^2, ...x_m^2\}$, a QPI model determines whether X_1 and X_2 are paraphrases of each other. During training, each input can be represented as (X_1, X_2, y), where y is a ground-truth label indicating the relationship between X_1 and X_2: either *Paraphrase* or *Non-paraphrase*. The goal of training is to optimize the QPI model $f(X_1, X_2)$ by minimizing prediction loss.

We develop two intention-aware QPI models using the variants of VAE, including CVAE and MVAE. The CVAE-based version utilizes the extracted intention class as an external condition to regulate VAE encoder during both the training and test phases. By contrast, the MVAE-based version regards the extracted intention classes as pseudo ground truth, and uses them to collaboratively regulate VAE encoder in an auxiliary task (intention prediction task) during training. During test, the invocation of the auxiliary task is cancelled for the purpose of increasing the efficiency.

4.1 CVAE-Based Intention-Aware QPI

We show the architecture of CVAE-based QPI in Fig. 3(a). It comprises three primary modules, including PLM-based embedding layer, intention-aware CVAE and binary classification for QPI.

PLM Embedding Layer. Given two questions X_1 and X_2, we use the intention extractor to obtain their intention class labels C_1 and C_2. On this basis, we concatenate the word sequences of questions and their intention class tags, in the form of $X_c=\{[CLS], X_1, C_1, [SEP], X_2, C_2, [SEP]\}$, where $[CLS]$ and $[SEP]$ serve as special tokens. Further, we feed X_c into a PLM to produce its initial embeddings H. PLM enables the information of C_1 and C_2 (i.e., signals of intentions) to be fused into the initial embeddings H.

In our experiments, we examine the effects of different PLMs, all of which are grounded on the transformer encoder. For English QPI, we apply BERT [7] and RoBERTa [20] as the embedding layer. For Chinese QPI, BERT [7], BERT-wwm, BERT-wwm-ext [6] and ERNIE [26] are adopted. Note that the special tokens of X_c is replaced with $<s>$ and $</s>$ when RoBERTa is used.

CVAE Module. We feed the initial representation H into the CVAE encoder $Q_\phi(Z|H,C)$ which serves to generate the latent variable Z ($Z=\{h_1^z, ..., h_t^z\}$). The variable will be taken by CVAE decoder $P_\theta(H|Z,C)$ for reconstructing H. The reconstruction loss and Kullback-Leible (KL) divergence [12] are computed for optimizing CVAE networks, where KL divergence is estimated by comparing to normal Gaussian distribution $P_{Gaus}(Z)$:

$$L_{CVAE}(\theta, \phi; H, C) = E[logP_\theta(H|Z,C)] - D_{KL}[Q_\phi(Z|H,C)||P_{Gaus}(Z)] \quad (1)$$

Classification for QPI. We adopt the variable Z as the intention-aware representation of the question pair. We feed Z into a two-layer fully-connected (FC) network which serves as the binary classifier. The FC network estimates the probabilities $p(\hat{y}|Z)$ that the questions are paraphrases of each other. Softmax is computed for binary classification over the probabilities $\{p(\hat{y}|Z), 1-p(\hat{y}|Z)\}$. The optimization objective of QPI is to minimize the cross-entropy loss as below:

$$\begin{aligned} L_{QPI} = &-(y * log(p(\hat{y}|Z)) \\ &+(1-y) * log(1 - p(\hat{y}|Z))). \end{aligned} \quad (2)$$

Training. We perform both CVAE and QPI for each batch of data during training. And both the losses L_{CVAE} and L_{QPI} are considered in the process of back propagation for optimization. Scale factors α and β are used to reconcile the losses in the synthetic optimization objective: $Loss = \alpha L_{CVAE} + \beta L_{QPI}$. During test, the CVAE decoder is pruned, which will not affect QPI.

4.2   MVAE-Based Intention-Aware QPI

The CVAE-based model suffers from the inconvenience that intention extraction needs to be conducted for each input, no matter whether QPI performs in the training or test phases. This leads to inefficiency for practical application.

Therefore, we additionally develop a user-friendly version using MVAE as the backbone. It regards the heuristically-extracted intention labels as pseudo ground truth for intention prediction. Multi-task learning is conducted during training to optimize the shareable VAE encoder, where the tasks of self-supervised VAE, intention prediction and QPI are considered. Figure 3(b) shows its architecture. During test, both VAE decoder and the channel of intention prediction is pruned. This model avoids the inconvenient run-time intention extraction. The computational details are as below.

PLM Embedding Layer. The embedding layer concatenates the word sequences of the questions: $X = \{[CLS], X_1, [SEP], X_2, [SEP]\}$. We feed X into a PLM to produce the initial representation H. The applied PLM models are the same with those in the CVAE-based version.

MVAE Module. We feed H into VAE encoder $Q_\phi(Z|H)$ to generate latent variable Z. Different from CVAE encoder $Q_\phi(Z|H, C)$, in this case, we exclude the condition C, and the signals of intentions are not directly fused into H and Z. The decoder network $P_\theta(H|Z)$ reconstructs H from Z. The optimization objective function is simplified as below:

$$L_{MVAE}(\theta, \phi; H) = E[log P_\theta(H|Z)] \\ -D_{KL}[Q_\phi(Z|H)\|P_{Gaus}(Z)] \tag{3}$$

Multi-task Learning. There are two tasks considered besides the self supervised VAE, including QPI and intention prediction. For QPI, the latent variable Z is fed into the two-layer FC network, and the probability $p(\hat{y}|Z)$ of mutual paraphrasing is estimated. We adopt the same optimization objective as Eq(2).

For intention prediction, we construct a simple neural intention classifier which merely possesses two FC layers with Softmax. We feed the latent variable Z into the classifier to perform 8-class intention classification (see the classes in Sect. 3.2). In this case, the heuristically-extracted intentions are used as pseudo ground truth to lead the supervised learning of intention prediction and VAE encoding. We hope that VAE encoder can learn to perceive intentions and highlight such signals in the representation Z. Cross-entropy loss L_{INT} is used for optimizing the predictor and shareable VAE encoder during back propagation.

Training. We perform MVAE for each batch of data during training, where the tasks are conducted in the order of VAE, intention predictor and QPI. Another scale factor γ is introduced into the reconciliation among the losses of different tasks. Correspondingly, the optimization objective is computed as: $Loss=\alpha L_{MVAE} + \beta L_{QPI} + \gamma L_{INT}$.

5   Experimentation

In this section, experimental settings and result analysis will be detailed introduced. Advanced methods of previous research are compared with our approach to verify the effectiveness of proposed methods.

Table 1. Statistics for LCQMC, BQ and QQP

Dataset	Size	Pos:Neg	Domain
LCQMC	260,068	1:0.7	Open-Domain
BQ	120,000	1:1	Bank
QQP	404,276	1:2	Open-Domain

5.1   Corpora and Evaluation Metrics

We evaluate our QPI models on three benchmark corpora, including English QQP [14], as well as Chinese LCQMC [19] and BQ [2].

Both QQP and LCQMC are large-scale open-domain corpora, where question pairs are natural as they are collected from QA websites (e.g., Zhihu and Quora) without rigorous selection. By contrast, BQ is a domain-specific corpus which contains various banking business questions. Each instance in all the three corpora is specified as a pair of sentence-level questions. The binary labels of *paraphrase* and *non-paraphrase* are provided. Table 1 shows the statistical information of the corpora. We strictly adhere to the canonical splitting method of training, validation and test sets for the corpora. Besides, we evaluate all the models in the experiments using accuracy (*ACC.*) and F1-score.

5.2   Hyperparameter Settings

We set our experiments on the HuggingFace's Transformers Library [29] with Pytorch version. We adopt an Adam optimizer with epsilon of 1e-8, and the number of BiGRU layers in both VAE/CVAE encoder and decoder is set to 2. For Chinese tasks, we set the learning rate to 1e-5 and batch size to 16 for LCQMC, and 2e-5 and 32 for BQ. The fine-tuning epoch is set to 5 for both datasets. For English QPI, we fine-tune the models for 50K steps, and checkpoints are evaluated every 1,000 steps. The learning rate is set to 2e-5 and batch size is 64. The others are set same as Chinese task. The hidden size of basic PLMs is 768, while that of large PLMs is 1,024. The hyper-parameters used to reconcile the loss are respectively set as below: $\alpha = 0.01, \beta = 0.99, \gamma = 0.1$. The hyperparameters settings are depending on our experience and intuition. We conduct simple experiments on adjusting them, it may still have better settings yet. All experiments are conducted on a single RTX 3090 GPU.

5.3   Main Results

Table 2 shows the main results on LCQMC, BQ and QQP. The reported performance of our models are average scores obtained in five random runs.

Chinese QPI. Our CVAE and MVAE-based QPI models yield significant improvements (p-value < 0.05 in statistical significance test) compared to the PLMs baselines. This demonstrates that our approaches generalize well when cooperating with different PLMs. In addition, our QPI models outperform all

Table 2. Comparison to the performance (%) of baselines and previous work on Chinese LCQMC, BQ and English QQP. The mark "◇" denotes the PLM-based QPI baselines, while "♣" denotes our CVAE and MVAE-based QPI models that obtain significant improvements (p-value < 0.05 in statistical significance test) over baselines.

Chinese Task					English Task		
Models	LCQMC		BQ		Models	QQP	
	ACC	F1	ACC	F1		ACC	F1
Text-CNN [11]	72.8	75.7	68.5	69.2	CENN [31]	80.7	\
BiLSTM [22]	76.1	78.9	73.5	72.7	L.D.C [28]	85.6	\
Lattice-CNN [18]	82.1	82.4	78.2	78.3	BiMPM [27]	88.2	\
BiMPM [27]	83.3	84.9	81.8	81.7	DIIN [8]	89.1	\
ESIM [4]	82.5	84.4	81.9	81.9	DRCN [15]	90.2	\
LET-BERT [21]	88.3	88.8	85.3	84.9	R^2-Net [32]	91.6	\
BERT [7]◇	85.7	86.8	84.5	84.0	BERT [7]◇	90.9	87.5
+ MVAE♣	88.8	88.9	85.6	85.5	+ MVAE♣	91.5	88.7
+ CVAE♣	**89.2**	89.3	**85.8**	85.5	+ CVAE♣	91.6	88.7
BERT-wwm [6]◇	86.8	87.7	84.8	84.2	BERT$_{large}$ [7]◇	91.0	87.7
+ MVAE♣	88.7	88.9	85.7	85.6	+ MVAE♣	91.8	89.0
+ CVAE♣	88.9	89.2	**85.8**	**85.8**	+ CVAE♣	91.8	89.0
BERT-wwm-ext [6]◇	86.6	87.7	84.7	83.9	RoBERTa [20]◇	91.4	88.4
+ MVAE♣	88.9	89.1	85.5	85.5	+ MVAE♣	91.8	89.1
+ CVAE♣	**89.2**	**89.4**	85.5	85.5	+ CVAE♣	91.8	89.2
ERNIE [26]◇	87.0	88.0	84.7	84.2	RoBERTa$_{large}$ [20]◇	91.9	89.1
+ MVAE♣	88.9	89.1	85.4	85.3	+ MVAE♣	**92.3**	**89.6**
+ CVAE♣	89.1	89.3	85.5	85.3	+ CVAE♣	**92.3**	**89.6**

the state-of-the-art models, including the recently-proposed LET-BERT [21]. The issues of complex Chinese word segmentation and ambiguity are successfully addressed to some extent by the use of external knowledge base *HowNet*. By contrast, our intention-aware approaches stably outperforms LET-BERT without using any external data. Therefore, we suggest that being aware of question intentions is as important as the contextual semantics for Chinese QPI.

English QPI. The previous work didn't report F1-scores on QQP. We evaluate our models with the F1 metric and report the performance to support the future comparative study. It can be observed that, similarly, CVAE and MVAE produce general improvements compared to the PLM-based baselines. The state-of-the-art approach R^2-Net [32] contains both CNN and BERT and is additionally trained in a scenario of congeniality recognition among multiple instances. Our approach achieves comparable performance to R^2-Net when BERT is used as the backbone. Besides, CVAE and MVAE-based QPI models outperform the advanced models when large PLMs (BERT$_{large}$, RoBERTa and RoBERTa$_{large}$) are used as backbones.

Comparing the performance between Chinese and English, we observe that the improvements we produce for English QPI are not as good as that for Chinese. It is most probably because that, different from English questions, the intentions of Chinese questions are generally obscure due to the lack of explicit question words. Therefore, the prior intention extraction and straight induction of intentions for encoding easily make positive effects on Chinese QPI.

Table 3. Ablation experiments on all benchmarks.

Models	LCQMC		BQ		QQP	
	ACC	F1	ACC	F1	ACC	F1
BERT	85.7	86.8	84.5	84.0	90.9	87.5
+ intention	87.3	88.1	85.1	84.6	91.2	88.2
+VAE	86.3	87.2	84.8	84.4	90.9	87.7
+CVAE+intention	**89.2**	**89.3**	85.8	**85.5**	**91.6**	**88.7**
+MVAE+intention	88.8	88.9	85.6	**85.5**	91.5	**88.7**

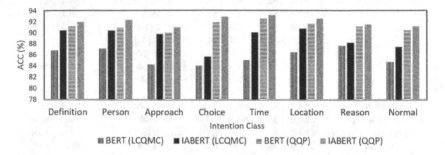

Fig. 4. Performance comparison between intention **Unaware** and intention **Aware** scenarios for all intention classes.

5.4   Ablation Experiments

We conduct ablation experiments using BERT as the baseline, which is connected with the two-layer FC-based classifier. There are four expanded models considered in the ablation experiments, including 1) "+intention" that merely concatenates the intention class with the input questions at the BERT-based embedding layer, 2) "+VAE" couples BERT with VAE before connecting it with the classifier, where intentions are disabled, 3) "+CVAE+intention" combines CVAE with BERT, where intentions are used for inducing the encoding process (as shown in Fig. 3.(a)), and 4) "+MVAE+intention" performs multi-task learning using intentions as pseudo ground truth (as shown in Fig. 3.(b)).

Table 3 shows the ablation performance. It can observed that merely connecting BERT with VAE produces insignificant improvements. By contrast, simply incorporating intention information into embeddings of questions yields substantial improvements. Besides, the utilization of CVAE and MVAE further improves the performance.

We also prove that the performance of MVAE is slightly worse than CVAE. It demonstrates that straight induction by intention class (CVAE) is more effective than indirect supervision from an auxiliary task (MVAE). Though, MVAE is more user-friendly for dispensing with intention extraction during test.

5.5 Effectiveness Analysis

We separate the LCQMC and QQP datasets into eight classes according to intention classes extracted by our automatic intention extraction. Then we compare our Intention-Aware BERT (CVAE) with the original fine-tuned BERT on different intention types of data to see the performance changes.

As shown in Fig. 4, our intention-aware model outperforms BERT in every category on both Chinese and English corpora. We can find that our strategy gains more benefits on Chinese tasks than English. For one thing, it contributes to reducing the difficulty in Chinese tasks. It is challenging for the model to notice the intent features in Chinese sentences due to the complex syntax and grammar, while the heuristic intention extraction helps alleviate this problem. For another, the semantic representing ability of Chinese PLMs are still need to be improved, which are deeply influenced by problems of Chinese word segmentation [3] and polysemy phenomenon [21]. Pre-training that are more in line with Chinese grammar and syntax are still under intensive research, such as BERT-wwm [6] and ERNIE [26]. In general, our experimental results show that query intention plays an important role in judging semantic relations.

5.6 Case Study

As shown in Table 4, we select several cases to reveal the effect of intention enhancement. We compare the Intention-Aware BERT (CVAE) with the Intention-Unaware model (fine-tuned BERT) on these cases. The fine-tuned BERT without integrating intentions fails to make the right predictions. In contrast, our model is able to notice the key intention information and works well on them.

Specifically, The No.1, No.2 and No.4 cases are remarkably similar in content but holding different intentions. In No.1 and No.2, models hardly capture the intent features due to diverse Chinese pragmatics, while it is easy for the

Table 4. Error analysis of Chinese QPI for fine-tuned BERT in the intention-**Unaware** and intention-**Aware** scenarios.

ID	Sentences	Label	Prediction	
			Unaware	Aware
1	怎样做一个好人? (*How* to be a good man?) 什么算一个好人? (*What* is a good man like?)	False	True	False
2	语文考试应该怎样复习? (*How* should I review for the Chinese test?) 语文考试应该复习什么? (*What* should I review for the Chinese test?)	False	True	False
3	什么时候可以发贷? (*What time* can I get a loan?) 何时能贷? (*When* can I get a loan?)	True	False	True
4	Who exactly is successful? What exactly is success?	False	True	False
5	What's the best way to forgive people? How do you forgive other people?	True	False	True

Intention-Aware model to perceive them through our heuristic intention extraction. Though the intentions are obviously revealed by the question words in No.4, as aforementioned, models hardly make interactive attention for question words. Thus, we suggest appropriately highlight the intention information.

We also show cases that are semantically equivalent but literally different, as No.3 and No.5, which makes models easily confused. We extract intent features through our designed heuristic approach, which will classify sentences with *"What's the best way"* pattern as an *"Approach"* type question. In this way, our Intention-Aware BERT is able to clearly capture the intent information instead of being confused by diverse sentence patterns with same intention.

6   Conclusion

In this work, we focus on the situation that advanced models are suffering from weak perception of intention. To alleviate this problem, we propose the Intention-aware approaches, which extracts intention classes and utilizes VAE-based models to integrate them with contextual representations. The proposed methods are evaluated on three benchmarks, including two Chinese and one English datasets. The experimental results demonstrate that our strategy is an effective way to alleviate the aforementioned problem, yielding significant improvements and outperforming state-of-the-art approaches.

Acknowledgements. The research is supported by National Key R&D Program of China (2020YFB1313601) and National Science Foundation of China (62076174, 62076175).

References

1. Cer, D., Diab, M., Agirre, E., Lopez-Gazpio, I., Specia, L.: SemEval-2017 task 1: semantic textual similarity multilingual and crosslingual focused evaluation. In: Proceedings of the 11th SemEval, pp. 1–14 (2017). https://doi.org/10.18653/v1/S17-2001. https://aclanthology.org/S17-2001
2. Chen, J., Chen, Q., Liu, X., Yang, H., Lu, D., Tang, B.: The BQ corpus: a large-scale domain-specific chinese corpus for sentence semantic equivalence identification. In: Proceedings of EMNLP 2018, pp. 4946–4951 (2018)
3. Chen, L., et al.: Neural graph matching networks for Chinese short text matching. In: Proceedings of the 58th ACL, pp. 6152–6158 (2020)
4. Chen, Q., Zhu, X., Ling, Z.H., Wei, S., Jiang, H., Inkpen, D.: Enhanced LSTM for natural language inference. In: Proceedings of the 55th ACL, pp. 1657–1668 (2017). https://doi.org/10.18653/v1/P17-1152. https://aclanthology.org/P17-1152
5. Chung, J., Gulcehre, C., Cho, K., Bengio, Y.: Empirical evaluation of gated recurrent neural networks on sequence modeling. arXiv preprint arXiv:1412.3555 (2014)
6. Cui, Y., Che, W., Liu, T., Qin, B., Yang, Z.: Pre-training with whole word masking for Chinese BERT. IEEE/ACM Trans. Audio Speech Language Process. **29**, 3504–3514 (2021)

7. Devlin, J., Chang, M.W., Lee, K., Toutanova, K.: BERT: pre-training of deep bidirectional transformers for language understanding. In: Proceedings of NAACL 2019, pp. 4171–4186 (2019). https://doi.org/10.18653/v1/N19-1423. https://aclanthology.org/N19-1423

8. Gong, Y., Luo, H., Zhang, J.: Natural language inference over interaction space. arXiv preprint arXiv:1709.04348 (2017)

9. He, H., Gimpel, K., Lin, J.: Multi-perspective sentence similarity modeling with convolutional neural networks. In: Proceedings of EMNLP 2015, pp. 1576–1586 (2015)

10. He, K., Zhang, X., Ren, S., Sun, J.: Deep residual learning for image recognition. In: Proceedings of CVPR 2016, pp. 770–778 (2016)

11. He, T., Huang, W., Qiao, Y., Yao, J.: Text-attentional convolutional neural network for scene text detection. IEEE TIP **25**(6), 2529–2541 (2016)

12. Hershey, J.R., Olsen, P.A.: Approximating the kullback leibler divergence between gaussian mixture models. In: 2007 IEEE International Conference on Acoustics, Speech and Signal Processing-ICASSP2007. vol. 4, pp. IV-317. IEEE (2007)

13. Hu, B., Lu, Z., Li, H., Chen, Q.: Convolutional neural network architectures for matching natural language sentences. In: NIPS 27 (2014)

14. Iyer, S., Dandekar, N., Csernai, K., et al.: First Quora dataset release: question pairs. data. quora. com (2017)

15. Kim, S., Kang, I., Kwak, N.: Semantic sentence matching with densely-connected recurrent and co-attentive information. In: Proceedings of AAAI, vol. 33, pp. 6586–6593 (2019)

16. Kingma, D.P., Welling, M.: Auto-encoding variational Bayes. arXiv preprint arXiv:1312.6114 (2013)

17. Kingma, D.P., Mohamed, S., Jimenez Rezende, D., Welling, M.: Semi-supervised learning with deep generative models. In: NIPS 27 (2014)

18. Lai, Y., Feng, Y., Yu, X., Wang, Z., Xu, K., Zhao, D.: Lattice CNNs for matching based chinese question answering. In: Proceedings of AAAI, vol. 33, pp. 6634–6641 (2019)

19. Liu, X., et al.: LCQMC: a large-scale Chinese question matching corpus. In: Proceedings of the 27th ACL, pp. 1952–1962 (2018)

20. Liu, Y., et al.: RoBERTa: a robustly optimized BERT pretraining approach. arXiv preprint arXiv:1907.11692 (2019)

21. Lyu, B., Chen, L., Zhu, S., Yu, K.: Let: linguistic knowledge enhanced graph transformer for chinese short text matching. In: Proceedings of the AAAI, vol. 35, pp. 13498–13506 (2021)

22. Mueller, J., Thyagarajan, A.: Siamese recurrent architectures for learning sentence similarity. In: Proceedings of AAAI, vol. 30 (2016)

23. Pang, L., Lan, Y., Cheng, X.: Match-ignition: Plugging pagerank into transformer for long-form text matching. In: Proceedings of the 30th CIKM, pp. 1396–1405 (2021)

24. Rücklé, A., Pfeiffer, J., Gurevych, I.: MultiCQA: zero-shot transfer of self-supervised text matching models on a massive scale. arXiv preprint arXiv:2010.00980 (2020)

25. Socher, R., Huang, E., Pennin, J., Manning, C.D., Ng, A.: Dynamic pooling and unfolding recursive autoencoders for paraphrase detection. In: NIPS 24 (2011)

26. Sun, Y., et al.: ERNIE: enhanced representation through knowledge integration. arXiv preprint arXiv:1904.09223 (2019)

27. Wang, Z., Hamza, W., Florian, R.: Bilateral multi-perspective matching for natural language sentences

28. Wang, Z., Mi, H., Ittycheriah, A.: Sentence similarity learning by lexical decomposition and composition. In: Proceedings of COLING 2016, the 26th International Conference on Computational Linguistics: Technical Papers, pp. 1340–1349 (2016)
29. Wolf, T., et al.: Transformers: state-of-the-art natural language processing. In: Proceedings of EMNLP 2020, pp. 38–45 (2020)
30. Yu, W., Wu, L., Zeng, Q., Tao, S., Deng, Y., Jiang, M.: Crossing variational autoencoders for answer retrieval. In: Proceedings of the 58th ACL, pp. 5635–5641 (2020). https://doi.org/10.18653/v1/2020.acl-main.498. https://aclanthology.org/2020.acl-main.498
31. Zhang, K., Chen, E., Liu, Q., Liu, C., Lv, G.: A context-enriched neural network method for recognizing lexical entailment. In: Proceedings of AAAI, vol. 31 (2017)
32. Zhang, K., Wu, L., Lv, G., Wang, M., Chen, E., Ruan, S.: Making the relation matters: relation of relation learning network for sentence semantic matching. In: Proceedings of AAAI, vol. 35, pp. 14411–14419 (2021)

Automatic and Analytical Field Weighting for Structured Document Retrieval

Tuomas Ketola[✉] and Thomas Roelleke

Queen Mary University of London, London, UK
t.j.h.ketola@qmul.ac.uk

Abstract. Probabilistic models such as BM25 and LM have established themselves as the standard in atomic retrieval. In structured document retrieval (SDR), BM25F could be considered the most established model. However, without optimization BM25F does not benefit from the document structure. The main contribution of this paper is a new field weighting method, denoted Information Content Field Weighting (ICFW). It applies weights over the structure without optimization and overcomes issues faced by some existing SDR models, most notably the issue of saturating term frequency across fields. ICFW is similar to BM25 and LM in its analytical grounding and transparency, making it a potential new candidate for a standard SDR model. For an optimised retrieval scenario ICFW does as well, or better than baselines. More interestingly, for a non-optimised retrieval scenario we observe a considerable increase in performance. Extensive analysis is performed to understand and explain the underlying reasons for this increase.

Keywords: Retrieval models · Structured documents

1 Introduction

The majority of data is inherently structured. Whether it is websites, product catalogues, or specific databases, the data has an underlying structure. Probabilistic models, such as the BM25 and LM have become the standard for non-structured (atomic) retrieval, especially if the use of learn-to-rank models is not warranted. However, no such widely accepted standard exists for structured document retrieval (SDR). One possible reason for this is that many of the existing SDR models do not work well without optimization. For example, even the most prominent candidate—BM25F—reverts back to considering the document as atomic if it is not optimized. This means that in some scenarios the rankings it produces are far from intuitive [14]. The main contribution of this paper is a field weighting method, denoted Information Content Field Weighting (ICFW). The method applies weights over the field-based scores produced by any atomic retrieval model (e.g. BM25, LM etc.) without optimization. By setting model parameters analytically, ICFW is able to overcome issues faced by some existing SDR models, most notably saturating term frequency across fields. ICFW is similar to the standard atomic models in its analytical grounding and transparency, making it a potential new candidate for a standard SDR model.

J. Kamps et al. (Eds.): ECIR 2023, LNCS 13980, pp. 489–503, 2023.
https://doi.org/10.1007/978-3-031-28244-7_31

A key aspect of many SDR models is the setting of weights over the document fields. Most of these models are characterized by one of two underlying aggregation functions: 1. **Field Score Aggregation** (FSA), where the weights are applied over field-based retrieval scores (meta-search) or 2. **Term Frequency Aggregation** (TFA), where the weights are applied to within-field term frequencies and the score is calculated over a flattened document representation, i.e. the document is seen as atomic (BM25F and MLM) [14,18,20]. In order to benefit properly from the structure, these models require the optimization of field weights and other parameters, either based on training data, or some other type of prior knowledge. ICFW is able to leverage the document structure without any such knowledge. Furthermore, by using the field based scores, as done by FSA-based models, whilst saturating term frequency across fields, as done by TFA-based models, ICFW is able to overcome some of the issues existing models face. By analytically setting the scale of term frequency saturation across fields, ICFW satisfies more of the SDR constraints introduced by [14] than existing models, thus producing more intuitive rankings.

The experimentation focuses on the BM25 as the underlying model for the baselines and the proposed model. However, ICFW could be used together with any atomic retrieval model (LM, DFR etc.). The experimentation considers two retrieval scenarios, one with training data and one without. For the former task ICFW does as well, or slightly better than the baselines (various versions of BM25F and FSA-BM25). For the latter—and arguably the more difficult one, as no user preferences are known—ICFW significantly outperforms the baselines. Significant increases in a accuracy, coupled with an in-depth analysis as to their underlying reasons, demonstrates in practical terms the contribution of this paper to establishing reliable standards for analytical SDR.

2   Background

There are many constraints for (atomic) retrieval [4–7]. This line of research allows for analytical evaluation of any retrieval model. In a similar fashion, [14] recently introduced formal constraints for SDR. They present three constraints with both formal and intuitive definitions. The intuitive definitions are presented below, we direct the reader to see the original paper for the formal definitions.

Field Importance: A model should be able to boost, or decrease the weight given to a field based on some notion of field importance.

Field Distinctiveness: Adding a query term to a new field should increase the retrieval score more than adding it to a field where it already occurs

Term Distinctiveness: Adding unseen query terms to a document should increase the retrieval score more than adding query terms already considered

[14] points out that none of the widely used analytical SDR models (BM25F, MLM [18], FSDM [29], PRMS [15], FSA) satisfy all three constraints. TFA-based models (BM25F, MLM) fail to satisfy the field distinctiveness constraint as they consider the document as atomic after applying the field weights and FSA-based

models (FSA, PRSM) fail to satisfy the term distinctiveness constraint because they do not saturate term frequency across fields (something also noted by [20]).

There are a multitude of SDR models other than those listed above. Older approaches—especially within the INEX initiative—tend to consider the structure and its semantics explicitly [2,12,16,24]. Some focus on what we might call **Term Importance**, i.e. taking into account that the importance of a term can vary across fields [22,26]. More recent work has focused on applying the lessons from atomic deep learn-to-rank models to SDR [3,27]. The most relevant approach to this work is the BM25-FIC, where the underlying idea is that more weight should be given to document fields with higher information content [13].

Definition 0 (BM25-FIC RSV). *Let q be a query, d be a document, t a term, m the number of fields, F_i a collection field (e.g. all titles in the collection), and f_i a document field (e.g. a title of a document). f_i is part of exactly one document. $|F_i|$ is the number of document fields for collection field type i. $\mathrm{df}(t, F_i)$ is the document (field) frequency, i.e. the number of fields (e.g. titles) in which the term occurs. The BM25-FIC RSV is defined as the weighted sum of field-based scores where the weight is proportional to the sum of field-based IDFs.*

$$P(t \in f_i | F_i) = \frac{\mathrm{df}(t, F_i)}{|F_i|} \tag{1}$$

$$w(f_i, q) := -\sum_{t \in q \cap f_i} \log(P(t \in f_i | F_i)) \quad \left(\propto \sum_{t \in q \cap f_i} \mathrm{IDF}(t, F_i) \right) \tag{2}$$

$$\mathrm{RSV}_{\mathrm{BM25\text{-}FIC}}(d, q, c) := \sum_{i=1}^{m} w(f_i, q) \ \mathrm{RSV}_{\mathrm{BM25}}(f_i, q, c) \tag{3}$$

Definition 0 is central to this paper as the proposed ICFW method has the same underlying idea of using information content (understood as the negative log of probability) for field weighting. However, unlike the BM25-FIC, ICFW saturates term frequency across fields, leading to better performance.

3   Information Content Field Weighting (ICFW)

ICFW leverages the information contained in the document structure without the need for optimization. Furthermore, due to its analytical nature and its relationship to the SDR constraints it is highly transparent. These are characteristics the ICFW shares with its counter parts in atomic retrieval (BM25, LM etc.) and what make it robust across different structure types and retrieval scenarios.

3.1   Model Specification

ICFW aggregates the field-based retrieval scores of a document by multiplying each by their information content-based field weight and summing these weighted scores together. The field weight is calculated as a combination of collection field-based information content and document field-based information content, where a scale parameter λ determines the weight given to the latter.

Definition 1 (Term Probabilities). *Let* $\mathrm{ff}(t, d)$ *be the field frequency; i.e. number of fields in d that contain term t.* $||F_i|| := \{f||f| > 0\}$ *is the number of non-empty document fields. Let* $m(d) := |\{f|f \in d\}|$ *denote the number of fields in document d. The probability of a term occurring in a document field* f_i *(of type i) given collection field* F_i *is denoted* $P(t \in f_i|F_i)$. *The probability of a term occurring in a document field* f_i *given document d is denoted* $P(t \in f_i|d)$.

$$P(t \in f_i|F_i) := \frac{\mathrm{df}(t, F_i)}{||F_i||} \tag{4}$$

$$P(t \in f_i|d) := \frac{\mathrm{ff}(t, d)}{m(d)} \tag{5}$$

Note that Eq.(4) corresponds to Eq.(1), except empty fields are considered.

Definition 2 (Field Probabilities). *The probability of q and* f_i *given collection field* F_i *is denoted* $P(q, f_i|F_i)$. *The probability of q and* f_i *given document d is denoted* $P(q, f_i|d)$.

$$P(q, f_i|F_i) = \prod_{t \in q \cap f_i} P(t \in f_i|F_i) \tag{6}$$

$$P(q, f_i|d) = \prod_{t \in q \cap f_i} P(t \in f_i|d) \tag{7}$$

Definition 3 (ICF and ICD). *The collection field-based information content of a document field* f_i *is denoted* $\mathrm{ICF}(q, f_i, F_i, d)$ *and the document-based information content of* f_i *is denoted* $\mathrm{ICD}(q, f_i, d)$. *The information content of an event is defined as its negative log probability as proposed by [10] and used previously in the DFR model by [1].*

$$\mathrm{ICF}(q, f_i, F_i, d) := -\log P(q, f_i|F_i) \tag{8}$$

$$\mathrm{ICD}(q, f_i, d) := -\log P(q, f_i|d) \tag{9}$$

If q is implicit and as F_i *follows from* f_i, $\mathrm{ICF}(q, f_i, F_i, d)$ *is shortened to* $\mathrm{ICF}(f_i, d)$ *and* $\mathrm{ICD}(q, f_i, d)$ *to* $\mathrm{ICD}(f_i, d)$.

Definition 4 (ICFW and Scale Parameter Lambda). *Let* $\vec{\lambda} = (\lambda_1, \ldots, \lambda_m)$ *be a vector of scaling parameters where each* λ_i *reflects the importance given to the document-based information content* ICD *for field* f_i. $\lambda_i >= 0$.

$$w_{\mathrm{icfw}, \lambda_i}(f_i, F_i, d, q) := \mathrm{ICF}(q, f_i, F_i, d) + \lambda_i \cdot \mathrm{ICD}(q, f_i, d) \tag{10}$$

where not ambiguous $w_{\mathrm{icfw}, \lambda_i}(f_j, d)$ *is short for* $w_{\mathrm{icfw}, \lambda_i}(f_i, F_i, d, q)$. *Note that if* $\forall i \lambda_i = 0$, *ICFW is equal to BM25-FIC (apart from the* $||F||$ *variable).*

Definition 5 (ICFW RSV).   *Let* $\text{RSV}_M(q, f_i, c)$ *a be retrieval score of a field where M is the retrieval model. Given document d, query q, $\vec{\lambda}$, collection c and retrieval model M, the score (retrieval status value) of d is denoted* $\text{RSV}_{ICFW,\vec{\lambda},M}(d, q, c)$.

$$\text{RSV}_{\text{ICFW},\vec{\lambda},M}(d, q, c) := \sum_{i=1}^{m} w_{\text{icfw},\lambda_i}(f_i, F_i, d, q) \sum_{t \in q} \text{RSV}_M(f_i, q, c) \quad (11)$$

The λ parameter scales the impact of the document-based information content. How to best set λ is one of the central research questions in this paper and will be discussed in the next section.

3.2   Setting the Scale Parameter Lambda

The parameter λ scales the impact of the document-based information content. If λ is set to 0, $w_{\text{icfw},\lambda_i}$ is defined only through information-content based on the collection field F_i, i.e. term occurrences would be considered independent between the fields as done in [13]. As discussed earlier in this paper and extensively by [20], this is not a good assumption and results in the term distinctiveness constraint not being satisfied.

However, as λ increases, term frequency is saturated more across fields: Higher λ puts more emphasis on ICD (Eq. (10)), meaning it gives more weight to document fields with distinct terms, rather than ones re-appearing. I.e. the second occurrence of a term increases the retrieval score less than the first one, no matter what field it is in (assuming similar IDFs across fields). The size of λ defines the scale of this **cross field term frequency saturation.**

The simplest way of setting lambda is to have it as a constant for the collection. In this way the term frequency saturation across fields is constant, same as for BM25F. Setting lambda this way will be considered in the experimentation. However, to find an appropriate value for lambda, optimization is needed. One of the main aims for this paper was to provide a field weighting method that does not need optimization. The following will describe an alternative approach to setting λ that analytically considers the scale of term frequency saturation with respect to the Term Distinctiveness constraint by [14].

The TFA-based models (BM25F etc.) satisfy the Term Distinctiveness constraint because they saturate the term frequency across fields, however in doing so they break the Field Distinctiveness constraint [14, 20, 28]. See [14] for a more in-depth discussion. ICFW does not have this same problem, as term frequency can be saturated across fields, without reverting back to considering the document as atomic, as done by TFA-based models. Lambda can be set for each query analytically, making sure the term frequency saturation is strong enough for the model to satisfy the term distinctiveness constraint. The following definitions and discussion will show how lambda is set to achieve this.

Definition 6 (Score Contribution of a Term). *Let f denote a document field with occurrences of t and $\bar{f}$ denote an amended version of document field f*

without occurrences of t. *The score contribution of a term* t *occurring in a field* f
is denoted as $S_{\mathrm{contr},M}(t,f,q,c)$. *Where not ambiguous* $S_{\mathrm{contr}}(t,f,d)$ *is short for*
$S_{\mathrm{contr},M}(t,f,q,c)$.

$$S_{\mathrm{contr},M}(t,f,q,c) := S_M(q,f,c) - S_M(q,\overline{f},c) \tag{12}$$

Definition 7 (Score Contribution Ratios). *Given terms* t_a *and* t_b *and document* d, *the cross-term score contribution ratio, i.e. the ratio of the score contributions* (S_{contr}) *of terms* t_a *and* t_b *is denoted* $Z(t_a,t_b,f_i,f_j,d)$. *Given term* t, *fields* f_i *and* f_j, *the cross-field score contribution ratio, i.e. the ratio of the score contributions of* t *in fields* f_i *and field* f_j, *is denoted* $x(t,f_i,f_j,d)$.

$$Z(t_a,t_b,f_i,f_j,d) := \frac{S_{\mathrm{contr}}(t_a,f_i,d)}{S_{\mathrm{contr}}(t_b,f_j,d)} \tag{13}$$

$$x(t,f_i,f_j,d) := \frac{S_{\mathrm{contr}}(t,f_i,d)}{S_{\mathrm{contr}}(t,f_j,d)} \tag{14}$$

Definition 8 (Scale Threshold - Two Terms). *Let* $q = \{t_1,\ldots,t_n\}$, d *a document with occurrences of term* t_a *in field* f_i *and occurrences of term* t_b *in field* $\overline{f}$. *Let* $\overline{d}$ *be an amended version of document* d, *where the occurrences of* t_b *in field* $\overline{f}$ *are replaced by further occurrences of term* t_a. *For presentation purposes* $Z(t_a,t_b,f_i,\overline{f},d)$ *is shortened to* Z *and* $x(t_a,\overline{f},f_i,d)$ *to* x.

$$\lambda_{threshold}(t_a,t_b,d,f_i) := \frac{\log \frac{\mathrm{df}(t_b,\overline{F})|\overline{F}|^{Zx}}{\mathrm{df}(t_a,\overline{F})^{Zx}|\overline{F}|}}{\log \frac{m^{Z+1}\,\mathrm{ff}(t_a,\overline{d})^{Z(x+1)}}{m^{Z(x+1)}\,\mathrm{ff}(t_a,d)^{Z+1}}} \tag{15}$$

$\lambda_{threshold}$ defines the λ value above which $\mathrm{score}(d) > \mathrm{score}(\overline{d})$. This is the λ above which the ICFW satisfies the term distinctiveness constraint with respect to t_a and t_b. See Appendix A for formal theorem and proof.

Definition 9 (Scale Threshold - Query). *In order to generalize Definition 8 to the entire query, rather than two query terms and* f_i, *we need to consider the rarest and most common query terms and the field with the smallest* $S_{\mathrm{contr}}(t_a,f_i,d)$. *Let* t_{ra} *be the rarest query term,* t_{co} *the most common query term and* $f_{\min}$ *the field with the smallest score contribution* $(S_{\mathrm{contr}}(t_{ra},f_i,d))$ *for term* t_{ra}.

$$\lambda_{threshold}(q,d) := \lambda_{threshold}(t_{ra},t_{co},d,f_{\min}) \tag{16}$$

Setting $t_a = t_{ra}$, $t_b = t_{co}$, $f_i = f_{\min}$ ensures $\mathrm{score}(d) > \mathrm{score}(\overline{d})$ for the entire document and query. This is because changing the most common term to the rarest term in $\overline{f}$ has the highest impact on the retrieval score, which needs to be offset by the ICD component and therefore a larger value for λ. If $\lambda > \lambda_{threshold}(q,d)$, ICFW satisfies the term distinctiveness constraint for any q and d. As ICFW already satisfies the field distinctiveness constraint, to the best of our knowledge it is the first SDR model to satisfy all three constraints by [14].

3.3   Approximating Appropriate Values for Lambda Threshold

If used directly Definition 9 is highly sensitive to rare terms. A single query term (t_{ra}) that is very rare in one of the fields defines λ for all documents. This is because Definition 9 sets λ so that an occurrence of the most common query term (t_{co}) would be enough to offset a second occurrence of t_{ra}, even if this is not feasible. Therefore, Definition 9 should not be viewed as an optimal lambda value, but as a good starting point with an intuitive explanation. With this in mind the calculation of $\lambda_{threshold}(q, d)$ is made less sensitive to large variations of score contributions in Eqs. (13) and (14): Firstly, we assume that $x = 1$, i.e. the score contribution of a term is assumed to be the same in all fields. This is done as we do not want λ to become too sensitive to variations in S_{contr} values for a single term across fields. Rather we are interested in satisfying the term distinctiveness constraint and thus care more about the ratio of S_{contr} values across terms. Secondly, we assume that the score contribution in Eq. (13) is calculated based on the first occurrence of a term in a field. With respect to BM25 this means is that we are only considering the contribution of the IDF to the score $(S_{contr,BM25}(t, f, d) = IDF(t, F))$. Finally, the effect of metrics based on singular terms and fields needs to be smoothed using the rest of the query terms and the collection. The three proposed methods for this smoothing approximate the df values in Definition 8 in terms of $df(t_{ra})$, $df(t_{co})$ and in terms of the IDF values in Z resulting in three proposed models.

Definition 10 (ICFW-Global (ICFW-G)).

$$df_{ICFW\text{-}G}(t_{ra}, F_i) := \min(\{df(t, c) : t \in q\}) \tag{17}$$

$$df_{ICFW\text{-}G}(t_{co}, F_i) := \max(\{df(t, c) : t \in q\}) \tag{18}$$

Definition 11 (ICFW-Global-Average (ICFW-GA)). *Let* t_{max} *be the most common query term in the collection*

$$df_{ICFW\text{-}GA}(t_{ra}, F_i) := \frac{\sum_{t \in q \setminus t_{max}} df(t, c)}{|\{t \in q \setminus t_{max}\}|} \tag{19}$$

$$df_{ICFW\text{-}GA}(t_{co}, F_i) := \max(\{df(t, c) : t \in q\}) \tag{20}$$

Definition 12 (ICFW-Local-Average (ICFW-LA)).

$$df_{ICFW\text{-}LA}(t_{ra}, F_i) := \frac{\sum_{t \in q \setminus t_{max}} df(t, F_i)}{|\{t \in q \setminus t_{max}\}|} \tag{21}$$

$$df_{ICFW\text{-}LA}(t_{co}, F_i) := \max(\{df(t, c) : t \in q\}) \tag{22}$$

ICFW-G uses the document frequency values over the whole collection. ICFW-GA further smooths the effect of rare query terms by estimating $df(t_{ra}, c)$ as the mean of collection level document frequencies of terms that are not the most common. ICFW-LA is similar to ICFW-GA, except the calculations are done at the collection field level, rather than the collection level (F vs. c). For the first two smoothing methods the value of lambda is the same for all fields, for the third one the value varies across fields.

3.4   Optimising ICFW

Even though the focus in this paper is on non-optimised SDR models, sometimes training data is available. In its current form Definition 5 does not offer many options for optimization, other than the parameters of the underlying model. Therefore we add an additional static field weight that can be optimised.

$$\mathrm{RSV}_{\mathrm{ICFW\text{-}opt},\vec{\lambda},M,\vec{w}_{\mathrm{stat}}}(d,q,c) :=$$
$$\sum_{i=1}^{m} w_{\mathrm{stat,i}} \cdot w_{\mathrm{icfw},\lambda_i}(f_i, F_i, d, q) \sum_{t \in q} \mathrm{RSV}_M(q, f_i, c) \qquad (23)$$

The training optimises the underlying model parameters (k_1 and b for BM25), the static fields weights $w_{\mathrm{stat},i}$ and lambda. We consider two methods for optimising lambda: **ICFW-Lambda-Const** (ICFW-LC) where lambda is optimised for each field as a constant and **ICFW-Lambda-Est** (ICFW-LE) where lambda is estimated based on the mean and variation of global query term idf values in a linear regression manner: $\lambda = B_0 + B_1 \mathrm{mean}(\mathrm{IDF}(q,c)) + B_2 \mathrm{var}(\mathrm{IDF}(q,c))$.

4   Experimentation and Analysis

RQ1: How does saturating term frequency affect ICFW performance?
RQ2: How well can lambda be estimated without optimization?
RQ3: How do the ICFW optimized candidates compare to baselines?

The experimentation considers two retrieval scenarios: A **non-optimised** one, where optimization using training data is not used and, an **optimised** one where it is. By not focusing solely on optimised versions of the baselines and proposed models this paper aims to provide a broader picture of the quality of the ICFW method. As the main question in this paper is about automatic weighting of fields for SDR, it makes sense to compare the non-optimised ICFW to other non-optimised SDR methods. Still, if training data is available models should be optimised. Therefore we also compare optimised versions of the baselines to an optimised version of ICFW. The experimentation is available on Github[1].

4.1   Data Collections

The collections in the experimentation are DBpedia [9], Homedepot[2] and Trec-8-Small-Web[3] (Trec-Web). A key objective of the experimentation is to demonstrate the robustness of ICFW on a variety of retrieval task types and a variety of document structure types. The retrieval task types are web search (Trec-Web), product search (Homedepot) and entity search (DBpedia). These collection structure types vary from the simple {*title*, *body*} of Trec-Web to the more

---

[1] https://github.com/TuomasKetola/icfw-for-SDR.
[2] https://www.kaggle.com/c/home-depot-product-search-relevance.
[3] https://trec.nist.gov/data/t8.web.html.

complex {*categories*, *related entities*, *similar entities*, *label* and *attributes*} of DBpedia. The collections sizes are 4.6 million for DBpedia, 200 k for Trec-Web and 55 k for Homedepot. More important than number of documents, is variety in structure complexity. In total Homedepot has 10 k+ queries. We have limited the number of queries by choosing 1000 with the most relevance judgements.

4.2   Baselines and Methodology

It is not our aim to demonstrate that ICFW combined with BM25 outperforms all SDR models. Instead we wish to show that ICFW is able to leverage the structure of the data in ways that existing analytical models are not. As we are comparing ICFW with existing field weighting methods, rather than existing SDR models, the experimentation will not include all the models in Sect. 2. Instead we will focus on the BM25 retrieval model and its various fielded versions. **FSA-BM25**: Linear sum of BM25 scores. **FSA-BM25-catchall**: Linear sum of BM25 scores with an additional catchall field that concatenates all fields together. **BM25F**: Fielded BM25 model where document length normalization is applied at a field level [28]. **BM25F-Simple**: A BM25 model where document length normalization is applied over the concatenated document [20]. No catchall field considered as BM25F already concatenates the fields. For the non-optimized task the field weights for all models are set as uniform. The BM25 hyperameters are set as $b = 0.8$ and $k_1 = 1.6$ (mid point in the recommended range [1.2–2.0] [21]). For the optimised task, models are optimised using coordinate ascent and 5-fold cross validation [17] for NDCG@100. The underlying BM25 model for all the approaches is the original one by Robertson et al. [11,19].

Table 1 shows the results. A significance test has been applied, even though there are different views on the methodology. [8,23] make the case that significance tests should not be used on multiple hypothesis (without correction) and that simple significance tests should not be applied to re-used test collections. Though the authors share similar views, significance tests are still often considered a must-have. Furthermore, some test collection are not reused (Homedepot) and the proposed models are similar, meaning as features they are correlated.

4.3   RQ1: The Effect of Term Frequency Saturation on Performance

Figure 1 shows the effect of lambda, i.e. cross field term frequency saturation on NDCG@100. For RQ1 it is worth only considering the top three graphs, as the bottom ones have inherent saturation due to the catchall field. ICFW-const-λ shows the accuracy for a ICFW-model where a single value for lambda is optimised. At $\lambda = 0$, the ICFW-const-λ model corresponds to the BM25-FIC [13]. We can see that for all the datasets (without catchall field), the optimal value of λ is greater than 0, albeit for Homedepot even at $\lambda = 0$, ICFW outperforms baselines clearly. This explains the results in [13]. So we can conclude that saturating term frequency across fields is indeed important. It is likely to be more important for data structures such as {title, body}, where there is greater dependence between term occurrences. This is evident in Fig. 1 from the high value of optimal lambda for Trec-Web.

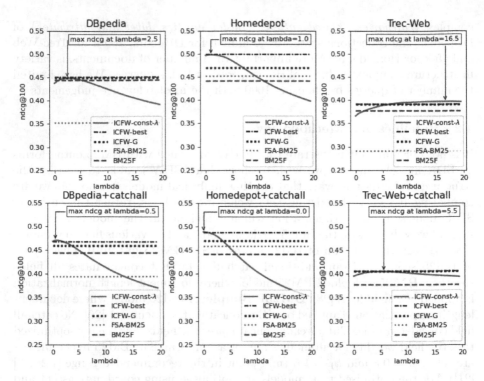

Fig. 1. Estimating λ analytically. +catchall denotes the use of a catchall field by the ICFW and FSA models. ICFW-const-λ shows change in accuracy for different levels of lambda. ICFW-best is the best non-optimised ICFW (Sect. 3.3) for a given dataset.

4.4   RQ2: Estimating Lambda Analytically

Figure 1 demonstrates that lambda can be estimated well analytically. We observe that the best performing smoothing method for each dataset from Sect. 3.3 (ICFW-best) is able to locate the maximum point of ICFW-const-λ well, even for very different values of lambda (0 for Homedepot+catchall vs. 16.5 for Trec-Web). The upper half of Table 1 shows the results for the non-optimized retrieval task. It is clear that ICFW generally outperforms all four baselines, independent of the smoothing method. ICFW-best is different for all the datasets: ICFW-LA, ICFW-G and ICFW-LA for DBpedia, Trec-Web and Homedepot respectively with no catchall field and ICFW-GA-all, ICFW-G-all and ICFW-LA-all with the catchall field. Out of the three smoothing methods for df, ICFW-G is the most straightforward one and its performance is therefore also reported in Fig. 1. It only falls significantly short of ICFW-best for Homedepot.

To put the performance increase in perspective consider the increase in accuracy gained from optimizing BM25F-Simple. The increases in MAP are approximately 0.05, 0.05 and 0.1 for DBpedia, Homedepot and Trec-Web respectively. The increases for moving from a non-optimised BM25F-Simple to the best ICFW version (ICFW-GA-all, ICFW-LA and ICFW-G-all respectively) are 0.03, 0.02 and 0.06, meaning we get around half of the accuracy gains by moving from BM25F-Simple to ICFW, compared to optimizing BM25F-Simple.

Table 1. Experimental results. The best performing baseline is underlined and the best performing model is in bold. * and ** denote significance at $p < 0.05$ and $p < 0.01$ respectively for a Wilcoxon signed ranks test. +all means the model considers a catchall field with all document fields concatenated together.

Dataset	DBpedia		Trec-Web		Homedepot	
Metric	map	ndcg@100	map	ndcg@100	map	ndcg@100
Non-optimized						
Baseline Models						
FSA-BM25	0.226	0.351	0.164	0.290	0.252	0.452
BM25F	0.295	0.444	0.229	0.377	0.249	0.440
BM25F-Simple	0.284	0.433	0.229	0.378	0.238	0.429
FSA-BM25+all	0.256	0.393	0.205	0.349	0.258	0.458
Proposed Models						
ICFW-G	0.299	0.448	0.243	0.391*	0.290**	0.486**
ICFW-GA	0.302	0.449	0.241	0.389	0.297**	0.496**
ICFW-LA	0.304*	0.453	0.233	0.378	**0.299****	**0.498****
ICFW-G+all	0.305*	0.459*	**0.251****	**0.406****	0.277**	0.470**
ICFW-GA+all	**0.313****	**0.468****	0.249**	0.403**	0.285**	0.482**
ICFW-LA+all	0.310**	0.464**	0.249**	0.402**	0.289**	0.487**
Optimized						
Baseline Models						
FSA-BM25	0.317	0.473	*0.286*	0.449	0.352	0.538
BM25F	0.338	0.494	0.279	0.444	0.354	0.544
BM25F-Simple	0.330	0.483	0.279	0.441	0.337	0.526
FSA-BM25+all	0.334	0.492	0.286	**0.451**	0.358	0.547
Proposed Models						
ICFW-LC	0.335	0.489	**0.287**	0.450	0.358	0.547
ICFW-LE	0.336	0.489	0.285	0.448	0.356	0.545
ICFW-LC+all	**0.344***	**0.500**	0.281	0.442	0.358	0.547
ICFW-LE+all	**0.344***	0.499	0.280	0.441	**0.360**	**0.548**

This is a significant finding since optimizing BM25F requires relevance information and optimization, whereas ICFW does not.

4.5   RQ3: Optimized ICFW Performance

The lower half of Table 1 presents the results for the optimised task. It is clear from the results that the differences between retrieval accuracy of the models, both between the baselines and the ICFW-models, are much smaller than they were for the non-optimised task. None of the baselines is clearly better than others across data collections. Interestingly, BM25F—which is usually considered state of the art for analytical SDR models—is the best only for DBpedia. Even there, the difference to FSA-BM25+all is marginal. There are few cases where we observe significant differences in retrieval behaviour between the baselines and ICFW. In general, the findings seems to correspond to the observation made by [25] for atomic retrieval, that once hyperparameter optimization is used, bag-of-words-based analytical models do not differ significantly in accuracy.

5   Conclusion

This paper has introduced ICFW: an information content-based field weighting method for SDR. The proposed method has two strengths compared to previous analytical SDR models: Firstly, it assigns weights automatically over the structure. Secondly, the model saturates term frequency across fields, without considering the document as atomic. The degree of this cross field term frequency saturation is controlled by the scale parameter λ. The experimentation first considers a constant level of λ, where the parameter was set to optimize NDCG@100. It is shown that a level of λ greater than 0 is beneficial and that the optimal level varies between datasets. However, as the underlying motivation for the paper was an automatic and analytical method, another approach to setting lambda was also considered: Sect. 3.2 introduces an analytical method (no optimization) for defining what we call lambda-threshold for any query. Lambda-threshold defines a level of λ above which the term distinctiveness constraint by [14] is satisfied. The analysis goes onto show that if λ is set in this way, retrieval performance is as good as if it was optimised and much better than the performance of established baselines (different fielded versions of the BM25). These results strongly suggest that ICFW leverages the structure in ways existing models do not.

Overall this paper contributes to the research area of proposing analytical standards for SDR. It has done so by introducing a method with the necessary transparency for advancing search systems used with structured data.

Acknowledgements. We would like to thank the reviewers for their comments, in particular regarding the presentation of the proposed models and candidate models, as well as the methodology of significance testing.

A   Scale Parameter Threshold

The underlying idea of the scale threshold theorem is that there exists a threshold for λ, above which the model satisfies the term distinctiveness constraint.

Let $q = \{t_1, \ldots, t_n\}$ be a query, d be a document with $n(t_a, f_i, d)$ occurrences of query term t_a in field f_i and $n(t_b, \overline{f}, d)$ occurrences of query term t_b in an average field $\overline{f}$. Let $\overline{d}$ be an amended version of document d where the occurrences of t_b are replaced with occurrences of t_a.

Theorem 1. *Given terms t_a and t_b, if $\lambda > \lambda_{threshold}$, then $RSV(d) > RSV(\overline{d})$.*

$$\forall (t_a, t_b) \in q: \ \lambda > \lambda_{threshold}(t_a, t_b, d, f_i) \implies \tag{24}$$
$$RSV_{ICFW,\vec{\lambda}}(q, d, c) > RSV_{ICFW,\vec{\lambda}}(q, \overline{d}, c)$$

Proof. Following Definition 8 for $\lambda_{\text{threshold}}$, the inequality becomes:

$$\lambda > \frac{\log \frac{\mathrm{df}(t_b, \overline{F}) |\overline{F}|^{Z_x}}{\mathrm{df}(t_a, \overline{F})^{Z_x} |\overline{F}|}}{\log \frac{m^{Z+1} \mathrm{ff}(t_a, \overline{d})^{Z(x+1)}}{m^{Z(x+1)} \mathrm{ff}(t_a, d)^{Z+1}}} \tag{25}$$

Considering the numerator first:

$$\log \frac{\mathrm{df}(t_b, \overline{F})|\overline{F}|^{Zx}}{\mathrm{df}(t_a, \overline{F})^{Zx}|\overline{F}|} = \log \frac{\frac{\mathrm{df}(t_b, \overline{F})}{|\overline{F}|}}{\frac{\mathrm{df}(t_a, \overline{F})^{Zx}}{|\overline{F}|^{Zx}}} \tag{26}$$

Following Eq. (4) for the definition of probabilities and Eq. (26) we obtain,

$$\log \frac{P(t_b, \overline{f}|\overline{F})}{P(t_a, \overline{f}|\overline{F})^{Zx}} = \log P(t_b, \overline{f}|\overline{F}) - Zx \log P(t_a, \overline{f}|\overline{F}) \tag{27}$$

Following Definition 3 we can re-write Eq. (27) to obtain,

$$\log P(t_b, \overline{f}|\overline{F}) - Zx \log P(t_a, \overline{f}|\overline{F}) = Zx\, \mathrm{ICF}(\overline{f}, \overline{d}) - \mathrm{ICF}(\overline{f}, d) \tag{28}$$

Moving onto the denominator,

$$\log \frac{m^{Z+1}\, \mathrm{ff}(t_a, \overline{d})^{Z(x+1)}}{m^{Z(x+1)}\, \mathrm{ff}(t_a, d)^{Z+1}} = \log \frac{\left[\frac{\mathrm{ff}(t_a, \overline{d})}{m}\right]^{Z(x+1)}}{\left[\frac{\mathrm{ff}(t_a, d)}{m}\right]^{Z+1}} \tag{29}$$

Inserting Eq. (5) to Eq. (29) and transforming the log expression we obtain,

$$\log \frac{P(f_i|\overline{d})^{Z(x+1)}}{P(f_i|d)^{Z+1}} = Z(x+1)\log P(t_a, f_i|\overline{d}) - Z\log P(t_a, f_i|d) - \log P(t_a, \overline{f}|d) \tag{30}$$

Following Definition 3 we can re-write Eq. (30) to obtain

$$\log \frac{P(f_i|\overline{d})^{Z(x+1)}}{P(f_i|d)^{Z+1}} = -Z(x+1)\,\mathrm{ICD}(f_i, \overline{d}) + Z\,\mathrm{ICD}(f_i, d) + \mathrm{ICD}(\overline{f}, d) \tag{31}$$

Inserting Eqs. (28) and (31) to Eq. (25) and solving for Z we obtain,

$$Z < \frac{\mathrm{ICF}(\overline{f}, d) + \lambda\,\mathrm{ICD}(\overline{f}, d)}{\lambda(x+1)\,\mathrm{ICD}(f_i, \overline{d}) - \lambda\,\mathrm{ICD}(f_i, d) + x\,\mathrm{ICF}(\overline{f}, \overline{d})} \tag{32}$$

Expanding the denominator we obtain,

$$Z < \frac{\mathrm{ICF}(\overline{f}, d) + \lambda\,\mathrm{ICD}(\overline{f}, d)}{\begin{array}{c}\mathrm{ICF}(f_i, \overline{d}) + \lambda\,\mathrm{ICD}(f_i, \overline{d}) + x\,\mathrm{ICF}(\overline{f}, \overline{d}) \\ + x\lambda\,\mathrm{ICD}(f_i, \overline{d}) - \mathrm{ICF}(f_i, d) - \lambda\,\mathrm{ICD}(f_i, d)\end{array}} \tag{33}$$

Following Eqs. (13), (14), (8) and (9) (33) is re-written:

$$\frac{S_{\mathrm{contr}}(t_a, f_i, d)}{S_{\mathrm{contr}}(t_b, \overline{f}, d)} < \frac{w_{\mathrm{icfw}}(\overline{f}, d)}{w_{\mathrm{icfw}}(f_i, \overline{d}) + xw_{\mathrm{icfw}}(\overline{f}, \overline{d}) - w_{\mathrm{icfw}}(f_i, d)} \tag{34}$$

Rearranging Eq. (34) we obtain,

$$w_{\mathrm{icfw}}(\overline{f}, d)\, S_{\mathrm{contr}}(t_{\mathrm{b}}, \overline{f}, d) + w_{\mathrm{icfw}}(f_i, d)\, S_{\mathrm{contr}}(t_{\mathrm{a}}, f_i, d) >$$
$$w_{\mathrm{icfw}}(f_i, \overline{d})\, S_{\mathrm{contr}}(t_{\mathrm{a}}, f_i, \overline{d}) + w_{\mathrm{icfw}}(\overline{f}, \overline{d})\, S_{\mathrm{contr}}(t_{\mathrm{a}}, \overline{f}, \overline{d}) \qquad (35)$$

Assuming the term frequencies from the theorem, the retrieval score difference is only dependent on the score contributions of term t_{a} in field f_i and term t_{b} in field $\overline{f}$. For $\overline{d}$ the same is true for the score contributions of term t_{a} in field f_i and term t_{a} in field $\overline{f}$. Following Definition 5 we rewrite Eq. (35) and obtain the implicated inequality from the theorem.

$$\square$$

References

1. Amati, G., Van Rijsbergen, C.J.: Probabilistic models of information retrieval based on measuring the divergence from randomness. ACM Trans. Inf. Syst. **20**(4), 357–389 (2002)
2. Amer-Yahia, S., Lalmas, M.: XML search: languages, INEX and scoring. ACM. SIGMOD Record **35**(4), 16–23 (2006)
3. Balaneshinkordan, S., Kotov, A., Nikolaev, F.: Attentive neural architecture for ad-hoc structured document retrieval. CIKM 2018, ACM, Torino, Italy (2018)
4. Fang, H., Tao, T., Zhai, C.: A formal study of information retrieval heuristics. SIGIR 2004, ACM, New York, NY, USA (2004)
5. Fang, H., Tao, T., Zhai, C.: diagnostic evaluation of information retrieval models. ACM Trans. Inf. Syst. **29**(2), 1–42 (2011)
6. Fang, H., Zhai, C.: An exploration of axiomatic approaches to information retrieval. SIGIR 2005 (2005)
7. Fang, H., Zhai, C.: Semantic term matching in axiomatic approaches to information retrieval. SIGIR 2006, ACM, New York, NY, USA (2006)
8. Fuhr, N.: Some common mistakes in IR evaluation, and how they can be avoided. ACM SIGIR Forum **51**(3), 32–41 (2018)
9. Hasibi, F., et al.: DBpedia-Entity v2: a test collection for entity search. SIGIR 2017 (2017)
10. Hintikka, J.: On semantic information. In: Yourgrau, W., Breck, A.D. (eds.) Physics, Logic, and History: Based on the First International Colloquium held at the University of Denver, pp. 147–172. Springer, Boston (1970). https://doi.org/10.1007/978-1-4684-1749-4_9
11. Kamphuis, C., de Vries, A.P., Boytsov, L., Lin, J.: Which BM25 Do You Mean? A Large-Scale Reproducibility Study of Scoring Variants. In: Jose, J.M., et al. (eds.) ECIR 2020. LNCS, vol. 12036, pp. 28–34. Springer, Cham (2020). https://doi.org/10.1007/978-3-030-45442-5_4
12. Kamps, J., Koolen, M., Geva, S., Schenkel, R., SanJuan, E., Bogers, T.: From XML retrieval to semantic search and beyond. In: Information Retrieval Evaluation in a Changing World: Lessons Learned from 20 Years of CLEF, pp. 415–437 (2019)
13. Ketola, T., Roelleke, T.: BM25-FIC: information content-based field weighting for BM25F. In: BIRDS@SIGIR (2020)
14. Ketola, T., Roelleke, T.: Formal constraints for structured document retrieval. In: Proceedings of the 2022 ACM SIGIR International Conference on Theory of Information Retrieval. ICTIR 2022, ACM, New York, NY, USA (2022)

15. Kim, J., Xue, X., Croft, W.B.: A probabilistic retrieval model for semistructured data. In: Boughanem, M., Berrut, C., Mothe, J., Soule-Dupuy, C. (eds.) ECIR 2009. LNCS, vol. 5478, pp. 228–239. Springer, Heidelberg (2009). https://doi.org/10.1007/978-3-642-00958-7_22
16. Malik, S., Lalmas, M., Fuhr, N.: Overview of INEX 2004. In: Fuhr, N., Lalmas, M., Malik, S., Szlávik, Z. (eds.) INEX 2004. LNCS, vol. 3493, pp. 1–15. Springer, Heidelberg (2005). https://doi.org/10.1007/11424550_1
17. Metzler, D., Croft, W.: Linear feature-based models for information retrieval. Inf. Retr. **16**, 1–23 (2007)
18. Ogilvie, P., Callan, J.: Combining document representations for known-item search. SIGIR 2003, ACM, New York, NY, USA (2003)
19. Robertson, S., Walker, S., Jones, S., Hancock-Beaulieu, M.M., Gatford, M.: Okapi at TREC-3, pp. 109–126 (1995)
20. Robertson, S., Zaragoza, H., Taylor, M.: Simple BM25 extension to multiple weighted fields, pp. 42–49. CIKM 2004, ACM, Washington, D.C., USA (2004)
21. Roelleke, T.: Information Retrieval Models: Foundations and Relationships. Morgan & Claypool Publishers (2013).https://doi.org/10.1007/978-3-031-02328-6
22. Roelleke, T., Lalmas, M., Kazai, G., Ruthven, I., Quicker, S.: The accessibility dimension for structured document retrieval. In: Crestani, F., Girolami, M., van Rijsbergen, C.J. (eds.) ECIR 2002. LNCS, vol. 2291, pp. 284–302. Springer, Heidelberg (2002). https://doi.org/10.1007/3-540-45886-7_19
23. Sakai, T.: On Fuhr's guideline for IR evaluation. ACM SIGIR Forum **54**(1), 1–8 (2021)
24. Schenkel, R., Theobald, M.: Structural feedback for keyword-based XML retrieval. In: Lalmas, M., MacFarlane, A., Rüger, S., Tombros, A., Tsikrika, T., Yavlinsky, A. (eds.) ECIR 2006. LNCS, vol. 3936, pp. 326–337. Springer, Heidelberg (2006). https://doi.org/10.1007/11735106_29
25. Trotman, A., Puurula, A., Burgess, B.: Improvements to BM25 and language models examined. In: Proceedings of the 2014 Australasian Document Computing Symposium. ADCS 2014, ACM, New York, NY, USA (2014)
26. Wang, J., Roelleke, T.: Context-specific frequencies and discriminativeness for the retrieval of structured documents. In: Lalmas, M., MacFarlane, A., Rüger, S., Tombros, A., Tsikrika, T., Yavlinsky, A. (eds.) ECIR 2006. LNCS, vol. 3936, pp. 579–582. Springer, Heidelberg (2006). https://doi.org/10.1007/11735106_69
27. Zamani, H., Mitra, B., Song, X., Craswell, N., Tiwary, S.: Neural ranking models with multiple document fields. WSDM 2018, ACM, New York, NY, USA (2018)
28. Zaragoza, H., Craswell, N., Taylor, M., Saria, S., Robertson, S.: microsoft Cambridge at TREC-13: web and hard tracks, p. 7 (2004)
29. Zhiltsov, N., Kotov, A., Nikolaev, F.: Fielded sequential dependence model for Ad-hoc entity retrieval in the web of data. SIGIR 2015 (2015)

An Experimental Study on Pretraining Transformers from Scratch for IR

Carlos Lassance, Hervé Dejean$^{(\boxtimes)}$, and Stéphane Clinchant

Naver Labs Europe, Meylan, France
{carlos.lassance,herve.dejean,stephane.clinchant}@naverlabs.com

Abstract. Finetuning Pretrained Language Models (PLM) for IR has been de facto the standard practice since their breakthrough effectiveness few years ago. But, is this approach well understood? In this paper, we study the impact of the pretraining collection on the final IR effectiveness. In particular, we challenge the current hypothesis that PLM shall be trained on a large enough generic collection and we show that pretraining from scratch on the collection of interest is surprisingly competitive with the current approach. We benchmark first-stage ranking rankers and cross-encoders for reranking on the task of general passage retrieval on MSMARCO, Mr-Tydi for Arabic, Japanese and Russian, and TripClick for specific domain. Contrary to popular belief, we show that, for finetuning first-stage rankers, models pretrained solely on their collection have equivalent or better effectiveness compared to more general models. However, there is a slight effectiveness drop for rerankers pretrained only on the target collection. Overall, our study sheds a new light on the role of the pretraining collection and should make our community ponder on building specialized models by pretraining from scratch. Last but not least, doing so could enable better control of efficiency, data bias and replicability, which are key research questions for the IR community.

Keywords: Pretrained language models · Transformers · IR

1 Introduction

Transformers models are the main breakthrough in artificial intelligence over the past five years. Pretraining transformers models with Masked Language Modeling (MLM), a form of self-supervision, as proposed in the seminal BERT model [11] led to major improvement in natural language processing, computer vision and other domains. Pretraining and self-supervision have then paved the way to a race on bigger foundation models. Information Retrieval (IR) followed the same trajectory, where Pretrained Language Models (PLM) have largely outperform previous neural model [13,15,23,30] but also traditional bag-of-words approaches such as BM25 [46]. These advances were all made possible by a combination of large datasets, PLMs such as BERT, but also priors coming from traditional IR methods such as BM25.

J. Kamps et al. (Eds.): ECIR 2023, LNCS 13980, pp. 504–520, 2023.
https://doi.org/10.1007/978-3-031-28244-7_32

However, these PLMs do not perform well out-of-the box for Information Retrieval (on the contrary to NLP[1]) as they require a significant fine-tuning procedure. In IR, there are two types of PLM: the cross-encoders for reranking a set of top k documents and the dual encoders to deal with an efficient first-stage retrieval [30]. However, the standard pretraining MLM task may not be the best task for IR as argued in [14]. More-so, a growing tendency is to introduce a "middle-training" step [14] to bridge this gap and adapt the PLM not only to the retrieval domain, but also to the way the sentences will be encoded. For example [14,15,34] adapt the Masked Language Modeling (MLM) loss, so that the model learns to condense information on the CLS token, which will be used as the de-facto sentence embedding during fine-tuning. Similarly, several middle training tasks have been proposed to better fit the IR tasks or by using weak supervision.

On the one hand, it seems that there is a widespread belief that the downstream effectiveness is essentially due to pretraining on a *large* external collection. For instance, the foundation models report [4] state that 'AI is undergoing a paradigm shift with the rise of models (e.g., BERT, DALL-E, GPT-3) that are trained on broad data at scale and are adaptable to a wide range of downstream tasks.' On the other hand, the middle training process seems to contradict the former. This is why, in this paper, we aim to investigate the following questions: do we actually need large scale pretraining for Neural IR? How much knowledge is actually encoded from the large pretraining collection? Besides, what is known about pretraining in IR is largely limited by the MSMARCO setting and therefore a related question is how one shall address pretraining language models for new languages or new domains when it comes to IR.

In this paper, we aim to verify if these preconceived notions are needed, or if we could just have combined the pretraining and middle training steps to generate PLMs that are already adapted to the problem at hand with a smaller cost than doing both separately. Overall, this paper makes the following contributions:

1. We study pretrained transformers from scratch on IR collections;
2. We show that first-stage rankers, pretrained on MSMARCO, are as effective or even better in-domain (MSMARCO), while out-of-domain those models generalize as well (sparse retrieval) or worse (dense retrieval);
3. We evaluate cross encoders that are pretrained from scratch and verify that they actually benefit from external pretraining;
4. We show that first-stage retrievers, trained from scratch on the target collection, are competitive or outperform domain specific models (e.g. SciBERT on TripClick) and multilingual models (e.g. MContriever on Mr. TiDy);
5. Variants of Transformers architectures, such as DeBERTa, alleged to be better in NLP benchmarks do not bring benefits to IR, even when trained from scratch.

---

[1] For instance, freezing the BERT encoding and learning an additional linear layer is sufficient to obtain good performance in NLP [11], while such approach is not as effective in IR.

2   Related Work

PLMs in IR: today, the standard practice of many IR researchers is to simply download an existing pretrained model in order to finetune it on their retrieval task. After their success on reranking, PLMs have been adopted for first-stage ranking with a bi-encoder network to tackle efficiency requirements [25,30,43] or with late interaction [26]. Several training strategies have been proposed to improve the effectiveness of bi-encoders, such as distillation [20,22,31,48] and hard negative mining [42,45,53]. Parallel to these developments, another research direction aimed at learning *sparse* representations from PLM to behave as lexical matching. COIL [16] (later improved in uniCOIL [29]) learns term-level dense representations to perform contextualized lexical match. SPLADE [13,28] directly learns high-dimensional sparse representation thanks to the MLM head of the PLM and the help of sparse regularization. Most notably, SPLADE achieved state-of-the art effectiveness on the zero-shot benchmark BEIR [51], being later surpassed by other methods with much more compute [36].

Rise of Middle Training: Several works recently proposed to perform an additional step of pretraining, before the final finetuning stage, a procedure that we will call here *middle training*. The rationale is that the PLM weights or its CLS pooling mechanism are not well-suited for retrieval or similarity tasks often used in IR. Two main ideas emerge from this literature: i) using a contrastive loss on different document spans, and ii) using an information bottleneck to better pre-condition the network to rely on its CLS representation to perform predictions [27]. In [6], the paper compares the Inverse Cloze Task, Wiki Link Prediction and Body First Selection. Their result show that a combination of all these tasks was beneficial compared to MLM pretraining only. In [35], hyperlinks are used for pretraining. Another pretraining relies on web page structure and their DOM in the WebFormer model [18]. Furthermore, Contriever [23] relies on contrastive loss from different text spans, similarly to Co-Condenser [15]. Co-Condenser extends the Condenser [14] idea which focuses on middle training the CLS token. Very recently, Retro-MAE [32] revisits the same idea, by masking twice an input passage so that the first masking produce a CLS representation reused for decoding the second masking of the passage. Pretraining for sparse models has been recently investigated in [28] to better condition the network with SPLADE finetuning. The idea is to reuse the FLOPS [41] regularization used during finetuning within the MLM middle training. In [1], 14 different pretraining tasks are compared when training BERT models, including predicting the *tfidf* scores of a document, which was shown to be beneficial. They then evaluate on several NLP tasks, including sentence similarity. In addition, [33,34] propose pretraining with representative word predictions: for each document a set of important word is defined by several heuristics and the model is pretrained to predict that set of words.

While pretraining for IR seems very trendy, the idea of pretraining representation for IR tasks can actually be traced before the advent of PLM (or Foundation Models). For instance, more than ten years ago, the supervised semantic indexing model [2], used hyperlinks anchor to build triplets in a contrastive task, which can be viewed as an ancestor to the pretrainings tasks on Wikipedia. Similarly, weak supervision coming from BM25 was used to pretrain neural IR models [10] before the use of PLM.

Foundational Models and Architectures: A loosely related line of work is the scaling laws literature for large pretrained language models [24]. The scaling law aims to understand how the architecture and model size influence the perplexity and accuracy of the language model. In [50], Tay et al. showed that perplexity was a poor predictive measure of downstream effectiveness and propose to favor depth rather than just width (i.e. hidden size) in a pattern they named *deep-narrow* architectures. While this question could be interesting to our work, most of the literature is focusing on the large data and model regime, while in IR we look to the other side of the spectrum with small or moderate size collections/models for efficiency purposes.

Finally, a very interesting work by Tay et al. [49] argued that pre-training and architectural advances have been conflated. In addition, they show that convolutional models are in fact competitive with transformers when they are pre-trained on the same collection and for tasks which do not require cross attention between two sentences (e.g. a bi-encoder network). Finally, they argued that the current approach is misguided and that both architecture and pretraining should be considered independently. Finally, [7] studies the impact of the pretraining collection for machine translation and in [12] for image related tasks with large models. Our work is related to those, as we study the impact of the pretraining collection for the final effectiveness of an IR system.

3   Pretraining from Scratch

All in all, the role of finetuning representation or performing a middle training seems important for IR systems. The PLM representations seem critical but for good effectiveness, the impact of the pretraining collection on the final effectiveness is unclear since these representations are then finetuned by diverse means. Is the effectiveness heavily influenced by the pretraining collection and its co-occurrence statistics? To investigate this question, we experiment with several PLM models **trained from scratch** on the target collection. This way, we will be able to measure the effectiveness gains obtained by pretraining on a larger external collection and answer the following questions: is there an advantage in pretraining directly and only on the target collection? When and what are the advantages of pretraining on a different larger collection?

On the one hand, pretraining from scratch on the target collection could have the advantage of better modelling the target collection by having more informative co-occurrence statistics between tokens. Moreover, 'smaller' sized models may be able to reach the same level of effectiveness, i.e. one can also include efficiency requirements when training these models from scratch. On the other hand, pretraining on a larger collection may lead to more robust/generic as the model has seen more domains, different token usages and could have 'more' knowledge.

Therefore, by comparing these two approaches, we hope to better understand how large external pretraining contributes to the final effectiveness. In this paper, our research question deals with general-purpose vs specific-purpose model. The mainstream approach is to adopt the general purpose model, by simply adapting it to an IR task. On contrary, this paper investigates specific purpose models to assess their effectiveness (cf. Fig. 1). In a nutshell, would pretraining from scratch work for IR models? More specifically, we look at the following research questions:

1. Do we need an external pretrained language model for Information Retrieval?
2. Do models pretrained on target collections generalize to other domains and tasks?
3. Can we take advantage of pretraining for specialized domains and non-English languages?
4. Does efficient pretraining allow us to use recent architectural advances of transformers?

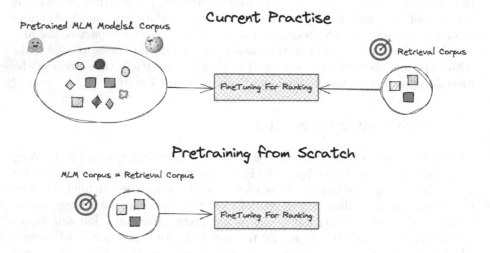

Fig. 1. Pretraining from Scratch: MLM is only performed on the retrieval collection.

To answer these questions, we compare the performance of standards models, such as BERT pretrained in Wikipedia and Book Corpus, to pretraining for scratch on the retrieval collections. For instance, we use the classical MSMARCO [39] dataset. By pretraining only on MSMARCO, we observe if there is indeed a benefit when pretraining on Wikipedia. Furthermore, we assess whether such models still generalise well on the BEIR dataset [51]. Indeed, one advantage of pretraining on a large collection could be that the neural retrievers

Table 1. Model comparison.

	BERT	DistilBERT	MLM 6L MLM 12L	
MLM collection	Wikipedia, BookCorpus	Wikipedia, BookCorpus	MSMARCO or Mr. TyDi or TripClick	
Pretraining size (words)	3,300M	3,300M	From 100M to 500M	
# params	110M	66M	67M	110M
# Layers	12	6	6	12
# number_heads	12	16	16	12

generalise better in zero shot settings. For specialized domains, such as health and biomedical data, we use the TripClick [44] dataset and compare pretraining from scratch to the performance of SciBERT [3] and PubMedBERT [17]. Similarly, by pretraining models only TripClick, we compare the performance of models trained on a much larger collection (i.e. PubMed). Finally, we will extend the previous experiments on non-English datasets, on Mr. TyDi [54] dataset with comparison with MContriever [23], which relied on a pretraining on large dataset with many languages. By conducting those experiments, we can assess whether an external pretrained language model is needed for Information Retrieval and if pretraining from scratch is an interesting alternative.

We will focus our study on BERT [11] and DistilBERT [47] architectures since they are the most popular ones in IR. Furthermore, we will measure the effectiveness of different architecture: *dense* bi-encoders [30,43], *SPLADE*, a state of the art sparse model [13,28] and cross-encoders for reranking [40].

3.1   Pretraining

In the following we always pretrain at least two types of models from scratch: a model with 12 layers based on the BERT architecture [11] and one with 6 layers based on DistilBERT [47]. We use BERT and DistilBERT as baselines all along the article. Table 1 summarises the main model characteristics we will consider. For pretraining from scratch, we always fix the vocabulary size to 32k (slightly larger than BERT's 30.5k), using wordpiece [52] to find the most common tokens of the target collection. We refer to those models as MLM 12L and MLM 6L.

We also use 2 models built on a setting called, MLM+FLOPS 12L and MLM+FLOPS 6L, by adding the FLOPS regularization [41], which helps to (pre)condition PLM for usage with SPLADE as proposed in [28]. The standard MLM loss is modified as follows: the MLM logits go through the SPLADE activation function (i.e. $\log(1 + ReLU(y_{\text{logits}}))$), which defines an MLM loss over a sparse set of logits. Finally, another term (FLOPS regularization) is added to force the logits to not only be nonnegative, but actually be sparse. As in SPLADE, a max pooling of the overall sentence is done to get a representation at the word level. On this final representation, the FLOPS regularization forces sparsification (and uniformity) over the overall vocabulary. The total loss is given by $\ell_{\text{MLM}} + \ell_{\text{MLM-SPLADE}} + \ell_{\text{FLOPS}}$.

Pretaining time is between 6 h to 1 day depending on the models and collections on 8 NVIDIA A100 80 Gb, compared to the original computational cost of

BERT (3 days using 64 TPUs), and the cost of finetuning (around 1 day on 4 V100 32 Gb), we consider our pretraining cost to be reasonable.

4   Experiments

Our research question is to assess whether models fully trained on the target collection perform as well as the generic DistilBERT or BERT models. First, we check the results on the general collection MSMARCO [39] (RQ1) and how the models trained on it generalize (RQ2) to a zero-shot scenario in BEIR [51]. We then verify if the results generalize to more specialized collections and non-English languages (RQ3). Finally, we take advantage of the fact that we are training models from scratch to test variants of transformer architectures (RQ4). Additional finetuning details are available at the end of the respective sections.

Experimental Setup. Pretraining is performed using 8 NVIDIAs A100 80 Gb either on MS-MARCO (RQ1 and RQ2), or TripClick (RQ3), or Mr. TyDi dataset (RQ3). We always use a learning rate of 1e-4. In the case of the MSMARCO collection, we pretrain using MLM and MLM+FLOPS using the entire passage collection (8.8M) combined with the training queries ($\tilde{8}$00k) for a total of (9.6M "documents"), while for TripClick we separate the documents into training and validation (90/10 split). For all collections, the batch size per GPU is either 150 (12L) or 200 (6L). We use an exponential warmup for FLOPS of 5k steps, a warmup of 1k steps for the logits and a learning rate warmup of 10k steps. The λ factor of FLOPS is set to $1e-3$, the max length before truncation to 256 tokens and the networks are pretrained for 125k steps. For Mr. TyDi, we pretrain 3 networks (Arabic, Russian, Japanese) using MLM+FLOPS on the entire language's passage collection (2M for Arabic, 7M for Japanese and 9.6M for Russian) with a batch size per GPU of 200. Finally for TripClick, the only difference is the number of epochs: 60 epochs, and the batch size per GPU is 256 for the 6L models, and 128 for the 12L models.

For finetuning, we used 4 V100 32 gb for a sparse model (SPLADE) and a dense neural bi-encoder model. Please note that **we do not use distillation** from a reranker as it would entail transferring information from an existing PLM. For SPLADE, we use the L1 regularization over queries (following [28]) with $\lambda_q = 1e-3$, and for documents a FLOPS regularisation with $\lambda_d = 1e-3$ ($\lambda_d = 5e-4$ on TripClick) following [28]. The learning rate is set up to $2e-5$. In all of our tables, **superscripts** denote **significant differences** according to a paired Student's t-test with Bonferroni's correction and $p \leq 0.05$ with the corresponding table row; MRR@10 and nDCG@10 have been multiplied by 100.

4.1   RQ1: Are Models Fully Trained on MSMARCO as Good as Models Pretrained on a Diverse Collection Set?

The first question we want to address is whether models that are solely trained on MSMARCO are as good as models that have used external corpora for

Table 2. Comparison on MSMARCO of first stage sparse neural (SPLADE) models. [†] indicates a method pretrained solely on MSMARCO.

#	Pretrained model	R-FLOPS	MSMARCO dev		TREC DL 19		TREC DL 20	
			MRR@10	R@1k	nDCG@10	R@1k	nDCG@10	R@1k
a	Distilbert	1.10	0.373	0.975^b	**0.732**	**0.853**	0.708	0.867^b
b	BERT	1.32	0.367	0.968	0.727	0.832	0.699	0.842
c	MLM 6L[†]	8.2	0.370	$\mathbf{0.982}^{ab}$	0.701	0.847	0.700	**0.877**
d	MLM+Flops 6L[†]	0.72	$\mathbf{0.382}^{abc}$	0.979^b	0.698	0.836	0.701	0.872
e	MLM+Flops 12L[†]	0.97	0.379^{bc}	0.980^{ab}	0.709	0.835	**0.709**	0.865

pretraining (i.e. BookCorpus and Wikipedia). We first investigate models trained as first stage rankers, either using a dense bi-encoder [25, 30] or sa SPLADE model.

We finetuned the models using negatives that were sampled from a previously trained SPLADE model. For each V100 we use a batch of the maximum size that does not exceed the total memory. Batches are constructed such that "one element" of the batch is composed of the query itself, a positive passage related to the query and 16 or 32 negatives, depending on the network size (the actual batch sizes per GPU varies from 54 to 102 depending on the pretrained model size). Finetuning is considered finished after two epochs (arbitrary decision based on initial experiments with a validation set). Note that the finetuning setting for both first stage and cross-encoders are almost the same, except for the fact that cross-encoders do not use in-batch negatives and use a learning rate of 1e-4.

We reports the retrieval flops (noted R-FLOPS) for SPLADE models, i.e., the number of floating point operations on the inverted index to return the list of documents for a given query. The R-FLOPS metric is defined by an estimation of the average number of floating-point operations between a query and a document which is defined as the expectation $\mathbb{E}_{q,d}\left[\sum_{j \in V} p_j^{(q)} p_j^{(d)}\right]$ where p_j is the activation probability for token j in a document d or a query q. It is empirically estimated from a set of approximately 100k development queries, on the MS MARCO collection. It is thus an indication of the inverted index sparsity and of the computational cost for a sparse model (which is different from the inference cost of the model).

First stage retrievers results are described in Table 2 for sparse models and Table 3 for dense models. Surprisingly, models trained solely on MSMARCO with MLM+FLOPS actually **can perform statistically significantly better** than their counterparts pretrained over larger collections, in both sparse and dense scenarios, while there's no statistically significant difference when considering just MLM on MSMARCO vs MLM on larger corpora. This shows that not only pretraining does not seem to care for a diverse collection, but that by focusing only on "off-the-shelf models" we could be losing possible performance gains of better initialized models. Also note that less computing was used to pretrain MLM+FLOPS 6L compared to its DistilBERT counterpart.

Table 3. Comparison on MSMARCO of first stage dense neural (DPR) models. [†] indicates a method pretrained solely on MSMARCO.

#	Pretrained model	MSMARCO dev		TREC DL 19		TREC DL 20	
		MRR@10	R@1k	nDCG@10	R@1k	nDCG@10	R@1k
a	Distilbert	0.342	0.961	0.673	0.774	0.670	0.816
b	BERT	0.347	0.961	**0.697**	0.785	**0.682**	0.809
c	MLM 6L[†]	0.346	0.968^{ab}	0.664	0.783	0.657	0.818
d	MLM+FLOPS 6L [†]	0.349	0.968^{ab}	0.670	0.781	0.668	0.837
e	MLM+FLOPS 12L[†]	$\mathbf{0.352}^{a}$	$\mathbf{0.969}^{ab}$	0.672	**0.800**	0.680	$\mathbf{0.848}^{bc}$

Table 4. Comparison of rerankers on MSMARCO. Models with † were pretrained solely on MSMARCO.

#	Pretrained model	MSMARCO dev	TREC DL 2019	TREC DL 2020
		MRR@10	nDCG@10	nDCG@10
a	Without reranking (First stage)	0.384	0.718	**0.737**
b	Distilbert	0.396^{af}	**0.764**	0.734
c	Bert	$\mathbf{0.404}^{abf}$	0.750	0.737
d	MLM 6L [†]	0.398^{af}	0.743	0.716
e	MLM+Flops 6L [†]	0.396^{af}	0.724	0.736
f	MLM+Flops 12L [†]	0.381	0.730	0.722

Finally, we also test in the case of reranking using cross-encoders. Results are available in Table 4. Overall there's no statistical significant gain on the models pretrained with external collections.

4.2   RQ2: Do Models Pretrained in MSMARCO Generalize Well on Other Collections?

In the previous section, we considered only in-domain results. While they have shown that we can outperform (or at least keep comparable effectiveness) using solely the target collection, they could be masking a possible gap in out-of-domain data. In order to verify that models solely pretrained and fine-tuned on MSMARCO do not lose effectiveness in out-of-domain data, we report results under the zero-shot BEIR benchmark in Table 5. We actually observe a small boost in effectiveness on sparse retrieval when using models solely trained on MSMARCO, while on Dense there's a more apparent decrease of performance. The biggest difference for dense models is on the TREC Covid dataset which is far from the MSMARCO collection. Given the nature of the BEIR benchmark (mean over 13 datasets), those differences may not be significative.[2]

---

[2] We could not find in the literature an easy/practical way to perform statistical significance testing over BEIR.

Table 5. Experiments on zero-shot retrieval on BEIR (nDCG@10) with models fine-tuned on MSMARCO. Models with † were pretrained solely on MSMARCO.

	SPLADE					Dense				
	Distilbert	Bert	M 6L†	M+F 6L†	M+F 12L†	Distilbert	Bert	M 6L†	M+F 6L†	M+F 12L†
arguana	45.30	44.30	**46.90**	42.90	45.40	34.00	37.30	40.20	37.40	40.70
climate-fever	14.90	15.30	14.50	15.70	15.40	16.50	15.90	15.40	15.30	**17.50**
dbpedia-entity	**39.10**	38.80	38.30	38.80	37.90	31.50	31.70	31.10	31.70	30.60
fever	**73.40**	71.20	68.60	72.30	70.70	70.80	70.20	59.10	59.90	56.10
fiqa	31.20	29.90	**33.20**	32.30	31.70	25.70	24.10	27.30	27.00	27.60
hotpotqa	66.90	65.50	64.00	67.30	**67.60**	49.70	50.20	47.30	46.40	47.60
nfcorpus	33.40	31.30	35.40	**35.90**	34.70	26.40	25.80	28.80	28.70	28.80
nq	**51.40**	50.60	48.50	49.20	49.40	46.00	47.10	41.90	41.50	42.90
quora	77.20	76.60	80.40	78.10	73.80	78.50	82.20	82.20	**83.30**	80.30
scidocs	14.90	14.90	14.40	**15.10**	14.50	11.40	11.80	10.70	11.20	11.00
scifact	66.00	64.70	**69.20**	69.20	69.00	52.00	54.80	56.70	57.30	56.50
trec-covid	67.60	**69.10**	65.60	64.70	68.00	66.10	65.70	56.00	48.90	49.90
webis-touche2020	27.60	27.00	24.70	**28.50**	26.20	22.20	23.50	23.20	19.70	19.30
Mean	**46.84**	46.09	46.43	**46.92**	46.48	40.83	41.56	39.99	39.10	39.14

4.3 RQ3: Can We Take Advantage of that Pretraining from Scratch in Collections of Specialized Domains/Languages

Domain Specific IR on TripClick. The TripClick collection [44] contains approximately 1.5 millions MEDLINE documents (title and abstract), and 692,000 queries. The test set is divided into three categories of queries: Head, Torso and Tail (decreasing frequency), which contain 1,175 queries each. For the Head queries a DCTR click model [9] was employed to created relevance signals, the other two sets use raw clicks. [21] showed that the original triplets were too noisy, and released a new training set which we use in this experiment (10 millions triplets).

As pretrained models, we use the off-the-shelf BERT, DistilBERT models, and similarly to [21] SciBERT [3] and PubMedBERT [17] which are both using a similar architecture to Bert (12 layers) and were pretrained using scientific documents (from where they extract their vocabulary). We consider them as off-the-shelf domain-specific models. While for finetuning, we use a batch size of 200 queries and only one negative per query, taking advantage solely of in-batch negatives for 90,000 iterations, which is equivalent to 1.8 epochs.

As Table 6 shows, models pretrained from scratch compete well against generic off-the-shelf models as well as against specialized ones. For this collection, the dense models perform better than the sparse ones. The conclusions depend on the model type: sparse or dense. For the sparse models, we see that at least one model based on pre-training from scratch outperforms off-the-shelf models such as BERT and DistilBERT, as well as both off-the-shelf domain-specific models (Scibert, PubMedBert). Regarding the dense architecture, pretrained models from scratch are on par with off-the-shelf domain-specific models which required much more training data (1.4B tokens for Scibert against 300M for TripClick). Pre-training from scratch allows for selecting the most suitable language model according to the fine-tuning architecture (sparse or dense). In our case a 6 layer

Table 6. Experiment on Tripclick with sparse (SPLADE) and dense models.† indicates a method pretrained solely on Tripclik.

#	Pretrained model	SPLADE				Dense			
		$Head_{dctr}$	Head	Torso	Tail	$Head_{dctr}$	Head	Torso	Tail
a	Distilbert	22.1^b	30.3^{bd}	23.9^b	24.8^{bd}	24.9	34.8	29.2	27.5
b	BERT	10.6	14.4	9.6	7.8	25.3	35.3	29.4	28.8^a
c	PubMedBert	22.5^{bd}	31.0^{bd}	24.5^b	24.3^{bd}	27.1^{abef}	37.6^{abef}	29.9^e	30.8^a
d	Scibert	21.1^b	28.4^b	23.0^b	22.2^b	27.8^{abef}	38.1^{abef}	29.2	30.3^a
e	MLM+F 6L†	$\mathbf{26.9}^{abcdfgh}$	$\mathbf{36.7}^{abcdfgh}$	$\mathbf{27.7}^{abcdfgh}$	$\mathbf{27.2}^{bcdfh}$	25.7	35.7	28.3	29.7
f	MLM 6L†	23.1^{bd}	31.9^{bd}	24.7^b	23.4^b	25.8	36.0	28.9	29.3
g	MLM+F 12L†	23.7^{abd}	32.5^{abd}	26.2^{abd}	26.6^{bdf}	26.7^{ab}	37.5^{abef}	29.9^e	30.1^a
h	MLM 12L†	24.0^{abcd}	32.6^{abcd}	24.8^{bd}	24.6^{bdf}	$\mathbf{28.0}^{abefg}$	$\mathbf{38.6}^{abef}$	$\mathbf{30.2}^e$	30.4^a
	Scibert [21]	–	–	–	–	24.3	–	–	–
	PubMebBert [21]	–	-	-	–	23.5	–	–	–
	BM25 [21]	–	–	–	–	14.0	-	-	–

language model is far better for SPLADE while a 12 layer better fits a dense model. We added at the bottom of Table 6 the results from [21], which show that our implementation choices are very competitive[3] for the sparse as well as for the dense models.

Mr. TyDi. is a multilingual dataset for monolingual retrieval composed of 11 typologically diverse languages [54]. For this study we focus on the three non-English languages with the most training data on Mr. TyDi: i) Arabic; ii) Russian; iii) Japanese. Note that, there is a large amount of data available for these languages (even outside of Mr. TyDi), but PLMs are not as well studied as in English. The main consensus seems to be that in this case one should focus on multi-lingual data, for which most is based on english as an anchor, as it is the case for many previous works [8,23,37,38,54]. We challenge this notion, by: a) using monolingual models; b) pretraining solely on Mr. TyDi.

We follow the fine-tuning protocol of MContriever [23], where they first fine-tune the model for retrieval on MMarco [5], a translated version of MSMARCO in multiple languages, for which we only use the target language for a given model (Arabic, Russian or Japanese) and then finally fine-tune on the Mr. TyDi collection. In our case we perform a three-step training, first step on MMarco using negatives sampled with the pretrained model (or MContriever for the baseline). We then perform two steps of training on Mr. TyDi, first using negatives extracted first from the MMarco finetuned model and second from the first stage of Mr.TyDi finetuning. Batch composition follows the previous MSMARCO fine-tuning, with a batch size of 3 queries 32 negatives (the actual batch sizes per GPU is thus 102). Finetuning stopped after two epochs.

---

[3] We were not able to find the parameters used in the experiments.

Results are available in Table 7. Compared to the previous state of the art [23][4], which is a dense retriever pretrained in a specific fashion on a much larger collection[5], we show statistically significant improvements on all languages while using solely the Mr. TyDi and MMarco collections on the target language. However, it is important to note that differently from [23] we did not actually test yet on all languages and thus can only evaluate the pretraining effect on these three languages, which are the three largest from Mr. TyDi.

Table 7. Comparison, on the Mr.TyDi dataset, of models trained from scratch against models pretrained in a large external collection with Contriever.[†] means a method pretrained solely on MrTiDy.

#	Pretrained model	Arabic		Russian		Japanese	
		MRR@100	R@100	MRR@100	R@100	MRR@100	R@100
	MContriever [23]	72.4	94.0	59.7	92.4	54.9	88.8
a	MContriever (reproduced)	72.7	93.6	59.9^b	91.6	49.9^b	85.2^b
b	MLM+FLOPS 6L DPR[†]	73.4	**94.9**	56.2	91.6	32.9	62.0
c	MLM+FLOPS 6L SPLADE[†]	$\mathbf{75.7}^{ab}$	94.2	$\mathbf{65.0}^{ab}$	$\mathbf{93.4}^{ab}$	$\mathbf{56.3}^{ab}$	$\mathbf{88.9}^{ab}$

4.4 RQ4: Impact of Architectures

Finally, one advantage of pretraining models from scratch is the fact that we can more easily experiment with different architectures. Indeed, considering that most IR works use a variant of BERT (either RoBERTA, DistilBERT or BERT) it raises a question whether variants of transformer architectures, benchmarked in NLP, could actually improve IR. To address this question, we use the sparse retrieval setting from RQ1, but this time also consider using the DeBERTa architecture [19] which beats BERT on many NLP tasks. Results are presented in Tables 8 and 9. Much to our dismay, we did not actually see major improvements using these architectural changes, we thus leave as future work how to better include these changes within first stage rankers.

Table 8. Comparison on MSMARCO of first stage sparse neural (SPLADE) models using different architectures. All methods are pretrained solely on MSMARCO.

#	Pretrained model	R-FLOPS	MSMARCO dev		TREC DL 19		TREC DL 20	
			MRR@10	R@1k	nDCG@10	R@1k	nDCG@10	R@1k
a	BERT MLM+FLOPS 6L	0.72	$\mathbf{0.382}^d$	0.979	0.698	0.836	0.701	**0.872**
b	BERT MLM+FLOPS 12L	0.97	0.379	**0.980**	0.709	0.835	0.709	0.865
c	DeBERTa MLM+FLOPS 6L	0.81	0.376	0.979	0.700	**0.845**	0.701	0.863
d	DeBERTa MLM+FLOPS 12L	0.79	0.373	0.978	**0.716**	0.839	**0.735**	0.871

---

[4] Note that MContriever TyDi (first row) is not available, statistical tests cannot be performed. We do our best to evaluate fairly under our training setting (second row).

[5] We suspect they use more compute, but could not find accurate compute information.

Table 9. Comparison on TripClick of first stage sparse (SPLADE) and dense models using different architectures. All methods are pretrained solely on TripClick.

#	Pretrained model	SPLADE				Dense			
		Head_{dctr}	Head	Torso	Tail	Head_{dctr}	Head	Torso	Tail
a	BERT MLM+FLOPS 6L	26.9^{bcdefh}	36.7^{bcdfh}	27.7^{bdfh}	27.2^{bdfh}	25.7^e	35.7^e	28.3	29.7^e
b	BERT MLM 6L	23.1	31.9	24.7	23.4	25.8^e	36.0^e	28.9	29.3
c	BERT MLM+FLOPS 12L	23.7^f	32.5^f	26.2^{fh}	26.6^{bfh}	26.7^{ef}	37.5^{abef}	29.9^{aef}	30.1^e
d	BERT MLM 12L	24.0^{fh}	32.6^f	24.8	24.6	$\mathbf{28.0}^{abcefgh}$	$\mathbf{38.6}^{abefgh}$	$\mathbf{30.2}^{aeh}$	$\mathbf{30.4}^e$
e	DeBERTa MLM FLOPS 6L	26.1^{bcdfh}	35.9^{bcdfh}	29.0^{bcdfh}	28.4^{bcdfh}	24.3	34.0	27.8	27.5
f	DeBERTa MLM 6L	22.2	30.7	23.3	23.3	24.9	34.7	28.7	28.9
g	DeBERTa MLM FLOPS 12L	$\mathbf{27.0}^{bcdfh}$	$\mathbf{37.5}^{bcdefh}$	$\mathbf{29.4}^{abcdfh}$	$\mathbf{28.7}^{abcdfh}$	26.6^f	36.7^{ef}	29.2	29.5
h	DeBERTa MLM 12L	22.6	31.3	23.8	23.3	26.4^{ef}	36.4^{ef}	28.6	29.0

5   Conclusion

Foundation models come with the promise to be highly general and modular. It is believed that they contain a wide "knowledge" due to their pretraining on a large collection, which is then believed to be the source of their improved performance. We have examined how this pretraining collection influence the performance of IR models. Our research question was to assess how much of this implicit knowledge, beneficial to the final performance, comes from pretraining on a large external collection. This is why we have experimented on a variety of collections, domains and languages to study how pretraining from scratch actually performed compared to their *de facto* approach of simple finetuning.

While we were expecting the standard pretrained models to work better, we surprisingly revealed that pretraining from scratch works better for first-stage retrieval on MSMARCO, TripClick and several non-English languages on the Mr. TyDi benchmark. In particular, the FLOPS regularization played a critical role in those results, suggesting that regularization or better pretraining techniques could further improve the results. Furthermore, pretrained models from scratch also behave well in the zero shot scenario for sparse models such as SPLADE. Nevertheless, pretraining from a large collection has a slight advantage when training rerankers.

Overall, these results, specific to IR, challenge the foundation model hypothesis for small models, i.e. that a more general model encapsulating the world knowledge would be better than a smaller one in a specific domain application. Furthermore, our study makes a contribution to the debate between general purpose and specific purpose models. In a way, our experiments showed that less is more. In addition, pretrained language models come also with many challenges such as the societal bias in the data they have been trained on. We hope that our study could convince practitioners, both from industry and academia, to reconsider specific purpose models by pretraining from scratch. Last but not least, doing so enable to better control efficiency, data bias and replicability, which are key research questions for the IR community.

References

1. Aroca-Ouellette, S., Rudzicz, F.: On losses for modern language models. In: Proceedings of the 2020 Conference on Empirical Methods in Natural Language Processing (EMNLP), pp. 4970–4981. Association for Computational Linguistics, Online, November 2020. https://doi.org/10.18653/v1/2020.emnlp-main.403, https://aclanthology.org/2020.emnlp-main.403

2. Bai, B., et al.: Supervised semantic indexing. In: Proceedings of the 18th ACM International Conference on Information and Knowledge Management, pp. 187–196. ACM (2009). https://doi.org/10.1145/1645953.1645979

3. Beltagy, I., Lo, K., Cohan, A.: SciBERT: a pretrained language model for scientific text (2019). https://doi.org/10.48550/ARXIV.1903.10676, https://arxiv.org/abs/1903.10676

4. Bommasani, R., et al.: On the opportunities and risks of foundation models (2021). https://doi.org/10.48550/ARXIV.2108.07258, https://arxiv.org/abs/2108.07258

5. Bonifacio, L.H., Campiotti, I., Jeronymo, V., Lotufo, R., Nogueira, R.: MMARCO: a multilingual version of the MS MARCO passage ranking dataset. arXiv preprint arXiv:2108.13897 (2021)

6. Chang, W.C., Yu, F.X., Chang, Y.W., Yang, Y., Kumar, S.: Pre-training tasks for embedding-based large-scale retrieval. In: International Conference on Learning Representations (2020). https://openreview.net/forum?id=rkg-mA4FDr

7. Clinchant, S., Jung, K.W., Nikoulina, V.: On the use of BERT for neural machine translation. In: Proceedings of the 3rd Workshop on Neural Generation and Translation, pp. 108–117. Association for Computational Linguistics, Hong Kong, November 2019. https://doi.org/10.18653/v1/D19-5611, https://aclanthology.org/D19-5611

8. Conneau, A., et al.: Unsupervised cross-lingual representation learning at scale. arXiv preprint arXiv:1911.02116 (2019)

9. Craswell, N., Zoeter, O., Taylor, M., Ramsey, B.: An experimental comparison of click position-bias models. In: Proceedings of the 2008 International Conference on Web Search and Data Mining, pp. 87–94 (2008)

10. Dehghani, M., Zamani, H., Severyn, A., Kamps, J., Croft, W.B.: Neural ranking models with weak supervision. In: Proceedings of the 40th International ACM SIGIR Conference on Research and Development in Information Retrieval. SIGIR 2017, pp. 65–74. Association for Computing Machinery, New York (2017). https://doi.org/10.1145/3077136.3080832

11. Devlin, J., Chang, M., Lee, K., Toutanova, K.: BERT: pre-training of deep bidirectional transformers for language understanding. CoRR abs/1810.04805 (2018), http://arxiv.org/abs/1810.04805

12. El-Nouby, A., Izacard, G., Touvron, H., Laptev, I., Jegou, H., Grave, E.: Are large-scale datasets necessary for self-supervised pre-training? arXiv preprint arXiv:2112.10740 (2021)

13. Formal, T., Lassance, C., Piwowarski, B., Clinchant, S.: From distillation to hard negative sampling: making sparse neural IR models more effective. In: Proceedings of the 45th International ACM SIGIR Conference on Research and Development in Information Retrieval. SIGIR 2022, pp. 2353–2359. Association for Computing Machinery, New York (2022). https://doi.org/10.1145/3477495.3531857

14. Gao, L., Callan, J.: Condenser: a pre-training architecture for dense retrieval. In: Proceedings of the 2021 Conference on Empirical Methods in Natural Language Processing, pp. 981–993. Association for Computational Linguistics, Online and Punta Cana, Dominican Republic, November 2021. https://doi.org/10.18653/v1/2021.emnlp-main.75, https://aclanthology.org/2021.emnlp-main.75

15. Gao, L., Callan, J.: Unsupervised corpus aware language model pre-training for dense passage retrieval. In: Proceedings of the 60th Annual Meeting of the Association for Computational Linguistics (Volume 1: Long Papers), pp. 2843–2853. Association for Computational Linguistics, Dublin, Ireland, May 2022. https://doi.org/10.18653/v1/2022.acl-long.203, https://aclanthology.org/2022.acl-long.203

16. Gao, L., Dai, Z., Callan, J.: COIL: Revisit exact lexical match in information retrieval with contextualized inverted list. In: Proceedings of the 2021 Conference of the North American Chapter of the Association for Computational Linguistics: Human Language Technologies, pp. 3030–3042. Association for Computational Linguistics, Online, June 2021. https://doi.org/10.18653/v1/2021.naacl-main.241, https://aclanthology.org/2021.naacl-main.241

17. Gu, Y., et al.: Domain-specific language model pretraining for biomedical natural language processing. ACM Trans. Comput. Healthc. 3(1), 1–23 (2022). https://doi.org/10.1145/3458754

18. Guo, Y., et al.: Webformer: pre-training with web pages for information retrieval. In: Proceedings of the 45th International ACM SIGIR Conference on Research and Development in Information Retrieval. SIGIR 2022, pp. 1502–1512. Association for Computing Machinery, New York (2022). https://doi.org/10.1145/3477495.3532086

19. He, P., Liu, X., Gao, J., Chen, W.: DeBERTa: decoding-enhanced BERT with disentangled attention. In: International Conference on Learning Representations (2021). https://openreview.net/forum?id=XPZIaotutsD

20. Hofstätter, S., Althammer, S., Schröder, M., Sertkan, M., Hanbury, A.: Improving efficient neural ranking models with cross-architecture knowledge distillation (2020)

21. Hofstätter, S., Althammer, S., Sertkan, M., Hanbury, A.: Establishing strong baselines for tripclick health retrieval (2022)

22. Hofstätter, S., Lin, S.C., Yang, J.H., Lin, J., Hanbury, A.: Efficiently teaching an effective dense retriever with balanced topic aware sampling. In: Proceedings of SIGIR (2021)

23. Izacard, G., et al.: Towards unsupervised dense information retrieval with contrastive learning (2021)

24. Kaplan, J., et al.: Scaling laws for neural language models. arXiv abs/2001.08361 (2020)

25. Karpukhin, V., et al.: Dense passage retrieval for open-domain question answering. In: Proceedings of the 2020 Conference on Empirical Methods in Natural Language Processing (EMNLP), pp. 6769–6781. Association for Computational Linguistics, Online, November 2020. https://doi.org/10.18653/v1/2020.emnlp-main.550, https://www.aclweb.org/anthology/2020.emnlp-main.550

26. Khattab, O., Zaharia, M.: ColBERT: efficient and effective passage search via contextualized late interaction over BERT. In: Proceedings of the 43rd International ACM SIGIR Conference on Research and Development in Information Retrieval. SIGIR 2020, pp. 39–48. Association for Computing Machinery, New York (2020). https://doi.org/10.1145/3397271.3401075

27. Kim, T., Yoo, K.M., Lee, S.G.: Self-guided contrastive learning for BERT sentence representations. In: Proceedings of the 59th Annual Meeting of the Association for Computational Linguistics and the 11th International Joint Conference on Natural Language Processing (Volume 1: Long Papers), pp. 2528–2540. Association for Computational Linguistics, Online, August 2021. https://doi.org/10.18653/v1/2021.acl-long.197, https://aclanthology.org/2021.acl-long.197

28. Lassance, C., Clinchant, S.: An efficiency study for splade models. In: Proceedings of the 45th International ACM SIGIR Conference on Research and Development in Information Retrieval. SIGIR 2022, pp. 2220–2226. Association for Computing Machinery, New York (2022). https://doi.org/10.1145/3477495.3531833

29. Lin, J., Ma, X.: A few brief notes on deepimpact, coil, and a conceptual framework for information retrieval techniques. CoRR abs/2106.14807 (2021). https://arxiv.org/abs/2106.14807

30. Lin, J., Nogueira, R., Yates, A.: Pretrained transformers for text ranking: BERT and beyond. arXiv:2010.06467 [cs] (Oct 2020), http://arxiv.org/abs/2010.06467, zSCC: NoCitationData[s0] arXiv: 2010.06467

31. Lin, S.C., Yang, J.H., Lin, J.: In-batch negatives for knowledge distillation with tightly-coupled teachers for dense retrieval. In: Proceedings of the 6th Workshop on Representation Learning for NLP (RepL4NLP-2021), pp. 163–173. Association for Computational Linguistics, Online, August 2021. https://doi.org/10.18653/v1/2021.repl4nlp-1.17, https://aclanthology.org/2021.repl4nlp-1.17

32. Liu, Z., Shao, Y.: Retromae: pre-training retrieval-oriented transformers via masked auto-encoder (2022). https://doi.org/10.48550/ARXIV.2205.12035, https://arxiv.org/abs/2205.12035

33. Ma, X., Guo, J., Zhang, R., Fan, Y., Ji, X., Cheng, X.: B-prop: bootstrapped pre-training with representative words prediction for ad-hoc retrieval. Proceedings of the 44th International ACM SIGIR Conference on Research and Development in Information Retrieval (2021)

34. Ma, X., Guo, J., Zhang, R., Fan, Y., Ji, X., Cheng, X.: Prop: pre-training with representative words prediction for ad-hoc retrieval. In: Proceedings of the 14th ACM International Conference on Web Search and Data Mining (2021)

35. Ma, Z., et al.: Pre-training for ad-hoc retrieval: hyperlink is also you need. In: Proceedings of the 30th ACM International Conference on Information and Knowledge Management (2021)

36. Muennighoff, N.: SGPT: GPT sentence embeddings for semantic search. arXiv preprint arXiv:2202.08904 (2022)

37. Nair, S., et al.: Transfer learning approaches for building cross-language dense retrieval models. In: Hagen, M., et al. (eds.) ECIR 2022. LNCS, vol. 13185, pp. 382–396. Springer, Cham (2022). https://doi.org/10.1007/978-3-030-99736-6_26

38. Nair, S., Yang, E., Lawrie, D., Mayfield, J., Oard, D.W.: Learning a sparse representation model for neural CLIR. In: Design of Experimental Search and Information REtrieval Systems (DESIRES) (2022)

39. Nguyen, T., et al.: MS MARCO: a human generated machine reading comprehension dataset. In: CoCo@ NIPs (2016)

40. Nogueira, R., Cho, K.: Passage re-ranking with BERT (2019)

41. Paria, B., Yeh, C.K., Yen, I.E.H., Xu, N., Ravikumar, P., Póczos, B.: Minimizing flops to learn efficient sparse representations (2020)

42. Qu, Y., et al: RocketQA: an optimized training approach to dense passage retrieval for open-domain question answering. In: In Proceedings of NAACL (2021)

43. Reimers, N., Gurevych, I.: Sentence-BERT: Sentence embeddings using Siamese BERT-networks. In: Proceedings of the 2019 Conference on Empirical Methods in Natural Language Processing. Association for Computational Linguistics, November 2019. http://arxiv.org/abs/1908.10084

44. Rekabsaz, N., Lesota, O., Schedl, M., Brassey, J., Eickhoff, C.: Tripclick: the log files of a large health web search engine. In: Proceedings of the 44th International ACM SIGIR Conference on Research and Development in Information Retrieval, pp. 2507–2513 (2021). https://doi.org/10.1145/3404835.3463242

45. Ren, R., et al.: RocketQAv2: a joint training method for dense passage retrieval and passage re-ranking. In: Proceedings of the 2021 Conference on Empirical Methods in Natural Language Processing, pp. 2825–2835. Association for Computational Linguistics, Online and Punta Cana, Dominican Republic, November 2021. https://doi.org/10.18653/v1/2021.emnlp-main.224, https://aclanthology.org/2021.emnlp-main.224

46. Robertson, S.E., Walker, S., Beaulieu, M., Gatford, M., Payne, A.: Okapi at TREC-4. Nist Special Publication Sp, pp. 73–96 (1996)

47. Sanh, V., Debut, L., Chaumond, J., Wolf, T.: DistilBERT, a distilled version of BERT: smaller, faster, cheaper and lighter. arXiv preprint arXiv:1910.01108 (2019)

48. Santhanam, K., Khattab, O., Saad-Falcon, J., Potts, C., Zaharia, M.: ColBERTv2: effective and efficient retrieval via lightweight late interaction (2021)

49. Tay, Y., et al.: Are pretrained convolutions better than pretrained transformers? In: Proceedings of the 59th Annual Meeting of the Association for Computational Linguistics and the 11th International Joint Conference on Natural Language Processing (Volume 1: Long Papers), pp. 4349–4359. Association for Computational Linguistics, Online, August 2021. https://doi.org/10.18653/v1/2021.acl-long.335, https://aclanthology.org/2021.acl-long.335

50. Tay, Y., et al.: Scale efficiently: insights from pre-training and fine-tuning transformers. arXiv abs/2109.10686 (2022)

51. Thakur, N., Reimers, N., Rücklé, A., Srivastava, A., Gurevych, I.: BEIR: a heterogenous benchmark for zero-shot evaluation of information retrieval models. CoRR abs/2104.08663 (2021). https://arxiv.org/abs/2104.08663

52. Wu, Y., et al.: Google's neural machine translation system: bridging the gap between human and machine translation (2016)

53. Xiong, L., et al.: Approximate nearest neighbor negative contrastive learning for dense text retrieval. In: International Conference on Learning Representations (2021). https://openreview.net/forum?id=zeFrfgyZln

54. Zhang, X., Ma, X., Shi, P., Lin, J.: Mr. Tydi: a multi-lingual benchmark for dense retrieval. In: Proceedings of the 1st Workshop on Multilingual Representation Learning, pp. 127–137 (2021)

Neural Approaches to Multilingual Information Retrieval

Dawn Lawrie[1]([✉]) [iD], Eugene Yang[1] [iD], Douglas W. Oard[1,2] [iD], and James Mayfield[1] [iD]

[1] HLTCOE. Johns Hopkins University, Baltimore, MD 21211, USA
{lawrie,eugene.yang,mayfield}@jhu.edu
[2] University of Maryland, College Park, MD 20742, USA
oard@umd.edu

Abstract. Providing access to information across languages has been a goal of Information Retrieval (IR) for decades. While progress has been made on Cross Language IR (CLIR) where queries are expressed in one language and documents in another, the multilingual (MLIR) task to create a single ranked list of documents across many languages is considerably more challenging. This paper investigates whether advances in neural document translation and pretrained multilingual neural language models enable improvements in the state of the art over earlier MLIR techniques. The results show that although combining neural document translation with neural ranking yields the best Mean Average Precision (MAP), 98% of that MAP score can be achieved with an 84% reduction in indexing time by using a pretrained XLM-R multilingual language model to index documents in their native language, and that 2% difference in effectiveness is not statistically significant. Key to achieving these results for MLIR is to fine-tune XLM-R using mixed-language batches from neural translations of MS MARCO passages.

Keywords: Multilingual ad-hoc retrieval · ColBERT-X · DPR-X · Multilingual training of MPLM

1 Introduction

With advances in neural models for machine translation (MT) and Information Retrieval (IR), it is time to revisit the problem of Multilingual IR (MLIR). Soon after Cross-Language IR (CLIR) was proposed as an information retrieval task, research began on MLIR [34]. MLIR seeks to produce a total ordering over retrieved documents, regardless of language, such that the most useful documents appear at the top of the ranking. Assuming a searcher can consume multilingual information (either directly or using MT), the search engine should be able to return useful information regardless of the language of the document.

Much prior work on MLIR has involved subsetting documents by language, performing CLIR on each document set, and merging the results [37]. The advent of neural machine translation and neural IR using Multilingual Pretrained Language Models (MPLMs) creates new opportunities for MLIR that we study here.

If MT were perfect, translating all documents into the query language and searching monolingually might suffice. Indeed, our experiments confirm that for the high-resource languages with which we have experimented (English, French, German, Italian, and

© The Author(s), under exclusive license to Springer Nature Switzerland AG 2023
J. Kamps et al. (Eds.): ECIR 2023, LNCS 13980, pp. 521–536, 2023.
https://doi.org/10.1007/978-3-031-28244-7_33

Spanish), using neural machine translation to convert each document into the query language is effective when used with neural ranking (in our experiments, ColBERT [26]) fine-tuned on MS MARCO [2]. However, using neural MT in that way incurs substantial indexing costs because a GPU is required first to translate the document and then again to encode it into dense vectors for neural IR. Alternatively, we can use translations of MS MARCO to fine-tune an MPLM; that approach is nearly as effective, not statistically different, and considerably faster at indexing time. Our use of MS MARCO makes English a natural choice as the query language, but our approach is extensible to any query language for which suitable fine-tuning data exists.

This paper makes the following contributions: (1) Using a collection containing five high-resource European languages, we show that neural MT with neural IR achieves higher MAP and Precision at 10 scores than any other known MLIR technique, but that reliance on neural MT greatly increases the time required to index a collection. (2) We show that extending the ColBERT-X [32] Translate-Train (TT) CLIR model to multiple languages achieves equivalent retrieval effectiveness with less than half the indexing time when used with mixed-language fine-tuning. (3) We show that some language bias in favor of query-language documents is present with all approaches, but that query-language bias is smaller with our Multilingual Translate-Train (MTT) implementation of ColBERT-X.

2   Background

We provide an overview of MLIR, followed by a brief review of traditional and neural IR. The term "multilingual" has been used in several ways in IR. Hull and Grefenstette [22], for example, note that it has been used to describe monolingual retrieval in multiple languages, as in Blloshmi et al. [5], and it has also been used to describe CLIR tasks that are run separately in several languages [7–9,27,31]. We adopt the Cross-Language Evaluation Forum (CLEF)'s meaning of MLIR: using a query to construct one ranked list in which each document is in one of several languages [36]. We note that this definition excludes mixed-language queries and mixed-language documents, which are yet other cases to which "multilingual" has been applied.

Five broad approaches to MLIR have been tried. Among the earliest, Rehder et al. [39] represented English, German and French documents in a learned trilingual embedding space, represented the query in the same embedding space, and then computed query-document similarity in the embedding space. The techniques and training data for creating multilingual embeddings were, however, too limited at the time to get good results from that technique. More recently, Sorg and Cimiano [44] garnered substantial attention by training embeddings on topically-related Wikipedia pages in English, German, French and Spanish. This paper extends this line of work.

A second approach by Nie and Jin [33] indexed terms from all documents in their original language then created queries containing translations of the query terms in all target languages. With many document languages, this can lead to long queries. A third approach is to translate indexed terms into the query language at indexing time; the original queries can then be used directly to find similar (translated) content [18,29,38]. We experiment with this approach as well. This approach is, however, only practical when just a few query languages are to be supported. To address that limitation, the second and third approaches can be combined to create a fourth approach in which documents and query terms are each converted into one of a small number of indexing languages. This has been called a "pivot language" approach, because in the limit all documents and queries can be translated into a single language.

The fifth, and most widely studied, approach is to first use monolingual or bilingual retrieval to create a ranked list for each document language, and then to merge those ranked lists to construct a single result list [37,43,45]. While this approach is architecturally similar to collection sharding, a widely-used approach to address efficiency, differences in collection statistics result in incompatible scores that require normalization prior to late fusion. Unfortunately, normalizing scores for collections across languages has been shown to be challenging [37].

Finally, one can simply show one ranked list per language to the user, as is done in the 2lingual search engine.[1] This approach does not scale well beyond a small number of languages, but it has the advantage of making it fairly clear to the searcher what the search engine has done.

Every MLIR ranked retrieval model must rank the indexed documents given a query. Traditional ranking methods such as computing inner products between the query and each indexed document containing a query term using sparse BM25 [40] term weights are fast, but neural IR methods yield better rankings [24,26,32] with more relevant documents earlier in the ranked list.

This paper focuses on tradeoffs between effectiveness and efficiency. Each technique described in this paper achieves ranking latency sufficient for interactive use (below 300 ms) on the collections that we experiment with, but the time required to index the documents varies. Indexing time consists of three components: text processing (e.g., casing and tokenization), machine translation, and representation (e.g., McCarley [30] and Magdy and Jones [29]). Of these, neural MT is the slowest, so IR methods that do not require neural MT at indexing time have a substantial indexing time advantage (e.g., Aljlayl and Frieder [1]). Our principal MLIR result is that MPLMs can achieve MAP close to the best results while producing substantial savings in indexing time.

We achieve this by extending the ColBERT-X [32] CLIR model to perform MLIR. ColBERT-X combines three key ideas. First, drawing insight from BERT [15], it represents documents using contextual embeddings, which better represent meaning than simple term occurrence. Second, using both multilinguality and improved pretraining from either multilingual BERT [47] or XLM-R [11], ColBERT-X generates similar contextual embeddings for terms with similar meaning, regardless of language. Third, drawing its structure from ColBERT [26], ColBERT-X limits ranking latency by separating query and document transformer networks, allowing offline indexing. ColBERT scores documents by focusing query term attention on the most similar contextual embedding in each document. Our experiments confirm that this approach yields better MLIR MAP than does computation of inner products between classification tokens for the query and each document, an approach known as Dense Passage Retrieval (DPR) [24].

---

[1] https://www.2lingual.com/.

3   Fine-Tuning MPLMs for MLIR

Following Nair et al. [32] we consider two high-level approaches to fine-tuning for generalizing neural retrieval models to MLIR. Both approaches use existing MPLMs such as XLM-R [11] to encode queries and documents in multiple languages. We adapt the MPLM to MLIR via task-specific fine-tuning. These approaches are applicable to any retrieval model that is able to encode text using an MPLM.

Consider a set of queries in a source language $\mathbf{L}_s$ and a set of documents in m target languages $\mathbf{L}_t = \cup_{i=i}^m \mathbf{L}_i$. We want to train a scoring function $\mathcal{M}_\Theta(q_{(s)}, d_{(t)}) \to \mathbb{R}$ for ranking documents with respect to a query. This paper denotes the language of an instance as a subscript $\bullet_{(l)}$.

3.1   English Training (ET)

Since MPLMs can encode text from many languages, we follow Nair et al. [32] and only fine-tune the model monolingually. When processing queries, we transfer the model to MLIR zero-shot. Specifically, consider a loss function $\mathcal{L}$ (for example, cross-entropy),

$$\Theta = \arg\min_\theta \sum_{q,d} \mathcal{L}_\theta(q_{(s)}, d_{(s)}, r_{q,d})$$

where $q_{(s)}$ and $d_{(s)}$ are representations of the queries and documents and $r_{q,d}$ is the relevance judgment of document d on query q, both in language $\mathbf{L}_s$, encoded by an MPLM. We use English as our query language because that is the language of MS MARCO. We refer to this approach as "English Training" or ET. However, this approach could equally well use any language for which similar extensive training data is available.

Despite only exposing the model to text in $\mathbf{L}_s$ during fine-tuning, the multilingual model can transfer its task model to other languages, as has been seen in prior CLIR work [32]. However, such zero-shot language transfer is suboptimal because of (1) the lack of alignment objectives between languages during pretraining [48]; and (2) differences in the representation of each language by the MPLM, which has been called *the curse of multilinguality* [11,46]. As we show in Sect. 6.1, such zero-shot transfer not only produces suboptimal retrieval effectiveness, it can also lead to language bias.

3.2   Multilingual Translate Training (MTT)

To mitigate those issues, we propose a Multilingual Translate-Train (MTT) approach that generalizes the CLIR Translate-Train (TT) approach to MLIR [32,42]. To expose target languages $\mathbf{L}_1...\mathbf{L}_m$ to the model, we translate the monolingual training documents into each target language using MT. Specifically, the training objective can be expressed as

$$\Theta = \arg\min_\theta \sum_{q,d} \sum_{l=1}^m \mathcal{L}_\theta(q_{(s)}, d_{(l)}, r_{q,d})$$

This objective exposes the retrieval model to language pairs that it might see when processing queries, resulting in a more effective, better-balanced model. We experiment with two batching approaches. In Mixed-language (MTT-M), each batch contains

Table 1. Dataset statistics of CLEF 2001, 2002, and 2003. CLEF 2001 and 2002 share the document collection but have different queries. Numbers in parentheses are the number of topics in each query set. We report the number of documents judged relevant over all the topics in a particular year.

Query set	English		German		Spanish		French		Italian		Total	
	# Rel.	# Docs	# Rel.	# Docs	# Rel.	# Docs	# Rel.	# Docs	# Rel.	# Docs	# Rel.	# Docs
2001 (50)	856	113,005	2,130	225,371	2,694	215,738	1,212	87,191	1,246	108,578	8,138	749,883
2002 (50)	821		1,938		2,854		1,383		1,072		8,068	
2003 (60)	1,006	169,477	1,825	294,809	2,367	454,045	946	129,806	–	–	6,144	1,048,137

documents in multiple languages, which encourages the model to learn similarity measures for all languages simultaneously.[2] With Single-language (MTT-S), each batch contains only documents in one language, helping the model to learn retrieval for one language pair at a time. We found that MTT-M yields better retrieval effectiveness; thus, we present MTT-M as our main result. Section 5.1 compares the two approaches. In Sect. 6.1, we also demonstrate that MTT-M reduces language bias in MLIR. Implementation details can be found in Appendix A

4   Experiments

One of the few test collections that currently supports MLIR evaluation with relevance judgments across multiple languages is from the Cross-Language Evaluation Forum (CLEF). Following Rahimi et al. [38] we use five document languages in the CLEF 2001–2002 collections [7,8] and four languages in the CLEF 2003 collection [9]. Table 1 shows collection statistics. We report performance for both title and title+description queries, also following Rahimi et al. [38]. Because the number of query elements (subwords) is limited when encoding a query for dense retrieval, we remove *stop structure* to ensure that no query exceeds the length limit. Stop structure includes phrases such as "Find documents" and a limited stop-word list including "on," "the," and "and."[3]

4.1   Neural Retrieval Models

We evaluate our proposed training approaches on two retrieval models – ColBERT-X [32] and DPR-X [48,49], which are multilingual variants of ColBERT [26] and DPR [24]. Nair et al. [32] generalized the ColBERT [26] model to CLIR, calling it ColBERT-X, by modifying the vocabulary space and replacing the monolingual pretrained language model with the MPLM XLM-RoBERTa (XLM-R) Large (550M parameters) [11]. With proper training, ColBERT-X achieves state-of-the-art effectiveness in CLIR. In this study, we integrate our proposed fine-tuning approaches with the ColBERT-X XLM-R implementation, which is based on the ColBERTv1 code base.

---

[2] Batches include the same query paired with document passages translated into each language.

[3] For a complete list: https://github.com/hltcoe/ColBERT-X/blob/main/scripts/stopstructure.txt.

We similarly adapted DPR [24, 48], a neural retrieval model that matches a single dense query vector to a single dense document vector. We name this model DPR-X. We use Tevatron [17], an open-source implementation of several neural end-to-end retrieval models in Python, for training, indexing, and retrieval.

For training data, we use MS MARCO-v1 [2], a commonly-used question-answering collection in English for fine-tuning neural retrieval models. For MTT, we use the publicly available mMARCO translations of MS MARCO [6], fine-tuning using the "small training triple" (query, positive and negative document) file released by mMARCO's creators. We trained all retrieval models with four GPUs (NVIDIA DGX and v100 with 32 GB Memory) with a per-GPU batch size of 32 triples for 200,000 update steps. All models are trained with half-precision floating points and optimized by the AdamW optimizer with a learning rate of 5×10^{-6}.

During indexing, documents are separated into overlapping spans of 180 tokens with a stride of 90 [32]. We aggregate by MaxP [3, 13], which takes the maximum score among the passages in a document as the document score.

4.2   Evaluation

We report previously published results for the state-of-the-art MULM [38] system as a baseline for models that do not perform MT on the full collection. MULM is essentially an MLIR version of Probabilistic Structured Queries (PSQ) [14]. PSQ maps term frequency vectors from document to query language using a matrix of translation probabilities generated using statistical machine translation. For MLIR, a translation matrix is created for each query-document language pair. The query likelihood model is used to score documents. Three key decisions led to good performance: (1) estimating collection statistics based on translation probabilities; (2) estimating document length based on the translation and using that for smoothing; and (3) truncating the translation list at three. As another baseline, we use BM25 ($b = 0.4$, $k_1 = 0.9$) as implemented in Patapsco [12] over neural machine translated documents (abbreviated ITD for Indexed Translated Documents). For BM25, English queries and documents are tokenized by spaCy [21] and stemmed by the NLTK [4] Porter stemmer (all supported by Patapsco).

For approaches that require document translation, we use directional MT models built on a transformer architecture (6-layer encoder/decoder) using Sockeye 2 [16, 19]. Measured by BLEU [35], Sockeye 2 achieves state-of-the-art effectiveness in each translation direction. Optimizations cut decoding time in half compared to Sockeye 1 [20]. We chose Sockeye 2 for its good trade-off between efficiency and effectiveness.

To evaluate effectiveness on multiple languages in CLEF 2001–2002 and CLEF 2003, we combine the relevance judgments (qrels) for all languages for each query. In general, different languages have different numbers of relevant documents for each query. To evaluate models trained with English training data, we also translate the document sets into English with MT for indexing. Our main effectiveness measures are Mean Average Precision (MAP) and Precision at 10 (P@10). Both measures focus on the top of the rankings, and both were used by Rahimi et al. [38], facilitating comparison between the neural approaches presented herein and prior state-of-the-art results.

To evaluate language bias, we count the number of relevant documents for a query across all languages, and calculate recall at that level. To compute the measure for a

Table 2. Configurations of experiments identifying the pre-trained language model when applicable, the fine tuning data and process, the retrieval model, and the language of the indexed documents. Under Fine-Tuning Data, MS MARCO refers to English MS MARCOv1, while mMARCO includes the translations into the various languages as well as the original English MS MARCOv1. A model that lists *either* under its Indexing Language can index either machine translated document (translation) or native documents in their various languages.

Name	Language model	Fine-tuning data	Fine-tuning process	Retrieval model	Indexing language
MULM	–	–	–	PSQ	Native
BM25-ITD	–	–	–	BM25	Translation
ColBERT-X(ET)	XLM-R	MS MARCO	ET	ColBERT-X	Either
ColBERT-X(MTT-M)	XLM-R	mMARCO	MTT-M	ColBERT-X	Either
ColBERT-X(MTT-S)	XLM-R	mMARCO	MTT-S	ColBERT-X	Either
DPR-X(ET)	XLM-R	MS MARCO	ET	DPR-X	Either
DPR-X(MTT-M)	XLM-R	mMARCO	MTT-M	DPR-X	Either
ColBERT(ET)	BERT	MS MARCO	ET	ColBERT	Translation

specific language, we keep this level constant, but ignore all documents in other languages (both in the MLIR results and in the relevance judgments). We call the mean of this measure over all queries *Recall@MLIR-Relevant*. When computing the mean, we omit from the calculation cases in which no relevant documents in that language are known (recall is undefined in such cases). This measure lies between 0 and 1, and values across that full range are achievable. We use the open source `ir-measures` [28][4] package to compute all effectiveness measures.

5   Results

We experiment with the Multilingual Translation Training (MTT) using two retrieval models and compare them to two strong baseline retrieval models: BM25-ITD indexing translated documents and MULM indexing native documents; these represent the state of the art on our test collections. Since per-query results for MULM have not been published we perform significance tests only between our systems and the BM25+ITD baseline (the stronger of the two baselines). Table 2 summarizes the experiments that facilitate this analysis. We first compare the effectiveness of our two batching strategies for MTT before examining their effectiveness relative to the baselines. Finally, we consider the trade-off between effectiveness and indexing time.

5.1   Multilingual Batching for Fine-Tuning

We compare two alternatives for fine-tuning the MTT condition and summarize the results with title+description queries in Table 3. In all cases, mixed-language batches (MTT-M) produce more effective retrieval models than single-language (MTT-S). This is likely because, in MLIR, the model must rank documents from different languages together instead of transferring trained models to other languages. The outcome might be different if our goal were to perform CLIR over monolingual document collections.

---

[4] https://ir-measur.es/.

Table 3. ColBERT-X MTT for Multiple or Single language training batches, indexing documents in their native language using title+description queries. † indicates significant improvement over MTT-S by paired t-test with 3-test Bonferroni correction ($p < 0.05$).

	MAP			P@10		
	2001	2002	2003	2001	2002	2003
MTT-M	**0.462**†	**0.462**†	**0.461**†	**0.704**	**0.752**	**0.653**
MTT-S	0.422	0.405	0.433	0.696	0.702	0.649

5.2   Effectiveness Relative to Baselines

Our main effectiveness results are shown in Table 4. For ColBERT-X and DPR-X, MTT-M consistently improves effectiveness when retrieving documents in their native language (i.e., *without document MT*) compared to English Training (ET). Such improvements are seen in all three query sets, and for both Title (T) and Title+Description (T+D) queries. Differences are larger for MAP than P@10, indicating that MTT-M affects more than just the top ranks.

ColBERT-X MTT-M numerically outperforms MULM for both query types and over all collections in MAP and nearly all collections in P@10. With longer, more fluent title+description queries, ColBERT-X MTT-M gives a larger improvement over MULM in both MAP and P@10, indicating that XLM-R favors queries with more context. Since DPR-X is less effective [48], MTT-M only brings its performance up to par with MULM.

With modern MT models, we can improve MLIR effectiveness. A common, yet strong, baseline of using BM25 to search over translated documents yields substantial improvement over MULM in both MAP and P@10 with both query types. We argue that BM25+ITD is a proper baseline to which future MLIR experiments should be compared.

We can also reduce neural IR to the monolingual case, training our retrieval model with English training and searching documents represented by English machine translations. For both ColBERT and DPR, an English-trained model (ET) indexing translated documents often yields better effectiveness than MTT-M indexing translated documents (ITD). Furthermore, an English trained model indexing translated documents yields better effectiveness than MTT-M indexing documents in their native language; however, these differences are only statistically significant for CLEF 2002 on Title queries using a paired t-test with 3-test Bonferroni correction ($p < 0.05$). We observe similar results with ColBERT using the BERT-Large pretrained LM trained under the same conditions except for using a learning rate of 3×10^{-6} (the value suggested by the authors). Compare Table 5 to ColBERT-X with English training, presented in Table 4.

5.3   Preprocessing and Indexing Time

Applying machine translation to entire document collections is expensive. Table 6 summarizes the cost for preprocessing and indexing the collection in GPU-hours for ColBERT-X and BM25. We omit consideration of query latency here since all of our

Table 4. MAP and P@10 on CLEF Title and Title+Description queries. Bold are best among a year; italics are best in a row (*i.e.*, with and without neural machine translation), † indicates significant difference from BM25+ITD by paired t-test with 16-test Bonferroni correction ($p < 0.05$).

Query set	ITD	MAP						P@10					
		MULM	BM25	ColBERT-X		DPR-X		MULM	BM25	ColBERT-X		DPR-X	
				MTT-M	ET	MTT-M	ET			MTT-M	ET	MTT-M	ET
Title queries													
2001	✓	–	*0.398*	0.377	0.391	0.338	0.344	–	*0.648*	0.612	0.596	0.548	0.584
	✗	0.349	–	*0.360*	0.322	0.327	0.298†	*0.650*	–	0.600	0.588	0.592	0.570
2002	✓	–	0.337	0.367	*0.389*	0.287	0.304	–	0.618	0.606	*0.670*	0.530	0.596
	✗	0.276	–	*0.352†*	0.333	0.282	0.277	0.592	–	0.614†	*0.622*	0.544	0.556
2003	✓	–	*0.349*	0.337	*0.349*	0.276†	0.266†	–	0.595	0.542	*0.573*	0.517	0.497†
	✗	0.305	–	*0.332†*	0.290	0.273†	0.247†	0.497	–	*0.546*	0.541	0.527	0.492†
All	✓	–	0.361	0.359	*0.375*	0.299†	0.302†	–	*0.619*	0.583	0.611	0.531†	0.554†
	✗	0.310	–	*0.347*	0.314†	0.293†	0.273†	0.575	–	*0.584†*	0.581	0.553†	0.536†
Title + Description queries													
2001	✓	–	0.436	0.472	*0.477*	0.365	0.356	–	0.704	0.718	*0.754*	0.658	0.650
	✗	0.387	–	*0.462*	0.405	0.358	0.324†	0.700	–	0.704†	*0.744*	0.658	0.644
2002	✓	–	0.398	0.470†	*0.480†*	0.347	0.332	–	0.696	*0.774*	0.770	0.664	0.620
	✗	0.347	–	*0.462*	0.410	0.335	0.310	0.666	–	*0.752*	0.720	0.672	0.640
2003	✓	–	0.394	*0.419*	0.410	0.343	0.328†	–	0.615	0.646	*0.661*	0.620	0.600
	✗	0.376	–	*0.409*	0.358	0.338	0.302†	0.563	–	*0.653*	0.637	0.622	0.575
All	✓	–	0.408	0.451†	*0.453†*	0.351†	0.338†	–	0.669	0.709	*0.725†*	0.646	0.622
	✗	0.368	–	*0.442*	0.390	0.343†	0.312†	0.643	–	*0.700†*	0.697	0.639	0.617

Table 5. Monolingual ColBERT model using BERT-Large trained with ET and evaluated with translated documents.

Queries	MAP				P@10			
	2001	2002	2003	All	2001	2002	2003	All
T	0.397	0.367	0.362	0.375	0.592	0.646	0.583	0.606
T+D	0.439	0.413	0.420	0.424	0.736	0.714	0.673	0.706

systems are sufficiently fast at query time for interactive use on collections of this size. We refer the interested reader to Santhanam et al. [41].

This table reveals that differences in total indexing time between searching native and translated documents range from four to 6.5 times depending on collection size and model.[5] Despite that searching translated documents with monolingual retrieval models is more effective, the computational cost of MT at indexing time is significantly higher; one might choose not to bear this cost in exchange for the small and not statistically significant numerical gain in measured effectiveness over searching documents in their native language with MTT-M fine-tuning for title+description queries.

---

[5] Although Marian [23] is faster than Sockeye 2, benchmark results from Sockeye 1 [20] and Sockeye 2 [19] confirm that Sockeye 2 is within a factor of 2 to 3 of Marian's speed, leaving our conclusions unchanged.

Table 6. ColBERT-X GPU hours for translating and indexing. BM25 does not use GPU.

Model	ITD	CLEF2001-2002			CLEF2003		
		Translation	Index	Total	Translation	Index	Total
BM25	✓	55.0	–	55.0	68.6	–	68.6
ET	✓	55.0	9.3	64.3	68.6	12.3	80.9
	✗	–	9.9	9.9	–	12.4	12.4
MTT-S	✓	55.0	16.9	71.9	68.6	19.0	87.6
	✗	–	16.7	16.7	–	21.9	21.9
MTT-M	✓	55.0	17.3	72.3	68.6	20.1	88.7
	✗	–	15.1	15.1	–	19.3	19.3

We also see this trade-off on a per-document basis. Figure 1 shows that ColBERT-X with English training searching translated documents (*ColBERT-X(ET)+ITD*) achieves the best effectiveness with both title (0.375 MAP) and title+description (0.453 MAP) queries. However, it has a high preprocessing cost of 0.32 s per document, whereas ColBERT-X trained with MTT-M searching documents in their native languages (*ColBERT-X(MTT-M)*) requires under 0.05 s per document. This is an 84% reduction in preprocessing cost at an apparent (but not statistically significant) cost of only 2% in MAP with title+description queries.

6 Analysis

This section investigates our experimental results by breaking down the collection in two ways – by document language, and by topic.

6.1 Language Bias

Since MPLMs are known to exhibit language biases [10,25], we investigate how retrieval models fine-tuned with our training schemes inherit or alleviate these biases. In MLIR

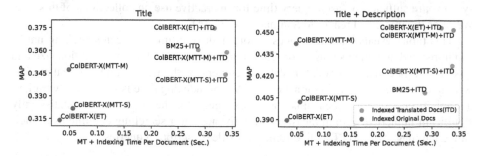

Fig. 1. Effectiveness (MAP) vs. efficiency (per-document GPU indexing time in seconds) trade-off on CLEF 2001–2003. MAP scores (y-axis) for Title and Title+Description queries are disjoint ranges. The upper left is the optimal part of the chart.

we consider a model biased if it ranks a language's documents systematically higher or lower than those of another language. While MLIR is not a new task, we are not aware of prior work that has examined language bias. Therefore we introduce two approaches to studying this phenomenon. The first approach examines rates of relevant documents. Since relevant documents are unevenly distributed across languages (e.g., Spanish has more than three times as many known relevant documents as English among the CLEF 2001 topics, averaging 54 vs. 17 relevant documents per topic, respectively), meaningful comparisons require us to focus on rates rather than on counts. In this analysis, we focus on Recall@MLIR-Relevant (see Sect. 4.2), illustrating our analysis using the 100 title+description queries in CLEF 2001–2002 topics to characterize the coverage of relevant documents in each language (results on CLEF 2003 topics are similar).

Figure 2 shows distributional statistics of Recall@MLIR-Relevant over topics by language and condition that have at least one known relevant document in that language (96 for German, 97 for Spanish, 94 for Italian, 90 for French, 73 for English). When transferring a ColBERT-X model fine-tuned zero-shot with English training (i.e., ColBERT-X(ET)) to other languages, the model favors English documents due to the fine-tuning condition. This results in a strong language bias in the retrieval results. Such biases can be ameliorated by fine-tuning with MTT. MTT-M appears to have more consistent behavior across languages compared to MTT-S, although the small apparent difference is not statistically significant. When indexing translated documents, Recall@MLIR-Relevant tends to be lower for English compared to other languages (though also not significantly). Since documents were translated sentence-by-sentence, we hypothesize that indexing translated documents provides more synonym variety when decoding similar terms, resulting in document expansion; this hypothesis requires more investigation, which we leave for future work.

An alternative approach to investigating language bias is to assume that in a bias-free approach to MLIR, the scores for relevant documents would be drawn from the same underlying distribution. Using the 2-sample Kolmogorov-Smirnov test, the null hypothesis is that the two samples are drawn from the same distribution. For this analysis, we chose English as a reference and tested each topic with at least three relevant

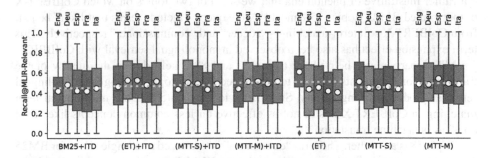

Fig. 2. R@MLIR-Relevant of BM25 and ColBERT-X variants for each language in CLEF2001-2002 with title+description queries. The yellow dashed line is the average over all languages, i.e., the R-Precision in MLIR. Outliers are defined as values beyond 1.5×interquartile range. Horizontal black bars indicate the median and white circles indicate the mean. (Color figure online)

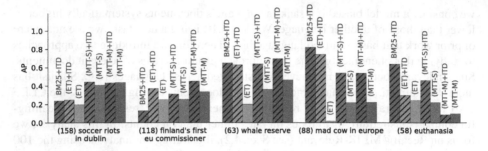

Fig. 3. Average Precision (AP) of BM25 and ColBERT-X on selected topics using title queries.

documents in each language. We then adjusted the p-values to account for multiple comparisons. We found that we could reject the null hypothesis for all languages and all configurations, indicating the document scores are not drawn from the same distribution based on language. Although some of this difference could result from differences in collection statistics (i.e., with some languages better supporting the queries than others based on the numbers of relevant documents), the differences we observe across retrieval models indicate that there are retrieval model effects as well. Notably, ColBERT-X(ET) retrieving documents in the native language has the largest percentage of topics with bias (from 15% to 30% depending on language pair), while all other configurations have no more than 12% of topics exhibiting biased scores. This confirms the qualitative analysis above, which revealed that ColBERT-X(ET) over the documents in their native language had the most skewed rates of relevant documents. Future research will need to address language bias in document scores.

6.2 Example Queries

For more insight into differences among the algorithms, we show effectiveness on individual queries in Fig. 3. Our query selection here is not meant to be representative, but rather illustrative of phenomena that we see. For two topics on which ColBERT-X outperformed BM25 (topics 158 and 118), the queries include terms that likely benefit from ColBERT-X soft term-matching – "soccer" and "commissioner" respectively. This term expansion effect has also been observed in monolingual retrieval with ColBERT.

MT is particularly helpful for topics 63 and 88, likely due to the quality of the translation for documents on these topics. Especially for topic 88, English monolingual retrieval produces strong results. Such behaviors indicate that the multilingual term matching in ColBERT-X is still not as effective on less common concepts like "mad cow" as is machine translation.

Topic 58 is an outlier. The term "euthanasia" is tokenized as a single token for BM25 but separated into _eu, thana, and sia by the XLM-R tokenizer; combined with the minimal context provided by a query, this prevents ColBERT-X from matching properly across languages. Such diverse behaviors suggest room for further MLIR improvements using system combination.

7   Conclusion and Future Work

This paper proposes the MTT training approach to MLIR that uses translated MS MARCO. When searching non-English documents, fine-tuning with MTT using mixed-language batches (MTT-M) enables neural models such as ColBERT and DPR to be more effective than if fine-tuned on English MS MARCO. ColBERT-X with MTT-M is not statistically different from monolingual English models applied to neural indexing-time translation of the collection into English, yet it achieves substantially better indexing time efficiency. These results may not hold for more diverse sets of languages or when MT is less effective; future work will examine the multilingual topics from the TREC 2022 the NeuCLIR track,[6] which judges the relevance of documents written in Chinese, Persian, and Russian. Our observation that the retrieval method that yields the best retrieval effectiveness is query-dependent suggests future work on system combination, but our focus on efficiency and on language bias also calls attention to issues beyond retrieval effectiveness that will merit consideration in such a study.

A   MTT Implementation Details

As described in Sect. 3.2, MTT-M consists of examples with different languages in the training batches. We implement it by mixing the translated MS-MARCO triples round-robin. Specifically, each triple consists of an English query and positive and negative passages translated into the target languages. We constructed such triples using the translated documents provided by mMARCO [6]. Each language results in a triple file of the same structure as `triples.train.small.tar.gz`.[7] The following Bash command creates a combined triple file that mixes all languages:

```
paste —d '\n' <(cat ./original_msmarco/triples.train.small.tsv) \
              <(cat ./mmarco/french/triples.train.small.tsv) \
              <(cat ./mmarco/german/triples.train.small.tsv) \
              <(cat ./mmarco/italian/triples.train.small.tsv) \
              <(cat ./mmarco/spanish/triples.train.small.tsv) \
| cat > combined.tsv
```

Training with four GPUs and a per-GPU batch size of 32 triples guarantees that each batch consists of examples in different languages based on ColBERT-X's[8] batching scheme.

For MTT-S, we modified the ColBERT-X batching mechanism to load multiple triple files and supply a batch of examples from only one source file whenever the training process requests one. After each request, we switch the source triple file to ensure all languages are presented equally to the model during training.

---

[6] https://neuclir.github.io/.

[7] https://msmarco.blob.core.windows.net/msmarcoranking/triples.train.small.tar.gz.

[8] https://github.com/hltcoe/ColBERT-X/blob/main/xlmr_colbert/training/lazy_batcher.py.

References

1. Aljlayl, M., Frieder, O.: Effective Arabic-English cross-language information retrieval via machine-readable dictionaries and machine translation. In: Proceedings of the Tenth International Conference on Information and Knowledge Management, pp. 295–302 (2001)
2. Bajaj, P., et al.: MS MARCO: a human generated machine reading comprehension dataset. arXiv preprint arXiv:1611.09268 (2016)
3. Bendersky, M., Kurland, O.: Utilizing passage-based language models for document retrieval. In: Macdonald, C., Ounis, I., Plachouras, V., Ruthven, I., White, R.W. (eds.) ECIR 2008. LNCS, vol. 4956, pp. 162–174. Springer, Heidelberg (2008). https://doi.org/10.1007/978-3-540-78646-7_17
4. Bird, S., Klein, E., Loper, E.: Natural Language Processing with Python: Analyzing Text with the Natural Language Toolkit. O'Reilly Media, Inc., Sebastopol (2009)
5. Blloshmi, R., Pasini, T., Campolungo, N., Banerjee, S., Navigli, R., Pasi, G.: IR like a SIR: sense-enhanced information retrieval for multiple languages. In: Proceedings of the 2021 Conference on Empirical Methods in Natural Language Processing, pp. 1030–1041, Association for Computational Linguistics, Online and Punta Cana, Dominican Republic, November 2021. https://doi.org/10.18653/v1/2021.emnlp-main.79, https://aclanthology.org/2021.emnlp-main.79
6. Bonifacio, L.H., et al.: mMARCO: a multilingual version of MS MARCO passage ranking dataset. arXiv preprint arXiv:2108.13897 (2021)
7. Braschler, M.: CLEF 2001 — overview of results. In: Peters, C., Braschler, M., Gonzalo, J., Kluck, M. (eds.) CLEF 2001. LNCS, vol. 2406, pp. 9–26. Springer, Heidelberg (2002). https://doi.org/10.1007/3-540-45691-0_2
8. Braschler, M.: CLEF 2002 — overview of results. In: Peters, C., Braschler, M., Gonzalo, J., Kluck, M. (eds.) CLEF 2002. LNCS, vol. 2785, pp. 9–27. Springer, Heidelberg (2003). https://doi.org/10.1007/978-3-540-45237-9_2
9. Braschler, M.: CLEF 2003 – overview of results. In: Peters, C., Gonzalo, J., Braschler, M., Kluck, M. (eds.) CLEF 2003. LNCS, vol. 3237, pp. 44–63. Springer, Heidelberg (2004). https://doi.org/10.1007/978-3-540-30222-3_5
10. Choudhury, M., Deshpande, A.: How linguistically fair are multilingual pre-trained language models? In: Proceedings of the AAAI Conference on Artificial Intelligence, vol. 35, pp. 12710–12718 (2021)
11. Conneau, A., et al.: Unsupervised cross-lingual representation learning at scale. In: Proceedings of the 58th Annual Meeting of the Association for Computational Linguistics, pp. 8440–8451. Association for Computational Linguistics, Online, July 2020. https://aclanthology.org/2020.acl-main.747
12. Costello, C., Yang, E., Lawrie, D., Mayfield, J.: Patapasco: a Python framework for cross-language information retrieval experiments. In: Hagen, M., et al. (eds.) ECIR 2022. LNCS, vol. 13186, pp. 276–280. Springer, Cham (2022). https://doi.org/10.1007/978-3-030-99739-7_33
13. Dai, Z., Callan, J.: Deeper text understanding for IR with contextual neural language modeling. In: Proceedings of the 42nd International ACM SIGIR Conference on Research and Development in Information Retrieval, pp. 985–988 (2019)
14. Darwish, K., Oard, D.W.: Probabilistic structured query methods. In: Proceedings of the 26th Annual International ACM SIGIR Conference on Research and Development in Information Retrieval, pp. 338–344 (2003)
15. Devlin, J., Chang, M.W., Lee, K., Toutanova, K.: BERT: pre-training of deep bidirectional transformers for language understanding. In: Proceedings of the 2019 Conference of the

North American Chapter of the Association for Computational Linguistics: Human Language Technologies, Volume 1 (Long and Short Papers), pp. 4171–4186, Association for Computational Linguistics, Minneapolis, June 2019. https://aclanthology.org/N19-1423

16. Domhan, T., Denkowski, M., Vilar, D., Niu, X., Hieber, F., Heafield, K.: The Sockeye 2 neural machine translation toolkit at AMTA 2020. In: Proceedings of the 14th Conference of the Association for Machine Translation in the Americas (Volume 1: Research Track), pp. 110–115, Association for Machine Translation in the Americas, Virtual, October 2020

17. Gao, L., Ma, X., Lin, J.J., Callan, J.: Tevatron: an efficient and flexible toolkit for dense retrieval. arXiv preprint arXiv:2203.05765 (2022)

18. Granell, X.: Multilingual Information Management: Information, Technology and Translators. Chandos Publishing, Cambridge (2014)

19. Hieber, F., Domhan, T., Denkowski, M., Vilar, D.: Sockeye 2: a toolkit for neural machine translation. In: EAMT 2020 (2020). https://www.amazon.science/publications/sockeye-2-a-toolkit-for-neural-machine-translation

20. Hieber, F., et al.: Sockeye: a toolkit for neural machine translation. arXiv preprint arXiv:1712.05690 (2017)

21. Honnibal, M., Montani, I., Van Landeghem, S., Boyd, A.: spaCy: industrial-strength natural language processing in Python. Technical report, Explosion (2020)

22. Hull, D.A., Grefenstette, G.: Querying across languages: a dictionary-based approach to multilingual information retrieval. In: Proceedings of the 19th Annual International ACM SIGIR Conference on Research and Development in Information Retrieval, pp. 49–57 (1996)

23. Junczys-Dowmunt, M., Heafield, K., Hoang, H., Grundkiewicz, R., Aue, A.: Marian: cost-effective high-quality neural machine translation in C++. arXiv preprint arXiv:1805.12096 (2018)

24. Karpukhin, V., et al.: Dense passage retrieval for open-domain question answering. In: Proceedings of the 2020 Conference on Empirical Methods in Natural Language Processing (EMNLP), pp. 6769–6781. Association for Computational Linguistics, Online, November 2020. https://aclanthology.org/2020.emnlp-main.550

25. Kassner, N., Dufter, P., Schütze, H.: Multilingual lama: investigating knowledge in multilingual pretrained language models. arXiv preprint arXiv:2102.00894 (2021)

26. Khattab, O., Zaharia, M.: ColBERT: efficient and effective passage search via contextualized late interaction over BERT. In: Proceedings of the 43rd International ACM SIGIR Conference on Research and Development in Information Retrieval, pp. 39–48 (2020)

27. Lawrie, D., Mayfield, J., Oard, D.W., Yang, E.: HC4: a new suite of test collections for ad hoc CLIR. In: Hagen, M., et al. (eds.) ECIR 2022. LNCS, vol. 13185, pp. 351–366. Springer, Cham (2022). https://doi.org/10.1007/978-3-030-99736-6_24

28. MacAvaney, S., Macdonald, C., Ounis, I.: Streamlining evaluation with ir-measures. In: Hagen, M., et al. (eds.) ECIR 2022. LNCS, vol. 13186, pp. 305–310. Springer, Cham (2022). https://doi.org/10.1007/978-3-030-99739-7_38

29. Magdy, W., Jones, G.J.F.: Should MT systems be used as black boxes in CLIR? In: Clough, P., et al. (eds.) ECIR 2011. LNCS, vol. 6611, pp. 683–686. Springer, Heidelberg (2011). https://doi.org/10.1007/978-3-642-20161-5_70

30. McCarley, J.S.: Should we translate the documents or the queries in cross-language information retrieval? In: Proceedings of the 37th Annual Meeting of the Association for Computational Linguistics, pp. 208–214 (1999)

31. Mitamura, T., et al.: Overview of the NTCIR-7 ACLIA tasks: advanced cross-lingual information access. In: NTCIR (2008)

32. Nair, S., et al.: Transfer learning approaches for building cross-language dense retrieval models. In: Hagen, M., et al. (eds.) ECIR 2022. LNCS, vol. 13185, pp. 382–396. Springer, Cham (2022). https://doi.org/10.1007/978-3-030-99736-6_26

33. Nie, J.-Y., Jin, F.: A multilingual approach to multilingual information retrieval. In: Peters, C., Braschler, M., Gonzalo, J., Kluck, M. (eds.) CLEF 2002. LNCS, vol. 2785, pp. 101–110. Springer, Heidelberg (2003). https://doi.org/10.1007/978-3-540-45237-9_8

34. Oard, D.W., Dorr, B.J.: A survey of multilingual text retrieval. Technical report, UMIACS-TR-96019 CS-TR-3615, UMIACS (1996)

35. Papineni, K., Roukos, S., Ward, T., Zhu, W.J.: BLEU: a method for automatic evaluation of machine translation. In: Proceedings of the 40th Annual Meeting of the Association for Computational Linguistics, pp. 311–318. Association for Computational Linguistics, Philadelphia , July 2002. https://doi.org/10.3115/1073083.1073135, https://aclanthology.org/P02-1040

36. Peters, C., Braschler, M.: The importance of evaluation for cross-language system development: the CLEF experience. In: LREC (2002)

37. Peters, C., Braschler, M., Clough, P.: Multilingual Information Retrieval: From Research to Practice. Springer, Heidelberg (2012). https://doi.org/10.1007/978-3-642-23008-0

38. Rahimi, R., Shakery, A., King, I.: Multilingual information retrieval in the language modeling framework. Inf. Retrieval J. **18**(3), 246–281 (2015). https://doi.org/10.1007/s10791-015-9255-1

39. Rehder, B., Littman, M.L., Dumais, S.T., Landauer, T.K.: Automatic 3-language cross-language information retrieval with latent semantic indexing. In: TREC, pp. 233–239. Citeseer (1997)

40. Robertson, S., Zaragoza, H., et al.: The probabilistic relevance framework: BM25 and beyond. Found. Trends® Inf. Retrieval **3**(4), 333–389 (2009)

41. Santhanam, K., Khattab, O., Potts, C., Zaharia, M.: PLAID: an efficient engine for late interaction retrieval. arXiv preprint arXiv:2205.09707 (2022)

42. Shi, P., Lin, J.: Cross-lingual relevance transfer for document retrieval. arXiv preprint arXiv:1911.02989 (2019)

43. Si, L., Callan, J., Cetintas, S., Yuan, H.: An effective and efficient results merging strategy for multilingual information retrieval in federated search environments. Inf. Retrieval **11**(1), 1–24 (2008)

44. Sorg, P., Cimiano, P.: Exploiting Wikipedia for cross-lingual and multilingual information retrieval. Data Knowl. Eng. **74**, 26–45 (2012). ISSN 0169-023X, https://www.sciencedirect.com/science/article/pii/S0169023X12000213, Appl. Nat. Lang. Inf. Syst

45. Tsai, M.F., Wang, Y.T., Chen, H.H.: A study of learning a merge model for multilingual information retrieval. In: Proceedings of the 31st Annual International ACM SIGIR Conference on Research and Development in Information Retrieval, pp. 195–202 (2008)

46. Xu, H., Van Durme, B., Murray, K.: BERT, mBERT, or BiBERT? A study on contextualized embeddings for neural machine translation. In: Proceedings of the 2021 Conference on Empirical Methods in Natural Language Processing, pp. 6663–6675. Association for Computational Linguistics, Online and Punta Cana, Dominican Republic, November 2021. https://aclanthology.org/2021.emnlp-main.534

47. Xu, Y.: Global divergence and local convergence of utterance semantic representations in dialogue. In: Proceedings of the Society for Computation in Linguistics 2021, pp. 116–124. Association for Computational Linguistics, Online, February 2021. https://aclanthology.org/2021.scil-1.11

48. Yang, E., Nair, S., Chandradevan, R., Iglesias-Flores, R., Oard, D.W.: C3: continued pretraining with contrastive weak supervision for cross language ad-hoc retrieval. In: Proceedings of the 45th International ACM SIGIR Conference on Research and Development in Information Retrieval (SIGIR) (2022). https://arxiv.org/abs/2204.11989

49. Zhang, X., Ma, X., Shi, P., Lin, J.: Mr. TyDi: a multi-lingual benchmark for dense retrieval. In: Proceedings of the 1st Workshop on Multilingual Representation Learning, pp. 127–137. Association for Computational Linguistics, Punta Cana, Dominican Republic, November 2021. https://aclanthology.org/2021.mrl-1.12

CoSPLADE: Contextualizing SPLADE for Conversational Information Retrieval

Nam Hai Le[1]([✉]) [iD], Thomas Gerald[2], Thibault Formal[1,3], Jian-Yun Nie[4],
Benjamin Piwowarski[1] [iD], and Laure Soulier[1,2] [iD]

[1] Sorbonne Université, CNRS, ISIR, 75005 Paris, France
{hai.le,thibault.formal,benjamin.piwowarski,
laure.soulier}@sorbonne-universite.fr
[2] Université Paris-Saclay, CNRS, SATT Paris Saclay, LISN, 91405 Orsay, France
{thomas.gerald,laure.soulier}@lisn.upsaclay.fr
[3] Naver Labs Europe, Meylan, France
thibault.formal@naverlabs.com
[4] University of Montreal, Montreal, Canada
nie@iro.umontreal.ca

Abstract. Conversational search is a difficult task as it aims at retrieving documents based not only on the current user query but also on the full conversation history. Most of the previous methods have focused on a multi-stage ranking approach relying on query reformulation, a critical intermediate step that might lead to a sub-optimal retrieval. Other approaches have tried to use a fully neural IR first-stage, but are either zero-shot or rely on full learning-to-rank based on a dataset with pseudo-labels. In this work, leveraging the CANARD dataset, we propose an innovative lightweight learning technique to train a first-stage ranker based on SPLADE. By relying on SPLADE sparse representations, we show that, when combined with a second-stage ranker based on T5Mono, the results are competitive on the TREC CAsT 2020 and 2021 tracks. The source code is available at https://github.com/nam685/cosplade.git.

Keywords: Information retrieval · Conversational search · First-stage ranking

1 Introduction

With the introduction of conversational assistants like Siri, Alexa or Cortana, conversational Information Retrieval, a variant of adhoc IR, has emerged as an important research domain [5,7]. In conversational IR, a search is conducted within a session, and the user's information need is expressed through a sequence of queries, similarly to natural conversations – thus introducing complex interdependencies between queries and responses.

Not surprisingly, neural IR models have been shown to perform the best on conversational IR [6,8]. Most prior works rely on a Historical Query Expansion step [39], i.e. a query expansion mechanism that takes into account all past queries and their associated answers. Such query expansion model is learned on the CANARD dataset [9], which is composed of a series of questions and their

© The Author(s), under exclusive license to Springer Nature Switzerland AG 2023
J. Kamps et al. (Eds.): ECIR 2023, LNCS 13980, pp. 537–552, 2023.
https://doi.org/10.1007/978-3-031-28244-7_34

associated answers, together with a disambiguated query, referred to as *gold query* in this paper. However, relying on a reformulation step is computationally costly and might be sub-optimal as underlined in [15,18]. Krasakis et al. [15] proposed to use ColBERT [14] in a zero-shot manner, replacing the query by the sequence of queries, without any training of the model. Lin et al. [18] proposed to learn a dense *contextualized* representation of the query history, optimizing a learning-to-rank loss over a dataset composed of weak labels. This makes the training process complex (labels are not reliable) and long.

In this work, we follow this direction of research but propose a much lighter training process for the first-stage ranker, where we focus on queries and do not make use of any passage – and thus of a learning-to-rank training. It moreover sidesteps the problem of having to derive weak labels from the CANARD dataset[1]. Given this strong supervision, we can consider more context – i.e. we use the answers provided by the system the user is interacting with, which allows to better contextualize the query, as shown in our experiments. The training loss we propose leverages the sparse representation of queries and documents provided by the SPLADE model [10]. In a nutshell, we require that the representation of the query matches that of the disambiguated query (i.e. the *gold query*). Our first-stage ranker achieves high performances, especially on recall – the most important measure in a multi-stage approach, comparable to the best systems in TREC CAsT [8], but also on precision-oriented measures – which shows the potential of our methodology.

Finally, to perform well, the second-stage ranker (i.e. re-ranker) needs to consider the conversation as well, which might require a set of heuristics to select some content and/or query reformulation such as those used in [20]. Leveraging the fact that our first-stage ranker outputs weights over the BERT vocabulary, we propose a simple mechanism that provides a conversational context to the re-ranker in the form of keywords selected by SPLADE.

In summary, our contributions are the following:

1. We propose the CoSPLADE (COntextualized SPLADE) model based on a new loss to optimize a first-stage ranker resulting in a lightweight training strategy and state-of-the-art results in terms of recall;
2. We show that, when combined with a second-stage ranker based on a context derived from the SPLADE query representation of the first stage, we obtain results on par with the best approaches in TREC CAsT 2020 and 2021.

2   Related Works

The first edition [6] of the TREC Conversational Assistance Track (*CAsT*) was implemented in 2019, providing a new challenge on Conversational Search. The

---

[1] Note that for the second stage, we rely on weak labels since our model is similar to previous works. Given that the gap between first-stage and second-stage rankers continues to decrease, training a second-stage ranker might not be necessary in the future.

principle is the following: a user queries the system with questions in natural language, and each time gets a response from the system. The challenge differs from classical search systems as involving previous utterances (either queries or answers) is key to better comprehending the user intent. In conversational IR, and in TREC *CAsT* [6–8] in particular, the sheer size of the document collection implies to design an efficient (and effective) search system.

Conversational IR is closely related to conversational Question-Answering [28–30] in the sense that they both include interaction turns in natural language. However, the objective is intrinsically different. While the topic or the context (i.e., the passage containing answers) is known in conversational QA, conversational IR aims to search among a huge collection of documents with potentially more exploratory topics. With this in mind, in the following, we focus on the literature review of conversational IR.

We can distinguish two lines of work in conversational search. The first one [4,34,35,37] focuses on a Contextual Query Reformulation (CQR) to produce a (plain or bag-of-words) query, representing ideally the information need free of context, which is fed into a search model. One strategy of CQR consists in selecting keywords from previous utterances by relying on a graph weighted by either word2vec similarity [34], term-based importance using BM25 [21], or classification models [35]. Other approaches [16,20,21,33,38] leverage the potential of generative language models (e.g., GPT2 or T5) to rewrite the query. Such approaches are particularly effective, reaching top performances in the TREC CAsT 2020 edition [6]. Query reformulation models also differ in the selected evidence sources. Models either focus on the early stage of the conversation [1], on a set of the queries filtered either heuristically [2] or by a classification model [23], or on both previous queries and documents [36]. Finally, to avoid the problem of generating a *single* query, [16,22] have proposed to use different generated queries and aggregate the returned documents.

The reformulation step is however a bottleneck since there is no guarantee that the "gold query" is optimal and thus generalizes well [15,18]. Moreover, generating text is time-consuming. To avoid these problems, the second line of work aims to directly integrate the conversation history into the retrieval model, bypassing the query reformulation step. As far as we know, only a few studies followed this path in conversational search. Qu et al. [26] compute a query representation using the k last queries in the dialogue [17]. Similarly, Lin et al. [18] average contextualized tokens embeddings over the whole query history. The representation is learned by optimizing a learning-to-rank loss over a collection with weak labels, which requires much care to ensure good generalization. Finally, Krasakis et al. [15] use a more lexical neural model, i.e. ColBERT [14], to encode the query with its context – but they do not finetune it at all. In this work, we go further by using a sparse model SPLADE [10], using a novel loss tailored to such sparse representations, and by using a lightweight training procedure that does not rely on passages, but only on a dataset containing reformulated queries.

3   Model

In TREC CAsT [6,8], each retrieval session contains around 10 turns of exchange. Each turn corresponds to a query and its associated canonical answer[2] is provided as context for future queries. Let us now introduce some notations that we use to describe our model. For each turn $n \leq N$, where N is the last turn of the conversation, we denote by q_n and a_n respectively the corresponding query and its response. Finally, the context of a query q_n at turn n corresponds to all the previous queries and answers, i.e. q_1, a_1, q_2, a_2, ..., q_{n-1}, a_{n-1}. The main objective of the *TREC CAsT* challenges is to retrieve, for each query q_n and its context, the relevant passages. In the next sections, we present our first-stage ranker and second-stage re-ranker, along with their training procedure, both based, directly or indirectly, on the SPLADE (v2) model described in [10]. SPLADE has shown results on par with dense approaches on in-domain collections while exhibiting stronger abilities to generalize in a zero-shot setting [10]. Moreover, it outputs a sparse representation of a document or a query in the BERT vocabulary, which is key to our model during training and inference. This explains why we did not consider interaction models such as ColBERT [32] or dense approaches [13]. The SPLADE model we use includes a contextual encoding function, followed by some aggregation steps: ReLU, log saturation, and max pooling over each token in the text. The output of SPLADE is a sparse vector with only positive or zero components in the BERT vocabulary space $\mathbb{R}^{|V|}$. In this work, we use several sets of parameters for the same SPLADE architecture and distinguish each version by its parameters θ, and the corresponding model by $SPLADE(\ldots;\theta)$.

3.1   First Stage

The original SPLADE model [10] scores a document using the dot product between the sparse representation of a document ($\hat{d}$) and of a query ($\hat{q}$):

$$s(\hat{q}, \hat{d}) = \hat{q} \cdot \hat{d} \tag{1}$$

In our work, like in [18], we suppose that the document representation has been sufficiently well-tuned on the standard ad-hoc IR task. The document embedding $\hat{d}$ is thus obtained using the pre-trained SPLADE model, i.e. $\hat{d} = SPLADE([\text{CLS}]\ d;\ \theta_{SPLADE})$ where θ_{SPLADE} are the original SPLADE parameters obtained from HuggingFace[3]. These parameters are not fine-tuned during the training process. We can thus use standard index built from the original SPLADE document representations to retrieve efficiently the top-k documents. In the following, we present how to contextualize the query representation using the conversation history. Then, we detail the training loss aiming at reducing the gap between the representation of the gold query and the contextualized representation.

---

[2] Selected by the organizer as the most relevant answer of a baseline system.

[3] The weights can be found at https://huggingface.co/naver/splade-cocondenser-ensembledistil.

Query Representation. Like state-of-the-art approaches for first-stage conversational ranking [15,18], we contextualize the query with the previous ones. Going further, we propose to include the answers in the query representation process, which is easier to do thanks to our lightweight training.

To leverage both contexts, we use a simple model where the contextual query representation at turn n, denoted by $\hat{q}_{n,k}$, is the combination of two representations, $\hat{q}_n^{queries}$ which encodes the current query in the context of all the previous queries, and $\hat{q}_{n,k}^{answers}$ which encodes the current query in the context of k the past answers[4]. Formally, the contextualized query representation $\hat{q}_{n,k}$ is:

$$\hat{q}_{n,k} = \hat{q}_n^{queries} + \hat{q}_{n,k}^{answers} \qquad (2)$$

where we use two versions of SPLADE parameterized by $\theta_{queries}$ for the full query history and $\theta_{answers,k}$ for the answers. These parameters are learned by optimizing the loss defined in Eq. (8).

Following [18], we define $\hat{q}_n^{queries}$ to be the query representation produced by encoding the concatenation of the current query and all the previous ones:

$$\hat{q}_n^{queries} = SPLADE([\text{CLS}]\ q_n\ [\text{SEP}]\ q_1\ [\text{SEP}]\ \dots\ [\text{SEP}]\ q_{n-1}; \theta_{queries}) \qquad (3)$$

using a set of specific parameters $\theta_{queries}$.

To take into account the answers that the user had access to, we need to include them in the representation. Following prior work [2], we can consider a various number of answers k, and in particular, we can either choose $k = 1$ (the last answer) or $k = n - 1$ (all the previous answers). Formally, the representation $\hat{q}_{n,k}^{answers}$ is computed as the mean of the representations of query-answer pairs. This allows to evade the token-length limits imposed by language models. Formally,

$$\hat{q}_{n,k}^{answers} = \frac{1}{k} \sum_{i=n-k}^{n-1} SPLADE(q_n\ [\text{SEP}]\ a_i; \theta_{answers,k}) \qquad (4)$$

Training. Based on the above, training aims at obtaining a good representation $\hat{q}_n$ for the last issued query q_n, i.e. to contextualize q_n using the previous queries and answers. To do so, we can leverage the gold query q_n^*, that is (hopefully), a contextualized and unambiguous query. We can compute the representation $\hat{q}_n^*$ of this query by using the original SPLADE model, i.e.

$$\hat{q}_n^* = SPLADE(q_n^*; \theta_{SPLADE}) \qquad (5)$$

For example, for a query "How old is he?" the matching gold query could be "How old is Obama?". The representation of the latter given by SPLADE would be as follows:

[("*Obama*", 1.5), ("*Barack*", 1.2), ("*age*", 1.2), ("*old*", 1.0), ("*president*", 0.8), ...]

---

[4] In the experiments, we also explore an alternative model where answers and queries are considered at once.

where the terms "Obama" and "Barack" clearly appear alongside other words related to the current query ("old" and the semantically related "age").

We can now define the goal of the training, which is to reduce the difference between the gold query representation $\hat{q}_n^*$ and the representation $\hat{q}_{n,k}$ computed by our model. An obvious choice of a loss function is to match the predicted and gold representations using cosine loss (since the ranking is invariant when scaling the query). However, as shown in the result section, we experimentally found better results with a modified MSE loss, whose first component is the standard MSE loss:

$$Loss_{MSE}(\hat{q}_{n,k}, \hat{q}_n^*) = MSE(\hat{q}_{n,k}, \hat{q}_n^*) \tag{6}$$

In our experiments, we observed that models trained with the direct MSE do not capture well words from the context, especially for words from the answers. The reason is that the manually reformulated gold query usually only contains a few additional words from the previous turns that are directly implied by the last query. Other potentially useful words from the answers may not be included. This is a conservative expansion strategy which may not be the best example to follow by an automatic query rewriting process. We thus added an asymmetric MSE, designed to encourage term expansion from past answers, but avoid introducing noise by restricting the terms to those present in the gold query q_n^*. Formally, our asymmetric loss is:

$$Loss_{asym}(\hat{q}_{n,k}^{answers}, \hat{q}_n^*) = \left(\max(\hat{q}_n^* - \hat{q}_{n,k}^{answers}, 0)\right)^2 \tag{7}$$

where the maximum is component-wise. This loss thus pushes the answer-biased representation $\hat{q}_{n,k}^{answers}$ to include tokens from the gold query. Contrarily to MSE, it does not impose (directly) an upper bound on the components of the $\hat{q}_{n,k}^{answers}$ representation – this is done indirectly through the final loss function described below.

The final loss we optimize is a simple linear combination of the losses defined above, and only relies on computing two query representations:

$$Loss(\hat{q}_{n,k}, \hat{q}_n^*) = Loss_{MSE}(\hat{q}_{n,k}, \hat{q}_n^*) + Loss_{asym}(\hat{q}_{n,k}^{answers}, \hat{q}_n^*) \tag{8}$$

There is an interplay between the two components of the global loss. More precisely, $Loss_{asym}$ pushes the $\hat{q}_{n,k}^{answers}$ representation to match the golden query representation $\hat{q}_n^*$ *if it can*, and $Loss_{MSE}$ pushes the queries-biased representation $\hat{q}_{n,k}$ to compensate *if not*. It thus puts a strong focus on extracting information from past answers, which is shown to be beneficial in our experiments.

Implementation Details. For the first-stage, we initialize both encoders (one encoding the queries, and the other encoding the previous answer) with pre-trained weights from SPLADE model for adhoc retrieval. We use the ADAM optimizer with train batch size 16, learning rate 2e-5 for the first encoder and 3e-5 for the second. We fine-tune for only 1 epoch over the CANARD dataset.

3.2 Reranking

We perform reranking using a T5Mono [24] approach, where we enrich the raw query q_n with keywords identified by the first-stage ranker. Our motivation is that these words should capture the information needed to contextualize the raw query. The enriched query q_n^+ for conversational turn n is as follows:

$$q_n^+ = q_n.\ Context: q_1\ q_2\ \ldots\ q_{n-1}.\ Keywords: w_1, w_2, ..., w_K \qquad (9)$$

where the w_i are the top-K most important words that we select by leveraging the first-stage ranker as follows. First, to reduce noise, we only consider words that appear either in any query q_i or in the associated answers a_i (for $i \le n-1$). Second, we order words by using the maximum SPLADE weight over tokens that compose the word.[5]

We denote the T5 model fine-tuned for this input as $T5^+$. As in the original paper [24], the relevance score of a document d for the query q_n is the probability of generating the token "$\texttt{true}$" given a prompt $pt(q_n^+, d) = $ "$\texttt{Query:}\ q_n^+.$ $\texttt{Document:}\ d.\ \texttt{Relevant:}$":

$$score(q_n^+, d; \theta) = \frac{p_{T5}(\texttt{true}|pt(q_n^+, d); \theta)}{p_{T5}(\texttt{true}|pt(q_n^+, d); \theta) + p_{T5}(\texttt{false}|pt(q_n^+, d); \theta)} \qquad (10)$$

where θ are the parameters of the T5Mono model.

Differently to the first stage training, we fine-tune the ranker by aligning the scores of the documents, and not the weight of a query (which is obviously not possible with the T5 model). Here the "gold" score of a document is computed using the original T5Mono with the gold query q_n^*. The T5 model is initialized with weights made public by the original authors[6], denoted as θ_{T5}. More precisely, we finetune the pre-trained T5Mono model using the MSE-Margin loss [12]. The loss function for the re-ranker (at conversation turn n, given documents d_1 and d_2) is computed as follows:

$$\mathcal{L}_R = \left[\left(s(q_n^+, d_1; \theta_{T5+}) - s(q_n^+, d_2; \theta_{T5+}) \right)\ \left(s(q_n^*, d_1; \theta_{T5}) - s(q_n^*, d_2; \theta_{T5}) \right) \right]^2$$

We optimize the θ_{T5+} parameters by keeping the original θ_{T5} to evaluate the score of gold queries.

Implementation details. We initialize θ_{T5+} as θ_{T5}, and fine-tune for 3 epochs, with a batch size of 8 and a learning rate 1e-4. We sample pairs (d_1, d_2) using the first-stage top-1000 documents: d_1 is sampled among the top-3, and d_2 among the remaining 997 to push the model to focus on important differences in scores.

---

[5] To improve coherence, we chose to make keywords follow their order of appearance in the context, but did not vary this experimental setting.
[6] We used the Huggingface checkpoint https://huggingface.co/castorini/monot5-base-msmarco.

4   Experimental Protocol

We designed the evaluation protocol to satisfy two main evaluation objectives: (i) Evaluating separately the effectiveness of the first-stage and the second-stage ranking components of our CoSPLADE model; (ii) Comparing the effectiveness of our CoSPLADE model with TREC CAsT 2020 and 2021 participants.

4.1   Datasets

To train our model, we used the CANARD corpus[7], a conversational dataset focusing on context-based query rewriting. More specifically, the CANARD dataset is a list of conversation histories, each being composed of a series of queries, short answers (human-written), and reformulated queries (contextualized). The training, development, and test sets include respectively 31.538, 3.418, and 5.571 contextual and reformulated queries.

To evaluate our model, we used the TREC CAsT 2020 and 2021 datasets which include respectively 25 and 26 information needs (topics) and a document collection composed of the MS MARCO dataset, an updated dump of Wikipedia from the KILT benchmark, and the Washington Post V4 collection. For each topic, a conversation is available, alternating questions and responses (manually selected passages from the collection, aka canonical answers). For each question (216 and 239 in total), the dataset provides its manually rewritten form as well as a set of about 20 relevant documents. We use the former to define an upperbound baseline (**Splade_GoldQuery**).

4.2   Metrics and Baselines

We used the official evaluation metrics considered in the TREC CAsT 2020 and 2021, namely nDCG@3, MRR, Recall@X, MAP@X, nDCG@X, where the cut-off is set to 1000 for the CAsT 2020 and 500 for the CAsT 2021. For each metric, we calculate the mean and variance of performance across the different queries in the dataset. With this in mind, we present below the different baselines and scenarios used to evaluate each component of our model.

First-Stage Ranking Scenarios. To evaluate the effectiveness of our first-stage ranking model (Sect. 3.1), we compare our approach CoSPLADE, based on the query representation of Eq. (2) with different variants (the document encoder is set to the original SPLADE encoder throughout our experiments): **SPLADE_rawQuery**(lower bound): SPLADE [11] using only the original and ambiguous user queries q_n; **SPLADE_goldQuery**(kind of upper bound): SPLADE using the manually rewritten query q_n^*; **CQE** [18], a state-of-the-art dense contextualized query representation learned using learning-to-rank on a dataset with pseudo-labels. While the two former aim at evaluating how much

---

[7] https://sites.google.com/view/qanta/projects/canard.

our model captures contextual information, the latter is a TREC CAsT participant closely related to our work.

To model answers when representing the query using $\hat{q}_{n,k}^{answers}$, we used two historical ranges ("**All**" with $k = n-1$ answers and "**Last**" where we use only the last one, i.e. $k = 1$) and three types of answer inputs: **Answer** in which answers are the canonical answers; **Answer-Short** in which sentences are filtered as in the best performing TREC CAsT approach [20]. This allows for consistent input length, at the expense of losing information; **Answer-Long** As answers from CANARD are short (a few sentences extracted from Wikipedia – contrarily to CAsT ones), we expand them to reduce the discrepancy between training and inference. For each sentence, we find the Wikipedia passage it appears in (if it exists in ORConvQA [25]), and sample a short snippet of 3 adjacent sentences from it.

Finally, we also conducted ablation studies (on the best of the above variants) by modifying either the way to use the historical context or the training loss: **flatContext** a one-encoder version of our SPLADE approach in which we concatenate all information of the context to apply SPLADE to obtain a single representation of the query (instead of two representations $\hat{q}_{n}^{queries}$ and $\hat{q}_{n,k}^{answers}$ as in Eqs. 2 and 3) trained using a MSE loss function (Eq. 6) since there are no more two representations. **MSE** the version of our SPLADE approach trained with the MSE loss (Eq. 6) instead of the proposed one (Eq. 8); **cosine** the version of our SPLADE approach trained with a cosine loss instead of the proposed loss (Eq. 8). The cosine loss is interesting because it is invariant to the scaling factor that preserves the document ordering (Eq. 1).

Second-Stage Ranking Scenarios. We consider different scenario for our second-stage ranking model: **T5Mono_RawQuery** the T5Mono ranking model [24] applied on raw queries; **T5Mono_GoldQuery** the T5Mono ranking model applied on gold queries; **T5Mono_CQR** the T5Mono ranking model applied on query reformulation generated with a pre-trained T5 (using the CANARD dataset); **CoSPLADE_[context]_[number]**: different versions of our second-stage ranking model input (Eq. 9), varying 1) the presence or absence of the past queries within the reformulation, and 2) the number K of keywords identified as relevant by the first-stage ranker: 5, 10, 20.

TREC Participant Baselines. For each evaluation campaign (2020 and 2021), we also compare our model with the best, the median and the lowest TREC CAsT participants presented in the two overviews [6,8], where participants are ranked according to the nDCG@3 metric. Please note that we are not able to assess if results are significant since we report the effectiveness metrics presented in the TREC overview [7,8].

Table 1. Effectiveness of different scenarios of our first-stage ranking model on the TREC CAsT 2021.

	Recall@500	MAP@500	MRR	nDCG@500	nDCG@5	nDCG@3
Baselines						
SPLADE_rawQuery	30.8 ± 2.7	5.5 ± 0.9	21.3 ± 2.9	17.8 ± 1.8	12.8 ± 1.9	13.1 ± 2.1
SPLADE_goldQuery	68.8 ± 2.0	16.1 ± 1.2	55.5 ± 3.3	42.8 ± 1.7	35.2 ± 2.4	38.3 ± 2.8
CQE [19] from [8]	79.1	28.9	60.3	55.7	–	43.8
Effect of answer processing: CoSPLADE_ ...						
AllAnswers	79.5 ± 2.2	28.8 ± 1.7	61.7 ± 3.1	55.3 ± 2.0	44.1 ± 2.6	46.5 ± 2.9
AllAnswers-short	72.8 ± 2.6	25.7 ± 1.9	54.4 ± 3.3	49.5 ± 2.3	38.6 ± 2.7	40.1 ± 3.0
AllAnswers-long	80.4 ± 2.1	29.3 ± 1.8	62.0 ± 3.2	55.6 ± 2.1	46.3 ± 2.7	**48.9 ± 3.0**
LastAnswer	83.4 ± 2.0	31.2 ± 1.8	61.8 ± 3.1	58.1 ± 2.0	46.0 ± 2.7	47.4 ± 3.0
LastAnswer-short	79.2 ± 2.2	28.1 ± 1.8	61.4 ± 3.3	54.3 ± 2.1	44.7 ± 2.7	46.4 ± 3.0
LastAnswer-long	**85.2 ± 1.8**	**32.0 ± 1.7**	**64.3 ± 03.0**	**59.4 ± 1.9**	47.7 ± 2.6	48.6 ± 3.0
CoSPLADE_LastAnswer-long variants						
flatContext	77.0 ± 2.0	26.0 ± 2.0	55.0 ± 3.0	52.0 ± 2.0	41.0 ± 3.0	42.0 ± 3.0
MSE loss	70.9 ± 2.4	21.6 ± 1.7	48.7 ± 3.4	45.2 ± 2.3	34.9 ± 2.8	36.9 ± 3.1
cosine loss	70.4 ± 2.5	22.6 ± 1.7	52.5 ± 3.3	46.9 ± 2.2	37.5 ± 2.7	39.0 ± 3.0

5   Results

5.1   First-Stage Ranking Effectiveness

In this section, we focus on the first-stage ranking component of our CoSPLADE model. To do so, we experiment different scenarios aiming at evaluating the impact of the designed loss (Eq. 8) and the modeling/utility of evidence sources (Eqs. 3 and 4). Results of these different baselines and scenarios on the TREC CAsT 2021 dataset are provided in Table 1 – similar trends are observed on CAsT 2020. We provide detailed results (at query level) in the GitHub repository.

In general, one can see that all variants of our approach (CoSPLADE_* models) outperform the scenario applying the initial version of SPLADE on raw and, more importantly, gold queries. This is very encouraging since this latter scenario might be considered an oracle, i.e. the query is manually disambiguated. Finally, we improve the results over CQE [18] for all the metrics – showing that our simple learning mechanism, combined with SPLADE, allows for achieving SOTA performance.

Leveraging Queries and Answers History Better Contextualizes the Current Query. The results of the flatContext scenario w.r.t. to the SPLADE_goldQuery allow comparing the impact of evidence sources related to the conversation since they both use the same architecture (SPLADE). We can observe that it obtains better results than SPLADE_goldQuery (e.g., 77 vs. 68.8 for the Recall@500 metric), highlighting the usefulness of context to better understand the information need.

More Detailed Answers Perform Better. Since answers are more verbose than questions, including them is more complex, and we need to study the different possibilities (CoSPLADE_AllAnswers* and CoSPLADE_LastAnswer*). One can see that: 1) trimming answers (*-short) into a few keywords is less effective than considering canonical answers, but 2) it might be somehow effective when combined with the associated Wikipedia passage (*-long). Moreover, it seems more effective to consider only the last answer rather than the whole set of answers in the conversation history[8]. Taking all together, these observations highlight the importance of the way to incorporate information from answers into the reformulation process.

Dual Query Representation with Asymmetric Loss Leverages Sparse Query Representations. The results of the flatContext scenario show that considering at once past queries and answers perform better (compared to the MSE loss scenario which is directly comparable). However, if we separate the representations *and* use an asymmetric loss function, the conclusion changes. Moreover, the comparison of our best scenario CoSPLADE_LastAnswer-long with a similar scenario trained by simply using a MSE or a cosine losses reveals the effectiveness of our asymmetric MSE (Eq. 7). Remember that this asymmetric loss encourages the consideration of previous answers in the query encoding. This reinforces our intuition that the conversation context, and particularly verbose answers, is important for the conversational search task. It also reveals that the context should be included at different levels in the architecture (input and loss).

5.2   Second-Stage Ranking Effectiveness

In this section, we rely on the CoSPLADE_LastAnswer-long model as a first stage ranker, and evaluate different variants of the second-stage ranking method relying on the T5Mono model. For fair comparison, we also mention results obtained by a T5Mono ranking model applied on raw and gold queries, as well as query reformulated using a T5 generative model. Results on the TREC CAsT 2021 dataset are presented in Table 2.

The analysis of the CoSPLADE model variants allows to highlight different observations regarding the usability of the context and the number of keywords added to the query. First, adding the previous questions to the current query in the prompt (i.e., "Context") seems to improve the query understanding and, therefore, positively impacts the retrieval effectiveness. For instance, when 5 keywords are added, the context allows reaching 51.5% for the nDCG@3 against 45.9% without context. Second, the performances tend to increase with the number of additional keywords, particularly for scenarios without context, which is sensible. This trend is less noticeable for the scenarios with context since the best scores are alternatively obtained considering either 10 or 20 keywords. Notice that adding 10 or 20 keywords is more effective than only considering 5 keywords (e.g. 54.4% vs. 51.5% for the nDCG@3 metric). Thus, It seems that keywords help to reformulate the initial information need but they can lead to saturation when they are too numerous and combined with other information.

---

[8] This might be due to the simple way to use past answers, i.e. Equation 4, but all the other variations we tried did not perform better.

Table 2. Effectiveness of different scenarios of our second-stage ranking model on TREC CAsT 2021.

	Recall@500	MAP@500	MRR	nDCG@500	nDCG@5	nDCG@3
Baselines						
T5Mono_RawQuery	78.4±2.3	21.0±1.8	39.6±3.2	45.9±2.1	29±2.9	28.4±3.0
T5Mono_GoldQuery	86.1±1.7	44.1±1.9	78.7±2.7	68.5±1.8	63.2±2.7	64.6±2.8
T5Mono_CQR	80.4±2.2	30.0±1.9	58.2±3.4	55.3±2.1	43.6±3.0	44.6±3.2
coSPLADE-based second stage variants						
CoSPLADE_NoContext_5	84.3±1.8	31.7±2.0	61.6±3.3	58.1±2.0	45.5±2.8	45.9±3.1
CoSPLADE_NoContext_10	83.1±1.9	32.0±1.7	66.0±3.1	59.1±1.9	48.5±2.6	49.8±2.9
CoSPLADE_NoContext_20	84.8±1.7	33.4±1.8	66.0±3.0	60.4±1.8	47.4±2.6	49.6±2.9
CoSPLADE_Context_5	**85.0±1.7**	35.0±1.8	68.4±3.0	61.7±1.9	51.5±2.6	51.5±02.9
CoSPLADE_Context_10	84.8±1.7	**36.5±1.9**	67.8±3.1	**63.0±1.9**	52.0±2.7	53.3±3.1
CoSPLADE_Context_20	84.9±1.7	35.5±1.8	**69.8±3.0**	62.2±1.9	51.9±2.6	**54.4±2.9**

Table 3. TREC CAsT 2020 and 2021 performances regarding participants

TREC CAsT 2020	Recall@1000	MAP@1000	MRR	nDCG@1000	nDCG@5	nDCG@3
TREC participant (best)	63.3	30.2	59.3	52.6	–	45.8
TREC participant (median)	52.1	15.1	42.2	36.4	–	30.4
TREC participant (low)	27.9	1.0	5.9	11.1	–	2.2
CoSPLADE	82.4±2.0	26.9±1.5	58.1±2.9	54.2±1.8	41.2±2.4	44.0±2.7
TREC CAsT 2021	Recall@500	MAP@500	MRR	nDCG@500	nDCG@5	nDCG@3
TREC participants 1 (best)	85.0	37.6	67.9	63.6	–	52.6
TREC participants 2 (median)	36.4	17.6	53.4	33.6	–	37.7
TREC participants 3 (low)	58.9	7.6	27.0	31.4	–	15.4
CoSPLADE	84.9±1.7	35.5±1.8	69.8±3	62.2±1.9	51.99±2.6	54.4±2.9

By comparing the best model scenarios with the more basic scenarios applying the T5Mono second-stage ranker on raw and gold queries, we can observe that our method allows improving the retrieval effectiveness regarding initial queries but is not sufficient for reaching the performance of T5Mono_GoldQuery. However, results obtained when applying T5Mono on queries reformulated by T5 highlight that the contextualization of an initial query is a difficult task. Indeed, the T5Mono_CQR scenario is less effective than the T5Mono_GoldQuery one, depending on the metrics, the scores differ from 6 to 20 points.

Moreover, it is interesting to notice that the SPLADE model applied on raw and gold queries (first-stage ranking in Table 1) obtains lower results than the T5Mono model on the same data (second-stage ranking in Table 2). It can be explained by the different purposes of those architectures: SPLADE is a sparse model focusing on query/document expansion while T5Mono is particularly devoted to increase precision. However, it is worth noting that combining SPLADE and T5Mono as first and second-stage rankers reaches the highest effectiveness results in our experimental evaluation. This demonstrates the effectiveness of the CoSPLADE approach to both contextualize queries and effectively rank documents.

5.3 Effectiveness Compared to TREC CAsT Participants

We finally compare our approach with TREC CAsT participants for the 2020 and 2021 evaluation campaigns (Table 3). For both years, we reached very close or better performances than the best participants. Indeed, CoSPLADE outperforms the best TREC participant for the 2020 evaluation campaign regarding Recall@1000 and nDCG@1000. For 2021, our model obtains better results than the best one for the MRR and nDCG@3 metrics. For both years, the best participant is the h2oloo team [8,20] where they use query reformulation techniques, using T5. Our results suggest that our approach focusing on a sparse first-stage ranking model allows combining the benefit of query expansion and document ranking in a single model that eventually helps the final reranking step. In other words, simply rewriting the query without performing a joint learning document ranking can hinder the overall performance of the search task.

5.4 Efficiency

The first stage ranker is based on SPLADE, a sparse retrieval model and is therefore efficient. Detailed analysis in terms of FLOPs can be found in [10]. As our loss does not include any regularization loss to preserve sparsity, it is interesting to look at how they evolve compared to SPLADE. The average number of non-zero entries increases only for our AllAnswers variants (from around 60 for gold/raw to 80), showing that in the future we might benefit from controlling sparsity to improve the CoSPLADE efficiency. Interestingly, the LastAnswers variants have a slightly higher sparsity and perform better.

6 Conclusion

In this paper, we have shown how a sparse retrieval neural IR model, namely SPLADE [10], could be leveraged together with a lightweight learning process to obtain a state-of-the-art first-stage ranker. We further showed that this first-stage ranker could be used to provide context to the second-stage ranker, leading to results comparable with the best-performing systems. Future work may explore strategies to better capture the information from the context or to explicitly treat user feedback present in the evaluation dataset. We also envision to evaluate our approach on other conversational QA datasets, such as CoQA [31], OR-ConvQA [27], or ConvMix [3].

Acknowledgements. We would like to thank ANR for supporting this work under the following grants: ANR JCJC SESAMS (ANR-18-CE23-0001) and ANR COST (ANR-18-CE23-0016-003).

References

1. Aliannejadi, M., Chakraborty, M., Ríssola, E.A., Crestani, F.: Harnessing evolution of multi-turn conversations for effective answer retrieval, pp. 33–42. https://doi.org/10.1145/3343413.3377968, http://arxiv.org/abs/1912.10554

2. Arabzadeh, N., Clarke, C.L.A.: Waterlooclarke at the TREC 2020 conversational assistant track (2020)

3. Christmann, P., Roy, R.S., Weikum, G.: Conversational question answering on heterogeneous sources. In: Amigó, E., Castells, P., Gonzalo, J., Carterette, B., Culpepper, J.S., Kazai, G. (eds.) SIGIR 2022: The 45th International ACM SIGIR Conference on Research and Development in Information Retrieval, Madrid, Spain, 11–15 July 2022, pp. 144–154. ACM (2022). https://doi.org/10.1145/3477495.3531815

4. Clarke, C.L.A.: Waterlooclarke at the TREC 2019 conversational assistant track. In: Voorhees, E.M., Ellis, A. (eds.) Proceedings of the Twenty-Eighth Text REtrieval Conference, TREC 2019, Gaithersburg, Maryland, USA, 13–15 November 2019. NIST Special Publication, vol. 1250. National Institute of Standards and Technology (NIST) (2019). https://trec.nist.gov/pubs/trec28/papers/WaterlooClarke.C.pdf

5. Culpepper, J.S., Diaz, F., Smucker, M.D.: Research frontiers in information retrieval: report from the third strategic workshop on information retrieval in Lorne (SWIRL 2018). SIGIR Forum **52**(1), 34–90 (2018). https://doi.org/10.1145/3274784.3274788

6. Dalton, J., Xiong, C., Callan, J.: CAsT 2020: The conversational assistance track overview, p. 10

7. Dalton, J., Xiong, C., Callan, J.: TREC CAsT 2019: The conversational assistance track overview. http://arxiv.org/abs/2003.13624

8. Dalton, J., Xiong, C., Callan, J.: TREC CAsT 2021: the conversational assistance track overview, p. 7 (2021)

9. Elgohary, A., Peskov, D., Boyd-Graber, J.: Can you unpack that? Learning to rewrite questions-in-context. In: Proceedings of the 2019 Conference on Empirical Methods in Natural Language Processing and the 9th International Joint Conference on Natural Language Processing (EMNLP-IJCNLP), pp. 5918–5924. Association for Computational Linguistics, Hong Kong, November 2019. https://doi.org/10.18653/v1/D19-1605, https://aclanthology.org/D19-1605

10. Formal, T., Lassance, C., Piwowarski, B., Clinchant, S.: From distillation to hard negative sampling: making sparse neural IR models more effective. In: Proceedings of the 45th International ACM SIGIR Conference on Research and Development in Information Retrieval. SIGIR 2022, pp. 2353–2359. Association for Computing Machinery, New York, July 2022. https://doi.org/10.1145/3477495.3531857

11. Formal, T., Piwowarski, B., Clinchant, S.: SPLADE: sparse lexical and expansion model for first stage ranking. In: Proceedings of the 44th International ACM SIGIR Conference on Research and Development in Information Retrieval. SIGIR 2021, pp. 2288–2292. Association for Computing Machinery, New York, July 2021. 10/gm2tf2, https://doi.org/10.1145/3404835.3463098

12. Hofstätter, S., Althammer, S., Schröder, M., Sertkan, M., Hanbury, A.: Improving efficient neural ranking models with cross-architecture knowledge distillation. arXiv abs/2010.02666 (2020)

13. Hofstätter, S., Lin, S.C., Yang, J.H., Lin, J., Hanbury, A.: Efficiently teaching an effective dense retriever with balanced topic aware sampling. In: Proceedings of the 44th International ACM SIGIR Conference on Research and Development in Information Retrieval. SIGIR 2021, pp. 113–122. Association for Computing Machinery, New York (2021). https://doi.org/10.1145/3404835.3462891
14. Khattab, O., Zaharia, M.: ColBERT: efficient and effective passage search via contextualized late interaction over BERT. http://arxiv.org/abs/2004.12832
15. Krasakis, A.M., Yates, A., Kanoulas, E.: Zero-shot Query Contextualization for Conversational Search. In: Proceedings of the 45th International ACM SIGIR Conference on Research and Development in Information Retrieval. SIGIR 2022, pp. 1880–1884. Association for Computing Machinery, New York, July 2022. https://doi.org/10.1145/3477495.3531769
16. Kumar, V., Callan, J.: Making information seeking easier: an improved pipeline for conversational search, p. 10
17. Lan, Z., Chen, M., Goodman, S., Gimpel, K., Sharma, P., Soricut, R.: ALBERT: a lite BERT for self-supervised learning of language representations. In: 8th International Conference on Learning Representations, ICLR 2020, Addis Ababa, Ethiopia, 26–30 April 2020. OpenReview.net (2020). https://openreview.net/forum?id=H1eA7AEtvS
18. Lin, S.C., Yang, J.H., Lin, J.: Contextualized query embeddings for conversational search. http://arxiv.org/abs/2104.08707
19. Lin, S.C., Yang, J.H., Lin, J.: In-batch negatives for knowledge distillation with tightly-coupled teachers for dense retrieval. In: Proceedings of the 6th Workshop on Representation Learning for NLP (RepL4NLP-2021), pp. 163–173. Association for Computational Linguistics. https://doi.org/10.18653/v1/2021.repl4nlp-1.17, https://aclanthology.org/2021.repl4nlp-1.17
20. Lin, S.C., Yang, J.H., Lin, J.: TREC 2020 notebook: CAsT track. Technical report, TREC, December 2021
21. Lin, S.C., Yang, J.H., Nogueira, R., Tsai, M.F., Wang, C.J., Lin, J.: Multi-stage conversational passage retrieval: an approach to fusing term importance estimation and neural query rewriting. http://arxiv.org/abs/2005.02230
22. Lin, S., Yang, J., Nogueira, R., Tsai, M., Wang, C., Lin, J.: Query reformulation using query history for passage retrieval in conversational search. CoRR abs/2005.02230 (2020). https://arxiv.org/abs/2005.02230
23. Mele, I., Muntean, C.I., Nardini, F.M., Perego, R., Tonellotto, N.: Finding context through utterance dependencies in search conversations. Technical report (2021)
24. Nogueira, R., Jiang, Z., Pradeep, R., Lin, J.: Document ranking with a pretrained sequence-to-sequence model. In: Findings of the Association for Computational Linguistics: EMNLP 2020, pp. 708–718. Association for Computational Linguistics. https://doi.org/10.18653/v1/2020.findings-emnlp.63, https://www.aclweb.org/anthology/2020.findings-emnlp.63
25. Qu, C., Yang, L., Chen, C., Qiu, M., Croft, W.B., Iyyer, M.: Open-retrieval conversational question answering, pp. 539–548. https://doi.org/10.1145/3397271.3401110, http://arxiv.org/abs/2005.11364
26. Qu, C., Yang, L., Chen, C., Qiu, M., Croft, W.B., Iyyer, M.: Open-retrieval conversational question answering. In: Proceedings of the 43rd International ACM SIGIR Conference on Research and Development in Information Retrieval. SIGIR 2020, pp. 539–548. Association for Computing Machinery, New York (2020). https://doi.org/10.1145/3397271.3401110

27. Qu, C., Yang, L., Chen, C., Qiu, M., Croft, W.B., Iyyer, M.: Open-retrieval conversational question answering. In: Huang, J.X., et al. (eds.) Proceedings of the 43rd International ACM SIGIR Conference on Research and Development in Information Retrieval, SIGIR 2020, Virtual Event, China, 25–30 July 2020, pp. 539–548. ACM (2020). https://doi.org/10.1145/3397271.3401110

28. Qu, C., Yang, L., Qiu, M., Croft, W.B., Zhang, Y., Iyyer, M.: BERT with history answer embedding for conversational question answering. In: Proceedings of the 42nd International ACM SIGIR Conference on Research and Development in Information Retrieval. SIGIR 2019, pp. 1133–1136. Association for Computing Machinery, New York (2019). https://doi.org/10.1145/3331184.3331341

29. Qu, C., et al.: Attentive history selection for conversational question answering. In: Proceedings of the 28th ACM International Conference on Information and Knowledge Management, pp. 1391–1400 (2019)

30. Reddy, S., Chen, D., Manning, C.D.: CoQA: a conversational question answering challenge. Trans. Assoc. Comput. Linguist. **7**, 249–266 (2019). https://doi.org/10.1162/tacl_a_00266, https://aclanthology.org/Q19-1016

31. Reddy, S., Chen, D., Manning, C.D.: COQA: a conversational question answering challenge. Trans. Assoc. Comput. Linguist. **7**, 249–266 (2019). https://doi.org/10.1162/tacl_a_00266

32. Santhanam, K., Khattab, O., Saad-Falcon, J., Potts, C., Zaharia, M.: ColBERTv2: effective and efficient retrieval via lightweight late interaction. In: Proceedings of the 2022 Conference of the North American Chapter of the Association for Computational Linguistics: Human Language Technologies, pp. 3715–3734. Association for Computational Linguistics, Seattle, United States, July 2022. https://doi.org/10.18653/v1/2022.naacl-main.272, https://aclanthology.org/2022.naacl-main.272

33. Vakulenko, S., Longpre, S., Tu, Z., Anantha, R.: Question rewriting for conversational question answering. In: Proceedings of the 14th ACM International Conference on Web Search and Data Mining, pp. 355–363. ACM. https://doi.org/10.1145/3437963.3441748, https://dl.acm.org/doi/10.1145/3437963.3441748

34. Voskarides, N., Li, D., Panteli, A., Ren, P.: ILPS at TREC 2019 conversational assistant track, p. 4

35. Voskarides, N., Li, D., Ren, P., Kanoulas, E., de Rijke, M.: Query resolution for conversational search with limited supervision, pp. 921–930. https://doi.org/10.1145/3397271.3401130, http://arxiv.org/abs/2005.11723

36. Yan, X., Clarke, C.L.A., Arabzadeh, N.: Waterlooclarke at the TREC 2021 conversational assistant track (2021)

37. Yang, J.H., Lin, S.C., Wang, C.J., Lin, J.J., Tsai, M.F.: Query and answer expansion from conversation history. In: TREC (2019)

38. Yu, S., et al.: Few-shot generative conversational query rewriting. http://arxiv.org/abs/2006.05009

39. Zamani, H., Trippas, J.R., Dalton, J., Radlinski, F.: Conversational Information Seeking, January 2022. https://doi.org/10.48550/arXiv.2201.08808, http://arxiv.org/abs/2201.08808, arXiv:2201.08808 [cs]

SR-CoMbEr: Heterogeneous Network Embedding Using Community Multi-view Enhanced Graph Convolutional Network for Automating Systematic Reviews

Eric W. Lee[✉] and Joyce C. Ho

Emory University, Atlanta, GA, USA
{ewlee4,joyce.c.ho}@emory.edu

Abstract. Systematic reviews (SRs) are a crucial component of evidence-based clinical practice. Unfortunately, SRs are labor-intensive and unscalable with the exponential growth in literature. Automating evidence synthesis using machine learning models has been proposed but solely focuses on the text and ignores additional features like citation information. Recent work demonstrated that citation embeddings can outperform the text itself, suggesting that better network representation may expedite SRs. Yet, how to utilize the rich information in heterogeneous information networks (HIN) for network embeddings is understudied. Existing HIN models fail to produce a high-quality embedding compared to simply running state-of-the-art homogeneous network models. To address existing HIN model limitations, we propose SR-CoMbEr, a community-based multi-view graph convolutional network for learning better embeddings for evidence synthesis. Our model automatically discovers article communities to learn robust embeddings that simultaneously encapsulate the rich semantics in HINs. We demonstrate the effectiveness of our model to automate 15 SRs.

Keywords: Systematic review · Network embedding · Heterogeneous information network · Multi-view learning · Graph convolution network

1 Introduction

Systematic reviews (SRs) serve as a cornerstone of evidence-based medicine and bridge the research-to-practice gap by ensuring all the available evidence is accessible to decision-makers. An excellent SR carefully synthesizes individual studies such as clinical trial results to guide and inform clinical practice. As a motivating example, a SR was used to synthesize findings from randomized intervention studies to determine the impact of angiotensin-converting-enzyme (ACE) inhibitors for treating high blood pressure [7]. As a result, ACE inhibitors now are commonly prescribed to treat hypertension, heart failure, and various other heart conditions. Unfortunately, conducting a SR is an extremely time-consuming and complex task [12]. Established methodologies for performing a

J. Kamps et al. (Eds.): ECIR 2023, LNCS 13980, pp. 553–568, 2023.
https://doi.org/10.1007/978-3-031-28244-7_35

Fig. 1. A simplified illustration of the SR screening process using "ACEInhibitors" from Cohen [7] dataset.

SR require a comprehensive search to identify all the relevant studies for inclusion [5]. Yet these broad searches yield imprecise search results (e.g., <2% relevant documents). Figure 1 provides an example of the laborious citation screening process for ACE inhibitors. Only 1.19% of the articles were selected for full-text review based on the title and abstract of which 1.61% were included (i.e., analyzed and evaluated) in the actual review itself. Thus current estimates for the average time to conduct a SR is 67 weeks from registration to publication [3]. Clearly, this process is unsustainable nor scalable, especially given the exponential growth of biomedical literature [2].

Given the importance of SR and the labor-intensive work it entails, research on machine learning and text mining methods to automate the evidence synthesis while maintaining the rigor of a traditional SR have been proposed [28]. In particular, semi-automation can help speed up the screening process, an extremely tedious endeavor due to a large number of articles [25]. The standard methodology for automating the screening process focuses predominately on the text itself using representations like bag-of-words or word embeddings [15,19,24]. Yet, recent work demonstrated that the rich citation structure can be utilized to improve the screening process [20]. Their work used a homogeneous network embedding technique, LINE, to learn the citation network representations and these representations were able to outperform the text itself on 10 of the 15 SRs. These promising results suggest that better network representation may expedite evidence synthesis.

Citation networks can be represented as a graph structure that includes articles (nodes) and references (edges). This representation is used across many application domains including social networks, the world wide web, and knowledge graphs. As real-world networks can be huge and complex, it is difficult to directly analyze the graph, thus learning meaningful low-dimensional vectors of the nodes and edges, or network embeddings have been proposed while preserving the features of the network [21]. Recently, there has been an emergence of deep learning-based models such as graph neural networks (GNN) to learn the network embeddings [11,22,31]. One popular method is Graph Convolutional Network (GCN) [17] which can efficiently learn the structural dependencies through convolutional operations on the graph. However, GCN is designed for a homogeneous network, whereas the biomedical citation graph contains multiple objects (nodes) and link types (edges) including author information, venue information, and Medical Subject Headings (MeSH) terms that are used for indexing articles.

Since many real-world networks are heterogeneous information networks (HIN) with multiple objects and link types, several variations of GNN and GCN models have been proposed for HIN embeddings. However, existing models have focused on preserving the meta-path structure (i.e., the path with various object types and edge types that captures the semantics of the network) by transforming the HIN into several homogeneous networks to learn the representations [9,33,42]. Unfortunately, the defined meta-path impacts the embedding quality. Thus, ie-HGCN [40] automatically evaluates all possible meta-paths and projects the representations of different types of neighbor objects into a common semantic space. Yet, ie-HGCN is susceptible to noise in the graph.

We propose SR-CoMbEr, a **C**ommunity **M**ulti-view **b**ased **E**nhanced Graph Convolutional Network for **S**ystematic **R**eview. SR-CoMbEr constructs multiple local GCNs, each centered around a community. To learn from the different object and link types, each community adopts a multi-view approach where a view-specific representation is learned to capture the complex structure information for each relation type. Moreover, we pose the multiple community GCN aggregation problem as a multi-modal problem to yield a robust final embedding that reflects the different community representations. Our main contributions of this work are:

- We pose the problem of HIN representation as a multi-view learning problem to avoid specification of the meta-path while automatically capturing the network semantics.
- We propose an innovative multiple, community-based multi-view GCN to capture the structural heterogeneity that is useful for downstream tasks.
- We conduct extensive experiments on SR screening to demonstrate the superior performance of SR-CoMbEr over HIN baselines.

2   Preliminaries

In this section, we introduce the *heterogeneous information network*, or HIN, and *Graph Convolutional Network*, or GCN, a state-of-the-art network embedding model.

2.1   Heterogeneous Information Network

A HIN contains multiple types of objects and links. Formally, such a network is defined as follows.

Definition 1. *HIN.* *A HIN is defined as $\mathcal{G} = (\mathcal{V}, \mathcal{E}, \phi, \psi)$, where $\mathcal{V}$ is the set of objects, $\mathcal{E}$ is the set of links, ϕ is the object type mapping function, and ψ is the link mapping function. ϕ is defined as $\phi : \mathcal{V} \to \mathcal{A}$, and ψ is defined as $\psi : \mathcal{E} \to \mathcal{R}$. $\mathcal{A}$ and $\mathcal{R}$ denotes predefined object and link types respectively where $|\mathcal{A}| + |\mathcal{R}| > 2$.*

A homogeneous network contains a single object and relation type such as a social network with User (U) as an object type and a single type of link U – U. On the other hand, HIN contains multiple types of objects such as a bibliographic network which has four types of objects (i.e., Author (A), Paper (P), Venue (V), and MeSH terms (M) and three link types, A – P, P – V, and P – M.

2.2   Graph Convolutional Networks

GCNs have been extensively studied and used for a wide range of tasks (see [17] for a survey). Formally, GCNs can be defined as follows. Suppose H^k is the feature representation of the k-th layer in GCN, the propagation becomes

$$H^k = \sigma(\tilde{D}^{-\frac{1}{2}} \tilde{A} \tilde{D}^{-\frac{1}{2}} H^{k-1} W^k) \tag{1}$$

where $\tilde{A} = A + I \in \boldsymbol{R}^{N \times N}$ is the adjacency matrix A with a self connection. $\tilde{D}$ is the degree matrix of $\tilde{A}$ which is formally defined as $\tilde{D}_{ii} = \sum_i \tilde{A}_{ij}$. And W^k is a trainable weight matrix. As shown in Eq. (1), the convolution operation is determined by the given graph structure and GCN only learns the node-wise linear transform $H^{k-1} W^k$. Thus, the convolution layer can be interpreted as the composition of a fixed convolution followed by an activation function σ on the graph after the node-wise linear transformation.

3   Related Works

Methods for semi-automating the citation screening step of SRs have been widely studied [28]. Most of these models use bag-of-words and their combinations as input representations to a supervised learning model (e.g., support vector machine or random forest) [7,15]. For example, Cohen *et al.* [6] proposed to use uni-grams and bi-grams to treat each of them as a single word, and Bannach-Brown *et al.* [1] used tri-gram and NLP tagger prior to extracting uni-grams.

However, articles contain rich information besides the text, such as citations, author, venue, and keywords. This information can be captured using a HIN where network embeddings can serve as the article representation. Several HIN network embedding methods have been proposed. Existing work focuses on preserving the meta-path structure which contains the semantic information of the graph. For example, ESim [33] uses multiple user-defined meta-path to learn representations in the user-preferred embedding space, and metapath2vec [9] is a skip-gram model that uses meta-path based random walk. Some works extend Graph Neural Networks (GNNs) for modeling HIN. For example, HAN [38] transforms the given HIN into a homogeneous network based on the meta-path and uses GNN based on hierarchical attention.

However, these models require manually selected meta-path or only accept one meta-path which may cause an information loss by not capturing all meaningful relations. Thus, some recent works proposed learning the meta-path. GTN [41] learns the meta-path to generate multiple new graphs based on the defined meta-path to apply GCN, and ie-HGCN [40] learns the weights of the meta-paths to select the best meta-path for their model. While HIN embedding methods are proposed by enhancing GCN by learning meta-path, some research attempts to use multi-view learning for HIN embedding. For example, Zhang *et al.* [43] propose to use a fusion of multiple GCNs modalities of brain images in relationship prediction, and Ma *et al.* [23] uses multi-view graph auto-encoder to capture the similarities of drug features.

4   SR-CoMbEr

SR-CoMbEr is inspired by the multiple-filtering local GCN model [39], which constructs multiple local versions of a homogeneous network to capture different aspects of the node attributes while providing robustness to noise. Yet, the local versions of the multiple GCN approach may fail to capture the complex neighborhood structure when solely focusing on a homogeneous network. Moreover, the model can be sensitive to the number of local filters. We address these limitations using three parts: (1) automatic identification of communities in HIN, (2) community multi-view learning to capture information from each link type, and (3) global consensus across the communities. Figure 2 depicts SR-CoMbEr's overall architecture, where the goal is to learn the representation of the target object α (i.e., circle node (P)).

4.1   Heterogeneous Community Detection

The ability to capture the neighborhood information is a crucial aspect of ensuring the quality of the network embedding. Many network embedding methods use random walks to capture the neighborhoods before passing them to a deep learning model. For example, the multiple-filtering local GCN model [39] uses random walk to construct $\mathcal{M}$ local networks are constructed. However, sampling of a single link type may not encapsulate the community structure via other link types while sampling multiple links may not be sufficient to capture the complicated structure [42]. However, utilizing the entire HIN can pose computational problems for large networks as well as limit their generalizability to unseen data [39]. Instead, we propose to utilize the community structure ubiquitous in networks, where a group of nodes exhibits more intra-connections than inter-connections with external nodes [10], to determine the construction of the local networks. Given a set of communities, a random walk is initiated using the nodes belonging to the community. Thus each local GCN version learns a better local embedding by integrating information found in the community structure. It is important to note that SR-CoMbEr does not restrict the random walk to just links between community nodes, therefore the local network may contain neighborhood information of nodes outside the community. Moreover, since a node may be part of multiple communities, the combination of multiple local GCNs will thereby reflect different neighborhood information for the same object.

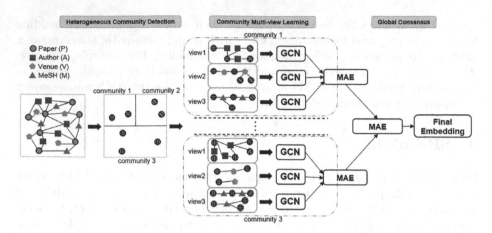

Fig. 2. The framework overview of the SR-CoMbEr. The input network is a toy example of a PubMed Network which contains four node types and three edge types. Four node types are Paper (P), Author (A), Venue (V), and MeSH Terms (M), and three edge types are P – A, P – V, and P – M. The target node is set to P which is used for the node classification task.

The community-based focus of each local GCN lends itself naturally to automatic detection of the "optimal" number of local filters, $\mathcal{M}$. While there are many types of community detection methods including clustering-based methods [26] and modularity-based methods [27], many of these models are developed for the homogeneous setting. Instead, SR-CoMbEr uses Tucker decomposition [35], a popular tensor factorization model, to identify the community structure and the number of optimal filters in the HIN setting. Tucker decomposition can be viewed as a generalization of singular value decomposition (SVD) which can detect communities in homogeneous networks [30]. The HIN tensor, $\mathcal{X}$, is a higher-order tensor where each object type serves as a mode of the tensor and the entries in the tensor capture the status of the links between the different modes of the tensor. For Fig. 2, a paper by author by venue by term tensor (4-mode tensor), can be constructed where each element captures who authored a paper, where it was published, and what terms were present in that paper. Thus, the tensor succinctly encapsulates the relations between different object types.

Formally, for a 3rd order tensor, $\mathcal{X} \in \mathbb{R}^{I \times J \times K}$, Tucker decomposition approximates the tensor into a core tensor, $\mathcal{H} \in \mathbb{R}^{P \times Q \times S}$ multiplied by a factor matrices along each mode, $A \in \mathbb{R}^{I \times P}, B \in \mathbb{R}^{J \times Q}, C \in \mathbb{R}^{K \times S}$:

$$\mathcal{X} \approx \mathcal{H} \times_1 A \times_2 B \times_3 C. \tag{2}$$

The core tensor, $\mathcal{H}$ captures the level of interactions between the different components, and the factor matrices, A, B, C, are often assumed to be column-wise orthonormal. We note that Tucker decomposition generalizes to any N-mode tensor, does not impose column-wise orthonormal factor matrices nor does the core tensor have decreasing Frobenius norm along each matrix slice. The general Tucker properties deviate from the SVD assumptions but can be integrated through the algorithmic choice for computing the decomposition. In addition, the column rank of each factor matrix can be different (i.e., $P \neq Q \neq S$) in the Tucker decomposition. We refer the reader to [18,29] for additional details.

Since each local filter encapsulates a community, the column rank of each factor matrix is set to be the same, $P = Q = S$. To compute the Tucker decomposition, we use the higher-order orthogonal iteration (HOOI) algorithm as it is one of the more efficient techniques. HOOI uses SVD to compute the orthonormal basis of each factor matrix [8]. Moreover, the resulting core tensor and factor matrices can be seen as the generalized counterparts of the matrix SVD. Thus, the superdiagonal entries of the core tensor ($H_{iii}, \forall i \in [1, R]$) is comparable to the singular values of SVD (i.e., diagonal entries in Σ). As a result, the number of communities can be calculated as the point in which the superdiagonal values converge, similar in fashion to using the Σ matrix in SVD to find the number of communities in a homogeneous network [30]. This eliminates the need for the user to grid search the number of filters $\mathcal{M}$.

The next step is to identify the nodes that belong to each community. Without loss of generality, we assume that the target object, α, corresponds to the first mode of the tensor. Each object can then be represented in a low-dimensional vector space (i.e., $P << I$) using the row vectors of the corresponding factor matrix A. Spectral clustering is performed on A to identify the community members using $\mathcal{M}$ for each node in the target object α. For simplicity of implementation, SR-CoMbEr uses the k-means algorithm to generate a hard cluster assignment but the framework can use any spectral clustering method. The graph for each community (local) filter is then obtained by performing a fixed-size random walk starting with only nodes within the community. Note that the community filters can contain not just nodes within the same community but also other nodes that are connected during the random walk process. The entire community detection process is summarized in Algorithm 1.

4.2   Community Multi-view Learning

Since random walk of $\mathcal{G}$ directly may fail to capture the complex structure, SR-CoMbEr treats each link type as a different view of the network. For each link type containing the target object α, a view of the community is created by performing the fixed-size random walk using only this link type. For each community GCN m, a view is constructed from each link type thus yielding $|\mathcal{R}|$ different representations, $X_1^m, \cdots, X_{|\mathcal{R}|}^m$. As an example, three views are constructed for Fig. 2 with a different link type (e.g., P–A, P–V, P–M). Thus, rather than having a single community GCN, each community will have multiple view-specific filters of the network.

Although each view contains a single link type, GCN still cannot be applied directly because the neighbors of an object are of different types. Moreover, the adjacency matrix is not a square matrix and thus cannot be fed into Equation (1),

---

Algorithm 1: Heterogeneous Community Detection in SR-CoMbEr.

Input: Graph $\mathcal{G} = (\mathcal{V}, \mathcal{E}, \phi, \psi)$, $\phi : \mathcal{V} \to \mathcal{A}$, $\psi : \mathcal{E} \to \mathcal{R}$
Output: Number of filters $\mathcal{M}$, Communities $C_1, ..., C_M$

1 Construct tensor $\mathcal{X}$ from $\mathcal{G}$;
2 Compute $\mathcal{X} \approx \mathcal{H} \times_1 A \times_2 B \times_3 C$ using HOOI;
3 Set $\mathcal{M}$ based on convergence of values in the superdiagonal entries of $\mathcal{H}$;
4 Detect communities of α, $C_1, C_2, \ldots, C_M$, using spectral clustering of A;
5 **return** $\mathcal{M}, C_1, C_2, \ldots, C_M$;

---

where $\tilde{A}$ is the square matrix. We thus use the idea of projection, introduced in ie-HGCN [40], to ensure both object types are in the same space. Suppose the view captures the link α–β, where $\mathcal{V}^\alpha$ and $\mathcal{V}^\beta$ represent the set of objects in the α and β node type, respectively. Let $A^{\alpha-\beta} \in \mathbf{R}^{|\mathcal{V}^\alpha| \times |\mathcal{V}^\beta|}$ denote the adjacency matrix between α and β and the degree matrix $D^{\alpha-\beta} = diag(\sum_j A_{ij}^{\alpha-\beta}) \in \mathbf{R}^{|\mathcal{V}^\alpha| \times |\mathcal{V}^\alpha|}$. Every object is then projected into the same space and passed to the GCN:

$$\tilde{A}^{\alpha-\beta} = (D^{\alpha-\beta})^{-1} \cdot A^{\alpha-\beta}$$
$$X_{\alpha-\beta} = \tilde{A}^{\alpha-\beta} \cdot W^{\alpha-\beta} \tag{3}$$

where $\tilde{A}^{\alpha-\beta}$ is the row-normalized matrix and $W^{\alpha-\beta}$ is the trainable convolution weight matrix of α–β relation.

The community embedding, X^m, should capture all the information from the $|\mathcal{R}|$ views while reducing information redundancy that may be present in the views. Moreover, certain views may learn better representations of the community. Thus, to summarize the different view modalities simultaneously, SR-CoMbEr adopts the multi-modal stacked autoencoder (MAE) [4]. MAE takes multiple input representations, concatenates the input together, and then passes this to an autoencoder to induce a succinct, shared representation from which to reconstruct the original (concatenated) inputs. Formally, the global consensus process for the shared representation in the mth community GCN is:

$$H^m = MAE(X_1^m, X_2^m, \ldots, X_{|\mathcal{R}|}^m). \tag{4}$$

4.3   Global Consensus

Since each community multi-view GCN representation H^m, captures community-specific information, the learned representation can differ. We formulate the aggregation of the community multi-view GCN representation as a multi-modal problem. Although the final shared representation can be computed as the average of the community representations, this assumes each community is equivalent. In practice, some community representations are of higher quality and thereby should have higher weights. MAE is used again to learn the final representation across the $\mathcal{M}$ communities:

$$H = MAE(H^1, H^2, \ldots, H^{\mathcal{M}}) \tag{5}$$

Algorithm 2: The pseudocode of SR-CoMbEr.

Input: Graph $\mathcal{G} = (\mathcal{V}, \mathcal{E}, \phi, \psi)$, $\phi : \mathcal{V} \to \mathcal{A}$, $\psi : \mathcal{E} \to \mathcal{R}$
Number of localized filters $\mathcal{M}$
Output: Final representation H

1   Compute $\mathcal{M}, C_1, C_2, \ldots, C_M$ using Algorithm 1;
    /* Loop through the communities */
2   **for** $i=1, \ldots, \mathcal{M}$ **do**
      /* Loop through the views */
3   **for** $\alpha - \beta \in \mathcal{R}$ **do**
4   Run random walk on objects $\in C_i^\alpha$ and $\in C_i^\beta$;
5   Compute $X_{\alpha - \beta}$ according to Eq. (3);
6   **end**
7   Compute H^i according to Eq. (4);
8   **end**
9   Compute loss and update parameters;
10 **return** H according to Eq. (5);

The final embedding representation, H, is then used for a variety of tasks such as classification, clustering, etc., where the loss function is tailored towards the specific task. For example, in a multi-class node classification task, H is passed to a fully connected layer with softmax activation, and the loss is defined as the cross-entropy over the object type. The weights are then learned using stochastic gradient descent with backpropagation. Algorithm 2 shows the overall training procedure of SR-CoMbEr.

5 Experimental Design

5.1 Dataset

We evaluate our model on the publicly available dataset provided by Cohen *et al.* [7]. The dataset includes 15 SRs (or topics) concerning different drug efficacies which were performed by members of evidence-based practice centers (EPCs). In the dataset, each SR topic contains a set of PubMed article identifiers (PMID) and their associated title/abstract screening status (i.e., whether or not the article passed the title/abstract screening stage). The PMID allows us to retrieve the metadata (citation, author, venue, and MeSH terms) from the PubMed database. There exist other SR datasets [32], however, the dataset does not contain the PMID. We extract a subset of articles from the PubMed database using Entrez API[1]. Including all the articles from the Cohen dataset and using Entrez API, we trace articles up to 2-hops based on the citation information and retrieve about 7.6M articles with the meta-data including author, venue, and MeSH terms. The number of articles screened ranged from 310 (Antihistamines) to 3465 (Statins)

---

[1] https://www.ncbi.nlm.nih.gov/books/NBK25501/.

Table 1. Comparison of baseline characteristics. The * symbol next to the model name denotes a homogeneous network model. The columns MP, SS, and MVF represent meta-path specification, subgraph sampling, and multi-view fusion, respectively.

	MP	SS	MVF	Module	Supervision
LINE*	✗	✗	✗	Skip-gram	✗
GCN*	✗	✗	✗	GCN	✓
HAN	✓	✗	✗	Transformer	✓
GAHNE	✗	✓	✓	GCN	✓
ie-HGCN	✗	✓	✓	GCN	✓
SR-CoMbEr	✗	✓	✓	GCN	✓

with anywhere from 2.07% (SkeletalMuscleRelaxants) to 32.49% (Triptans) passing the abstract screening process. This demonstrates a relatively large degree of imbalance.

5.2 Baselines

We compare with five baselines spanning both homogeneous and HIN embedding methods in the SR task. Table 1 compares the characteristics of baseline models.

- **LINE** [34]. A conventional network embedding method that is using first- and second-proximity. Since it is designed for a homogeneous network, we transform the HIN by considering collapsing the object and link types as a single type and use LINE to learn the representation of the whole HIN.
- **GCN** [17]. A semi-supervised graph convolutional network that is designed for a homogeneous network. Similar to LINE, we ignore the heterogeneity of the network and collapse it into a homogeneous network..
- **HAN** [38]. A model to learn representations for HIN. It transforms the HIN into several homogeneous sub-networks by user-defined meta-paths. For object-level aggregation, it uses GAT [36], then uses an attention mechanism to fuse object representations from each sub-networks.
- **GAHNE** [21]. A model to learn representations for HIN. It converts the network into a series of homogeneous sub-networks to capture the semantic information. Then an aggregation mechanism fuses the sub-networks with supplemental information from the whole network.
- **ie-HGCN** [40]. A HIN embedding model that evaluates all possible meta-paths and projects the representations of different types of neighbor objects into a common semantic space using object- and type-level aggregation.

5.3 Evaluation Metrics

The recent trend for evaluating a SR task uses the area under the receiver operating curve (AUC) for predicting whether or not the abstract was screened or not to report the results [6,24]. Thus we evaluate the models using AUC.

5.4   Implementation Details

Our method is implemented in Keras and the source code is publicly available[2]. The source codes of the other baselines are provided by their authors and are implemented in either PyTorch or TensorFlow. All experiments are conducted on a machine with 1 Nvidia GeForce GTX 1080Ti and 11GB GPU memory. For each SR task, we randomly split the articles in the SR into train-validation-test as 50%−25%−25%, and use the validation set for the hyperparameter tuning. Articles not in the target SR task are marked as irrelevant in the training process.

For the baseline models, we adopt the same hyperparameter settings introduced in their respective papers. For LINE [34], we use a dimension of 128 for each first- and second-order proximity resulting a dimension of 256 for the final embedding. For GCN [17], we use the learning rate of 0.01, the dropout rate of 0.5, and the L2 penalty weight decay of 0.0005. For HAN [38], the number of attention heads is set to 8, and the meta-paths PAP, PMP, and APVPA are used (P: Paper, A: Author, M: MeSH terms, and V: Venue). For GAHNE [21], we used a learning rate of 0.005, a dropout of 0.5, an L2 penalty of 0.001, and a dimension of 128. For ie-HGCN [40], the number of layers is set to 5, and the dimension for the four hidden layers starting from the second layer is set to 64, 32, 16, and 8. For SR-CoMbEr, we use $\mathcal{M} = 12$, set the random walk length to 20, and the embedding dimension to 128. The Adam optimizer [16] is used with a learning rate of 0.01 and all parameters are initialized randomly. Dropout is used for all layers except the output layer with a dropout rate of 0.5.

6   Experimental Results

6.1   Systematic Review

The AUC on the Cohen dataset is reported in Table 2 for each SR. The best results are bolded and the second-best results are underlined. The results show that HIN embedding outperforms homogeneous network embedding (LINE and GCN). This demonstrates citation information and other node types (author, venue, and MeSH terms) help to improve the performance of the SR task.

From the table, we observe SR-CoMbEr outperforms all other baselines from 0.002 to 0.018 by comparing with the second-best AUC score. This indicates the importance of effectively modeling the HIN and demonstrates the effectiveness of SR-CoMbEr in the SR task. Between the existing HIN models, HAN shows the limitation of the user-defined meta-path. The results suggest that there are more hidden but important paths that are difficult for users to define. In contrast, the performance between GAHNE and ie-HGCN is similar. GAHNE performs better when there are more papers excluded from the abstract screening process. For example, the "SkeletalMuscleRelaxants" dataset has a total of 1643 articles in the beginning but only 34 articles are selected from the abstract screening which is only 2%. While GAHNE performs better in cases when fewer articles are

---

[2] https://github.com/ewhlee/SR-CoMbEr.

Table 2. Performance results (AUC score) for the SR task. The best score for each SR is bolded and the second highest is underlined.

SR	LINE	GCN	HAN	GAHNE	ie-HGCN	SR-CoMbEr
ACEInhibitors	0.622	0.627	0.649	0.662	<u>0.667</u>	**0.672**
ADHD	0.597	0.605	0.621	0.644	<u>0.646</u>	**0.659**
Antihistamines	0.541	0.544	0.567	<u>0.588</u>	0.586	**0.593**
AtypicalAntipsychotics	0.601	0.607	0.617	<u>0.638</u>	0.636	**0.641**
BetaBlockers	0.629	0.632	0.658	0.671	<u>0.677</u>	**0.684**
CalciumChannelBlockers	0.636	0.64	0.662	<u>0.67</u>	0.666	**0.688**
Estrogens	0.577	0.583	0.607	<u>0.629</u>	0.626	**0.631**
NSAIDs	0.637	0.639	0.662	<u>0.691</u>	0.685	**0.697**
Opioids	0.632	0.635	0.654	0.667	<u>0.671</u>	**0.686**
OralHypoglycemics	0.555	0.559	0.582	<u>0.591</u>	0.583	**0.598**
ProtonPumpInhibitors	0.638	0.641	0.664	0.677	<u>0.681</u>	**0.687**
SkeletalMuscleRelaxants	0.64	0.643	0.658	0.672	<u>0.677</u>	**0.684**
Statins	0.606	0.609	0.633	0.653	<u>0.659</u>	**0.665**
Triptans	0.617	0.624	0.64	0.652	<u>0.66</u>	**0.671**
UrinaryIncontinence	0.633	0.639	0.658	<u>0.678</u>	0.675	**0.683**

selected, ie-HGCN performs better in cases when more papers are selected. For example, "AtypicalAntipsychotics" has a total of 1120 articles in the beginning and 363 articles passed the screening which is 32%.

6.2   Ablation Study

We assess the importance of each component in SR-CoMbEr for the final embedding. *LMV* is a localized, multi-view model that does not use the heterogeneous community detection component (i.e., Sect. 4.1). Each localized, multi-view filter is subsampled using a random walk of all the nodes in the graph. Then the local representations are aggregated using an average function. *CoAvg* extends *LMV* by using the community detection module to construct the localized, multi-view filters. However, unlike the SR-CoMbEr, it does not use the MAE to learn the shared representation from the community filters (i.e., Eq. (5) is replaced with $H = AVG(H^1, H^2, ..., H^M)$). Table 3 summarizes the AUC scores on the test set of the two different multi-view learning techniques on the ACEInhibitors SR task. As shown in the table, incorporating the community information improves the performance (see CoAvg versus LMV). By leveraging the community structure, the embedding model can capture different neighborhood information to learn a better representation. While the overall results suggest that although the performance boost is less compared to the community detection component, MAE is beneficial to automatically learn the weights from each of the community representations for the final embedding.

To better understand the importance of the community detection algorithm, we compared the performance using SVD to identify the communities using just

Table 3. Comparison of the AUC score using different community detection algorithms on ACEInhibitors from the SR task.

LMV	CoAvg	CP	SVD	SR-CoMbEr
0.658	0.665	0.668	0.662	**0.672**

one view of the network [30] and CANDECOMP-PARAFAC (CP), a special case of Tucker decomposition where the core tensor only has values along the superdiagonal entries [18]. For SVD, let $F \in \mathbb{R}^{m \times n}$ denote the adjacency matrix of the link type with the largest number of nodes and the target node α. Under SVD, $F = U\Sigma V^*$, where $U \in \mathbb{R}^{m \times p}$, $V \in \mathbb{R}^{n \times p}$ matrix, and $\Sigma \in \mathbb{R}^{p \times p}$. Spectral clustering is then performed in a similar fashion using $\mathcal{M}$ as the number of clusters on the target object, α, and U as the low-dimensional embedding. For CP decomposition, the alternating least square method is used to find the leading left singular values [13]. As shown in the table, SR-CoMbEr (using HOOI algorithm) for community detection outperforms other techniques (see CP and SVD). While we identify 12 local filters for SR-CoMbEr using HOOI, SVD identifies 9 and CP identifies 14. This shows the importance of identifying the optimal number of filters as too many or too few filters can degrade the performance.

7   Conclusion

In this paper, we propose SR-CoMbEr to learn citation network representations for SRs. To avoid defining the meta-path, we formulate the problem using multi-view learning to automatically capture the semantics of HIN. To encode the structural heterogeneity and neighborhood information, we use community detection and multiple community-based views of the network and fuse the representations to obtain the final representation. We also introduce the use of HOOI to compute the optimal number of filters in concert with community detection. The experiments on 15 SRs show that SR-CoMbEr outperforms several state-of-the-art HIN embedding models.

There are several limitations to our study. First, the improvements in results are not substantially better even after all the extensive modeling. This is typical in SR automation as the evaluation measures may be ill-suited for capturing major improvements due to the dominance of irrelevant documents. Second, our evaluation was limited in the number of topics considered. There are other evaluation resources such as the CLEF eHealth TAR data, clinical outcomes [37], SIGIR 2017 SysRev Query Collection [32], and the SWIFT-review dataset [14] that can be explored for future work. Other promising future directions include the incorporation of the article text as well as the structure of the PubMed HIN.

Acknowledgements. We thank the reviewers for their insightful suggestions and comments. This work was supported by the National Science Foundation award IIS-1838200 and IIS-2145411.

References

1. Bannach-Brown, A., et al.: Machine learning algorithms for systematic review: reducing workload in a preclinical review of animal studies and reducing human screening error. Systematic Rev. **8**(1), 23 (2019)
2. Bastian, H., Glasziou, P., Chalmers, I.: Seventy-five trials and eleven systematic reviews a day: How will we ever keep up? PLOS Med. **7**(9), e1000326 (2010)
3. Borah, R., Brown, A.W., Capers, P.L., Kaiser, K.A.: Analysis of the time and workers needed to conduct systematic reviews of medical interventions using data from the prospero registry. BMJ open **7**(2), e012545 (2017)
4. Cadena, C., Dick, A.R., Reid, I.D.: Multi-modal auto-encoders as joint estimators for robotics scene understanding. In: Robotics: Science and Systems (2016)
5. Chandler, J., Churchill, R., Higgins, J., Lasserson, T., Tovey, D., et al.: Methodological standards for the conduct of new cochrane intervention reviews. Cochrane Collaboration, Sl (2013)
6. Cohen, A.M.: Optimizing feature representation for automated systematic review work prioritization. In: AMIA Annu. Symp. Proceed. **2008**, 121–125 (2008). American Medical Informatics Association (2008)
7. Cohen, A.M., Hersh, W.R., Peterson, K., Yen, P.Y.: Reducing workload in systematic review preparation using automated citation classification. J. Am. Med. Inf. Assoc. **13**(2), 206–219 (2006)
8. De, L., De-Moor, B., Vandewalle, J.: On the best rank-1 and rank-(r1 r2...rn) approximation of higher-order tensors. SIAM J. Matrix Anal. Appl. **21**(4), 1324–1342 (2000)
9. Dong, Y., Chawla, N.V., Swami, A.: metapath2vec: scalable representation learning for heterogeneous networks. In: Proceedings of the 23rd ACM SIGKDD International Conference on Knowledge Discovery and Data Mining, Halifax, NS, Canada, 13–17 August 2017 (2017)
10. Girvan, M., Newman, M.E.: Community structure in social and biological networks. Proceedings of the National Academy Of Sciences (12) (2002)
11. Gori, M., Monfardini, G., Scarselli, F.: A new model for learning in graph domains. In: Proceedings. 2005 IEEE International Joint Conference on Neural Networks, 2005. IEEE (2005)
12. Haddaway, N.R., Westgate, M.J.: Predicting the time needed for environmental systematic reviews and systematic maps. Conserv. Biol. **33**, 434–443 (2018)
13. Harshman, R.A., et al.: Foundations of the parafac procedure: models and conditions for an "explanatory" multimodal factor analysis (1970)
14. Howard, B.E., et al.: Swift-review: a text-mining workbench for systematic review. Syst. Control Found. Appl. **5**(1), 1–16 (2016)
15. Khabsa, M., Elmagarmid, A., Ilyas, I., Hammady, H., Ouzzani, M.: Learning to identify relevant studies for systematic reviews using random forest and external information. Mach. Learn. **102**(3), 465–482 (2015). https://doi.org/10.1007/s10994-015-5535-7
16. Kingma, D.P., Ba, J.: Adam: a method for stochastic optimization. In: Proceedings of ICLR (2015)
17. Kipf, T.N., Welling, M.: Semi-supervised classification with graph convolutional networks. In: Proceedings of ICLR (2017)
18. Kolda, T.G., Bader, B.W.: Tensor decompositions and applications. SIAM Rev. **51**(3), 455–500 (2009)

19. Kontonatsios, G., et al.: A semi-supervised approach using label propagation to support citation screening. J. Biomed. Inf. **72**, 67–76 (2017)

20. Lee, E.W., Wallace, B.C., Galaviz, K.I., Ho, J.C.: MMiDaS-AE: multi-modal missing data aware stacked autoencoder for biomedical abstract screening. In: Proceedings of the ACM Conference on Health, Inference, and Learning (2020)

21. Li, X., Wen, L., Qian, C., Wang, J.: GAHNE: graph-aggregated heterogeneous network embedding. In: 2020 IEEE 32nd International Conference on Tools with Artificial Intelligence (ICTAI). IEEE (2020)

22. Li, Y., Tarlow, D., Brockschmidt, M., Zemel, R.S.: Gated graph sequence neural networks. In: Proceedings of ICLR (2016)

23. Ma, T., Xiao, C., Zhou, J., Wang, F.: Drug similarity integration through attentive multi-view graph auto-encoders. In: Proceedings of the Twenty-Seventh IJCAI 2018, Stockholm, Sweden (2018)

24. Miwa, M., Thomas, J., O'Mara-Eves, A., Ananiadou, S.: Reducing systematic review workload through certainty-based screening. J. Biomed. Inf. **51**, 242–253 (2014)

25. Morris, Z.S., Wooding, S., Grant, J.: The answer is 17 years, what is the question: understanding time lags in translational research. J. Royal Soc. Med. **104**(12), 510–520 (2011)

26. Newman, M.E.J.: Detecting community structure in networks. Eur. Phys. J. B **38**(2), 321–330 (2004). https://doi.org/10.1140/epjb/e2004-00124-y

27. Newman, M.E.: Modularity and community structure in networks. Proceed. Nat. Acad. Sci. **103**(23), 8577–8582 (2006)

28. O'Mara-Eves, A., Thomas, J., McNaught, J., Miwa, M., Ananiadou, S.: Using text mining for study identification in systematic reviews: a systematic review of current approaches. Syst. Rev. **4**(1), 5 (2015). https://doi.org/10.1186/2046-4053-4-5

29. Papalexakis, E.E., Faloutsos, C., Sidiropoulos, N.D.: Tensors for data mining and data fusion: Models, applications, and scalable algorithms. ACM Trans. Intell. Syst. Technol. (TIST) **8**(2), 2915921 (2016)

30. Sarkar, S., Dong, A.: Community detection in graphs using singular value decomposition. Phys. Rev. E **83**(4), 046114 (2011)

31. Scarselli, F., Gori, M., Tsoi, A.C., Hagenbuchner, M., Monfardini, G.: The graph neural network model. IEEE Trans. Neural Netw. **20**(1), 61–80 (2008)

32. Scells, H., Zuccon, G., Koopman, B., Deacon, A., Azzopardi, L., Geva, S.: A test collection for evaluating retrieval of studies for inclusion in systematic reviews. In: Proceedings of the 40th International ACM SIGIR Conference on Research and Development in Information Retrieval, pp. 1237–1240 (2017)

33. Shang, J., Qu, M., Liu, J., Kaplan, L.M., Han, J., Peng, J.: Meta-path guided embedding for similarity search in large-scale heterogeneous information networks. ArXiv preprint (2016)

34. Tang, J., Qu, M., Wang, M., Zhang, M., Yan, J., Mei, Q.: LINE: large-scale information network embedding. In: Proceedings of the 24th International Conference on WWW, Florence, Italy, 18–22 May 2015 (2015)

35. Tucker, L.R.: Some mathematical notes on three-mode factor analysis. Psychometrika **31**(3), 279–311 (1966). https://doi.org/10.1007/BF02289464

36. Velickovic, P., Cucurull, G., Casanova, A., Romero, A., Liò, P., Bengio, Y.: Graph attention networks. In: Proceedings of ICLR (2018)

37. Wallace, B.C., Trikalinos, T.A., Lau, J., Brodley, C., Schmid, C.H.: Semi-automated screening of biomedical citations for systematic reviews. BMC Bioinf. **11**(1), 1–11 (2010)

38. Wang, X., et al.: Heterogeneous graph attention network. In: The World Wide Web Conference, WWW 2019, San Francisco, CA, USA, 13–17 May 2019 (2019)
39. Wanyan, T., Zhang, C., Azad, A., Liang, X., Li, D., Ding, Y.: Attribute2vec: Deep network embedding through multi-filtering GCN. ArXiv preprint (2020)
40. Yang, Y., Guan, Z., Li, J., Huang, J., Zhao, W.: Interpretable and efficient heterogeneous graph convolutional network. ArXiv preprint (2020)
41. Yun, S., Jeong, M., Kim, R., Kang, J., Kim, H.J.: Graph transformer networks. In: Advances in Neural Information Processing Systems 32: Annual Conference on Neural Information Processing Systems 2019, NeurIPS 2019, 8–14 Dec 2019, Vancouver, BC, Canada (2019)
42. Zhang, C., Song, D., Huang, C., Swami, A., Chawla, N.V.: Heterogeneous graph neural network. In: Proceedings of the 25th ACM SIGKDD International Conference on Knowledge Discovery & Data Mining, KDD 2019, Anchorage, AK, USA, 4–8 Aug 2019 (2019)
43. Zhang, X., He, L., Chen, K., Luo, Y., Zhou, J., Wang, F.: Multi-view graph convolutional network and its applications on neuroimage analysis for Parkinson's disease. In: AMIA Annu. Symp. Proceed. **2018**, 1147–1156 (2018). American Medical Informatics Association (2018)

Multimodal Inverse Cloze Task for Knowledge-Based Visual Question Answering

Paul Lerner[1]([✉]) [ID], Olivier Ferret[2] [ID], and Camille Guinaudeau[1] [ID]

[1] Université Paris-Saclay, CNRS, LISN, 91400 Orsay, France
{paul.lerner,camille.guinaudeau}@lisn.upsaclay.fr
[2] Université Paris-Saclay, CEA, List, 91120 Palaiseau, France
olivier.ferret@cea.fr

Abstract. We present a new pre-training method, Multimodal Inverse Cloze Task, for Knowledge-based Visual Question Answering about named Entities (KVQAE). KVQAE is a recently introduced task that consists in answering questions about named entities grounded in a visual context using a Knowledge Base. Therefore, the interaction between the modalities is paramount to retrieve information and must be captured with complex fusion models. As these models require a lot of training data, we design this pre-training task, which leverages contextualized images in multimodal documents to generate visual pseudo-questions. Our method is applicable to different neural network architectures and leads to a 9% relative-MRR and 15% relative-F1 gain for retrieval and reading comprehension, respectively, over a no-pre-training baseline.

Keywords: Visual question answering · Pre-training · Multimodal fusion

1 Introduction

Knowledge-based Visual Question Answering about named Entities (KVQAE) is a challenging task recently introduced in [50]. It consists in answering questions about named entities grounded in a visual context using a Knowledge Base (KB). Figure 1 provides two examples of visual questions along with relevant visual passages from a KB. To address the task, one must thus *retrieve* relevant information from a KB. This contrasts with standard Visual Question Answering (VQA [1]), where questions target the content of the image (e.g. the color of an object or the number of objects) or Knowledge-based VQA (about coarse-grained object categories) [40], where one can rely on off-the-shelf object detection [17].

This work was supported by the ANR-19-CE23-0028 MEERQAT project. This work was granted access to the HPC resources of IDRIS under the allocation 2021-AD011012846 made by GENCI. We thank the anonymous reviewers for their helpful comments.

J. Kamps et al. (Eds.): ECIR 2023, LNCS 13980, pp. 569–587, 2023.
https://doi.org/10.1007/978-3-031-28244-7_36

Visual Question (input)	Relevant visual passage in the Knowledge Base
"Which constituency did this man represent when he was Prime Minister?"	Macmillan indeed lost Stockton in the landslide Labour victory of 1945, but returned to Parliament in the November 1945 by-election in Bromley.
"In which year did this ocean liner make her maiden voyage?"	Queen Elizabeth 2, often referred to simply as QE2, is a floating hotel and retired ocean liner built for the Cunard Line which was operated by Cunard as both a transatlantic liner and a cruise ship from 1969 to 2008.

Fig. 1. Example of visual questions about named entities from the ViQuAE dataset along with relevant visual passages from its Knowledge Base [32].

In KVQAE, both text and image modalities bring useful information that must be combined. Therefore, the task is more broadly related to *Multimodal Information Retrieval* (IR) and *Multimodal Fusion.*

There are two paradigms for multimodal IR and for multimodal learning more generally: early fusion (data- and feature-level) and late fusion (score- and decision-level) [30]. On the one hand, late fusion is more straightforward as both Natural Language Processing and Computer Vision techniques can be applied independently. However, it neglects interactions between modalities. For example, in Fig. 1, attempting to recognize Harold Macmillan without accounting for him being a Prime Minister is suboptimal. On the other hand, the richness of early fusion often comes at the cost of increasing complexity and model parameters. This adds an extra challenge for KVQAE, where the two existing datasets are either small, because of a costly annotation process (ViQuAE [32]), or generated automatically (KVQA [50]), which leads to several limitations discussed in [32]. To address this challenge, we propose the multimodal Inverse Cloze Task (ICT) for pre-training the two early fusion models we define in this article for tackling KVQAE. Multimodal ICT consists in considering a sentence paired with a nearby image as a visual pseudo-question and its multimodal context as a relevant visual passage. It is related to the visual cloze task proposed by [61], a downstream task that requires modeling temporal events. Textual ICT was first introduced in [31] to pre-train a neural retriever for textual Question Answering (QA) and can be seen as a generalization of the skip-gram objective [41].

Our main contributions are: (i) Multimodal ICT, a new pre-training method that allows tackling KVQAE, even for small datasets as ViQuAE; (ii) a multimodal IR framework for KVQAE; (iii) experiments with different neural network architectures, including recently proposed multimodal BERTs.

2 Related Work

Dense Retrieval. Dense Retrieval is a rapidly evolving field, surveyed in [11,36], with new pre-training tasks, optimizing methods, and variants of the Transformer architecture emerging [14,15,23,47]. [31] were the first to outperform sparse bag-of-words representations such as BM25 with dense representations for QA. Their approach relies on three components: (i) pre-trained language models such as BERT [10], which allow to encode the semantic of a sentence in a dense vector; (ii) a contrastive learning objective that optimizes the similarities between questions' and text passages' embeddings (see Sect. 3); (iii) an unsupervised training task, ICT (see Sect. 1). [27] criticize the latter for being computationally intensive[1] and argue that regular sentences are not good surrogates of questions. Instead, they propose DPR, which takes advantage of (i) the heuristic of whether the passage contains the answer to the question to deem it relevant; (ii) unsupervised IR methods such as BM25 to mine hard negatives examples, which proved to be the key of their method's success. We aim to take advantage of both approaches by (i) pre-training our model on text QA datasets like DPR; (ii) incorporating multimodality into this hopefully-well-initialized model by adapting the ICT of [31] to multimodal documents.

Multimodal Fusion and Pre-training. The success of BERT in NLP [10], which relies on the easily-parallelizable Transformer architecture [57], an unsupervised training objective, and a task-agnostic architecture, has concurrently inspired many works in the VQA and cross-modal retrieval fields [7,34,35,38,55, 56]. These models are unified under a single framework in [5] and partly reviewed in [28]. All of these models rely on the Transformer architecture, often initialized with a pre-trained BERT, in order to fuse image and text. The training is weakly supervised, based upon image caption datasets such as COCO [37] or Conceptual Captions [51], and pre-trained object detectors like Faster R-CNN [48]. [22] show that these models learn nontrivial interactions between the modalities for VQA. Multimodal BERTs can be broadly categorized into *single-stream* and *multi-stream*. Single-stream models feed both text tokens' embeddings and image regions' embeddings to the same Transformer model, relying on the *self-attention* mechanism to fuse them. Instead, in the multi-stream architecture, text and image are first processed by two independent Transformers before using *cross-attention* to fuse the modalities. Both architectures have been shown to perform equally well in [5]. In this work, we use a single-stream model to take advantage of pre-training on text-only (on QA datasets). Also note that, while inspired by these work, we do not use the same training objectives or data, which are arguably unsuited for named entities' representations, as explained in the next section.

---

[1] [31] use a batch size of over 4K questions.

Multimodal Information Retrieval and KVQAE. Multimodal IR has largely been addressed using late fusion techniques (see [9] for a survey) but we are mostly interested in early fusion techniques in this work.

[9] review first attempts at early fusion. It was then systematically done by concatenating the features of both modalities in a single vector, with a focus on the feature weighting scheme. Concatenation is confronted with the curse of dimensionality as the resulting feature space equals the sum of the dimensions of each modality's features.

[44] and [39] concurrently proposed an approach quite similar to ours for Knowledge-based VQA. They adapt DPR [27] by replacing the question encoder with LXMERT [56], which allows to fuse the question and image. However, unlike us, they keep the passage encoder based on text-only and use the same pre-training objectives as [56], namely Masked Language Modeling, Masked Region Modeling, and Image-Text Matching. We expect that these objectives are suited to learn representations of coarse-grained object categories but not named entities. In other words, they are suited for standard VQA but not KVQAE. For example, Masked Region Modeling relies on an object detector, which is not applicable to KVQAE. While both [44] and [39] experiment on OK-VQA [40], their results are inconsistent: [44] show that their model is competitive with a BM25 baseline that takes as input the question and the *human-written* caption of the image while the model of [39] is outperformed by BM25 with an *automatically-generated* caption. The discrepancies between these works can be explained because they use neither the same KB nor the same evaluation metrics. [19] also experiment with different multimodal BERTs but dispense passage-level annotation for an end-to-end training of the retriever and answer classifier[2].

Although they experiment with KVQA [50], we do not consider the work of [16,21] as their systems take a *human-written* caption as input, which makes the role of the image content unclear. [50] follow a late fusion approach at the decision-level. First, they detect and disambiguate the named entity mentions in the question. Then, they rely on a face recognition step as their dataset, KVQA, is restricted to questions about person named entities. Facts from both textually- and visually-detected entities are retrieved from Wikidata[3] and processed by a memory network [59]. In contrast, our work is in line with [32], who use unstructured text from Wikipedia as KB. Unlike the late fusion approach of [32], which considers the question and the image independently, we aim at a unified representation of the text and image, both on the visual question and KB sides.

3   Methods

In this section, we first formalize our KVQAE framework, then describe the models before diving into the three training stages: (i) DPR for textual Question Answering; (ii) Multimodal Inverse Cloze Task, our main contribution;

---

[2] Standard (Knowledge-based) VQA is often treated as a classification task.

[3] https://www.wikidata.org/.

(iii) Fine-tuning for KVQAE. Finally, we discuss the inference mechanism and implementation details.

3.1   Information Retrieval Framework

In our multimodal setting, both visual questions (from the dataset) and visual passages (from the KB) consist of a text-image pair (t, i), as in Fig. 1. Our goal is to find the optimal model E to encode adequate representations $\mathbf{q} = E(t_q, i_q)$ and $\mathbf{p} = E(t_p, i_p)$ such that they are close if (t_p, i_p) is relevant for (t_q, i_q) (denoted with the superscripts $(^+)$ and $(^-)$). Search then boils down to retrieving the K closest visual passages to the visual question. When computing the similarity between two vectors, here with the dot product, the objective used throughout all the training stages (Sect. 3.3) is to minimize the following negative log-likelihood loss for all visual questions in the dataset (see Fig. 2) [27,31].

$$- \log \frac{\exp\left(\mathbf{q} \cdot \mathbf{p}^+\right)}{\exp\left(\mathbf{q} \cdot \mathbf{p}^+\right) + \sum_j \exp\left(\mathbf{q} \cdot \mathbf{p}_j^-\right)} \tag{1}$$

This contrastive objective allows to efficiently utilize passages relevant to other questions in the batch as *in-batch negatives*, since computing the similarity between two vectors is rather inexpensive compared to the forward pass of the whole model. We present two different models E in the next section according to their fusion mechanism.

3.2   Models

All of the models described in this section take advantage of CLIP[4] [46] to represent images and BERT[5] [10] to represent either text or multimodal data. BERT is trained for masked language modeling and next sentence prediction on Wikipedia and BooksCorpus [62]. CLIP has been trained with a contrastive objective in a weakly-supervised manner over 400M image and caption pairs. It has demonstrated better generalization capacities than fully-supervised models and is efficient for KVQAE, as empirically demonstrated in [32]. We experiment with two different fusion techniques: ECA and ILF.

Early Cross-Attention fusion (ECA) is carried out by a single-stream Transformer model like the multimodal BERTs described above (e.g. UNITER [7]). However, instead of relying on a fixed object detector such as Faster R-CNN, we take advantage of CLIP, as motivated above. To enable early fusion, the visual embedding produced by CLIP is projected in the same space as the text using a linear layer with $\mathbf{W}_c \in \mathbb{R}^{c \times d}$ parameters trained from scratch: $\mathbf{e}_c = \text{CLIP}(i) \cdot \mathbf{W}_c$. The resulting embedding is then concatenated with the word embeddings of the text, acting as an additional "visual token". Those embeddings, noted $[t; \mathbf{e}_c]$, are then fed to the Transformer model, where the attention mechanism should

---

[4] With a ResNet-50 backbone [20].

[5] Uncased "base" 12-layers version available at https://huggingface.co.

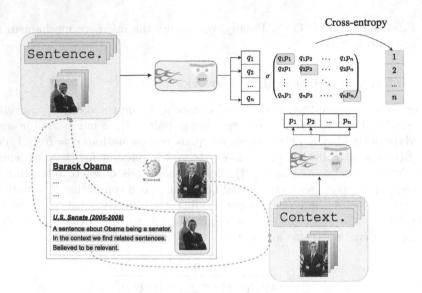

Fig. 2. Overview of Multimodal Inverse Cloze Task via Wikipedia/WIT.

enable interaction between the modalities. The final embedding corresponds to the special [CLS] token: $\mathrm{ECA}(t, i) = \mathrm{BERT}([t; \mathbf{e}_c])_{[\mathrm{CLS}]}$. The Transformer model is first initialized from BERT.

Intermediate Linear Fusion (ILF) introduces an additional $\mathbf{W}_t \in \mathbb{R}^{d \times d}$ parameters trained from scratch used to simply project the representation of the [CLS] token in the same space as the CLIP embedding before summing the two[6]: $\mathrm{ILF}(t, i) = \mathrm{BERT}(t)_{[\mathrm{CLS}]} \cdot \mathbf{W}_t + \mathbf{e}_c$.

Because both ECA and ILF produce multimodal representations $\mathbf{q}$ and $\mathbf{p}$, ranking is done directly using their similarity $\mathbf{q} \cdot \mathbf{p}$. As a baseline, we follow [32] and linearly combine text and image similarities after zero-mean and unit-variance normalization (omitted in the following equation):

$$\alpha \times \mathrm{BERT}(t_q)_{[\mathrm{CLS}]} \cdot \mathrm{BERT}(t_p)_{[\mathrm{CLS}]} + (1 - \alpha) \times \cos(\mathrm{CLIP}(i_q), \mathrm{CLIP}(i_p)) \quad (2)$$

The interpolation hyperparameter α is optimized on the validation set using grid search to maximize Mean Reciprocal Rank. The left term (text similarity) is referred to as DPR in the rest of the paper.

3.3   Training Stages

The models are trained sequentially in three stages. The first two stages correspond to pre-training stages: the first one is dedicated to IR for Question Answering while the second one focuses on multimodal IR. The last stage corresponds to the training for our target task of IR for KVQAE.

---

[6] Note that this is equivalent to concatenating both before projecting like $[\mathrm{BERT}(t)_{[\mathrm{CLS}]}; \mathrm{CLIP}(i)] \cdot [\mathbf{W}_t; \mathbf{W}_c]$.

Stage 1: DPR for Textual Question Answering. Leaving visual representations aside, a DPR model is trained starting from the BERT initialization [27].
DPR consists of two BERT encoders: one for the question t_q and one for the
text passage t_p. We use the model pre-trained by [32] on TriviaQA, filtered of all
questions used in their dataset, ViQuAE. They use the KILT [43] version of TriviaQA and Wikipedia, which serves as KB at this stage. Each article is then split
into disjoint passages of 100 words for text retrieval, while preserving sentence
boundaries, and the title of the article is appended to the beginning of each passage. This yields 32M passages, that is ≈ 5.4 passages per article. Following [27],
irrelevant passages (i.e. hard negatives) are mined using BM25 [49].

Stage 2: Multimodal Inverse Cloze Task. This is the main contribution
of the paper. We propose to extend the ICT of [31] to multimodal documents.
ICT consists in considering a sentence as a pseudo-question t_q and its context
as a relevant passage t_p^+. Note that the title of the article is appended to the
beginning of each passage t_p (as in Stage 1). We extend it using the contextual
images of Wikipedia paragraphs for the pseudo-question and the *infobox* image
for the passage (see Fig. 2). [31] empirically demonstrated that a key success of
their approach was to leave the pseudo-question in the relevant passage in 10%
of the training samples so that the model will learn to perform word matching,
as lexical overlap is ultimately a very useful feature for retrieval. In our case,
however, we argue that it is neither necessary, as the model should be strongly
initialized from Stage 1 training on TriviaQA, nor beneficial, as the model could
then ignore the image modality. Question and passage encoders pre-trained in
Stage 1 are used to initialize the visual question and visual passage encoders,
respectively.

The process is eased thanks to the WIT dataset [53]. WIT consists of millions of images with associated text from Wikipedia and Wikimedia Commons
in 108 different languages. We are, however, only interested in English for this
work. While [53] have multiple strategies to find text related to a given Wikipedia
image, such as its Commons' caption, we use only the contextual paragraph as text
source in order to mimic the downstream KVQAE setting. The resulting English
subset of WIT yields 400K *infobox* images/articles that correspond to 1.2M para-
graphs/images. Those 1.2M paragraphs consist of 13.6M sentences, i.e. potential
pseudo-questions, which are 26 words long on average. Therefore, to stick as close
as possible to stages 1 and 3, where passages are up to 100 *words* long, passages
consist of *four* sentences. This slightly differs from [31] who consider passages of
up to 288 *wordpieces*, *prior* to the pseudo-question masking.

Because both ViQuAE and WIT images are taken from Wikimedia Commons[7], we can estimate from the image URLs that 14% of ViQuAE images
overlap with WIT. This might lead to a bias that we analyze in Sect. 4.1.

Inspired by [2], to prevent *catastrophic forgetting* and enforce a *modality-
invariant* representation of the entities, the last l layers of BERT are frozen
during this stage. In this way, we tune only the first, modality-specific layers

---

[7] https://commons.wikimedia.org/.

of ECA, the intuition being to "replace" the text-named entities learned during Stage 1 with the "visual" entities present in the images. ILF fully freezes BERT during this stage, relying only on the $\mathbf{W}_t$ parameters to tune the text representation. Furthermore, CLIP is systematically frozen throughout all stages.

We do not have a straightforward way of mining irrelevant visual passages at this stage. In early experiments, we tried to synthesize them by permuting images in the batch: $(t_p^+, i_p^+) \leftarrow (t_p^+, i_p^-)$, but it did not improve the results.

After filtering corrupted images or images with inappropriate image formats (e.g. .svg) and paragraphs with a single sentence, we end up with 975K paragraphs/images. We refer to it as WIT in the rest of the paper. It is split into train (878K), validation (48K, to tune hyperparameters), and test (48K, as a sanity check) subsets such that there is no overlap between articles.

Stage 3: Knowledge-Based Visual Question Answering About Named Entities. This stage consists in fine-tuning the model on a downstream KVQAE dataset, which provides visual questions (t_q, i_q) and relevant visual passages (t_p^+, i_p^+). Following [2], all layers of the model are tuned during this stage.

A subtlety of this stage is the selection of irrelevant visual passages (t_p^-, i_p^-). As mentioned in Sect. 2, it was shown to be essential to DPR [27], and it is more generally important for contrastive learning [26]. In [32], irrelevant passages are mined with BM25 to train DPR. However, we suppose that this is suboptimal for ECA and ILF as BM25 will only mine textually-plausible passages but not visually-plausible ones. Therefore, we use the system provided by [32] to mine irrelevant passages. It is a late-fusion of DPR, ArcFace [8], CLIP, and ImageNet-ResNet [20]. This leads to different training setups between DPR (used as a baseline) and our models. However, we have experimented both for DPR and found no significant differences[8].

We use the same KB as [32], which is based upon KILT's Wikipedia and Wikidata images of the corresponding entities. It consists of 1.5M articles (thus images/entities) split into 12M passages of at most 100 words as in Stage 1.

Visual questions in ViQuAE are split into train (1,190), validation (1,250), and test (1,257) without overlap between images' URLs [32]. We do not experiment with KVQA [50] for the following reasons: (i) it is generated automatically from Wikidata so our text-based KB has a poor coverage of the answers; (ii) it comprises yes/no questions for which passage relevance cannot be assessed automatically.

3.4 Inference

For efficient retrieval, every passage in the KB is embedded along with its corresponding image by the visual passage encoder beforehand. Given a question grounded in an image, both are embedded by the visual question encoder. Search is then carried out with maximum inner product search using Faiss [25].

---

[8] Evaluation methods are detailed in Sect. 4.

3.5 Implementation Details

Our code is built upon PyTorch [42], Hugging Face's `transformers` [60] and `datasets` [33] (itself wrapping Faiss). It is freely available along with the data and trained models[9].

To train ECA, we use the same hyperparameters as [32] for DPR, themselves based upon [27]. In particular, we use a learning rate of 2×10^{-5} along with the Adam optimizer [29]. It is scheduled linearly with 100 and 4 warm-up steps for stages 2 and 3, respectively. However, for ILF, we found, based on the validation set, that it converged faster with a learning rate of 2×10^{-3} and a constant scheduler during Stage 2. We believe this is because ILF fully freezes BERT in Stage 2, so it does not require careful scheduling or a small learning rate. Dropout [54] is applied in BERT and after projecting embedding with $\mathbf{W}_c$ and $\mathbf{W}_t$ with a probability of 0.1 (as in the standard BERT configuration). Likewise, layer normalization [3] is applied in BERT and after summing the two embeddings in ILF. Gradients' norms are clipped at 2.

Models in stages 2 and 3 are trained with a batch size of 512 and 298 visual questions, respectively. The success of contrastive learning partly relies on a large number of *in-batch negatives* and, therefore, a large batch size [45]. We found that gradient checkpointing [6] enables the use of much larger batch sizes. Instead of [32] who use *four* NVIDIA V100 GPUs with 32 GB of RAM each for a total batch size of 128 questions, we are able to fit a batch of 298 questions (as stated above) in a *single* V100 GPU. Stage 2 takes most of the compute budget, with most models converging after $\approx$ 8K steps, which takes around three days[10]. Checkpoint selection is made based on the validation *in-batch* Mean Reciprocal Rank, for all stages. In-batch means that only the other visual passages in the batch are ranked and that each visual question is paired with only one relevant visual passage (as during training).

4 Results

The retrieval models are evaluated in two different ways: (i) by computing standard IR metrics on visual passage retrieval; (ii) by feeding retrieved visual passages to a reader module that is tasked with extracting the concise answer to the question, thus achieving KVQAE. Put differently, either evaluate whether the system is able to retrieve a *relevant passage* for the question or whether it is able to *answer* the question. We find both metrics to correlate. Ablation studies are carried out with IR metrics.

ViQuAE is based upon TriviaQA, so it is only distantly supervised: the answer is considered correct if it string-matches the ground truth and, likewise, a passage

---

[9] https://github.com/PaulLerner/ViQuAE.

[10] Jean Zay GPUs consume 0.482 kW (or 0.259 kW after heat recovery) in France, which has an average grid emission factor of 0.0569 kgCO2e/kWh according to https://bilans-ges.ademe.fr/en.

Table 1. IR evaluation on ViQuAE. l: Number of frozen layers during Multimodal ICT. Superscripts denote significant differences in Fisher's randomization test with $p \leq 0.01$. Hits@1 is omitted as it is equivalent to P@1.

#	Model	Multimodal ICT	MRR@100	P@1	P@20	Hits@20
a	DPR	NA	32.8	22.8	16.4	61.2
b	DPR + CLIP	NA	34.5^a	24.8^a	15.8	61.8
c	ECA (l = NA)	✗	34.6	25.9^a	17.2^{ab}	61.6
d	ECA (l = 6)	✓	$\mathbf{37.8}^{abce}$	26.7^a	$\mathbf{19.5}^{abce}$	$\mathbf{67.6}^{abce}$
e	ECA (l = 0)	✓	35.1	24.7	17.6^b	63.7
f	ILF (l = 12)	✓	37.3^a	$\mathbf{26.8}^a$	19.1^{abce}	66.9^{abc}

is deemed relevant if it contains the ground truth[11]. Moreover, Wikipedia aliases of the ground truth are considered to be valid answers.

4.1 Information Retrieval

Because of the setting of ViQuAE, it is impossible to get complete coverage of relevant passages. Therefore we do not use any metric based on recall (e.g. R-Precision, mAP, etc.). Instead, we evaluate the models with Precision@K (P@K), Mean Reciprocal Rank (MRR), and Hits@K. Hits@K is the proportion of questions for which IR retrieves *at least one* relevant passage in top-K. Statistical significance tests are conducted using Fisher's randomization test [12,52]. Metrics and statistical tests are computed with ranx [4] and are reported in Table 1.

The best models pre-trained with Multimodal ICT (d and f) outperform the text-only (a) and late-fusion (b) baselines on all metrics. Some qualitative examples are shown in Fig. 3. In the first row, we can see evidence of *cross-input* cross-modal interactions between the image depicting Winston Churchill and the passage that mentions him (while being illustrated by a totally different image). In contrast, the late fusion baseline exhibits textual bias by returning a passage that mentions several English palaces (highlighted in red). The same observation can be made for the second row, where St Paul's Cathedral is only mentioned in the relevant passage but not depicted in the contextual image. Cross-modal interactions prove useful in this case because of the heterogeneity of visual depictions: Winston Churchill is depicted by a statue in the visual question but by a photograph in the KB.

We can see that Multimodal ICT is essential to ECA (c vs. d). Without it, it performs on par with late fusion. We believe this is because of overfitting on the small training set of ViQuAE. However, we find that fine-tuning on ViQuAE is also essential to ECA, which exhibits catastrophic forgetting because of the sequential learning setup: indeed, after Stage 2, it falls behind DPR (not shown in the table). We see that the freezing technique of [2] helps to prevent catastrophic forgetting to some extent (d vs. e). It is also visible in the upstream WIT pre-training where ECA achieves 91.6 and 92.9 in-batch MRR on WIT's test set

---

[11] After standard preprocessing (lowercasing, stripping articles, and punctuation).

Visual Question	ECA top-1	DPR + CLIP top-1
"In which English palace was this man born?"	Blenheim Palace was the birthplace of the 1st Duke's famous descendant, *Winston Churchill* [...]	In 1762, George purchased Buckingham House (on the site now occupied by Buckingham Palace) for use as a family retreat. His other residences were Kew and Windsor Castle. St James's Palace was retained for official use.
"Who designed this cathedral?"	He was appointed [...] Surveyor of the Fabric of *St Paul's Cathedral*, where he was responsible for maintaining the building designed by Sir Christopher Wren.	Sir George Gilbert Scott led the restoration of Salisbury Cathedral between 1863 – 1878. It was during this time that Skidmore created the cathedral's choir screen.

Fig. 3. Qualitative examples where ECA ($l = 6$) finds a relevant visual passage in top-1 but late fusion falls behind (column on the right). We can see evidence of *cross-input* cross-modal interactions with ECA.

Table 2. In-batch results (re-ranking 1024 visual passages) on the upstream WIT test set.

#	Model	In-batch MRR	In-batch P@1
d	ECA ($l = 6$)	91.6	86.6
e	ECA ($l = 0$)	**92.9**	**88.3**
f	ILF ($l = 12$)	87.1	79.9

with $l = 6$ and $l = 0$, respectively: fitting WIT better leads to further forgetting (cf. Table 2).

Unlike what is suggested by related work (Sect. 2), we find that the linear fusion model performs on par with the more early, cross-attention based, fusion model (f vs. d in Table 1). This suggests that the improvement over the late fusion baseline indeed comes from the Multimodal ICT pre-training, which is not very sensitive to the model's architecture. Interestingly, *cross-input* cross-modal interactions (as shown in Fig. 3) are possible with both ECA and ILF. So they may be the primary reason for performance improvement. Moreover, the architecture of ILF allows to fully freeze BERT during Stage 2, which circumvents catastrophic forgetting[12]. We leave other training strategies (e.g. multi-tasking, using adapters [24]) for future work.

---

[12] ILF only achieves 87.1 in-batch MRR on WIT's test set because of the freezing.

Table 3. Reading Comprehension evaluation on ViQuAE, averaged over 5 runs of the *reader*. *l*: Number of frozen layers during Multimodal ICT.

#	IR Model	Multimodal ICT	Exact Match	F1
a	DPR	NA	16.9 ± 0.4	20.1 ± 0.5
b	DPR + CLIP	NA	19.0 ± 0.4	22.3 ± 0.4
c	ECA (l = NA)	✗	17.7 ± 0.6	21.2 ± 0.8
d	ECA (l = 6)	✓	20.6 ± 0.3	24.4 ± 0.2
e	ECA (l = 0)	✓	20.8 ± 0.8	24.3 ± 0.9
f	ILF (l = 12)	✓	$\mathbf{21.3 \pm 0.6}$	$\mathbf{25.4 \pm 0.3}$

Nothing suggests that ECA is better on the 14% of ViQuAE images that overlap with WIT. ECA is better on the out-of-WIT subset (38.0 vs. 36.5 MRR), but it is the other way around for DPR and late fusion.

4.2   Reading Comprehension

To extract the answers from the retrieved passages, we keep the same model as [32]. It uses the Multi-passage BERT architecture [58] and is thus based on text only because *once the relevant passage has been retrieved*, the question may be answered without looking at the image. To limit the variations due to training and the number of experiments, we use the model trained by [32] off-the-shelf and simply change its input passages. It takes the top-24 passages as input. The model was first trained on TriviaQA (filtered of all questions used in ViQuAE), then fine-tuned on ViQuAE, much like stages 1 and 3. The authors provide *five* different versions of the model that correspond to different random seeds.

We use Exact Match and F1-score (at the bag-of-words level) to evaluate the extracted answers. In Table 3 we can verify that more relevant passages indeed lead to better downstream answers. The only difference with the IR evaluation is the role of the freezing technique of [2] (d vs. e), which is less clear here.

5   Generic vs. Specialized Image Representations

Numbers reported in the previous section are actually on par with the best results of [32]. This is because the latter is based on ArcFace and ImageNet-ResNet, in addition to DPR and CLIP. In particular, [32] have a heuristic for taking advantage of the face representations provided by ArcFace: they use ArcFace if faces are detected and a combination of CLIP and ImageNet-ResNet otherwise. They show that this method improves retrieval precision for questions about persons (for which face representations are relevant). However, this approach is not scalable for two reasons: (i) there are near 1,000 different entity types in ViQuAE (according to Wikidata's ontology), and not all can benefit from specialized representations; (ii) combining several representations (e.g. CLIP and

ImageNet-ResNet) for the same entity type is computationally expensive and quickly saturates. To provide a comparable system to the late fusion of [32], we have tried integrating ArcFace and ImageNet-ResNet in ECA. However, we have failed to outperform the CLIP-only version of ECA. Intuitively, we think that ECA dilutes the specialized representations of ArcFace and is unable to preserve them throughout all twelve layers of BERT. Therefore, in this setting, ECA is overall on par with late fusion (37.7 vs. 37.9 MRR, not significant) but better on questions about non-persons (39.3 vs. 35.7 MRR), which again suggests that it is unable to exploit ArcFace's representations.

6 Conclusion and Perspectives

We have presented a new pre-training method, Multimodal Inverse Cloze Task, for Knowledge-based Visual Question Answering about Named Entities. Multimodal ICT leverages contextual images in multimodal documents to generate visual pseudo-questions. It enables the use of more complex multimodal fusion models than previously proposed late fusion methods. Consequently, our method improves retrieval accuracy over the latter by 10% relative-MRR, leading to a 9% relative-F1 improvement in downstream reading comprehension (i.e. answer extraction), on the recently introduced ViQuAE dataset. We believe it is thanks to cross-modal interactions, which are prohibited by late fusion. More precisely, we qualitatively observed that *cross-input* interactions occurred between the image of the visual question and the text of the KB, which counteracts the heterogeneity of visual depictions.

We have experimented our pre-training method with two different neural networks architectures: (i) ECA, which follows recently proposed Multimodal BERTs by fusing modalities Early via Cross-Attention; (ii) ILF, a more standard model that fuses modalities through a linear projection. We found that both perform equally well, unlike in standard VQA and cross-modal retrieval. We argue that it might be because *cross-input* cross-modal interactions are the most important and may be captured by both models. However, further investigations are required, it might also be because of their different training settings, which leads ECA to catastrophic forgetting.

While aiming for generic multimodal representations of named entities, we found that integrating specialized representations in our models, such as ArcFace for faces, was not beneficial. We hypothesize that the studied architectures may be inappropriate but we leave this issue for future studies.

For future work, we think that generalizing Multimodal ICT for re-ranking (processing (t_q, i_q) and (t_p, i_p) simultaneously) and reading comprehension (generating or extracting the answer from (t_p, i_p)) is an exciting research lead. Indeed, there is evidence that sharing the same model for IR and reading comprehension, or IR and re-ranking, is beneficial for textual QA [13] and cross-modal retrieval [18], respectively: two tasks that closely relate to KVQAE.

References

1. Antol, S., et al.: VQA: visual question answering. In: 2015 IEEE International Conference on Computer Vision (ICCV), pp. 2425–2433. IEEE, Santiago, December 2015. https://doi.org/10.1109/ICCV.2015.279, http://ieeexplore.ieee.org/document/7410636/

2. Aytar, Y., Castrejon, L., Vondrick, C., Pirsiavash, H., Torralba, A.: Cross-modal scene networks. IEEE Trans. Pattern Anal. Mach. Intell. **40**(10), 2303–2314 (2017). Publisher: IEEE

3. Ba, J.L., Kiros, J.R., Hinton, G.E.: Layer normalization. arXiv:1607.06450 [cs, stat], July 2016. http://arxiv.org/abs/1607.06450, arXiv: 1607.06450

4. Bassani, E.: `ranx`: A blazing-fast Python library for ranking evaluation and comparison. In: Hagen, M., et al. (eds.) ECIR 2022. LNCS, vol. 13186, pp. 259–264. Springer, Cham (2022). https://doi.org/10.1007/978-3-030-99739-7_30

5. Bugliarello, E., Cotterell, R., Okazaki, N., Elliott, D.: Multimodal pretraining unmasked: a meta-analysis and a unified framework of vision-and-language BERTs. Trans. Assoc. Comput. Linguist. **9**, 978–994 (2021). https://doi.org/10.1162/tacl_a_00408

6. Chen, T., Xu, B., Zhang, C., Guestrin, C.: Training deep nets with sublinear memory cost, April 2016 https://doi.org/10.48550/arXiv.1604.06174, http://arxiv.org/abs/1604.06174, number: arXiv:1604.06174 [cs]

7. Chen, Y.-C., et al.: UNITER: UNiversal image-TExt representation learning. In: Vedaldi, A., Bischof, H., Brox, T., Frahm, J.-M. (eds.) ECCV 2020. LNCS, vol. 12375, pp. 104–120. Springer, Cham (2020). https://doi.org/10.1007/978-3-030-58577-8_7

8. Deng, J., Guo, J., Xue, N., Zafeiriou, S.: Arcface: additive angular margin loss for deep face recognition. In: Proceedings of the IEEE/CVF Conference on Computer Vision and Pattern Recognition (CVPR), June 2019. https://openaccess.thecvf.com/content_CVPR_2019/html/Deng_ArcFace_Additive_Angular_Margin_Loss_for_Deep_Face_Recognition_CVPR_2019_paper.html

9. Depeursinge, A., Müller, H.: Fusion techniques for combining textual and visual information retrieval. In: Müller, H., Clough, P., Deselaers, T., Caputo, B. (eds.) ImageCLEF: Experimental Evaluation in Visual Information Retrieval. The Information Retrieval Series, pp. 95–114, Springer, Heidelberg (2010). https://doi.org/10.1007/978-3-642-15181-1_6

10. Devlin, J., Chang, M.W., Lee, K., Toutanova, K.: BERT: pre-training of deep bidirectional transformers for language understanding. In: Proceedings of the 2019 Conference of the North American Chapter of the Association for Computational Linguistics: Human Language Technologies, Volume 1 (Long and Short Papers), pp. 4171–4186. Association for Computational Linguistics, Minneapolis, June 2019. https://doi.org/10.18653/v1/N19-1423, https://aclanthology.org/N19-1423

11. Fan, Y., et al.: Pre-training methods in information retrieval. Found. Trends® Inf. Retrieval **16**(3), 178–317 (2022). https://doi.org/10.1561/1500000100

12. Fisher, R.A.: The Design of Experiments, 2nd edn. Oliver & Boyd, Edinburgh & London (1937). https://www.cabdirect.org/cabdirect/abstract/19371601600

13. Fun, H., Gandhi, S., Ravi, S.: Efficient retrieval optimized multi-task learning. arXiv:2104.10129 [cs], April 2021. http://arxiv.org/abs/2104.10129, arXiv: 2104.10129

14. Gao, L., Callan, J.: Condenser: a pre-training architecture for dense retrieval. In: Proceedings of the 2021 Conference on Empirical Methods in Natural Language Processing, pp. 981–993. Association for Computational Linguistics, Online and Punta Cana, Dominican Republic, November 2021. https://aclanthology.org/2021.emnlp-main.75

15. Gao, L., Callan, J.: Unsupervised corpus aware language model pre-training for dense passage retrieval. In: Proceedings of the 60th Annual Meeting of the Association for Computational Linguistics (Volume 1: Long Papers), pp. 2843–2853. Association for Computational Linguistics, Dublin, May 2022. https://doi.org/10.18653/v1/2022.acl-long.203, https://aclanthology.org/2022.acl-long.203

16. Garcia-Olano, D., Onoe, Y., Ghosh, J.: Improving and diagnosing knowledge-based visual question answering via entity enhanced knowledge injection. In: Companion Proceedings of the Web Conference 2022. WWW 2022, pp. 705–715. Association for Computing Machinery, New York (2022). https://doi.org/10.1145/3487553.3524648

17. Gardères, F., Ziaeefard, M.: ConceptBert: concept-aware representation for visual question answering. In: Findings of the Association for Computational Linguistics: EMNLP 2020, p. 10 (2020). https://aclanthology.org/2020.findings-emnlp.44/

18. Geigle, G., Pfeiffer, J., Reimers, N., Vulić, I., Gurevych, I.: Retrieve fast, rerank smart: cooperative and joint approaches for improved cross-modal retrieval. Trans. Assoc. Comput. Linguist. **10**, 503–521 (2022). https://doi.org/10.1162/tacl_a_00473

19. Guo, Y., Nie, L., Wong, Y., Liu, Y., Cheng, Z., Kankanhalli, M.: A unified end-to-end retriever-reader framework for knowledge-based VQA. In: Proceedings of the 30th ACM International Conference on Multimedia. MM 2022, pp. 2061–2069. Association for Computing Machinery, New York (2022). https://doi.org/10.1145/3503161.3547870

20. He, K., Zhang, X., Ren, S., Sun, J.: Deep residual learning for image recognition. In: Proceedings of the IEEE Conference on Computer Vision and Pattern Recognition, pp. 770–778 (2016). https://openaccess.thecvf.com/content_cvpr_2016/papers/He_Deep_Residual_Learning_CVPR_2016_paper.pdf

21. Heo, Y.J., Kim, E.S., Choi, W.S., Zhang, B.T.: Hypergraph transformer: weakly-supervised multi-hop reasoning for knowledge-based visual question answering. In: Proceedings of the 60th Annual Meeting of the Association for Computational Linguistics (Volume 1: Long Papers), pp. 373–390. Association for Computational Linguistics, Dublin, May 2022. https://doi.org/10.18653/v1/2022.acl-long.29, https://aclanthology.org/2022.acl-long.29

22. Hessel, J., Lee, L.: Does my multimodal model learn cross-modal interactions? It's harder to tell than you might think! In: Proceedings of the 2020 Conference on Empirical Methods in Natural Language Processing (EMNLP), pp. 861–877. Association for Computational Linguistics, Online, November 2020. https://doi.org/10.18653/v1/2020.emnlp-main.62, https://aclanthology.org/2020.emnlp-main.62

23. Hofstätter, S., Lin, S.C., Yang, J.H., Lin, J., Hanbury, A.: Efficiently teaching an effective dense retriever with balanced topic aware sampling. In: Proceedings of the 44th International ACM SIGIR Conference on Research and Development in Information Retrieval. SIGIR 2021, pp. 113–122. Association for Computing Machinery, New York (2021). https://doi.org/10.1145/3404835.3462891

24. Houlsby, N., et al.: Parameter-efficient transfer learning for NLP. In: Proceedings of the 36th International Conference on Machine Learning, pp. 2790–2799. PMLR, May 2019. https://proceedings.mlr.press/v97/houlsby19a.html, ISSN 2640-3498

25. Johnson, J., Douze, M., Jégou, H.: Billion-scale similarity search with GPUs. IEEE Trans. Big Data **7**(3), 535–547 (2019). https://doi.org/10.1109/TBDATA.2019.2921572

26. Kalantidis, Y., Sariyildiz, M.B., Pion, N., Weinzaepfel, P., Larlus, D.: Hard negative mixing for contrastive learning. In: Advances in Neural Information Processing Systems, vol. 33, pp. 21798–21809. Curran Associates, Inc. (2020). https://proceedings.neurips.cc/paper/2020/hash/f7cade80b7cc92b991cf4d2806d6bd78-Abstract.html

27. Karpukhin, V., et al.: Dense passage retrieval for open-domain question answering. In: Proceedings of the 2020 Conference on Empirical Methods in Natural Language Processing (EMNLP), pp. 6769–6781. Association for Computational Linguistics, Online, November 2020. https://www.aclweb.org/anthology/2020.emnlp-main.550

28. Khan, S., Naseer, M., Hayat, M., Zamir, S.W., Khan, F.S., Shah, M.: Transformers in vision: a survey. ACM Comput. Surv. (2021). https://doi.org/10.1145/3505244, just Accepted

29. Kingma, D.P., Ba, J.: Adam: a method for stochastic optimization. In: ICLR (Poster) (2015). http://arxiv.org/abs/1412.6980

30. Kludas, J., Bruno, E., Marchand-Maillet, S.: Information fusion in multimedia information retrieval. In: Boujemaa, N., Detyniecki, M., Nürnberger, A. (eds.) AMR 2007. LNCS, vol. 4918, pp. 147–159. Springer, Heidelberg (2008). https://doi.org/10.1007/978-3-540-79860-6_12

31. Lee, K., Chang, M.W., Toutanova, K.: Latent retrieval for weakly supervised open domain question answering. In: Proceedings of the 57th Annual Meeting of the Association for Computational Linguistics, pp. 6086–6096. Association for Computational Linguistics, Florence, Italy, July 2019. https://doi.org/10.18653/v1/P19-1612, https://aclanthology.org/P19-1612

32. Lerner, P., et al.: ViQuAE, a dataset for knowledge-based visual question answering about named entities. In: Proceedings of The 45th International ACM SIGIR Conference on Research and Development in Information Retrieval. SIGIR 2022. Association for Computing Machinery, New York (2022). https://doi.org/10.1145/3477495.3531753, https://hal.archives-ouvertes.fr/hal-03650618

33. Lhoest, Q., et al.: Datasets: a community library for natural language processing. In: Proceedings of the 2021 Conference on Empirical Methods in Natural Language Processing: System Demonstrations, pp. 175–184. Association for Computational Linguistics, Online and Punta Cana, Dominican Republic, November 2021. https://aclanthology.org/2021.emnlp-demo.21

34. Li, G., Duan, N., Fang, Y., Gong, M., Jiang, D.: Unicoder-VL: a universal encoder for vision and language by cross-modal pre-training. In: Proceedings of the AAAI Conference on Artificial Intelligence, vol. 34, no. 07, pp. 11336–11344, April 2020. https://doi.org/10.1609/aaai.v34i07.6795, https://ojs.aaai.org/index.php/AAAI/article/view/6795, number: 07

35. Li, L.H., Yatskar, M., Yin, D., Hsieh, C.J., Chang, K.W.: Visualbert: A simple and performant baseline for vision and language (2019). https://doi.org/10.48550/ARXIV.1908.03557, https://arxiv.org/abs/1908.03557

36. Lin, J., Nogueira, R., Yates, A.: Pretrained transformers for text ranking: BERT and beyond. Synth. Lect. Hum. Lang. Technol. **14**(4), 1–325 (2021). https://doi.org/10.2200/S01123ED1V01Y202108HLT053

37. Lin, T.-Y., et al.: Microsoft COCO: common objects in context. In: Fleet, D., Pajdla, T., Schiele, B., Tuytelaars, T. (eds.) ECCV 2014. LNCS, vol. 8693, pp. 740–755. Springer, Cham (2014). https://doi.org/10.1007/978-3-319-10602-1_48

38. Lu, J., Batra, D., Parikh, D., Lee, S.: ViLBERT: pretraining task-agnostic visiolinguistic representations for vision-and-language tasks. In: Advances in Neural Information Processing Systems, vol. 32, pp. 13–23 (2019). https://proceedings.neurips.cc/paper/2019/hash/c74d97b01eae257e44aa9d5bade97baf-Abstract.html

39. Luo, M., Zeng, Y., Banerjee, P., Baral, C.: Weakly-supervised visual-retriever-reader for knowledge-based question answering. In: Proceedings of the 2021 Conference on Empirical Methods in Natural Language Processing, pp. 6417–6431. Association for Computational Linguistics, Online and Punta Cana, Dominican Republic, November 2021. https://doi.org/10.18653/v1/2021.emnlp-main.517, https://aclanthology.org/2021.emnlp-main.517

40. Marino, K., Rastegari, M., Farhadi, A., Mottaghi, R.: OK-VQA: a visual question answering benchmark requiring external knowledge. In: Proceedings of the IEEE Conference on Computer Vision and Pattern Recognition, pp. 3195–3204 (2019). https://ieeexplore.ieee.org/document/8953725/

41. Mikolov, T., Sutskever, I., Chen, K., Corrado, G.S., Dean, J.: Distributed representations of words and phrases and their compositionality. In: Advances in Neural Information Processing Systems, vol. 26 (2013). https://papers.neurips.cc/paper/2013/hash/9aa42b31882ec039965f3c4923ce901b-Abstract.html

42. Paszke, A., et al.: PyTorch: an imperative style, high-performance deep learning library. In: Advances in Neural Information Processing Systems, vol. 32 (2019). https://papers.nips.cc/paper/2019/hash/bdbca288fee7f92f2bfa9f7012727740-Abstract.html

43. Petroni, F., et al.: KILT: a benchmark for knowledge intensive language tasks. In: Proceedings of the 2021 Conference of the North American Chapter of the Association for Computational Linguistics: Human Language Technologies, pp. 2523–2544. Association for Computational Linguistics, Online, June 2021. https://doi.org/10.18653/v1/2021.naacl-main.200, https://aclanthology.org/2021.naacl-main.200

44. Qu, C., Zamani, H., Yang, L., Croft, W.B., Learned-Miller, E.: Passage retrieval for outside-knowledge visual question answering. In: Proceedings of the 44th International ACM SIGIR Conference on Research and Development in Information Retrieval. SIGIR 2021, pp. 1753–1757. Association for Computing Machinery, New York (2021). https://doi.org/10.1145/3404835.3462987

45. Qu, Y., et al.: RocketQA: an optimized training approach to dense passage retrieval for open-domain question answering. In: Proceedings of the 2021 Conference of the North American Chapter of the Association for Computational Linguistics: Human Language Technologies, pp. 5835–5847. Association for Computational Linguistics, Online, June 2021. https://doi.org/10.18653/v1/2021.naacl-main.466, https://aclanthology.org/2021.naacl-main.466

46. Radford, A., et al.: Learning transferable visual models from natural language supervision. In: International Conference on Machine Learning, pp. 8748–8763. PMLR (2021)

47. Ram, O., Shachaf, G., Levy, O., Berant, J., Globerson, A.: Learning to retrieve passages without supervision. In: Proceedings of the 2022 Conference of the North American Chapter of the Association for Computational Linguistics: Human Language Technologies, pp. 2687–2700. Association for Computational Linguistics, Seattle, July 2022. https://aclanthology.org/2022.naacl-main.193

48. Ren, S., He, K., Girshick, R., Sun, J.: Faster R-CNN: towards real-time object detection with region proposal networks. In: Advances in Neural Information Processing Systems, vol. 28, pp. 91–99 (2015). https://proceedings.neurips.cc/paper/2015/hash/14bfa6bb14875e45bba028a21ed38046-Abstract.html

49. Robertson, S.E., Walker, S., Jones, S., Hancock-Beaulieu, M.M., Gatford, M.: Okapi at TREC-3. In: Harman, D.K. (ed.) Third Text REtrieval Conference (TREC-3). NIST Special Publication, vol. 500–225, pp. 109–126. National Institute of Standards and Technology (NIST) (1995)

50. Shah, S., Mishra, A., Yadati, N., Talukdar, P.P.: KVQA: knowledge-aware visual question answering. In: Proceedings of the AAAI Conference on Artificial Intelligence, vol. 33, pp. 8876–8884 (2019). https://144.208.67.177/ojs/index.php/AAAI/article/view/4915

51. Sharma, P., Ding, N., Goodman, S., Soricut, R.: Conceptual captions: a cleaned, hypernymed, image alt-text dataset for automatic image captioning. In: Proceedings of the 56th Annual Meeting of the Association for Computational Linguistics (Volume 1: Long Papers), pp. 2556–2565. Association for Computational Linguistics, Melbourne, Australia, July 2018. https://doi.org/10.18653/v1/P18-1238, https://aclanthology.org/P18-1238

52. Smucker, M.D., Allan, J., Carterette, B.: A comparison of statistical significance tests for information retrieval evaluation. In: Proceedings of the Sixteenth ACM Conference On Conference on Information and Knowledge Management. CIKM 2007, pp. 623–632. Association for Computing Machinery, New York, November 2007. https://doi.org/10.1145/1321440.1321528

53. Srinivasan, K., Raman, K., Chen, J., Bendersky, M., Najork, M.: Wit: Wikipedia-based image text dataset for multimodal multilingual machine learning. In: Proceedings of the 44th International ACM SIGIR Conference on Research and Development in Information Retrieval. SIGIR 2021, pp. 2443–2449. Association for Computing Machinery, New York (2021). https://doi.org/10.1145/3404835.3463257

54. Srivastava, N., Hinton, G., Krizhevsky, A., Sutskever, I., Salakhutdinov, R.: Dropout: a simple way to prevent neural networks from overfitting. J. Mach. Learn. Res. 15(1), 1929–1958 (2014). Publisher: JMLR.org

55. Su, W., et al.: Vl-BERT: pre-training of generic visual-linguistic representations. In: International Conference on Learning Representations (2020). https://openreview.net/forum?id=SygXPaEYvH

56. Tan, H., Bansal, M.: LXMERT: learning cross-modality encoder representations from transformers. In: Proceedings of the 2019 Conference on Empirical Methods in Natural Language Processing and the 9th International Joint Conference on Natural Language Processing (EMNLP-IJCNLP), pp. 5100–5111. Association for Computational Linguistics, Hong Kong, November 2019. https://doi.org/10.18653/v1/D19-1514, https://www.aclweb.org/anthology/D19-1514

57. Vaswani, A., et al.: Attention is all you need. In: Advances in Neural Information Processing Systems, pp. 5998–6008 (2017)

58. Wang, Z., Li, L., Li, Q., Zeng, D.: Multimodal data enhanced representation learning for knowledge graphs. In: 2019 International Joint Conference on Neural Networks (IJCNN), pp. 1–8, July 2019). https://doi.org/10.1109/IJCNN.2019.8852079, Issn 2161-4407

59. Weston, J., Chopra, S., Bordes, A.: Memory networks (2014). https://doi.org/10.48550/ARXIV.1410.3916, https://arxiv.org/abs/1410.3916

60. Wolf, T., et al.: HuggingFace's transformers: state-of-the-art natural language processing. arXiv:1910.03771 [cs], July 2020. http://arxiv.org/abs/1910.03771

61. Zhang, H., et al.: Modeling temporal-modal entity graph for procedural multimodal machine comprehension. In: Proceedings of the 60th Annual Meeting of the Association for Computational Linguistics (Volume 1: Long Papers), pp. 1179–1189. Association for Computational Linguistics, Dublin, May 2022. https://doi.org/10.18653/v1/2022.acl-long.84, https://aclanthology.org/2022.acl-long.84

62. Zhu, Y., et al.: Aligning books and movies: towards story-like visual explanations by watching movies and reading books. In: Proceedings of the IEEE International Conference on Computer Vision (ICCV), December 2015. https://www.cv-foundation.org/openaccess/content_iccv_2015/html/Zhu_Aligning_Books_and_ICCV_2015_paper.html

A Transformer-Based Framework
for POI-Level Social Post Geolocation

Menglin Li[1], Kwan Hui Lim[1(✉)], Teng Guo[2], and Junhua Liu[1,3]

[1] Singapore University of Technology and Design, Singapore, Singapore
menglin_li@mymail.sutd.edu.sg, kwanhui_lim@sutd.edu.sg
[2] Dalian University of Technology, Dalian, China
[3] Forth AI, Singapore, Singapore
j@forth.ai

Abstract. POI-level geo-information of social posts is critical to many location-based applications and services. However, the multi-modality, complexity, and diverse nature of social media data and their platforms limit the performance of inferring such fine-grained locations and their subsequent applications. To address this issue, we present a transformer-based general framework, which builds upon pre-trained language models and considers non-textual data, for social post geolocation at the POI level. To this end, inputs are categorized to handle different social data, and an optimal combination strategy is provided for feature representations. Moreover, a uniform representation of hierarchy is proposed to learn temporal information, and a concatenated version of encodings is employed to capture feature-wise positions better. Experimental results on various social media datasets demonstrate that the three variants of our proposed framework outperform multiple state-of-art baselines by a large margin in terms of accuracy and distance error metrics.

Keywords: Location prediction · Geolocation · Social media · Twitter · Transformer

1 Introduction

Knowing the posting location of social media data is important for many useful applications, including local event/place recommendations [8,24], location-based advertisements [6,11], emergency location identification and disaster response [23,44]. However, geotagged social posts are very limited as less than 1% of tweets are labeled with geo-coordinates [1]. This constraint motivates our research on geolocation, which is a topic that has received significant attention in the past decade. However, most prior studies concentrate on user geolocation, which is estimating the home location of users [34,39,45,46]. This type of geo-information is insufficient for applications like emergency location identification and natural disaster response [21], which require the location of individual posts. Hence, in this paper, we focus on the problem of social post geolocation to infer the locations of individual posts.

© The Author(s), under exclusive license to Springer Nature Switzerland AG 2023
J. Kamps et al. (Eds.): ECIR 2023, LNCS 13980, pp. 588–604, 2023.
https://doi.org/10.1007/978-3-031-28244-7_37

For social post geolocation, previous efforts typically aim at inferring locations at the city level [2,21,43]. Although there is good performance at the city level, location information at such a coarse-grained level is still insufficient for the various applications mentioned earlier. While some researchers studied the task of geo-coordinates estimation, it is challenging to achieve high accuracy [28,31]. In real-life scenarios, semantic toponyms are more practical and understandable compared to numerical latitude and longitude [42]. Therefore, we study the problem of social post geolocation at the Point-Of-Interest (POI) level, a fine-grained semantic level.

However, Social Post Geolocation at the POI level is a challenging problem due to the complexity, multi-modality, and diverse nature of social media data and their platforms. Firstly, the user-generated textual content is short, freeform, and often noisy, containing acronyms, misspellings, and special tokens. It is non-trivial to understand such complex text precisely for location estimation. Secondly, there are other non-textual contents such as time, social networks, images, and videos, which can be used for this task but also lead to the multi-modality issue. The ability to represent and fuse different data types is vital for geolocation. Lastly, it is increasingly important to develop a geolocation framework with a generalization ability to deal with the emergence of diverse social platforms, like photo-sharing and micro-blogging platforms. Many works focus on a single social platform with specific inputs, thus limiting their performance on other social platforms due to the difference in data fields. For better generalizability across platforms, some approaches utilize text content solely for geolocation but at the expense of missing out on other non-textual content and limiting performance.

To address these limitations, we present a transformer-based model, named transTagger, for POI-level social post geolocation, which is a general framework that builds upon the Bidirectional Encoder Representations from Transformers (BERT) model with good generalization ability across different social platforms for accurate fine-grained location inference. The main contributions of this work can be summarized as follows:

- We design a general categorization to tackle the multi-modality and diverse nature of social media data and their platforms and provide four datasets with ground truth covering two cities and two platforms.
- We fuse features and learn their correlations using transformer encoders with a concatenated version of positional encodings, along with a novel temporal representation to provide an optimal combination strategy of representations for multi-modality fusion. We denote this model, transTagger.
- We construct two additional variants, hierTagger and mtlTagger, by incorporating the hierarchy of locations into transTagger, and experimental results demonstrate that our models outperform state-of-the-art baselines by a considerate margin in terms of accuracy and distance error metrics.[1]

---

[1] Our code and dataset are made publicly available at https://github.com/lazylml/transTagger.

The rest of the paper is organized as follows. In Sect. 2, we review the critical related work in the geolocation field and briefly introduce hierarchical classification techniques. In Sect. 3, we first present the problem formulation and then describe our proposed model transTagger and two variants in detail. Then Sect. 4 introduces the experimental setting, while Sect. 5 presents and discusses our experimental results. Following that, we summarize and conclude this paper in Sect. 6.

2 Related Work

In this section, we review two main categories of work that are related to our research, namely social post geolocation and hierarchical geolocation works.

2.1 Post Geolocation

Post geolocation focuses on estimating the originating locations of social posts. Unlike user geolocation, which leverages a user's entire posting history, post geolocation considers only an individual post or tweet and uses that as input. For example, the work [13] uses the convolutional mixture density network for location estimation with single tweet content. Term co-occurrences in tweets, which exhibit spatial clustering or dispersion tendency, are detected and used to extend feature space in probabilistic language models [32]. For location prediction during disaster events, Ouaret et al. [31] present an iterative Random Forest fitting-prediction framework to learn semi-supervised models. A name entity recognizer [28] is developed for geolocating tweets with the help of GeoNames gazetteer. Kulkarni et al. [19] present a multi-level geocoding model that learns to associate texts with geographical locations and represent locations using S2 hierarchy. Others propose to locate tweets based on BERT architecture with different tokenization settings, like vocabulary sizes [36]. In special cases, historical locations of users are involved to boost location inference performance, like using the Markov model to formalize tweet geolocation in a flood-related disaster based on history tweets [38].

Many researchers consider metadata to infer tweet locations [2,17,20]. Pliakos and Kotropoulos construct a hypergraph based on images, users, geotags and tags of Flickr, which is further used for simultaneous image tagging and geolocation prediction [33]. A refined language model that is learned from massive corpora of social content, including tags, titles, descriptions, user ids, and image ids, is proposed to estimate the location of a post [16]. Miura et al. [29] propose a simple neural network structure with fully-connected layers and an average pooling process based on message text and user metadata for geolocation prediction. To classify the microblogs of WeiBo into 8 semantic categories, the work [42] explores the effect of user attributes and designs a neural network-based architecture with 4 feature fusion strategies.

2.2   Hierarchical Geolocation

Although the class hierarchy has been shown to be effective in closely relevant fields, like text classification [12,18,27,48], this problem has not thus far received the attention it deserves. Only a handful of existing works estimate the locations of tweets and explore geolocation performance using hierarchical locations. Previous efforts [26,43] represent locations as a tree and construct a local classifier for each parent node to infer locations, which corresponds to a typical hierarchical classification technique, Local Classifier per Parent Node (LCPN) [37]. Multi-Task Learning (MTL) is incorporated to combine losses across multiple levels and predict locations at each level simultaneously [9,19]. Most of these works aim at user location inference, whereas we study post geolocation.

Similar to our work, some research has attempted to infer fine-grained locations of tweets [3,4,30]. By investigating two properties, spatial focus and spatial homophily, a learning-to-rank framework [3,4] is designed by ranking candidate venues. The work [30] extracts semantic similarities between tweets and POI reviews locally and globally to provide a Spatially-aware Geotext Matching model building upon MLP. Both methods need to compute similarity features explicitly with additional datasets, like check-in data or POI reviews from Foursquare, which is non-trivial and time-consuming to collect. While these works advance the task of tweet geolocation, our work differs from these earlier works in various ways, which we discuss next. Our method takes in tweet content and metadata of the Twitter dataset directly as inputs, building upon BERT and using transformer encoders to learn correlations among features. Additionally, we employ a uniform representation of decomposed hierarchical time elements to further boost performance as the importance of temporal features is highlighted by many studies [21,25,30,38]. Moreover, we explore the effect of location hierarchy on the post geolocation performance by leveraging LCPN and MTL in our proposed models.

3   Method

3.1   Problem Formulation

The **Social Post Geolocation** problem is defined as estimating the originating location of tweets. In the same spirit as prior studies [21,42,43], the task is formulated as a classification problem where the predicted target is a location. Unlike these earlier works, which classify posts into countries or cities, we aim at inferring locations at a finer-granularity level, that is at the landmark or POI level. More specifically, the social post geolocation problem is represented as inferring POIs, given text and metadata of social media as input.

3.2   Method Overview

The overall structure of our proposed model, transTagger, is shown in Fig. 1. To tackle the inconsistency of different social platforms, we classify the inputs of

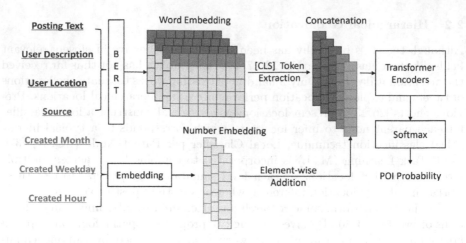

Fig. 1. The architecture of our proposed model transTagger

social media data into three categories. Information contained in social media data can be divided into user-generated and system-generated according to sources. The user-generated content is of free form and could be very noisy. Besides posting text, the user-generated content also includes user locations, user descriptions and so on. They form the first category of inputs and we denote it **Text**. System-generated content comprises textual fields and numerical fields. The former is mostly categorical text, like source (indicating whether the tweet is posted from the phone or web platform), which falls into the second category of inputs: **Categorical Text (CT)**. For the latter, numerical fields, a typical one is the time (when the post is created), and others are less explored and employed in post geolocation and we leave them for future research. The third category is **Time** and we discuss the various representation techniques used in later sections. **Text**, **CT**, and **Time** are depicted in orange, green, and blue, respectively in Fig. 1.

Our model applies BERT to learn semantic information and contextual information of **Text** and **CT** and maps features into a word embedding space. Following that, the representations of [CLS] tokens are extracted from all textual features and combined with embeddings of **Time**. Then we use several layers of transformer encoders to learn the correlation of all features. The POI probability of each post is calculated using a fully-connected layer with softmax as the activation function.

3.3   Feature Representation

We apply the pre-trained model by plugging in the post geolocation task-specific inputs and outputs into BERT. At the architecture level, BERT is an L-layer bi-directional transformer encoder [5]. The hidden size and the number of self-attention heads for each component are denoted as H and A, respectively.

Text, including posting texts, user locations, user descriptions, and **CT**, like sources, are all used as inputs. Here a degenerate text-$\varnothing$ pair corresponds to sentence A and sentence B since we formulate the post geolocation task as a classification problem and there is no "sentence" pair. An input sample is regarded as a sentence in this paper although it may actually contain multiple sentences. During tokenization, each sentence is converted into a sequence of tokens and a special classification token, [CLS], is injected in front of every input sample [5]. Then the first token becomes [CLS]. Apart from the above token embedding, other embeddings are utilized to take the position information inside sentences or between sentence pairs into consideration. Position embedding represents the position of each token in a sentence. In contrast, segment embedding is used to distinguish sentences A and B and thus is set to all zero in our case. The element-wise addition of token, position and segment embeddings forms the input representation [5].

We denote the learned embedding in the final hidden layer of each input sample as $E \in \mathbb{R}^{N \times H}$ where N is the sentence length. The corresponding embedding of the [CLS] token is represented as $C \in \mathbb{R}^H$. This token embedding can be seen as the aggregation of sentence representation, which is used for subsequent applications. Note that all the parameters are fine-tuned in an end-to-end manner based on our task, post geolocation.

Time is a vital factor in relation to human mobility and thus, of great importance for location inference. However, most works simply represent it as one-hot encoding based on the timestamp, which does not capture the full extent of temporal information and ignores the hierarchy of time elements, like hours and months. Inspired by this work [47], we propose a uniform representation of hierarchical time elements, UniHier, to learn temporal information. Hierarchical time elements are extracted from **Time**, including hours, weekdays, and months. Then each element is represented as a learnable embedding vector with dimension H and limited vocab size. A uniform representation of time is constructed by the element-wise addition of all embedding vectors.

3.4   Feature Fusion

Assuming that **Text** contains m fields, **CT** contains n fields, we extract [CLS] token vectors of **Text** and **CT**, and concatenate them with the UniHier representation of **Time**, then a feature matrix $F \in \mathbb{R}^{(m+n+1) \times H}$ is generated.

To learn the correlation of all features, we employ a multi-layer transformer encoder as described in the work [41]. Positional encodings are represented using sine and cosine functions of different frequencies as below and pos is the position, i is the dimension:

$$PE_{(pos, 2i)} = sin(pos/10000^{2i/H}) \tag{1}$$

$$PE_{(pos, 2i+1)} = cos(pos/10000^{2i/H}) \tag{2}$$

These positional encodings are fixed during training and with dimension H. In contrast to the now ubiquitous transformer encoder that sums feature representations and the corresponding positional encodings, we concatenate them and term it the concatenated version of positional encodings. Experiments demonstrate that this approach improves performance.

After concatenating with positional encodings, this feature matrix is utilized to calculate POI probabilities with a softmax layer. This model is then trained using the Adam update rule as the optimizer.

3.5   Hierarchical Prediction

The hierarchy of locations enables the application of hierarchical prediction and thus improves the performance of post geolocation. We incorporate LCPN, a typical hierarchical classification approach, with transTagger, and construct a variant, hierTagger. By combining the class hierarchy with MTL, we build upon our earlier described transTagger and propose another variant, mtlTagger. Due to space constraints, we briefly describe how to build these two variants and refer interested readers to our released source code for the implementation details.

HierTagger. The LCPN approach aims to train a multi-class classifier for each parent node in the class hierarchy, to distinguish between its child nodes [37]. The class hierarchy is typically a tree or a Direct Acyclic Graph (DAG), which is represented as a tree in our case. We build the tree of toponyms at different scales, from coarse to fine, starting from a root node that covers the whole research area. For every parent node in this tree, we employ transTagger to construct a local classifier, which is trained independently. Then a top-down class prediction approach is applied during the testing phase.

mTLTagger. MTL provides models with better generalization ability by sharing representations between related tasks [35]. The predictions of post location at coarser levels are designed as auxiliary tasks. We incorporate transTagger with hard parameter sharing, a commonly used approach with MTL in neural networks, to predict post location at different scales, from coarse to fine. The prediction result for the coarser level, denoted as q, is further utilized to constrain the finer level prediction by adding q to the loss function of the finer level. A correlation matrix between the two levels is employed to help the loss function of the finer level better understand the coarser level's prediction result.

4   Experimental Setting

4.1   Datasets

We perform our experiments using datasets from two different social media platforms, Flickr and Twitter, for two cities of Melbourne and Singapore.

Twitter. We collected 266,614 geotagged tweets that were posted in Melbourne from 2010 to 2018, and 482,765 geotagged tweets that were posted in Singapore from 2018 to 2022. We also combined tweets from Melbourne and Singapore for experiments to test the robustness of our models. The Twitter datasets of Melbourne, and Singapore, and their combination are denoted as Twitter-Mel, Twitter-SG, and Twitter-SM, respectively.

Flickr. The Flickr dataset comprises 78,131 geotagged images that were posted in Melbourne from 2004 to 2020, extracted using the Flickr API or from the Yahoo! Flickr Creative Commons 100M (YFCC-100M) [40]. We further augmented this dataset by collecting the metadata of Flickr users. This dataset is denoted as Flickr-Mel.[2]

A list of POIs and their categories are obtained using the Google Place API.[3] For Singapore, our research area is the whole country/city and there are 9,666 POIs. For Melbourne, we concentrate on the central city area and there are 242 POIs. To implement hierarchical prediction, POI themes and POI sub-themes are involved as labels to construct the class hierarchy. Specifically, there are 16 POI themes (e.g., Leisure/Recreation), 49 POI sub-themes (e.g., Park/Garden), and 242 POIs (e.g., Batman Park).

Our work aims to predict the specific POI where a post is sent from, in contrast to existing efforts that focus on coarse-level predictions at the city, country, or even continent level. To this end, we label a tweet tw in the Twitter dataset (or image im in the case of the Flickr dataset) as one and only one POI. Following the proximity principle [22], we compare the distance between tw (or im) and the POI location using their latitude and longitude coordinates, and label it with the POI if their distance differs by less than 100 m. Any tw (or im) that is not assigned a POI label is then filtered out. Note that the above statistics of the Twitter and Flickr datasets are computed after POI-labelling preprocessing.

Our two variants involve the use of class hierarchy of POIs. For example, hierTagger utilizes POI-theme level and POI-level labels, while mtlTagger contains three loss functions that are designed for POI theme, POI sub-theme, and POI predictions, respectively.

4.2 Evaluation Metrics

We use two evaluation metrics that are frequently used in geolocation tasks, namely accuracy and distance error. Accuracy, denoted as $acc@k$, reflects the proportion of correct predictions based on the top-k results and we evaluate with k as 1, 5, 10, and 20. Mean distance error, represented as $mean$, measures the mean distance between the predicted location and actual POI location. We also experimented using median distance error and observe that our models

---

[2] We also collected a Flickr dataset for Singapore but excluded it for further experimentation due to a low number of data points.

[3] https://developers.google.com/maps/documentation/places/web-service/overview.

achieve 0 error, thus we do not report the results for concision. Unless otherwise specified, all results reported in this paper are at the POI level to make our models comparable.

4.3   Parameter Setting

In our experiments, the max sequence length for text and other textual features is 100. To represent **Time** inputs using UniHier, they are randomly initialized from a uniform distribution $U(-1.0, 1.0)$ with dimension 128 (this value corresponds to the dimension of word embeddings) and vocab size is limited to 60 since the finest granularity is a minute. These embeddings are then learned during training.

The hyperparameter tuning is conducted using Bayesian optimization on the learning rate, the number of encoder layers, the number of heads, hidden size, and batch size. The number of layers, the number of attention heads, and the hidden size of the transformer encoder before the softmax layer are set as 3, 48, and 1300, respectively. The training of our model is performed using Adam with an initial learning rate of 3e–4 and a batch size of 128. We train the model with 4 epochs. Additionally, the block threshold for hierTagger is set as 0.01 and the loss weights for mtlTagger are 0.1, 0.1, and 1.

4.4   Baselines

We compare our proposed model and two variants with various popular geolocation models, including **MNB-Ngrams** (Multinomial Naive Bayes with Uni/Bi/Tri-grams) [2,4,7,26,32], **CNN-TT** (Convolutional Neural Network with Text and Time) [22], and **HLPNN** (Hierarchical Location Prediction Neural Network) [9]. The CNN text classification model [15] is widely used for geolocation [10,13,25], which we include as a baseline **CNN** in our experiments, along with its variant that uses one-hot encoding **CNN-1Hot** [14]. Besides HLPNN, another hierarchical classification model, **HDLTex** (Hierarchical Deep Learning for Text Classification) [18] is utilized as one of baselines. Our two proposed variants, hierTagger and mtlTagger, are also involved in comparisons.

Table 1. Baseline comparison on Flickr-Mel and Twitter-Mel

	Flickr-Mel					Twitter-Mel				
	Acc@1↑	Acc@5↑	Acc@10↑	Acc@20↑	Mean (m)↓	Acc@1↑	Acc@5↑	Acc@10↑	Acc@20↑	Mean (m)↓
HLPNN	68.68	83.62	88.95	93.87	247.6	61.45	76.7	81.85	87.05	433.2
HDLTex	56.89	64.71	66.49	70.14	604	56.2	64.67	66.33	67.69	512.5
CNN-TT	75.49	87.63	90.83	94.14	241	67.85	80.69	84.93	89.19	351.9
CNN	59.4	74.19	81.16	88.43	528	60.45	77.27	83.45	88.54	408.5
CNN-1Hot	59.91	76.69	83.25	90.14	697.7	63.08	76.89	80.92	85.43	362
MNB-Ngrams	54.35	71.71	79.93	88.61	1071	49.82	73.6	79.05	84.62	500.7
transTagger	**77.88**	89.85	**93.05**	93.05	**175.8**	**71.96**	84.64	**88.2**	88.2	**303.3**
hierTagger	77.59	**90.13**	92.91	**95.87**	183.5	71.42	84.34	88.12	**91.49**	319.5
mtlTagger	77.22	89.44	92.86	95.73	182.9	71.84	**84.67**	88.03	91.44	317.9

Table 2. Baseline comparison on Twitter-SG and Twitter-SM

	Twitter-SG					Twitter-SM				
	Acc@1↑	Acc@5↑	Acc@10↑	Acc@20↑	Mean (km)↓	Acc@1↑	Acc@5↑	Acc@10↑	Acc@20↑	Mean (km)↓
CNN-TT	53.76	67.27	70.29	72.81	2.617	54.8	68.64	72.41	75.7	154.3
CNN	49.77	62.04	64.72	67.66	2.949	48.98	62.11	65.5	69.09	542.7
CNN-1Hot	38.34	50.11	52.96	55.86	3.536	38.85	52.05	55.54	59.21	882.9
transTagger	**61.94**	**73.75**	**76.75**	**76.75**	**2.215**	**64.88**	**76.8**	**80.08**	**80.08**	**3.69**

5 Experimental Results

5.1 Baseline Comparison

To verify the effectiveness of our proposed models, experiments are designed to compare the performance of three variants and various baselines on the Flickr-Mel and Twitter-Mel datasets, as shown in Table 1. Similar experiments are conducted on the Twitter-SG and Twitter-SM datasets as well as to further examine the robustness of geolocation performance, as presented in Table 2. We only report results for transTagger and three baselines as the hierarchical labels are not available for the latter two datasets.

Overall, transTagger, hierTagger, and mtlTagger outperform all baselines, including the hierarchical ones, across all four datasets. Compared with a strong baseline like CNN-TT, transTagger outperforms by a substantial margin, obtaining an improvement of 2.39%, 4.11%, 8.18% and 10.08% in accuracy (acc@1) on Flickr-Mel, Twitter-Mel, Twitter-SG, and Twitter-SM, respectively. The latter two datasets contain many more POIs and the improvement of transTagger over the baselines is even larger. This indicates that our model is versatile enough to handle a large number of classes (POIs) well. In addition to accuracy, the mean distance error is also greatly reduced. To be specific, transTagger reduces the mean distance error by 65.2, 48.6, 402, and 1174 m, compared with CNN-TT. In contrast to Table 1, we use kilometers (km) to denote distance in Table 2 because Twitter-SG and Twitter-SM cover much larger areas and thus values of mean distance error are relatively higher. In addition, the distance calculation of Twitter-SM involves two cities and thus is quite sensitive to prediction accuracy as this dataset is a mixture of Twitter-SG and Twitter-Mel. Therefore, the distance errors would increase greatly in comparison to the corresponding accuracy that decreases slightly, as shown in Table 2 and Table 4.

The overall results show that our proposed models provide superior performance for POI-level post geolocation across all cities and platforms, compared to the various baselines.

5.2 Representation Combination Selection

Taking generalization into consideration, we categorize inputs into three types: **Text**, **CT**, and **Time**, and performed a representation for each type, as previously described in Sect. 3. However, there are multiple ways to represent each input type. For **CT**, one way is to treat categorical texts as normal texts and

Table 3. Representation combination selection on Flickr-Mel and Twitter-Mel

	Flickr-Mel					Twitter-Mel				
	Acc@1↑	Acc@5↑	Acc@10↑	Acc@20↑	Mean (m)↓	Acc@1↑	Acc@5↑	Acc@10↑	Acc@20↑	Mean (m)↓
transTagger										
Text-Text	77.88	89.85	93.05	93.05	175.8	**71.96**	**84.64**	88.2	88.2	**303.3**
1Hot-Text	77.88	89.85	93.05	93.05	175.8	71.69	84.6	**88.29**	**88.29**	322.1
Text-UniHier	**78.04**	**90.16**	**93.28**	**93.28**	**171.6**	70.03	84.32	88.09	88.09	313.8
1Hot-UniHier	**78.04**	**90.16**	**93.28**	**93.28**	**171.6**	69.69	84.25	87.87	87.87	308.3
Text-1Hot	77.49	89.66	92.99	92.99	184.1	69.91	84.13	87.71	87.71	316.9
1Hot-1Hot	77.49	89.66	92.99	92.99	184.1	69.5	84.26	88.04	88.04	321.6
hierTagger										
Text-Text	77.59	**90.13**	92.91	**95.87**	183.5	71.42	84.34	88.12	91.49	319.5
1Hot-Text	77.59	**90.13**	92.91	**95.87**	183.5	**71.49**	**84.45**	**88.15**	**91.56**	324.5
Text-UniHier	**78.18**	89.94	**93.15**	95.83	**169.5**	70.03	84.29	88.03	91.52	314.4
1Hot-UniHier	**78.18**	89.94	**93.15**	95.83	**169.5**	69.6	84.18	87.78	91.07	**308.8**
Text-1Hot	77.23	89.43	92.82	95.56	190.2	69.82	84.04	87.57	91.16	316.7
1Hot-1Hot	77.23	89.43	92.82	95.56	190.2	69.35	84.19	87.96	91.46	321.9
mtlTagger										
Text-Text	77.22	89.44	92.86	95.73	182.9	**71.84**	**84.67**	88.03	91.44	317.9
1Hot-Text	77.22	89.44	92.86	95.73	182.9	71.48	84.45	**88.04**	**91.64**	315.1
Text-UniHier	**78.93**	**90.18**	93.31	95.97	**168.3**	69.91	84.3	87.94	91.52	**312.9**
1Hot-UniHier	**78.93**	**90.18**	93.31	95.97	**168.3**	69.16	84.06	88.01	91.39	314.2
Text-1Hot	77.84	89.9	**93.36**	**96.26**	179.1	69.62	84.18	87.83	91.35	317.3
1Hot-1Hot	77.84	89.9	**93.36**	**96.26**	179.1	69.39	84	87.72	91.33	314.9

use BERT or other language models to generate representations, and we call this Text embedding. Another commonly used approach is one-hot encoding. For **Time**, one way is to treat date/time as a standard text and generate temporal embedding using language models. Hence, there are two ways to represent **CT**: text and one-hot, and three ways for **Time**: text, one-hot, and UniHier. This results in six combinations of these representation methods, which we further experiment to find an optimal representation combination strategy. The results are illustrated in Tables 3 and 4, where Text denotes Text embedding, and Text-UniHier refers to using Text embedding for **CT** and UniHier representation for **Time**, and so forth. Note that the results of Text-Text and 1Hot-Text are duplicated for Flickr since there are no **CT** fields. Similarly for Text-UniHier and 1Hot-UniHier, Text-1Hot and 1Hot-1Hot.

The results show that Text-Text delivers the overall best performance across all Twitter datasets. However, Text-UniHier (or 1Hot-UniHier) outperforms others for the Flickr dataset. One possible reason is that Flickr contains more

Table 4. Representation combination selection of transTagger on Twitter-SG and Twitter-SM

	Twitter-SG					Twitter-SM				
	Acc@1↑	Acc@5↑	Acc@10↑	Acc@20↑	Mean (km)↓	Acc@1↑	Acc@5↑	Acc@10↑	Acc@20↑	Mean (km)↓
Text-Text	**61.94**	**73.75**	**76.75**	**76.75**	**2.215**	**64.88**	76.8	**80.08**	**80.08**	**3.69**
1Hot-Text	61.37	73.26	76.36	76.36	2.292	64.84	**76.88**	80.06	80.06	3.263
Text-UniHier	58.1	72.71	75.92	75.92	2.318	61.9	76.1	79.55	79.55	56.63
1Hot-UniHier	57.82	72.63	75.94	75.94	2.332	61.53	76.13	79.48	79.48	69.64
Text-1Hot	58.13	72.71	75.92	75.92	2.334	61.74	76	79.33	79.33	67.52
1Hot-1Hot	57.3	72.43	75.65	75.65	2.349	61.21	75.8	79.24	79.24	58.33

time fields, including photo taken time and photo posted time, compared to Twitter that only contains tweet created time. Therefore, the best representation combination is Text-Text. In the event where multiple time inputs are involved, it is recommended to represent temporal inputs using UniHier.

We further compare the performance of three variants. Contrary to our expectations, hierTagger and mtlTagger show no distinct advantage, except for acc@20. Hence, these two variants are recommended when this specific metric is important. The intuition of utilizing hierarchical locations is that the prediction results at coarser level can help guide the geolocation at target level. However, this process might involve error propagation and thus impair the expressive power of the whole architecture. An effective mechanism for correcting these prediction errors is a promising direction to boost geolocation performance, and we leave this for future work.

Table 5. Ablation study on Twitter-SG and Twitter-SM

	Twitter-SG					Twitter-SM				
	Acc@1↑	Acc@5↑	Acc@10↑	Acc@20↑	Mean (km)↓	Acc@1↑	Acc@5↑	Acc@10↑	Acc@20↑	Mean (km)↓
transTagger	**61.94**	**73.75**	**76.75**	**76.75**	**2.215**	**64.88**	**76.8**	80.08	80.08	**3.69**
w/o transformer	60.3	72.44	75.64	75.64	2.338	63.92	76.7	**80.1**	**80.1**	6.108
w/o position	61.28	73.01	75.96	75.96	2.292	64.39	76.43	79.75	79.75	4.917

5.3 Ablation Study

We compare transTagger with two ablations to examine the effectiveness of two model components, namely transformer encoders and position encodings. Table 5 shows the performance breakdown on Twitter-SG and Twitter-SM. For w/o position, we replace the concatenation version of positional encodings with the commonly used add-on version. The w/o transformer ablation removes the transformer encoders which are used to learn the correlation of features. The results demonstrate that all components contribute to improving the post geolocation performance of transTagger. Among all components, encoders have the greatest effect as shown by how it increases accuracy (including acc@1, acc@5, acc@10, and acc@20) and reduces the mean distance error by the largest margin.

Table 6. Coarse-Level Geolocation

	Flickr-Mel(POI-Theme)				Flickr-Mel(POI)			
	Acc@1↑	Acc@5↑	Acc@10↑	Acc@20↑	Acc@1↑	Acc@5↑	Acc@10↑	Acc@20↑
HLPNN	79.92	97.16	**100**	**100**	68.68	83.62	88.95	93.87
hierTagger	**83.22**	**97.93**	**100**	**100**	**77.59**	**90.13**	**92.91**	**95.87**
mtlTagger	81.57	97.49	99.97	**100**	77.22	89.44	92.86	95.73

5.4   Coarse-Level Geolocation

We now study the prediction results of coarse-level geolocation since our two hierarchical variants both incorporate the toponym hierarchy. Although mtlT-agger is capable of inferring locations at three levels, only the results of POI theme and POI are listed in Table 6 to make mtlTagger consistent and comparable with hierTagger. We observed that our models not only outperform at the target level (POI) by a large margin but also present outstanding coarse-level (POI theme) performance, even when compared with the competitive hierarchical geolocation algorithm HLPNN [9]. Furthermore, hierTagger obtained an absolute improvement of almost 2 points compared to mtlTagger (acc@1) for POI-theme geolocation even though the two have a similar capability of estimating POI-level locations. To force the model to focus more on our target task, POI geolocation, we set the weights of mtlTagger as 0.1, 0.1, and 1, for the loss functions of POI theme, POI sub-theme, and POI, respectively. In turn, this might be the cause of a negative impact on coarse-level prediction.

6   Conclusion

In this paper, we propose a transformer-based general framework, transTagger, for POI-level post geolocation. The inputs are categorized into three types: **Text**, **CT**, and **Time** to handle different social data, and the optimal representation combination, Text-Text, is provided by experimenting with all combinations. A novel representation of time, UniHier, is presented and verified to be useful in the case of multiple temporal inputs. Transformer encoders are employed to enhance geolocation performance and a concatenated version of encodings is incorporated to capture feature-wise positions. The effectiveness and robustness of our model are demonstrated on four datasets, covering two cities and two social platforms. Two variants, hierTagger and mtlTagger, by incorporating respective LCPN and MTL with transTagger, are shown to lift acc@20 effectively.

While these results are encouraging, we believe our approach can be further improved via two future directions. Firstly, we can explore more representation methods for different inputs, like numeral embeddings to extract time entities accurately. Secondly, we can also incorporate other modalities in addition to text and numbers, such as images and videos to provide more comprehensive knowledge.

Acknowledgments. This research / project is supported by the National Research Foundation, Singapore, and Ministry of National Development, Singapore under its Cities of Tomorrow R&D Programme (CoT Award COT-V2-2020-1). Any opinions, findings and conclusions or recommendations expressed in this material are those of the author(s) and do not reflect the views of National Research Foundation, Singapore and Ministry of National Development, Singapore. The computational work was partially performed on resources of the National Supercomputing Centre, Singapore.

References

1. Cheng, Z., Caverlee, J., Lee, K.: You are where you tweet: a content-based approach to geo-locating twitter users. In: Proceedings of the 19th ACM international conference on Information and knowledge management, pp. 759–768 (2010)
2. Chi, L., Lim, K.H., Alam, N., Butler, C.J.: Geolocation prediction in twitter using location indicative words and textual features. In: Proceedings of the 2nd Workshop on Noisy User-generated Text (WNUT), pp. 227–234 (2016)
3. Chong, W.H., Lim, E.P.: Exploiting contextual information for fine-grained tweet geolocation. In: Proceedings of the International AAAI Conference on Web and Social Media, vol. 11 (2017)
4. Chong, W.H., Lim, E.P.: Exploiting user and venue characteristics for fine-grained tweet geolocation. ACM Trans. Inf. Syst. (TOIS) 36(3), 1–34 (2018)
5. Devlin, J., Chang, M.W., Lee, K., Toutanova, K.: BERT: Pre-training of deep bidirectional transformers for language understanding. In: Proceedings of the 2019 Conference of the North American Chapter of the Association for Computational Linguistics: Human Language Technologies, Volume 1 (Long and Short Papers), pp. 4171–4186. Association for Computational Linguistics, Minneapolis, Minnesota (2019). https://doi.org/10.18653/v1/N19-1423. https://aclanthology.org/N19-1423
6. Evans, C., Moore, P., Thomas, A.: An intelligent mobile advertising system (imas): Location-based advertising to individuals and business. In: 2012 Sixth International Conference on Complex, Intelligent, and Software Intensive Systems, pp. 959–964. IEEE (2012)
7. Han, B., Cook, P., Baldwin, T.: Text-based twitter user geolocation prediction. J. Artif. Intell. Res. 49, 451–500 (2014)
8. Ho, N.L., Lim, K.H.: POIBERT: a transformer-based model for the tour recommendation problem. In: Proceedings of the 2022 IEEE International Conference on Big Data (2022)
9. Huang, B., Carley, K.: A hierarchical location prediction neural network for Twitter user geolocation. In: Proceedings of the 2019 Conference on Empirical Methods in Natural Language Processing and the 9th International Joint Conference on Natural Language Processing (EMNLP-IJCNLP), pp. 4732–4742. Association for Computational Linguistics, Hong Kong, China (2019). https://doi.org/10.18653/v1/D19-1480. https://aclanthology.org/D19-1480
10. Huang, B., Carley, K.M.: On predicting geolocation of tweets using convolutional neural networks. In: Lee, D., Lin, Y.-R., Osgood, N., Thomson, R. (eds.) SBP-BRiMS 2017. LNCS, vol. 10354, pp. 281–291. Springer, Cham (2017). https://doi.org/10.1007/978-3-319-60240-0_34
11. Huang, H., Gartner, G., Krisp, J.M., Raubal, M., Van de Weghe, N.: Location based services: ongoing evolution and research agenda. J. Location Based Serv. 12(2), 63–93 (2018)
12. Huang, W., et al.: Hierarchical multi-label text classification: an attention-based recurrent network approach. In: Proceedings of the 28th ACM International Conference on Information And Knowledge Management, pp. 1051–1060 (2019)
13. Iso, H., Wakamiya, S., Aramaki, E.: Density estimation for geolocation via convolutional mixture density network. CoRR abs/1705.02750 (2017). http://arxiv.org/abs/1705.02750
14. Johnson, R., Zhang, T.: Effective use of word order for text categorization with convolutional neural networks. In: NAACL (2015)

15. Kim, Y.: Convolutional neural networks for sentence classification. In: Proceedings of the 2014 Conference on Empirical Methods in Natural Language Processing (EMNLP), pp. 1746–1751. Association for Computational Linguistics, Doha, Qatar (2014). https://doi.org/10.3115/v1/D14-1181. https://www.aclweb.org/anthology/D14-1181

16. Kordopatis-Zilos, G., Papadopoulos, S., Kompatsiaris, Y.: Geotagging social media content with a refined language modelling approach. In: Chau, M., Wang, G.A., Chen, H. (eds.) PAISI 2015. LNCS, vol. 9074, pp. 21–40. Springer, Cham (2015). https://doi.org/10.1007/978-3-319-18455-5_2

17. Kordopatis-Zilos, G., Popescu, A., Papadopoulos, S., Kompatsiaris, Y.: Placing images with refined language models and similarity search with PCA-reduced VGGfeatures. In: MediaEval (2016)

18. Kowsari, K., Brown, D.E., Heidarysafa, M., Meimandi, K.J., Gerber, M.S., Barnes, L.E.: HDLTex: hierarchical deep learning for text classification. In: 2017 16th IEEE International Conference on Machine Learning And Applications (ICMLA), pp. 364–371. IEEE (2017)

19. Kulkarni, S., Jain, S., Hosseini, M.J., Baldridge, J., Ie, E., Zhang, L.: Spatial language representation with multi-level geocoding. arXiv preprint arXiv:2008.09236 (2020)

20. Li, M., Lim, K.H.: Geotagging social media posts to landmarks using hierarchical BERT (student abstract). In: Proceedings of the Thirty-Sixth AAAI Conference on Artificial Intelligence (AAAI2022) (2022)

21. Li, P., Lu, H., Kanhabua, N., Zhao, S., Pan, G.: Location inference for non-geotagged tweets in user timelines. IEEE Trans. Knowl. Data Eng. 31(6), 1150–1165 (2018)

22. Lim, K.H., Karunasekera, S., Harwood, A., George, Y.: Geotagging tweets to landmarks using convolutional neural networks with text and posting time. In: Proceedings of the 24th International Conference on Intelligent User Interfaces: Companion, pp. 61–62 (2019)

23. Liu, J., Singhal, T., Blessing, L.T., Wood, K.L., Lim, K.H.: CrisisBERT: a robust transformer for crisis classification and contextual crisis embedding. In: Proceedings of the 32nd ACM Conference on Hypertext and Social Media, pp. 133–141 (2021)

24. Liu, J., Wood, K.L., Lim, K.H.: Strategic and crowd-aware itinerary recommendation. In: Proceedings of the 2020 European Conference on Machine Learning and Knowledge Discovery in Databases (ECML-PKDD2020) (2020)

25. Liu, R., Cong, G., Zheng, B., Zheng, K., Su, H.: Location prediction in social networks. In: Cai, Y., Ishikawa, Y., Xu, J. (eds.) APWeb-WAIM 2018. LNCS, vol. 10988, pp. 151–165. Springer, Cham (2018). https://doi.org/10.1007/978-3-319-96893-3_12

26. Mahmud, J., Nichols, J., Drews, C.: Where is this tweet from? inferring home locations of twitter users. In: Proceedings of the International AAAI Conference on Web and Social Media, vol. 6 (2012)

27. Meng, Y., Shen, J., Zhang, C., Han, J.: Weakly-supervised hierarchical text classification. In: Proceedings of the AAAI conference on artificial intelligence, vol. 33, pp. 6826–6833 (2019)

28. Mircea, A.: Real-time classification, geolocation and interactive visualization of covid-19 information shared on social media to better understand global developments. In: Proceedings of the 1st Workshop on NLP for COVID-19 (Part 2) at EMNLP 2020 (2020)

29. Miura, Y., Taniguchi, M., Taniguchi, T., Ohkuma, T.: A simple scalable neural networks based model for geolocation prediction in twitter. In: Proceedings of the 2nd Workshop on Noisy User-generated Text (WNUT), pp. 235–239 (2016)
30. Mousset, P., Pitarch, Y., Tamine, L.: End-to-end neural matching for semantic location prediction of tweets. ACM Trans. Inf. Syst. (TOIS) **39**(1), 1–35 (2020)
31. Ouaret, R., Birregah, B., Soulier, E., Auclair, S., Boulahya, F.: Random forest location prediction from social networks during disaster events. In: 2019 Sixth International Conference on Social Networks Analysis, Management and Security (SNAMS), pp. 535–540. IEEE (2019)
32. Ozdikis, O., Ramampiaro, H., Nørvåg, K.: Spatial statistics of term co-occurrences for location prediction of tweets. In: Pasi, G., Piwowarski, B., Azzopardi, L., Hanbury, A. (eds.) ECIR 2018. LNCS, vol. 10772, pp. 494–506. Springer, Cham (2018). https://doi.org/10.1007/978-3-319-76941-7_37
33. Pliakos, K., Kotropoulos, C.: Simultaneous image tagging and geo-location prediction within hypergraph ranking framework. In: 2014 IEEE International Conference on Acoustics, Speech and Signal Processing (ICASSP), pp. 6894–6898. IEEE (2014)
34. Qian, Y., Tang, J., Yang, Z., Huang, B., Wei, W., Carley, K.M.: A probabilistic framework for location inference from social media. arXiv preprint arXiv:1702.07281 (2017)
35. Ruder, S.: An overview of multi-task learning in deep neural networks. arXiv preprint arXiv:1706.05098 (2017)
36. Scherrer, Y., Ljubešić, N.: Social media variety geolocation with geoBERT. In: Proceedings of the Eighth Workshop on NLP for Similar Languages, Varieties and Dialects. The Association for Computational Linguistics (2021)
37. Silla, C.N., Freitas, A.A.: A survey of hierarchical classification across different application domains. Data Min. Knowl. Disc. **22**(1), 31–72 (2011)
38. Singh, J.P., Dwivedi, Y.K., Rana, N.P., Kumar, A., Kapoor, K.K.: Event classification and location prediction from tweets during disasters. Ann. Oper. Res. **283**(1), 737–757 (2019)
39. Tao, H., Gao, Y., Wang, Z., Khan, L., Thuraisingham, B.: An episodic learning based geolocation detection framework for imbalanced data. In: 2021 International Joint Conference on Neural Networks (IJCNN), pp. 1–8. IEEE (2021)
40. Thomee, B., et al.: YFCC100M: the new data in multimedia research. Commun. ACM **59**(2), 64–73 (2016)
41. Vaswani, A., Shazeer, N., Parmar, N., Uszkoreit, J., Jones, L., Gomez, A.N., Kaiser, Ł., Polosukhin, I.: Attention is all you need. In: Advances in Neural Information Processing Systems 30 (2017)
42. Wang, N., et al.: Semantic place prediction with user attribute in social media. IEEE Multimedia **28**(4), 29–37 (2021)
43. Wing, B., Baldridge, J.: Hierarchical discriminative classification for text-based geolocation. In: Proceedings of the 2014 conference on empirical methods in natural language processing (EMNLP), pp. 336–348 (2014)
44. Zheng, X., Han, J., Sun, A.: A survey of location prediction on twitter. IEEE Trans. Knowl. Data Eng. **30**(9), 1652–1671 (2018)
45. Zhong, T., Wang, T., Zhou, F., Trajcevski, G., Zhang, K., Yang, Y.: Interpreting twitter user geolocation. In: Proceedings of the 58th Annual Meeting of the Association for Computational Linguistics, pp. 853–859 (2020)
46. Zhou, F., Qi, X., Zhang, K., Trajcevski, G., Zhong, T.: MetaGeo: a general framework for social user geolocation identification with few-shot learning. IEEE Transactions on Neural Networks and Learning Systems (2022)

47. Zhou, H., et al.: Informer: beyond efficient transformer for long sequence time-series forecasting. In: Proceedings of the AAAI Conference on Artificial Intelligence, vol. 35, pp. 11106–11115 (2021)
48. Zhou, J., et al.: Hierarchy-aware global model for hierarchical text classification. In: Proceedings of the 58th Annual Meeting of the Association for Computational Linguistics, pp. 1106–1117 (2020)

Document-Level Relation Extraction with Distance-Dependent Bias Network and Neighbors Enhanced Loss

Hao Liang[1,2] and Qifeng Zhou[1,2](✉)

[1] Department of Automation, Xiamen University, Xiamen 361005, China
`lianghao6@stu.xmu.edu.cn`, `zhouqf@xmu.edu.cn`
[2] Xiamen Key Laboratory of Big Data Intelligent Analysis and Decision-making,
Xiamen 361005, China

Abstract. Document-level relation extraction (DocRE), in contrast to sentence-level, requires additional context to be considered. Recent studies, when extracting contextual information about entities, treat information about the whole document equally, which inevitably suffers from irrelevant information. This has been demonstrated to make the model not robust: it predicts correctly when an entire document is fed but errs when non-evidence sentences are removed. In this work, we propose three novel components to improve the robustness of the model by selectively considering the context of the entities. Firstly, we propose a new method for computing the distance between tokens that reduces the distance between evidence sentences and entities. Secondly, we add a distance-dependent bias network to each self-attention building block to exploit the distance information between tokens. Finally, we design an auxiliary loss for entities with higher attention to close tokens in the attention mechanism. Experimental results on three DocRE benchmark datasets show that our model not only outperforms existing models but also has strong robustness.

Keywords: Relation extraction · Self attention · Pre-training model

1 Introduction

Relationship extraction (RE) aims to find predefined relations between entities from the texts. It is the fundamental of knowledge graph [2,38], question answering [7], and information extraction [17,28]. Early RE focused on the sentence-level [18,39], but realistic application scenarios often span across sentences, thus it is natural to shift to Document-level RE (DocRE) [31,32,43]. DocRE requires finding useful information from a large amount of context for logical reasoning, which is highly challenging.

Identifying the relation of an entity pair within a document by focusing on only a portion of the document is quite intuitive. Huang *et al.* [6] proposed a heuristic method to select three sentences for each pair of entities instead of inputting the whole document, which is effective but will delete a lot of helpful

J. Kamps et al. (Eds.): ECIR 2023, LNCS 13980, pp. 605–621, 2023.
https://doi.org/10.1007/978-3-031-28244-7_38

| **William B. Maclay** |
| [1] *William* was a United States Representative from *New York* . |
| ... |
| [4] Born in *New York City* , he received private instruction and was graduated from the *New York University* in 1836 ... |

Entity Pair: (*William*, *New York*)	(*William*, *New York University*)
Relation: place of birth	educated at
Evidence set: $\{1, 4\}$	$\{1, 4\}$
Mentions location set: $\{1, 4\}$	$\{1, 4\}$

Fig. 1. An illustration on the DocRED dataset. New York and New York City are mention of the same entity.

contexts. Xu *et al.* [32] gives a weight to each sentence when identifying entity pairs, but the model requires two forward propagations to identify each pair of entities, which is not efficient. Nevertheless, these works point in a direction to improve the robustness and performance of the DocRE model, that is, the model should focus more on evidence sentences rather than non-evidence sentences.

The DocRED dataset [34] is labeled with the evidence sentences of an entity pair and all mentions of entities, Fig. 1 is an example. The entity pair (*William, New York*) has relation *a place of birth*. A human identifies this relation only by the sentences ⟨1⟩ and ⟨4⟩ (evidence sentences), that is, the Evidence set $\{1, 4\}$, denoted as set E. The Mentions location set is a set of sentences where all the mentions of this entity pair are located, denoted as set M. In this case, $E \subseteq M$, does this mean that we only need to consider the sentences in which the entities mention are located? In this regard, we made a statistic for DocRED dataset, see Table 1. "0" in the table means that for an entity pair, if we consider only the sentences in M, the percentage of $E \subseteq M$ is 90.9%. Now we extend one sentence outward with M as the center, that is, if M is $\{3\}$, it will be extended to $\{2, 3, 4\}$, then the percentage of $E \subseteq M$ is 96.4% as shown in "1" in the Table 1, and the rest of the values in the table have the same meaning in turn. From the Table 1, we may observe that the percentage growth is not significant as the number of considered sentences becomes larger. This represents that for an entity pair, the more distant token, the more noise and the less helpful information. Therefore, we assume that the DocRE model should pay differential attention to tokens at

Table 1. The percentage that $E \subseteq M$ as M expands in the DocRED dataset.

Expand number	0	1	2	3	4	5	6	7	...	13
Probability	90.9%	96.4%	98.1%	98.9%	99.3%	99.6%	99.7%	99.8%	...	100%

different distances when identifying the relation of entity pairs, and the closer the token is, the more valuable it will be.

To this end, we propose **S**elf-Attention with **D**istance-dependent **B**ias **N**etwork (SDBN) and **N**eighbors **E**nhanced **L**oss (NEL). Specifically, we design a Bias Network that can improve the self-attention mechanism by using the mutual distance between tokens within a document as prior knowledge. Meanwhile, we design an auxiliary NEL to encourage the model to have a higher attention score for tokens that are closer to an entity pair. In addition, we use crossing-distance in our model, that is, the distance of an entity from other tokens is determined by the closest distance between the token and the entity mentions. Take Fig. 1 as an illustration, the distance between *New York City* in sentence $\langle 4 \rangle$ and *William* in sentence $\langle 1 \rangle$ is **3** (three sentences apart), but since *New York* in sentence $\langle 1 \rangle$ and *New York City* belong to the same entity, we define the crossing-distance between *New York City* and *William* as **0**, which can be seen as the context of *New York City* is enhanced from $\langle 4 \rangle$ to $\langle 1, 4 \rangle$.

2 Methodology

2.1 Crossing-Distance Calculation

In identifying the relation of an entity pair, tokens with different distances have different bias network parameters to generate attention preferences. Thus we create an adjacency matrix to record the distance between tokens according to the following distance calculation.

Given a document $D = \{s_1, s_2, ..., s_N\}$, containing a set of entities $\{e_i\}_{i=1}^n$, where each sentence s_i is a sequence words. We convert D into a sequence of tokens $x = (x_1, x_2, ..., x_n)$, and a sequence $l = (l_1, l_2, ..., l_n), l_i \in \{1, 2, ..., N\}$ that records the sentence in which the token is located. Each entity may have multiple mentions, and each mention may consist of many tokens. Thus for each entity, we use a set $\left\{l_k^i\right\}_{k=1}^{N_{e_i}}$ to note in which sentences this entity is in. We classify all tokens into the following two categories.

- *Entity token*: Token that belongs to an entity.
- *Non-entity token*: Token that does not belong to any entity.

For set operations, we define the function $F(A, B)$, which works to determine the absolute minimum of the difference between the elements of A and B. There are three distance calculation methods derived between the two types of tokens:
Entity token and *Entity token*:

$$min(F(\{l_k^i\}_{k=1}^{N_{e_i}}, \left\{l_k^j\right\}_{k=1}^{N_{e_j}}), \mu) \qquad (1)$$

where e_i and e_j represent the entities to which the two tokens belong, respectively, and μ is a hyper-parameter, the distance beyond μ is considered as μ. Shaw *et al.* [19] and Lee *et al.* [9] have done similar clipping in the calculation

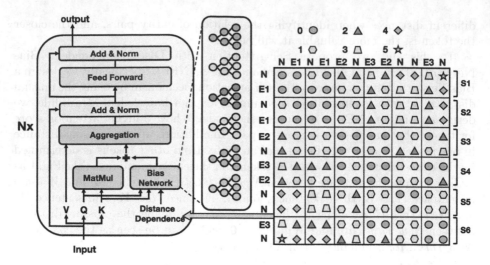

Fig. 2. The overall architecture of SDBN, when $\mu = 5$. Left illustrates structured self-attention as its basic building block. Right explains our bias network. This example consists of six sentences: S1, S2, S3, S4, S5, S6, and three entities: E1, E2, and E3. N denotes non-entity tokens. Element in row i and column j represents the distance of query token x_i to key token x_j, we use different graphic shapes to distinguish different distances.

of relative distances, which allows the model to be generalized to any sequence length.

Entity token and *Non-entity token*:

$$min(F(\{l_k^i\}_{k=1}^{N_{e_i}}, \{l_j\}), \mu) \tag{2}$$

Non-entity token and *Non-entity token*:

$$min(F(\{l_i\}, \{l_j\}), \mu) \tag{3}$$

For explanation, we assume that $\mu = 5$. Using (1)(2)(3), we construct the distance information of the entire document from each other as an adjacency matrix with elements from a finite set: $\{0, 1, 2, 3, 4, 5\}$.

2.2   SDBN

SDBN inherits the architecture of Transformer [23] encoder, which is a stack of the identical building block. As its core part, we added a bias network that can utilize the distance information in the self-attention mechanism. It makes the model generate entity representation with attention preferences for contexts of different distances.

A token sequence $x = (x_1, x_2, \ldots, x_n)$ is provided as input, following the calculate of Sect. 2.1, we introduce $A = \{a_{ij}\}$ to represent adjacency matrix, where $i, j \in \{1, 2, \ldots, n\}$ and $a_{ij} \in \{0, 1, 2, 3, 4, 5\}$ is a discrete variable denotes the distance from x_i to x_j.

The input representation $x_i^m \in \mathbb{R}^{d_{in}}$ is first projected into the query/key/ value vector respectively in each layer m.

$$\mathbf{q}_i^m = x_i^m W_m^Q, \mathbf{k}_i^m = x_i^m W_m^K, \mathbf{v}_i^m = x_i^m W_m^V \tag{4}$$

where $W_m^Q, W_m^K, W_m^V \in \mathbb{R}^{d_{in} \times d_{out}}$. Based on these inputs and the distance adjacency matrix A, we compute the raw attention scores and distance-dependent attention biases, then aggregate them together as the final attention scores to engage in the self-attention mechanism.

The raw attention score is produced by query-key product as in standard self-attention:

$$e_{ij}^m = \frac{\mathbf{q}_i^m \mathbf{k}_j^m}{\sqrt{d}} \tag{5}$$

To model the distance dependency based on their contextualized query / key representations, we use an additional module in parallel to it. We parameterize it as bias network that transforms a_{ij}, together with the query and key vectors $\mathbf{q}_i^m$ and $\mathbf{k}_j^m$, into an attentive bias, then apply it to e_{ij}^m:

$$\tilde{e}_{ij}^m = e_{ij}^m + \frac{bias\,network\,(\mathbf{q}_i^m, \mathbf{k}_j^m, a_{ij})}{\sqrt{d}} \tag{6}$$

The proposed bias network regulates the attention flow from x_i to x_j. As a result, the model gains from the information provided by distance dependencies.

After obtaining the regulated attention scores $\tilde{e}_{ij}^m$, a softmax operation is used to aggregate the value vectors.

$$\mathbf{z}_i^{m+1} = \sum_{j=1}^{n} \frac{\exp \tilde{e}_{ij}^m}{\sum_{k=1}^{n} \exp \tilde{e}_{ik}^m} \mathbf{v}_j^m \tag{7}$$

here $\mathbf{z}_i^{m+1} \in \mathbb{R}^{d_{out}}$ is the updated contextual representation of x_i^m. Figure 2 gives the overview of SDBN. In the next section, we describe the bias network.

2.3   Bias Network

We instantiate each discrete distance a_{ij} as neural layers with particular parameters, train, then apply them in a compositional manner to include them into an end-to-end trainable deep model. As a consequence, we get a structured model composed of corresponding layer parameters for each input adjacency matrix A made up of a_{ij}. Regarding the specific layout of these neural layers, we apply two options: Bilinear Transformation and Decomposed Linear Transformation.

$$\text{bias}_{ij}^m = \text{Bilinear}\,(a_{ij}, \mathbf{q}_i^m, \mathbf{k}_j^m)$$

$$\text{or} \tag{8}$$

$$= \text{Decomp}\,(a_{ij}, \mathbf{q}_i^m, \mathbf{k}_j^m)$$

Bilinear Transformation. Lin *et al.* [12] proposed bilinear to combine the features of two images, which simplifies the gradient calculation and can be directly applied to our bias network.

$$\text{bias }_{ij}^m = \mathbf{q}_i^m \mathbf{W}_{m,a_{ij}}^B \mathbf{k}_j^{m^T} + b_{m,a_{ij}} \tag{9}$$

here, a_{ij} is parameterized as a trainable neural layer $\mathbf{W}_{l,a_{ij}}^B \in \mathbb{R}^{d_{out} \times 1 \times d_{out}}$, which projects the query and key vector into a single-dimensional bias. Regarding the second term, we directly represent prior bias for each distance, regardless of its context, in $b_{m,a_{ij}}$.

Decomposed Linear Transformation. We introduce bias on the query and key vector respectively, according to Xu *et al.* [31], thus the bias is decomposed into:

$$bias_{ij}^l = \mathbf{q}_i^m \mathbf{K}_{m,a_{ij}}^T + \mathbf{Q}_{m,a_{ij}} \mathbf{k}_j^m + b_{m,a_{ij}} \tag{10}$$

where $\mathbf{K}_{m,a_{ij}}, \mathbf{Q}_{m,a_{ij}} \in \mathbb{R}^d$ are also trainable neural layers.

So the overall computation of ditance-dependent self-attention is:

$$
\begin{aligned}
\tilde{e}_{ij}^m &= \frac{\mathbf{q}_i^m \mathbf{k}_j^{m^T} + bias\,network\,(\mathbf{q}_i^m, \mathbf{k}_j^m, a_{ij})}{\sqrt{d}} \\
&= \frac{\mathbf{q}_i^m \mathbf{k}_j^{m^T} + \mathbf{q}_i^m \mathbf{W}_{m,a_{ij}}^B \mathbf{k}_j^m + b_{m,a_{ij}}}{\sqrt{d}} \\
or \\
&= \frac{\mathbf{q}_i^m \mathbf{k}_j^{m^T} + \mathbf{q}_i^m \mathbf{K}_{m,a_{ij}}^T + \mathbf{Q}_{m,a_{ij}} \mathbf{K}_j^{m^T} + b_{m,a_{ij}}}{\sqrt{d}}
\end{aligned} \tag{11}
$$

Adaptive distance dependencies aren't shared between different layers or attention heads since Bias Network model them based on context.

2.4   Neighbors Enhanced Loss

The proposed SDBN model takes the document text D as input and builds its contextual representation within and throughout the encoding stage, guided by the distance. We create a fixed dimensional representation for each target entity using average pooling after the encoding stage. The entity representation is denoted as $e_i \in \mathbb{R}^{d_e}$. Thus for the ith entity pair in D, to determine whether there is a relation j, we have the following equation:

$$P_{ij} = sigmoid(\mathbf{e}_{ih} \mathbf{W}_j \mathbf{e}_{it}) \tag{12}$$

where $\mathbf{W}_j \in \mathbb{R}^{d_e \times d_e}$, $\mathbf{e}_{ih}$ is the head entity and $\mathbf{e}_{it}$ is the tail entity. The model is trained using a loss consisting of two components: Binary Cross-Entropy Loss and Neighbors Enhanced Loss.

Binary Cross-Entropy Loss. This loss function is widely adopted and we also use it as part of the loss function.

$$\mathcal{L}_1 = - \sum_{i_h \neq i_t} \sum_{j \in \mathcal{R}} (r_{ij} \log P_{ij}$$
$$+ (1 - r_{ij}) \log(1 - P_{ij})) \tag{13}$$

where $\mathcal{R}$ denotes the set of relation types and $r_{ij} \in \{0,1\}$ is the groundtruth label regarding entity pair i and relation j.

Neighbors Enhanced Loss. Based on our findings in Sect. 1, the closer the context is to an entity, the greater its value to that entity, especially after crossing-distance was used to enhance the entity context. To give the model this tendency, we designed Neighbors Enhanced loss. Since we use SDBN as an encoder, which has learned the crossing-distance dependencies at the token level, directly using their attention heads to enhance the attention to neighbors is appropriate.

Specifically, given a pre-trained multi-head attention matrix $\mathbf{G} \in \mathbb{R}^{H \times n \times n}$, where $\mathbf{G}_{kij}$ indicates attention score from token i to token j in the kth attention head; n is the length of the input token sequence. Previously we have recorded the distance between token i and token j using a_{ij}. Suppose we want token i to pay more attention to contexts within the distance $\beta (\beta \leq \mu, \beta$ is a hyper-parameter.), we only need to use $a_{ij} \leq \beta$ to find the attention score of token i to these contexts. Therefore we use the following loss:

$$\mathcal{L}_2 = -log(\frac{\sum_{k \in H} \sum_{a_{ij} \leq \beta} \mathbf{G}}{\sum_{k \in H} \sum_{a_{ij} \in A} \mathbf{G}}) \tag{14}$$

$\mathcal{L}_2$ will push the attention score of $a_{ij} \leq \beta$ higher than the other attention scores. In our experiments, we use the attention matrix of the last SDBN layer, which avoids damaging the attentional flow and serves as a guide. Thus the total loss is:

$$\mathcal{L} = \mathcal{L}_1 + \lambda \mathcal{L}_2 \tag{15}$$

where λ is a hyper-parameter to control the attention flow.

3   Experiment

3.1   Datasets

DocRED. DocRED [34] is a relation extraction dataset created from Wikipedia and Wikidata. The dataset's documents are each annotated by humans with references to named entities, coreference data, intra- and inter-sentence relations, and supporting documentation. The dataset also offers vastly distantly supervised data in addition to human-annotated data. The two major metrics for evaluation are Ign F1 and F1 score according to Yao *et al.* [34], where Ign F1 is the F1 score that excludes triples from the annotated training data. The test results need to be submitted to the official Codalab.[1]

CDR. CDR [11] is a human-annotated chemical-disease relation extraction dataset in biomedicine, consisting of 500 documents, which is tasked with predicting binary interactions between chemical and disease concepts.

GDA. GDA [29] is a large-scale dataset in the biomedical domain and its task is to predict the binary interactions between genes and disease concepts. It contains 29192 documents for training, and we took 20% of the training set as the development set.

3.2 Implementation Details

Our model is implemented based on the Huggingface Transformers [27], using the pre-trained models Roberta-large [14] and Bert-large [4] and SciBERT-base [1]. For the DocRED dataset, we first train using the distantly supervised dataset to initialize the bias network. All hyper-parameters are grid searched and the best performers are used on the test set. The optimizer used for all experiments is Adam. The model is trained on a single NVIDIA V100 GPU with 32 GB memory.

3.3 Compared Methods

Due to the effectiveness of the pre-trained model, many models are constructed based on it, including ours. We mainly compare with these Transformer-based models.

- Wang *et al.* [26] built an enhanced Bert baseline using a two-stage prediction approach.
- Ye *et al.* [35] proposed a CorefBERT, which can help pre-trained models to better exploit co-reference relations in the context.
- Xu *et al.* [33] explicitly created paths such as logical reasoning and co-reference disambiguation in DocRE.
- Zhou *et al.* [43] proposed a method to enhance the entity context and introduce an adaptive threshold to solve the multi-label problem.
- Xu *et al.* [32] was designed with a loss function that allows the model to focus more on evidence sentences, which can reduce noise and improve the robustness of the model.
- Xie *et al.* [30] designed a lightweight model for extracting evidence sentences to be trained jointly with the RE model to help the RE model focus on evidence sentences.

---

[1] https://competitions.codalab.org/competitions/20717. Our model is named SDBN-DF.

- Zhang *et al.* [37] used the U-shaped Network in computer vision on DocRE to get global information at the entity-level.
- Xu *et al.* [31] used prior knowledge of entity structure to help model inference, which inspired our work.

Table 2. Results on DocRED. Subscript DL and BL refer to Decomposed Linear Transformation and Bilinear Transformation.

Models	Dev		Test	
	Ign F1	F1	Ign F1	F1
Coref-Roberta large [35]	57.35	59.43	57.90	60.25
BERT Two-stage [26]	–	54.42	–	53.92
GAIN+SIEF [32]	59.82	62.24	59.87	62.29
ATLOP-Roberta large [43]	61.46	63.37	61.39	64.40
DRN-Bert base [33]	59.33	61.39	59.15	61.37
EIDER-Roberta large [30]	62.48	64.37	62.85	64.79
DocuNet-Roberta large [37]	62.35	64.26	62.39	64.55
SSAN+Adaptation [31]	63.76	65.69	63.78	65.92
SDBN$_{DL}$+Bert large	63.32	65.38	63.38	65.42
SDBN$_{BL}$+Bert large	63.47	65.34	63.58	65.46
SDBN$_{DL}$+Roberta large	64.38	66.92	64.62	67.03
SDBN$_{BL}$+Roberta large	**64.72**	**67.26**	**64.88**	**67.12**

3.4 Main Results

Table 2 shows the main results on the DocRED dataset. We used Bert large and Roberta large pre-trained models on DF and BL, respectively. Similar to the results of other papers, Roberta large gives a significantly better result than Bert large. BL gives a slightly better result than DL, indicating that the former is better adapted to the introduced crossing-distance.

We compared the model EIDER-Roberta large [30], which also utilizes evidence sentences, and obtained a 2.03/2.33 boost on the test set of lgnF1/F1. Meanwhile, we compared SSAN [31], which also introduces attention bias, and obtained a 1.1/1.2 improvement on lgnF1/F1 of the test set. The comparison with other models demonstrates the effectiveness of our SDBN. Table 3 shows the main results of CDR and GDA datasets. We use the SciBERT base as a pre-trained model, which has a better performance in the biomedical domain. On the CDR and GDA test set, we improved by 2.10/1.72 over CGM2IR-SciBERT [41] on F1, which is already outperforming all existing work. These results demonstrate that our approach is highly applicable and generalizable.

3.5 Ablation Study

On DocRED, we conducted an ablation study of the best-performing method, SDBN$_{BL}$+Roberta large, by turning off one component at a time. The results of the model ablation study are shown in Table 4. It is obvious that the results of the model decrease significantly when there is no Bias Network in the model. However, when the NEL is removed, the Bias Network alone also gets better results. It implies that our proposed crossing-distance is a better guide to the attention flow and the model benefits from it.

The NEL enhancement is not apparent, because the context at close range, although helpful, also contains distracting information. However, NEL can influence the tendency of attention bias in the Bias Network, and the two complement each other to achieve better results.

Table 3. Results on CDR and GDA.

Model	CDR	GDA
ATLOP-SciBERT [43]	69.4	83.90
EIDER-SciBERT [30]	70.6	84.54
CGM2IR-SciBERT [41]	73.8	84.70
SSAN-SciBERT [31]	68.7	83.70
$SDBN_{DL}$+SciBERT base	**75.9**	86.08
$SDBN_{BL}$+SciBERT base	75.2	**86.42**

Table 4. Ablation Study of SDBN on DocRED. We turn off different components of the model one at a time.

Model	Ign F1	F1
$SDBN_{BL}$+Roberta large	64.88	67.12
- Bias Network	61.82	63.38
- Neighbors Enhanced Loss	63.62	66.24
- both	59.47	61.42

3.6 Robustness Analysis

We designed an experiment to verify the robustness of our model. In the training stage, we make no changes and input the whole document. During the testing stage, we randomly remove two of the non-evidence sentences that are not evidence of any entities from each document in the development set. Every 5 epochs, the result of F1 is compared to the performance of the model without deleting sentences, and the absolute value of the difference is recorded.

Our approach is compared with Xie et al. [30] and Xu et al. [32], both of which share the central notion of making the model concentrate more on the evidence sentences. The comparison results are shown in Fig. 3. Obviously, our SDBN changes the least after randomly deleting two non-evidence sentences, since this does not change the crossing-distance of the evidence sentences from the entity. This shows the robustness of our model, as it is less sensitive to non-evidence sentences for DocRE.

3.7 Hyper-parameter Analysis

Our model has two hyper-parameters μ and β. μ is a maximum distance cut when computing the crossing-distance, and distance exceeding μ will be cut to μ. The meaning of β is that the token needs to be more focused on the context within β distance.

The best result of our grid search is $\mu = 5$ and $\beta = 3$. Therefore, our experiments on F1 score fix one of the hyperparameters and then change the other one. The experimental results are in Fig. 4. Since $\beta \leq \mu$, the part of $\beta > \mu$ in

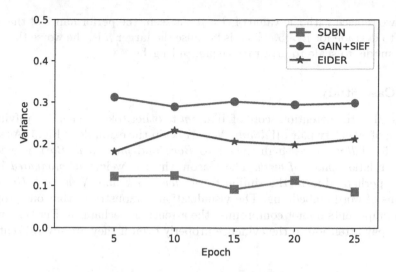

Fig. 3. Robustness of DocRE models.

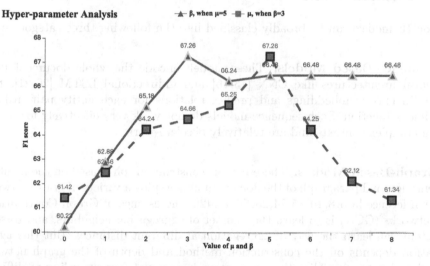

Fig. 4. Performances of the classification (in F1 score) on the development set of different hyper-parameter β and μ.

William B. Maclay

[1] *William* was a United States Representative from *New York.*

...

[4] Born in *New York City* , he received private instruction and was graduated from the *New York University* in 1836 ...

Fig. 5. Case study on DocRED dataset.

Fig. 4 we replace with the value of $\beta = \mu$. As seen, the performance of the model starts to degrade for $\mu > 5$, which is because the larger μ is, the worse the ability of our model to scale to arbitrary sequence lengths.

3.8   Case Study

We visualize the attention score of *William* to other tokens when identifying the relation of an entity pair (*William, New York*) in the example of Fig. 1. As shown in Fig. 5, *William* gives high weight to *Born* and *graduated*. *Born* is evidence for the relation *place of birth*. The reason why the weight of *graduated* is also high is perhaps due to the similarity of *New York* and *New York University* in terms of word embedding. The visualization demonstrates that our proposed three components do not compromise the attention mechanism. Entities are still able to pay attention to the evidence properly even if they are several sentences apart.

4   Related Work

DocRE models can be broadly classified into the following three categories:

Sequence-Based Models. These models encode the whole document using neural architectures like CNN [13,16] and bidirectional LSTM [20,21], then obtain entity embeddings and predict relations for each entity pair utilizing bilinear function. Such sequence models do not work very effectively for modeling complex contexts and are relatively obsolete work.

Graph-Based Models. These models construct graphs based on the mention, sentence, and paragraph of the document and employ a variety of graph networks for inference [3,5,8,10,15,24,25,36,40,42]. The essence of Graph Convolutional Networks (GCN) is to learn the manner of aggregating neighbors, and to some extent can learn the way of aggregation at different distances. But this aggregation depends on the construction method and depth of the graph network, and will not personalize the aggregation in various ways according to different nodes. The Graph Attention Networks (GAT) will be personalized to calculate aggregation weights based on node information, but the distance information is weakened. Our work can be seen as a neutralization of GCN and GAT, calculating the attention of one token to another token considering both the personalized embedding of the two as well as the distance between them.

Transformer-Based Models. Without using graph structures, these models adapt pre-trained language models directly to DocRE [22,26,33,35,37,43]. These models achieve great performance based on the strong adaptability of the pre-trained models.

Xu *et al.* [31] used a priori information on entity structure to guide the attention flow of the pre-trained model, which inspired our work. Unlike it, we propose a new distance calculation method to guide the attention flow and design an auxiliary loss to help the model make better utilization of the attention bias.

5 Conclusion

In this work, we propose three novel techniques SDBN and NEL, as well as a new way of computing the distance between tokens. The new distance enhances the context of entities; SDBN allows tokens to have different attention preferences for contexts of different distances; NFL allows tokens to pay more attention to closer contexts. The three can be perfectly combined to enable the model to pay more attention to useful contexts. This will reduce the interference of irrelevant information when identifying the relation of entity pairs.

Pronouns do not belong to any entity, hence they cannot benefit from crossing-distance. The model can only rely on the powerful encoding capabilities of self-attention to implicitly exploit pronoun information. In the future, we intend to refer to Ye *et al.* [35], which explicitly exploits pronoun information.

Acknowledgement. We would like to thank the anonymous reviewers for their insightful feedback and comments. This work is partially supported by China Natural Science Foundation under grant (No. 62171391) and the Natural Science Foundation of Fujian Province of China under grant (No. 2020J01053).

References

1. Beltagy, I., Lo, K., Cohan, A.: SciBERT: a pretrained language model for scientific text. In: Proceedings of the 2019 Conference on Empirical Methods in Natural Language Processing and the 9th International Joint Conference on Natural Language Processing (EMNLP-IJCNLP), pp. 3615–3620. Association for Computational Linguistics, Hong Kong, China (2019). https://doi.org/10.18653/v1/D19-1371. https://aclanthology.org/D19-1371

2. Cao, Y., Ji, X., Lv, X., Li, J., Wen, Y., Zhang, H.: Are missing links predictable? an inferential benchmark for knowledge graph completion. In: Proceedings of the 59th Annual Meeting of the Association for Computational Linguistics and the 11th International Joint Conference on Natural Language Processing (Volume 1: Long Papers), pp. 6855–6865. Association for Computational Linguistics, Online (2021). https://doi.org/10.18653/v1/2021.acl-long.534. https://aclanthology.org/2021.acl-long.534

3. Christopoulou, F., Miwa, M., Ananiadou, S.: Connecting the dots: document-level neural relation extraction with edge-oriented graphs. In: Proceedings of the 2019 Conference on Empirical Methods in Natural Language Processing and the 9th International Joint Conference on Natural Language Processing (EMNLP-IJCNLP), pp. 4925–4936. Association for Computational Linguistics, Hong Kong, China (2019). https://doi.org/10.18653/v1/D19-1498. https://aclanthology.org/D19-1498

4. Devlin, J., Chang, M.W., Lee, K., Toutanova, K.: BERT: pre-training of deep bidirectional transformers for language understanding. arXiv preprint arXiv:1810.04805 (2018). https://arxiv.org/abs/1810.04805

5. Guo, Z., Zhang, Y., Lu, W.: Attention guided graph convolutional networks for relation extraction. In: Proceedings of the 57th Annual Meeting of the Association for Computational Linguistics, pp. 241–251. Association for Computational Linguistics, Florence, Italy (2019). https://doi.org/10.18653/v1/P19-1024. https://aclanthology.org/P19-1024

6. Huang, Q., Zhu, S., Feng, Y., Ye, Y., Lai, Y., Zhao, D.: Three sentences are all you need: local path enhanced document relation extraction. In: Proceedings of the 59th Annual Meeting of the Association for Computational Linguistics and the 11th International Joint Conference on Natural Language Processing (Volume 2: Short Papers). pp. 998–1004. Association for Computational Linguistics, Online (2021). https://doi.org/10.18653/v1/2021.acl-short.126. https://aclanthology.org/2021.acl-short.126

7. Jia, R., Lewis, M., Zettlemoyer, L.: Question answering infused pre-training of general-purpose contextualized representations. In: Findings of the Association for Computational Linguistics: ACL 2022, pp. 711–728. Association for Computational Linguistics, Dublin, Ireland (May 2022). https://doi.org/10.18653/v1/2022.findings-acl.59. https://aclanthology.org/2022.findings-acl.59

8. Kipf, T.N., Welling, M.: Semi-supervised classification with graph convolutional networks. arXiv preprint arXiv:1609.02907 (2016). https://arxiv.org/abs/1609.02907

9. Lee, B.K., Lessler, J., Stuart, E.A.: Weight trimming and propensity score weighting. PLoS ONE 6(3), e18174 (2011)

10. Li, B., Ye, W., Sheng, Z., Xie, R., Xi, X., Zhang, S.: Graph enhanced dual attention network for document-level relation extraction. In: Proceedings of the 28th International Conference on Computational Linguistics. pp. 1551–1560. International Committee on Computational Linguistics, Barcelona, Spain (Online) (2020). https://doi.org/10.18653/v1/2020.coling-main.136. https://aclanthology.org/2020.coling-main.136

11. Li, J., Sun, Y., et al.: BioCreative V CDR task corpus: a resource for chemical disease relation extraction. Database 2016, baw068 (2016)

12. Lin, T.Y., RoyChowdhury, A., Maji, S.: Bilinear CNN models for fine-grained visual recognition. In: Proceedings of the IEEE International Conference on Computer Vision, pp. 1449–1457 (2015)

13. Liu, C.Y., Sun, W.B., Chao, W.H., Che, W.X.: Convolution neural network for relation extraction. In: Motoda, H., Wu, Z., Cao, L., Zaiane, O., Yao, M., Wang, W. (eds.) ADMA 2013. LNCS (LNAI), vol. 8347, pp. 231–242. Springer, Heidelberg (2013). https://doi.org/10.1007/978-3-642-53917-6_21

14. Liu, Y., et al.: Roberta: a robustly optimized bert pretraining approach. arXiv preprint arXiv:1907.11692 (2019). https://arxiv.org/abs/1907.11692

15. Nan, G., Guo, Z., Sekulic, I., Lu, W.: Reasoning with latent structure refinement for document-level relation extraction. In: Proceedings of the 58th Annual Meeting of the Association for Computational Linguistics, pp. 1546–1557. Association for Computational Linguistics, Online (2020). https://doi.org/10.18653/v1/2020.acl-main.141. https://aclanthology.org/2020.acl-main.141

16. Nguyen, T.H., Grishman, R.: Relation extraction: Perspective from convolutional neural networks. In: Proceedings of the 1st Workshop on Vector Space Modeling for Natural Language Processing, pp. 39–48. Association for Computational Linguistics, Denver, Colorado (2015). https://doi.org/10.3115/v1/W15-1506. https://aclanthology.org/W15-1506

17. Papanikolaou, Y., Staib, M., Grace, J.J., Bennett, F.: Slot filling for biomedical information extraction. In: Proceedings of the 21st Workshop on Biomedical Language Processing, pp. 82–90. Association for Computational Linguistics, Dublin, Ireland (2022). https://doi.org/10.18653/v1/2022.bionlp-1.7. https://aclanthology.org/2022.bionlp-1.7

18. Park, S., Kim, H.: Improving sentence-level relation extraction through curriculum learning. arXiv preprint arXiv:2107.09332 (2021), https://arxiv.org/abs/2107.09332

19. Shaw, P., Uszkoreit, J., Vaswani, A.: Self-attention with relative position representations. In: Proceedings of the 2018 Conference of the North American Chapter of the Association for Computational Linguistics: Human Language Technologies, Volume 2 (Short Papers), pp. 464–468. Association for Computational Linguistics, New Orleans, Louisiana (2018). https://doi.org/10.18653/v1/N18-2074. https://aclanthology.org/N18-2074

20. Song, L., Zhang, Y., Wang, Z., Gildea, D.: N-ary relation extraction using graphstate LSTM. In: Proceedings of the 2018 Conference on Empirical Methods in Natural Language Processing, pp. 2226–2235. Association for Computational Linguistics, Brussels, Belgium (2018). https://doi.org/10.18653/v1/D18-1246. https://aclanthology.org/D18-1246

21. Sorokin, D., Gurevych, I.: Context-aware representations for knowledge base relation extraction. In: Proceedings of the 2017 Conference on Empirical Methods in Natural Language Processing, pp. 1784–1789. Association for Computational Linguistics, Copenhagen, Denmark (2017). https://doi.org/10.18653/v1/D17-1188. https://aclanthology.org/D17-1188

22. Tang, H., et al.: HIN: hierarchical inference network for document-level relation extraction. In: Lauw, H.W., Wong, R.C.-W., Ntoulas, A., Lim, E.-P., Ng, S.-K., Pan, S.J. (eds.) PAKDD 2020. LNCS (LNAI), vol. 12084, pp. 197–209. Springer, Cham (2020). https://doi.org/10.1007/978-3-030-47426-3_16

23. Vaswani, A., et al.: Attention is all you need. In: Advances in Neural Information Processing Systems 30 (2017)

24. Veličković, P., Cucurull, G., Casanova, A., Romero, A., Lio, P., Bengio, Y.: Graph attention networks. arXiv preprint arXiv:1710.10903 (2017)

25. Wang, D., Hu, W., Cao, E., Sun, W.: Global-to-local neural networks for document-level relation extraction. In: Proceedings of the 2020 Conference on Empirical Methods in Natural Language Processing (EMNLP), pp. 3711–3721. Association for Computational Linguistics, Online (2020). https://doi.org/10.18653/v1/2020.emnlp-main.303. https://aclanthology.org/2020.emnlp-main.303

26. Wang, H., Focke, C., Sylvester, R., Mishra, N., Wang, W.: Fine-tune bert for DocRED with two-step process. arXiv preprint arXiv:1909.11898 (2019)

27. Wolf, T., et al.: Transformers: State-of-the-art natural language processing. In: Proceedings of the 2020 Conference on Empirical Methods in Natural Language Processing: System Demonstrations, pp. 38–45. Association for Computational Linguistics, Online (2020). https://doi.org/10.18653/v1/2020.emnlp-demos.6. https://aclanthology.org/2020.emnlp-demos.6

28. Wu, X., Zhang, J., Li, H.: Text-to-table: a new way of information extraction. In: Proceedings of the 60th Annual Meeting of the Association for Computational Linguistics (Volume 1: Long Papers), pp. 2518–2533. Association for Computational Linguistics, Dublin, Ireland (2022). https://doi.org/10.18653/v1/2022.acl-long.180. https://aclanthology.org/2022.acl-long.180

29. Wu, Y., Luo, R., Leung, H.C.M., Ting, H.-F., Lam, T.-W.: RENET: a deep learning approach for extracting gene-disease associations from literature. In: Cowen, L.J. (ed.) RECOMB 2019. LNCS, vol. 11467, pp. 272–284. Springer, Cham (2019). https://doi.org/10.1007/978-3-030-17083-7_17

30. Xie, Y., Shen, J., Li, S., Mao, Y., Han, J.: Eider: Empowering document-level relation extraction with efficient evidence extraction and inference-stage fusion. In: Findings of the Association for Computational Linguistics: ACL 2022, pp. 257–268. Association for Computational Linguistics, Dublin, Ireland (2022). https://doi.org/10.18653/v1/2022.findings-acl.23. https://aclanthology.org/2022.findings-acl.23

31. Xu, B., Wang, Q., Lyu, Y., Zhu, Y., Mao, Z.: Entity structure within and throughout: modeling mention dependencies for document-level relation extraction. In: Proceedings of the AAAI Conference on Artificial Intelligence, vol. 35, pp. 14149–14157 (2021). https://arxiv.org/abs/2102.10249

32. Xu, W., Chen, K., Mou, L., Zhao, T.: Document-level relation extraction with sentences importance estimation and focusing. In: Proceedings of the 2022 Conference of the North American Chapter of the Association for Computational Linguistics: Human Language Technologies, pp. 2920–2929. Association for Computational Linguistics, Seattle, United States (2022). https://aclanthology.org/2022.naacl-main.212

33. Xu, W., Chen, K., Zhao, T.: Discriminative reasoning for document-level relation extraction. In: Findings of the Association for Computational Linguistics: ACL-IJCNLP 2021, pp. 1653–1663. Association for Computational Linguistics, Online (2021). https://doi.org/10.18653/v1/2021.findings-acl.144

34. Yao, Y., et al.: DocRED: a large-scale document-level relation extraction dataset.In: Proceedings of the 57th Annual Meeting of the Association for Computational Linguistics, pp. 764–777. Association for Computational Linguistics, Florence, Italy (2019). https://doi.org/10.18653/v1/P19-1074. https://aclanthology.org/P19-1074

35. Ye, D., Lin, Y., Du, J., Liu, Z., Li, P., Sun, M., Liu, Z.: Coreferential Reasoning Learning for Language Representation. In: Proceedings of the 2020 Conference on Empirical Methods in Natural Language Processing (EMNLP), pp. 7170–7186. Association for Computational Linguistics, Online (2020). https://doi.org/10.18653/v1/2020.emnlp-main.582. https://aclanthology.org/2020.emnlp-main.582

36. Zeng, S., Xu, R., Chang, B., Li, L.: Double graph based reasoning for document-level relation extraction. In: Proceedings of the 2020 Conference on Empirical Methods in Natural Language Processing (EMNLP), pp. 1630–1640. Association for Computational Linguistics, Online (2020). https://doi.org/10.18653/v1/2020.emnlp-main.127. https://aclanthology.org/2020.emnlp-main.127

37. Zhang, N., et al.: Document-level relation extraction as semantic segmentation. arXiv preprint arXiv:2106.03618 (2021). https://www.ijcai.org/proceedings/2021/0551.pdf

38. Zhang, Y., Li, P., Liang, H., Jatowt, A., Yang, Z.: Fact-tree reasoning for N-ary question answering over knowledge graphs. In: Findings of the Association for Computational Linguistics: ACL 2022, pp. 788–802. Association for Computational Linguistics, Dublin, Ireland (2022). https://doi.org/10.18653/v1/2022.findings-acl.66. https://aclanthology.org/2022.findings-acl.66

39. Zhang, Y., Qi, P., Manning, C.D.: Graph convolution over pruned dependency trees improves relation extraction. In: Proceedings of the 2018 Conference on Empirical Methods in Natural Language Processing, pp. 2205–2215. Association for Computational Linguistics, Brussels, Belgium (2018). https://doi.org/10.18653/v1/D18-. https://aclanthology.org/D18-1244

40. Zhang, Z., Yu, B., Shu, X., Liu, T., Tang, H., Yubin, W., Guo, L.: Document-level relation extraction with dual-tier heterogeneous graph. In: Proceedings of the 28th International Conference on Computational Linguistics, pp. 1630–1641. International Committee on Computational Linguistics, Barcelona, Spain (Online) (2020). https://doi.org/10.18653/v1/2020.coling-main.143. https://aclanthology.org/2020.coling-main.143

41. Zhao, C., Zeng, D., Xu, L., Dai, J.: Document-level relation extraction with context guided mention integration and inter-pair reasoning. arXiv preprint arXiv:2201.04826 (2022)

42. Zhou, H., Xu, Y., Yao, W., Liu, Z., Lang, C., Jiang, H.: Global context-enhanced graph convolutional networks for document-level relation extraction. In: Proceedings of the 28th International Conference on Computational Linguistics, pp. 5259–5270. International Committee on Computational Linguistics, Barcelona, Spain (Online) (2020). https://doi.org/10.18653/v1/2020.coling-main.461. https://aclanthology.org/2020.coling-main.461

43. Zhou, W., Huang, K., Ma, T., Huang, J.: Document-level relation extraction with adaptive thresholding and localized context pooling. In: Proceedings of the AAAI Conference on Artificial Intelligence vol. 35, pp. 14612–14620 (2021). https://arxiv.org/abs/2010.11304

Investigating Conversational Agent Action in Legal Case Retrieval

Bulou Liu[1], Yiran Hu[2], Yueyue Wu[1], Yiqun Liu[1(✉)], Fan Zhang[3], Chenliang Li[4], Min Zhang[1], Shaoping Ma[1], and Weixing Shen[2]

[1] Quan Cheng Laboratory, Department of Computer Science and Technology, Institute for Internet Judiciary, Tsinghua University, Beijing, China
`yiqunliu@tsinghua.edu.cn`
[2] School of Law, Tsinghua University,Beijing, China
[3] School of Information Management, Wuhan University,Wuhan, China
[4] School of Cyber Science and Engineering, Wuhan University,Wuhan, China

Abstract. Legal case retrieval is a specialized IR task aiming to retrieve supporting cases given a query case. Existing work has shown that the conversational search paradigm can improve users' search experience in legal case retrieval with humans as intermediary agents. To move further towards a practical system, it is essential to decide what action a computer agent should take in conversational legal case retrieval. Existing works try to finish this task through Transformer-based models based on semantic information in open-domain scenarios. However, these methods ignore search behavioral information, which is one of the most important signals for understanding the information-seeking process and improving legal case retrieval systems. Therefore, we investigate the conversational agent action in legal case retrieval from the behavioral perspective. Specifically, we conducted a lab-based user study to collect user and agent search behavior while using agent-mediated conversational legal case retrieval systems. Based on the collected data, we analyze the relationship between historical search interaction behaviors and current agent actions in conversational legal case retrieval. We find that, with the increase of agent-user interaction behavioral indicators, agents are increasingly inclined to return results rather than clarify users' intent, but the probability of collecting candidates does not change significantly. With the increase of the interactions between the agent and the system, agents are more inclined to collect candidates than clarify users' intent and are more inclined to return results than collect candidates. We also show that the agent action prediction performance can be improved with both semantic and behavioral features. We believe that this work can contribute to a better understanding of agent action and useful guidance for developing practical systems for conversational legal case retrieval.

Keywords: Conversational search · Agent action · Legal case retrieval

---

B. Liu and Y. Hu—Equal contributions from both authors.

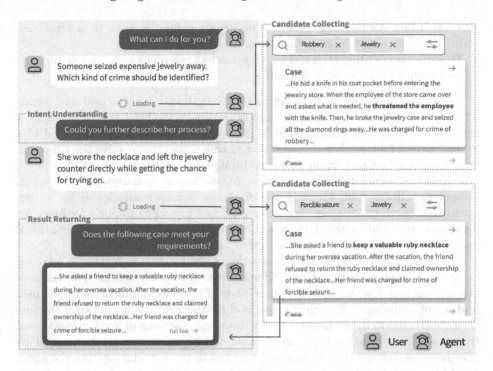

Fig. 1. An example of the three kinds of agent actions in conversational legal case retrieval.

1   Introduction

In recent years, legal case retrieval has attracted much attention in the IR research community. It aims to retrieve supporting cases for a given query case and contributes to better legal systems. Existing works show that an automatic system not only performs the legal case retrieval tasks with higher performance than lawyers, but also performs more efficiently [15]. Under traditional legal case retrieval systems, users need to issue queries to express their complex information needs [6,14], which requires sufficient domain knowledge [12,19,23]. Liu et al. [8,9] show that conversational search paradigm, where human experts play the role of intermediary conversational agents, can be adopted to improve users' legal case retrieval experience in terms of query formulation, result examination, and users' satisfaction and search success.

The conversational agent action prediction task aims to decide what action the agents will take based on the context of the conversation and helps provide useful and meaningful responses in conversational search systems [2]. As shown in Fig. 1, there are three kinds of conversational agent actions in legal case retrieval:

- **Intent Understanding (IU):** The agents ask clarifying questions to understand users' search intent better.

- **Candidate Collecting (CC):** The agents submit queries to a legal case ad-hoc search system and collect candidate cases.
- **Result Returning (RR):** The agents select relevant cases from candidates and return them to users as results.

It is essential to understand human conversational agent action and decide what action to take automatically before we move further towards a practical conversational search system (that is, to use an automated agent instead of a human expert) for legal case retrieval. Existing works try to solve the conversational agent action prediction problem through Transformer-based models [5,10] which exploit semantic information in open-domain scenarios [22]. However, behavioral information, which is one of the most important signals for understanding the information-seeking process [3] and providing implicit feedback for legal case retrieval system [19], has not been incorporated into conversational agent action prediction in legal case retrieval.

This paper investigates the conversational agent action in legal case retrieval from the behavioral perspective. Different from traditional search systems, conversational search systems contain two kinds of behavioral information (i.e., user and agent behaviors). Specifically, we analyze the relationship between the historical interaction behaviors and the current agent actions in legal case retrieval from two aspects: agent-user interactions and agent-system interactions. Furthermore, we try to utilize behavioral features to predict which action the agent would like to take. Our research questions are as follows:

- **RQ1:** What is the relationship between the historical behaviors and the current agent actions in legal case retrieval?
- **RQ2:** Can we improve the conversational agent action prediction performance with behavioral features involved in legal case retrieval?

To shed light on these research questions, we conducted a lab-based user study with 106 tasks to collect user and agent search behavior using agent-mediated conversational legal case retrieval systems. It's worth noting that no available conversational legal case retrieval system exists currently. Therefore, we recruit legal experts as intermediary agents to complete the procedure in a wizard-of-oz fashion. To answer RQ1, we compare the differences in user and agent historical behaviors w.r.t. different conversational agents' actions. Furthermore, we define the conversational agent action prediction as a classification task and demonstrate the effectiveness of features extracted from the user and agent behaviors for RQ2.

2 Related Work

Legal case retrieval is a specialized IR task that differs from general web search in various aspects, such as the needs for data authority [1] and the definition of relevance [6,7,11,18]. Behavior information plays an important role in legal case retrieval. [19] investigated user behavior in legal case retrieval. They found that users of legal case retrieval devote more search effort and appear to be more patient and cautious. They further applied the behavioral features to relevance prediction. [8] have shown that conversational search paradigm can be adopted to improve users' legal case retrieval experience in terms of query formulation, result examination, and users' satisfaction and search success. They revealed that it is necessary to develop conversational legal case retrieval systems.

And agent action prediction is important for develop practical conversational systems. [16] proposed a hierarchical deep reinforcement learning approach to learning the dialogue policy at different temporal scales. [21] presented an agent that efficiently learns dialogue policy from demonstrations through policy shaping and reward shaping. [22] proposed a Transformer-based model to predict agent action in conversational search systems in open-domain scenarios.

Compared to these studies, our work focuses on the behavioral perspective of the conversational agent action in legal case retrieval.

3   User Study

To investigate the relationship between the historical behaviors and the current agent actions in conversational legal case retrieval, we conducted a lab-based user study with 106 tasks. We show the details in this section.

3.1   Conversational Legal Case Retrieval

It's worth noting that no available conversational legal case retrieval system exists currently. We add an intermediary agent to complete the procedure in a wizard-of-oz fashion. The agent needs to understand users' intents via conversations, construct queries and pick cases from SERPs for the user. Specifically, the procedure contains the following steps:

1. The user submits a legal issue question to the agent in natural language.
2. The agent asks clarifying questions until the background information of the search issue is sufficient.
3. The agent submits queries to the legal case retrieval system. She then selects cases from the SERPs and responds to the user with the selected ones.

In particular, the conversational legal case retrieval procedure contains rich logs of behavioral information. On the one hand, it contains agents' interactions with users, such as search questions, clarifying questions, and the cases returned by the agent. On the other hand, it includes agents' interactions with the system, such as queries and clicks. Therefore, we extracted ten behavioral features from two aspects: agent-user interactions and agent-system interactions, which are shown in Table 1. In detail, as for agent-user interaction behaviors, we focus on conversational input behaviors and agent answering behaviors. As for agent-system interactions, we concentrate on query formulation and the search engine result page (SERP) examination behaviors, especially the examination behaviors in the SERPs from the last query. Note that there were no users' interactions with the system because the user study dataset was collected in a wizard-of-oz approach. And we just kept behavioral information before the current action for analysis, i.e., the same setting for the action prediction task.

Table 1. The list of 10 behavioral features extracted from the agent-user interactions and agent-system interactions.

Group	Behavioral Features
Agent-User	Number of utterances/words in conversations Number of returned results/returned cases
Agent-System	Number of queries/query words Number of clicks in all queries/in last query Avg./Max. click rank in last query

3.2 Tasks and Participants

We collected 106 search tasks from legal practitioners' real information need via online forums and social networks, covering 3 legal domains: 34 civil tasks (involving 10 topics, such as "Inheritance", "Personality rights", "Contracts" and "Marriage"), 35 criminal tasks (involving 7 topics, such as "Robbery", "Fraud", "Bribery", "Forcible Rape" and "Traffic accident") and 37 commercial tasks (involving 9 topics, such as "Company", "Expertise Bankruptcy" and "Insurance"). Compared with existing user studies for legal case retrieval or conversational search [19,20], we believe that the number of tasks is enough for a between-subjects analysis. Each task contained a query case description and a legal issue. Users were expected to retrieve legal cases which may help to answer the issue question.

There were two kinds of participants: users and agents. As for users, we recruited 30 participants (12 males and 18 females) via online forums and social networks. They were all native Chinese speakers and college law students. All users had no previous experience with conversational search systems. No users conducted two tasks in the same topic, which also can avoid the task learning effects on the results. Note that the tasks have negligible or no learning effects on each other even in the same domain if they are not in the same topics.

We recruited 15 graduate students from law school (5 for civil law, 5 for criminal law and 5 for commercial law) to be agents. They were all native Chinese speakers and qualified in legal practice[1]. To ensure an adequate level of domain expertise, they only participated in the task related to their research fields. In addition, they all achieved a score of 95 or more in the courses corresponding to their experimental topics. This can reduce the effect of individual variability. And they were trained with 5 auxiliary search tasks beforehand to familiarize with the query construction skills in the legal case retrieval system, guaranteeing an adequate level of search expertise.

---

[1] They had passed the "National Uniform Legal Profession Qualification Examination".

Table 2. Statistics of agent actions in the user study dataset.

#Tasks	#Intent Understanding	#Candidate Collecting	#Result Returning
106	437	385	208

As for the legal case retrieval system, we choose a leading commercial legal search engine[2] in China. Users and agents had a conversation (just in text form) via Zoom[3].

3.3   Procedure

Before the experiments, we firstly requested each participant to complete a warm-up search task. We then introduce the details of the procedure as follows:

Query Case and Issue Reading. In the first step, the user read the query case description and the legal issue carefully. She could refer to the query case at any time during the session, so she did not need to memorize the case description at this step.

Pre-task Questionnaire. Next, the user was asked to finish a pre-search questionnaire, including: domain knowledge level, task difficulty level, and prior interest level of the task with a 5-point Likert scale (1: not at all, 2: slightly, 3: somewhat, 4: moderately, 5: very).

Task Completion. After that, the user started performing searches with the agent. At this step, we collected the agent's interactions with the system, including queries, clicks, etc. Moreover, we recorded the conversation contents, including users' legal questions, agents' clarifying questions, the cases returned by the agent.

Post-task Questionnaire. After examining the supporting cases returned by the agent, the user was required to complete a post-task questionnaire. At this step, we collected explicit feedback signals with respect to the search experience, including five-grade workload and satisfaction.

Result Assessment. After completing the post-task questionnaire, the user was further asked to annotate the cases that agents clicked in the SERPs. That is, a relevance score is annotated to each case (1: irrelevant, 2: relevant). As for the cases that weren't clicked, we simply regarded them as irrelevant.

To drive the conversational legal case retrieval process, the intermediary agents can take three kinds of actions (i.e., Intent Understanding, Candidate Collecting, and Result Returning). Through these actions, the agents understand user intent by clarifying questions, collect candidate cases from the traditional legal search system by submitting queries, and return relevant cases from candidates to users, respectively. Table 2 shows distribution of each agent action in the user study dataset.

---

[2] https://ydzk.chineselaw.com/case.
[3] https://zoom.us/.

4   Results

4.1   Analysis on Conversational Agent Action

To address **RQ1**, we report the relationship between the historical interaction features and the current agent action using box plots. Specifically, we compare the differences in users' and agents' historical behaviors given different conversational agents' actions from two aspects: agent-user interactions and agent-system interactions. We also perform a series of one-way ANOVA tests [4] and pairwise t-tests [17] to determine the significance.

Comparison of Agent-User Interactions. Firstly, we compare historical agent-user interaction behaviors w.r.t. different agent actions. Here, we focus on conversational input behaviors and agent answering behaviors. The results of ANOVA tests (ANOVA-p) are reported in Fig. 2. We can make the following observations.

From the first line in Fig. 2, we can observe that the conversational input behavioral indicators (i.e., the number of utterances and words) show significant differences between the three agent actions (ANOVA-$p < 0.05$). Moreover, we find that the number of utterances and words under "Result Returning" action is significantly more than that under "Intent Understanding" action ($p < 0.05$). However, there are no significant differences according to pairwise t-tests in conversational input behaviors between the "Candidate Collecting" action and the other actions. This illustrates that the agent tends to adopt the "Intent Understanding" action when the conversation length is short and tends to adopt the "Result Returning" action when the conversation contains sufficient information. Furthermore, the agents may take the "Candidate Collecting" action regardless of the length of the conversation.

We further investigate the agent answering behaviors (i.e., the number of returned results and returned cases), and the results of ANOVA tests are shown in the second line in Fig. 2. There are also significant differences in these two indicators before agents adopt the three actions (ANOVA-$p < 0.01$). Furthermore, we find that the number of returned results and returned cases before agents take the "Intent Understanding" action are significantly less than that before agents take the other two actions ($p < 0.01$). And these two indicators do not show significant differences before the "Candidate Collecting" action and the "Result Returning" action. These indicate that as the agent answering behavioral indicators increases, the agent will decrease the probability of taking the "Intent Understanding" action and prefer to take the other two actions.

Comparison of Agent-System Interactions. Then we compare historical agent-system interaction behaviors w.r.t. different agent actions. Specifically, we concentrate on query formulation and SERP examination behaviors. The results of ANOVA tests (ANOVA-p) are reported in Fig. 3. We can make the following observations.

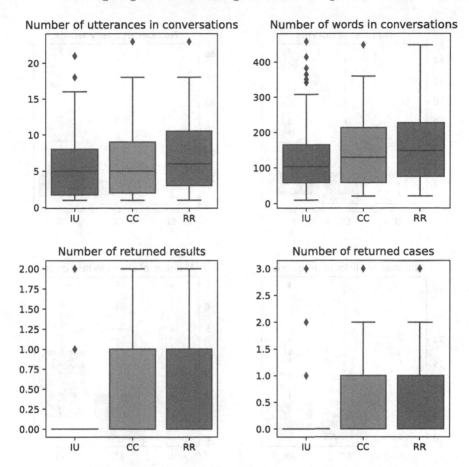

Fig. 2. Comparison of historical agent-user interaction behavioral measures given different current agent actions.

Overall, we find that the historical agent-system interaction behavioral indicators before the three conversational agent actions follow the following relative order: "Intent Understanding" < "Candidate Collecting" < "Result Returning" (shown in Fig. 3). We can observe that agents submitted fewer queries and query words before "Intent Understanding" actions than those before another action ($p < 0.001$). And the number of queries and query words before taking "Candidate Collecting" actions are less than those before taking "Result Returning" actions ($p < 0.001$). The above phenomenons also exist for the number of clicks, especially in the last query. This suggests that with the increase of the interactions between the agent and the system, agents are more inclined to collect candidates than clarify users' intent and are more inclined to return results than collect candidates.

Furthermore, we investigate two indicators related to the examination behaviors in the last SERP: the average/maximum click rank in the last query.

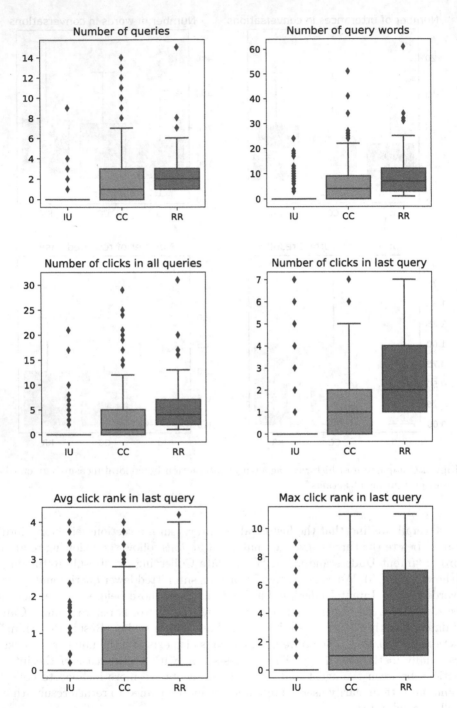

Fig. 3. Comparison of historical agent-system interaction behavioral measures given different current agent actions.

The results of ANOVA tests are shown in the last line in Fig. 3. Similarly, these two indicators show significant differences between the three actions (ANOVA-$p < 0.001$). And we can observe that the click rank is larger before the"Result Returning" action than that before the other two actions significantly ($p < 0.001$). It shows that the agents are more inclined to return results when they examine and click on the results with a larger rank. Because they are more convinced that they have understood user intent and submitted accurate queries.

Summary. To answer **RQ1**, our findings are as follows: 1) With the increase of agent-user interaction behavioral indicators, agents are increasingly inclined to return results rather than clarify users' intent, but the probability of collecting candidates does not change significantly; 2) With the increase of the interactions between the agent and the system, agents are more inclined to collect candidates than clarify users' intent and are more inclined to return results than collect candidates.

4.2   Conversational Agent Action Prediction

To address **RQ2**, we try to improve the conversational agent action prediction with behavioral features. Two groups of features are adopted in the experiments: agent-user behaviors and agent-system behaviors (ref. Table 1). We define the prediction task as a multi-class classification task and use Macro-F1 for evaluation. Furthermore, we further analyze the effect of these features through three binary classification tasks: IU vs. CC, IU vs. RR and CC vs. RR. We use the F1-score to evaluate the three classification tasks.

We first investigate the prediction performance without semantic features. Random is the baseline which decides actions randomly. As this task can be treated as a multi-class classification problem, we apply a gradient boosting classifier [13] and perform 5-fold cross-validation. The results are shown in Table 3. We can observe that using both groups of features achieves the best performance on all classification tasks and using agent-system interaction behavioral features also outperforms Random significantly. These suggest that both groups of features are useful for the conversational agent action prediction in legal case

Table 3. Performance comparison of conversational agent action prediction task. AU and AS denotes that the method incorporates agent-user and agent-system behavioral features, respectively. Best results are in boldface. † indicates that the difference to Random is statistically significant at 0.05 level from the student t-test.

Method	Overall	IU vs. CC	IU vs. RR	CC vs. RR
Random	0.3607	0.5259	0.5158	0.5112
AU	0.4018	0.5625	$0.8087^{\dagger}$	0.6182
AS	$0.5142^{\dagger}$	$0.7775^{\dagger}$	$0.8171^{\dagger}$	$0.6971^{\dagger}$
AU+AS	$\mathbf{0.5265^{\dagger}}$	$\mathbf{0.8058^{\dagger}}$	$\mathbf{0.8260^{\dagger}}$	$\mathbf{0.7043^{\dagger}}$

retrieval. And agent-user behaviors only significantly improve the performance of IU vs. RR task, suggesting that they do not provide much information to distinguish whether to take the Candidate Collecting action.

Existing works try to solve the conversational agent action prediction problem through Transformer-based models based on semantic information. To further investigate the prediction performance with semantic features, we concatenated all the utterances in the conversation together and used LawFormer [24] as the encoder. Here LawFormer is a Longformer-based pre-trained language model for Chinese legal long documents understanding. Then we fed the [CLS] embedding into full connected layers and fine-tuned LawFormer for each tasks as the baseline. The model is optimized by the cross-entropy loss. We performed 5-fold cross-validation and the results are shown in Table 4. We can find Law-Former outperforms all the methods without involving semantic features (shown in Table 3). It illustrates that semantic information is also very useful for the conversational agent action prediction in legal case retrieval.

Then we regarded the probability distribution (i.e., the probabilities of taking the three actions, 3-dimensional in total) of the full connected layers' output as semantic features. We combine the semantic features with behavioral features together and also apply a gradient boosting classifier to obtain the final agent action prediction results. Note that we just utilize the LawFormer fine-tuned in the baseline and do not take further fine-tuning strategies. And the model frame-

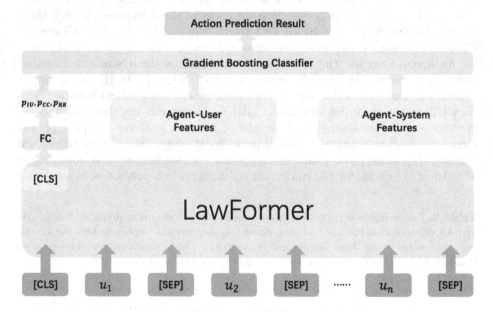

Fig. 4. Combining semantic and behavioral features for the conversational agent action prediction. $u_1, u_2, ..., u_n$ denote all the utterances in the conversation. FC denotes the full connected layers. p_{IU}, p_{CC}, p_{RR} denote the probabilities of taking the three actions predicted by the LawFormer baseline.

work is shown in Fig. 4. As for the three binary classification tasks: IU vs. CC, IU vs. RR and CC vs. RR, the semantic features change from three-dimensional to two-dimensional as the output of the fully connected layer changes. Other experimental settings remain the same and the results are shown in Table 4. AU and AS denotes that the method incorporates agent-user and agent-system behavioral features, respectively. We find that combining behavioral features with semantic features achieves significantly better classification performance than LawFormer, especially in the IU vs. CC task and the CC vs. RR task. This illustrates that historical behaviors are useful supplementary information for semantic features to distinguish whether to take the Candidate Collecting action.

Table 4. Performance comparison of conversational agent action prediction task with using semantic features. AU and AS denotes that the method incorporates agent-user and agent-system behavioral features, respectively. Best results are in boldface. † indicates that the difference to LawFormer is statistically significant at 0.05 level from the student t-test.

Method	Overall	IU vs. CC	IU vs. RR	CC vs. RR
LawFormer	0.5425	0.8232	0.8739	0.7335
LawFormer+AU	0.5675	0.8318	0.8844	0.7398
LawFormer+AS	0.5828	0.8428	0.8462	0.7564
LawFormer+AU+AS	**0.6177**†	**0.8669**†	**0.8870**	**0.8015**†

Concerning **RQ2**, we find that behavioral features can improve the conversational agent actions prediction performance in legal case retrieval whether semantic features are involved or not.

5 Conclusion

In this paper, we investigate three kinds of conversational agent actions (i.e., Intent Understanding, Candidate Collecting, and Result Returning) in legal case retrieval from a behavioral perspective. We find that with the increase of agent-user interaction behavioral indicators, agents are increasingly inclined to return results rather than clarify users' intent, but the probability of collecting candidates does not change significantly. Moreover, with the increase of the interactions between the agent and the system, agents are more inclined to collect candidates than clarify users' intent and are more inclined to return results than collect candidates. We further show that the agent action prediction performance can be improved with both semantic and behavioral features in legal case retrieval. We believe that this work can contribute to a better understanding of agent action and useful guidance for developing practical systems for conversational legal case retrieval.

As for future work, we firstly plan to utilize more sophisticated algorithms (e.g., reinforcement learning) to incorporate behavioral information into the legal conversational agent action prediction task more effectively. Secondly, we prepare to take more fine-grained behavioral information (e.g., mouse movements, hovers and so on) into consideration. At last, we also try to improve retrieval performance through more accurate action prediction in conversational legal case retrieval.

Acknowledgement. This work is supported by the Natural Science Foundation of China (Grant No. 61732008, 62002194) and Tsinghua University Guoqiang Research Institute.

References

1. Arewa, O.B.: Open access in a closed universe: Lexis, Westlaw, law schools, and the legal information market. Lewis Clark L. Rev. **10**, 797 (2006)
2. Azzopardi, L., Dubiel, M., Halvey, M., Dalton, J.: Conceptualizing agent-human interactions during the conversational search process. In: The Second International Workshop on Conversational Approaches to Information Retrieval (2018)
3. Buscher, G., Dengel, A., Biedert, R., Elst, L.V.: Attentive documents: eye tracking as implicit feedback for information retrieval and beyond. ACM Trans. Interact. Intell. Syst. (TiiS) **1**(2), 1–30 (2012)
4. Cuevas, A., Febrero, M., Fraiman, R.: An anova test for functional data. Comput. Statist. Data Anal. **47**(1), 111–122 (2004)
5. Devlin, J., Chang, M.W., Lee, K., Toutanova, K.: BERT: pre-training of deep bidirectional transformers for language understanding. arXiv preprint arXiv:1810.04805 (2018)
6. Ferrer, A.S., Hernández, C.F., Boulat, P.: Legal search: foundations, evolution and next challenges. the Wolters kKuwer experience. Revista Democracia Digital e Governo Eletrônico **1**(10), 120–132 (2014)
7. Fleiss, J.L.: Measuring nominal scale agreement among many raters. Psychol. Bull. **76**(5), 378 (1971)
8. Liu, B., et al.: Conversational vs traditional: comparing search behavior and outcome in legal case retrieval. In: Proceedings of the 44th International ACM SIGIR Conference on Research and Development in Information Retrieval, pp. 1622–1626 (2021)
9. Liu, B., et al.: Query generation and buffer mechanism: Towards a better conversational agent for legal case retrieval. Inf. Process. Manage. **59**(5), 103051 (2022)
10. Liu, Y., et al.: RoBERTa: a robustly optimized BERT pretraining approach. arXiv preprint arXiv:1907.11692 (2019)
11. Ma, Y., Shao, Y., Liu, B., Liu, Y., Zhang, M., Ma, S.: Retrieving legal cases from a large-scale candidate corpus. In: Proceedings of the Eighth International Competition on Legal Information Extraction/Entailment, COLIEE2021 (2021)
12. Mao, J., Liu, Y., Kando, N., Zhang, M., Ma, S.: How does domain expertise affect users' search interaction and outcome in exploratory search? ACM Trans. Inf. Syst. (TOIS) **36**(4), 1–30 (2018)
13. Mason, L., Baxter, J., Bartlett, P., Frean, M.: Boosting algorithms as gradient descent. In: Advances in Neural Information Processing Systems 12 (1999)
14. McGinnis, J.O., Pearce, R.G.: The great disruption: how machine intelligence will transform the role of lawyers in the delivery of legal services. Actual Probs. Econ. L. **82**, 1230 (2019)

15. McGinnis, J.O., Wasick, S.: Law's algorithm. Fla. L. Rev. **66**, 991 (2014)
16. Peng, B., et al.: Composite task-completion dialogue policy learning via hierarchical deep reinforcement learning. arXiv preprint arXiv:1704.03084 (2017)
17. Semenick, D.: Tests and measurements: the t-test. Strength Condition. J. **12**(1), 36–37 (1990)
18. Shao, Y., Liu, B., Mao, J., Liu, Y., Zhang, M., Ma, S.: Thuir@ coliee-2020: leveraging semantic understanding and exact matching for legal case retrieval and entailment. arXiv preprint arXiv:2012.13102 (2020)
19. Shao, Y., Wu, Y., Liu, Y., Mao, J., Zhang, M., Ma, S.: Investigating user behavior in legal case retrieval. In: Proceedings of the 44th International ACM SIGIR Conference on Research and Development in Information Retrieval, pp. 962–972 (2021)
20. Vtyurina, A., Savenkov, D., Agichtein, E., Clarke, C.L.: Exploring conversational search with humans, assistants, and wizards. In: Proceedings of the 2017 CHI Conference Extended Abstracts on Human Factors in Computing Systems, pp. 2187–2193 (2017)
21. Wang, H., Peng, B., Wong, K.F.: Learning efficient dialogue policy from demonstrations through shaping. In: Proceedings of the 58th Annual Meeting of the Association for Computational Linguistics, pp. 6355–6365 (2020)
22. Wang, Z., Ai, Q.: Controlling the risk of conversational search via reinforcement learning. In: Proceedings of the Web Conference 2021, pp. 1968–1977 (2021)
23. White, R.W., Dumais, S.T., Teevan, J.: Characterizing the influence of domain expertise on web search behavior. In: Proceedings of the second ACM International Conference on Web Search and Data Mining, pp. 132–141 (2009)
24. Xiao, C., Hu, X., Liu, Z., Tu, C., Sun, M.: Lawformer: a pre-trained language model for Chinese legal long documents. AI Open **2**, 79–84 (2021)

MS-Shift: An Analysis of MS MARCO Distribution Shifts on Neural Retrieval

Simon Lupart[1(✉)], Thibault Formal[1,2], and Stéphane Clinchant[1]

[1] Naver Labs Europe, Meylan, France
{simon.lupart,thibault.formal,stephane.clinchant}@naverlabs.com
[2] Sorbonne Université, ISIR, Paris, France

Abstract. Pre-trained Language Models have recently emerged in Information Retrieval as providing the backbone of a new generation of neural systems that outperform traditional methods on a variety of tasks. However, it is still unclear to what extent such approaches generalize in zero-shot conditions. The recent BEIR benchmark provides partial answers to this question by comparing models on datasets and tasks that differ from the training conditions. We aim to address the same question by comparing models under more *explicit* distribution shifts. To this end, we build three query-based distribution shifts within MS MARCO (query-semantic, query-intent, query-length), which are used to evaluate the three main families of neural retrievers based on BERT: sparse, dense, and late-interaction – as well as a monoBERT re-ranker. We further analyse the performance drops between the train and test query distributions. In particular, we experiment with two generalization indicators: the first one based on train/test query vocabulary overlap, and the second based on representations of a trained bi-encoder. Intuitively, those indicators verify that the further away the test set is from the train one, the worse the drop in performance. We also show that models respond differently to the shifts – dense approaches being the most impacted. Overall, our study demonstrates that it is possible to design more *controllable* distribution shifts as a tool to better understand generalization of IR models. Finally, we release the MS MARCO query subsets, which provide an additional resource to benchmark zero-shot transfer in Information Retrieval.

Keywords: Neural IR · Zero-shot retrieval · Distribution shift

1 Introduction

The ability of machine learning models to generalize to unseen cases under distribution shifts remains a major challenge and concern for systems deployed in the real world [8]. In Information Retrieval (IR), the question of generalization has often been eluded, due to the robust and long-standing performance of term-based approaches [51]. However, with the recent advent of neural IR based on Pre-trained Language Models (PLM) like BERT [10], the generalization issue has become as relevant as ever, as recently shown in the zero-shot

© The Author(s), under exclusive license to Springer Nature Switzerland AG 2023
J. Kamps et al. (Eds.): ECIR 2023, LNCS 13980, pp. 636–652, 2023.
https://doi.org/10.1007/978-3-031-28244-7_40

BEIR benchmark [43]. By comparing various types of BERT-based models on different domains and tasks, Thakur et al. show how cross-encoders, as well as retrieval models with lexical priors such as doc2queryT5 [32] or ColBERT [24], tend to be more robust, while dense bi-encoders such as DPR [23] or TAS-B [20] seem to suffer more from domain shifts – with performance lower than BM25 overall. Knowing that many production systems use (or will use) models based on PLM (e.g. [50]), while being exposed to new documents and queries every day, robustness is thus a critical aspect that must be assessed.

Outside of the IR field, Wiles et al. [47] propose a framework to evaluate computer vision models under various distribution shifts, in order to assess the important aspects for which robustness is required, and which models are effectively robust. Intuitively, a dataset is composed of samples with various attributes (for instance, color, shape, or lightning), where some attribute values would be seen at training time – and others not. The objective is then to be able to learn representations that are invariant to such variations, to better transfer to unseen attributes. Our objective echoes the same research question, from the IR point of view, where *terms* constitute the unit of variation – their frequency in a training dataset having a potential impact on model effectiveness. Even if the BEIR benchmark implicitly defines various distribution shifts, we would like to *explicitly* control the shifts to understand which of those are critical to evaluate robustness. In particular, we show that within MS MARCO, we can construct several distribution shift experiments that we believe will ease the study of these phenomena. Our main contributions are as follows:

- We design and release multiple distribution shifts based on MS MARCO query attributes to help analysing the robustness to unseen attributes[1];
- We compare the three main families of first-stage retrievers based on BERT (dense or sparse bi-encoders, and late interaction), as well as a cross-encoder in those controlled shifts, and show that dense models are the most impacted;
- We analyse how the drops in effectiveness can be linked to the "distance" between train and test query distributions, in particular with two possible generalization indicators (term- and model-based).

The structure of the paper is organized as follows: Sect. 2 outlines prior works on generalization in IR. Section 3 details our methodology, while Sect. 4 contains the experimental setting. Results and analyses are reported in Sect. 5. Section 6 summarizes the main conclusions and future research directions.

2   Related Work

Pre-trained Language Models (PLM) have impacted IR at its very core, owing to their ability to model complex semantic relevance signals, which makes them appealing to replace traditional term-based approaches in modern search engines.

---

[1] Splits of MS MARCO queries available at https://github.com/naver/ms-marco-shift.

From re-rankers like monoBERT [31] to models that directly tackle first-stage ranking – including dense [20,23,25,35,50] and sparse [12,14] bi-encoders, as well as late-interaction models [24,39], the effectiveness gains offered by such approaches is quite impressive. Initially evaluated on in-domain settings (like MS MARCO [4]), where train and test queries follow the same distribution, conclusions became more contrasted when Thakur et al. released the zero-shot BEIR benchmark [43] – in which some models like DPR [23] achieve lower overall performance than (unsupervised) BM25. In more detail, the BEIR benchmark consists of a test suite of 18 datasets – each containing documents, queries, and corresponding *qrels* – that are used to evaluate models in zero-shot, i.e., without any sort of training based on those datasets. The selected datasets were chosen by three factors: diversity in tasks, domains, and difficulty. This makes BEIR really challenging as, differently from classical evaluation settings where the collection usually stays unchanged, here both queries and documents are "new". Furthermore, the task may also vary from the initial training objective (e.g., Question-Answering or Fact-Checking). By measuring the similarity between datasets/domains – relying on the weighted Jaccard similarity [22] between the document collections – the authors argued that BEIR indeed contains a diverse set of tasks, and is perfectly suited for zero-shot evaluation of neural IR models.

In the meantime, other works have investigated various aspects of robustness or generalization capabilities of re-ranker models, in order to uncover their weaknesses or failure cases. First, there have been several works on the systematic testing of transformer-based rankers [6,29,34,44]. More specific to the IR field, the impact of shifting trends in search engines was studied on neural re-rankers as well (pre- and post-BERT), under the lens of catastrophic forgetting [27] and lifelong learning [18]. Penha et al. [33] also analysed the robustness of various re-ranking models to typos (i.e. variations without changes of semantic) – as search engines directly interact with users and may be exposed to such issues. Following works further complement the findings for dense bi-encoders against misspellings and paraphrasing [40,55–57]. From a different perspective, various neural IR models (mostly re-rankers) have been shown to be vulnerable under adversarial attacks – usually by substituting words in documents or queries [26,41,46,48,49]. Overall, such studies usually focus on a single type of model, and lack the comparison between the various architectures proposed to tackle *first-stage* ranking, for which robustness might even be more critical.

A few works recently analysed the generalization (or zero-shot properties) of BERT-based first-stage rankers [37]. Lexical matching has been shown to be architecture-dependent, especially in zero-shot [15]. Using the two train sets of MS MARCO and Natural Question, Ren et al. [36] identify key factors that affect the zero-shot properties of dense models – including the overlap between the source and target query sets, as well as the query type distribution. We similarly study the role of such overlap, but we rely on our *explicit* shifts built within MS MARCO. Zhan et al. [54] provide a thorough analysis on MS MARCO by (*i*) identifying a strong train/test overlap within the dataset, that plagues the current evaluation of neural rankers, (*ii*) fixing this issue, relying on two new

Table 1. Examples of queries from the topic clusters. We identified the following topics: Names and Public Figures, Dated Events, Pricing/Units, Medical Treatments and Biology/Physics.

Topic	Queries
C_0	+what does the name brooke mean ; Camel Two Humps called ; How Did George Peppard Die ; How Much is Bobby Brown Worth
C_1	+when is mardi grai ; +which president has Living grandsons ; 23 is What day of 2016 ; 2015 ncca footbal rankings
C_2	1 cm is how many millimeters ; . what is the major Difference between a treaty and an executive agreement? ; 1 point perspective definition
C_3	ECT is a treatment that is used for ; The ABO blood Types are examples of ; The vitamin that prevents beriberi is ; 1.5 g of sodium per day
C_4	Ebolavirus is an enveloped virus, which means ; % of earths crust is dysprosium ; +what is forbs as a food for animals?

resampling strategies that allow to accurately compare zero-shot properties of several retrieval architectures, on datasets without train/test overlaps, (iii)showing that bi-encoders fail to properly generalize compared to cross-encoders. While also being related, our work differs in that we aim to create *controllable* subsets that would not only avoid train/test overlaps, but would also help to analyse the link between effectiveness drops and train/test similarity. Similar findings on leakages between Robust04 and MS MARCO have been recently outlined [16], which further motivates the need to build datasets with controlled shifts that do not contain such leakages. Other related works also investigated particularities of the MS MARCO collection, and possible bias the dataset could contain [3,19,30]. Finally, several methods have been proposed to adapt neural rankers to new domains or tasks – usually without requiring supervised annotations on the targeted distributions [9,42,45,52,53]. Such works which try to fix – rather than understand the causes of – generalization issues, are out of scope for our work, even though they can indirectly help on the understanding of generalization.

3 Methodology

3.1 Distribution Shifts

To investigate the behavior of models facing different types of shifts, we start from the MS MARCO passage dataset [4], which contains approximately $8.8M$ passages and $500k$ training queries. We then build upon it three shifts – defined as a change of attributes distribution between train and test distribution – from queries. Our goal is to cover all three *lexical* (inferred from terms), *semantic* (connected to meaning), and *syntactic* shifts (related to word structure). Additionally, note that shifts on training queries implicitly entail a shift in documents,

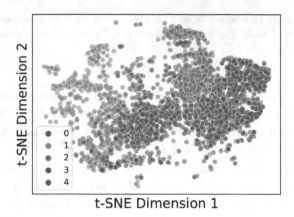

Fig. 1. t-SNE on the query topics, where each cluster contains both training and evaluation queries. Clusters 3 and 4 are close to each other but correspond to resp. Medical Treatments (red) and Biology/Physics (purple), the other cluster are well separated. (Color figure online)

as both relevant and negative-sampled documents used for training won't follow the same distribution.

Query Topics. We first propose to separate queries into five semantic clusters, referred to as $(C_i)_{i \in [\![0,4]\!]}$, alongside their complements $\overline{C_i} = \{C_j | j \in [\![0,4]\!]; j \neq i\}$. Formally, we proceed in three steps: (i) use a k-means algorithm on the [CLS] DistilBERT representations of MS MARCO queries to build 100 initial clusters[2], (ii) select the five clusters that maximize the sum of pairwise ℓ_2 distances between their respective centroids, (iii) expand those five native clusters by joining nearest clusters, until we have groups of $\approx$25k queries (without overlap). This process differs from a classical k-mean algorithm as the final clusters do not form a partition of the entire set of queries ($|\bigcup_i C_i| \simeq 125k \ll 500k$), resulting in larger boundaries between clusters compared to the works from [54]. Additionally, starting from 100 clusters enables us to define more specific topics from the start (we also experimented with larger values for k, but didn't notice any improvements in the quality of the clustering). Finally, we split each cluster into **train** and **test** sets, allowing us to compare both in-domain and out-of-domain performance. Training is done on the train set of each $\overline{C_i}$, referred to as $\overline{C_i}^t$. It contains $\approx$100k queries, for which we sample 100 negatives using BM25, resulting in approximately 10M triplets in total. Models are evaluated on the test sets of the C_i (referred to as C_i^e, containing 6200 queries each). We provide a t-SNE [28] visualization of the resulting clusters in Fig. 1: we observe that clusters are clearly distinct from each other in the embedding space. We additionally provide examples from each cluster in Table 1: clusters correspond to different topics, so the distribution shift here is more semantic.

---

[2] We consider a pre-trained DistilBERT [38] that has not been fine-tuned.

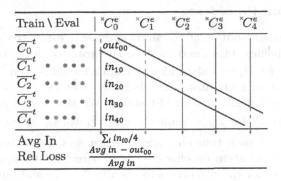

Train \ Eval	$^x C_0^e$	$^x C_1^e$	$^x C_2^e$	$^x C_3^e$	$^x C_4^e$
$\overline{C_0}^t$	$\bullet\bullet\bullet\bullet$	out_{00}			
$\overline{C_1}^t$	$\bullet\ \bullet\bullet\bullet$	in_{10}			
$\overline{C_2}^t$	$\bullet\bullet\ \bullet\bullet$	in_{20}			
$\overline{C_3}^t$	$\bullet\bullet\bullet\ \bullet$	in_{30}			
$\overline{C_4}^t$	$\bullet\bullet\bullet\bullet$	in_{40}			
Avg In	$\sum_i in_{i0}/4$				
Rel Loss	$\dfrac{Avg\ in - out_{00}}{Avg\ in}$				

Fig. 2. Zero-shot evaluation procedure. Lines correspond to models trained on different clusters, and columns to evaluation sets. For C_0, we define *Avg In* as the mean test performance on C_0^e for models trained on $\overline{C_1}^t, \overline{C_2}^t, \overline{C_3}^t$ and $\overline{C_4}^t$, and *Rel Loss* as the relative difference between the *Avg In* and the zero-shot performance (i.e. trained on $\overline{C_0}^t$ and evaluated on C_0^e).

WH-Words Queries. Besides query topics, we investigate queries styles and goals, through an analysis of question words [36,58]. In comparison to Natural Language Processing, question words in IR are a much stronger signal, as they define the query intent recently studied in [5], such as: *instruction, reason, evidence-based, comparison, experience* or *debate*. In order to evaluate models on such shifts, we manually build three clusters (*wha, how, who*, also referred to as $(W_i)_{i=0,1,2}$), for queries respectively related to definitions ("what", "definition"), instructions ("how") and finally more general questions linked to persons, locations or context ("who", "when", "where", "which"). This shift is entirely rule-based, as we separate queries on the above lists of fixed terms. Similarly, we split each cluster into **train/test** splits, and we train models on the training sets of the complements (containing $10M$ triplets in total). We evaluate models on the test W_i, each containing 6500 queries.

Short and Long Queries. Query length is known to greatly impact retrieval – for instance, aggregating information from long queries has been deemed difficult for both traditional and neural IR methods [43]. We thus define the last shift on this attribute. To analyse this effect, we split the train set into groups of short and long queries, from the median query length at the word level (6 for MS MARCO). **Train/test** sets contain respectively $10M$ training triplets and 3500 queries for evaluation.

3.2 Evaluation Procedure

We perform leave-one-out on all the shifts in order to evaluate the in-domain and zero-shot effectiveness of various models. For instance, let's consider the topic clusters. For $i \in [\![0, 4]\!]$, we independently train a model $\mathcal{M}_i$ on $\overline{C_i}^t$, and evaluate it on the test set of each cluster. It, therefore, creates a zero-shot experiment on C_i^e, as its distribution was out-of-training. We also have access to in-domain evaluation on the C_j^e for $j \neq i$. Similarly with wh-words, we train on $\overline{wha}^t$, $\overline{how}^t$ and $\overline{who}^t$, and test each time on respective test sets. Note that, in addition to the leave-one-out rotation on clusters, we also rely on train/test splits inside each cluster, in order to have access to both in-domain and out-of-domain performances. We, therefore, report *Avg In* as the average performance measure, when the distribution of the evaluated cluster is seen at training time, and *Rel Loss* as the relative loss between the above average measure, and the zero-shot performance, i.e. when the evaluated cluster is out of the training distribution. Figure 2 illustrates the overall evaluation procedure. Note that, as some clusters may intrinsically contain harder queries, what we are really interested in here is the comparison of the performance inside a column (i.e., on the same evaluation set, with different training sets). In general, the lowest performance is achieved on the diagonal (as it corresponds to zero-shot evaluation).

4 Experimental Setup

We compare three first-stage ranking models: (i) a standard dense `bi-encoder` [23] which represents queries and documents in a dense low dimensional space by the means of the `[CLS]` embedding, (ii) the late-interaction `ColBERT` [24], (iii) and the sparse bi-encoder `SPLADE-max` [13] which represents queries and documents as sparse high-dimensional bag-of-words vectors. Those models represent the three main families of representation-based models in IR, and we are thus interested in their behavior with respect to zero-shot generalization. Finally, we also include BM25 as a reference point, as well as a `monoBERT` cross-encoder [31] re-ranking BM25 top-1000 documents. All models rely on a pre-trained DistilBERT [38] backbone model, and are fine-tuned on a particular query subset. We limit ourselves to a standard training procedure, relying on contrastive learning and BM25/in-batch negatives [23], without further improvements such as distillation [21], hard negative mining [50] or middle-training [17], as those techniques are more general, and may be applied to any baseline model. For evaluation, we report MRR@10 – the official MS MARCO performance measure. We also report the Atomized Search Length[3] (ASL) [1], as it complements MRR@10 by not focusing on top of the ranking. In a nutshell, it is defined, for a given query, as the average number of irrelevant documents ranked before a relevant one. The dense and sparse bi-encoders are fine-tuned on 4 V100 GPUs, for $100k$ iterations, with a batch size of 128, using the MS MARCO triplets

---

[3] We used a 100 bounded ASL.

Table 2. Comparison of the average performance and relative loss (MRR@10) from seen to unseen clusters. In bold are the best on each cluster (in terms of performance and loss). All losses between the *Avg In* and *Out*, for each model, are statistically significant with p-value< 0.05 for paired t-test.

Models		C_0^e	C_1^e	C_2^e	C_3^e	C_4^e
BM25		19.2	25.9	16.4	18.1	17.5
Bi-encoder	Avg in	33.2	37.2	28.5	21.5	21.4
	Out	30.4	30.5	25.9	19.0	19.6
	Rel loss	8.3%	18.0%	9.0%	11.5%	8.6%
SPLADE	Avg in	36.8	38.7	31.2	26.2	25.0
	Out	34.5	34.0	30.2	24.5	24.7
	Rel loss	6.3%	12.2%	**3.2%**	6.4%	**1.4%**
ColBERT	Avg in	**39.7**	42.3	**34.6**	**28.8**	**27.7**
	Out	**38.6**	**38.7**	**33.4**	**27.7**	**27.1**
	Rel loss	2.7%	**8.5%**	3.4%	**3.7%**	2.2%
monoBERT	Avg in	39.4	**42.7**	33.4	27.1	26.3
	Out	**38.6**	38.4	31.8	25.9	25.7
	Rel loss	**2.1%**	10.2%	4.8%	4.5%	2.4%

from our shifts. ColBERT and monoBERT are both trained for $150k$ iterations with a default batch size of 32, on 2 V100 GPUs. For the bi-encoders, best checkpoints are selected using an approximated early stopping [20] relying on a validation set composed of 1600 queries, which is *not* subject to the shifts. From our observations, the best checkpoints would generally correspond to $40k$ iterations, resulting in about $5M$ training triplets (using $bs = 128$). ColBERT and monoBERT, on the other hand, do not rely on early stopping, but overall see the same number of training samples ($5M$, in $150k$ iterations with $bs = 32$). Thus, both training procedures are very similar. For all other parameters, we adopt the default values reported in the original papers.

5 Results and Analysis

In this section, we first compare the drops in performance of different architectures when subject to the shifts. We then link those results with two indicators we define in Sect. 5.2, which can be used to analyse the behavior of different models regarding their zero-shot capability.

5.1 Performance Evaluation on Distribution Shifts

Query Topics. Table 2 reports the performance of models on both in-domain (*Avg In*) and out-of-domain (*Out*), as well as the relative (*Rel Loss*) due to the shift, on each topic cluster. We first notice that drops in performance can

Table 3. Comparison of the avg. MRR@10 and relative losses in zero-shot for wh-words and query length. All losses have p-value<0.05 for paired t-test.

Models		wha^e	how^e	who^e	$short^e$	$long^e$
BM25		18.2	14.7	22.3	19.0	18.5
Bi-encoder	Avg in	27.8	26.0	33.1	34.0	27.1
	Out	23.4	19.6	27.9	29.8	25.2
	Rel loss	15.8%	24.8%	15.8%	12.5%	7.0%
SPLADE	Avg in	30.3	28.9	37.7	34.9	30.3
	Out	28.6	21.2	32.5	33.5	27.1
	Rel loss	5.5%	26.8%	13.7%	**3.9%**	10.4%
ColBERT	Avg in	33.5	**31.7**	40.0	**38.4**	32.5
	Out	**31.8**	**27.3**	36.6	**36.4**	**31.6**
	Rel loss	**5.2%**	14.0%	8.6%	5.1%	**2.7%**
monoBERT	Avg in	**33.9**	30.5	**40.1**	37.7	**32.6**
	Out	31.1	26.4	**37.1**	32.3	28.9
	Rel loss	8.3%	**13.5%**	**7.5%**	14.3%	11.3%

be significant (up to 18% for the dense bi-encoder). Overall, we observe that dense models are the most impacted by the shifts across clusters, followed by SPLADE, and finally ColBERT and monoBERT. Interaction approaches also demonstrate both better performance on in-domain and out-of-domain – compared to representation-based models. All in all, the results are in line with the ones reported in [43]. However, contrary to the results reported in [54], SPLADE seems here to be substantially more robust than dense bi-encoders, and this on every cluster. Note that, the average query/document sizes for SPLADE (which correspond to the number of non-zero dimensions in the sparse representations) are respectively 24 and 130, which is already way below the dense representations ([CLS] vector of size 768) – despite its better performance overall. Moreover, as opposed to the BEIR benchmark, no model here performs lower than BM25 under zero-shot evaluation. This phenomenon is likely due to the fact that we "stay" within MS MARCO, tackling the same retrieval task. BM25 performance also indicates to some extent the difficulty of each cluster – some of them (e.g. C_1^e) supposedly relying more on word matching than others. Finally, note that all the drops (between in- and out-of-domain) are statistically significant, with p-values<0.05.

WH-Words Queries. Table 3 shows the results on wh-words and query length shifts. On the former (left columns), drops in effectiveness are on average much more important compared to topic clusters – with up to 26.8% on the how^e cluster for SPLADE. However, the overall comparison between architectures remains the same. We also notice that models with higher *Avg In* are also the ones with the lowest relative losses – ruling out overfitting as the cause of better in-domain performance. Alexander et al. [2] conduct a study on query taxonomies. In particular, they refer to *how* queries as having an *instrumental* intent ("how", "how to", "how do"), in contrast to usual queries which have *factual intents* (e.g. "who", "when", "where", "which", "what", "definition"). We notice in our results the same pattern on those queries, with both higher drops, as well as lower BM25 results. Those queries tend to rely less on word matching, and more on a general understanding of the given situation – it is thus harder for models to accurately perform on such a cluster in zero-shot.

Length. Concerning the length-based shift (Table 3, right columns), we first see that short queries are easier than longer ones: a model trained on long queries will be better on the *short* than on the *long* evaluation set. Interestingly, we notice that short and long queries are somewhat complementary in a training set – a model cannot be trained with long queries only. Besides, we also observe that models trained on long queries tend to have a higher recall (not reported), while models trained on short queries a higher precision.

5.2   Train/Test Distribution Similarity

From our previous experiments, it is difficult to estimate *a priori* the strength of a shift on the downstream performance. We thus consider in the following two measures of similarities between train and test queries, that partially correlate with the zero-shot performance drop – and so the strength of the shift – and that can easily be interpreted: (i) a term-based similarity, which measures the vocabulary overlap between out-of-domain and in-domain query sets, (ii) a model-based similarity, which takes advantage of internal representations of trained models.

Jaccard Similarity. The weighted Jaccard [22] can be used to measure the vocabulary overlap between two sets of documents or queries [11,43]. More formally, given a source and a target collection, it is defined as:

$$J(S,T) = \frac{\sum_V \min(S_k, T_k)}{\sum_V \max(S_k, T_k)}$$

where S_k and T_k are the normalized frequencies of word k in source and target datasets respectively, and V is the union vocabulary. In our case, we aim to measure the similarity $J(C_i, \overline{C_i})$ between queries from cluster C_i (out-of-domain) and its complement $\overline{C_i}$ (in-domain)[4]. Intuitively, we are trying to quantity to which extent a given out-of-domain query set is *far* for the training set, based on term statistics. This metric is computed at the cluster level, such that we can associate it with the overall out-of-domain loss.

---

[4] Formally, we should compute $J(C_i^e, \overline{C_i}^t)$, but terms statistics at the cluster level for train and eval sets are very similar.

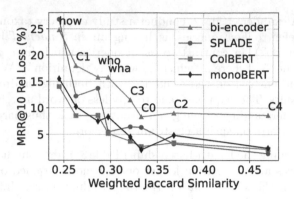

Fig. 3. Relative loss on zero-shot clusters with respect to weighted Jaccard similarities between train and test query distributions. Leftmost part: low similarities and high losses. Rightmost part: high similarities and small losses.

We plot in Fig. 3 the relative loss on each of the out-of-domain cluster, with respect to the weighted Jaccard $J(C_i, \overline{C_i})$ and $J(W_i, \overline{W_i})$. From the figure, we observe that, as the Jaccard similarity increases – and so, the terms overlap – the relative loss in performance diminishes. Such a measure is thus indicative when it comes to predicting generalization capabilities of neural models. An important point is that the behavior is common for both Query Topics and WH-Words Queries, suggesting that this pattern may be independent of the nature of the shifts. With the hypothesis that all models still partly rely on term matching, it also shows that they tend to have issues when learning the general pattern of word matching, independently from the terms themselves. This extends the observation made for dense models in [36], and supports [15] about generalization to unseen words. We further notice that, for the clusters with the highest Jaccard (C_2^e and C_4^e), the relative loss is lower for SPLADE, compared to ColBERT. We hypothesize that, as sparse models rely more on lexical components (through the BoW representation), they are better able to transfer when the query vocabulary distributions are closer. Finally, the same relation can be observed for ASL (instead MRR@10) – although not reported here – indicating the quality of the indicator for recall-oriented metrics.

Model-Based Similarity. To complement the above lexical indicator, we additionally introduce a semantic measure of similarity, which relies on the internal representations of a trained bi-encoder. Intuitively, for a model trained on a given set of queries, we compute the distances between those training queries and a targeted test query. More formally, we compute the mean retrieval score of a dense bi-encoder $\mathcal{M}_i$ between a test query $q^e \in C_i^e$ and the training queries from $\overline{C_i}^t$, as follows:

$$R(q^e, \overline{C_i}^t) = \frac{1}{|\overline{C_i}^t|} \sum_{q^t \in \overline{C_i}^t} s_{\mathcal{M}_i}(q^e, q^t) \qquad (1)$$

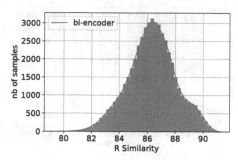 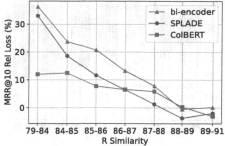

Fig. 4. Distributions of query scores. R similarity gives an estimation of how close each query is to the training set.

Fig. 5. Relative MRR@10 loss on zero-shot queries with respect to different intervals of R similarities.

where $s_{\mathcal{M}_i}$ is the output score of the dense bi-encoder $\mathcal{M}_i$ trained on $\overline{C_i}^t$ (in our case, $s_{\mathcal{M}_i}$ is a dot product between query embeddings). We use *trained* dense bi-encoders as the baseline for the similarity computation, as their symmetric nature makes it more natural to compute similarities between queries. This representation-based similarity thus enables to quantify how *far* is a test query from the training set. Contrary to the Jaccard similarity, this indicator is, however, defined *at the query level* – such that it is possible to link it to the loss associated with each query. Moving away from cluster-based to query-level indicators is interesting from a practical standpoint. Given a query from a new domain, we would like to be able to infer model performance. We note the interesting parallel that can be made with the vast literature around Query Performance Prediction, where the goal is to estimate the performance of an IR system without relevance judgements [7].

We show on Fig. 4 the distributions of the above similarity $R(q^e, \overline{C_i}^t)$ and $R(q^e, \overline{W_i}^t)$ for all $\{(q^e, \overline{C_i}^t) | i \in [\![0, 4]\!], q^e \in C_i^e\}$ and $\{(q^e, \overline{W_i}^t) | i \in [\![0, 2]\!], q^e \in W_i^e\}$, for the dense bi-encoder. Then, in Fig. 5, we plot the average relative loss in terms of MRR@10 with respect to different intervals of R similarities – corresponding to zero-shot queries from topics and wh-words clusters. Overall, those intervals represent the spectrum of queries that are the further away from the train set (leftmost part, *low similarity*) to the closest ones (rightmost part, *high similarity*). For the three architectures, the farther the out-of-domain test query is, the highest the loss. When comparing models, ColBERT seems to generalize better on the most distant queries compared to SPLADE and the dense bi-encoder. Results are aligned with the ones observed for the Jaccard – but for a semantic notion of distance. It is also interesting to see that the trend holds for SPLADE and ColBERT, given that the similarity is entirely based on the dense bi-encoder representations. We can thus infer that the knowledge accumulated by dense bi-encoders at training time could be used as a potential signal to predict performance, without additional supervision.

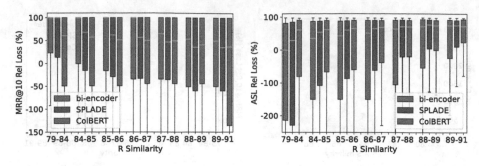

Fig. 6. Boxplots of the MRR@10 and ASL relative loss with respect to R similarity (Median indicated in orange). Left: MRR@10 ($\uparrow$). Right: For ASL ($\downarrow$), the lower the better, so the higher the relative losses, the better. Both metrics indicate better performances on queries close to the training distribution (rightmost parts of both graphs). (Color figure online)

In addition, in Fig. 6, we represent the same intervals with boxplots, to analyse the variance of the MRR@10 and ASL (as those metrics are initially defined at the query level). Looking at the variance of the MRR@10, we see that it decreases as we go further away from the training set, implying that the losses on the top of the ranking will be high for distant queries. On the other hand, with ASL, the variance behaves differently: we have high confidence in the closest queries, on which the ASL will improve (decrease), while performance on distant queries is uncertain. ASL being a more recall-oriented metric, both results are thus complementary and give an overview of the behavior at both the very top of the ranking with MRR@10, and at deeper ranks with ASL.

6   Conclusion

In this work, we focus on zero-shot evaluation of neural IR models. We propose to benchmark neural retrievers based on PLM against three controlled distribution shifts, created by partitioning MS MARCO training queries – based on different characteristics (semantic, intent, and length). Overall, we observe that interaction approaches are more robust than representation-based models, and this across all shifts. We further link the observed drops in performance to two indicators which verify that, the further away the test set is from the train one, the worse the drop in performance. Our analysis seems to suggest that a model-based similarity could possibly be used as *unsupervised* predictors of performance. We believe it opens the path to future research directions that need to be investigated in, and foster open questions for the IR community on how to measure model robustness. Furthermore, the effect of techniques such as distillation or pre-training remains yet to be analysed, as they could possibly correct model biases and lead to better generalization.

Acknowledgements. We first would like to thank Carlos Lassance, Hervé Déjean and Jean-Michel Renders for the valuable discussions and feedback on the paper. We also would like to thank Guglielmo Faggioli, Stefano Marchesin, Nicola Ferro and Benjamin Piwowarski a lot, for the knowledge they provided us on Query Performance Prediction.

References

1. Alex, J., Hall, K., Metzler, D.: Atomized search length: beyond user models (2022). https://doi.org/10.48550/ARXIV.2201.01745, https://arxiv.org/abs/2201.01745
2. Alexander, D., Kusa, W., de Vries, A.P.: ORCAS-I: queries annotated with intent using weak supervision. arXiv e-prints arXiv:2205.00926, May 2022
3. Arabzadeh, N., Vtyurina, A., Yan, X., Clarke, C.L.A.: Shallow pooling for sparse labels (2021). https://doi.org/10.48550/ARXIV.2109.00062, https://arxiv.org/abs/2109.00062
4. Bajaj, P., et al.: MS marco: a human generated machine reading comprehension dataset (2018)
5. Bolotova, V., Blinov, V., Scholer, F., Croft, W.B., Sanderson, M.: A non-factoid question-answering taxonomy. In: Proceedings of the 45th International ACM SIGIR Conference on Research and Development in Information Retrieval. SIGIR 2022, pp. 1196–1207. Association for Computing Machinery, New York (2022). https://doi.org/10.1145/3477495.3531926
6. Câmara, A., Hauff, C.: Diagnosing BERT with retrieval heuristics. Adv. Inf. Retrieval **12035**, 605–618 (2020)
7. Carmel, D., Yom-Tov, E.: Estimating the query difficulty for information retrieval. In: Proceedings of the 33rd International ACM SIGIR Conference on Research and Development in Information Retrieval. SIGIR 2010, p. 911. Association for Computing Machinery, New York (2010). https://doi.org/10.1145/1835449.1835683
8. Csurka, G., Volpi, R., Chidlovskii, B.: Unsupervised domain adaptation for semantic image segmentation: a comprehensive survey (2021)
9. Dai, Z., et al.: Promptagator: few-shot dense retrieval from 8 examples (2022). https://doi.org/10.48550/ARXIV.2209.11755, https://arxiv.org/abs/2209.11755
10. Devlin, J., Chang, M., Lee, K., Toutanova, K.: BERT: pre-training of deep bidirectional transformers for language understanding. CoRR abs/1810.04805 (2018), http://arxiv.org/abs/1810.04805
11. Ferro, N., Sanderson, M.: How do you test a test? A multifaceted examination of significance tests. In: Proceedings of the Fifteenth ACM International Conference on Web Search and Data Mining. WSDM 2022, pp. 280–288. Association for Computing Machinery, New York (2022). https://doi.org/10.1145/3488560.3498406
12. Formal, T., Lassance, C., Piwowarski, B., Clinchant, S.: From distillation to hard negative sampling: making sparse neural IR models more effective. In: Proceedings of the 45th International ACM SIGIR Conference on Research and Development in Information Retrieval. SIGIR 2022, pp. 2353–2359. Association for Computing Machinery, New York (2022). https://doi.org/10.1145/3477495.3531857
13. Formal, T., Lassance, C., Piwowarski, B., Clinchant, S.: Splade v2: sparse lexical and expansion model for information retrieval (2021)
14. Formal, T., Piwowarski, B., Clinchant, S.: Splade: sparse lexical and expansion model for first stage ranking. In: Proceedings of the 44th International ACM SIGIR Conference on Research and Development in Information Retrieval. SIGIR 2021, pp. 2288–2292. Association for Computing Machinery, New York (2021). https://doi.org/10.1145/3404835.3463098

15. Formal, T., Piwowarski, B., Clinchant, S.: Match your words! A study of lexical matching in neural information retrieval (2021)
16. Fröbe, M., Akiki, C., Potthast, M., Hagen, M.: How train-test leakage affects zero-shot retrieval (2022). https://doi.org/10.48550/ARXIV.2206.14759, https://arxiv.org/abs/2206.14759
17. Gao, L., Callan, J.: Unsupervised corpus aware language model pre-training for dense passage retrieval. In: Proceedings of the 60th Annual Meeting of the Association for Computational Linguistics (Volume 1: Long Papers), pp. 2843–2853. Association for Computational Linguistics, Dublin, May 2022. https://doi.org/10.18653/v1/2022.acl-long.203, https://aclanthology.org/2022.acl-long.203
18. Gerald, T., Soulier, L.: Continual learning of long topic sequences in neural information retrieval (2022). https://doi.org/10.48550/ARXIV.2201.03356, https://arxiv.org/abs/2201.03356
19. Gupta, P., MacAvaney, S.: On survivorship bias in MS MARCO. In: Proceedings of the 45th International ACM SIGIR Conference on Research and Development in Information Retrieval. ACM, July 2022. https://doi.org/10.1145/3477495.3531832
20. Hofstätter, S., Lin, S.C., Yang, J.H., Lin, J., Hanbury, A.: Efficiently teaching an effective dense retriever with balanced topic aware sampling. In: Proceedings of SIGIR (2021)
21. Hofstätter, S., Althammer, S., Schröder, M., Sertkan, M., Hanbury, A.: Improving efficient neural ranking models with cross-architecture knowledge distillation (2020). https://doi.org/10.48550/ARXIV.2010.02666, https://arxiv.org/abs/2010.02666
22. Ioffe, S.: Improved consistent sampling, weighted minhash and l1 sketching. In: Proceedings of the 2010 IEEE International Conference on Data Mining. ICDM 2010, pp. 246–255. IEEE Computer Society, USA (2010). https://doi.org/10.1109/ICDM.2010.80
23. Karpukhin, V., et al.: Dense passage retrieval for open-domain question answering (2020)
24. Khattab, O., Zaharia, M.: Colbert: efficient and effective passage search via contextualized late interaction over BERT. In: Proceedings of the 43rd International ACM SIGIR Conference on Research and Development in Information Retrieval. SIGIR 2020, pp. 39–48. Association for Computing Machinery, New York (2020). https://doi.org/10.1145/3397271.3401075
25. Lin, S.C., Yang, J.H., Lin, J.: In-batch negatives for knowledge distillation with tightly-coupled teachers for dense retrieval. In: Proceedings of the 6th Workshop on Representation Learning for NLP (RepL4NLP-2021), pp. 163–173. Association for Computational Linguistics, Online, August 2021. https://doi.org/10.18653/v1/2021.repl4nlp-1.17, https://aclanthology.org/2021.repl4nlp-1.17
26. Liu, J., et al.: Order-disorder: imitation adversarial attacks for black-box neural ranking models (2022)
27. Lovón-Melgarejo, J., Soulier, L., Pinel-Sauvagnat, K., Tamine, L.: Studying catastrophic forgetting in neural ranking models. In: 43rd European Conference on Information Retrieval - ECIR 2021. Lucca (online), Italy, April 2021. https://hal.archives-ouvertes.fr/hal-03156630
28. van der Maaten, L., Hinton, G.: Visualizing data using t-SNE. J. Mach. Learn. Res. **9**, 2579–2605 (2008). http://www.jmlr.org/papers/v9/vandermaaten08a.html
29. MacAvaney, S., Feldman, S., Goharian, N., Downey, D., Cohan, A.: AbniRML: analyzing the behavior of neural IR models (2020). https://doi.org/10.48550/ARXIV.2011.00696, https://arxiv.org/abs/2011.00696

30. Mackenzie, J., Petri, M., Moffat, A.: A sensitivity analysis of the MSmarco passage collection (2021). https://doi.org/10.48550/ARXIV.2112.03396, https://arxiv.org/abs/2112.03396
31. Nogueira, R., Cho, K.: Passage re-ranking with BERT (2019)
32. Nogueira, R., Lin, J.: From doc2query to docttttttquery (2019)
33. Penha, G., Câmara, A., Hauff, C.: Evaluating the robustness of retrieval pipelines with query variation generators. In: Hagen, M., et al. (eds.) ECIR 2022. LNCS, vol. 13185, pp. 397–412. Springer, Cham (2022). https://doi.org/10.1007/978-3-030-99736-6_27
34. Rau, D., Kamps, J.: How different are pre-trained transformers for text ranking? (2022). https://doi.org/10.48550/ARXIV.2204.07233, https://arxiv.org/abs/2204.07233
35. Ren, R., et al.: RocketQAv2: a joint training method for dense passage retrieval and passage re-ranking. In: Proceedings of the 2021 Conference on Empirical Methods in Natural Language Processing, pp. 2825–2835. Association for Computational Linguistics, Online and Punta Cana, Dominican Republic, November 2021. https://doi.org/10.18653/v1/2021.emnlp-main.224, https://aclanthology.org/2021.emnlp-main.224
36. Ren, R., et al.: A thorough examination on zero-shot dense retrieval (2022). https://doi.org/10.48550/ARXIV.2204.12755, https://arxiv.org/abs/2204.12755
37. Rosa, G.M., et al.: No parameter left behind: how distillation and model size affect zero-shot retrieval (2022). https://doi.org/10.48550/ARXIV.2206.02873, https://arxiv.org/abs/2206.02873
38. Sanh, V., Debut, L., Chaumond, J., Wolf, T.: DistilBERT, a distilled version of BERT: smaller, faster, cheaper and lighter. In: 5th Workshop on Energy Efficient Machine Learning and Cognitive Computing @ NeurIPS 2019 (2019). http://arxiv.org/abs/1910.01108
39. Santhanam, K., Khattab, O., Saad-Falcon, J., Potts, C., Zaharia, M.: ColBERTv2: effective and efficient retrieval via lightweight late interaction. In: Proceedings of the 2022 Conference of the North American Chapter of the Association for Computational Linguistics: Human Language Technologies, pp. 3715–3734. Association for Computational Linguistics, Seattle, July 2022. https://doi.org/10.18653/v1/2022.naacl-main.272, https://aclanthology.org/2022.naacl-main.272
40. Sidiropoulos, G., Kanoulas, E.: Analysing the robustness of dual encoders for dense retrieval against misspellings. In: Amigó, E., Castells, P., Gonzalo, J., Carterette, B., Culpepper, J.S., Kazai, G. (eds.) SIGIR 2022: The 45th International ACM SIGIR Conference on Research and Development in Information Retrieval, Madrid, Spain, 11–15 July 2022, pp. 2132–2136. ACM (2022). https://doi.org/10.1145/3477495.3531818
41. Song, J., Zhang, J., Zhu, J., Tang, M., Yang, Y.: TRAttack: text rewriting attack against text retrieval. In: Proceedings of the 7th Workshop on Representation Learning for NLP, pp. 191–203. Association for Computational Linguistics, Dublin, Ireland, May 2022. https://doi.org/10.18653/v1/2022.repl4nlp-1.20, https://aclanthology.org/2022.repl4nlp-1.20
42. Thakur, N., Reimers, N., Lin, J.: Domain adaptation for memory-efficient dense retrieval (2022). https://doi.org/10.48550/ARXIV.2205.11498, https://arxiv.org/abs/2205.11498
43. Thakur, N., Reimers, N., Rücklé, A., Srivastava, A., Gurevych, I.: BEIR: a heterogenous benchmark for zero-shot evaluation of information retrieval models. CoRR abs/2104.08663 (2021). https://arxiv.org/abs/2104.08663

44. Völske, M., et al.: Towards axiomatic explanations for neural ranking models. In: Proceedings of the 2021 ACM SIGIR International Conference on Theory of Information Retrieval, pp. 13–22 (2021)
45. Wang, K., Thakur, N., Reimers, N., Gurevych, I.: GPL: generative pseudo labeling for unsupervised domain adaptation of dense retrieval (2021). https://doi.org/10.48550/ARXIV.2112.07577, https://arxiv.org/abs/2112.07577
46. Wang, Y., Lyu, L., Anand, A.: BERT rankers are brittle: a study using adversarial document perturbations (2022). https://doi.org/10.48550/ARXIV.2206.11724, https://arxiv.org/abs/2206.11724
47. Wiles, O., et al.: A fine-grained analysis on distribution shift. CoRR abs/2110.11328 (2021). https://arxiv.org/abs/2110.11328
48. Wu, C., Zhang, R., Guo, J., Fan, Y., Cheng, X.: Are neural ranking models robust? (2021)
49. Wu, C., Zhang, R., Guo, J., de Rijke, M., Fan, Y., Cheng, X.: Prada: practical black-box adversarial attacks against neural ranking models (2022)
50. Xiong, L., et al.: Approximate nearest neighbor negative contrastive learning for dense text retrieval. In: International Conference on Learning Representations (2021). https://openreview.net/forum?id=zeFrfgyZln
51. Yang, W., Lu, K., Yang, P., Lin, J.: Critically examining the "neural hype": weak baselines and the additivity of effectiveness gains from neural ranking models. In: Proceedings of the 42nd International ACM SIGIR Conference on Research and Development in Information Retrieval. SIGIR 2019, pp. 1129–1132. Association for Computing Machinery, New York (2019). https://doi.org/10.1145/3331184.3331340
52. Yu, Y., Xiong, C., Sun, S., Zhang, C., Overwijk, A.: COCO-DR: combating distribution shifts in zero-shot dense retrieval with contrastive and distributionally robust learning (2022). https://doi.org/10.48550/ARXIV.2210.15212, https://arxiv.org/abs/2210.15212
53. Zhan, J., et al.: Disentangled modeling of domain and relevance for adaptable dense retrieval. arXiv preprint arXiv:2208.05753 (2022)
54. Zhan, J., Xie, X., Mao, J., Liu, Y., Zhang, M., Ma, S.: Evaluating extrapolation performance of dense retrieval (2022). https://doi.org/10.48550/ARXIV.2204.11447, https://arxiv.org/abs/2204.11447
55. Zhou, Y., et al.: Towards robust ranker for text retrieval, June 2022. https://doi.org/10.48550/arXiv.2206.08063
56. Zhuang, S., Zuccon, G.: Dealing with typos for BERT-based passage retrieval and ranking. In: Conference on Empirical Methods in Natural Language Processing (2021)
57. Zhuang, S., Zuccon, G.: Characterbert and self-teaching for improving the robustness of dense retrievers on queries with typos. In: Proceedings of the 45th International ACM SIGIR Conference on Research and Development in Information Retrieval. p. 1444–1454. SIGIR '22, Association for Computing Machinery, New York, NY, USA (2022). https://doi.org/10.1145/3477495.3531951,https://doi.org/10.1145/3477495.3531951
58. Zukerman, I., Horvitz, E.: Using machine learning techniques to interpret WH-questions. In: Proceedings of the 39th Annual Meeting on Association for Computational Linguistics. ACL 2001, pp. 547–554. Association for Computational Linguistics, USA (2001). https://doi.org/10.3115/1073012.1073082

Listwise Explanations for Ranking Models Using Multiple Explainers

Lijun Lyu[1,2]([envelope])[iD] and Avishek Anand[2][iD]

[1] L3S Research Center, Leibniz University Hannover, Hannover, Germany
[2] Delft University of Technology, Delft, The Netherlands
{L.Lyu,Avishek.Anand}@tudelft.nl

Abstract. This paper proposes a novel approach towards better interpretability of a trained text-based ranking model in a post-hoc manner. A popular approach for post-hoc interpretability text ranking models are based on *locally approximating* the model behavior using a simple ranker. Since rankings have multiple relevance factors and are aggregations of predictions, existing approaches that use a single ranker might not be sufficient to approximate a complex model, resulting in low fidelity. In this paper, we overcome this problem by considering multiple simple rankers to better approximate the entire ranking list from a black-box ranking model. We pose the problem of local approximation as a GENERALIZED PREFERENCE COVERAGE (GPC) problem that incorporates multiple simple rankers towards the listwise explanation of ranking models. Our method MULTIPLEX uses a linear programming approach to judiciously extract the explanation terms, so that to explain the entire ranking list. We conduct extensive experiments on a variety of ranking models and report fidelity improvements of 37%–54% over existing competitors. We finally compare explanations in terms of multiple relevance factors and topic aspects to better understand the logic of ranking decisions, showcasing our explainers' practical utility.

Keywords: Explanation · Neural · Ranking · Post-hoc · List-wise

1 Introduction

Recent approaches for ranking text documents have focused heavily on neural models [12,16,17]. Neural rankers learn the complex and often non-linear relationships between the query and document that are difficult to encode using closed-form analytical ranking functions like BM25 [2]. However, the superior ranking performance of such models comes at the expense of reduced interpretability, thus increasing the risk of encoding spurious correlations and undesirable biases [25,32]. In parallel to developing better rankers, there has been an increased focus on interpreting neural ranking models [7,23–25] that specifically aim at explaining the rationale behind the ranking decisions.

This paper aims to propose post-hoc approaches to interpret neural text rankers. Post-hoc methods explain *already-trained* models and do not compromise on the accuracy of the learned model, hence making them popular choices for interpreting machine learning models. One prevalent strategy in post-hoc interpretability is to *locally approximate* a trained model with a *simple and interpretable proxy or a surrogate model*.

J. Kamps et al. (Eds.): ECIR 2023, LNCS 13980, pp. 653–668, 2023.
https://doi.org/10.1007/978-3-031-28244-7_41

Explainers	Explanation Terms
Term Matching	charlotte, north, sales, 2008
Position Aware	basketball, north, states, learn
Semantic Similarity	felidae, carnivorous, boko, extinction, deserts, iucn
MULTIPLEX	felidae, carnivorous, boko, extinction, deserts, gvwr, north

Fig. 1. Explaining the query bobcat with multiple relevance factors – (i) "charlotte-bobcat basketball club"; (ii) "learn to hunt bobcat"; (iii) "animal bobcat" and (iv) "bobcat mechanical retailer". MULTIPLEX carefully chooses from multiple relevance factors to explain a ranking. See Fig. 6 for more examples.

The degree of approximation is called *fidelity* and the objective is to maximize the fidelity between the proxy model and the underlying black-box model. Post-hoc methods for rankings entail using simple rankers to locally approximate (on a per-query basis) complex rankers such that the simple ranker has a high rank correlation (or high fidelity) with the complex ranking. Adapting this general post-hoc framework to ranking models has two specific challenges – *how do we aggregate multiple decisions inherent in a single ranking?* And *how do we explain ranking decisions with different inherent relevance factors?*

Rankings as Aggregations of Decisions. Text ranking models output a ranked list of documents for a given query. Unlike other learning tasks (e.g. regression and classification) that deal with a single decision, the ranking task can be viewed as an *aggregation of multiple pointwise or pairwise decisions* [1]. Any interpretability approach or explainer should therefore explain the reasoning behind the ranking list, or multiple-preference pair predictions. Therefore existing explanation techniques such as feature-attribution methods [21,22,28] that explain a single decision (pointwise) cannot be seamlessly used for rankings. Instead, a *listwise explanation* method that intends to cover all individual decisions in the entire ranking list is needed for rankings.

Different Explanations for Different Relevance Factors. Secondly, it is well-known that when ranking text, multiple relevance factors (also called ranking heuristics or axioms) determine the relevance of a document to a query, e.g., *lexical matching*, *semantic similarity*, *term proximity* etc. Unlike traditional models that explicitly encode each of these relevance factors, neural rankers automatically learn them from data. The next challenge in explaining rankings is ascertaining the relevance factor that best explains a given decision. Informally, there might not exist a single relevance factor that explains or satisfies all preferences $d_i \succ d_j$ in a given ranking. Therefore trying to approximate a ranking with a single relevance factor might result in low fidelity. A notable example is the listwise explanation approach [25] that considers covering multiple ranking decisions, but uses a single explainer which captures only one relevance factor (i.e., term matching), resulting in low-fidelity explanations due to the mismatch of exact terms.

In this paper, we define an explanation to be a combination of the underlying *relevance factors* along with the actual *machine intent*. In this paper, we **firstly** consider multiple simple rankers or explainers(formally defined in Sect. 3.1), which rely on different *well-known* and *human-understandable* (to system designers, or IR practitioners) relevance heuristics. **Secondly**, we explain the *machine intent* in terms of *expansion* *terms* (in addition to the query terms) such that the simple ranker explains a complex black-box model by inducing a similar ranking list. Thus a combination of *simple rankers* that represents a relevance factor, along with its *expanded query terms* (also called explanation terms) is the listwise explanation of the reasoning behind the ranking.

Approach wise, we carefully select a small set of explanation terms sourced from the documents of the ranked list to maximize the explanation's approximation ability (i.e. fidelity). Specifically, we define the GENERALIZED PREFERENCE COVERAGE (GPC) framework, on which we optimize the preference coverage using approximated integer linear programming. Our method MULTIPLEX is shown to be able to improve the fidelity, and more interestingly combine terms from multiple explainers, implicitly covering multiple topics for an ambiguous query. Figure 1 shows an example of explanation terms extracted by each single explainer and MULTIPLEX can cover terms of multiple aspects. Note the aspects of terms are specified by manual observation.

We conduct extensive experiments using datasets from the TREC test collections – TREC-DL and Clueweb09 with three neural rankers to evaluate MULTIPLEX. We report fidelity improvements of 37%–54% over existing competitors. We also present anecdotal examples that showcase the practical utility of MULTIPLEX in understanding neural rankers. The datasets and source code are publicly available[1].

2 Related Work

Feature Attribution for Ranking Models. The earliest works of interpreting ranking models were simple extensions to existing pointwise explanation techniques – explain a single instance given a trained ML model for general machine learning tasks in vision and language. [24, 29] adapted the popular surrogate-based LIME [20] to generate terms as the explanation for a trained black-box ranker. On the other hand, [7] applied a game-theory feature attribution method [15] to interpret the relevance score of a document given a query. Alternatively, other prevalent gradient-based feature attribution methods [21,22,28] can be adapted in the same way to attribute the relevance prediction to the textual input elements. All these methods provide pointwise explanations (why is doc_i relevant?) or pairwise explanations (why is doc_i ranked higher than doc_j?). We instead focus on listwise explanations or explaining the entire ranked list.

Listwise Explanations for Ranking Models. There is limited work on listwise explanations, i.e., explaining the entire ranking list. LiEGe [33] tackles the task as text generation. Specifically, LiEGe employs a Transformer-style model to generate terms for each document in a ranked list, and the explanation contains all generated terms. However, this method presupposes documents with labeled explanation terms, which is unrealistic in most application scenarios. Additionally, the explanation generator is not human-understandable, hindering understanding of the explanation generation process. In contrast, GreedyLM [25] uses a simple ranker to replicate the ranking list of a complex black-box model by expanding the query terms. The *simple ranker* and *expanded query*

---

[1] https://github.com/GarfieldLyu/RankingExplanation.

terms constitute the explanation for the complex model. We follow the same philosophy that the *explanation terms* along with the explanation generation process should be human interpretable. However, a limitation of [25] is that it assumes that a single relevance factor (modeled by a simple surrogate ranker) is adequate to explain an entire ranking. We challenge this assumption in this paper and use multiple simple explainers instead.

Axioms as Explanations. Another line of work uses IR axioms (or ranking heuristics) to ground the decisions of complex models. Axioms are well-understood, interpretable, and deterministic sets of rules that lay down the fundamental relevance factors of documents given a query. Recent works [4, 19] diagnosed a group of ad-hoc neural rankers with a set of axioms and found out that neural models only to a limited extent adhere to the IR axioms. Similarly, [30] also found it hard to characterize BERT models in terms of IR axioms. The hypothesis is axiomatic approaches are limited to using just the query terms, resulting in low fidelity. In this work, we consider a much larger vocabulary of explanation terms to optimize the fidelity of our explanations.

In parallel, there are other works dealing with explaining learning-to-rank (LTR) [23, 26], probing contextual ranking models [27, 31], and intrinsic methods for extractive explanations [10, 14, 34]. We point the readers to a recent survey [3] in explainable information retrieval for a more detailed overview. In this work, we operate on text rankers and generate term-based explanations in a post-hoc manner.

3 Background and Preliminaries

We start with the notion of a ranker Φ that takes as input a keyword query $\mathcal{Q}$ to output an ordering π over a set of documents $\pi = (d_1 \succ d_2 \succ \ldots \succ d_n)$ based on the relevance of the documents to the query, i.e., $\Phi(\mathcal{Q}) \rightarrow \pi$. We aim to interpret Φ in a model-agnostic manner, using simple proxy rankers (called explainers Ψ). Note that the output of a ranker can be viewed as a set of preferences over the documents, or w.l.o.g $\pi = \{(d_i \succ d_j)\}$. Therefore explaining a ranking π is akin to explaining all or most of the preference pair decisions in π. An example of a single decision is whether the preference pair $(d_i \succ d_j)$ is true/false.

3.1 Explainers for Ranking

The explainer Ψ mimicking a black-box ranking model is essentially a simple ranker operating based on human-understandable closed form formulae (i.e. ranking heuristics). A popular example of such interpretable rankers is BM25 [2] model, which ranks documents for a given query by measuring the *term-matching* frequency of query terms in each document. Apart from term matching, there are also other factors or heuristics that might affect the relevance judgment such as the *term position*. Specifically, in news articles, the title and the introductory paragraphs are regarded to be more important. A ranking model should then weigh the term matching that occurred in the earlier paragraphs more than the rest. Additionally, *semantic similarity* is known to be crucial to address the exact mismatch problem. This is particularly true in neural models with embedding vectors as input. However, the semantic meaning of a term is less interpretable as it can vary if the context changes due to different training procedures or datasets. In this regard, we draw the line of choosing the commonly-used context-free embeddings (i.e. GloVe [18]) as human-understandable input representation, instead of other contextualized embeddings (i.e., generated by BERT language model).

This set of simple ranking heuristics can be large given different granularities [4, 19]. In this work we start from **three** explainers to encode the above three ranking heuristics. Note that our framework allows a flexible amount of explainers, and thus more heuristics can be added if necessary. In summary, the explainers rank a document (d) based on its relevance to a query (q) by:

Term Matching or Ψ_{lm}: $\Psi_{lm}(q, d) = \frac{1}{|d|} \sum_{t \in q} \text{tf}(t, d)$, where $\text{tf}(t, d)$ denotes the term frequency of t in d.

Position Aware or Ψ_{pa}: a position-aware term-matching model [8], $\Psi_{pa}(q, d) = \sum_{t \in d} \frac{1}{|d|} \sum_{p \in d} \text{tf}(t, p)^{\frac{1}{p}}$, where p denotes the p_{th} paragraph in d.

Semantic Similarity or Ψ_{emb}: $\Psi_{emb}(q, d) = \frac{1}{|q| \times |d|} \sum_{t \in q, w \in d} \text{cosine}(t, w)$, where t and w are represented by the pre-trained GloVe embedding vectors [18].

3.2 Explanations to a Ranking Model

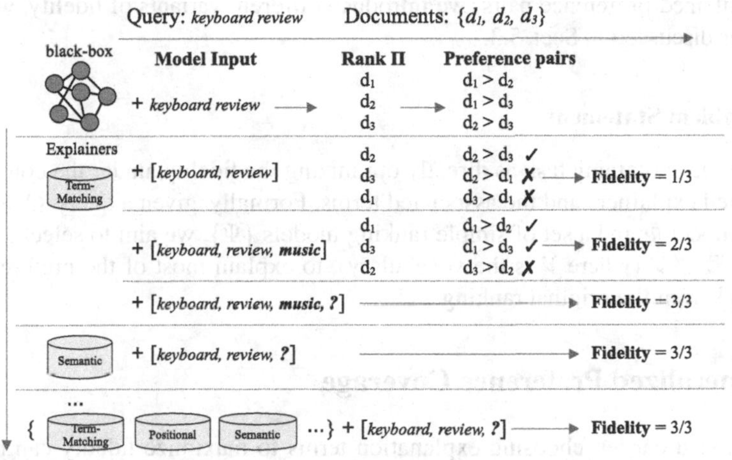

Fig. 2. Explaining black-box model with simple rankers and query terms.

The output of an interpretability procedure is an explanation, which should be *simple*, *human-understandable*, and *faithful* to the behavior of Φ. For the ranking task, the explanation can be decomposed into **two parts**: (1) a simple ranker whose decision-making process is fully transparent; (2) the machine intent of Φ in terms of an expanded query. The quality or fidelity(in XAI parlance) of the explanation can be evaluated by comparing the ranked lists induced by Φ and Ψ by standard rank-correlation metrics, e.g., Kendall's tau or just counting concordant preference pairs.

Take Fig. 2 as an example of interpreting the ranking induced by a black-box model. The simple Term Matching explainer with the input terms ("keyboard" and "review") can be regarded as an explanation, with a fidelity of $1/3$, as only one out of three preference pairs agrees with the original ranking. It is common that the query term is under-specified, and thus the simple ranker fails to extract the exact query intent. One solution is to use *query expansions* (e.g., RM3 [11]) to improve ranking performance. For instance, when adding "music" to the query, the explainer is aware of the musical preference of the black-box ranker and improves the explanation fidelity to $2/3$. The questions we ask are: (1) *which terms can be added to the query to maximize fidelity?*, and if more than one explainer is applied, (2) *how can we combine multiple simple explainers to cover as many pairs as possible?*

Fidelity Variants. Note that rankings can be misleading because they do not show the magnitude of the relevance difference. Sometimes the relevance scores of a preference pair can be very close, and explaining such pair is challenging even to humans. Therefore, to avoid uncertainty due to small score differences, we obtain a set of *important* preference pairs after excluding the similar pairs whose prediction difference is below some threshold. As Fig. 2 shows, suppose the black-box ranker predicts similar scores for d_2 and d_3, then $d_2 \succ d_3$ is not considered for evaluation. As a result, the Term Matching explainer, along with the input terms ("keyboard", "review" and "music"), can faithfully cover all pairs and get 100% fidelity. Given different choices of selecting to-be-explained preference pairs, we introduce different variants of fidelity, which will be further discussed in Sect. 5.3.

3.3 Problem Statement

We solve the explaining task as directly optimizing the fidelity, under the constraints of pre-defined explainers and the associated terms. Formally, given a query $\mathcal{Q}$, a complex ranking model Φ and a set of simple ranking models $\{\Psi\}$, we aim to select a small set of terms $\mathbb{E} \in \mathcal{V}$ (where $\mathcal{V}$ is the vocabulary), to explain most of the preference pairs $\{d_i \succ d_j\}$ from the original ranking π.

4   Generalized Preference Coverage

As mentioned earlier, choosing explanation terms to maximize fidelity can be formulated as a coverage problem of the preference pairs. We briefly describe the preference coverage (PC) framework as introduced in [25], using a single explainer as a precursor to introducing the generalized PC problem.

4.1 The Preference Coverage Framework

Similar to [25], the PC framework operates on a preference matrix constructed with a single Ψ. First, a set of n potentially important candidate terms $\mathcal{X}(\mathcal{X} \subseteq \mathcal{V}, |\mathcal{X}| = n)$ are extracted from the list of documents using simple statistics (e.g., *tf-idf*). Then, m preference pairs are sampled from π to create the preference matrix $\mathbf{M} \in \mathbb{R}^{n \times m}$. Each

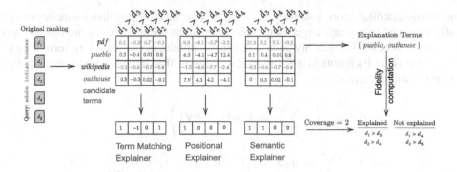

Fig. 3. Approach overview of MULTIPLEX using multiple explainers.

cell in $\mathbf{M}$ represents the utility or degree of Ψ in explaining the preference $d_{\pi(i)} \succ d_{\pi(j)}$ with t as input, by computing a preference score $f_{ij}^t = \Psi(t, d_{\pi(i)}) - \Psi(t, d_{\pi(j)})$. A positive f score means with t, the Ψ can explain or cover this pair, otherwise cannot. Each t can now be viewed as an m-dimensional vector $\mathbf{f}$, where each element represents how well it explains a specific pair. The PC framework using a single Ψ aims to choose a subset of rows $\mathbb{E} \subseteq \mathcal{X}$ (equivalent to selecting terms) from $\mathbf{M}$ so as to maximize the number of non-zero values in the aggregated vector. Since choosing or not choosing the row/term is a boolean decision, we can formulate the PC objective as an Integer Linear Program (ILP):

$$maximize \sum_{i=1}^{m} \left(\mathtt{sign}(\mathbf{x}^\top \mathbf{M})\right)_i, \quad \text{s.t.} \ \ \mathbf{x} = [x_1, \cdots, x_n]; \ \ x_i \in \{0, 1\} \qquad \text{(PC)}$$

$\mathbf{x}$ is a selection vector with boolean values where $x_i = 1$ indicates selecting term $\mathcal{X}_i$, and $x_i = 0$ otherwise. The $\mathtt{sign}$ is an element-wise operation. Namely, $\mathbb{E} = \{i | x_i == 1\}$. This equation however is NP-hard and not solvable by the prevalent convex programming solvers (e.g., supported by CVXPY [6]) due to the non-differentiable $\mathtt{sign}$ function. Next, we present an improved formulation of the PC problem followed by a generalization to accommodate multiple explainers called the GENERALIZED PREFERENCE COVERAGE problem.

4.2 Optimizing PC for Multiple Explainers

Compared to PC, our proposal should be (i) practically solvable, (ii) ensuring sparse output $\mathbf{x}$ so that the explanation is human-understandable, and (iii) flexible to combine multiple explainers or $\mathbf{M}$.

Correspondingly, the first change we introduce is using $\mathtt{tanh}$ to approximate the non-convex $\mathtt{sign}$ operator. Secondly, we add a ℓ_1-regularization $\|\mathbf{x}\|$ to enforce sparsity constraints on the number of terms to be selected. A straightforward way to combine all explainers is to sum up their scores, i.e., $\Psi_{multi}(t, d) = \sum \Psi(t, d)$. However, different explainers can have different output ranges and exhibit high variance. For instance, the term-matching score usually lies in $[0, 1]$, whereas the position-aware score typically operates in a much larger range. Normalization these scores in the optimization procedure is central to flexibly adding multiple explainers. We therefore formulate the GENERALIZED PREFERENCE COVERAGE problem that intends to optimize multiple matrices simultaneously as:

$$minimize \quad \left(-\sum_{i=1}^{m} (\texttt{tanh}(\mathbf{v}))_i + \|\mathbf{x}\| \right) \tag{GPC}$$

$$\text{s.t. } \mathbf{v} = \sum_{j=1}^{p} \texttt{tanh}(\mathbf{x}^{\top}\mathbf{M}_j), \quad 0 \le x_i \le 1, \quad a \le \sum_{i=1}^{m} x_i \le b$$

Like in PC, GPC also maximizes the number of positive elements in the aggregated vector $\mathbf{v}$, computed by summing up multiple vectors transposed from multiple $\mathbf{M}$. $\mathbf{M}_j$ denotes the matrix constructed by the jth explainer from the total p explainers. Note that $\texttt{tanh}$ is also element-wise. The sparsity constraint is ensured by a and b, namely the lower/upper bound of the term-selection budget. The current formulation can now be solved by the latest proposed solver GENO [13] that handles constraints with the *augmented lagrangian algorithm*.

Picking the ith term will choose all i^{th} row vectors simultaneously. Before summing them up, each vector element is already transformed to the same range by $\texttt{tanh}$ activation. This accounts for the variable range problem. Figure 3 briefly shows the coverage computing when selecting "pueblo" and "outhouse" during optimization.

5 Experimental Setup

5.1 Datasets and Ranking Models

We choose two datasets: (1) **Clueweb09** collection (category B), for all ranking models, we use 120/40/40 splits for train/dev/test, and the explanation experiments are conducted on the test queries. (2) 40 randomly selected queries from **Trec-DL** 2019 passage ranking test set, and the ranking models are trained on the MS MARCO passage ranking dataset. We focus on the following three neural ranking models:

DRMM [9] computes the term-document similarity histograms beforehand and then jointly learns a matching and a term gate layer from the query and matching histograms. We take the implementation from MatchZoo[2].

BERT [5] takes the query and document separated by [SEP] as input and computes the pooled ([CLS]) representation, on which a feed-forward layer predicts the final relevance score. Both DRMM and BERT models are trained to optimize the margin between the scores of a relevant/non-relevant input pair.

DPR [12] encodes the query and document by two separate BERT models. The relevance is simply measured by the cosine similarity of the two pooled representations. We use the pretrained checkpoints directly without fine-tuning.

---

[2] https://github.com/NTMC-Community/MatchZoo.

5.2   Baseline and Competitors

We compare our approach named MULTIPLEX with the following methods:

QUERY-TERMS serves as the baseline by feeding only the query terms to our explainers. By comparing this baseline, we argue that only the original query is insufficient to discover the underlying ranking logic.

DEEPLIFT [21] is a popular white-box feature attribution method. To adapt it to ranking, we first compute the importance of a word in a document using Captum[3], then we take the average across all documents and extract important terms as a listwise explanation for a query. Note that we omit this baseline for DRMM since its input is a histogram, thus the importance cannot be attributed to the word level.

GREEDY-LM [25] uses a term-matching explainer to approximate neural rankers. It optimizes the preference coverage greedily. Our approach shares a similar pipeline of generating candidate terms and preference matrices. By comparing this baseline, we show the improvements of combining multiple explainers and approximated linear programming optimization.

5.3   Metrics

Since multiple explainers are applied, a preference pair from the original ranking is counted as explained as long as a single explainer can explain it. This evaluation does not apply to GREEDY-LM as it generates explanation terms based on a single explainer. For both GREEDY-LM and MULTIPLEX, we fix 200 candidate terms and 500 sampled pairs for preference matrix construction. We also fix a maximum of 10 explanation terms for all methods except QUERY-TERMS. For both datasets, we consider a ranking depth (k) of 100.

Similar to [19], we measure fidelity by computing the fraction of the maintained preference pairs by the explainers given the explanation terms. In other words, the fidelity measures the coverage over the *feasible* preference pairs. As mentioned in Sect. 3, depending on the choice of feasible preference pairs, we consider the following three variants of fidelity:

Fidelity-global ($\mathcal{F}_{\text{global}}$) includes all $\binom{k}{2}$ pairs induced by a k-length ranking list.

Fidelity-sampled ($\mathcal{F}_{\text{sampled}}$) considers the sampled pairs from the matrix construction.

Fidelity-diff ($\mathcal{F}_{\text{diff}}$) discards all pairs whose relevance score difference $< g$. The magnitude of g is chosen based on the relevance score distribution of a particular model. For BERT we set $g = 2$ as the prediction margin appears to be larger than the rest two models, for which $g = 0.05$.

6   Evaluation Results

To show the effectiveness of our approach, we first present the quality of our approach in terms of fidelity on all datasets and models compared to other competitors in Table 1.

---

[3] https://github.com/pytorch/captum.

Table 1. Fidelity ($\mathcal{F}$) results. The best results are in bold.

Model	Method	Clueweb09			Trec-DL		
		$\mathcal{F}_{global}$	$\mathcal{F}_{diff}$	$\mathcal{F}_{sampled}$	$\mathcal{F}_{global}$	$\mathcal{F}_{diff}$	$\mathcal{F}_{sampled}$
BERT	QUERY-TERMS	0.81	0.88	0.76	0.81	0.82	0.63
	DEEPLIFT [21]	0.77	0.81	0.67	0.70	0.75	0.62
	GREEDY-LM [25]	0.63	0.77	0.69	0.59	0.69	0.84
	MULTIPLEX	**0.88**	**0.97**	**0.93**	**0.86**	**0.93**	**0.97**
DPR	QUERY-TERMS	0.81	0.86	0.71	0.82	0.84	0.64
	DEEPLIFT [21]	0.68	0.71	0.57	0.60	0.63	0.58
	GREEDY-LM [25]	0.61	0.68	**0.88**	0.63	0.70	0.75
	MULTIPLEX	0.87	**0.93**	0.87	**0.87**	**0.92**	**0.96**
DRMM	QUERY-TERMS	0.82	0.85	0.72	0.80	0.81	0.59
	DEEPLIFT [21]	–	–	–	–	–	–
	GREEDY-LM [25]	0.57	0.60	0.72	0.53	0.54	0.34
	MULTIPLEX	**0.88**	**0.92**	**0.84**	**0.85**	**0.88**	**0.95**

Then we show the improvements of adding multiple explainers by an ablation study presented in Fig. 4. Finally, we discuss how our explanations can be used to explain a specific preference pair, as well as other potential use cases.

6.1 Effectiveness of Explanations

In terms of fidelity (cf. Table 1), our method consistently outperforms other competitors. Besides, for all methods the global fidelity ($\mathcal{F}_{global}$) scores are always lower than $\mathcal{F}_{diff}$ where close, hence potentially noisy pairs are all excluded. This shows that all methods and prominently MULTIPLEX can better explain document pairs with larger differences in relevance scores.

Ranking Heuristics vs. Query Expansion. Though both factors constitute the explanation of ranking, which one is more crucial? Take QUERY-TERMS and GREEDY-LM as a comparison, note that QUERY-TERMS includes the given query terms but three ranking heuristics, while GREEDY-LM on the contrary only relies on one term-matching but richer query information. Their fidelity results show QUERY-TERMS outperforms GREEDY-LM by a large margin, strongly suggesting that ranking heuristics particularly semantic similarity, are more effective in explaining neural models.

The Importance of Explanation Aggregation. Applying simple aggregation strategies (i.e. *average*) on the prevalent pointwise feature attribution methods is shown to be less effective by the results of DEEPLIFT. Compared to QUERY-TERMS, the extra expanded query terms extracted by DEEPLIFT seem unhelpful in enhancing fidelity but introducing noise. On the other hand, methods directly optimizing fidelity (i.e. GREEDY-LM and MULTIPLEX) explicitly include the aggregation in the optimization loop. The $\mathcal{F}_{sampled}$ results of DEEPLIFT and GREEDY-LM further confirm the importance of aggregation.

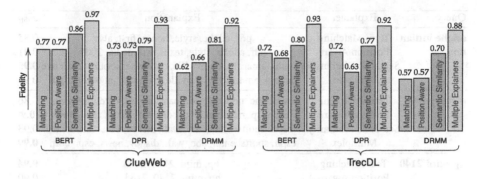

Fig. 4. *Fidelity-diff* results of each single and combined explainer using our method.

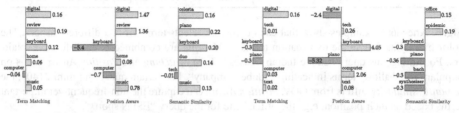

(a) BERT: music-related terms are *pos*.   (b) DPR: music-related terms are *neg*.

Fig. 5. Query: keyboard review. Document pair: clueweb09-en0008-49-09140 (musical keyboard) *vs.* clueweb09-en0010-56-37788 (technical keyboard). BERT prefers the former whereas DPR prefers the latter, resulting in opposite explanations.

The Benefits of Our Optimization Solution. We also experimented with every single Ψ to extract explanation terms with our approximated ILP objective shown in Fig. 4. Comparing the fidelity results of term-matching (orange bar) with the $\mathcal{F}_{\mathrm{diff}}$ of GREEDY-LM (using the same explainer) in Table 1, we show the superiority of our optimizing strategy over the greedy-algorithm.

The Benefits of Combining Explainers. As Fig. 4 indicates, semantic explainer overall generates the most faithful explanations than the rest. However, combining all explainers can further improve the preference coverage and in turn increase the fidelity results. When one explainer fails to explain a pair, it is still possible to be covered by other explainers. Moreover, we also notice that combining multiple explainers in optimization can generate explanation terms exhibiting multiple topic aspects, especially for short and ambiguous queries. More examples are presented in Fig. 1 and Fig. 6.

6.2 Utility of Explanations

Explaining Document Preference. We now show how to explain a single preference pair using MULTIPLEX, i.e., why does a model prefer d_i over d_j? We start by constructing preference scores for each candidate term as described in Sect. 4.1. Next, we select

Query	Explainer	Explanation	$\mathcal{F}_{\text{diff}}$
adobe indian houses	Term Matching	pdf, adobe, style, house, first, also	0.85
	Position Aware	pdf, adobe, style, texas, wikipedia, 2009	0.81
	Semantic Similarity	pueblo, amarillo, castroville, outhouse, abourezk, alcove	**0.95**
	Multiplex	pueblo, amarillo, castroville, outhouse, abourezk, pdf	0.91
espn sports	Term Matching	espn, abc, network, company, award, entertainment,	0.86
	Position Aware	espn, sportscenter, abc, company, news, espn.com	**0.99**
	Semantic Similarity	espn, sportscenter, abc, walt, disney, entertainment,	0.93
	Multiplex	espn, sportscenter, abc, walt, disney, news, espn.com	**0.99**
hp mini 2140	Text Matching	hp, mini, 2140, 2133	**0.94**
	Position Aware	hp, mini, 2140, 2133	0.90
	Semantic Similarity	hp, touchpad, overview, hdd,	0.71
	Multiplex	hp, mini, 2140, 2133, touchpad, overview	0.91

Fig. 6. Anecdotal examples show that each explainer selects terms from a different aspect. The color highlights denote the explanation terms in *Multiple* are combined from different explainers. For ambiguous query "adobe Indian houses", *Term Matching* and *Position Aware* focus on popular but 'shallow' terms indicating "adobe company". For certain query "hp mini 2140", the *semantic similarity* suffers from OOV. *Position Aware* can capture the non-frequent yet important terms based on their position, e.g., the official site for the query "ESPN sports".

the important terms with significant scores. Figure 5 illustrates the explanation terms of two opposing decisions by BERT and DPR respectively, for `keyboard review`.

Discovering Model Preference and Spurious Correlations. We believe that explanation terms encode relevance factors that rank relevant documents over others. Based on this assumption, we create a perturbed document by adding explanation terms to a potentially *non-relevant document* (e.g. at the lowest rank). We then feed this modified document to the black-box model and measure the rank improvement. Unsurprisingly, the terms extracted by MULTIPLEX result in the maximum rank increase (cf. Fig. 7), meaning our method can better identify the black-box model's preference. Moreover, we manually selected some ambiguous queries, and our initial observation of their explanation terms suggests the ranking model shows some topic preference when ranking the documents, while the explanation terms representing the preferred topics are also shown dominant quantitively. Thus, it helps understand the model's topic preference more easily by analyzing the explanations instead of going through hundreds of documents.

Another possible usage is model debugging, or finding spurious correlations in models or datasets, by analyzing explanation terms. One simple example is "Wikipedia" which appears as an explanation term for many different queries. This is not surprising as the Wikipedia entity pages are usually labeled as relevant. We leave a more systematic exploration of making use of ranking explanations to future work.

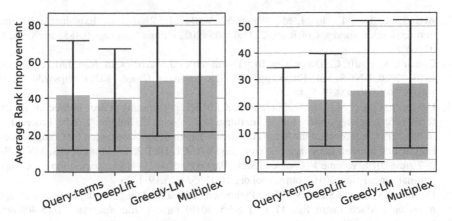

Fig. 7. Average rank improvements. Left: on all test queries; Right: on hand-picked ambiguous queries. Note that for each query the document size ≤ 100.

7   Conclusion and Outlook

This paper proposes a post-hoc model-agnostic framework to explain text ranking models using multiple explainers. Our method MULTIPLEX systematically combines multiple explainers to capture different relevance factors encoded in the ranking decisions. The extensive experiments show that our method can generate high-fidelity explanations for over-parameterized models like BERT, delivering up to 54% fidelity improvements. Our method explains a ranking by a set of terms attributed to a union of multiple explainers. It is interesting to examine which explainer (or ranking heuristic) contributes to which extent using which particular terms for future work. We also plan to extend our framework to account for n-grams and to make our explanation generation procedure efficient enough to be used during query processing. Moreover, it is well known that validating explanations is challenging, especially in the absence of ground-truth data. We measure fidelity in this work, however, the fidelity might not reflect the real underline logic of a complex model. Therefore, incorporating human perspectives into the evaluation and meanwhile, balancing the cost of annotating numerous decisions in a ranking are also worth exploring in future work.

Acknowledgements. This work is partially supported by German Research Foundation (DFG), under the Project IREM with grant No. AN 996/1-1.

References

1. Ailon, N., Charikar, M., Newman, A.: Aggregating inconsistent information: ranking and clustering. J. ACM **55**(5), 23:1–23:27 (2008). https://doi.org/10.1145/1411509.1411513
2. Amati, G., Van Rijsbergen, C.J.: Probabilistic models of information retrieval based on measuring the divergence from randomness. ACM Trans. Inf. Syst. **20**(4), 357–389 (2002). https://doi.org/10.1145/582415.582416, http://doi.acm.org/10.1145/582415.582416

3. Anand, A., Lyu, L., Idahl, M., Wang, Y., Wallat, J., Zhang, Z.: Explainable information retrieval: a survey. CoRR abs/2211.02405 (2022). https://doi.org/10.48550/arXiv.2211.02405

4. Câmara, A., Hauff, C.: Diagnosing BERT with retrieval heuristics. In: Jose, J.M., et al. (eds.) ECIR 2020. LNCS, vol. 12035, pp. 605–618. Springer, Cham (2020). https://doi.org/10.1007/978-3-030-45439-5_40

5. Devlin, J., Chang, M., Lee, K., Toutanova, K.: BERT: pre-training of deep bidirectional transformers for language understanding. In: Burstein, J., Doran, C., Solorio, T. (eds.) Proceedings of the 2019 Conference of the North American Chapter of the Association for Computational Linguistics: Human Language Technologies, NAACL-HLT 2019, Minneapolis, MN, USA, 2–7 June 2019, Volume 1 (Long and Short Papers), pp. 4171–4186. Association for Computational Linguistics (2019). https://doi.org/10.18653/v1/n19-1423

6. Diamond, S., Boyd, S.P.: CVXPY: a Python-embedded modeling language for convex optimization. J. Mach. Learn. Res. **17**, 83:1–83:5 (2016). http://jmlr.org/papers/v17/15-408.html

7. Fernando, Z.T., Singh, J., Anand, A.: A study on the interpretability of neural retrieval models using deepshap. In: Proceedings of the 42Nd International ACM SIGIR Conference on Research and Development in Information Retrieval. SIGIR 2019, pp. 1005–1008. ACM, New York (2019). https://doi.org/10.1145/3331184.3331312, http://doi.acm.org/10.1145/3331184.3331312

8. Fetahu, B., Markert, K., Anand, A.: Automated news suggestions for populating Wikipedia entity pages. In: Bailey, J., et al. (eds.) Proceedings of the 24th ACM International Conference on Information and Knowledge Management. CIKM 2015, Melbourne, VIC, Australia, 19–23 October 2015, pp. 323–332. ACM (2015). https://doi.org/10.1145/2806416.2806531

9. Guo, J., Fan, Y., Ai, Q., Croft, W.B.: A deep relevance matching model for ad-hoc retrieval. In: Mukhopadhyay, S., et al. (eds.) Proceedings of the 25th ACM International Conference on Information and Knowledge Management. CIKM 2016, Indianapolis, IN, USA, 24–28 October 2016, pp. 55–64. ACM (2016). https://doi.org/10.1145/2983323.2983769

10. Hofstätter, S., Mitra, B., Zamani, H., Craswell, N., Hanbury, A.: Intra-document cascading: learning to select passages for neural document ranking. In: Proceedings of the 44th International ACM SIGIR Conference on Research and Development in Information Retrieval. SIGIR 2021, pp. 1349–1358. Association for Computing Machinery, New York (2021). https://doi.org/10.1145/3404835.3462889

11. Jaleel, N.A., et al.: Umass at TREC 2004: novelty and HARD. In: Voorhees, E.M., Buckland, L.P. (eds.) Proceedings of the Thirteenth Text REtrieval Conference, TREC 2004, Gaithersburg, Maryland, USA, 16–19 November 2004. NIST Special Publication, vol. 500-261. National Institute of Standards and Technology (NIST) (2004). http://trec.nist.gov/pubs/trec13/papers/umass.novelty.hard.pdf

12. Karpukhin, V., et al.: Dense passage retrieval for open-domain question answering. In: Webber, B., Cohn, T., He, Y., Liu, Y. (eds.) Proceedings of the 2020 Conference on Empirical Methods in Natural Language Processing, EMNLP 2020, Online, 16–20 November 2020, pp. 6769–6781. Association for Computational Linguistics (2020). https://doi.org/10.18653/v1/2020.emnlp-main.550

13. Laue, S., Mitterreiter, M., Giesen, J.: GENO - generic optimization for classical machine learning. In: Wallach, H.M., Larochelle, H., Beygelzimer, A., d'Alché-Buc, F., Fox, E.B., Garnett, R. (eds.) Advances in Neural Information Processing Systems 32: Annual Conference on Neural Information Processing Systems 2019. NeurIPS 2019, 8–14 December 2019. Vancouver, BC, Canada, pp. 2187–2198 (2019). https://proceedings.neurips.cc/paper/2019/hash/84438b7aae55a0638073ef798e50b4ef-Abstract.html

14. Leonhardt, J., Rudra, K., Anand, A.: Extractive explanations for interpretable text ranking. ACM Trans. Inf. Syst. (2022). https://doi.org/10.1145/3576924

15. Lundberg, S.M., Lee, S.: A unified approach to interpreting model predictions. In: Guyon, I., et al. (eds.) Advances in Neural Information Processing Systems 30: Annual Conference on Neural Information Processing Systems 2017, 4–9 December 2017, Long Beach, CA, USA, pp. 4765–4774 (2017). https://proceedings.neurips.cc/paper/2017/hash/8a20a8621978632d76c43dfd28b67767-Abstract.html
16. McDonald, R.T., Brokos, G., Androutsopoulos, I.: Deep relevance ranking using enhanced document-query interactions. In: Riloff, E., Chiang, D., Hockenmaier, J., Tsujii, J. (eds.) Proceedings of the 2018 Conference on Empirical Methods in Natural Language Processing, Brussels, Belgium, 31 October–4 November 2018, pp. 1849–1860. Association for Computational Linguistics (2018). https://doi.org/10.18653/v1/d18-1211
17. Nogueira, R.F., Cho, K.: Passage re-ranking with BERT. CoRR abs/1901.04085 (2019). http://arxiv.org/abs/1901.04085
18. Pennington, J., Socher, R., Manning, C.D.: Glove: global vectors for word representation. In: Moschitti, A., Pang, B., Daelemans, W. (eds.) Proceedings of the 2014 Conference on Empirical Methods in Natural Language Processing, EMNLP 2014, 25–29 October 2014, Doha, Qatar, A meeting of SIGDAT, a Special Interest Group of the ACL, pp. 1532–1543. ACL (2014). https://doi.org/10.3115/v1/d14-1162
19. Rennings, D., Moraes, F., Hauff, C.: An axiomatic approach to diagnosing neural IR models. In: Azzopardi, L., Stein, B., Fuhr, N., Mayr, P., Hauff, C., Hiemstra, D. (eds.) ECIR 2019. LNCS, vol. 11437, pp. 489–503. Springer, Cham (2019). https://doi.org/10.1007/978-3-030-15712-8_32
20. Ribeiro, M.T., Singh, S., Guestrin, C.: "why should I trust you?": Explaining the predictions of any classifier. In: Krishnapuram, B., Shah, M., Smola, A.J., Aggarwal, C.C., Shen, D., Rastogi, R. (eds.) Proceedings of the 22nd ACM SIGKDD International Conference on Knowledge Discovery and Data Mining, San Francisco, CA, USA, 13–17 August 2016, pp. 1135–1144. ACM (2016). https://doi.org/10.1145/2939672.2939778
21. Shrikumar, A., Greenside, P., Kundaje, A.: Learning important features through propagating activation differences. In: Precup, D., Teh, Y.W. (eds.) Proceedings of the 34th International Conference on Machine Learning, ICML 2017, Sydney, NSW, Australia, 6–11 August 2017. Proceedings of Machine Learning Research, vol. 70, pp. 3145–3153. PMLR (2017), http://proceedings.mlr.press/v70/shrikumar17a.html
22. Simonyan, K., Vedaldi, A., Zisserman, A.: Deep inside convolutional networks: visualising image classification models and saliency maps. In: Bengio, Y., LeCun, Y. (eds.) 2nd International Conference on Learning Representations, ICLR 2014, Banff, AB, Canada, 14–16 April 2014, Workshop Track Proceedings (2014). http://arxiv.org/abs/1312.6034
23. Singh, J., Anand, A.: Posthoc interpretability of learning to rank models using secondary training data. In: Workshop on ExplainAble Recommendation and Search (EARS 2018) at SIGIR 2018 (2018). https://ears2018.github.io/ears18-singh.pdf
24. Singh, J., Anand, A.: Exs: explainable search using local model agnostic interpretability. In: Proceedings of the Twelfth ACM International Conference on Web Search and Data Mining. WSDM 2019, pp. 770–773. ACM, New York (2019). https://doi.org/10.1145/3289600.3290620, http://doi.acm.org/10.1145/3289600.3290620
25. Singh, J., Anand, A.: Model agnostic interpretability of rankers via intent modelling. In: Hildebrandt, M., Castillo, C., Celis, L.E., Ruggieri, S., Taylor, L., Zanfir-Fortuna, G. (eds.) FAT* '20: Conference on Fairness, Accountability, and Transparency, Barcelona, Spain, 27–30 January 2020, pp. 618–628. ACM (2020). https://doi.org/10.1145/3351095.3375234
26. Singh, J., Khosla, M., Wang, Z., Anand, A.: Extracting per query valid explanations for blackbox learning-to-rank models. In: Hasibi, F., Fang, Y., Aizawa, A. (eds.) ICTIR 2021: The 2021 ACM SIGIR International Conference on the Theory of Information Retrieval, Virtual Event, Canada, 11 July 2021, pp. 203–210. ACM (2021). https://doi.org/10.1145/3471158.3472241

27. Singh, J., Wallat, J., Anand, A.: BERTnesia: investigating the capture and forgetting of knowledge in BERT. In: Alishahi, A., Belinkov, Y., Chrupala, G., Hupkes, D., Pinter, Y., Sajjad, H. (eds.) Proceedings of the Third BlackboxNLP Workshop on Analyzing and Interpreting Neural Networks for NLP, BlackboxNLP@EMNLP 2020, Online, November 2020, pp. 174–183. Association for Computational Linguistics (2020). https://doi.org/10.18653/v1/2020.blackboxnlp-1.17

28. Sundararajan, M., Taly, A., Yan, Q.: Axiomatic attribution for deep networks. In: Precup, D., Teh, Y.W. (eds.) Proceedings of the 34th International Conference on Machine Learning, ICML 2017, Sydney, NSW, Australia, 6–11 August 2017. Proceedings of Machine Learning Research, vol. 70, pp. 3319–3328. PMLR (2017). http://proceedings.mlr.press/v70/sundararajan17a.html

29. Verma, M., Ganguly, D.: LIRME: locally interpretable ranking model explanation. In: Piwowarski, B., Chevalier, M., Gaussier, É., Maarek, Y., Nie, J., Scholer, F. (eds.) Proceedings of the 42nd International ACM SIGIR Conference on Research and Development in Information Retrieval, SIGIR 2019, Paris, France, 21–25 July 2019, pp. 1281–1284. ACM (2019). https://doi.org/10.1145/3331184.3331377

30. Völske, M., et al.: Towards axiomatic explanations for neural ranking models. In: Hasibi, F., Fang, Y., Aizawa, A. (eds.) ICTIR 2021: The 2021 ACM SIGIR International Conference on the Theory of Information Retrieval, Virtual Event, Canada, 11 July 2021, pp. 13–22. ACM (2021). https://doi.org/10.1145/3471158.3472256

31. Wallat, J., Beringer, F., Anand, A., Anand, A.: Probing BERT for ranking abilities. In: Advances in Information Retrieval - 45th European Conference on IR Research, ECIR 2023, Dublin, Ireland. LNCS. Springer, Cham (2023)

32. Wang, Y., Lyu, L., Anand, A.: BERT rankers are brittle: a study using adversarial document perturbations. In: Crestani, F., Pasi, G., Gaussier, É. (eds.) ICTIR 2022: The 2022 ACM SIGIR International Conference on the Theory of Information Retrieval, Madrid, Spain, 11–12 July 2022, pp. 115–120. ACM (2022). https://doi.org/10.1145/3539813.3545122

33. Yu, P., Rahimi, R., Allan, J.: Towards explainable search results: a listwise explanation generator. In: Amigó, E., Castells, P., Gonzalo, J., Carterette, B., Culpepper, J.S., Kazai, G. (eds.) SIGIR 2022: The 45th International ACM SIGIR Conference on Research and Development in Information Retrieval, Madrid, Spain, 11–15 July 2022, pp. 669–680. ACM (2022). https://doi.org/10.1145/3477495.3532067

34. Zhang, Z., Rudra, K., Anand, A.: Explain and predict, and then predict again. In: Proceedings of the 14th ACM International Conference on Web Search and Data Mining, pp. 418–426 (2021)

Improving Video Retrieval Using Multilingual Knowledge Transfer

Avinash Madasu[1,2], Estelle Aflalo[1], Gabriela Ben Melech Stan[1], Shao-Yen Tseng[1], Gedas Bertasius[2], and Vasudev Lal[1(✉)]

[1] Cognitive Computing Research, Intel Labs, Santa Clara, USA
vasudev.lal@intel.com
[2] University of North Carolina at Chapel Hill, Chapel Hill, USA

Abstract. Video retrieval has seen tremendous progress with the development of vision-language models. However, further improving these models require additional labelled data which is a huge manual effort. In this paper, we propose a framework MKTVR, that utilizes knowledge transfer from a multilingual model to boost the performance of video retrieval. We first use state-of-the-art machine translation models to construct pseudo ground-truth multilingual video-text pairs. We then use this data to learn a video-text representation where English and non-English text queries are represented in a common embedding space based on pretrained multilingual models. We evaluate our proposed approach on four English video retrieval datasets such as MSRVTT, MSVD, DiDeMo and Charades. Experimental results demonstrate that our approach achieves state-of-the-art results on all datasets outperforming previous models. Finally, we also evaluate our model on a multilingual video-retrieval dataset encompassing six languages and show that our model outperforms previous multilingual video retrieval models in a zero-shot setting.

Keywords: Video-retrieval · Multi-lingual · Multi-modal

1 Introduction

The task of text-to-video retrieval aims to retrieve videos that are semantically similar to a given text query. In the last-few years, there has been a significant progress in the area of video retrieval. These works were developed in two parallel directions. The first line of work [3,15,22] focused on video-text pretraining on large-scale datasets like Howto100M [31] and WebVid-2M [3]. While the second line of work [29] focused on using pretrained image features like CLIP [33] for video retrieval often surpassing the models pretrained on video datasets.

In Natural Language Processing, some works [9] explored the idea of using multilingual data to improve the performance on monolingual English datasets. Conneau et al. introduced a new pretraining objective: Translation Language

---

A. Madasu—Contribution during Avinash's internship at Intel.

J. Kamps et al. (Eds.): ECIR 2023, LNCS 13980, pp. 669–684, 2023.
https://doi.org/10.1007/978-3-031-28244-7_42

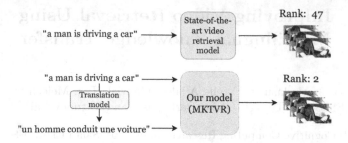

Fig. 1. Illustration of the improved video retrieval ranking for a test sample in MSRVTT dataset. For the current state-of-the-art video retrieval model, the rank of the ground truth video is 47. When multilingual data is used as knowledge transfer, the ranking of the ground truth video improved significantly from 47 to 2.

Modeling (TLM) in which random words were masked in the concatenated sentences of English and multilingual data and the model predicts the masked words. There, the objective was to use multilingual context to predict masked English words if the English context was not sufficient and vice-versa.

Most of the current video retrieval datasets typically contain around 10k videos and captions of maximum lengths ranging from 30 to 60. In video retrieval, the text encoder projects the input text caption and videos into a common embedding space. With longer captions, the text embeddings might lose required contextual information resulting in incorrect video retrievals. One could address this by incorporating more structured knowledge (e.g., parts of speech, dependency graphs) in the text encoder [6]. However, a drop in performance is observed [6] with the addition of structural knowledge in a text-to-video retrieval setting. One could argue that the reason might be the smaller video retrieval datasets and creating meaningful structural knowledge becomes a challenging task.

In this work, we are interested in **improving the performance of video retrieval using multilingual knowledge transfer**. Multilingual data serves as a powerful knowledge augmentation for monolingual models [9]. Nevertheless, creating multilingual data requires huge human effort. To overcome this, we use state-of-the-art machine translation models [37] to convert English text captions into other languages. Specifically, we choose languages whose performance on XNLI benchmark [10] is comparable to that of English (i.e., French, German, Spanish). With this, we create high quality multilingual data without requiring human labelling. To the best of our knowledge, this is the first work that uses multilingual knowledge transfer to improve video retrieval.

We propose a model based on CLIP [33] to effectively utilize and adapt the multilingual knowledge transfer. Our model takes a video, English text caption and multilingual text caption as inputs and extracts joint video-text representations. The multilingual text representations should act as a knowledge augmentation to the English text representations aiding in video-retrieval. For this purpose, we introduce a dual cross-modal (DCM) encoder block which learns the similarity between English text representations and video representations.

In addition, the DCM encoder block also associates the video representations with the multilingual text representations. In the common embedding space, our model learns the important contextual information from multilingual representations which is otherwise missing from the English text representations effectively serving as knowledge transfer.

We validate our proposed model on five English video retrieval datasets: MSRVTT-9k [44], MSRVTT-7k [44], MSVD [7], DiDeMo [2] and Charades [35]. We show that our approach achieves state-of-the-art results, outperforming previous models on most datasets. In addition to the evaluation on monolingual video retrieval datasets, we also compare the performance of our model on multilingual MSRVTT dataset [18] in a zero-shot setting and demonstrate that our framework surpasses the previous works by a large margin.

To summarize, our contributions are as follows: (i) We generate multilingual data using external state-of-the-art machine translation models. (ii) We propose a model that is capable of knowledge transfer from multilingual data to improve the performance of video retrieval. (iii) We evaluate the proposed framework on five English video retrieval benchmarks and achieve state-of-the-art results in both text-to-video and video-to-text retrieval settings. (iv) Finally, we demonstrate that our model significantly outperforms previous approaches on multilingual video retrieval datasets in a zero-shot setting.

2 Related Work

2.1 Video Retrieval

The task of video retrieval has seen tremendous progress in the recent years. This is partly due to the availability of large-scale video datasets like HowTo100M [31] and WebVid-2M [3]. Besides the adaption of transformers to image tasks like image classification [11] spurred the development of models based on transformers. However, videos require computationally more memory and compute power and can be infeasible to compute self-attention matrices. With the introduction of more efficient architectures [4] large-scale pretraining on videos became a possibility. In this direction, several transformer based architectures [3,15,30] were proposed and pretrained on large video datasets which achieved state-of-the-art results on downstream video retrieval datasets in both zero-shot and fine-tuning settings.

In a parallel direction, a few works [29] have adopted image level features pretrained on large scale image-text pairs to perform video retrieval. Surprisingly, these works have performed significantly better than the models that are pretrained from scratch on large scale video datasets. Compared to these models, our approach completely differs in the architecture and the training methodology.

2.2 Multilingual Training

The recent success of multimodal image-text models on a variety of tasks, such as retrieval and question-answering, has been mostly limited to monolingual models trained on English text. This is mainly due to the availability and high-quality of English-based multimodal datasets. Recent work indicates that incorporating a second language or a multilingual encoder, thus creating a shared multilingual token embedding space, can improve monolingual pure-NLP downstream tasks [9]. This concept was rapidly embraced for training multimodal models. Previous works had used images as a bridge for translating between two languages, without using a language-to-language shared dataset for training [7,34,36].

Recent work has focused on multimodal tasks, such as image retrieval, aiming to add multilingual capabilities to multimodal models [5,17]. The work often indicates that incorporating a second language during training of multimodal models, improves performance on single-language multimodal tasks such as image retrieval, compared to multimodal models that were trained on a single language [17,19,41]. MULE [19], which is a multilingual universal language encoder trained on image-multilingual text pairs, showed an improvement on image-sentence retrieval tasks of up to 20% compared to monolingual models. Nevertheless, all these previous works focus on designing a universal model for image and video retrieval. Our objective is to use multilingual knowledge transfer to improve the performance on current video retrieval datasets.

3   MKTVR: Multilingual Knowledge Transfer for Video Retrieval

In this section, we introduce our framework MKTVR: Multilingual Knowledge Transfer for Video Retrieval. We first describe the problem statement, then the multilingual data augmentation strategy and finally go over the proposed approach that enables knowledge transfer from multilingual data for video retrieval.

3.1   Problem Statement

Given a set of videos V, their corresponding English text captions E and related multilingual text captions M, our goal is to learn similarity functions $s_1(v_i, e_i)$ and $s_2(v_i, m_i)$ ($v_i \in V$, $e_i \in E$ and $m_i \in M$). In other words, we propose a framework MKTVR that enables end-to-end learning on a tuple of video, English text caption and multilingual text caption by bringing closer the joint representations of those three elements. Specifically, for each video V and the English text captions E, we generate the multilingual translations M using external state-of-the-art machine translation models [37]. Next, we present the proposed approach that facilitates the end-to-end learning using multilingual data.

3.2   Approach

Our model, illustrated in Fig. 2, is comprised of three components: (i) video encoder (ii) text encoder (iii) dual cross modal encoder. Next, we describe the framework in detail.

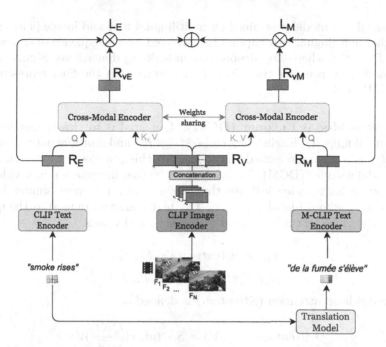

Fig. 2. Illustration of the proposed MKTVR model. The model takes as input a video, a corresponding English text query and a translated multilingual query. The multilingual text query is obtained using the off-the-shelf machine translation model. It is used only for the inference and is not part of the architecture. The video and English text features are extracted using CLIP model whereas multilingual text features are extracted using M-CLIP model. The features are then passed onto a cross-model encoder to learn the association in a common embedding space. Cross-entropy loss is then applied to measure the similarity between text features R_E and R_{vE}, R_M and R_{vM}. The final loss is the sum of both the losses.

Video Encoder. Given a video V, we consider uniformly sampled clips $C \in R^{N_v \times H \times W \times 3}$ where N_v is the number of frames, H and W are the spatial dimensions of a RGB frame. We then use a pretrained CLIP-ViT image encoder [33] to extract the frame embeddings $F_v \in R^{N_v \times D_v}$ where D_v denotes the dimensions of the frame embeddings. The frame embeddings are concatenated to obtain the final representation for the video V.

Text Encoder. Let the inputs English text caption be E and multilingual text caption M of lengths p and q respectively. We use a pretrained CLIP-ViT text encoder to convert the English text caption into a sequence of embeddings $R_E = R^{E_p \times D_E}$ where D_E denote the embedding dimensions. We consider the representation of the token $[EOS]$ as the final representation of English text caption. To encode multilingual text caption M, we use a M-CLIP[1] model which is a

---
[1] https://github.com/FreddeFrallan/Multilingual-CLIP

multilingual clip model pretrained on multilingual text and image pairs. Specifically, the multilingual text caption is converted into a sequence of embeddings $R_M = R^{M_q \times D_M}$ where D_M denote the embedding dimensions. Similar to the CLIP model, we consider the $[EOS]$ representation as the final representation of M-CLIP model.

Dual Cross-Modal Encoder (DCM). Our goal is to closely associate the video embeddings R_v, English text embeddings R_E and multilingual text embeddings R_M in a common embedding space. For this purpose, we propose a dual cross-modal encoder (DCM). To incorporate textual information into video features and to learn video features that are semantically most similar to text features, we use multi-head attention. The text features are used as the queries whereas the video features are used as the keys and values.

$$r_{vE} = Attention(T_E, F_v, F_v) \tag{1}$$

$$r_{vM} = Attention(M_E, F_v, F_v) \tag{2}$$

where multi-head attention (*Attention*) is defined as:

$$Attention(Q, K, V) = Softmax(\frac{QK^T}{\sqrt{d}})V \tag{3}$$

Here Q, K and V are same as the original multi-head attention matrices in the transformer encoder. We then apply a fully connected layer on the attention outputs and finally layer normalization to obtain R_{vE} and R_{vM}.

$$R_{vE} = LN(FC(r_{vE}) + r_{vE}) \tag{4}$$

$$R_{vM} = LN(FC(r_{vM}) + r_{vM}) \tag{5}$$

where FC is the fully connected layer and LN is the layer normalization layer.

Loss. We use the standard video-text matching loss [42] to train the model. It is measured as the dot product similarity between matching text embeddings and video embeddings in a batch. First, we compute the loss L_E between R_{vE} and R_E and then compute the loss L_M between R_{vM} and R_M. The final loss is the sum of losses L_E and L_M.

$$L = L_E + L_M. \tag{6}$$

where $L_E = L_E^{t2v} + L_E^{v2t}$ and $L_M = L_M^{t2v} + L_M^{v2t}$

$$\mathcal{L}_E^{v2t} = -\frac{1}{B} \sum_{i=1}^{B} \log \frac{\exp(R_E^{(i)} \cdot R_{vE}^{(i)})}{\sum_{j=1}^{B} \exp(R_E^{(i)} \cdot R_{vE}^{(j)})}, \tag{7}$$

$$\mathcal{L}_E^{t2v} = -\frac{1}{B} \sum_{i=1}^{B} \log \frac{\exp(R_{vE}^{(i)} \cdot R_E^{(i)})}{\sum_{j=1}^{B} \exp(R_{vE}^{(i)} \cdot R_E^{(j)})}. \tag{8}$$

$$\mathcal{L}_M^{\text{v2t}} = -\frac{1}{B} \sum_{i=1}^{B} \log \frac{\exp(R_M^{(i)} \cdot R_{vM}^{(i)})}{\sum_{j=1}^{B} \exp(R_M^{(i)} \cdot R_{vM}^{(j)})}, \tag{9}$$

$$\mathcal{L}_M^{\text{t2v}} = -\frac{1}{B} \sum_{i=1}^{B} \log \frac{\exp(R_{vM}^{(i)} \cdot R_M^{(i)})}{\sum_{j=1}^{B} \exp(R_{vM}^{(i)} \cdot R_M^{(j)})}. \tag{10}$$

Inference. During inference for English video retrieval datasets, we freeze the multilingual text encoder and measure the retrieval performance only using R_E and R_{vE}. Similarly, for multilingual datasets, we freeze the English text encoder and calculate the retrieval score using R_M and R_{vM}.

4 Experiments

4.1 Datasets

We perform experiments on five standard text-video retrieval datasets: MSRVTT-9k and MSRVTT-7k splits [44], MSVD [7], DiDeMo [2] and Charades [35].

MSRVTT contains 10K videos with each video ranging from 10 to 32 s and 200K captions. We report the results both on MSRVTT-9k and MSRVTT-7k datasets following [29].

MSVD consists of 1970 videos and 80K descriptions. We use the standard training, validation and testing splits following [29].

DiDeMo is made up of 10K videos and 40K localized descriptions of the videos. We concatenate all the sentences for each video and evaluate the paragraph-to-video retrieval following [24,29].

Charades contains of 9848 videos and each video is associated with a caption. We use the standard training and test splits following [24].

MSRVTT multilingual is a multilingual version of MSRVTT in which the English captions are translated into nine different languages. We use the standard splits following [18].

4.2 Metrics

For evaluating the performance of models, we use recall at rank K ($R@1$, $R@5$, $R@10$), median rank (MedR) and mean rank (MnR). Unless specified, the values reported are the mean of three runs with different seeds.

Table 1. Text-to-video and video-to-text retrieval results on MSR-VTT dataset 9k split. Recall at rank 1 (R@1)↑, rank 5 (R@5)↑, rank 10 (R@10)↑, Median Rank (MdR)↓ and Mean Rank (MnR)↓ are reported. Results of other methods taken from mentioned references. Our model surpasses previous state-of-the-art performance. In video-to-text retrieval, our model achieved 1.6 points boost in performance.

Type	Model	Text-to-video retrieval					Video-to-text retrieval				
		R@1	R@5	R@10	MdR	MnR	R@1	R@5	R@10	MdR	MnR
Others	JsFusion [45]	10.2	31.2	43.2	13.0	–	–	–	–	–	–
	HT [31]	14.9	40.2	52.8	9.0	–	–	–	–	–	–
	HERO [23]	20.5	46.8	60.9	–	–	–	–	–	–	–
	CE [26]	20.9	48.8	62.4	6.0	28.2	20.6	50.3	64.0	5.3	25.1
	ClipBERT [21]	22.0	46.8	59.9	–	–	–	–	–	–	–
	SupportSET [32]	27.4	56.3	67.7	3.0	–	–	–	–	–	–
	VideoCLIP [43]	30.9	55.4	66.8	4.0	–	–	–	–	–	–
	FrozenInTime [3]	31	59.5	70.5	3.0	–	–	–	–	–	–
	CLIP [33]	31.2	53.7	2.6	4.0	–	–	–	–	–	–
	HIT [25]	30.7	60.9	73.2	2.6	-	32.1	62.7	74.1	3.0	–
	AlignPrompt [22]	33.9	60.7	73.2	–	–	–	–	–	–	–
	All-in-one [39]	34.4	65.4	75.8	–	–	–	–	–	–	–
	MDMMT [12]	38.9	69.0	79.7	**2.0**	–	–	–	–	–	–
CLIP based	CLIP4Clip [29]	44.5	71.4	81.6	–	15.3	43.1	70.5	81.2	**2.0**	12.4
	VCM [6]	43.8	71.0	80.9	**2.0**	14.3	45.1	72.3	82.3	**2.0**	10.7
	MCQ [15]	44.9	71.9	80.3	**2.0**	15.3	–	–	–	–	–
	MILES [16]	44.3	71.1	80.8	**2.0**	14.7	–	–	–	–	–
	CAMoE [8]	44.6	**72.6**	81.8	**2.0**	**13.3**	45.1	72.4	83.1	**2.0**	10.0
	CLIP2Video [13]	45.6	**72.6**	81.7	**2.0**	14.6	43.5	72.3	82.1	**2.0**	10.2
	CLIP2TV [14]	46.1	72.5	**82.9**	**2.0**	15.2	43.9	70.9	82.2	**2.0**	12.0
Ours	**MKTVR**	**46.6**	**72.6**	82.2	**2.0**	13.9	**45.5**	**73.4**	**84.7**	**2.0**	**8.07**

4.3 Implementation Details

We use translations of French for the multilingual inputs to train the MKTVR model. The video encoder and the English text encoder are initialized with CLIP-ViT-B-32. The multilingual text encoder is initialized with M-CLIP-ViT-B-32. The dimension size of the video, English caption and multilingual caption representations is 512. The dual cross-model encoder is initialized randomly and trained from scratch. The dimension size of the key, query and value projection layers is 512. The fully connected layer in the transformer has a size of 512 and a dropout of 0.4 is applied on this layer. We use 16 frames for MSRVTT-9k, MSRVTT-7k and MSVD datasets, 42 frames for DiDeMo and Charades datasets. The maximum sequence length is set to 32 for MSRVTT-9k and MSRVTT-7k, 64 for DiDeMo and 30 for charades dataset. The model is trained using AdamW [28] a learning rate of 1e-4 and a cosine decay of 1e-6. The MSRVTT-9k and MSRVTT-7k datasets are trained with a batch size of 32 and for 15 epochs. The MSVD dataset is trained with a batch size of 32 and for 5 epochs. The DiDeMo and charades datasets are trained with a batch size of 16 for 12 and 15 epochs respectively.

5 Results and Discussion

5.1 Evaluation on English Video Retrieval Datasets

In Table 1 we report the results of our proposed approach on MSVRTT-9k dataset. It can be observed that the difference between CLIP based models and other models is very significant ($> 5\%$). Therefore, it explains the incentive to build our model using CLIP features. On MSRVTT-9k split, our model significantly outperforms CLIP4Clip model on all the metrics in both text-to-video and video-to-text retrieval settings. VCM employs a knowledge graph between video and text modalities making its performance superior to other models in a video-to-text retrieval task. Our model surpasses VCM significantly in all the metrics elucidating that the multilingual representations serve as a powerful knowledge transfer. Moreover, our approach outperforms MCQ and MILES which are pretrained on WebVid-2M data, initialized with CLIP features, employing additional semantic information like parts-of-speech. This validates that our model doesn't require any pretraining on videos and structural knowledge injection. The multilingual text representations in our model effectively serves this purpose.

In Tables 2, 3, 4 and 5 we report the results on MSRVTT-7k, MSVD, DiDeMo and Charades datasets respectively. Our model outperforms all the previous approaches across all the metrics on all the datasets. For the MSRVTT-7k split, our model achieves a significant boost of 2.1%, 2.5% and 4.4% in R@1, R@5 and R@10 respectively compared to the previous baselines. For the MSVD dataset, we notice an improvement of 0.2%, 0.5% and 0.3% in R@1, R@5 and R@10 respectively. MSVD is a relatively smaller dataset with test size of 670 videos and hence, the improvements are relatively marginal.

Table 2. Text-to-video retrieval results on MSR-VTT - 7k split. Recall at rank-1 (R@1), rank-5 (R@5), rank-10 (R@10), Median Rank (MdR) are reported. Results of other methods taken from mentioned references.

Model	R@1 (↑)	R@5 (↑)	R@10(↑)	MdR (↓)
HowTo100M [31]	10.2	31.2	43.2	13.0
ActBERT [47]	8.6	23.4	33.1	36.0
NoiseE [1]	17.4	41.6	53.6	8.0
ClipBERT [21]	22.0	46.8	59.9	6.0
CLIP4clip- [29]	42.1	71.9	81.4	**2.0**
Singularity [20]	42.7	69.5	78.1	**2.0**
MKTVR	**44.8**	**72.0**	**82.5**	**2.0**

For the DiDeMo dataset, our model showed a marginal boost of 0.2% in R@1 but a significant boost of 4.3% and 2.9% in R@5 and R@10 respectively compared to the previous approaches. For the Charades dataset, our model outperformed previous approaches by 0.9% in R@1 and by a significant margin of 4.6%, 7.6% and 6.0% in R@5, R@10 and MedianR respectively. ECLIPSE uses audio as additional information for video retrieval. We showed that multilingual text acts as a better knowledge transfer input.

Table 3. Text-to-video retrieval results on MSVD dataset. Recall at rank-1 (R@1), rank-5 (R@5), rank-10 (R@10), Median Rank (MdR) are reported. Results of other methods taken from mentioned references.

Model	R@1 (↑)	R@5 (↑)	R@10(↑)	MdR (↓)
VSE [13]	12.3	30.1	42.3	14.0
CE [26]	19.8	49.0	63.8	6.0
SSML [1]	20.3	49.0	63.3	6.0
SUPPORT-SET [32]	28.4	60.0	72.9	4.0
FROZEN [3]	33.7	64.7	76.3	3.0
CLIP [33]	37.0	64.1	73.8	3.0
CLIP4Clip [29]	46.2	76.1	84.6	2.0
CLIP2Video [13]	47.0	76.8	85.9	2.0
MKTVR	**47.2**	**77.3**	**86.2**	**2.0**

Table 4. Text-to-video retrieval result on DiDeMo dataset. Recall at rank-1 (R@1), rank-5 (R@5), rank-10 (R@10), Median Rank (MdR) are reported. Results of other methods taken from mentioned references.

Model	R@1 (↑)	R@5 (↑)	R@10 (↑)	MdR (↓)
S2VT [38]	11.9	33.6	–	13.0
FSE [46]	13.9	36	–	11.0
CE [26]	16.1	41.1	–	8.3
ClipBERT [21]	20.4	48.0	60.8	6.0
FrozenInTime [3]	31.0	59.8	72.4	3.0
OA-Trans [40]	34.8	64.4	75.1	3.0
CLIP4clip [29]	43.4	70.2	80.6	**2.0**
CLIP2TV [14]	43.9	70.5	79.8	**2.0**
TS2-Net [27]	41.8	71.6	82.0	**2.0**
ECLIPSE [24]	44.2	70.0	80.2	**2.0**
MKTVR	**44.4**	**74.3**	**83.1**	**2.0**

Table 5. Text-to-video retrieval result on charades dataset. Recall at rank-1 (R@1), rank-5 (R@5), rank-10 (R@10), Median Rank (MdR) are reported. Results reported are taken from [24].

Model	R@1 (↑)	R@5 (↑)	R@10 (↑)	MdR	MnR
ClipBERT [21]	6.7	17.3	25.2	32.0	149.7
FrozenInTime [3]	11.9	28.3	35.1	17.0	103.8
CLIP4clip [29]	13.9	30.4	37.1	14.0	98.0
ECLIPSE [24]	15.7	32.9	42.4	16.0	84.9
MKTVR	**16.6**	**37.5**	**50.0**	**10.0**	**52.7**

5.2 Evaluation on Multilingual Video Retrieval Datasets

In addition to the monolingual datasets, we also evaluate the proposed approach on multilingual video retrieval datasets. Specifically, we use the model trained only using French captions and test on 6 languages such as German (de), Czech (cs), Chinese (zh), Swahili (sw), Russian (ru) and Spanish (es) in a zero-shot setting. Table 6 shows the results on MSRVTT-multilingual dataset. Our model achieved a significant boost of 8.2% (average) in R@1 in a zero-shot setting. It is worth noting that our model in a zero-short evaluation outperformed the previous approaches fine-tuned on these languages by a huge margin of 6.1% (average). MMP [18] is pretrained on the large scale multilingual dataset HowTo100M on 9 languages. However, our model trained on just 1 language outperformed MMP. This shows that our dual cross-modal (DCM) encoder block can effectively learn the association among video, English and multilingual representations even when large video pretraining is not involved.

Table 6. Text-to-video retrieval (R@1 metric) results on MSR-VTT - multilingual [18]. Results of other methods taken from [18]. Our model is trained on Charades dataset and using only french language and evaluated in a zero-shot setting on MSRVTT multilingual dataset. In zero-shot evaluation on other languages, our model significantly outperforms previous models trained in both zero-shot and fine-tuning setting.

Model	de	cs	zh	ru	sw	es
m-BERT (zero-shot)	11.1	8.2	6.9	7.9	1.4	12
m-BERT MMP (zero-shot)	15	11.2	8.4	11	3.4	15.1
XLM-R (zero-shot)	16.3	16	14.9	15.4	7.7	17.3
XLM-MMP (zero-shot)	19.4	19.3	18.2	19.1	8.4	20.4
m-BERT (fine-tune)	18.2	16.9	16.2	16.5	13	18.5
XLM- R + MMP (fine-tune)	21.1	20.7	20	20.5	14.4	21.9
MKTVR - fr (zero-shot)	**27.4**	**28.2**	**24.1**	**26.6**	**22.5**	**26.5**

5.3   Ablation Studies

Effect of Multilingual Knowledge Transfer. We investigate the effect of multilingual knowledge transfer on the video-retrieval performance. Precisely, we train a model without the multilingual text encoder keeping the rest of the architecture intact. As shown in Fig. 3, using multilingual data as knowledge transfer significantly improved the performance on DiDeMo and Charades datasets. The improvement is 2.4% for DiDeMo and 3.62% for Charades datasets.

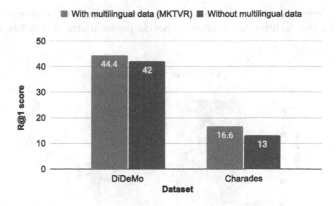

Fig. 3. Comparison of models with and without using multilingual data as input. The first model takes as input only video and English text captions whereas the second model takes video, English text and multilingual text captions as input. As shown in the figure, using multilingual text data as knowledge transfer significantly improved the performance.

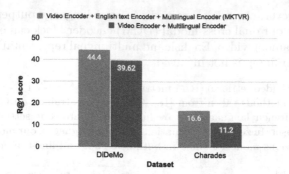

Fig. 4. Comparison of models consisting of only multilingual text encoder and multilingual text encoder + English text encoder. Using a separate English text encoder for encoding English text captions outperforms the model using multilingual text encoder to encode English text captions

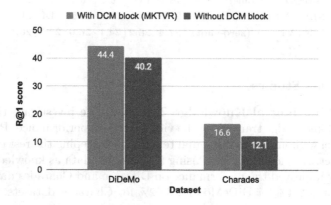

Fig. 5. Comparison of models with and without DCM block in the architecture. Using DCM block in the architecture showed superior performance to models without the DCM block.

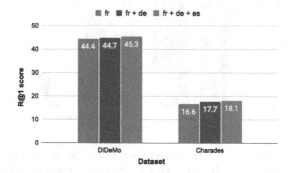

Fig. 6. Comparison of MKTVR trained with different multilingual caption data. It is evident from the figure that training with more languages improved the performance.

Using only Multilingual Text Encoder. Next, we ablate the choice of using an English text encoder. We validated previously that multilingual data improves the performance of video retrieval. This raises the question: *Why a separate English text encoder is required if multilingual text encoder can be used for both English text and multilingual text representations?* In Fig. 4, we show the results of two different model variants. The first model uses a separate English text encoder to encode English text captions whereas in the second model, both the English text and multilingual text are encoded using the same multilingual text encoder. Results show that encoding English text captions using a separate English text encoder surpasses the model using multilingual text encoder to encode both English text and multilingual text. Multilingual pretraining employs a part of English data whereas the English text encoder is pretrained on a comparatively larger English data. Hence, leveraging a separate English text encoder showed much superior performance to using multilingual text encoder for English text.

Effectiveness of Dual Cross Encoder Block. Next, we ablate the effectiveness of dual cross encoder block. We train a model without the DCM block and directly compute the loss between video representations and English text representations and video representations and multilingual text representations. From Fig. 5, we can see that the model using DCM block achieves better performance than the model without the encoder block. This justifies our motivation to use DCM block in our model.

Training with More Languages. Next, we ablate training our model with more than one language. Concretely we train our model with German (de) and Spanish (es) captions. These languages are chosen because their performance on XNLI dataset [10] is comparable to English. The results are shown in Fig. 6 and it is seen that training with more languages improved the performance on video retrieval. These results validate that multilingual data act as an effective knowledge transfer mechanism for improving video retrieval.

6   Conclusion

In this paper we introduced MKTVR, a multilingual knowledge transfer framework to improve the performance of video retrieval. We constructed multilingual captions using off-the-shelf state-of-the-art machine translation models. We then proposed a CLIP-based model that enables multilingual knowledge transfer using a dual cross-modal encoder block. Experiment results on five standard video retrieval datasets showed that our framework achieved state-of-the-art results on all the datasets. Finally, our model also showed superior performance to previous approaches on multilingual video retrieval datasets in a zero-shot setting. In the future, we will focus on more efficient ways of multilingual knowledge transfer for video retrieval.

References

1. Amrani, E., Ben-Ari, R., Rotman, D., Bronstein, A.: Noise estimation using density estimation for self-supervised multimodal learning. In: Proceedings of the AAAI Conference on Artificial Intelligence. vol. 35, pp. 6644–6652 (2021)
2. Anne Hendricks, L., Wang, O., Shechtman, E., Sivic, J., Darrell, T., Russell, B.: Localizing moments in video with natural language. In: Proceedings of the IEEE International Conference on Computer Vision, pp. 5803–5812 (2017)
3. Bain, M., Nagrani, A., Varol, G., Zisserman, A.: Frozen in time: A joint video and image encoder for end-to-end retrieval. In: Proceedings of the IEEE/CVF International Conference on Computer Vision, pp. 1728–1738 (2021)
4. Bertasius, G., Wang, H., Torresani, L.: Is space-time attention all you need for video understanding? In: ICML. vol. 2, p. 4 (2021)
5. Burns, A., Kim, D., Wijaya, D., Saenko, K., Plummer, B.A.: Learning to scale multilingual representations for vision-language tasks. In: European Conference on Computer Vision, pp. 197–213. Springer (2020)
6. Cao, S., Wang, B., Zhang, W., Ma, L.: Visual consensus modeling for video-text retrieval (2022)
7. Chen, D., Dolan, W.B.: Collecting highly parallel data for paraphrase evaluation. In: Proceedings of the 49th Annual Meeting of the Association for Computational Linguistics: Human Language Technologies, pp. 190–200 (2011)
8. Cheng, X., Lin, H., Wu, X., Yang, F., Shen, D.: Improving video-text retrieval by multi-stream corpus alignment and dual softmax loss. arXiv preprint arXiv:2109.04290 (2021)
9. Conneau, A., Lample, G.: Cross-lingual language model pretraining. In: Advances in Neural Information Processing Systems, vol. 32 (2019)
10. Conneau, A., et al.: Xnli: Evaluating cross-lingual sentence representations. In: Proceedings of the 2018 Conference on Empirical Methods in Natural Language Processing, pp. 2475–2485 (2018)
11. Dosovitskiy, A., et al.: An image is worth 16x16 words: Transformers for image recognition at scale. In: International Conference on Learning Representations (2020)
12. Dzabraev, M., Kalashnikov, M., Komkov, S., Petiushko, A.: Mdmmt: Multidomain multimodal transformer for video retrieval. In: Proceedings of the IEEE/CVF Conference on Computer Vision and Pattern Recognition, pp. 3354–3363 (2021)
13. Fang, H., Xiong, P., Xu, L., Chen, Y.: Clip2video: Mastering video-text retrieval via image clip. arXiv preprint arXiv:2106.11097 (2021)
14. Gao, Z., Liu, J., Chen, S., Chang, D., Zhang, H., Yuan, J.: Clip2tv: An empirical study on transformer-based methods for video-text retrieval. arXiv preprint arXiv:2111.05610 (2021)
15. Ge, Y., et al.: Bridging video-text retrieval with multiple choice questions. In: Proceedings of the IEEE/CVF Conference on Computer Vision and Pattern Recognition, pp. 16167–16176 (2022)
16. Ge, Y., et al.: Miles: Visual bert pre-training with injected language semantics for video-text retrieval. arXiv preprint arXiv:2204.12408 (2022)
17. Gella, S., Sennrich, R., Keller, F., Lapata, M.: Image pivoting for learning multilingual multimodal representations. In: Proceedings of the 2017 Conference on Empirical Methods in Natural Language Processing, pp. 2839–2845 (2017)

18. Huang, P.Y., Patrick, M., Hu, J., Neubig, G., Metze, F., Hauptmann, A.G.: Multilingual multimodal pre-training for zero-shot cross-lingual transfer of vision-language models. In: Proceedings of the 2021 Conference of the North American Chapter of the Association for Computational Linguistics: Human Language Technologies, pp. 2443–2459 (2021)
19. Kim, D., Saito, K., Saenko, K., Sclaroff, S., Plummer, B.: Mule: Multimodal universal language embedding. In: Proceedings of the AAAI Conference on Artificial Intelligence. vol. 34, pp. 11254–11261 (2020)
20. Lei, J., Berg, T.L., Bansal, M.: Revealing single frame bias for video-and-language learning. arXiv preprint arXiv:2206.03428 (2022)
21. Lei, J., et al.: Less is more: Clipbert for video-and-language learning via sparse sampling. In: Proceedings of the IEEE/CVF Conference on Computer Vision and Pattern Recognition, pp. 7331–7341 (2021)
22. Li, D., Li, J., Li, H., Niebles, J.C., Hoi, S.C.: Align and prompt: Video-and-language pre-training with entity prompts. In: Proceedings of the IEEE/CVF Conference on Computer Vision and Pattern Recognition, pp. 4953–4963 (2022)
23. Li, L., Chen, Y.C., Cheng, Y., Gan, Z., Yu, L., Liu, J.: Hero: Hierarchical encoder for video+ language omni-representation pre-training. In: Proceedings of the 2020 Conference on Empirical Methods in Natural Language Processing (EMNLP), pp. 2046–2065 (2020)
24. Lin, Y.B., Lei, J., Bansal, M., Bertasius, G.: Eclipse: Efficient long-range video retrieval using sight and sound. arXiv preprint arXiv:2204.02874 (2022)
25. Liu, S., Fan, H., Qian, S., Chen, Y., Ding, W., Wang, Z.: Hit: Hierarchical transformer with momentum contrast for video-text retrieval. In: Proceedings of the IEEE/CVF International Conference on Computer Vision, pp. 11915–11925 (2021)
26. Liu, Y., Albanie, S., Nagrani, A., Zisserman, A.: Use what you have: Video retrieval using representations from collaborative experts. arXiv preprint arXiv:1907.13487 (2019)
27. Liu, Y., Xiong, P., Xu, L., Cao, S., Jin, Q.: Ts2-net: Token shift and selection transformer for text-video retrieval. arXiv preprint arXiv:2207.07852 (2022)
28. Loshchilov, I., Hutter, F.: Decoupled weight decay regularization. In: International Conference on Learning Representations (2018)
29. Luo, H., et al.: Clip4clip: An empirical study of clip for end to end video clip retrieval. arXiv preprint arXiv:2104.08860 (2021)
30. Madasu, A., Oliva, J., Bertasius, G.: Learning to retrieve videos by asking questions. arXiv preprint arXiv:2205.05739 (2022)
31. Miech, A., Zhukov, D., Alayrac, J.B., Tapaswi, M., Laptev, I., Sivic, J.: Howto100m: Learning a text-video embedding by watching hundred million narrated video clips. In: Proceedings of the IEEE/CVF International Conference on Computer Vision, pp. 2630–2640 (2019)
32. Patrick, M., Huang, P.Y., Asano, Y., Metze, F., Hauptmann, A.G., Henriques, J.F., Vedaldi, A.: Support-set bottlenecks for video-text representation learning. In: International Conference on Learning Representations (2020)
33. Radford, A., et al.: Learning transferable visual models from natural language supervision. In: International Conference on Machine Learning, pp. 8748–8763. PMLR (2021)
34. Sigurdsson, G.A., et al.: Visual grounding in video for unsupervised word translation. In: Proceedings of the IEEE/CVF Conference on Computer Vision and Pattern Recognition, pp. 10850–10859 (2020)

35. Sigurdsson, G.A., Varol, G., Wang, X., Farhadi, A., Laptev, I., Gupta, A.: Hollywood in homes: crowdsourcing data collection for activity understanding. In: Leibe, B., Matas, J., Sebe, N., Welling, M. (eds.) ECCV 2016. LNCS, vol. 9905, pp. 510–526. Springer, Cham (2016). https://doi.org/10.1007/978-3-319-46448-0_31

36. Surís, D., Epstein, D., Vondrick, C.: Globetrotter: Connecting languages by connecting images. In: Proceedings of the IEEE/CVF Conference on Computer Vision and Pattern Recognition, pp. 16474–16484 (2022)

37. Tang, Y., et al.: Multilingual translation with extensible multilingual pretraining and finetuning. arXiv preprint arXiv:2008.00401 (2020)

38. Venugopalan, S., Rohrbach, M., Donahue, J., Mooney, R., Darrell, T., Saenko, K.: Sequence to sequence-video to text. In: Proceedings of the IEEE International Conference on Computer Vision, pp. 4534–4542 (2015)

39. Wang, A.J., et al.: All in one: Exploring unified video-language pre-training. arXiv preprint arXiv:2203.07303 (2022)

40. Wang, J., et al.: Object-aware video-language pre-training for retrieval. In: Proceedings of the IEEE/CVF Conference on Computer Vision and Pattern Recognition, pp. 3313–3322 (2022)

41. Wehrmann, J., Souza, D.M., Lopes, M.A., Barros, R.C.: Language-agnostic visual-semantic embeddings. In: Proceedings of the IEEE/CVF International Conference on Computer Vision, pp. 5804–5813 (2019)

42. Wu, H.Y., Zhai, A.: Classification is a strong baseline for deep metric learning. In: Sidorov, K., Hicks, Y. (eds.) Proceedings of the British Machine Vision Conference (BMVC), pp. 224.1-224.12. BMVA Press (September 2019). https://doi.org/10.5244/C.33.224

43. Xu, H., et al.: Videoclip: Contrastive pre-training for zero-shot video-text understanding. In: Proceedings of the 2021 Conference on Empirical Methods in Natural Language Processing, pp. 6787–6800 (2021)

44. Xu, J., Mei, T., Yao, T., Rui, Y.: Msr-vtt: A large video description dataset for bridging video and language. In: Proceedings of the IEEE Conference on Computer Vision and Pattern Recognition, pp. 5288–5296 (2016)

45. Yu, Y., Kim, J., Kim, G.: A joint sequence fusion model for video question answering and retrieval. In: Proceedings of the European Conference on Computer Vision (ECCV), pp. 471–487 (2018)

46. Zhang, B., Hu, H., Sha, F.: Cross-modal and hierarchical modeling of video and text. In: Proceedings of the European Conference on Computer Vision (ECCV), pp. 374–390 (2018)

47. Zhu, L., Yang, Y.: Actbert: Learning global-local video-text representations. In: Proceedings of the IEEE/CVF Conference on Computer Vision and Pattern Recognition, pp. 8746–8755 (2020)

Service Is Good, Very Good or Excellent? Towards Aspect Based Sentiment Intensity Analysis

Mamta$^{(\boxtimes)}$ and Asif Ekbal

Department of Computer Science and Engineering, IIT Patna, Patna, India
{mamta_1921cs11,asif}@iitp.ac.in

Abstract. Aspect-based sentiment analysis (ABSA) is a fast-growing research area in natural language processing (NLP) that provides more fine-grained information, considering the aspect as the fundamental item. The ABSA primarily measures sentiment towards a given aspect, but does not quantify the intensity of that sentiment. For example, intensity of positive sentiment expressed for *service* in *service is good* is comparatively weaker than in *service is excellent*. Thus, aspect sentiment intensity will assist the stakeholders in mining user preferences more precisely. Our current work introduces a novel task called aspect based sentiment intensity analysis (ABSIA) that facilitates research in this direction. An annotated review corpus for ABSIA is introduced by labelling the benchmark SemEval ABSA restaurant dataset with the seven (7) classes in a semi-supervised way. To demonstrate the effective usage of corpus, we cast ABSIA as a natural language generation task, where a natural sentence is generated to represent the output in order to utilize the pre-trained language models effectively. Further, we propose an effective technique for the joint learning where ABSA is used as a secondary task to assist the primary task, i.e. ABSIA. An improvement of 2 points is observed over the single task intensity model. To explain the actual decision process of the proposed framework, model explainability technique is employed that extracts the important opinion terms responsible for generation (Source code and the dataset has been made available on https://www.iitp.ac.in/~ai-nlp-ml/resources.html#ABSIA, https://github.com/20118/ABSIA)

Keywords: Sentiment analysis · Absa · Aspect sentiment intensity · Explainability · Generation · Joint learning

1 Introduction

Aspect based Sentiment Analysis (ABSA) is an important research area in natural language processing (NLP), which aims to associate sentiments to the fine-grained aspect terms [28]. The ABSA has mitigated the shortcomings of document/sentence level sentiment analysis, which only provide coarse-grained sentiment information by ignoring the aspects in the text [10,20–22]. Aspects are the attributes of an entity, for example, food of restaurant, ambience of restaurant,

© The Author(s), under exclusive license to Springer Nature Switzerland AG 2023
J. Kamps et al. (Eds.): ECIR 2023, LNCS 13980, pp. 685–700, 2023.
https://doi.org/10.1007/978-3-031-28244-7_43

etc. In recognition of its tremendous potential in recommendation systems, a number of ABSA resources and systems have been developed covering a variety of domains, including reviews of electronic products, such as laptops, digital cameras, etc., and restaurants [2,8,28]. An attempt has also been made in legal domain [24].

The very first benchmark setup of ABSA was proposed in SemEval 2014 [28], which primarily focused on English language. In the subsequent attempt, this has been extended to eight (8) more languages [26,27]. A few other challenges have been introduced for ABSA, *e.g.*, GermanEval for German ABSA [39] and TASS for Spanish language [30]. Authors in [1–3] have explored ABSA in the Indian context. Authors in [18] proposed a task in financial opinion mining to predict the scores for each predefined targets. Sentiment scores are defined in continuous numeric value ranging from -1 (negative) to +1 (positive). The dataset contains limited number of news headlines or financial microblogs. There are other attempts to measure the sentiment intensity [23,31,36]. However, these are focused on document level sentiment analysis. Traditionally, neural based models have shown promising results for ABSA, which models the relations of aspects and their contextual words with the help of target connection LSTMs (Long Short Term Memory), target dependent LSTMs [34], memory networks [4,19], attention [5,17,37,38], and graph based models [11,13,25,35,42] etc. Recently, transformers based models have also achieved promising results in almost every of area of NLP, including ABSA [7,40]. These studies model the relations between aspect and sentences by concatenating the aspect information at the end of the sentence. Further, a new direction is introduced to cast different NLP problems as the sequence generation tasks [6,14,29]. These studies leverage the pre-trained knowledge present in the language models without adding additional parameters to the model.

Existing ABSA systems in review domains classify the given aspect into one of the sentiment classes, *viz.* positive, negative, neutral, or conflict. However, it does not convey the fine-grained information such as strength or intensity of sentiment expressed. The user may not only express the sentiment towards a given aspect, but also the level of that sentiment. For example, sentence *service is good and food is excellent* have positive sentiments for the aspects, *service* and *food*, but both express different levels of sentiments for the aspects *service* and *food*. The level of sentiment expressed for *service* is comparatively weaker than *food*. Similarly, the level of positive sentiment expressed for aspect *service* in *service is wonderful* is more than in *service is nice*. Thus, measuring the intensity of sentiment is of foremost importance in analyzing the finer-level details of the sentiments expressed towards a given aspect. Such analysis will be incredibly helpful in mining user preferences while designing recommendation systems.

To mitigate the limitations of the existing literature, in this work we introduce the novel task of aspect based sentiment intensity analysis (ABSIA). Further, an ABSIA corpus is created by extending the benchmark SemEval-2014 restaurant domain dataset [28]. For this, we follow a two-step approach: the document level sentiment intensity dataset [23] is utilized to weakly label the

ABSA dataset for the seven classes, *viz.* low negative (-1), moderate negative (-2), high negative (-3), neutral (0), low positive (1), moderate positive (2), and high positive (3). This step is followed by manual verification by the linguists. Motivated by the ability of pre-trained language models to solve various NLP problems as well as ABSA subtasks [12,15,41], we cast ABSIA into a natural language generation task. Our approach focuses on joint modeling for both the ABSA and ABSIA tasks to share the knowledge between them. These tasks are treated as sequence-to-sequence (Seq2Seq) tasks, where input to the encoder is the review sentence, and output from the decoder is a natural language sentence describing the sentiment or the sentiment intensity of the given aspect term. We utilize the BART (Bi-directional Autoencoder Representation from Transformers) pre-trained model, which consists of a bi-directional encoder and a decoder [12]. Solving classification tasks as Seq2Seq tasks has following advantages: (i). to solve classification task, neural network is added on top of pre-trained representation, with separate network parameters; (ii). the addition of aspect terms renders the aspect-specific input representation not exactly a natural language sentence, which makes it different from the pre-training scenario. Intuitively, linking pre-training and ABSA at a task level instead of at the representation level would utilize the pre-trained knowledge effectively.

However, these models lack transparency, which makes it difficult to understand their actual decision process. No attempt has been made so far to explain the behavior of these models. We exploit the model explainability technique, SHAP (SHapley Additive exPlanations) [16] to understand the behavior of our proposed model by extracting the opinion terms responsible for generating the output label. To the best of our knowledge, this is the very first attempt to introduce intensity in aspect based sentiment analysis for the reviews. We believe that our current work will attract the attention from community to deep dive into aspect based sentiment intensity analysis (ABSIA).

We summarize the key contributions of our current work as follows:

- We propose a novel task named ABSIA, which quantifies the intensity of sentiment expressed for a given aspect.
- We create a benchmark dataset for ABSIA following a semi-supervised approach, which contains aspect terms, their corresponding sentiment and sentiment intensity labels.
- We propose a generative pre-trained language model to jointly learn two tasks, where ABSIA is treated as the primary task and ABSA is the auxiliary task.
- To explain the decision process of our model, we harness the model explainability technique SHAP. To the best of our knowledge, this is the very first attempt towards explaining template based generation models.

2 Resource Creation

To construct the ABSIA corpus, we utilize the benchmark SemEval 2014 ABSA dataset [28], annotated for aspect terms and their corresponding polarities. We follow a two-step approach to annotate this dataset for the ABSIA task to reduce the manual annotation efforts. First, the existing document level sentiment intensity dataset [23] is utilized to train the classifier, which is then used to assign weak labels to the ABSA dataset.

Weak Labelling: SemEval 2018 shared task [23] considers document sentiment intensity as an ordinal classification problem and provides the dataset for seven intensity classes ranging from -3 to +3. The intensity value of -3 signifies the high negative and +3 signifies the high positive sentiment intensity. There are a total of 2567 instances in the dataset, out of which 1181 are for training, 449 for development, and 937 for testing. We opt to annotate sentiment intensities as ordinal classes to reduce the error rate in weak labeling.

To train the sentiment intensity classifier, we leverage the pre-trained language model BERT [7]. We fine-tune it by adding a fully connected layer on top of it. Input to BERT model is a sequence of tokens present in the sentence $S = w_1, w_2, w_3, ..., w_n$. BERT inserts a special token $[CLS]$ at the beginning and $[SEP]$ at the end of the sentence. The representation at $[CLS]$ is considered as final contextual representation of the S and is further passed to a fully connected layer containing 7 neurons. This output layer returns the probability for each class, and the class with the highest probability is selected as the final class. The trained system is then used to assign weak labels to the SemEval 2014 ABSA dataset. ABSA dataset contains sentences, aspects and corresponding sentiment polarities. It contains total of 2892 positive, 1001 negative, 829 neutral, and 105 conflict instances. We ignore the conflict instances due to its relatively low number. In order to assign weak labels to each aspect of a sentence S, we create a context window of size s around it, i.e., s words from the left and s words from the right of the aspect. It is based on the intuition that opinion words related to a given aspect are present nearby. For a review sentence S with n words and m aspect terms, we create m context windows, each containing $2 * s + 1$ words. These m context windows are passed to the trained BERT classifier which assigns a sentiment intensity label to each aspect of S.

Manual Correction: To ensure the annotation quality of the dataset, we engaged three linguists with sufficient subject knowledge and having experience on the construction of supervised corpora. Two of them have doctoral degrees, and one among them has masters. Linguists are asked to manually verify the annotations of every aspect sentence pair. Every aspect is accompanied by a sentiment intensity class, *viz.* -1, -2, -3, 0, 1, 2, and 3. Guidelines along with some examples were explained to the linguists before starting the verification process [23]. They were provided with a few gold label samples to get an idea about the actual annotations. Linguists are advised to annotate aspect asp_i based on the opinion terms associated with it. To aggregate the annotations from different linguists, majority voting technique is used. We attained an overall Fleiss' kappa [32] score of 0.75 among the three linguists, which can be considered as reliable. The use of weak labeling prior to actual annotation reduced the annotation time and also helped in achieving the correct intensity labels. A few samples along with their aspect sentiment intensity labels are shown in Table 1.

Table 1. Samples from ABSIA corpus

Text	Aspects	Label
Largest and freshest pieces of sushi, and delicious!	Sushi	3
Food was decent, but not great	Food	1
Our waiter was horrible; so rude and disinterested	Waiter	-3
The sashimi is cut a little thinly	Sashimi	-1

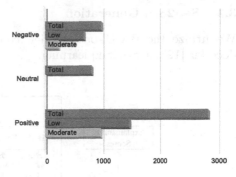

Fig. 1. Class-wise distribution

Data Distribution: ABSIA dataset contains 1478 low positive, 981 moderate positive, 384 high positive, 682 low negative, 228 moderate negative, and 80 high negative aspects instances. Class-wise distribution of aspects for sentiment and sentiment intensity is depicted in Fig. 1. There is a strong dominance of the low positive class among the other two intensity classes of positive sentiment. Similar to the positive class, the low negative class dominates the high negative and moderate negative classes.

3   Methodology

We present a natural language generation based framework to classify the aspect sentiment intensity into 7 classes, *viz.* -1, -2, -3, 0, 1, 2, and 3. The detailed architecture is shown in Fig. 2. There are 2 components, *viz.*, Seq2Seq generation using BART and model explainability. In this section, we first formulate ABSIA as a generation task, followed by presenting the details of model components.

Task Formulation: The ABSIA task aims to categorize the sentiment intensity of a sentence $S = w_1, w_2, w_3, ..., w_n$ towards a given aspect asp_i into 7 classes, *viz.* -1, -2, -3, 0, 1, 2, and 3. ABSIA is formulated as a language model ranking problem under a Seq2Seq framework, where we generate a natural sentence to represent the output class. Output sentence is a pre-defined template filled with given aspect and corresponding class label. For a given input sequence S with aspect asp_i, template $T_{int} = t_1, t_2, ..., t_{m1}$ is the target sequence, with $m1$ words. T_{int} is filled by given aspect asp_i and corresponding aspect sentiment intensity l_r to obtain gold sequence $T_{int_{asp_i, l_r}}$, where r is the list of all sentiment intensity labels. To effectively share the knowledge between ABSA and ABSIA tasks, we propose joint modelling of ABSA and ABSIA tasks. Hence, ABSA task is also formulated as a language model ranking problem. Aspect asp_i and the corresponding sentiment label p_c are used to fill the template $T_{pol} = t_1, t_2, ..., t_{m2}$ to obtain a gold sequence $T_{pol_{asp_i, p_c}}$. Here, c is the list of all sentiment labels and $m2$ is the number of words in T_{pol}.

3.1   Seq2Seq Generation

We utilize the BART-base model, consisting of 6-layer encoder and 6-layer decoder [12] for Seq2Seq learning.

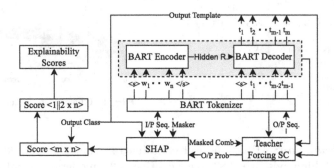

Fig. 2. Proposed Framework

Template Creation: The output template contains the given aspect asp_i and its associated sentiment p_c or sentiment intensity label l_r. The template T_{int} for ABSIA task is: *The aspect $< asp_i >$ has $< l_r >$ intensity.* As an example, the output template for the sentence *service is excellent,* is *The aspect **service** has **high positive** intensity.* Similarly, the template T_{pol} for ABSA task is *The sentiment polarity of aspect $< asp_i >$ is $< p_c >$.* Accordingly, the template for the same example is: *The sentiment polarity of aspect **service** is **positive**.*

Model Training: Seq2Seq framework requires parallel data for training. Firstly, we obtain the parallel data by generating the gold sequences as discussed above. We create input output pair (S, T_{pol}) and (S, T_{int}). Suppose there are a aspects for S. Then, for both (*pol* and *int*) tasks, a gold sequences will be created for S, one for each aspect. Given a sequence pair (S, T) for task q (q can be *pol* or *int*), the input sentence S is fed to BART tokenizer. $< s >$ and $< /s >$ tokens will be added to the start and end of S, respectively, as shown in Fig. 2. The tokenized input sequence is passed to BART encoder, which generates the hidden representations of the sentence S as shown in Eq. 1.

$$h^{enc} = BART\_ENC(S) \tag{1}$$

This encoded representation is further fed to a decoder to generate the output tokens. The decoder takes the encoder representation h^{enc} and previous output tokens from the template T as input to generate the current output token. At the kth time step of the decoder, the decoder yields a representation h_k^{dec} by applying attention over encoder hidden state h^{enc} and previous output tokens $t_{1:k-1}$ as shown in Eq. 2.

$$h_k^{dec} = BART\_DEC(h^{enc}, t_{1:k-1}) \tag{2}$$

Finally, Softmax is applied over the decoder representation to obtain the probability vector over the whole vocabulary, as described in Eq. 3. The Eq. 3 gives the conditional probability of word t_k.

$$P(t_k|t_{1:k-1}) = Softmax(h_k^{dec}W + b) \qquad (3)$$

where, W and b are weight and bias metrices having dimensions $(d_h, |V|)$ and $(|V|)$ respectively. Here, $|V|$ is the vocabulary size of the pre-trained BART model. The highest probability word is selected as the final word. Our model is trained to minimize the cross-entropy loss between the output of the decoder and the original template, as shown in Eq. 4.

$$Loss_q = \sum_{k=1}^{m} logP(t_k|t_{1,k-1}, S) \qquad (4)$$

Here, m can be either $m1$ or $m2$, i.e., the number of output tokens either in the template T_{int} or T_{pol}, respectively, based on the task q. Joint learning of ABSA and ABSIA tasks update the parameters of the model for the loss produced by both the tasks, $Loss_{pol}$ and $Loss_{int}$. Every mini-batch comprises of a different task that the model learns and parameters are updated for each mini-batch. As a result, both the tasks can share the information among themselves and help ABSIA to improve its performance.

Model Inference: During model inference, we first obtain all the label values for both the tasks. For a sentence S with aspect asp_i, task int, we create r templates by filling the asp_i and sentiment intensity label l_r to the pre-defined template T_{int}. Then the trained BART model is used to assign a score to each template $T_{int_{asp_i, l_r}}$ as shown in Eq. 5.

$$f(T_{int_{asp_i, l_r}}) = \sum_{k=1}^{m1} logP(t_k|t_{1-1, S}) \qquad (5)$$

We calculate $f(T_{int_{asp_i, l_r}})$ for each ABSIA template and then choose the final template having the largest score. Similarly, for ABSA task, we obtain c templates for a given sentence S and aspect asp_i. We obtain score $f(T_{pol_{asp_i, p_c}})$ for each sentiment template similar to Eq. 5 and choose the final sentiment template with the largest score.

3.2   Model Explainability

Model explainability is introduced at the inferencing time to understand the reason behind predictions. To achieve this goal, our first step towards it is to find the contribution of each word for the output template generation. In general, the word importance is computed as the difference between a prediction for a given sentence S (with n words) and the expected prediction when the word w_j is not present in S and replaced by $[MASK]$. It is described in Eq. 6.

$$\phi_j(S) = M(w_1, \ldots, w_n) - E[M(w_1, \ldots, w_{j-1}, [MASK], w_{j+1} \ldots, w_n)] \qquad (6)$$

We use the Shapely algorithm, inspired by coalitional game theory, to determine the relevance of each word in a given sentence [16]. Shapley calculates the relevance score for each word based on possible coalitions for a particular prediction. Equation 7 explains the computation using a value function, which calculates the feature importance over the difference in prediction with or without w_j, over all combinations.

$$\phi_j(w) = \sum_{Q \subseteq S \setminus j} \frac{|Q|!(|S|-|Q|-1)!}{|S|!} \left(v_{(Q \cup \{j\})}(w) - v_Q(w) \right) \tag{7}$$

$v_Q(w)$ (value function) is the payout function for coalitions of players (feature values), which denotes the influence of a subset of feature values. It generalizes (Eq. 6), in the following form

$$v_Q(w) = \mathbb{E}\left[M \mid W_i = w_i, \forall i \in Q \right] - \mathbb{E}[M] \tag{8}$$

where M provides the prediction over the set of features provided, S is the complete set of features, $Q \in S$ is a subset of features, and $|\cdot|$ is the size of feature set [33].

To adapt SHAP for the BART based generation model, we first obtain the score for each template and choose the final output template for a given input sentence S with aspect asp_i as discussed above. We create an explicit word masker to tokenize the sentence S into sentence fragments consisting of words, which serves as a basis for word masking in SHAP (here mask refers to hiding a particular word from the sentence). To explain the likelihood of generating this output template for S, we wrap the BART model along with tokenizer and output template with a teacher forcing scoring class (SC). Teacher forcing SC forces the model to generate the provided output template. The input sentence along with the designed masker and predicted output template is passed to SHAP ex-plainer, which generates various masked combinations of the input sentence. These masked combinations are passed to the teacher forcing scoring class, generating output probabilities for the output template. These output probabilities are returned to SHAP explainer, which returns the contribution of each word (from S) for generating each word of the output template. Output template for the ABSIA task contains two words (e.g. low positive) to describe the final class l_r (except neutral). We extract the relevance scores of each word from S for generating these two output words. We then take the average of both the score vectors to obtain the final relevance scores. ABSA task has only one word in the output template to describe the output class. We directly extract the relevance scores of the input sequence S, contributing to the generation of output class p_c.

4   Experiments and Analysis

We develop our models in Pytorch, a python based deep learning library[1]. All the experiments were carried out on an NVIDIA GeForce GTX 1080 Ti GPU.

---

[1] https://pytorch.org/.

We split the dataset tuples as: train (70%), validation (10%) and test-set (20%). Detailed data statistics are shown in Table 2. The batch size is set to 16 for all the models. The number of epochs is set to 30 and learning rate is set to 3e-5. Model parameters are optimized by Adam [9]. We use grid search to find the best set of hyperparameters on validation set. Since ABSIA is our primary task, we save the best model according to the development set accuracy of the ABSIA task. We experimented with multiple loss weight values and got the best results when loss weight values are 1 for both the tasks.

Baselines: We consider ABSIA as our primary task and try to enhance its performance with joint learning of ABSA and ABSIA. So, we define the baselines for the ABSIA task as mentioned below:

- TD LSTM: It uses two LSTM networks (left LSTM and right LSTM) to model the contexts preceding and following the aspect term [34].
- TC LSTM: It adds a target connection component to TD-LSTM, to incorporate semantic relatedness between aspect terms and context words [34].
- Inter-aspect relation modeling with memory networks (IARM) [19]: IARM learns the dependency of a given aspect with the other aspects in the sentence using memory network.
- Single task BERT: The BERT-base model is fine-tuned on the ABSIA task by adding a fully connected layer on top of it.
- Multi-task BERT: Multi-task system utilizes BERT as a shared encoder with two task-specific layers on top of it to solve ABSA and ABSIA tasks jointly.
- Single task BART: BART model consisting of encoder and decoder is fine-tuned for ABSIA task by adding a fully connected layer at decoder [12].
- Multi-task BART: This corresponds to the BART based multi-task system for ABSA and ABSIA tasks. Two task specific layers are added on the top of the decoder, one for each task.
- BART generation for ABSIA: Single task generation based model for intensity.
- BART generation model joint learning: Joint learning of ABSA and ABSIA tasks in generation framework.

4.1 Experimental Results

We evaluate the performance of our model using accuracy measure. Table 3 shows the results of our primary task ABSIA. TC LSTM reports an accuracy of 58.12% and TD LSTM outperforms TC LSTM. IARM reports improved performance, indicating the effectiveness of considering the interrelation between aspects of the same sentence. Further, single task BERT and BART outperform these models, illustrating the effectiveness of transfer learning and contextual representations; while, multi-task learning based frameworks outperform all the other models. The improved performance demonstrates the role of the ABSA task in enhancing the performance of the ABSIA task. Our proposed intensity generation single task model outperforms the single task BERT and BART based classification models, which shows that it can detect multiple sentiment intensities in a

Table 2. Data Statistics of ABSIA dataset

Type	-3	-2	-1	0	1	2	3
Train	51	170	477	0	1065	697	277
Dev	11	21	68	58	110	79	25
Test	18	37	137	0	303	205	82
Total	80	228	682	817	1478	981	384

Table 3. Results

Model	Accuracy
TD LSTM	58.12
TC LSTM	59.04
IARM	60.17
ST BERT	61.93
ST BART	62.45
MT BERT	63.31
MT BART	63.89
Intensity generation	63.87
Joint generation	**66.02**

sentence towards different aspect terms. Joint modelling of ABSA and ABSIA results in improved performance than all other baselines, showing the potential of both joint learning and utilization of pre-trained knowledge at the task level rather than at the representation level. We follow the paired T-test (significance test), which validates the performance gain over all the baselines is significant with 95% confidence (p-value<0.05).

Table 4. Impact of templates on ABSIA task

Template	Accuracy
The sentiment intensity of $< asp_i >$ is l_r	65.32%
The aspect $< asp_i >$ has sentiment intensity of l_r	66.29%
The aspect $< asp_i >$ has $< l_r >$ sentiment intensity .	67.74%

Impact of Templates: We experiment with several templates to observe their impact on the model performance. Table 4 shows the results of using different templates on the development dataset. For example, the template $< The\ aspect < asp_i >$ has sentiment intensity of $< l_r >$, reports accuracy value of 66.29% on development set. We select the final template with the highest accuracy on the development set.

4.2   Detailed Analysis

This section presents a detailed analysis of our results. The confusion matrix for the ABSIA task for single task and joint learning framework is shown in Fig. 3. We can see that joint model has improved the number of correct classifications for almost every class. In joint model, it is observed that the majority of misclassifications from low negative class are into neutral classes and from moderate negative class into low negative class. Moderate and high positive classes are

.	1	2	3	0	-1	-2	-3
1	231	30	8	20	14	0	0
2	67	112	16	5	4	1	0
3	16	19	47	0	0	0	0
0	26	3	1	100	14	4	0
-1	13	1	0	21	86	16	0
-2	3	0	0	5	19	9	1
-3	0	0	0	0	2	7	9

.	1	2	3	0	-1	-2	-3
1	236	29	8	18	10	2	0
2	55	117	14	9	5	5	0
3	15	16	49	0	2	0	0
0	29	4	0	95	18	2	0
-1	8	0	0	22	95	10	2
-2	3	0	0	3	21	10	0
-3	0	0	0	0	3	5	10

Fig. 3. Confusion matrix for single task ABSIA and joint model

often confused with low positive classes. This could be due to the close resemblance between different levels of a class and the higher number of instances of low positive class.

Explaining Generation Framework: This section describes the actual decision process of proposed BART based template generation framework by extracting the opinion terms responsible for generating the final output label. We first explain how the intensity generation model derives its final predictions, followed by explanation of the joint model. Consider the following example, *service is slow*. The intensity generation model generates low negative sentiment intensity for aspect *service*. However, the model predicts high negative intensity in case of the sentence *service is severely slow*. Figures 4 and 5 explain the predictions of the model. Tokens with red colour signify the terms which are responsible for the final label generation (positive SHAP scores). In contrast, the words with blue colour negatively influence the final generation (negative SHAP scores). More intense colour signifies the greater influence of the term for final label generation. Figure 4 reveals that model generates intensity label with the help of term *slow*. On the other hand, Fig. 5 illustrates the importance of *severely* over *slow* to generate the high positive intensity label. Thus, our model is capable of distinguishing between different levels of intensities, which are not explored in the previous studies. Furthermore, we exploit the behavior of our proposed model. We observe that ABSIA task benefits from the ABSA task to perform better. Example cases are shown in Fig. 6 to demonstrate this observation. The left side explains the SHAP predictions for the single task intensity model and the right side describes the predictions of the joint learning framework. Example 1 carry low negative sentiment intensity for aspect *delivery*. The single task model generates the intensity as neutral. We can see that model's focus is on *delivery* followed by *also ordered for*. However, with joint learning of both tasks, ABSA helps the model to shift its focus to *forgot*, which helps the ABSIA to generate the correct class. Similarly, for example 2, the single task model predicts

Fig. 4. SHAP output: low intensity   **Fig. 5.** SHAP output: high intensity

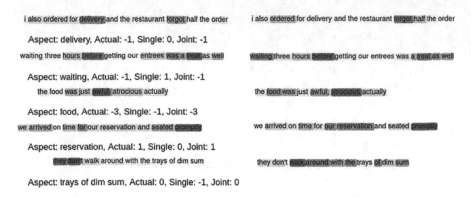

Fig. 6. Explaining behavior of single task and joint learning for ABSIA task

intensity as low positive because of its less focus on *atrocious*. However, the joint learning framework has shifted the model's focus to *atrocious* and *awful*, which helps for correct prediction. In example 4, for aspect reservation, the single task model made predictions based on *we, arrived, and for*; hence it predicts intensity class as neutral. However, the joint learning framework helps the model to focus on *promptly*, leading to correct prediction. Similarly, for example 5, the actual label is neutral for aspect *dimsum*. However, the single task model pays more attention to the negative term *don't*, which leads to incorrect intensity prediction. The ABSA task helps the ABSIA task to shift its focus to neutral terms (*walk, around, and with*). To explain the limitations of our proposed framework, we show samples misclassified by the joint learning framework in Fig. 7. SHAP analysis shows that samples with implicit opinions are misclassified by our framework (examples 1, 2, and 3). Sometimes the model cannot link correct opinion terms to the aspect when there is stronger sentiment intensity expressed for the other aspect (example 4). In example 4, model is focusing on word *excellent* rather than words associated with aspect *mayo*.

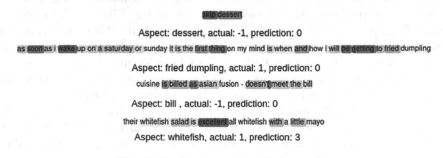

Fig. 7. Explaining misclassified samples

5   Conclusion

In this paper, we have introduced a new task of aspect based sentiment intensity analysis, which would immensely benefit the organizations and enterprises to mine the user preferences more precisely. We have developed a benchmark setup to create the annotated corpus for the ABSIA task using a semi-supervised way. To demonstrate the effective usage of the dataset, we cast ABSIA as a generation task and proposed a template based generation framework for intensity prediction. Further, we have proposed a joint framework to improve the performance of ABSIA task with the help of ABSA. We further utilized the explainability technique to explain the predictions of single task and joint learning frameworks. SHAP explains the decision process of our models by extracting the opinion terms responsible for generating the output class. In the future we shall investigate the usage of term importance information provided by SHAP to reduce the misclassification rate. We would also focus how to augment this information to minimise misclassifications without hampering the correct classifications.

References

1. Akhtar, M.D., Ekbal, A., Bhattacharyya, P.: Aspect Based Sentiment Analysis: Category Detection and Sentiment Classification for Hindi. In: Gelbukh, A. (ed.) CICLing 2016. LNCS, vol. 9624, pp. 246–257. Springer, Cham (2018). https://doi.org/10.1007/978-3-319-75487-1_19
2. Akhtar, M.S., Ekbal, A., Bhattacharyya, P.: Aspect based sentiment analysis in Hindi: resource creation and evaluation. In: Proceedings of the Tenth International Conference on Language Resources and Evaluation (LREC'16), pp. 2703–2709 (2016)
3. Akhtar, M.S., Kumar, A., Ekbal, A., Bhattacharyya, P.: A hybrid deep learning architecture for sentiment analysis. In: Proceedings of COLING 2016, the 26th International Conference on Computational Linguistics, Tech. Pap, pp. 482–493 (2016)
4. Chen, P., Sun, Z., Bing, L., Yang, W.: Recurrent attention network on memory for aspect sentiment analysis. In: Proceedings of the 2017 conference on empirical methods in natural language processing, pp. 452–461 (2017)
5. Cheng, J., Zhao, S., Zhang, J., King, I., Zhang, X., Wang, H.: Aspect-level sentiment classification with heat (hierarchical attention) network. In: Proceedings of the 2017 ACM on Conference on Information and Knowledge Management, pp. 97–106 (2017)
6. Daza, A., Frank, A.: A sequence-to-sequence model for semantic role labeling. arXiv preprint arXiv:1807.03006 (2018)
7. Devlin, J., Chang, M.W., Lee, K., Toutanova, K.: Bert: pre-training of deep bidirectional transformers for language understanding. arXiv preprint arXiv:1810.04805 (2018)
8. Jiang, Q., Chen, L., Xu, R., Ao, X., Yang, M.: A challenge dataset and effective models for aspect-based sentiment analysis. In: Proceedings of the 2019 Conference on Empirical Methods in Natural Language Processing and the 9th International Joint Conference on Natural Language Processing (EMNLP-IJCNLP), pp. 6280–6285 (2019)

9. Kingma, D.P., Ba, J.: Adam: A method for stochastic optimization. In: Bengio, Y., LeCun, Y. (eds.) 3rd International Conference on Learning Representations, ICLR 2015, San Diego, CA, USA, May 7–9, 2015, Conference Track Proceedings (2015). https://arxiv.org/abs/1412.6980

10. Kumar, S., De, K., Roy, P.P.: Movie recommendation system using sentiment analysis from microblogging data. IEEE Trans. Comput. Soc. Syst. **7**(4), 915–923 (2020)

11. Lei, Z., Yang, Y., Yang, M., Zhao, W., Guo, J., Liu, Y.: A human-like semantic cognition network for aspect-level sentiment classification. In: Proceedings of the AAAI Conference on Artificial Intelligence. vol. 33, pp. 6650–6657 (2019)

12. Lewis, M., Liu, Y., Goyal, N., Ghazvininejad, M., Mohamed, A., Levy, O., Stoyanov, V., Zettlemoyer, L.: BART: denoising sequence-to-sequence pre-training for natural language generation, translation, and comprehension. In: Proceedings of the 58th Annual Meeting of the Association for Computational Linguistics, Assoc. Comput. Linguist, pp. 7871–7880. (2020). 10.18653/v1/2020.acl-main.703. https://aclanthology.org/2020.acl-main.703

13. Li, R., Chen, H., Feng, F., Ma, Z., Wang, X., Hovy, E.: Dual graph convolutional networks for aspect-based sentiment analysis. In: Proceedings of the 59th Annual Meeting of the Association for Computational Linguistics and the 11th International Joint Conference on Natural Language Processing (Long Papers). 1 pp. 6319–6329. Assoc. Comput. Linguist. (2021). 10.18653/v1/2021.acl-long.494. https://aclanthology.org/2021.acl-long.494

14. Li, X., et al.: Entity-relation extraction as multi-turn question answering. arXiv preprint arXiv:1905.05529 (2019)

15. Liu, J., Teng, Z., Cui, L., Liu, H., Zhang, Y.: Solving aspect category sentiment analysis as a text generation task. In: Proceedings of the 2021 Conference on Empirical Methods in Natural Language Processing. Assoc. Comput. Linguist. Punta Cana, Dominican Republic , pp. 4406–4416(2021). 10.18653/v1/2021.emnlp-main.361, https://aclanthology.org/2021.emnlp-main.361

16. Lundberg, S.M., Lee, S.I.: A unified approach to interpreting model predictions. In: Proceedings of the 31st international conference on neural information processing systems, pp. 4768–4777 (2017)

17. Ma, D., Li, S., Zhang, X., Wang, H.: Interactive attention networks for aspect-level sentiment classification. arXiv preprint arXiv:1709.00893 (2017)

18. Maia, M., et al.: Www'18 open challenge: financial opinion mining and question answering. In: Companion proceedings of the the web conference 2018, pp. 1941–1942 (2018)

19. Majumder, N., Poria, S., Gelbukh, A., Akhtar, M.S., Cambria, E., Ekbal, A.: Iarm: Inter-aspect relation modeling with memory networks in aspect-based sentiment analysis. In: Proceedings of the 2018 conference on empirical methods in natural language processing, pp. 3402–3411 (2018)

20. Mamta, Ekbal, A., Bhattacharyya, P.: Exploring multi-lingual, multi-task and adversarial learning for low-resource sentiment analysis. Transactions on Asian and Low-Resource Language Information Processing (2022)

21. Mamta, Ekbal, A., Bhattacharyya, P., Srivastava, S., Kumar, A., Saha, T.: Multi-domain tweet corpora for sentiment analysis: Resource creation and evaluation. In: Calzolari, N., et al., (eds.) Proceedings of The 12th Language Resources and Evaluation Conference, LREC 2020, Marseille, France, May 11–16, 2020, European Lang. Resour. Assoc, pp. 5046–5054. (2020). https://aclanthology.org/2020.lrec-1.621/

22. Meena, A., Prabhakar, T.V.: Sentence Level Sentiment Analysis in the Presence of Conjuncts Using Linguistic Analysis. In: Amati, G., Carpineto, C., Romano, G. (eds.) ECIR 2007. LNCS, vol. 4425, pp. 573–580. Springer, Heidelberg (2007). https://doi.org/10.1007/978-3-540-71496-5_53

23. Mohammad, S., Bravo-Marquez, F., Salameh, M., Kiritchenko, S.: SemEval-2018 task 1: Affect in tweets. In: Proceedings of The 12th International Workshop on Semantic Evaluation. Assoc. Comput. Linguist. New Orleans, Louisiana, pp. 1–17. (2018). 10.18653/v1/S18-1001. https://aclanthology.org/S18-1001

24. Mudalige, C.R et al.: SIGMALAW-ABSA: Dataset for aspect-based sentiment analysis in legal opinion texts. In: 2020 IEEE 15th International Conference on Industrial and Information Systems (ICIIS), pp. 488–493. IEEE (2020)

25. Oh, S., Lee, D., Whang, T., Park, I., Gaeun, S., Kim, E., Kim, H.: Deep context- and relation-aware learning for aspect-based sentiment analysis. In: Proceedings of the 59th Annual Meeting of the Association for Computational Linguistics and the 11th International Joint Conference on Natural Language Processing (Short Papers) 2, pp. 495–503. Assoc. Comput. Linguist. (Aug 2021). 10.18653/v1/2021.acl-short.63, https://aclanthology.org/2021.acl-short.63

26. Pontiki, M., et al.: Semeval-2016 task 5: Aspect based sentiment analysis. In: International workshop on semantic evaluation, pp. 19–30 (2016)

27. Pontiki, M., Galanis, D., Papageorgiou, H., Manandhar, S., Androutsopoulos, I.: Semeval-2015 task 12: Aspect based sentiment analysis. In: Proceedings of the 9th international workshop on semantic evaluation (SemEval 2015), pp. 486–495 (2015)

28. Pontiki, M., Galanis, D., Pavlopoulos, J., Papageorgiou, H., Androutsopoulos, I., Manandhar, S.: Semeval-2014 task 4: Aspect based sentiment analysis. In: Nakov, P., Zesch, T. (eds.) Proceedings of the 8th International Workshop on Semantic Evaluation, SemEval@COLING 2014, Dublin, Ireland, August 23–24, 2014, Assoc. Comput. Linguist, pp. 27–35. (2014). https://doi.org/10.3115/v1/s14-2004, https://doi.org/10.3115/v1/s14-2004

29. Raffel, C., Shazeer, N., Roberts, A., Lee, K., Narang, S., Matena, M., Zhou, Y., Li, W., Liu, P.J., et al.: Exploring the limits of transfer learning with a unified text-to-text transformer. J. Mach. Learn. Res. **21**(140), 1–67 (2020)

30. Román, J.V., Cámara, E.M., Morera, J.G., Zafra, S.M.J.: Tass 2014-the challenge of aspect-based sentiment analysis. Procesamiento del Lenguaje Natural **54**, 61–68 (2015)

31. Socher, R., et al.: Recursive deep models for semantic compositionality over a sentiment treebank. In: Proceedings of the 2013 conference on empirical methods in natural language processing, pp. 1631–1642 (2013)

32. Spitzer, R.L., Cohen, J., Fleiss, J.L., Endicott, J.: Quantification of agreement in psychiatric diagnosis: A new approach. Arch. Gen. Psychiatry **17**(1), 83–87 (1967)

33. Strumbelj, E., Kononenko, I.: Explaining prediction models and individual predictions with feature contributions. Knowl. Inf. syst. **41**(3), 647–665 (2014)

34. Tang, D., Qin, B., Feng, X., Liu, T.: Effective lSTMs for target-dependent sentiment classification. arXiv preprint arXiv:1512.01100 (2015)

35. Tian, H., et al.: Skep: Sentiment knowledge enhanced pre-training for sentiment analysis. arXiv preprint arXiv:2005.05635 (2020)

36. Tian, L., Lai, C., Moore, J.D.: Polarity and intensity: the two aspects of sentiment analysis. arXiv preprint arXiv:1807.01466 (2018)

37. Wang, W., Pan, S.J.: Recursive neural structural correspondence network for cross-domain aspect and opinion co-extraction. In: Proceedings of the 56th Annual Meeting of the Association for Computational Linguistics (Long Papers) 1, pp. 2171–2181 (2018)

38. Wang, Y., Huang, M., Zhu, X., Zhao, L.: Attention-based LSTM for aspect-level sentiment classification. In: Proceedings of the 2016 conference on empirical methods in natural language processing, pp. 606–615 (2016)
39. Wojatzki, M., Ruppert, E., Holschneider, S., Zesch, T., Biemann, C.: Germeval 2017: Shared task on aspect-based sentiment in social media customer feedback. Proceedings of the GermEval, pp. 1–12 (2017)
40. Xu, H., Liu, B., Shu, L., Yu, P.S.: Bert post-training for review reading comprehension and aspect-based sentiment analysis. arXiv preprint arXiv:1904.02232 (2019)
41. Yan, H., Dai, J., Ji, T., Qiu, X., Zhang, Z.: A unified generative framework for aspect-based sentiment analysis. In: Proceedings of the 59th Annual Meeting of the Association for Computational Linguistics and the 11th International Joint Conference on Natural Language Processing (Long Papers) 1, pp. 2416–2429. Assoc. Comput. Linguist. (2021). 10.18653/v1/2021.acl-long.188, https://aclanthology.org/2021.acl-long.188
42. Yang, M., Jiang, Q., Shen, Y., Wu, Q., Zhao, Z., Zhou, W.: Hierarchical human-like strategy for aspect-level sentiment classification with sentiment linguistic knowledge and reinforcement learning. Neural Networks **117**, 240–248 (2019)

Effective Hierarchical Information Threading Using Network Community Detection

Hitarth Narvala[✉], Graham McDonald, and Iadh Ounis

University of Glasgow, Glasgow, UK
h.narvala.1@research.gla.ac.uk,
{graham.mcdonald,iadh.ounis}@glasgow.ac.uk

Abstract. With the tremendous growth in the volume of information produced online every day (e.g. news articles), there is a need for automatic methods to identify related information about events as the events evolve over time (i.e., information threads). In this work, we propose a novel unsupervised approach, called *HINT*, which identifies coherent **H**ierarchical **I**nformation **T**hreads. These threads can enable users to easily interpret a hierarchical association of diverse evolving information about an event or discussion. In particular, HINT deploys a scalable architecture based on network community detection to effectively identify hierarchical links between documents based on their chronological relatedness and answers to the 5W1H questions (i.e., who, what, where, when, why & how). On the NewSHead collection, we show that HINT markedly outperforms existing state-of-the-art approaches in terms of the quality of the identified threads. We also conducted a user study that shows that our proposed network-based hierarchical threads are significantly ($p < 0.05$) preferred by users compared to cluster-based sequential threads.

1 Introduction

In the digital age, the rise of online platforms such as news portals have led to a tremendous growth in the amount of information that is produced every day. The volume of such information can make it difficult for the users of online platforms to quickly find related and evolving information about an event, activity or discussion. However, presenting this information to the users as a hierarchical list of articles, where each branch of the hierarchy contains a chronologically evolving sequence of articles that describe a story relating to the event, would enable the users to easily interpret large amounts of information about an event's evolution. For example, Fig. 1(a) presents different stories that are related to the event "Lira, rand and peso crash" as separate branches of a hierarchical list. We refer to this structure of information as a *Hierarchical Information Thread*. Figure 1(a) illustrates the following three characteristics of hierarchical threads: (1) all of the articles in the thread present coherent information that relates to the same event, (2) different stories (i.e., branches) capture diverse information relating to the event, and (3) the articles that discuss a story are chronologically ordered.

Compared to hierarchical threads, a sequential thread cannot simultaneously capture both the chronology and the logical division of diverse information about

© The Author(s), under exclusive license to Springer Nature Switzerland AG 2023
J. Kamps et al. (Eds.): ECIR 2023, LNCS 13980, pp. 701–716, 2023.
https://doi.org/10.1007/978-3-031-28244-7_44

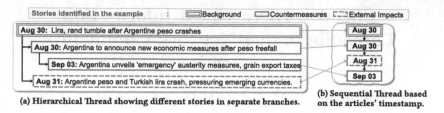

(a) Hierarchical Thread showing different stories in separate branches.

(b) Sequential Thread based on the articles' timestamp.

Fig. 1. Comparative example of Hierarchical & Sequential Information Threads.

an event. For example in Fig. 1(b), a simple chronological order of the articles cannot represent the articles about "Countermeasures" as a coherent story in the thread. In contrast, hierarchical threads (Fig. 1(a)) can enable the users to find diverse stories about the event's evolution in an easily interpretable structure.

We propose a novel unsupervised approach, *HINT*,[1] for identifying **H**ierarchical **In**formation **T**hreads by analysing the network of related articles in a collection. In particular, we leverage article timestamps and the 5W1H questions (Who, What, Where, When, Why and How) [8] to identify related articles about an event or discussion. We then construct a network representation of the articles, and identify threads as strongly connected hierarchical network communities.

We evaluate the effectiveness of HINT on the NewSHead collection [7], in both an offline setting and a user study. In our offline evaluation, we show that HINT markedly improves the quality of the threads in terms of Normalised Mutual Information (NMI) and Homogeneity (h) (up to +232.08% NMI & +400.71% h) compared to different established families of related methods in the literature, i.e., document threading [6] and event extraction [12] approaches. We also compare the effectiveness of our hierarchical information threading approach with a recent work on cluster-based sequential information threading [14], which we refer to as SeqINT. In terms of thread quality, we show that HINT is more effective in generating quality threads than SeqINT (+10.08% NMI and +19.26% h). We further conduct a user study to evaluate the effectiveness of HINT's hierarchical threads compared to SeqINT's sequential threads. Our user study shows that the users significantly ($p < 0.05$) preferred the HINT threads in terms of the event's description, interpretability, structure and chronological correctness than the SeqINT threads. We also analyse the scalability of HINT's architecture by simulating a chronologically incremental stream of NewSHead articles. We show that the growth in the execution time of HINT is slower compared to the growth in the number of articles over time.

2   Related Work

Existing tasks such as topic detection and tracking (TDT) [2] and event threading (ET) [13] broadly relate to the problem of identifying information about events. TDT and ET tasks typically focus on identifying *clusters* of events to capture related information about evolving topics or dependent events. However, unlike hierarchical information threads, these clusters of events do not provide a

---

[1] HINT's code is available at: https://github.com/hitt08/HINT.

finer-grained view of an event's evolving stories, which makes it difficult for the users to find relevant stories based on their interests.

Topic Detection and Tracking (TDT) [2] is the task of identifying threads of documents that discuss a topic, i.e., *topic-based* threads about the chronological evolution of a topic. TDT approaches (e.g. [1,5,18,24]) focus on identifying topical relationships between the documents to automatically detect topical clusters (i.e., topics), and to track follow-up documents that are related to such topics. These topics are often a group of many related events [24]. Differently from topic-based threads about many related events, hierarchical information threads describe evolving information about different *stories* that relate to a *specific* event.

Event threading approaches (e.g., [12,13,20]) first extract events as clusters of related documents, and then identify threads of the event clusters. Differently from event threading, our focus is to identify hierarchical information threads of documents that describe different stories about a single event, activity or discussion. We used the EventX [12] event extraction approach as a baseline in our experiments, since it also identifies related documents about specific events.

Another related task is to identify a few *specific* document threads in a collection such as threads about the most important events [6] or threads that connect any two given documents in the collection [19]. Our work on hierarchical information threading is different in multiple aspects from the aforementioned document threading approaches that aim to identify specific threads in a collection. First, we focus on identifying threads about all of the events in a collection. Second, unlike document threading approaches that use document term features, we focus on the 5W1H questions and chronological relationships between documents to identify evolving information about events. Lastly, unlike existing document threading approaches that generate sequential threads, we propose hierarchical threading to describe various aspects about an event (e.g. different stories).

Recently, Narvala et al. [14] introduced an information threading approach. They deploy clustering to identify *sequential* threads using 5W1H questions and the documents' timestamps (we refer to this approach as SeqINT). Unlike the cluster-based SeqINT approach, in this work, we focus on identifying threads of *hierarchically* associated documents using network community detection methods to capture the evolving stories of an event. Moreover, the SeqINT approach only supports static collections, whereas, our proposed network-based approach can also be deployed to generate information threads in dynamic collections.

3 Proposed Approach: HINT

In this section, we present our proposed approach for identifying Hierarchical Information Threads (HINT). Our approach leverages the chronological relationships between documents, 5W1H questions' answers along with the entities that are mentioned in multiple documents in a collection, to define a directed graph structure of the collection (i.e., a network of documents). We then deploy a community detection algorithm to identify coherent threads by identifying hierarchical links in the network of documents. Figure 2 shows the components of HINT, which we describe in this section, i.e., (1) 5W1H Extraction, (2) Constructing a

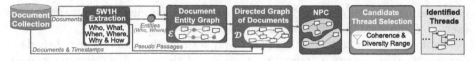

Fig. 2. Components of HINT

Document-Entity Graph, (3) Constructing a Directed Graph of the Documents, (4) Nearest Parent Community Detection, and (5) Candidate Thread Selection.

5W1H Extraction: We first extract the phrases of text that answers the 5W1H questions from each document in the collection using the Giveme5W1H approach [8]. We then concatenate all of the 5W1H questions' answers for each document as a pseudo-passage (i.e., one pseudo-passage per document). To vectorise the pseudo-passages, we use transformer-based [23] contextual embeddings to capture the context of the events described by the pseudo-passages. We use these embeddings when constructing the Directed Graph of Documents.

Constructing Document-Entity Graph: After 5W1H extraction, we construct an undirected document-entity graph, $\mathcal{E}$, to identify the common entities between the documents in the collection. The graph $\mathcal{E}$ comprises two types of nodes, i.e., the entities and documents in the collection. We first identify the key entities associated with an event by leveraging the 5W1H questions' answers. In particular, we re-use the available answers to the "who" and "where" questions, which directly correspond to named-entities, i.e., "person/organisation" (who) and "place" (where). In other words, we re-purpose the available named-entity information from the 5W1H extraction to avoid needing an additional named-entity recogniser. We then create an edge between the documents and their corresponding entities, i.e., at most two edges per document node (who and/or where).

Constructing a Directed Graph of Documents: We then use the 5W1H questions' answers, the document-entity graph $\mathcal{E}$ along with the creation timestamps of the documents to construct a document graph, $\mathcal{D}$, from which we identify candidate hierarchical threads. In the graph $\mathcal{D}$, the nodes are the documents in the collections. We define directed edges between documents in $\mathcal{D}$ based on the document timestamp such that the edges between two documents go forward in time. In addition, we define weights for the edges based on the relatedness of the child node to the parent node in a directed edge between two documents. In particular, to effectively capture the relatedness of nodes based on the event they describe, the weight of each edge is defined as: (1) the similarity between the 5W1H pseudo-passages of the documents, (2) the chronological relationship between the documents, and (3) the number of entities mentioned in both of the documents.

To calculate the edge weights of the graph, we first compute the cosine similarity ($cos(p_x, p_y)$) of the 5W1H pseudo-passage embeddings, p_x & p_y (for documents x & y respectively). To capture the chronological relationship between x & y, we compute the documents' time-decay (inspired by Nallapati et al. [13]), i.e., the normalised time difference between the creation times of x & y, defined as:

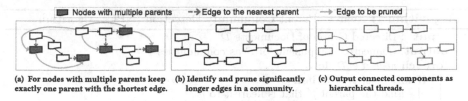

(a) For nodes with multiple parents keep exactly one parent with the shortest edge.
(b) Identify and prune significantly longer edges in a community.
(c) Output connected components as hierarchical threads.

Fig. 3. Nearest Parent Community Detection

$$td(x,y) = e^{-\alpha \frac{|t_x - t_y|}{T}} \tag{1}$$

where t_x & t_y are the creation timestamps of documents x & y respectively, T is the time difference between the oldest and latest document in the collection and α is a parameter to factor in time decay. In a dynamic collection, the value of T can be dynamically estimated based on the maximum time difference between articles in the existing threads identified from the historical articles.

We then use the document-entity graph $\mathcal{E}$ to calculate an entity similarity score for each pair of documents in the graph $\mathcal{D}$. To compute an entity similarity score for a pair of documents, x & y, we first identify the number of paths ($|\mathbb{P}_{xy}|$) that connect x & y in the graph $\mathcal{E}$ through exactly one entity node. Second, if $|\mathbb{P}_{xy}| = 0$, we identify the length of the shortest path ($|s_{xy}|$) that connects x & y through multiple entities or other document nodes in $\mathcal{E}$. Intuitively, for documents that have common entities, a higher value of $|\mathbb{P}_{xy}|$ denotes a higher similarity between documents x & y, with respect to the entities that are mentioned in the documents. In contrast, for documents that do not have any common entities (i.e., $|\mathbb{P}_{xy}| = 0$), a longer length of the shortest path, $|s_{xy}|$, denotes less similarity between x & y. Based on the aforementioned description of $|\mathbb{P}_{xy}|$ and $|s_{xy}|$, we define the overall entity similarity score between documents x & y as follows:

$$es(x,y) = \frac{\lambda}{2}(1 + (1 - e^{-\gamma \frac{|\mathbb{P}_{xy}|}{M}})) + \frac{(1-\lambda)}{2}e^{-\gamma \frac{|s_{xy}|}{N}}, \lambda = \begin{cases} 1, & \text{if } |\mathbb{P}_{xy}| > 0 \\ 0, & \text{otherwise} \end{cases} \tag{2}$$

where, M is the largest number of common entities between any two documents in the collection, N is the largest shortest path in the collection, and γ is a parameter to control the relative weights of the number of common entities or the length of the shortest path between x & y.

Lastly, we define the edge weights in the document graph $\mathcal{D}$ (i.e., the distance between x & y) using Eqs. (1) and (2), and the 5W1H cosine similarity, as:

$$w(x,y) = 1 - cos(p_x, p_y) \cdot td(x,y) \cdot es(x,y) \tag{3}$$

Algorithm 1: Nearest Parent Community Detection (NPC) Algorithm

---

    input : Directed Graph of Documents $\mathcal{D}$
    output: Connected components of $\mathcal{D}$ as communities
    foreach *node* $n \in \mathcal{D}$ **do**
        if inDegree(n) > 1 **then**
            Find the parent p' that is nearest to n
            foreach $p \in$ parents(n) **do**
                if $p \neq p'$ **then**
                    Remove edge ($p \to n$)
    foreach *connected component* $c \in \mathcal{D}$ **do**
        Compute outlier weight threshold for c using Equation (4).
        foreach edge $e \in c$ **do**
            if weight(e) > *threshold* **and** outDegree(childNode(e)) > 1 **then**
                Remove e from $\mathcal{D}$

Nearest Parent Community Detection (NPC): From the Directed Graph $\mathcal{D}$, we identify hierarchically connected communities for thread generation. We propose a Nearest Parent Community Detection (NPC) method that identifies strongly connected components of graph $\mathcal{D}$ as communities of hierarchically linked documents. The NPC algorithm is presented in Algorithm 1 and is illustrated in Fig. 3. To identify hierarchical links between document nodes, as shown in Fig. 3(a), NPC first identifies the nodes that have multiple parents, and follows a greedy approach to keep only the edge that corresponds to the nearest parent (i.e., the edge with the lowest weight; shown with dashed green arrow in Fig. 3(a)). This selection of only the nearest parent node results in various hierarchically connected components of graph $\mathcal{D}$, as shown in Fig. 3(b). However, the connected graph components may still have some weakly connected parent and child nodes (i.e., edges with high weights). Therefore, to remove such weak connections, we split the connected graph components by identifying edges that have significantly higher weights based on the outlier detection method [22]. In particular, within a connected graph component, we determine a threshold edge weight. This threshold value corresponds to the outliers in the distribution of the edge weights within a connected graph component defined as follows [22]:

$$threshold = P_3 + 1.5 * (P_3 - P_1) \qquad (4)$$

where P_1 and P_3 are respectively the values for the first and third quartiles (i.e. 25 and 75 percentile) of the edge weight distribution, and $(P_3 - P_1)$ is the interquartile range. We compute this threshold for each connected graph component. While pruning the outlier edges, we do not prune edges where the child nodes do not have any outward edges so that the graph does not contain any isolated nodes. Finally, as shown in Fig. 3(c), NPC outputs the connected graph components (i.e., strongly connected communities) as candidate hierarchical threads.

Candidate Thread Selection: From the candidate threads identified by NPC, we select the output threads based on thread coherence and diversity of information. Our focus is on selecting a maximum number of threads from the candidates that are coherent and providing diverse information about their respective events. However, popular metrics (e.g. Cv [16]) for directly computing coherence for all threads in a large collection can be computationally expensive. Therefore, following Narvala et al. [14], to efficiently select candidate threads, we define an estimate of coherence and diversity using the following three measures: (1) The number of documents in a thread $\mathbb{T}$ (i.e., the thread length $|\mathbb{T}|$), (2) The time period, $\mathbb{T}_{span}$, between the timestamps of the first and last documents in a candidate thread, and (3) The mean pairwise document cosine similarity, $\mathbb{T}_{MPDCS}$, of a candidate thread, $\mathbb{T}$, calculated over all pairs of consecutive documents in the candidate thread.

Following [14], we optimise a minimum and maximum threshold range of the aforementioned measures based on coherence, diversity and the total number of selected threads using a smaller sample of NewSHead articles. To compute coherence and diversity, we use the C_v metric [16] and KL Divergence [10] respectively.

4 Experimental Setup

We now describe our experimental setup for the offline evaluation where we evaluate the threads quality (Sect. 5), and for the user study where we evaluate the effectiveness of hierarchical and sequential threads with real users (Sect. 6).

Dataset: There are very limited datasets available for evaluating information threads. In particular, previous work (e.g. [6, 13]) use manually annotated datasets which are not publicly available. Moreover, classical text clustering datasets such as 20 Newsgroups [11] only contain topic labels and not event labels, which are needed to evaluate event-based information threads.

Therefore, we use the publicly available NewSHead [7] test collection, which contains news story labels and URLs to news articles. Each of the NewSHead story label corresponds to a group of 3–5 articles about a story of an event. For our experiments, we crawled 112,794 NewSHead articles that are associated with 95,786 story labels. We combine the articles from multiple stories about an event into a single set, and refer to these sets as the true thread labels. In particular, since the NewSHead stories often share common articles (i.e., overlapping sets), we perform a union of these overlapping story sets, to create the true thread labels. This resulted in 27,681 true thread labels for the NewSHead articles (average of 4.07 articles per thread). In addition, considering the scalability limits of some of the baseline approaches that we evaluate, similar to Gillenwater et al. [6], we split the collection based on the article creation time into three test sets (37,598 articles each). We execute the threading approaches on these test sets separately, and evaluate their effectiveness collectively on all the three test sets.

Baselines: We compare the effectiveness of HINT to the following baselines:

- **k-SDPP** [6]: We first evaluate the k-SDPP document threading approach, using the publicly available implementation of SDPP sampling [9]. Since the length of k-SDPP threads are fixed, we specify $k=4$, based on the mean length of the NewSHead threads. Moreover, k-SDPP samples a fixed number of threads. We perform 200 k-SDPP runs with sample size 50 from each of the three test sets (i.e., $200 * 50 * 3 = 30,000$ threads, based on 27,681 NewSHead threads).

- **EventX** [12]: Second, we evaluate the EventX event extraction approach, using its publicly available implementation.
- **SeqINT** [14]: Third, we evaluate SeqINT to compare the effectiveness of cluster-based sequential threading with our hierarchical information threading approach (HINT). We use the edge weight function defined in Equation (3) as the distance function for clustering in SeqINT. Unlike HINT, SeqINT requires an estimate of the number of clusters. For our experiments, we use the number of true thread labels in each of the three test sets as the number of clusters in SeqINT.

HINT: We now present HINT's implementation details and configurations.

- **Pseudo-Passage Embedding**: We evaluate two contextual embedding models [15] for representing the 5W1H pseudo-passages namely: *all-miniLM-L6-v2* and *all-distilRoBERTa-v1*. We denote the aforementioned two embedding models as *mLM* and *dRoB*, respectively, when discussing our results in Sect. 5.1.
- **Community Detection**: We evaluate the effectiveness of NPC compared to two widely-used community detection methods: Louvain [3] and Leiden [21].
- **Parameters**: We tune HINT's parameters based on thread coherence and diversity on a small sample of NewSHead (Sect. 3), using the following values:

  ○ $\alpha; \gamma \Rightarrow \{10^i \ \forall \ -3 \leq i \leq 3; \text{step} = 1\}\}$
  ○ $x \leq |\mathbb{T}| \leq y \Rightarrow \{x, y\} \in \{\{3, i\} \ \forall \ 10 \leq i \leq 100; \text{step} = 10\}$,
  ○ $x \leq \mathbb{T}_{span} \leq y \Rightarrow \{x, y\} \in \{\{0, i\} \ \forall \ 30 \leq i \leq 360; \text{step} = 30\}$,
  ○ $x \leq \mathbb{T}_{MPDCS} \leq y \Rightarrow \{x, y\} \in \{\{0 + i, 1 - i\} \ \forall \ 0 \leq i \leq 0.4; \text{step} = 0.1\}$.

5   Offline Evaluation

Our offline evaluation compares the effectiveness of HINT in terms of the quality of generated threads, compared to the baselines discussed in Sect. 4. We aim to answer the following two research questions:

- **RQ1**: Is HINT more effective for identifying good quality threads than the existing document threading and event extraction approaches?
- **RQ2**: Is our NPC component more effective at identifying communities for thread generation than existing general community detection methods?

Evaluation Metrics: We evaluate thread quality based on the agreement of articles in the generated threads with the NewSHead thread labels. However, we note that thread quality cannot indicate whether the sequence of articles in a thread is correct, which we evaluate later in our user study (Sect. 6). Intuitively, our offline evaluation considers threads as small clusters of articles. We use the following popular cluster quality metrics to measure the thread quality: Homogeneity Score (h) [17] and Normalised Mutual Information (NMI) [4].

Since all of the NewSHead articles have an associated thread label, we compute h and NMI using all of the articles in the collection to measure the thread quality. Moreover, for each of the evaluated approaches, it is possible that the approach will not include all of the NewSHead articles in the generated threads. Therefore, we also report the number of generated threads along with the total and mean of the number of articles (mean $|\mathbb{T}|$) in each of the generated threads.

5.1   Results

Table 1 presents the number, length and quality of the generated threads. Firstly addressing **RQ1**, we observe from Table 1 that the NPC configurations for HINT markedly outperform the k-SDPP and EventX approaches from the literature along with the SeqINT approach in terms of h and NMI (e.g. NMI; mLM-NPC: 0.797 vs k-SDPP: 0.190 vs EventX: 0.240 vs SeqINT: 0.724). Even though both HINT and SeqINT use 5W1H questions, HINT's NPC community detection and graph construction using time decay and entity similarity contributes to its higher effectiveness over SeqINT. Moreover, since we measure h and NMI on the entire collection, the number of articles identified as threads (e.g. mLM-NPC: 74.67% articles) is an important factor in HINT's effectiveness compared to existing methods. For example, EventX identified only 16.58% articles as threads, which affects its overall effectiveness. To investigate this, we evaluate EventX and HINT using only the NewSHead articles that are identified as threads (16.58% & 74.67% respectively). Even for this criteria, HINT outperforms EventX (e.g. 0.927 vs 0.883 NMI). We further observe that the number of threads identified are markedly higher for HINT (e.g. mLM-NPC: 18,340) compared to k-SDPP (4,599), EventX (7,149), and SeqINT (13,690). Furthermore, we observe that the mean number of articles per thread (mean $|\mathbb{T}|$) for HINT (4.59) is the closest to the true threads (4.07) in NewSHead. Therefore, for RQ1, we conclude that HINT is indeed effective for generating quality information threads compared to existing document threading (k-SDPP) and event extraction (EventX) approaches as well as cluster-based information threading (SeqINT).

Moving on to **RQ2**, from Table 1 we observe that the Louvain and Leiden configurations of HINT are the least effective. Upon further investigations, we found that these general community detection methods identify comparatively larger communities than NPC, which can affect the coherence of the generated threads. Therefore, the candidate selection component in HINT when using Louvain or Leiden selects a very small number of threads (e.g., mLM-Louvain: 20, mLM-Leiden: 17, compared to mLM-NPC: 18,340). Therefore, in response to RQ2, we conclude that our proposed NPC is the most suitable method to identify the strongly connected communities for effective thread generation.

Table 1. Results for the Thread Quality of the evaluated approaches. (True #articles=112,794; #threads=27,681 and mean $|\mathbb{T}|$=4.07).

Configuration	h	NMI	#Articles	#Threads	mean $|\mathbb{T}|$
K-SDPP	0.107	0.190	13,076	4,599	2.84
EventX	0.141	0.240	18,698	7,149	2.62
SeqINT$_{mLM}$	0.592	0.724	69,430	13,690	5.07
SeqINT$_{dRoB}$	0.541	0.684	63,336	12,522	5.06
HINT$_{mLM-Louvain}$	0.001	0.003	207	20	10.35
HINT$_{dRoB-Louvain}$	0.001	0.003	202	15	13.47
HINT$_{mLM-Leiden}$	0.001	0.001	78	17	**4.59**
HINT$_{dRoB-Leiden}$	0.001	0.001	69	14	4.93
HINT$_{mLM-NPC}$	**0.706**	**0.797**	**84,228**	**18,340**	**4.59**
HINT$_{dRoB-NPC}$	0.686	0.783	81,770	17,819	4.59

5.2   Ablation Study

We now present an analysis of the effectiveness of different components of HINT.
• **Effect of Time-Decay and Entity Similarity**: We first analyse the effectiveness of the time-decay and entity similarity scores to compute the weights of the edges in the Document Graph ($\mathcal{D}$). In particular, we evaluate HINT in two additional settings to compute the edge weights: (1) cosine similarity of the 5W1H pseudo-passages (i.e., by setting $td(x,y) = es(x,y) = 1$ in Equation (3)), and (2) cosine similarity and time-decay (TD) (i.e., $es(x,y) = 1$). From Table 2, we observe that our proposed configuration to compute the edge weights with both time-decay and entity similarity (e.g. mLM-TD-ENT: 0.797 NMI) outperforms other configurations that include only cosine similarity (e.g. mLM: 0.759 NMI) or cosine and time-decay similarity (e.g. mLM-TD: 0.796 NMI). However, we also observe that the improvements from including the time-decay similarity are larger compared to including both time-decay and entity similarity. As future work, we plan to investigate whether including an entity recognition component in addition to the 5W1H extraction can further improve the thread quality.

Table 2. Effect of Time-Decay and Entity Similarity on thread quality.

Configuration	h	NMI
mLM	0.657	0.759
dRoB	0.642	0.747
mLM-TD	0.705	0.796
dRoB-TD	0.686	0.783
mLM-TD-ENT	**0.706**	**0.797**
dRoB-TD-ENT	0.686	0.783

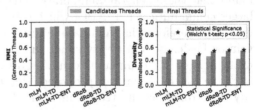

Fig. 4. Effect of Candidate Selection on NMI and Diversity.

• **Effect of Candidate Thread Selection**: We also analyse the effect of the candidate selection on the quality and diversity of the generated threads. We use KL divergence [10] to measure the threads' diversity of information. For a given thread, we hold-out each document from the thread and compute the KL divergence between the probability distributions of the words in the held-out document and the words in the rest of the documents in the thread. Since in this analysis we are focused on the quality of the generated threads, we compute h and NMI using only the articles that are identified as part of the *generated* threads.

Figure 4 shows the thread quality (NMI) and information diversity of the candidate threads identified by the NPC and the final output threads from the candidate selection component. We first observe that the quality of the candidate threads and the final threads are comparable. However, the final threads are significantly (Welch's t-test; $p < 0.05$) more diverse than the candidate threads. Therefore, our proposed candidate selection component can effectively select quality information threads that describe diverse information about an event.

6   User Study

As we described in Sect. 3, our proposed HINT approach captures hierarchical links between documents. These hierarchical links can present chronological hierarchies and a logical division of diverse information, e.g. different stories that are each related to the same event. However, unlike HINT's hierarchical threads, sequential threads (such as from SeqINT) may not be able to capture such logical division of diverse information. Therefore, it is important to know which of these presentation strategies (i.e., hierarchical or sequential) is preferred by users. We conducted a user study that evaluates whether hierarchical information threads are more descriptive and more interpretable to users than sequential threads. In particular, we compared HINT with the best performing baseline from our offline evaluation, i.e., SeqINT. We selected the best configurations of HINT and SeqINT from our offline evaluation (i.e., $\text{HINT}_{\text{mLM-NPC}}$ & $\text{SeqINT}_{\text{mLM}}$).

Our user study aims to answer the following two research questions:

• **RQ3**: Do users prefer the hierarchical threads that are generated by HINT compared to the cluster-based sequential SeqINT threads?
• **RQ4**: Do the hierarchical links between articles in HINT threads effectively present a logical division of diverse information about an event?

Experiment Design: We follow a within-subject design for our user study, i.e., we perform a pairwise evaluation of the threads generated by the HINT and SeqINT approaches. In other words, each user in the user study evaluates pairs of threads, where each pair of threads is about the same event, but the threads are generated from different threading approaches (i.e., HINT and SeqINT). When selecting the threads to present to the users, we select the best pairs of threads based on the threads' precision scores calculated over both of the threads in a pair, i.e., the ratio of the number of articles associated with a single true thread label to the total number of articles in a thread. In addition, we select threads that have exactly 4 articles based on the mean number of articles in the NewSHead thread labels. Overall, we selected 16 pairs of threads.We then distributed the selected pairs into 4 unique sets (i.e., 4 pairs per set), such that each of our study participants evaluates the pairs of threads from a particular set.

The user study participants were asked to select their preferred thread from each of the pairs with respect to each of the following criteria: (1) the description of an event in the thread, (2) the interpretability of the thread, (3) the structure of the thread, and (4) the explanation of the event's evolution in the thread. We also asked participants to rate each of the threads with respect to the thread's: (1) coherence, (2) diversity of information, and (3) chronology of the presented articles. We deployed a 4-point likert scale to capture the participants' ratings. The choice of a 4-point scale was based on the number of articles that we fix (i.e. 4) in each of the threads. Additionally, participants were asked to rate the HINT threads with respect to the logical division of the information in the branches of the thread (i.e., the logical hierarchies). The participants were presented with the title of the articles in each of the threads, as illustrated in the example in Fig. 1.

Participant Recruitment: We recruited 32 participants using the MTurk (www.mturk.com) crowdsourcing platform. The recruited participants were all 18+ years of age and from countries where English is their first language. From the 32 participants, we first assigned 8 participants to each of the 4 sets of thread pairs. We further created 4 participant groups for each of the sets (i.e., 2 participants per group-set combination), using balanced Latin square counterbalancing by permuting the 4 pairs of threads in each set.

6.1   Results

Figure 5 shows our user study's results in terms of the participants' preferences and ratings. We use the chi-square goodness-of-fit test to measure statistical significance between the participants' preferring the HINT or SeqINT threads, as shown in Table 3. We also use a paired-samples t-test to measure the statistical significance between the participants' ratings for HINT and SeqINT (Table 4).

First, addressing **RQ3**, from Fig. 5(a) and Table 3, we observe that participants significantly (chi-square test; $p < 0.05$) prefer our proposed HINT approach compared to SeqINT, for all four criteria, i.e. description, interpretability, structure and evolution. Further, from Fig. 5(b), we observe that the participants rate the HINT threads higher for all of the three criteria, i.e., coherence, diversity and chronology. Moreover, as shown in Table 4, the participants' ratings for HINT are significantly (t-test; $p < 0.05$) higher with respect to diversity and chronology. However, the improvement in coherence ratings for HINT is not significant compared to SeqINT. This shows that both HINT and SeqINT threads can identify related articles about an event. However, HINT threads provide significantly more diverse information about the event, as shown in Fig. 5(b). Overall, for RQ3, we conclude that the participants significantly preferred hierarchical HINT threads over the sequential SeqINT threads. Moreover, the participants' ratings show that the HINT threads provide significantly more diverse and chronologically correct information about an event than the SeqINT threads.

Table 3. Chi-square goodness-of-fit test results.

Table 4. Paired Samples t-test results.

Criteria	$\chi^2(1)$	Cohen's w	p	Power
Description	13.781	0.328	< 0.001	96.00%
Interpretability	15.125	0.344	< 0.001	97.33%
Structure	11.281	0.297	0.001	91.93%
Evolution	12.500	0.313	< 0.001	94.30%

Criteria	Cohen's d	p	Power
Coherence	0.117	0.187	25.96%
Diversity	0.294	0.001	91.08%
Chronology	0.251	0.005	80.46%

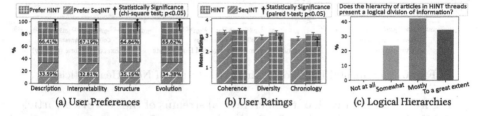

(a) User Preferences   (b) User Ratings   (c) Logical Hierarchies

Fig. 5. Pairwise participants' preferences and ratings of the threading methods.

Moving on to **RQ4**, Fig. 5(c) shows the participants' ratings for the logical division of information by the different hierarchies in the HINT threads. From Fig. 5(c), we observe that the majority of participants (44%) said that the hierarchies in the HINT threads are *mostly* logical. Moreover, none of the participants said that the hierarchies in the HINT threads are *not at all* logical. Therefore, for RQ4, we conclude that HINT threads present a logical presentation of diverse information (i.e. distinct stories) about an event through the hierarchical association between related articles.

Overall, our user study shows that HINT's hierarchical threads are significantly preferred by users compared to sequential threads. Moreover, the study shows that HINT can effectively present a logical hierarchical view of aspects (e.g. stories) about the evolution of an event.

7 Identifying Incremental Threads

We now present an analysis on the scalability of the HINT's architecture. This analysis focuses on the overall efficiency of HINT's novel components, i.e., the document-entity graph ($\mathcal{E}$), document graph ($\mathcal{D}$), NPC, and candidate selection.

We deploy HINT to generate threads incrementally by simulating a chronological stream of NewSHead articles. NewSHead articles were published between May 2018 and May 2019, i.e., over a period of 394 days [7]. We first generate threads from the articles that were published in the first 30 days in the collection (i.e., historical run). From the historical run, we store the NPC communities as a single graph of hierarchically connected document nodes ($\mathcal{D}'$), as illustrated in Fig. 3(c). We then simulate three incremental article streams such that, in each stream, documents from different sequential time intervals are input to HINT to be added to existing threads or generate new threads, i.e., daily (every 1 day), weekly (every 7 days), and monthly (every 30 days). For each incremental run, we extend the document graph $\mathcal{D}'$ by computing the similarity between the new articles in the stream and the existing articles in $\mathcal{D}'$ using Equation (3). We then perform community detection on $\mathcal{D}'$ using NPC, followed by candidate selection of the newly identified or extended threads.

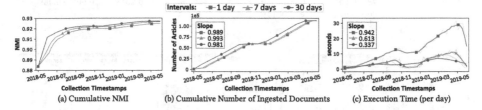

(a) Cumulative NMI   (b) Cumulative Number of Ingested Documents   (c) Execution Time (per day)

Fig. 6. Incremental HINT on a simulated stream of NewSHead articles.

Figure 6 shows, for each of the incremental streams of the NewSHead articles, the NMI of the generated threads, the total number of ingested documents and HINT's execution time. From Fig. 6(a), we observe that the quality (NMI) of the HINT threads quickly increases during the initial 2 months of the incremental runs (i.e., between May and July 2018) and remains comparable thereafter. This shows that HINT is still effective when there are only a small number of articles. Furthermore, Figs. 6(b) and (c) show that there is a linear increase in the execution time of HINT as the number of ingested articles increases. Most importantly, we observe that the rate of increase in HINT's execution time is slower than the increase in ingested articles (e.g. 0.981 slope as the number of monthly ingested documents increases vs 0.337 slope for the execution time in seconds). Additionally from Fig. 6(c), we observe that the rate of increase in the daily execution times is the highest, followed by the weekly and monthly execution times. This suggests that the time taken for incremental executions of HINT can be reduced by increasing the frequency of days between the incremental executions.

Overall, this analysis shows that HINT can effectively and efficiently identify threads in a dynamic collection. Moreover, HINT's architecture is scalable, as the rate of increase in HINT's execution time is slower compared to the increase in the number of ingested articles (Figs. 6(b) and (c)).

8   Conclusions

We proposed a novel unsupervised approach, HINT, for hierarchical information threading. The hierarchical threads generated by HINT can help users to easily interpret evolving information about stories related to an event, activity or discussion. Our offline evaluation showed that HINT can effectively generate quality information threads compared to approaches from the literature. In addition, our user study showed that HINT's hierarchical information threads are significantly preferred by users compared to cluster-based sequential threads. Moreover, with its scalable network community-based architecture, HINT can efficiently identify threads in a dynamic collection to capture and track evolving information.

References

1. Allan, J.: Topic detection and tracking: event-based information organization, The Information Retrieval Series, vol. 12. Springer, US (2012). https://doi.org/10.1007/978-1-4615-0933-2
2. Allan, J., Carbonell, J.G., Doddington, G., Yamron, J., Yang, Y.: Topic detection and tracking pilot study final report. In: Proceedings of the DARPA Broadcast News Transcription and Understanding Workshop (1998)
3. Blondel, V.D., Guillaume, J.L., Lambiotte, R., Lefebvre, E.: Fast unfolding of communities in large networks. J. Statist. Mech. Theory Exper. 2008(10), P10008 (2008)
4. Cai, D., He, X., Han, J.: Document clustering using locality preserving indexing. IEEE Trans. Knowl. Data Eng. **17**(12), 1624–1637 (2005)
5. Fan, W., Guo, Z., Bouguila, N., Hou, W.: Clustering-based online news topic detection and tracking through hierarchical Bayesian nonparametric models. In: Proceedings of the 44th International ACM SIGIR Conference on Research and Development in Information Retrieval (2021)
6. Gillenwater, J., Kulesza, A., Taskar, B.: Discovering diverse and salient threads in document collections. In: Proceedings of the 2012 Joint Conference on Empirical Methods in Natural Language Processing and Computational Natural Language Learning (2012)
7. Gu, : Generating representative headlines for news stories. In: Proceedings of The Web Conference (2020)
8. Hamborg, F., Breitinger, C., Gipp, B.: Giveme5W1H: a universal system for extracting main events from news articles. In: Proceedings of the 13th ACM Conference on Recommender Systems, 7th International Workshop on News Recommendation and Analytics (2019)
9. Kulesza, A., Taskar, B.: Structured determinantal point processes. In: Proceedings of the Advances in Neural Information Processing Systems (2010)
10. Kullback, S., Leibler, R.A.: On information and sufficiency. Annal. Math. Statist. **22**(1), 79–86 (1951)
11. Lang, K.: NewsWeeder: Learning to filter Netnews. In: Proceedings of the 12th International Conference on Machine Learning (1995)
12. Liu, B., Han, F.X., Niu, D., Kong, L., Lai, K., Xu, Y.: Story forest: extracting events and telling stories from breaking news. ACM Trans. Knowl. Discov. Data **14**(3), 31 (2020)
13. Nallapati, R., Feng, A., Peng, F., Allan, J.: Event threading within news topics. In: Proceedings of the 13th ACM International Conference on Information and Knowledge Management (2004)
14. Narvala, H., McDonald, G., Ounis, I.: Identifying chronological and coherent information threads using 5W1H questions and temporal relationships. Inf. Process. Manage. **60**(3), 103274 (2023)
15. Reimers, N., Gurevych, I.: Sentence-BERT: sentence embeddings using Siamese BERT-networks. In: Proceedings of the 2019 Conference on Empirical Methods in Natural Language Processing and the 9th International Joint Conference on Natural Language Processing (2019)
16. Röder, M., Both, A., Hinneburg, A.: Exploring the space of topic coherence measures. In: Proceedings of the 8th ACM International Conference on Web Search and Data Mining (2015)

17. Rosenberg, A., Hirschberg, J.: V-Measure: A conditional entropy-based external cluster evaluation measure. In: Proceedings of the 2007 Joint Conference on Empirical Methods in Natural Language Processing and Computational Natural Language Learning (2007)
18. Saravanakumar, K.K., Ballesteros, M., Chandrasekaran, M.K., McKeown, K.: Event-Driven news stream clustering using entity-aware contextual embeddings. In: Proceedings of the 16th Conference of the European Chapter of the Association for Computational Linguistics: Main Volume (2021)
19. Shahaf, D., Guestrin, C.: Connecting two (or less) dots: Discovering structure in news articles. ACM Trans. Knowl. Discov. Data 5(4), 24 (2012)
20. Shahaf, D., Yang, J., Suen, C., Jacobs, J., Wang, H., Leskovec, J.: Information cartography: Creating zoomable, large-scale maps of information. In: Proceedings of the 19th ACM SIGKDD International Conference on Knowledge Discovery and Data Mining (2013)
21. Traag, V.A., Waltman, L., Van Eck, N.J.: From Louvain to Leiden: guaranteeing well-connected communities. Sci. Rep. 9, 5233 (2019)
22. Tukey, J.W., et al.: Exploratory data analysis, vol. 2. Reading, MA (1977)
23. Vaswani, A., et al.: Attention is all you need. In: Proceedings of Advances in Neural Information Processing Systems (2017)
24. Zong, C., Xia, R., Zhang, J.: Topic detection and tracking. In: Text Data Mining, pp. 201–225. Springer, Singapore (2021). https://doi.org/10.1007/978-981-16-0100-2_9

HADA: A Graph-Based Amalgamation Framework in Image-Text Retrieval

Manh-Duy Nguyen[1](✉) [iD], Binh T. Nguyen[2,3,4] [iD], and Cathal Gurrin[1] [iD]

[1] School of Computing, Dublin City University, Dublin, Ireland
manh.nguyen5@mail.dcu.ie
[2] VNU-HCM, University of Science, Ho Chi Minh City, Vietnam
[3] Vietnam National University, Ho Chi Minh City, Vietnam
[4] AISIA Lab, Ho Chi Minh City, Vietnam

Abstract. Many models have been proposed for vision and language tasks, especially the image-text retrieval task. State-of-the-art (SOTA) models in this challenge contain hundreds of millions of parameters. They also were pretrained on large external datasets that have been proven to significantly improve overall performance. However, it is not easy to propose a new model with a novel architecture and intensively train it on a massive dataset with many GPUs to surpass many SOTA models already available to use on the Internet. In this paper, we propose a compact graph-based framework named HADA, which can combine pretrained models to produce a better result rather than starting from scratch. Firstly, we created a graph structure in which the nodes were the features extracted from the pretrained models and the edges connecting them. The graph structure was employed to capture and fuse the information from every pretrained model. Then a graph neural network was applied to update the connection between the nodes to get the representative embedding vector for an image and text. Finally, we employed cosine similarity to match images with their relevant texts and vice versa to ensure a low inference time. Our experiments show that, although HADA contained a tiny number of trainable parameters, it could increase baseline performance by more than 3.6% in terms of evaluation metrics on the Flickr30k dataset. Additionally, the proposed model did not train on any external dataset and only required a single GPU to train due to the small number of parameters required. The source code is available at https://github.com/m2man/HADA.

Keywords: Image-text retrieval · Graph neural network · Fusion model

1 Introduction

Image-text retrieval is one of the most popular challenges in vision and language tasks, with many state-of-the-art (SOTA) models recently introduced [3,10,17–19,25,28]. This challenge includes two subtasks, which are image-to-text retrieval and text-to-image retrieval. The former subtask utilises an image

© The Author(s), under exclusive license to Springer Nature Switzerland AG 2023
J. Kamps et al. (Eds.): ECIR 2023, LNCS 13980, pp. 717–731, 2023.
https://doi.org/10.1007/978-3-031-28244-7_45

query to retrieve relevant texts in a multimodal dataset, while the latter is concerned with text queries for ranked videos.

Most of the SOTA models in this research field share two things in common: (1) they were built on transformer-based cross-modality attention architectures [3,19] and (2) they were pretrained on the large-scale multimodal data crawled from the Internet [13,17–19,28]. However, these things have their own disadvantages. The attention structure between two modalities could achieve an accurate result, but it costs a large amount of inference time due to the massive computation required. For instance, UNITER [3] contained roughly 303 million parameters, and it took a decent amount of time (more than 12 s for each query on a dataset with 30000 images [31]) to perform the retrieval in real-time. Many recent works have resolved this model-related problem by introducing joint-encoding learning methods. They can learn visual and semantic information from both modalities without using any cross-attention modules, which can be applied later to rerank the initial result [18,25,31]. Figure 1 illustrates the architecture of these pipelines. Regarding the data perspective, the large collected data usually comes with noisy annotations, which could impact on to the models trained on it. Several techniques have been proposed to mitigate this issue [17–19]. However, training on a massive dataset still burdens computation, such as the number of GPUs required to train the model successfully and efficiently [28].

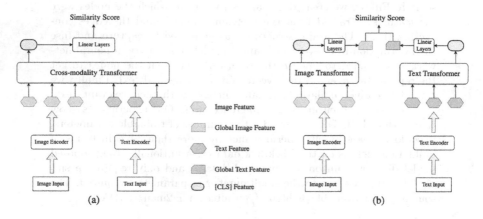

Fig. 1. Two most popular pipelines of the SOTA for image-text retrieval challenge. (a) A cross-modality transformer network is applied to measure the similarity between an image and a text based on their features. (b) Each modality used their own transformer network to get its global embedding.

It has motivated us to answer the question: *Can we combine many SOTA models, which are currently available to use, to get a better-unified model without intensive training using many GPUs?* In this paper, we introduce a graph-based amalgamation framework, called **HADA**, which utilises a graph-based structure to fuse the features produced by other pretrained models. We did not use any time-consuming cross-modality attention network to ensure fast retrieval speed. A graph neural network was employed to extract visual and textual embedded vectors from fused graph-based structures of images and texts, where we can measure their cosine similarity. To the best of our knowledge, the graph structure has been widely applied in the image-text retrieval challenge [7,21,26,27,35]. Nevertheless, it was utilized to capture the interaction between objects or align local and global information within images. HADA is the first approach that applies this data structure to combine SOTA pretrained models by fusing their features in each modality. We trained HADA on the Flickr30k dataset without using any large-scale datasets. Then, we applied the Momentum Distillation technique [18], which has been shown to mitigate not only the harmful effect of noise annotation, but also improve accuracy on a clean dataset. Our experiments showed that HADA, with the tiny extra number of training parameters, could improve total recall by 3.64% compared to the input SOTA, without training with millions of additional image-text pairs as other models require. This is the most crucial contribution since it is expensive to utilise multiple GPU, especially for small and medium businesses or start-up companies. Therefore, we believe that HADA can be applied in both academic and industrial domains.

Our main contribution can be summarised as follows: (1) We introduced HADA, a compact pipeline that can combine two or many SOTA pretrained models to address the image-text retrieval challenge. (2) We proposed a way to fuse the information between input pretrained models by using graph structures. (3) We evaluated the performance of HADA on the well-known Flickr30k dataset [37] and MSCOCO dataset [20] without using any other large-scale dataset but still improved the accuracy compared to the baseline input models.

2   Related Work

A typical vision-and-language model, including an image-text retrieval task, was built using transformer-based encoders. In specific, OSCAR [19], UNITER [3], and VILLA [10] firstly employed Faster-RCNN [29], and BERT [6] to extract visual and text features from images and texts. These features were then fed into a cross-modality transformer block to learn the contextualized embedding that captured the relations between regional features from images and word pieces from texts. An additional fully connected layer was used to classify whether the images and texts were relevant to each other or not based on the embedding vectors. Although achieving superior results, these approaches had a drawback in applying them to real-time use cases. It required a huge amount of time to perform the online retrieval, since models had to process the intensive cross-attention transformer architecture many times for each query [31].

Recently, some works have proposed an approach to resolve that problem by utilizing two distinct encoders for images and text. Data from each modality can now be embedded offline and hence improve retrieval speed [13,17,18,25,28,31]. In terms of architecture, all approaches used the similar BERT-based encoder for semantic data but different image encoders. While LightningDOT [31] encoded images with detected objects extracted by the Faster-RCNN model, FastnSlow [25] applied the conventional Resnet network to embed images. On the other side, ALBEF [18] and BLIP [17] employed the Vision Transformer backbone [8] to get the visual features corresponding to their patches. Because these SOTA did not use the cross-attention structure, which was a critical point to achieve high accuracy, they applied different strategies to increase performance. Specifically, pretraining a model on a large dataset can significantly improve the result [13, 18,19]. For instance, CLIP [28] and ALIGN [13] were pretrained on 400 million and 1.8 billion image-text pairs, respectively. Another way was that they ran another cross-modality image-text retrieval model to rerank the initial output and get a more accurate result [18,31].

Regarding graph structures, SGM [35] introduced a visual graph encoder and a textual graph encoder to capture the interaction between objects appearing in images and between the entities in text. LGSGM [26] proposed a graph embedding network on top of SGM to learn both local and global information about the graphs. Similarly, GSMN [21] presented a novel technique to assess the correspondence of nodes and edges of graphs extracted from images and texts separately. SGRAF [7] built a reasoning and filtration graph network to refine and remove irrelevant interactions between objects in both modalities.

Although there are many SOTAs with different approaches for image-text retrieval problems, there is no work that tries combining these models, rather they introduce a new architecture and pretrain on massive datasets instead. Training an entirely new model from scratch on the dataset is a challenging task since it will create a burden on the computation facilities such as GPUs. In this paper, we introduced a simple method that combined the features extracted from the pretrained SOTA by applying graph structures. Unlike other methods that also used this data structure, we employed graphs to fuse the information between the input features, which was then fed into a conventional graph neural network to obtain the embedding for each modality. Our HADA consisted of a small number of trainable parameters, hence can be easily trained on a small dataset but still obtained higher results than the input models.

3   Methodology

This section will describe how our HADA addressed the retrieval challenge by combining any available pretrained models. Figure 2 depicted the workflow of HADA. We started with only two models ($N_{models} = 2$) as illustrated in Fig. 2 for simplicity. Nevertheless, HADA can be extended with a larger N_{models}. HADA began using some pretrained models to extract the features from each modality. We then built a graph structure to connect the extracted features together, which were fed into a graph neural network (GNN) later to update them. The outputs of the GNN were concatenated with the original global features produced by the pretrained models. Finally, simple linear layers were employed to get the final representation embedding features for images and texts, which can be used to measure similarity to perform the retrieval. For evaluation, we could extract our representation features offline to guarantee high-speed inference time.

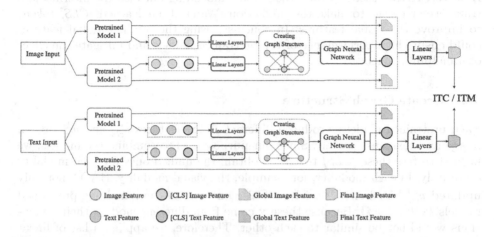

Fig. 2. The pipeline of the proposed HADA. The red borders indicated trainable components. The ITM and ITC infered the training tasks which will be discussed later. (Color figure online)

3.1 Revisit State-of-the-Art Models

We only used the pretrained models without using the cross-modality transformer structure to extract features as depicted in Fig. 1 in order to reduce the number of computations and ensure the high-speed inference time. Basically, they used a unimodal encoder to get the features of an image or a text followed by a transformer network to embed them and obtain the *[CLS]* embedding. This *[CLS]* token was updated by one or many fully connected layers to become a representative global feature that can be compared with that of the remaining modality to get the similarity score.

HADA began with the output of the transformer·layer from the pretrained models. In detail, for an input image I, we obtained the sequence of patch tokens from each model i denoted as $\mathbf{v}^{(i)} = \{v_{cls}^{(i)}, v_1^{(i)}, v_2^{(i)}, ..., v_{N_i}^{(i)}\}$, where $v_j^{(i)} \in \mathbb{R}^{d_v^{(i)}}$ and N_i was the length of the sequence. This length depended on the architecture of the image encoder network employed in the pretrained model. For example, it could be the number of patches if the image encoder was a Vision Transformer (ViT) network [8], or the number of detected objects or regions of interest if the encoder was a Faster-RCNN model [29]. Additionally, we also extracted the global visual representation feature $h_v^{(i)} \in \mathbb{R}^{d_h^{(i)}}$ from $v_{cls}^{(i)}$ as illustrated in Fig. 1. Regarding the semantic modality, we used the same process as that of the visual modality. Specifically, we extracted the sequence of patch tokens $\mathbf{w}^{(i)} = \{w_{cls}^{(i)}, w_1^{(i)}, w_2^{(i)}, ..., w_L^{(i)}\}$ where $w_j^{(i)} \in \mathbb{R}^{d_w^{(i)}}$ and L was the length of the text, and the global textual representation embedding $h_w^{(i)} \in \mathbb{R}^{d_h^{(i)}}$ for an input text T using the pretrained model i. The input model i matched a pair of an image I and a text T by calculating the dot product $\langle h_v^{(i)}, h_w^{(i)} \rangle$ of their global features. However, HADA used not only the global embedding but also the intermediate transformer tokens to make the prediction. We used our learned *[CLS]* tokens to improve the global features. In contrast, using the original global features could ensure high performance of the pretrained models and mitigate the effect of unhelpful tokens.

3.2   Create Graph Structure

Each pretrained model i produced different *[CLS]* features $v_{cls}^{(i)}$ and $w_{cls}^{(i)}$ for an image and text, respectively. Since our purpose was to combine the models, we needed to fuse these *[CLS]* tokens to obtain the unified ones for each modality separately. In each modality, for example, the visual modality, HADA not only updated $v_{cls}^{(i)}$ based on $\mathbf{v}^{(i)}$ solely but also on those of the remaining pretrained models $\{\mathbf{v}^{(j)} \mid j \neq i\}$. Because these $\mathbf{v}$ came from different models, their dimensions would not be similar to each other. Therefore, we applied a list of linear layers $f_v^{(i)} : \mathbb{R}^{d_v^{(i)}} \to \mathbb{R}^{d_p}$ to map them in the same dimensional space:

$$\mathbf{p}^{(i)} = \{f_v^{(i)}(x) | x \in \mathbf{v}^{(i)}\} = \{p_{cls}^{(i)}, p_1^{(i)}, p_2^{(i)}, ..., p_{N_i}^{(i)}\}$$

We performed a similar process for the textual modality to obtain:

$$\mathbf{s}^{(i)} = \{f_w^{(i)}(x) | x \in \mathbf{w}^{(i)}\} = \{s_{cls}^{(i)}, s_1^{(i)}, s_2^{(i)}, ..., s_L^{(i)}\}, \text{where } f_w^{(i)} : \mathbb{R}^{d_w^{(i)}} \to \mathbb{R}^{d_s}$$

We then used graph structures $\mathcal{G}p = \{\mathcal{V}_p, \mathcal{E}_p\}$ and $\mathcal{G}_s = \{\mathcal{V}_s, \mathcal{E}_s\}$ to connect these mapped features together, where $\mathcal{V}$ and $\mathcal{E}$ denoted the list of nodes and edges in the graph $\mathcal{G}$ accordingly. In our HADA, nodes indicated the mapped features. Specifically, $\mathcal{V}_p = \{\mathbf{p}^{(i)}\}$ and $\mathcal{V}_s = \{\mathbf{s}^{(i)}\}$ for all $i \in [1, N_{models}]$. Regarding edges, we symbolized $e_{a \to b}$ as a directed edge from node a to node b in the graph, thus the set of edges of the visual graph $\mathcal{E}_p$ and the textual graph $\mathcal{E}_s$ were:

$$\mathcal{E}_p = \{e_{x \to p_{cls}^{(j)}} \mid x \in \mathbf{p}^{(i)} \text{ and } i, j \in [1, N_{models}]\}$$

$$\mathcal{E}_s = \{e_{x \to s_{cls}^{(j)}} \mid x \in \mathbf{s}^{(i)} \text{ and } i, j \in [1, N_{models}]\}$$

To be more detailed, we created directed edges that went from every patch feature to the *[CLS]* feature, including from the *[CLS]* itself, for all pretrained models but not in the reverse direction, as shown in Fig. 2. The reason was that *[CLS]* was originally introduced as a representation of all input data, so it would summarize all patch tokens [2,6,8]. Therefore, it would be the node that received information from other nodes in the graph. This connection structure ensured that HADA could update the *[CLS]* tokens based on the patch tokens from all pretrained models in a fine-grained manner.

3.3   Graph Neural Network

Graph neural networks (GNN) have witnessed an increase in popularity over the past few years, with many GNN structures having been introduced recently [1,5, 11,15,30,34]. HADA applied the modified Graph Attention Network (GATv2), which was recommended to be used as a baseline whenever employing a GNN [1], to fuse the patch features from different pretrained models together to get the unified [CLS] features. Let $\mathcal{N}_k = \{x \in \mathcal{V} \mid e_{x \to k} \in \mathcal{E}\}$ be the set of neighbor nodes from which there was an edge connecting to node k in the graph $\mathcal{G}$. GATv2 used a scoring function se to weight every edge indicating the importance of the neighbor nodes x in $\mathcal{N}_k$ before updating the node $k \in \mathbb{R}^d$:

$$se(e_{x \to k}) = \mathbf{A}^\top \text{LeakyRELU}(\mathbf{W}_1 x + \mathbf{W}_2 k])$$

where $\mathbf{A} \in \mathbb{R}^{d'}$, $\mathbf{W}_1 \in \mathbb{R}^{d' \times d}$, and $\mathbf{W}_2 \in \mathbb{R}^{d' \times d}$ were learnable parameters. These weights were then normalized across all neighbor nodes in $\mathcal{N}_k$ by using a softmax function to get the attention scores:

$$\alpha_{e_{x \to k}} = \frac{\exp(se(e_{x \to k}))}{\sum_{y \in \mathcal{N}_k} \exp(se(e_{y \to k}))}$$

The updated node $k' \in \mathbb{R}^{d'}$ was then calculated based on its neighbors in $\mathcal{N}_k$, including k if we add an edge connect it to itself:

$$k' = \sigma \Big(\sum_{x \in \mathcal{N}_k} \alpha_{e_{x \to k}} \cdot \mathbf{W}_1 x \Big),$$

where σ was a nonlinear activate function. Furthermore, this GATv2 network could be enlarged by applying a multi-head attention structure, and improved performance [34]. The output now was a concatenation of each head output, which was similar to Transformer architecture [33]. An extra linear layer was used at the end to convert these concatenated nodes to the desired dimensions.

We used distinct GATv2 structures with H attention heads for each modality in this stage, as illustrated in Fig. 2. HADA took the input graphs $\mathcal{G}_p$ and $\mathcal{G}_s$ with nodes $\mathcal{V}_p$ and $\mathcal{V}_s$ in the vector space of d_p and d_s dimensions and updated them to $\mathcal{V'}_p = \{\mathbf{p}'^{(i)}\}$ and $\mathcal{V'}_s = \{\mathbf{s}'^{(i)}\}$ with dimensions of d'_p and d'_s. We then concatenated the updated [CLS] nodes p'_{cls} and s'_{cls} from all pretrained models with their corresponding original global embedding h_v and h_w. Finally, we fed them into a list of linear layers to get our normalized global representation $h_p \in \mathbb{R}^{d_h}$ and $h_s \in \mathbb{R}^{d_h}$.

3.4   Training Tasks

Image-Text Contrastive Learning. HADA encoded the input image I and text T to h_p and h_s, accordingly. We used a similarity function that was a dot product $S(I, T) = \langle h_p, h_s \rangle = h_p^\top h_s$ to ensure that a pair of relevant image-text (positive pair) would have a higher similar representation compared to irrelevant pairs (negative pairs). The contrastive loss for image-to-text (i2t) retrieval and text-to-image (t2i) retrieval for the mini-batch of M relevant pairs $(\boldsymbol{I}_m, \boldsymbol{T}_m)$ were:

$$\mathcal{L}_{i2t}(\boldsymbol{I}_m) = -\log\frac{\exp(S(\boldsymbol{I}_m, \boldsymbol{T}_m)/\tau)}{\sum_{i=1}^{M} \exp(S(\boldsymbol{I}_m, \boldsymbol{T}_i)/\tau)}$$

$$\mathcal{L}_{t2i}(\boldsymbol{T}_m) = -\log\frac{\exp(S(\boldsymbol{T}_m, \boldsymbol{I}_m)/\tau)}{\sum_{i=1}^{M} \exp(S(\boldsymbol{T}_m, \boldsymbol{I}_i)/\tau)}$$

where τ was a temperature parameter that could be learned during training. Such contrastive learning has been used in many vision-and-language models and has been proven to be effective [17,18,28,31]. In our experiment, we trained HADA with the loss that optimized both subtasks:

$$\mathcal{L}_{ITC} = \frac{1}{M}\sum_{m=1}^{M}(\mathcal{L}_{i2t}(\boldsymbol{I}_m) + \mathcal{L}_{t2i}(\boldsymbol{T}_m))$$

Inspired by ALBEF [18], we also applied momentum contrast (MoCo) [12] and their momentum distillation strategy for this unsupervised representation learning to cope with the problem of noisy information in the dataset and improve accuracy.

Image-Text Matching. This objective was a binary classification task to distinguish irrelevant image-text pairs that had similar representations. This task would ensure that they were different in fine-grained details. We implemented an additional disciminator layer $dc : \mathbb{R}^{4d_h} \rightarrow \mathbb{R}$ on top of the final embedding features h_p and h_s to classify whether the image $\boldsymbol{I}$ and the text $\boldsymbol{T}$ is a positive pair or not:

$$dc(h_p, h_s) = \text{sigmoid}(\mathbf{W}^\top[h_p\|h_s\|\text{abs}(h_p - h_s)\|h_p \odot h_s])$$

where $\mathbf{W} \in \mathbb{R}^{4d_h}$ was trainable parameters, $\|$ indicated the concatenation, abs(.) was the absolute value, and $\odot$ denoted elementwise multiplication. We used binary cross-entropy loss for this ordinary classification task:

$$\mathcal{L}_{itm}(\boldsymbol{I}, \boldsymbol{T}) = y\log(dc(h_p, d_s)) + (1 - y)\log(1 - dc(h_p, d_s))$$

where y was the one-hot vector representing the ground truth label of the pair.

For each positive pair in the minibatch of M positive pairs, we sampled 1 hard negative text for the image and 1 hard negative image for the text. These negative samples were chosen from the current mini-batch in which they were not relevant based on the ground-truth labels, but have the highest similarity dot product score. Therefore, the objective for this task was:

$$\mathcal{L}_{ITM} = \frac{1}{3M}\sum_{m=1}^{M}(\mathcal{L}_{itm}(\boldsymbol{I}_m, \boldsymbol{T}_m) + \mathcal{L}_{itm}(\boldsymbol{I}_m, \boldsymbol{T}'_m) + \mathcal{L}_{itm}(\boldsymbol{I}'_m, \boldsymbol{T}_m))$$

where $\boldsymbol{T}'_m$ and $\boldsymbol{I}'_m$ were the hard negative text and image samples in the mini-batch that were corresponding with the $\boldsymbol{I}_m$ and $\boldsymbol{T}_m$, respectively. The final loss function in HADA was:

$$\mathcal{L} = \mathcal{L}_{ITC} + \mathcal{L}_{ITM}$$

4 Experiment

4.1 Dataset and Evaluation Metrics

We trained and evaluated HADA on two different common datasets in the image-text retrieval task, which are Flickr30k [37] and MSCOCO [20]. The Flickr30k dataset consists of 31K images collected on the Flickr website, while MSCOCO comprises 123K images. Each image contains five relevant texts or captions that describe the image. We used Karpathy's split [14], which has been widely applied by all models in the image-text retrieval task, to split each dataset into train/evaluate/test on 29K/1K/1K and 113K/5K/5K images on Flickr30k and MSCOCO, respectively.

The common evaluation metric in this task was the Recall at K ($R@K$) because many SOTA works used this metric [3,10,13,17–19,28,31]. This metric scores the proportion of the number of queries that we found the correct relevant output in the top K of the retrieved ranked list:

$$R@K = \frac{1}{N_q} \sum_{q=1}^{N_q} \mathbf{1}(q, K)$$

where N_q is the number of queries and $\mathbf{1}(q, K)$ is a binary function returning 1 if the model finds the correct answer of the query q in the top K of the retrieved output. In particular, for the image-to-text subtask, $R@K$ is the percentage of the number of images where we found relevant texts in the top K of the output result. In our experiment, we used R@1, R@5, R@10, and RSum, which was the sum of them.

4.2 Implementation Details

In our experiment, we combined two SOTA models that had available pretrained weights fine-tuned on the Flickr30k dataset: ALBEF[1] and LightningDOT[2]. None of them used the cross-modality transformer structure when retrieved to ensure the fast inference speed[3]. Although they used the same BERT architecture to encode a text, the former model employed the ViT network to encode an image, while the latter model applied the Faster-RCNN model. We chose these two

---

[1] https://github.com/salesforce/ALBEF.

[2] https://github.com/intersun/LightningDOT.

[3] Indeed, these two models applied the cross-modality transformer network to rerank the initial result in the subsequent step. However, we did not focus on this stage.

models because we wanted to combine different models with distinct embedding backbones to utilize the advantages of each of them.

Regarding ALBEF, their ViT network encoded an image to 577 patch tokens including the *[CLS]* one ($N_{ALB} = 576$ and $d_v^{(ALB)} = 768$). This *[CLS]* was projected to the lower dimension to obtain the global feature ($d_h^{(ALB)} = 256$). Because LightningDOT encoded an image based on the detected objects produced by the Faster-RCNN model, its N_{DOT} varied depending on the number of objects in the image. The graph neural network, unlike other conventional CNNs, can address this inconsistent number of inputs due to the flexible graph structure with nodes and edges. Unlike ALBEF, the dimensions of image features and global features from LightningDOT were the same with $d_v^{(DOT)} = d_h^{(DOT)} = 768$. In terms of text encoder, the output of both models was similar since they used the same BERT network: $d_w^{(ALB)} = d_w^{(DOT)} = 768$. We projected these features to a latent space where $d_p = d_s = 512$, which was the average of their original dimensions. We used a 1-layer GATv2 network with $H = 4$ multi-head attention to update the graph features while still keeping the input dimensions of $d_p' = d_s' = 512$. We also applied Dropout with $p = 0.7$ in linear layers and graph neural networks. In total, our HADA contained roughly 10M trainable parameters.

The input pretrained models were pretrained on several large external datasets. For example, ALBEF was pretrained on 14M images compared to only 29K images on Flickr30k that we used to train HADA. We used this advantage in our prediction instead of training HADA in millions of samples. We modified the similarity score to a weighted sum of our predictions and the original prediction of the input models. Therefore, the weighted similarity score that we used was:

$$S(\boldsymbol{I}, \boldsymbol{T}) = (1 - \alpha)\langle h_p, h_s \rangle + \alpha \langle h_v^{(ALB)}, h_w^{(ALB)} \rangle$$

where α was a trainable parameter. We did not include the original result of the LightningDOT model since its result was lower than ALBEF by a large margin and, therefore, could have a negative impact on overall performance[4].

We trained HADA for 50 epochs (early stopping[5] was implemented) using the batch size of 20 on one NVIDIA RTX3080Ti GPU. We used the AdamW [23] optimizer with a weight decay of 0.02. The learning rate was set at $1e^{-4}$ and decayed to $5e^{-6}$ following cosine annealing [22]. Similarly to ALBEF, we also applied RandAugment [4] for data augmentation. The initial temperature parameter was 0.07 [36], and we kept it in the range of $[0.001, 0.5]$ during training. To mitigate the dominant effect of ALBEF global features on our weighted similarity score, we first trained HADA with $\alpha = 0$. After the model had converged, we continued to train but initially set $\alpha = 0.5$ and kept it in the range of $[0.1, 0.9]$.

---

[4] We tried including the LightningDOT in the weighted similarity score, but the result was lower than using only ALBEF.

[5] In our experiment, it converged after roughly 20 epochs.

4.3 Baselines

We built two baselines that also integrated ALBEF and LightningDOT as input to show the advantages of using graph structures to fuse these input models.

Baseline B1. We calculated the average of the original ranking results obtained from ALBEF and LightningDOT and considered them as the distance between images and text. This meant that the relevant pairs should be ranked at the top, whilst irrelevant pairs would rank lower.

Baseline B2. Instead of using a graph structure to fuse the features extracted from the pretrained models, we only concatenated their global embedding and fed them into the last linear layers to obtain the unified features. We trained this baseline B2 following the same strategy as described in Sect. 4.2 using the weighted similarity score.

4.4 Comparison to Baseline

Table 1 illustrated the evaluation metrics of the different models in the Flickr30k dataset. Similarly to LightningDOT, our main target was to introduce an image-text retrieval model that did not implement a cross-modality transformer module to ensure that it can perform in real time without any delay. Thus, we only reported the result from LightningDOT and ALBEF that did not use the time-consuming compartment to rerank in the subsequent step. If the model has a better initial result, it can have a better-reranked result by using the cross-modality transformer later. We also added UNITER [3], and VILLA [10] to our comparison. These approaches both applied cross-modality transformer architecture.

Table 1. Performance of models on Flickr30k Dataset. The symbol † indicated the results were originally reported in their research, while others were from our re-implementation using their public pretrained checkpoints. The column ∆R showed the difference compared to ALBEF.

Methods	Image-to-Text				Text-to-Image				Total	∆R
	R@1	R@5	R@10	RSum	R@1	R@5	R@10	RSum	RSum	
UNITER†	87.3	98	99.2	284.5	75.56	94.08	96.76	266.4	550.9	↓13.68
VILLA†	87.9	97.2	98.8	283.9	76.26	94.24	96.84	267.34	551.24	↓13.34
LightningDOT	83.6	96	98.2	277.8	69.2	90.72	94.54	254.46	532.26	↓32.32
LightningDOT†	83.9	97.2	98.6	279.7	69.9	91.1	95.2	256.2	535.9	↓28.68
ALBEF	92.6	99.3	99.9	291.8	79.76	95.3	97.72	272.78	564.58	0
B1	90.7	99	99.6	289.3	79.08	94.5	96.94	270.52	559.82	↓4.76
B2	91.4	99.5	99.7	290.6	79.64	95.34	97.46	272.44	563.04	↓1.54
HADA	**93.3**	**99.6**	**100**	**292.9**	**81.36**	**95.94**	**98.02**	**275.32**	**568.22**	**↑3.64**

It was clear that our HADA obtained the highest metrics at all recall values compared to others. HADA achieved a slightly better R@5 and R@10 in Image-to-Text (I2T) and Text-to-Image (T2I) subtasks than ALBEF. However, the gap became more significant at R@1. We improved the R@1 of I2T by 0.7% (92.96 → 93.3) and the R@1 of T2I by 1.6% (79.76 → 81.36). In total, our RSum was 3.64% higher than that of ALBEF (564.58 → 568.22).

The experiment also showed that LightningDOT, which encoded images using Faster-RCNN, performed worse than ALBEF when its total RSum was lower than that of ALBEF by approximately 30%. The reason might be that the object detector was not as powerful as the ViT network, and LightningDOT was pretrained on 4M images compared to 14M images used to train ALBEF. Although also using object detectors as the backbone but applying a cross-modality network, UNITER and VILLA surpassed LightningDOT by a large margin at 15%. It proved that this intensive architecture made a large impact on multimodal retrieval.

Regarding our two baselines, B1 and B2, both of them failed to get better results than the input model ALBEF. Model B1, using the simple strategy of taking the average ranking results and having no learnable parameters, performed worse than model B2, which used a trainable linear layer to fuse the pretrained features. Nevertheless, the RSum of B2 was lower than HADA by 5.18%. It showed the advantages of using a graph structure to fuse the information between models to obtain a better result.

4.5   HADA with Other Input Models

To show the stable performance of HADA, we used it to combine two other different pretrained models, including BLIP [17] and CLIP [28]. While CLIP is well-known for its application in many retrieval challenges [9,24,31,32], BLIP is the enhanced version of ALBEF with the bootstrapping technique in the training process. We used the same configuration as described in 4.2 to train and evaluate HADA in Flickr30k and MSCOCO datasets. We used the pretrained BLIP and CLIP from the LAVIS library [16]. It was noted that the CLIP we used in this experiment was the zero-shot model since the fine-tuned CLIP for these datasets is not available yet.

Table 2. Performance of models on the test set in Flickr30k and MSCOCO datasets. The column ΔR showed the difference compared to BLIP in that dataset.

Dataset	Methods	Image-to-Text				Text-to-Image				Total	ΔR
		R@1	R@5	R@10	RSum	R@1	R@5	R@10	RSum	RSum	
Flickr30k	BLIP	94.3	99.5	99.9	293.7	83.54	96.66	98.32	278.52	572.22	0
	CLIP	88	98.7	99.4	286.1	68.7	90.6	95.2	254.5	540.6	↓31.62
	HADA	**95.2**	**99.7**	**100**	**294.9**	**85.3**	**97.24**	**98.72**	**281.26**	**576.16**	↑3.94
MSCOCO	BLIP	**75.76**	**93.8**	**96.62**	**266.18**	57.32	81.84	88.92	228.08	494.26	0
	CLIP	57.84	81.22	87.78	226.84	37.02	61.66	71.5	170.18	397.02	↓97.24
	HADA	75.36	92.98	96.44	264.78	**58.46**	**82.85**	**89.66**	**230.97**	**495.75**	↑1.49

Table 2 showed the comparison between HADA and the input models. CLIP performed worst on both Flickr30k and MSCOCO with huge differences compared to BLIP and HADA because CLIP was not fine-tuned for these datasets. Regarding the Flickr30k dataset, HADA managed to improve the RSum by more than 3.9% compared to that of BLIP. Additionally, HADA obtained the highest scores in all metrics for both subtasks. Our proposed framework also increased the RSum of BLIP by 1.49% in the MSCOCO dataset. However, BLIP performed slightly better HADA in the I2T subtask, while HADA achieved higher performance in the T2I subtask.

5   Conclusion

In this research, we proposed a simple graph-based framework, called HADA, to combine two pretrained models to address the image-text retrieval problem. We created a graph structure to fuse the extracted features obtained from the pretrained models, followed by the GATv2 network to update them. Our proposed HADA only contained roughly 10M learnable parameters, helping it become easy to train using only one GPU. Our experiments showed the promise of the proposed method. Compared to input models, we managed to increase total recall by more than 3.6%. Additionally, we implemented two other simple baselines to show the advantage of using the graph structures. This result helped us to make two contributions: (1) to increase the performance of SOTA models in image-text retrieval tasks and (2), to not require many GPUs to train on any large-scale external dataset. It has opened the possibility of applying HADA in the industry where large-scale GPU utilisaiton may be considered too costly in financial or environmental terms.

Although we achieved a better result compared to the baselines, there are still rooms to improve the performance of HADA. Firstly, it can be extended not only by two pretrained models as proposed in this research but can be used with more than that number. Secondly, the use of different graph neural networks, such as the graph transformer [30], can be investigated in future work. Third, the edge feature in the graph is also considered. Currently, HADA did not implement the edge feature in our experiment, but they can be learnable parameters in graph neural networks. Last but not least, pretraining HADA on a large-scale external dataset as other SOTA have done might enhance its performance.

Acknowledgement. This publication has emanated from research supported in part by research grants from Science Foundation Ireland under grant numbers SFI/12/RC/2289, SFI/13/RC/2106, and 18/CRT/6223.

References

1. Brody, S., Alon, U., Yahav, E.: How attentive are graph attention networks? arXiv preprint arXiv:2105.14491 (2021)
2. Chen, C.F.R., Fan, Q., Panda, R.: Crossvit: cross-attention multi-scale vision transformer for image classification. In: Proceedings of the IEEE/CVF International Conference on Computer VisionM pp. 357–366 (2021)

3. Chen, Y.-C., et al.: UNITER: UNiversal image-TExt representation learning. In: Vedaldi, A., Bischof, H., Brox, T., Frahm, J.-M. (eds.) ECCV 2020. LNCS, vol. 12375, pp. 104–120. Springer, Cham (2020). https://doi.org/10.1007/978-3-030-58577-8_7

4. Cubuk, E.D., Zoph, B., Shlens, J., Le, Q.V.: Randaugment: practical automated data augmentation with a reduced search space. In: Proceedings of the IEEE/CVF Conference on Computer Vision and Pattern Recognition Workshops, pp. 702–703 (2020)

5. Defferrard, M., Bresson, X., Vandergheynst, P.: Convolutional neural networks on graphs with fast localized spectral filtering. Adv. Neural Inf. Process. Syst. **29** (2016)

6. Devlin, J., Chang, M.W., Lee, K., Toutanova, K.: Bert: pre-training of deep bidirectional transformers for language understanding. arXiv preprint arXiv:1810.04805 (2018)

7. Diao, H., Zhang, Y., Ma, L., Lu, H.: Similarity reasoning and filtration for image-text matching. In: Proceedings of the AAAI Conference on Artificial Intelligence, vol. 35, pp. 1218–1226 (2021)

8. Dosovitskiy, A., et al.: An image is worth 16×16 words: transformers for image recognition at scale. arXiv preprint arXiv:2010.11929 (2020)

9. Dzabraev, M., Kalashnikov, M., Komkov, S., Petiushko, A.: Mdmmt: multidomain multimodal transformer for video retrieval. In: Proceedings of the IEEE/CVF Conference on Computer Vision and Pattern Recognition, pp. 3354–3363 (2021)

10. Gan, Z., Chen, Y.C., Li, L., Zhu, C., Cheng, Y., Liu, J.: Large-scale adversarial training for vision-and-language representation learning. Adv. Neural Inf. Process. Syst. **33**, 6616–6628 (2020)

11. Hamilton, W., Ying, Z., Leskovec, J.: Inductive representation learning on large graphs. Adv. Neural Inf. Process. Syst. **30**, 1–11 (2017)

12. He, K., Fan, H., Wu, Y., Xie, S., Girshick, R.: Momentum contrast for unsupervised visual representation learning. In: Proceedings of the IEEE/CVF Conference on Computer Vision and Pattern Recognition, pp. 9729–9738 (2020)

13. Jia, C., et al.: Scaling up visual and vision-language representation learning with noisy text supervision. In: International Conference on Machine Learning, pp. 4904–4916. PMLR (2021)

14. Karpathy, A., Fei-Fei, L.: Deep visual-semantic alignments for generating image descriptions. In: Proceedings of the IEEE Conference on Computer Vision and Pattern Recognition, pp. 3128–3137 (2015)

15. Kipf, T.N., Welling, M.: Semi-supervised classification with graph convolutional networks. arXiv preprint arXiv:1609.02907 (2016)

16. Li, D., Li, J., Le, H., Wang, G., Savarese, S., Hoi, S.C.H.: Lavis: a library for language-vision intelligence. arXiv preprint arXiv:2209.09019 (2022)

17. Li, J., Li, D., Xiong, C., Hoi, S.: Blip: bootstrapping language-image pre-training for unified vision-language understanding and generation. arXiv preprint arXiv:2201.12086 (2022)

18. Li, J., Selvaraju, R., Gotmare, A., Joty, S., Xiong, C., Hoi, S.C.H.: Align before fuse: vision and language representation learning with momentum distillation. Adv. Neural Inf. Process. Syst. **34**, 9694–9705 (2021)

19. Li, X., et al.: OSCAR: object-semantics aligned pre-training for vision-language tasks. In: Vedaldi, A., Bischof, H., Brox, T., Frahm, J.-M. (eds.) ECCV 2020. LNCS, vol. 12375, pp. 121–137. Springer, Cham (2020). https://doi.org/10.1007/978-3-030-58577-8_8

20. Lin, T.-Y., et al.: Microsoft COCO: common objects in context. In: Fleet, D., Pajdla, T., Schiele, B., Tuytelaars, T. (eds.) ECCV 2014. LNCS, vol. 8693, pp. 740–755. Springer, Cham (2014). https://doi.org/10.1007/978-3-319-10602-1_48

21. Liu, C., Mao, Z., Zhang, T., Xie, H., Wang, B., Zhang, Y.: Graph structured network for image-text matching. In: Proceedings of the IEEE/CVF Conference on Computer Vision and Pattern Recognition, pp. 10921–10930 (2020)

22. Loshchilov, I., Hutter, F.: SGDR: stochastic gradient descent with warm restarts. arXiv preprint arXiv:1608.03983 (2016)

23. Loshchilov, I., Hutter, F.: Decoupled weight decay regularization. arXiv preprint arXiv:1711.05101 (2017)

24. Luo, H., et al.: Clip4clip: an empirical study of clip for end to end video clip retrieval. arXiv preprint arXiv:2104.08860 (2021)

25. Miech, A., Alayrac, J.B., Laptev, I., Sivic, J., Zisserman, A.: Thinking fast and slow: efficient text-to-visual retrieval with transformers. In: Proceedings of the IEEE/CVF Conference on Computer Vision and Pattern Recognition, pp. 9826–9836 (2021)

26. Nguyen, M.D., Nguyen, B.T., Gurrin, C.: A deep local and global scene-graph matching for image-text retrieval. arXiv preprint arXiv:2106.02400 (2021)

27. Nguyen, M.-D., Nguyen, B.T., Gurrin, C.: Graph-based indexing and retrieval of lifelog data. In: Lokoč, J., et al. (eds.) MMM 2021. LNCS, vol. 12573, pp. 256–267. Springer, Cham (2021). https://doi.org/10.1007/978-3-030-67835-7_22

28. Radford, A., et al.: Learning transferable visual models from natural language supervision. In: International Conference on Machine Learning, pp. 8748–8763. PMLR (2021)

29. Ren, S., He, K., Girshick, R., Sun, J.: Faster r-cnn: towards real-time object detection with region proposal networks. Adv. Neural Inf. Process. Syst. **28** (2015)

30. Shi, Y., Huang, Z., Feng, S., Zhong, H., Wang, W., Sun, Y.: Masked label prediction: unified message passing model for semi-supervised classification. arXiv preprint arXiv:2009.03509 (2020)

31. Sun, S., Chen, Y.C., Li, L., Wang, S., Fang, Y., Liu, J.: Lightningdot: pre-training visual-semantic embeddings for real-time image-text retrieval. In: Proceedings of the 2021 Conference of the North American Chapter of the Association for Computational Linguistics: Human Language Technologies, pp. 982–997 (2021)

32. Tran, L.D., Nguyen, M.D., Nguyen, B., Lee, H., Zhou, L., Gurrin, C.: E-myscéal: embedding-based interactive lifelog retrieval system for lsc'22. In: Proceedings of the 5th Annual on Lifelog Search Challenge, pp. 32–37 (2022)

33. Vaswani, A., et al.: Attention is all you need. Adv. Neural Inf. Process. Syst. **30** (2017)

34. Veličković, P., Cucurull, G., Casanova, A., Romero, A., Lio, P., Bengio, Y.: Graph attention networks. arXiv preprint arXiv:1710.10903 (2017)

35. Wang, S., Wang, R., Yao, Z., Shan, S., Chen, X.: Cross-modal scene graph matching for relationship-aware image-text retrieval. In: Proceedings of the IEEE/CVF Winter Conference on Applications of Computer Vision, pp. 1508–1517 (2020)

36. Wu, Z., Xiong, Y., Yu, S.X., Lin, D.: Unsupervised feature learning via non-parametric instance discrimination. In: Proceedings of the IEEE Conference on Computer Vision and Pattern Recognition, pp. 3733–3742 (2018)

37. Young, P., Lai, A., Hodosh, M., Hockenmaier, J.: From image descriptions to visual denotations: new similarity metrics for semantic inference over event descriptions. Trans. Assoc. Comput. Linguisti. **2**, 67–78 (2014)

Author Index

A

Abacha, Asma Ben III-557
Abdollah Pour, Mohammad Mahdi I-3
Abolghasemi, Amin I-66
Adams, Griffin III-557
Adhya, Suman II-321
Aflalo, Estelle I-669
Afzal, Anum II-608
Afzal, Zubair II-341
Agarwal, Anmol II-331
Agarwal, Sahaj I-150
Agichtein, Eugene II-664
Agrawal, Puneet III-341
Aidos, Helena III-491
Aji, Stergious III-281
Alam, Firoj III-506
Alcaraz, Benoît I-18
Alexander, Daria II-512
Aliannejadi, Mohammad I-134, II-522
Alkhalifa, Rabab III-499
Amato, Giuseppe II-110
Amigó, Enrique III-593
Anand, Abhijit II-255
Anand, Avishek I-653, II-255
Ananyeva, Marina II-502
Andrei, Alexandra III-557
Anelli, Vito Walter I-33
Ao, Shuang III-423
Arabzadeh, Negar II-350, II-589, II-599,
    II-655, III-315
Ares, M. Eduardo I-346
Ariza-Casabona, Alejandro I-49
Arslanova, Elena III-51
Artemova, Ekaterina II-571
Askari, Arian I-66
Augereau, Olivier III-536
Azarbonyad, Hosein II-341, III-536
Azizov, Dilshod III-506

B

Bagheri, Ebrahim II-350, II-589, II-599,
    II-655, III-315
Balloccu, Giacomo III-3
Balog, Krisztian III-177
Bandyopadhyay, Dibyanayan I-101
Barreiro, Álvaro III-300
Barriere, Valentin III-527
Barrón-Cedeño, Alberto III-506
Bast, Hannah III-324
Ben Melech Stan, Gabriela I-669
Benamara, Farah II-367
Benedetti, Alessandro III-20
Bergamaschi, Roberto III-491
Beringer, Fabian II-255
Bertasius, Gedas I-669
Besançon, Romaric II-637
Bevendorff, Janek III-518
Bharadwaj, Manasa I-3
Bhargav, Samarth I-134
Bhatia, Rohan II-141
Bhatia, Sumit III-377
Bhatt, Sahil Manoj I-150
Bhattacharya, Paheli III-331
Bhattacharyya, Pushpak II-156
Bhattarai, Bimal I-167
Biasini, Mirko I-182
Biessmann, Felix III-230
Bigdeli, Amin II-350, II-589, III-315
Bilal, Iman III-499
Bloch, Louise III-557
BN, Vinutha I-101
Bobic, Aleksandar III-195
Bock, Philipp II-627
Bogomasov, Kirill III-201
Bonagiri, Vamshi II-331
Bondarenko, Alexander II-571, III-527
Bongard-Blanchy, Kerstin I-18

Bonnet, Pierre   III-568
Boratto, Ludovico   III-3, III-373
Borkakoty, Hsuvas   III-499
Boros, Emanuela   I-377
Bosser, Anne-Gwenn   III-546
Botella, Christophe   III-568
Bouadjenek, Mohamed Reda   I-394
Boualili, Lila   II-359
Bourgeade, Tom   II-367
Boytsov, Leonid   III-51
Braslavski, Pavel   II-571, III-51
Braun, Daniel   II-608
Breitsohl, Jan   II-676
Bridge, Derek   I-330
Brockmeyer, Jason   III-242
Brüngel, Raphael   III-557
Bruun, Simone Borg   I-182
Bucur, Ana-Maria   I-200
Bulín, Martin   III-206

C
Cabanac, Guillaume   III-392
Callan, Jamie   I-298
Camacho-Collados, Jose   III-499
Campos, Ricardo   III-211, III-217, III-248,
     III-377
Cancedda, Christian   III-3
Carmignani, Vittorio   I-182
Carrara, Fabio   II-110
Carrillo-de-Albornoz, Jorge   III-593
Caselli, Tommaso   III-506
Castro, Mafalda   III-217
Cavalla, Paola   III-491
Caverlee, James   II-685
Chatterjee, Arindam   I-101
Chatterjee, Shubham   III-324
Chatzakou, Despoina   II-172
Cheema, Gullal S.   III-506
Chen, Haonan   III-148
Chen, Hsien-Hao   II-448
Chen, Lingwei   II-439
Chen, Xiaobin   I-216
Chinea-Ríos, Mara   III-518
Chiò, Adriano   III-491
Chiril, Patricia   II-367
Choi, Hyunjin   II-377

Choudhry, Arjun   II-386
Clinchant, Stéphane   I-232, I-504, I-636,
     II-16, II-484
Cock, Martine De   II-188
Cole, Elijah   III-568
Collins, Marcus D.   II-664
Coman, Ioan   III-557
Conrad, Jack G.   III-331
Conrad, Stefan   III-201
Constantin, Mihai Gabriel   III-557
Correia, Diogo   III-211
Correia, Gonçalo M.   II-406
Cosma, Adrian   I-200
Craps, Jeroen   II-646
Crestani, Fabio   III-585
Cross, Sebastian   III-35

D
D'Amico, Edoardo   I-249
Da San Martino, Giovanni   III-506
Dagliati, Arianna   III-491
Dalton, Jeff   III-324
de Carvalho, Mamede Alves   III-491
de Jesus, Gabriel   III-429
de Lange, Thomas   III-557
De Luca, Ernesto William   III-294
de Rijke, Maarten   I-409, II-94, III-68
de Vries, Arjen P.   II-512, III-224, III-307,
     III-412
de Vries, Arjen   III-324
Dejean, Hervé   I-504
Déjean, Hervé   II-16
Deldjoo, Yashar   I-33
Denton, Tom   III-568
Dercksen, Koen   III-224
Deshayes, Jérôme   III-557
Deveaud, Romain   III-499
Di Camillo, Barbara   III-491
Di Noia, Tommaso   I-33
Di Nunzio, Giorgio Maria   III-384
Dietz, Laura   III-117, III-324
Dinu, Liviu P.   I-200
Dogariu, Mihai   III-557
Dominguez, Jose Manuel García   III-491
Donabauer, Gregor   II-396
Dong, Mengxing   I-264

Dou, Zhicheng II-79
Doucet, Antoine I-377
Drăgulinescu, Ana Maria III-557
Draws, Tim I-279
Duricic, Tomislav III-255
Durso, Andrew III-568

E

Efimov, Pavel III-51
Eggel, Ivan III-568
Ekbal, Asif I-101, I-685, II-156
El-Ebshihy, Alaa III-436, III-499
Elsayed, Tamer II-430, III-506
Elstner, Theresa III-236
Ermakova, Liana III-536, III-546
Espinosa-Anke, Luis III-499
Ewerth, Ralph II-204

F

Fadljevic, Leon III-255
Faggioli, Guglielmo I-232, III-388
Fan, Zhen I-298
Faralli, Stefano III-373
Färber, Michael III-288
Farinneya, Parsa I-3
Fariselli, Piero III-491
Farnadi, Golnoosh II-188
Farre-Maduell, Eulalia III-577
Fedorova, Natalia III-357
Fedulov, Daniil III-262
Fenu, Gianni III-3
Ferret, Olivier I-569, II-637
Ferro, Nicola I-232, III-388, III-491
Fidalgo, Robson II-534
Figueira, João II-406
Filianos, Panagiotis I-182
Filipovich, Ihar III-557
Flek, Lucie I-118
Flick, Alexander III-230
Formal, Thibault I-232, I-537, I-636
Foster, Jennifer II-493
Franco-Salvador, Marc III-518
Friedrich, Christoph M. III-557
Fröbe, Maik I-313, III-236, III-527
Frommholz, Ingo III-392

G

Gabbolini, Giovanni I-330
Gabín, Jorge I-346

Galassi, Andrea III-506
Galuščáková, Petra III-499
Ganguly, Debasis III-281, III-331
Gangwar, Vivek II-125
Gao, Jun II-304
Gao, Luyu I-298
García Seco de Herrera, Alba II-221, III-557
Gaur, Manas II-331
Gerald, Thomas I-537
Geuer, Tim III-201
Ghosh, Kripabandhu III-331
Ghosh, Saptarshi III-331
Ghosh, Soumitra II-156
Giamphy, Edward I-377
Gienapp, Lukas I-313
Gkotsis, George II-3
Glotin, Hervé III-568
Goëau, Hervé III-568
Goethals, Bart II-646
Goeuriot, Lorraine III-499
Goldsack, Tomas I-361
Gollub, Tim III-242
Gonçalves, Francisco III-248
González-Gallardo, Carlos-Emiliano I-377
Gonzalez-Saez, Gabriela III-499
Gonzalo, Julio III-593
Gospodinov, Mitko II-414
Goyal, Pawan III-331
Grabmair, Matthias II-627
Grahm, Bastian III-236
Granmo, Ole-Christoffer I-167
Gu, Lingwei III-148
Guinaudeau, Camille I-569
Guo, Teng I-588
Gupta, Aaryan II-386
Gupta, Manish I-150, III-341
Gupta, Pankaj II-386
Gupta, Shashank I-394
Gupta, Shrey II-331
Gurjar, Omkar I-150
Gurrin, Cathal I-717, II-493
Gusain, Vaibhav II-423
Gütl, Christian III-195
Gwon, Youngjune II-377

H

Haak, Fabian III-443
Hada, Rishav I-279
Hagen, Matthias I-313, III-236, III-527
Hager, Philipp I-409

Hai Le, Nam I-537
Hakimov, Sherzod II-204
Halvorsen, Pål III-557
Hamdi, Ahmed I-377, III-600
Haouari, Fatima II-430, III-506
Harel, Itay I-426
Hauff, Claudia III-132
Heini, Annina III-518
Hemamou, Léo III-527
Hendriksen, Mariya III-68
Hicks, Steven III-557
Ho, Joyce C. I-553
Hong, Yu I-264, I-474, II-32
Hosokawa, Taishi I-441
Hosseini-Kivanani, Nina I-18
Hou, Shifu II-439
Hrúz, Marek III-568
Hu, Xuke III-398
Hu, Yingjie III-398
Hu, Yiran I-622
Huang, Yu-Ting II-448
Huet, Stéphane III-536
Huibers, Theo II-522
Hurley, Neil I-249
Hwang, Seung-won II-466

I
Idrissi-Yaghir, Ahmad III-557
Ignatov, Dmitry I. II-502
Iizuka, Kojiro I-459
Inel, Oana I-279
Ioannidis, George III-557
Ionescu, Bogdan III-557
Ircing, Pavel III-206

J
Jaenich, Thomas II-457
Jäger, Sebastian III-230
Jameel, Shoaib II-221
Jansen, Thera Habben III-307
Jatowt, Adam I-441, II-544, III-211, III-288, III-377, III-546
Jha, Prince II-141
Jha, Rohan I-298
Jiao, Lei I-167
Jin, Zhiling I-474, II-32
Joe, Seongho II-377
Joly, Alexis III-568
Jorge, Alípio III-217, III-248, III-377

Jose, Joemon M. II-676
Joshi, Meghana III-341
Ju, Mingxuan II-439

K
Kahl, Stefan III-568
Kalinsky, Oren II-617
Kamps, Jaap III-269, III-536
Kanoulas, Evangelos I-134, III-384, III-412
Kato, Makoto P. I-459
Kellenberger, Benjamin III-568
Kersten, Jens III-398
Ketola, Tuomas I-489
Khatri, Inder II-386
Kiesel, Johannes III-527
Kim, Jongho II-466
Klinck, Holger III-568
Kochkina, Elena III-499
Kolesnikov, Sergey II-502
Kolyada, Nikolay III-236
Kompatsiaris, Ioannis II-172
Korikov, Anton I-3
Körner, Erik III-518
Kovalev, Vassili III-557
Kowald, Dominik III-255
Kozlovski, Serge III-557
Kraaij, Wessel I-66
Krallinger, Martin III-577
Kredens, Krzysztof III-518
Krithara, Anastasia III-577
Kruschwitz, Udo II-396
Kuiper, Ernst III-68
Kumaraguru, Ponnurangam II-331
Kumari, Gitanjali I-101
Kurland, Oren I-426
Kutlu, Mucahid II-239
Kuuramo, Crista II-62

L
Lacic, Emanuel III-255
Laine, Markku II-62
Lajewska, Weronika III-177
Lal, Vasudev I-669
Landoni, Monica II-522
Larooij, Maik III-269
Lashinin, Oleg II-502
Lassance, Carlos I-504, II-16
Lawlor, Aonghus I-249
Lawrie, Dawn I-521

Le Goff, Jean-Marie III-195
Lease, Matthew II-239
Lee, Dohyeon II-466
Lee, Eric W. I-553
Lee, Hyunjae II-377
Lee, Jaeseong II-466
Leidner, Jochen L. II-3, III-275
Leith, Douglas II-423
Lerner, Paul I-569
Leśniak, Kacper Kenji I-182
Li, Chenliang I-622
Li, Menglin I-588
Li, Qiuchi II-475
Li, Yanling I-264
Li, Zhao II-304
Li, Zhenpeng II-304
Liakata, Maria III-499
Liang, Hao I-605
Libov, Alex II-617
Liew, Xin Yu III-450
Likhobaba, Daniil III-262
Lim, Kwan Hui I-588
Lima-Lopez, Salvador III-577
Lin, Chenghua I-361
Lin, Jimmy III-148, III-163
Lioma, Christina I-182, II-475
Litvak, Marina III-377
Liu, Bulou I-622
Liu, Junhua I-588
Liu, Yan II-694
Liu, Yiqun I-622
Loebe, Frank III-236
Lorieul, Titouan III-568
Losada, David E. III-585
Lotufo, Roberto II-534
Lou, Jing-Kai II-448
Loudcher, Sabine II-47, II-562
Loureiro, Daniel III-499
Luck, Simon III-527
Lupart, Simon I-636, II-484
Lyu, Chenyang II-493
Lyu, Lijun I-653

M

Ma, Shaoping I-622
MacAvaney, Sean II-414, III-101
Macdonald, Craig II-288, II-414
Mackenzie, Joel III-86
Madasu, Avinash I-669
Madeira, Sara C. III-491

Maistro, Maria I-182
Majumder, Prasenjit III-384
Makrehchi, Masoud III-362
Malitesta, Daniele I-33
Mamta, I-685
Manguinhas, Hugo III-557
Marchesin, Stefano I-232
Marcos, Diego III-568
Marras, Mirko III-3, III-373
Martín-Rodilla, Patricia III-585
Martins, Bruno I-84
Marx, Maarten III-269
Matthes, Florian II-608
Mavrin, Borislav I-3
Mayerl, Maximilian III-518
Mayfield, James I-521
Mayr, Philipp III-392
McDonald, Graham I-701, II-457
Meij, Edgar III-324
Mendes, Afonso II-406
Mendes, Gonçalo Azevedo I-84
Menzner, Tim III-275
Meurs, Marie-Jean II-386
Michiels, Lien II-646
Miller, Tristan III-546
Mitra, Bhaskar II-350
Mittag, Florian III-275
Moffat, Alistair III-86
Morante, Roser III-593
Moreno, José G. I-377
Moreno, Jose G. II-580
Moriceau, Véronique II-367
Morita, Hajime I-459
Mothe, Josiane III-388
Mourad, Ahmed III-35
Moussi, Sara Si III-568
Muhammad, Khalil I-249
Mulhem, Philippe III-499
Müller, Henning III-557, III-568
Müller-Budack, Eric II-204
Müllner, Peter III-456
Murgia, Emiliana II-522

N

Najjar, Amro I-18
Nakov, Preslav III-506
Nandi, Rabindra Nath III-506
Narvala, Hitarth I-701
Naumov, Sergey II-502
Nentidis, Anastasios III-577

Nguyen, Binh T.   I-717
Nguyen, Manh-Duy   I-717, II-493
Nguyen, Thong   III-101
Nicol, Maxime   II-386
Nie, Jian-Yun   I-537
Nigam, Shubham Kumar   III-331
Ninh, Van-Tu   II-493
Nioche, Aurélien   II-62
Nogueira, Rodrigo   II-534
Nugent, Tim   II-3

O

Oard, Douglas W.   I-521
Oulasvirta, Antti   II-62
Ounis, Iadh   I-701, II-288, II-457
Oza, Pooja   III-117

P

Pal, Santanu   I-101
Pal, Vaishali   II-16
Paliouras, Georgios   III-577
Palma Preciado, Victor Manuel   III-546
Papachrysos, Nikolaos   III-557
Paparella, Vincenzo   I-33
Parapar, Javier   I-346, III-300, III-585
Parker, Andrew   III-281
Pasi, Gabriella   I-66, II-512
Patel, Yash   III-600
Paul, Shounak   III-331
Pavlichenko, Nikita   III-357
Peikos, Georgios   II-512
Peng, Rui   I-474, II-32
Penha, Gustavo   III-132
Pentyala, Sikha   II-188
Pera, Maria Soledad   II-522
Perasedillo, Filipp   III-269
Pereira, Jayr   II-534
Pérez, Anxo   III-300
Pesaranghader, Ali   I-3
Petrocchi, Marinella   III-405
Pęzik, Piotr   III-518
Pfahl, Bela   II-544
Pham, Ba   II-655
Picek, Lukáš   III-568
Pickelmann, Florian   III-288
Pielka, Maren   II-553
Piot-Pérez-Abadín, Paloma   III-300
Piroi, Florina   III-412, III-499
Piwowarski, Benjamin   I-232, I-537

Planqué, Robert   III-568
Plaza, Laura   III-593
Pomo, Claudio   I-33
Popel, Martin   III-499
Popescu, Adrian   III-557
Potthast, Martin   I-313, III-236, III-242,
         III-518, III-527
Potyagalova, Anastasia   III-462
Poux-Médard, Gaël   II-47, II-562
Pradeep, Ronak   III-148, III-163
Pucknat, Lisa   II-553
Pugachev, Alexander   II-571
Purificato, Erasmo   III-294
Putkonen, Aini   II-62

Q

Qian, Hongjin   II-79

R

Raiber, Fiana   III-388
Rajapakse, Thilina C.   II-94
Rangel, Francisco   III-518
Ravenet, Brian   III-527
Reagle, Joseph   II-331
Reimer, Jan Heinrich   III-527
Resch, Bernd   III-398
Rieger, Alisa   I-279
Riegler, Michael A.   III-557
Robles-Kelly, Antonio   I-394
Roelleke, Thomas   I-489
Roha, Vishal Singh   II-580
Rokhlenko, Oleg   II-664
Rosso, Paolo   I-118, I-200, III-518, III-593
Roy, Nirmal   I-279
Rückert, Johannes   III-557
Ruggeri, Federico   III-506
Ruggero, Anna   III-20

S

Saha, Sriparna   II-125, II-141, II-580, III-349
Saha, Tulika   III-349
Saini, Naveen   II-580
Sajed, Touqir   I-3
Salamat, Sara   II-589, II-599, II-655
SanJuan, Eric   III-536
Sanner, Scott   I-3
Santosh, T. Y. S. S   II-627

Sanyal, Debarshi Kumar II-321
Scarton, Carolina I-361
Schäfer, Henning III-557
Schlatt, Ferdinand III-527
Schlicht, Ipek Baris I-118
Schmidt, Svetlana II-553
Schneider, Phillip II-608
Schöler, Johanna III-557
Schütz, Mina III-468
Sedmidubsky, Jan II-110
Sensoy, Murat III-362
Servajean, Maximilien III-568
Servan, Christophe III-499
Seyedsalehi, Shirin II-350, II-589, III-315
Shahania, Saijal III-294
Shani, Chen II-617
Shapira, Natalie II-617
Sharma, Karishma II-694
Sharma, Shubham II-125
Shen, Weixing I-622
Shrivastava, Manish I-150
Sidorov, Grigori III-546
Sifa, Rafet II-553
Šimsa, Štěpán III-600
Singh, Apoorva II-125, II-141
Singh, Gopendra Vikram II-156
Skalický, Matyáš III-600
Smirnova, Alisa III-357
Smyth, Barry I-249
Snider, Neal III-557
Soulier, Laure I-537
Spina, Damiano III-593
Stamatatos, Efstathios III-518
Stan, Alexandru III-557
Ştefan, Liviu-Daniel III-557
Stein, Benno III-236, III-242, III-518,
    III-527
Stilo, Giovanni III-373
Storås, Andrea M. III-557
Struß, Julia Maria III-506
Strzyz, Michalina II-406
Stylianou, Nikolaos II-172
Sugiyama, Kazunari I-441
Šulc, Milan III-568, III-600

Sun, Jia Ao II-188
Švec, Jan III-206
Szpektor, Idan I-426

T
Tahmasebzadeh, Golsa II-204
Taitelbaum, Hagai I-426
Talebpour, Mozhgan II-221
Tamber, Manveer Singh III-148, III-163
Taneva-Popova, Bilyana III-362
Tavazzi, Eleonora III-491
Tayyar Madabushi, Harish III-499
Thambawita, Vajira III-557
Theiler, Dieter III-255
Thiel, Marcus III-294
Timmermans, Benjamin I-279
Tintarev, Nava I-279
Tiwari, Abhishek III-349
Tolmach, Sofia II-617
Toroghi, Armin I-3
Tourille, Julien II-637
Tragos, Elias I-249
Trajanovska, Ivana III-230
Tsai, Ming-Feng II-448
Tsatsaronis, George II-341
Tseng, Shao-Yen I-669
Tsikrika, Theodora II-172
Tuo, Aboubacar II-637
Türkmen, Mehmet Deniz II-239
Twardowski, Bartlomiej I-49

U
Ustalov, Dmitry III-262, III-357

V
Vakulenko, Svitlana III-68
van Ginneken, Bram III-224
van Leijenhorst, Luke III-307
Velcin, Julien II-47, II-562
Vellinga, Willem-Pier III-568
Verachtert, Robin II-646
Verberne, Suzan I-66, III-392, III-412
Vishwakarma, Dinesh Kumar II-386
Viviani, Marco III-405

Vladika, Juraj   II-608
Vo, Duc-Thuan   II-655
Volokhin, Sergey   II-664
Vrochidis, Stefanos   II-172

W
Wallat, Jonas   II-255
Wang, Chuan-Ju   II-448
Wang, Qing   II-274
Warke, Oliver   II-676, III-476
Wertheim, Heiman   III-307
Wiegmann, Matti   III-236, III-518
Wiggers, Gineke   III-412
Wijaya, Tri Kurniawan   I-49
Wilson, Marianne   III-482
Wolska, Magdalena   III-518
Wu, Jia   II-304
Wu, Tung-Lin   II-448
Wu, Yueyue   I-622

X
Xiao, Nanfeng   I-216

Y
Yalcin, Mehmet Orcun   I-279
Yang, Eugene   I-521
Yang, Shen   II-274
Yao, Jianmin   I-474, II-32

Yates, Andrew   II-359, III-101
Ye, Yanfang   II-439
Yeh, Chia-Yu   II-448
Yetisgen, Meliha   III-557
Yi, Zixuan   II-288
Yim, Wen-Wai   III-557

Z
Zangerle, Eva   III-518
Zarrinkalam, Fattane   II-599, II-655
Zhang, Dell   III-362
Zhang, Fan   I-622
Zhang, Han   II-685
Zhang, Min   I-622
Zhang, Yizhou   II-694
Zhang, Zhihao   I-361
Zheng, Li   II-304
Zhou, Chuan   II-304
Zhou, Guodong   I-474, II-32
Zhou, Liting   II-493
Zhou, Qifeng   I-605, II-274
Zhu, Ziwei   II-685
Zihayat, Morteza   II-350, II-589, II-599,
                        III-315
Zoeter, Onno   I-409
Zou, Bowei   I-264
Zubiaga, Arkaitz   III-499
Zuccon, Guido   III-35

Printed in the United States
by Baker & Taylor Publisher Services

Printed in the United States
by Baker & Taylor Publisher Services